Rick Knoernschild

827-4226

VERTEBRATE NATURAL HISTORY

VERTEBRATE NATURAL HISTORY

Mary F. Willson
UNIVERSITY OF ILLINOIS
AT URBANA-CHAMPAIGN

Saunders College Publishing
Harcourt Brace Jovanovich College Publishers
Fort Worth Philadelphia San Diego
New York Orlando Austin San Antonio
Toronto Montreal London Sydney Tokyo

Text Typeface: Palatino
Compositor: Bi-Comp, Incorporated
Acquisitions Editor: Michael Brown
Project Editor: Patrice L. Smith
Copyeditor: Diane Ramanauskas
Managing Editor & Art Director: Richard L. Moore
Art/Design Assistant: Virginia A. Bollard
Text Design: Emily Harste
Cover Design: Lawrence R. Didona
Text Artwork: Judy A. Johnson, Patti L. Katusic, & Robert A. von Neumann
Text Artwork Camera Preparation: Tom Mallon
Production Manager: Tim Frelick
Assistant Production Manager: Maureen Iannuzzi

Cover Credit: High Perch—Redwing Blackbird. Copyright 1983 by E. R. Degginger, FPSA.

Library of Congress Cataloging in Publication Data

Willson, Mary F.
Vertebrate natural history.

Includes bibliographies and index.
1. Vertebrates. I. Title.
QL605.W54 1984 596 83-10114
ISBN 0-03-061804-5

VERTEBRATE NATURAL HISTORY ISBN 0-03-061804-5

Printed in the United States of America.
Library of Congress catalog card number 83-10114.

234 032 9876543

Harcourt Brace Jovanovich, Inc.
The Dryden Press
Saunders College Publishing

Preface

Many of my students are perplexed to learn that science is not just an ordered collection of facts but a process of discovery. The gaps in our knowledge distress some of them but excite others, who like to think about unanswered questions and ways of finding good answers. The latter are usually the students who learn the most and have the most fun doing it. This book presents what we regard as facts as well as reasoned speculation in an effort to encourage students to think more about this subject and initiate the process of discovery.

I have not provided references in the text itself for every assertion made in this book. This is particularly true in the first 10 chapters; much of this material is readily available in standard textbooks of physiology or anatomy and in surveys indicated in the reference sections for each chapter.

This book has been a long time coming. It has matured and improved as it sat on the shelf or passed through periodic revisions. I am sure that more improvement will be appropriate and I will welcome constructive criticism. Over the years I have accumulated a debt of gratitude to a number of colleagues and friends, who have explained, commented, corrected, found references, encouraged, and provided help of various other sorts: E. M. Banks, J. L. Brown, J. S. Findley, T. H. Frazzetta, C. Gans, N. H. Goldberg, H. W. Greene, J. R. Karr, J. H. Kaufmann, J. R. King, L. J. Miller, L. M. Page, P. W. Price, C. L. Prosser, B. J. Rathcke, R. R. Roth, D. W. Schemske, C. C. Smith, R. M. Storm, R. L. Trivers, J. Verner, G. P. Waldbauer. I have not wittingly left out the names of any who have helped over the years; I hope that any omissions will be ascribed to the passage of years and forgiven. I am also grateful to those who provided photographs and permissions, as credited *in situ,* and to the three artists, whose talents, hard work, and patience are vastly appreciated. R. C. Snyder and E. E. Provost provided constructive overviews of the manuscript. K. P. and H. W. Ambrose III gave staunch assistance in the early days. The librarians of the Biology Library (University of Illinois) and the Illinois Natural History Survey were, as usual, uncommonly helpful. P. L. Katusic was indispensible in the preparation of the final manuscript. Completion of this task would not have been possible without such assistance.

I dedicate this endeavor to my family—all of them, but especially to my mother and the memory of my father.

Mary F. Willson

Contents

PART 1

PERSPECTIVE

1 *Introduction and Principles*

What is "natural history"? My dictionary defines "history" as a systematic written account of events that is usually connected with a philosophical explanation of their causes. The word "history" is derived from a Greek word meaning a process of learning by inquiry. "Natural history," therefore, refers to the recording and explaining of events in nature; in its present usage, the term usually applies to biological (more than chemical, physical, astronomical, or geological) events. Although chemistry, physics, and sometimes geology are often involved in the explanation of biological features, biology is the focus of attention. The subjects of natural history are whole organisms, not the genes, cells, or tissues that compose the organisms, even though the functioning of genes and tissues is closely related to the behavior and functioning of the whole organism in its natural setting. Natural history involves the study of not only anatomy, physiology, systematics, distribution, but also behavior and ecology. This book is about the natural history of vertebrates (animals with backbones); in this text, I try to place vertebrate form and function in an evolutionary and ecological context.

Humans have long been interested in the natural history of other vertebrates. At first, this interest centered on how to capture them, how to escape from them, and how to maintain or diminish their populations. Recently, the usefulness of animals has again claimed our attention. Animals are useful as agents of population control for species that humans think are desirable or undesirable, as regulators of the balance of nature, as alternatives to domesticated protein sources, as sources of genetic variability for the development of new domestic strains, and as aesthetic phenomena. From such utilitarian motives, our observations of other animals lead to the development of general principles about the relationships of animals with each other and with their environment.

Much human activity is centered on imposing order upon our observations; we do this by trying to *explain* what we observe. Primitive attempts to create order result in myth; more sophisticated orderings represent significant steps toward rationally comprehensible patterns. The first generalizations of natural historians stem from observations and experience. For example, a series of observations could tell us that the long toes of jacanas (*Jacana spinosa*) provide adequate support for these aquatic birds as they walk across lily pads and floating debris (Fig. 1–1). Because the engineering principles involved are already well known, however, an engineer could design a "lily-trotter" of some sort (although it might not be very birdlike)

Figure 1–1 The American jaçana, *Jacana spinosa,* total length about 25 cm. This species lives in Central and South America. There are seven members of the family Jacanidae distributed around the world in tropical marshes where lily pads and other floating vegetation provide a walking surface for the lily trotters. Their name comes from a Spanish version of a Tupi Indian name for the bird in the Amazon basin.

from data concerning the weight of the bird, the buoyancy and flexibility of lily pads, and the customary speeds of the bird. In this example, some elements of jacana anatomy could be predicted from basic principles of physics.

In a similar way, by making many observations of the habitats in which field sparrows (*Spizella pusilla*) are commonly seen, we can learn to expect where, within their geographic range, they will be found. But, unlike the physical principles involved in lilytrotting, the mechanisms that determine patterns of habitat selection are less well understood, and we cannot yet make predictions from basic principles alone. Clearly, any species that survives must be able to exploit food resources, utilize nest or den sites, and escape from predators in some piece of habitat; and they must do these things better than any other organism in those particular circumstances. Some of the physiological and morphological principles involved (temperature tolerance, tooth structure, vision, for example) are understood, but how they combine to produce an animal capable of exploiting a given habitat is still a matter of observation and correlation rather than of prediction. Much of modern natural history is concerned with the progress from observational pattern to predictable pattern.

EVOLUTION

Evolution is the fundamental principle of biology. Comprehension of evolutionary processes is essential for full understanding of any aspect of biology, from medicine (the evolution of fever) to sociology (the evolution of altruism) to the technology of "pest" control (the evolution of resistance). Similarly, for the modern study of natural history, it is inadequate to construct a catalog of cute little tricks employed by organisms and call that "natural history." In some cases, our specific knowledge extends only as far as observing that a certain trick exists, and we lack sufficient knowledge to explain the evolution of that feature. Closer study will undoubtedly reveal something of that evolution, however, as it has so many times before. Conversely, our lack of knowledge regarding certain specific features cannot prevent us from understanding the general process of evolution and its relevance to many characteristics of living organisms.

Evolution is defined as a change in the frequencies of genes in the gene pool of a population. Evolution depends on genetic variation (the existence of more than one allele at a locus); obviously, there could be no change in the gene pool without it. The original source of genetic variation is mutation, supplemented by chromosomal rearrangements that place different genes in different orders that may influence their activity. Evolution is a result of a variety of processes (to be discussed below) that affect the individuals composing a population.

Each organism possesses a complement of genes that is its genotype. During meiósis and gamete production in sexually reproducing organisms, chromosome sets are broken up and chromosomes are randomly assorted into haploid gametes. When gametes unite to form a zygote, then, that zygote contains a different collection of chromosomes than either of its parents. This process, taken alone, does not change gene frequencies. It does, however, create different genotypes. Because different genotypes usually are outwardly expressed in different phenotypes, existing genetic variation is exposed to the environment. Some phenotypes are better suited to certain environments than others. As a result, they survive better and reproduce more successfully; that is, they leave more descendants than others. Because of this differential effect of the environment on the array of phenotypes, genotypes differentially and nonrandomly contribute genes to future generations. This is called "natural selection." Natural

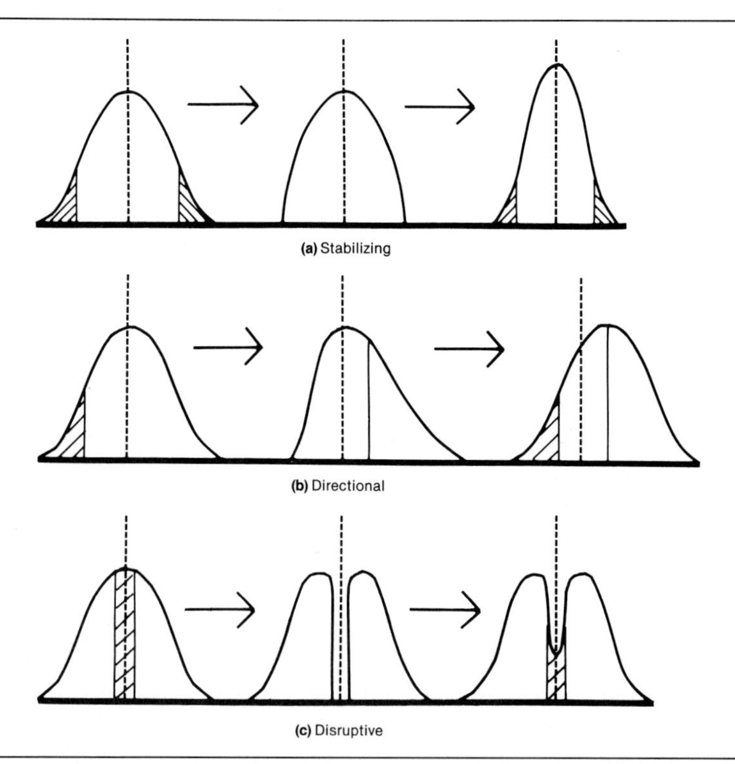

Figure 1–2 The three modes of selection. On the horizontal axis is the degree of development of a phenotypic character; the vertical axis shows the frequency of each phenotype in the population. The shaded areas indicate the phenotypes that are selected against. For each mode, the first curve shows the frequency distribution of phenotypes in the original population; the second suggests the distribution after the indicated mode of selection has eliminated some phenotypes; and the third illustrates the condition after reproduction and recombination have produced additional variability. The dotted vertical lines indicate the initial mean for each distribution.

selection generally produces adaptations that suit the animal's environment in some way. Selection changes the array of genetic variation present in a population and, therefore, is one of the major processes of evolution.

Not all changes in the gene pool may involve adaptation. First of all, genes are linked together on chromosomes, and although bits of chromosomes may break off and attach to other chromosomes, chromosomes (or parts of them) form linkage groups that are usually transmitted as intact entities. Therefore, certain groups of genes tend to be found together consistently, and selection that favors one gene will automatically and indirectly favor others in the linkage group. If the linked genes all confer selective advantage to the owners, the linkage group may persist. If some of them are disadvantageous, selection will favor unlinking those genes or the evolution of modifier genes (which alter the effects of other genes). In the meantime, however, disadvantageous genes may be temporarily and indirectly favored because of their physical association with other, advantageous genes.

Furthermore, in populations in which breeding is sometimes accomplished by just a few random members, changes in the gene pool may take place as a result of "sampling error" or genetic drift. Not all genes in the pool are represented in zygotes and the ensuing generation, because not all individuals bred. Since breeding was at random and not related to suitability, such changes in the gene pool are not related to adaptation. Drift may be far more common and important than is generally believed. Migration into or out of a population (gene flow) may also result in changes in gene frequency. The effect of migration depends on the size of the population, the frequency of migration, and whether or not it is random with respect to the genotype of the migrants.

Another source of change in gene frequency derives from a number of "anomalies" that may take place during cell division and result in the loss of certain chromosomes. Although these anomalies have the appearance of mistakes, they actually may have some as yet unknown adaptive value.

Variation and Natural Selection

Much variation in a population is continuous variation; that is, all individuals exhibit a characteristic, but to varying degrees. If we were to choose some characteristic—length of the jawbone, for example—we would probably find that most individuals have jawbones of medium length, but members of the population can be found with extremely long or short ones. Given such a distribution of phenotypic frequencies, which we will suppose has a genetic basis, selection can act on that distribution in three basic ways (Fig. 1–2).

Individuals at the tails of the distribution (those with very long or very short jawbones, in this case) may be less successful than those in the middle. As a result, such phenotypes are continually eliminated from the population more rapidly than those in the middle. In this condition, the

Figure 1–3 Directional selection by differential predation on color patterns in the water snake *Nerodia sipedon* living on islands in Lake Erie. Dark-colored snakes tend to be eliminated from the population because they are more conspicuous against the light-colored backgrounds found on the islands. Dark adults, migrating to the islands from the mainland, apparently reintroduce genes for dark color into the island gene pool. (From *The Process of Evolution* by P. Ehrlich and R. Holm. Copyright © 1965 McGraw-Hill Book Company. Used with the permission of McGraw-Hill Book Company.)

peak of the distribution (the mean of the characteristic) stays at the same value from generation to generation. This is called stabilizing selection.

Selection may also discriminate against only one tail of the distribution (for example, individuals with short jaws). Individuals genetically constituted to produce longer jaws then provide more genes to future generations, and, gradually, the distribution of jawbone lengths in our example shifts toward longer and longer jaws. This process is called directional selection.

In a third case, selection may discriminate against individuals close to the mean and favor those at either extreme. This results in a bimodal frequency distribution of phenotypes and is called disruptive selection.

These three modes of selection can be exemplified by cases from the real world. The now classic case of stabilizing selection is that of house sparrows (*Passer domesticus*) brought down by a severe storm in New England. Of 136 stunned birds, 72 recovered and 64 died. Bumpus (1898) made many skeletal measurements on both groups and found that individuals whose measurements were near the average had a higher probability of recovery than those either larger or smaller than the average. Grant (1972), and others, reanalyzed these data and showed that the results applied to female sparrows but not to males. He suggested that female sparrows are subordinate to males in most social interactions; they may be kept from food resources by the dominant males and, therefore, have a higher risk of death in severe conditions. Small females probably have small energy reserves and are unable to replace them when they are depleted; large females perhaps can mobilize stored energy less rapidly in response to metabolic stress. These circumstances may result in selection against both size extremes.

One of the best examples of directional selection comes from Camin and Ehrlich's (1958) study of water snakes (*Nerodia sipedon*) on islands in Lake Erie. All *N. sipedon* on the surrounding mainland and some of those on the islands are patterned with dark bands, but the island populations also include numerous lightly banded or completely unbanded individuals (Fig. 1–3). Litters of young snakes comprise both banded and unbanded forms, but unbanded individuals occur significantly more frequently among adult snakes than among young snakes. This change in pattern frequency is not due to age-related changes in color of individuals but

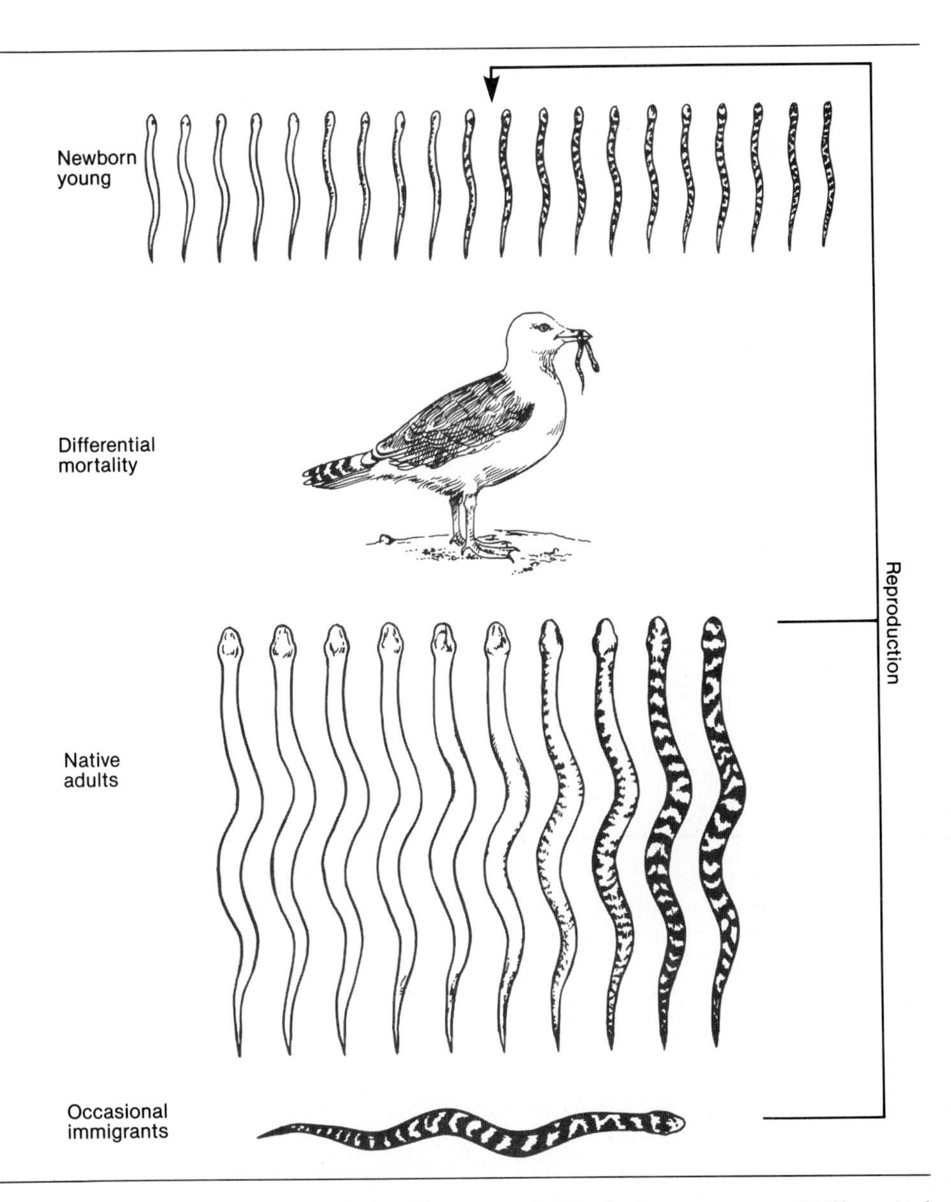

rather to the differential elimination of the darker patterns. Differential survival of light-colored individuals occurs consistently on several islands and is undoubtedly related to differential removal by predators such as gulls. Most island snakes inhabit limestone rocks and beaches, where light color is a better camouflage than dark bands (which are valuable as camouflage in the typical swamp habitat of this species on the mainland). Selection against banded patterns on the islands is very strong and would eventually eradicate the dark forms entirely were it not for the immigration of banded individuals from the mainland.

Good examples of disruptive selection among vertebrates are hard to find, although it must have occurred many times. Disruptive selection must be involved in the evolution of sexual differences in size, color, or pattern that are observed in many species of vertebrate. Small male and large female house sparrows were especially vulnerable in Bumpus' winter-storm sample (Johnston et al., 1972). This differential in survival maintains the sexual differences in body size.

Opposing selection forces favor two color types in some populations of the three-spine stickleback (*Gasterosteus aculeatus*). By implication, intermediate color forms are at a selective disadvantage (Fig. 1–4). Three-spine stickleback males typically develop a bright reddish throat during their breeding season. Males of populations in many areas are monomorphic ("one form") for this characteristic—that is, all males develop red throats.

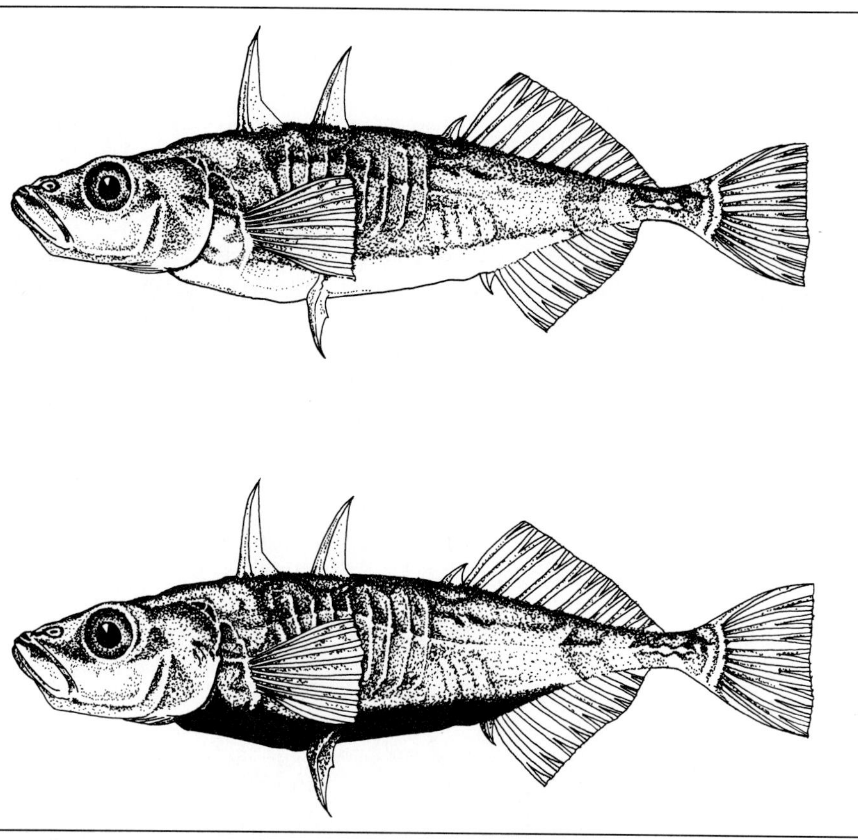

Figure 1–4 A male three-spine stickleback, *Gasterosteus aculeatus*. The common name derives from the three spines located anterior to the dorsal fin proper. The formal name of the genus denotes the series of bony plates found along the flanks, at least in some populations. Two color morphs are shown here.

But in certain streams in western North America, several populations have polymorphic ("several forms") males: Some have red throats, others (the majority) have silver or black throats. Females prefer to mate with red-throated males, thus giving this morph a reproductive advantage (Semler, 1971). Red-throated males also may have a higher success in defending their eggs and young against cannibalism by other sticklebacks, especially other males. On the other hand, the prevalence of nonred males in certain populations suggests that something in those areas must be selecting against red males. Cutthroat trout (*Salmo clarkii*) attacked red males more frequently than they did nonred males, and trout predation was particularly successful in well-lit waters where the red color was conspicuous. Thus, we would expect to find nonred males primarily in clear waters with an abundance of predatory trout. Males with pinkish (intermediate-colored) throats were apparently unobserved; presumably, they would have neither the reproductive nor the antipredator advantage and thus be disfavored by selection. This example is complicated by the existence of at least two nonred morphs, the probable existence of other conditions affecting the maintenance of the polymorphism (Hagen et al., 1980; Hagen and Moodie, 1979), and the fact that selection against the intermediate form is inferred and not demonstrated directly. Nevertheless, the idea of disruptive selection and the resulting population of two (or more) major forms is adequately illustrated.

Both stabilizing and directional selection tend to reduce existing variation by eliminating certain types. On the other hand, disruptive selection tends to increase the total amount of variation in a population, although variation around each peak may be reduced in each generation. Variation is continually reconstituted in sexual organisms by mutation and by recombination of chromosomes, however.

Fitness

We use the term "fitness" to describe the relative genetic contribution of different individuals to future generations. Individuals producing the most offspring that in turn reproduce successfully are the most fit, and the fitness of other members of the population is measured relative to the most fit. Thus, if individual *A* produces 100 surviving offspring and individual *B* produces only 90, clearly the second kind has produced only 90 percent as well as the first and is at a 10 percent disadvantage. It is customary to set the fitness of *A* at 1.00; then the fitness of *B* relative to *A* is 0.90. The difference in the fitness of *A* and *B* (10 percent in this case) is called the selection coefficient, which measures the relative disadvantage of the type disfavored by selection or the intensity of selection against the inferior kind. It is important to realize that fitness is defined in this relative way. A certain constellation of phenotypic characteristics may be most fit in one set of circumstances, but a new constellation in the next generation may then be more fit, even if the first constellation is still there and no environmental change has occurred.

Total fitness must also include some measure of genes passed on by the relatives of an individual, because relatives, by definition, carry a number of genes in common. Members of a family (a genetic lineage) share a higher proportion of genes with each other than with the rest of the population. It is possible, therefore, for an individual to contribute indirectly to its fitness by enhancing the fitness of its relatives. Fitness derived from both individual reproduction and that of relatives is called inclusive fitness. This fitness is far more difficult to measure in natural situations than the individual component of fitness (which itself is not easy to measure accurately) and few studies of adaptation have accounted for inclusive fitness. Selection that depends on inclusive fitness is known as kin selection (Kurland, 1980). We will see some examples in which inclusive fitness and kin selection apparently have played an important role in the evolution of certain kinds of social behavior.

Adaptation

The language that we sometimes use to discuss adaptation needs to be clearly understood. If some trait is described as being "for" a particular function (a leg for running, a tongue for lapping), this is a verbal shorthand indicating that the shape and size of the legs or tongues gave their owners some advantage that allowed them to survive and reproduce better than other individuals with somewhat different legs or tongues; that is, selection favored one conformation over another. Note that the view is retrospective: Selection pressures acted on available genetic variation and only certain variants left descendents. We never mean that a particular trait was purposefully invented in anticipation of need. Likewise, when we speak of the "design" of a morphological trait, we are concerned with the arrangement of its components and how well its form matches its function. Again, the mechanism of achieving a good design is natural selection of better genotypes and not some cognitive or supernatural plan. No goals are implied; there are only results, and the forms that work better are perpetuated.

It is also important to realize that many adaptive characteristics have multiple effects, some of which confer selective advantage to the organism and some of which are just secondary effects or by-products. Two examples will illustrate this point.

Consider the legs of a modern horse. They are used for many activities—pawing snow, swimming, stomping snakes, scratching—but it is clear that none of these is their primary function. Selection on these activities did not determine the form of horses' legs. These other activities are secondary, useful enough, perhaps, but many kinds of legs can and do accomplish these results. Horses' legs have clearly evolved in response to selection for improved running ability. Various principles of physics can be used to show that the "design" of horses' legs is appropriate for a running animal. Contrasts with the legs of other animals, such as bats or moles, reinforce this interpretation, for the arrangement of the parts of a mole's leg clearly indicates that there has not been selection for a running function. Running is an important element of adaptation in horses. Wild horses and their relatives, the zebras and asses, typically live in open (grassy or barren) habitats in which good running abilities are undoubtedly advantageous in escaping predators (flying and digging are not viable options for largish mammals) and perhaps in seeking pasture and water at widely separated points.

To many people, this example may be a silly one, so incredibly obvious as to be a ludicrous waste of time. It is quite analogous to the next example, however, one that has confused many ecologists. Many animals occupy a territory at certain times during their life cycles. A territory is a piece of real estate defended by its occupant against other individuals of the same species, sex, or feeding habits; thus, a territory is a rather exclusive property. Territorial behavior has evolved as a means of defending some important resource (food or nest sites, for example). Because it requires the outlay of time and energy for defense on the part of the owner, this behavior must provide to the owner some advantage it would not otherwise have. Two possible consequences of territoriality may be the regulation of population density (by the enforced spacing of individuals in a habitat) or population size (if some individuals are excluded from holding a territory and are then more subject to starvation, predation, or disease). This example is discussed again in Chapter 11, but, for the time being, note that these last two results of territoriality are population consequences (results measured at the population level rather than at the individual level) and by-products of selection on individuals for territorial behavior. Population consequences are not the results that determine the evolution of this behavior pattern; the evolution of this behavior has depended on the resulting advantages that accrue to the individual territory owner.

Adaptations must be discussed chiefly in terms of advantages to individuals (including their kin). "Evolution for the benefit of the species" (or population, or community) is a doubtful proposition. "Group selection" has acquired many meanings and is a subject of much debate. However the debate is resolved, the attempt to understand animal behavior should begin at the level of the individual organism.

We can answer the question of why a certain animal does something in two ways. Take the question "Why do most of the birds in Wisconsin breed in the spring?" We could answer that the pituitary gland responds to increasing day length as perceived by the eye and increases the production of sex hormones. This response induces mating behavior and reproduction. Such an answer really describes *how* these birds can breed in spring; it is not, fundamentally, an answer to *why*. The birds respond to what are called proximate factors, which are certain environmental cues that, in themselves, may or may not confer adaptive value except as signals.

We could also answer the question by saying that spring is a time of good food production, which is necessary for obtaining the requisite energy and nutrients to make eggs and feed young. Furthermore, springtime breeding allows young birds time to grow and achieve independence from their parents before winter comes. In this case, we have answered in terms of so-called "ultimate factors," which are usually ecological conditions that give adaptive value to a characteristic. Any environmental cue that is sufficiently correlated with such ultimate factors to be a good predictor of ecological conditions may be used as a proximate factor. Day length is often a proximate factor in the regulation of breeding seasons because it is often closely correlated with seasonal weather conditions and, thus, with prey populations. Proximate factors exist only because of ultimate factors; they are only the means to an end and become important to adaptation only by association with ultimate factors.

The study of adaptation is often far from easy (Clutton-Brock and Harvey, 1979). Not only are many kinds of detailed information required, but adaptations also are seldom "perfect" even when we can ascribe major biological significance to them. An engineer could probably design a better running machine than a horse and propose a better lily-trotter than a jacana.

It is, in fact, the imperfection of adaptation that helped convince Darwin and many others of the reality of evolution (Gould, 1978). A fetal whale (in certain species) has teeth it never uses. They are resorbed and replaced by a whalebone filter used for sifting tiny organisms from the water. Persistence of such seemingly functionless characteristics is evidence that the modern whalebone whales are descended from fully toothed ancestors.

Limitations to the perfection of adaptation are many. Selection acts on whole animals, not on isolated characteristics; thus, all characteristics evolve in a context of other features. Our exemplary horse is obviously not just a running machine; it must eat, mate, carry young (if it is a female), fight, swat flies, and do many other things. Features important to these other necessary activities may prevent the evolution of a perfect running machine. In fact, probably all adaptations are compromises between conflicting selective pressures.

An example of such a compromise is provided by the male dickcissel (*Spiza americana*). Dickcissels nest in grassy, weedy fields in eastern North America (Fig. 1–5). Males are more brightly colored than females and often mate with more than one female. Males usually spend considerable time advertising for and courting females, and females do virtually all the incubating and feeding of young. The time spent by breeding males in various activities has been recorded and rough estimates of the energy utilized in each activity have been made. A typical time and energy budget is given in Table 1–1.

Figure 1–5 A male dickcissel, *Spiza americana*. The common name is an onomatopoetic rendition of its song.

Observations were made on how these birds responded to an environmental change in the form of severe hot weather. High ambient temperatures increase metabolic demands related to ventilation and cooling off. Above 34°C (94°F), especially at high humidities, the males were forced to rest more than usual. Foraging, courtship, and pair-bond maintenance changed relatively little, but singing and territory defense decreased noticeably. As a result, males obtained fewer mates and their reproductive output was decreased. If they had not increased their resting time, however, it is likely that the risk of death or disability would have been significantly increased and the probability of future reproduction reduced. From this observation, one can infer that male dickcissels that "trade" a slight reduction of present reproductive output in return for enhanced future

Table 1–1 Time and energy budget of breeding male dickcissels. Percentages are based on daytime activities throughout a nesting season.

Activity	*Time spent (%)*	*Energy expended (%)*
Foraging	17–20	15
Resting	8–10	6
Singing	50–60	34
Courtship	<1	~1
Pair-bond maintenance	6–10	8
Interspecific aggression	<1	≪1
Territory defense	<1	<1
Miscellaneous flight	9–14	35

(From Schartz and Zimmerman, 1971.)

reproduction are selectively favored. The behavioral indication of this trade is the temperature sensitivity of the time and energy budget. This is one case in which time and energy budgets may be directly related to fitness.

The nature of the trade-offs among different activities will obviously vary with the specific requirements of each kind of organism. Some organisms sacrifice number of offspring to size and quality of offspring. For example, monkeys and carnivorous mice (*Onychomys*) expend proportionately more energy producing and raising each young than do skunks (*Mephitis, Spilogale*) or voles (*Microtus*), and their litter sizes are smaller. Some animals sacrifice self-maintenance to reproduction. Yellow-billed magpies (*Pica nuttalli*) may lose weight or tatter their plumage, and Pacific salmon (*Oncorhynchus*) characteristically lose their lives in the process. Some lizards may sacrifice color-matching of the background and camouflage protection from predation for the ability to use color changes in temperature regulation or vice versa.

Limitations are also historical. It is unlikely, for instance, that buffalo will ever fly or eat fish because they have a constellation of general features that render such changes improbable. Although mechanisms exist for accomplishing seemingly major changes in certain circumstances, basically the evolution of species is limited by the present accumulation of genes and combinations of genes and, therefore, by past selection pressures. Some bats and owls, for example, may often occupy similar habitats and eat similar foods, but they do these things in different ways, using different morphologies, because they arrived at this point from different starting places.

Giant pandas (*Ailuropoda melanoleuca*), which live in the high mountains of western China, have two "thumbs" in their forepaws (Gould, 1978). One is lined up with the other four ordinary digits and betrays the panda's descent from carnivores that used all five digits in running and digging. The second "thumb" develops from a wrist bone and angles out from the paw as a proper thumb should. It is operated by muscles that work the ordinary thumb in other bears. Pandas are principally vegetarians and use this secondary, opposable, thumb to grasp and manipulate their favorite food, bamboo shoots. Apparently, the original thumb had become so modified that it could no longer be restructured for this purpose and the somewhat clumsy secondary thumb was developed instead.

Furthermore, the direction of selection on a given population may shift from time to time. The population observed at any one time may present some characteristics of selective value before the last shift, depending on the time that has passed and the effectiveness of the new selective pressures. In addition, adaptation may be constrained by the other sources of evolutionary change.

Speciation and Adaptive Radiation

Because different populations of the same species may encounter different evolutionary forces, they gradually accumulate different assortments of genes. In itself, this process does not result in the formation of new species, because gene flow may still occur between populations. Speciation is possible only when gene flow is interrupted (or at least vastly reduced) and populations become effectively isolated from each other. Only then can different populations assemble assortments of characteristics sufficiently different that we call them different species. Technically, the usual definition of a biological species is a population of organisms that, at least potentially, can interbreed and do not, in nature, usually breed with members of other such populations. We do not actually know who breeds with whom for many kinds of animals, but it is the job of taxonomists to make inferences about this matter from morphological or other data. Most species are really defined by such indirect evidence. What exists in nature is an array of populations in different states of isolation; we find it convenient to label some of them as species.

In real populations, of course, many individuals never even meet each other, much less breed together. A song sparrow (*Melospiza melodia*) from Alaska is not likely to encounter one from Virginia, but we call them the same species because populations of song sparrows can be found across North America and gene flow very likely occurs among them. Although some geographic differentiation of populations is evident (body size, song, and other characteristics vary regionally), we infer that all these populations are conspecific because they have many characteristics in common. "Many" here is defined as what bird taxonomists have decided is "enough" to infer interpopulation gene flow. In some cases, we have evidence, usually from morphological intermediates but sometimes also from observing hybrid matings, that gene flow may occur between a series of adjacent populations (*A* and *B*, *B* and *C*, *C* and *D*, *D* and *E*, and so on), although populations at the ends of such series, if brought together, are not capable of interbreeding. Because of gene flow all along the line, however, we usually refer to this set of populations as a single species.

Because the accumulation of genetic differences in different populations is commonly believed to be gradual, it is often inferred that the phenotypic changes accompanying the genetic differentiations must be gradual also; however, this need not be the case. If a gene specifies a change relatively early in development or at a critical stage of developmental integration, that alteration can profoundly affect the shape and size of the adult trait or suite of traits. Thus, a single genetic difference sometimes may produce a prodigious shift in adult phenotype, and the morphological changes accompanying population differentiation may be more stepwise than gradual (see Stebbins and Ayala, 1981).

The classic view of animal speciation has been that populations must undergo some form of geographic isolation from each other for a sufficiently long time that, should they come together again, interbreeding between them would be impossible. This model of speciation is called allopatric ("different countries"), referring to a period of existence in different locales.

In recent years, however, increasing evidence has accumulated that other models of animal speciation may have validity. Reproductive isolation of different populations may be accomplished by means other than geographic isolation. Differences in breeding season or habitat and behavioral differences, if strong enough, can lead to speciation within the same geographic range—namely, sympatric ("same country") speciation. Most of the evidence for this speciation model comes from insects, but it may eventually be applied to vertebrates as well. Current research on frogs and on fishes, especially, may document this possibility.

We have long known that speciation in plants can occur via polyploidy, which is a multiplication of the normal number of chromosome sets such that some individuals have three or more sets of chromosomes instead of the usual two. If polyploidy is accompanied by the ability to reproduce without collaboration with another individual, speciation may occur instantly because polyploids often have difficulty mating successfully with normal diploids. (The gametes of polyploids have unusual numbers of chromosomes, which makes zygote formation and subsequent cell division difficult.) Because differences in ploidy levels are uncommon among closely related species of modern vertebrates, and because so few vertebrates are capable of unisexual reproduction, this mode of speciation has always been considered trivial for them. However, considerable evidence for significant changes in ploidy during the adaptive radiation of the vertebrates exists (Ohno, 1970; Ferris and Whitt, 1977), so the classic view may need to be modified.

Speciation results in the formation of separate gene pools with the possibility of independent evolutionary changes. New adaptive designs emerge in each species, and a multiple divergence of new adaptive types from the ancestral type is observed. This process is called adaptive radiation. Usually, new species are at first very similar to their antecedents, but sometimes a new species differs in some small way that has a disproportionately large effect, and, suddenly, a novel capacity exists. Such deceptively small novelties ultimately allowed vertebrates to begin to emerge onto land, to begin to truly fly, and to begin metabolic regulation of body temperature. We often call these broad ways of life adaptive zones.

Once the initial step is made, adaptive radiation may begin to fill such zones with species that specialize within them. Invasion of a new adaptive zone, therefore, can be followed by a burst of speciation as the zone is progressively subdivided by increasing numbers of species, each exploiting some part of the zone. Eventually, all possible means of specializing and subdividing may be used, and then, of course, formation of more species ceases until another "invention" opens another adaptive zone. Sometimes environmental changes erase a zone, or part of one, and some species may become extinct.

Many examples of adaptive radiation could be presented, but two will suffice. Horses, tapirs, and rhinoceroses all belong to a single order (Perissodactyla) of placental mammals. They are distantly related to deer, antelope, and pigs (order Artiodactyla), elephants (order Proboscidia), and manatees (order Sirenia). All four orders are inferred, largely from fossil evidence, to be derived from a single early group of vegetarian mammals. The adaptive zone could be characterized as one of medium- to large-sized vegetarians. All modern species are primarily vegetarian, too, but have radiated from the primitive stock in many directions. Manatees are slow, aquatic forms with reduced appendages; the others are basically terrestrial. Elephants (and their many extinct relatives) seem to have originated as small animals with tusks used for digging; body size gradually increased and tusks were used for other purposes. Modern perissodactyls and artiodactyls are distinguished mainly by the structure of the lower legs: In perissodactyls, the normal five digits have been reduced, usually to one or three; in artiodactyls they have been reduced to two. Furthermore, differences in the digestive tracts give evidence of differences in food-processing—the artiodactyl stomach is often highly compartmentalized and very specialized.

Within the perissodactyls, for instance, the tapirs of Latin America, Malaya, and the East Indies are nocturnal forest creatures that seem quite similar to the primitive representatives of the order. The hind foot has three toes (although the forefoot has four). Rhinoceroses, now found only in Africa, the East Indies, and East Asia, are typically large, heavy-bodied forms with three toes on each foot. The third extant branch of the perissodactyl radiation is the horses, specialized to running and eating grass (see

Janis, 1976). The number of digits has been reduced to only one during the course of horse evolution, and body size has increased. All existing types of horses are characteristic of open plains.

A second example is chosen on a smaller taxonomic scale and has become a classic. The Hawaiian honeycreepers of the family Drepanididae are small birds found only in the Hawaiian archipelago. Their ancestors are uncertain, but they are probably derived from one of the American groups of songbirds. Once in Hawaii, honeycreepers radiated (probably from an ancestral, short-billed finchlike bird) to form a number of species of different ecological types, including long- and short-billed insect-eaters, long- and short-billed nectar-eaters, and thick-billed seed- and fruit-eaters (see Fig. 1–6).

This adaptive radiation is simpler to analyze than those of the mammals cited previously, because it appears to have been based largely on dietary factors and, hence, is less complex than the mammalian example. Two subfamilies have been distinguished. The subfamily Psittarostrinae contained about 17 species, including insect-eaters that foraged in different styles: Four species of the genus *Loxops* gleaned insects from leaves and twigs or extracted them from buds and seedpods, and one hunted by scurrying over the bark of trees and gleaning insects; species of the two genera *Hemignathus* and *Pseudonestor* foraged by hammering on trunks and branches with the lower bill or striking off flakes of bark. The genus *Psittarostra* had six species of thick-billed seed- and fruit-eaters. Members of the subfamily Drepanidinae, on the other hand, were primarily nectar-feeders with brushy, fluid-lapping tongues, although bill shape (and probably foraging behavior) varied greatly. One drepanidine, *Ciridops anna,* was probably a fruit-eater. It is appropriate to refer to many of these species in the past tense because many of them are now extinct (Olson and James, 1982).

Figure 1–6 The adaptive radiation of the Hawaiian honeycreepers (Drepanididae). The presumed relationships are indicated by the upward-branching lines. All genera, but not all species of each genus, are depicted here. This hypothetical scheme is based on the assumption that a single finch-like ancestor colonized the Hawaiian archipelago and subsequently radiated, with species foraging in many different ways. The relationships of some genera, such as *Ciridops* and the newly discovered *Melamprosops,* are quite uncertain, and even the basic scheme of radiation has been rendered in other ways (Amadon, 1950; Richards and Bock, 1973). Nevertheless, the fundamental idea of adaptive radiation from a single (or a few?) basic types into many is amply illustrated. (From Raikow, R.J., 1976. The origin and evolution of the Hawaiian honeycreepers [Drepanididae]. *Living Bird, 15*:95–117.)

BODY SIZE

Size and Metabolism

All animals require energy to maintain themselves as integrated, functioning units. Energy is obtained by oxidation of food and transformed by various physiological processes of metabolism. Metabolic rate (the rate at which energy is used by an organism) is usually measured by oxygen (O_2) consumption or food (calories) utilized per unit time.

Metabolic rate is closely related to body size for all animals. As a rule, metabolic rate increases with increasing size, but somewhat more slowly.

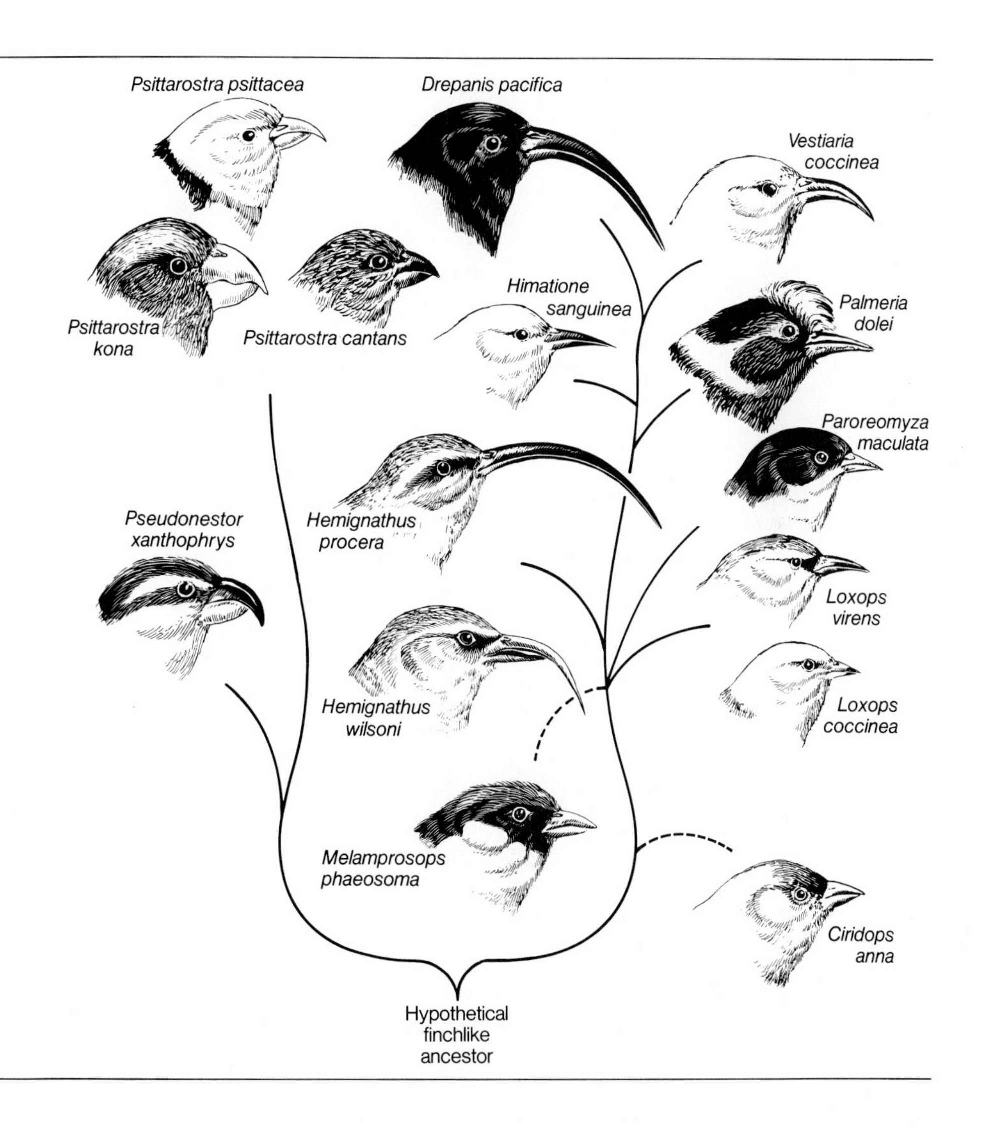

Thus, a 9-g harvest mouse may use about 22 cm^3 O_2/h, while a 96-g ground squirrel may use about 99 cm^3 O_2/h and a 2500-g domestic cat may use about 1700 cm^3 O_2/h. The general form of the relationship is $M = aW^b$ (where M = metabolic rate, a is a proportionality constant that varies among taxa, W is body weight, and b is an exponent less than or occasionally equal to 1). If both metabolic rate and body weight are expressed as logarithms or plotted on logarithmic axes on a graph, then b is the slope of the line that shows the relationship between metabolism and size (Fig. 1–

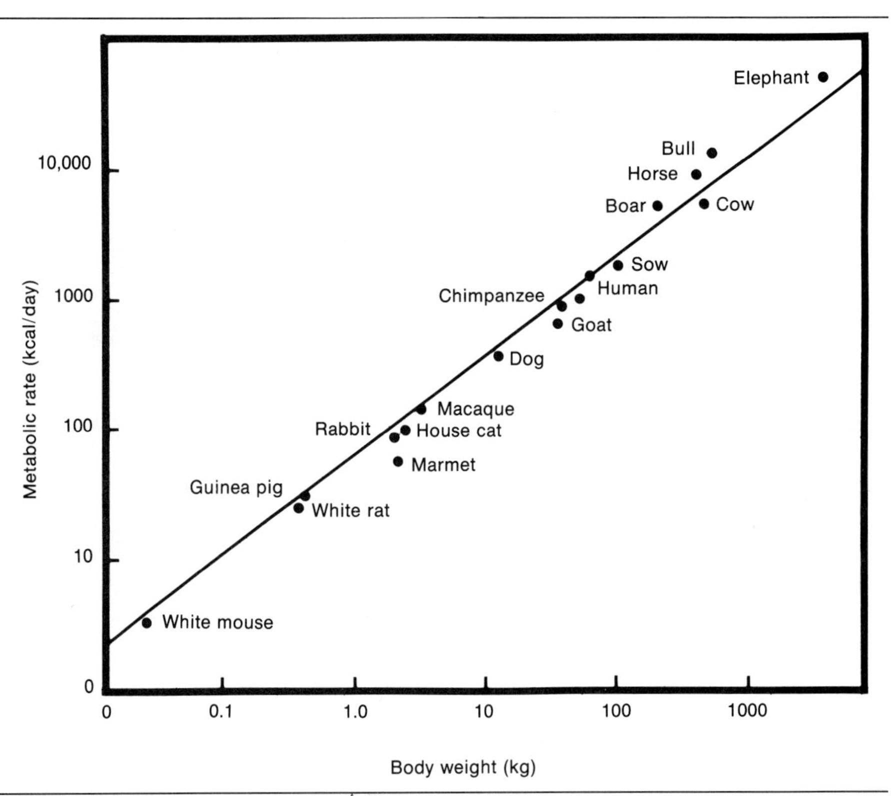

Figure 1–7 The relationship between total metabolic rate and body weight in mammals. Note the double logarithmic scales. The vertical axis on this graph is in units of kilocalories of energy per day. A common conversion factor that translates kilocalories into oxygen consumption is 4.7 kcal/L of O_2 for an inactive animal. Similar graphs can be drawn for birds, which have a higher intercept, and reptiles, and fishes, both of which have lower intercepts. (Modified from M.S. Gordon and M. Kleiber. Reprinted with permission of Macmillan Publishing Co., Inc. from *Animal Physiology: Principles and Functions* by Malcolm S. Gordon. Copyright © 1977 by Malcolm S. Gordon. Originally from M. Kleiber, 1932. Body size and metabolism. *Hilgardia,* 6:315–353. Courtesy of Division of Agricultural Sciences, University of California.)

7). For almost all animals, b falls in the range of 0.60 to 0.90 and is often about 0.75 (Fig. 1–8), although some interesting variations are known (Chapter 5). The value of a indicates the y intercept or the elevation of the line. This value differs greatly among animals and is considerably higher in birds (especially those of the order Passeriformes) and in mammals than in fishes or reptiles.

Because large animals require more food and more oxygen than do small ones, large bodies cost more to maintain than smaller ones. Because the

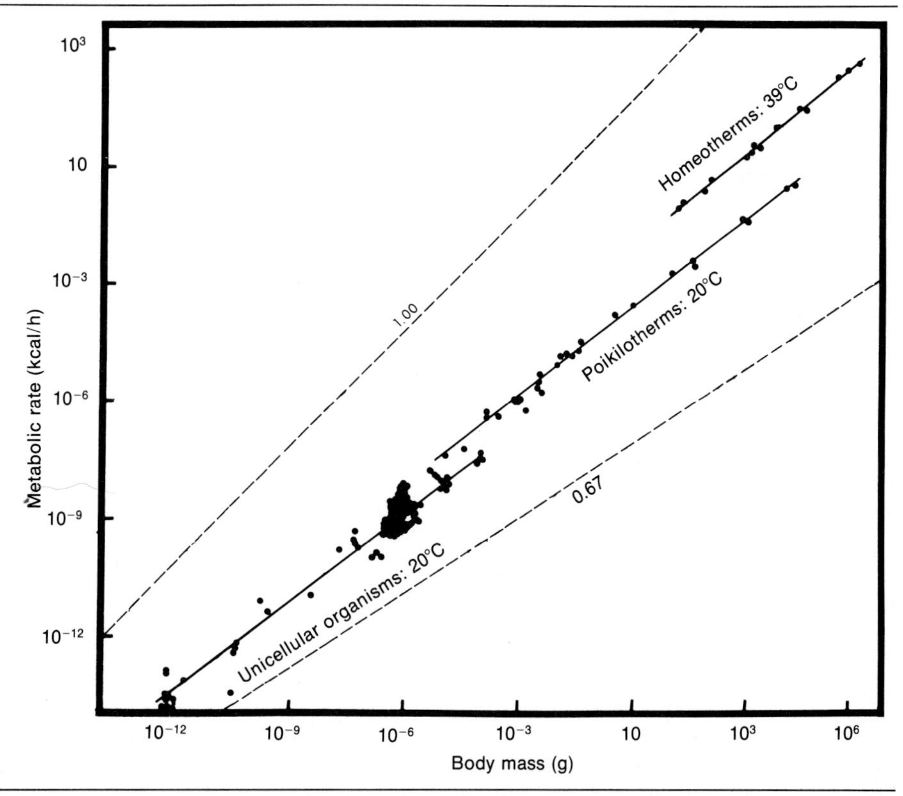

Figure 1–8 The relationship between total metabolic rate and body weight for different kinds of vertebrates, and some simpler organisms for comparison, again on a log-log plot. Note that the metabolic rate for most birds and mammals lies above that of most fishes and amphibians (and some reptiles). The upper dashed line has a slope of 1.00, meaning that the metabolic rate is directly proportional to body weight: An animal 10 times as big has a metabolic rate 10 times as high. The lower dashed line is near the lower limit of observed values for the exponent b in the equation $M = aW^b$. (Modified from K. Schmidt-Nielsen, 1979. *Animal Physiology.* 2nd ed. Cambridge University Press, Cambridge, England.)

observed relationship between metabolism and size is less than 1:1 (usually $b < 1$), however, each gram of body weight in a large animal is less costly than each gram of a small one. The harvest mouse may use 2.5 cm^3 O_2/g/h, the ground squirrel 1.03 cm^3 O_2/g/h, and the cat 0.68 cm^3 O_2/g/h (Fig. 1–9*a,b* are typical graphs of this kind of relationship). Consequently, small animals must eat more food and breathe more oxygen relative to their body size. The rate of delivery of oxygen and nutrients to the tissues must therefore be greater in small animals than in large ones, and blood supply to tissucs and the heart rate are correspondingly greater. The volume of the stomach and lungs of vertebrates is usually proportional to body weight, so that those organs in the ground squirrel might be about 10½ times as big as those of the mouse. The total amount of O_2 and food used per hour is only 4½ times as great for the larger animal, so the frequency at which lungs and stomach must be refilled is much less than that for the smaller animal. As an approximation, our model mouse must eat and breathe a little more than two times as often as the model ground squirrel (assuming, for simplicity, that the animals are similar in all respects but size). Given this situation, clearly there must exist a lower size limit for vertebrates. If a vertebrate is "too small," it must spend all its time eating (which is biologically impossible) and its food must be exceeding abundant (which is biologically improbable). The smallest known bird and mammal weigh about 2 g each.

Physiologists do not completely understand the reason for the seeming consistency of the correlation between body size and metabolic rate. Animals vary enormously in surface area, insulation, fat content, behavior, and proportion of metabolically less active tissues such as bone, and many such factors combine to determine metabolic rates. Nevertheless, the general correlation is often useful in predicting metabolic rates for animals of known size. Deviations from the predicted values are interesting because they suggest adaptive differences in metabolism. Some of these differences are mentioned later in this text.

Figure 1–9 Two graphs of metabolic rate *per unit* body weight. (*a*) This graph is a semilogarithmic plot (only the horizontal axis has log scale; the vertical axis is arithmetic) of the per gram metabolic rate of mammals. (Modified from K. Schmidt-Nielsen, 1979. *Animal Physiology*. 2nd ed. Cambridge University Press, Cambridge, England.) (*b*) This graph presents a log-log plot of per gram metabolic rate of lizards at 30°C. The slope is steeper and the weight specific rate is less than for mammals. Also, the logarithmic transformation has straightened the line, making comparisons much easier. Code: *Uta* (2, 5); *Sceloporus* (1, 7); *Crotaphytus* (10); *Dipsosaurus* (12); *Iguana* (21); *Amphibolurus* (18); *Uromastix* (22); *Eumeces* (11); *Tiliqua* (17, 19); *Lacerta* (3, 4, 6, 9, 14); *Varanus* (8, 13, 15, 16, 20, 23). (Modified from G. Bartholomew and V.A. Tucker, 1964. Size, body temperature, thermal conductance, oxygen consumption and heart rate in Australian varanid lizards. *Physiol. Zool.*, *37*:341–354.)

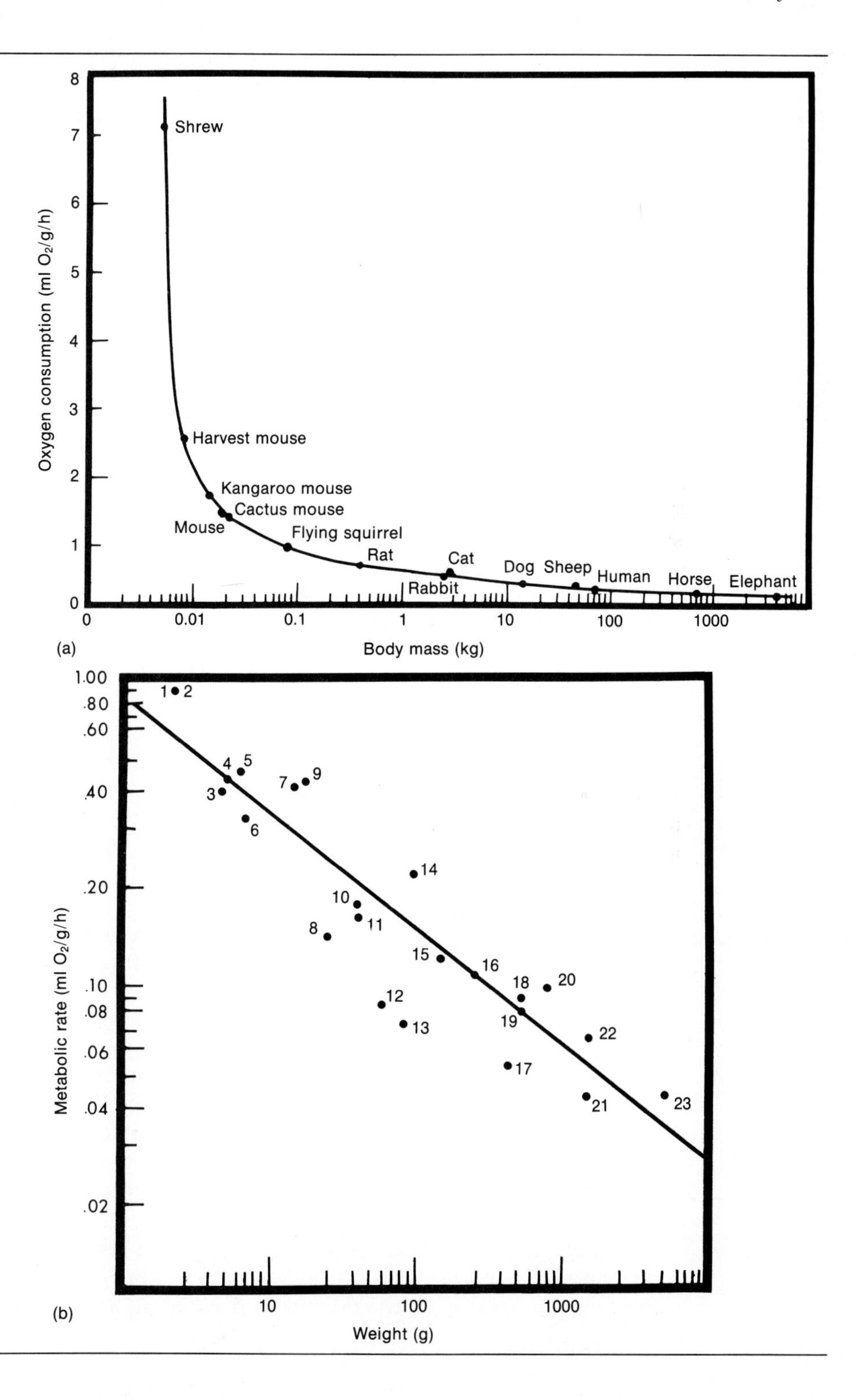
Oxygen consumption (ml O2/g/h)
8
7
6
5
4
3
2
1
0
Shrew
Harvest mouse
Kangaroo mouse
Cactus mouse
Mouse
Flying squirrel
Rat
Cat
Rabbit
Dog
Sheep
Human
Horse
Elephant
0
0.01
0.1
1
10
100
1000
(a)
Body mass (kg)
Metabolic rate (ml O2/g/h)
1.00
.80
.60
.40
.20
.10
.08
.06
.04
.02
1
2
3
4
5
6
7
8
9
10
11
12
13
14
15
16
17
18
19
20
21
22
23
10
100
1000
(b)
Weight (g)

The discussion so far has dealt with metabolism as if it had a single value for a given creature. However, metabolism obviously must vary with an animal's level of activity. An animal at minimal metabolism does not eat or move or grow or breed but only keeps its physiological machinery going. Metabolic scope is the energy available to an animal for engaging in activities of any sort; it is measured by the difference between minimum and maximum metabolism under given conditions. Metabolic scope varies with body size: a 5-g salmon (*Oncorhynchus*) can increase its metabolism less than fivefold at maximum activity, but a 2.5 kg individual can produce an increase of more than sixteenfold. Larger salmon, therefore, should be capable of more intense and more sustained activity than are smaller ones.

Size and Proportion

Different kinds of animals seldom, if ever, differ in size alone; they also differ in relative size or proportions of different body parts. No animals of mouselike proportions have achieved the size of an elephant, and, conversely, no animals of elephantine proportions are the size of a mouse. This is true because of specific, incontrovertible, geometric and physical relationships. In fact, changes in size often necessitate changes in shape and proportion. Changes in shape as a function of changes in size constitute the relationship of allometry.

The geometric principle at the foundation of allometric relationships is the way in which surface area and volume change with changes in absolute size. Imagine two cubes, one with edges four times as long as the other. The surface area of the larger cube is then 16 times greater than that of the small one, and the volume is 64 times greater (Fig. 1–10*a*). The linear dimensions are in the ratio of $1:L$ (where L is the length of one side, which is analogous to body length). The surface areas are in the ratio of $1:L^2$, and the volumes are in the ratio of $1:L^3$. If the cubes are constructed of the same materials, the weights are in the same ratios as the volumes. If we started with measures of volume or weight (W), then the linear dimension can be calculated by $W^{1/3}$ and surface areas by $W^{2/3}$ (that is, the square of the linear dimensions).

As demonstration of the biological problems imposed by this relationship, imagine two terrestrial animals of similar proportions but different sizes. A basic problem concerns the ability of the skeleton to support the animal's weight. The ability of a bone to support a load is proportional to its cross-sectional area. Therefore, if one of our imaginary animals were 4 times as tall as the other, its limbs could support 16 times (that is, L^2) the weight. However, the weight that needs to be supported is 64 times (that is, L^3) greater in the larger beast than in the smaller one. Obviously, this relationship places a limit on the maximum size that can be achieved by a terrestrial vertebrate. It also implies that increases in size necessitate disproportionate increases in bone diameter if the skeleton is to support the

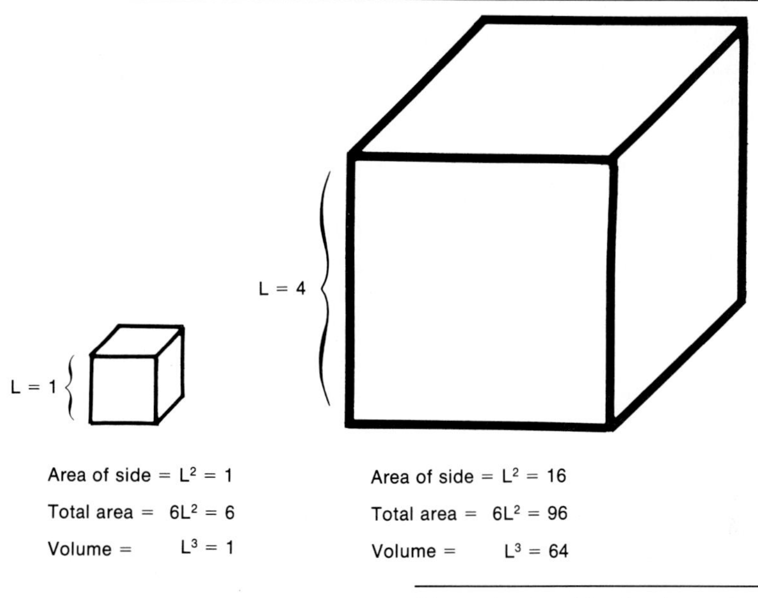

(a)

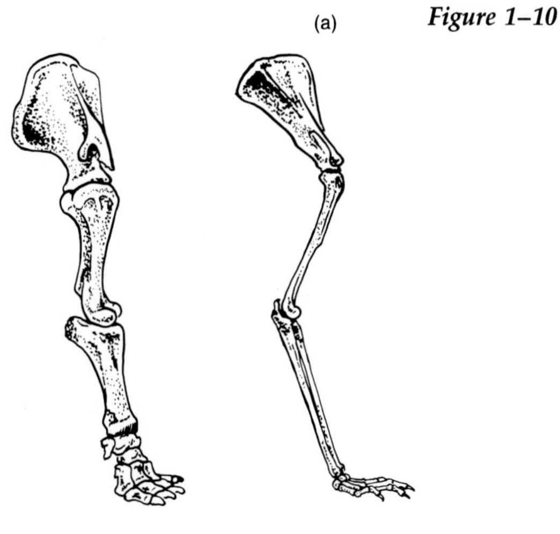

(b)

Figure 1–10 (*a*) Surface:volume relationships illustrated by cubes. (*b*) Legs of an Indian elephant (*Elephas maximus*) and a deer mouse (*Peromyscus maniculatus*) drawn to the same length. Note the great differences in diameter necessitated by differences in body size. Elephants may reach weights of 5900 kg, and deer mice about 20 g. (Elephant from R. Owen, F.R.S., 1886–1888. *On the Anatomy of Vertebrates*. Vol. 2. *Birds and Mammals.* Longmans, Green, London.) (*c*) Change of shape maintaining equal volume but greatly increasing surface area.

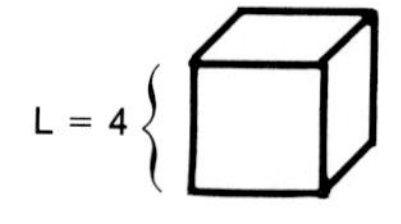

L = 64

L = 1

Area of side = $L^2 = 16$

Total area = $6L^2 = 96$

Volume = $L^3 = 64$

Area = $4(64 \times 1) + 2(1) = 258$

Volume = $64 \times 1 \times 1 = 64$

(c)

weight of the body. Consequently, elephant legs must be much stouter relative to their length than are the legs of mice (Fig. 1–10*b*), and identically shaped animals of different sizes are quite imaginary.

An example from real animals illustrates this relationship. Large mammals commonly have relatively larger teeth than smaller ones. This difference is clearly shown in vegetarian species such as horses or cattle that grind their plant food with large molar (cheek) teeth. The metabolic requirements vary approximately as the ¾-power of W or $(L^3)^{3/4} = L^{9/4}$, but the grinding surface of a tooth varies only as L^2. The rate of food preparation by the teeth thus lags behind metabolic requirements, providing selection for differentially enhanced tooth size in larger mammals. Large tooth size has been accomplished in part by expanding the grinding surface of molars and incorporating more anterior teeth (premolars) into the molar pattern or by increasing the height of the tooth crown (reducing the risk of wearing the tooth down to uselessness). Several researchers have inferred that increased body size thus necessitated a differential increase of tooth size in these mammals.

Surface area and volume can vary independently of each other only by means of changes in shape. Large cubes and small ones are still cubic in shape. If the larger cube were deformed into a new shape with the same volume as before, we could get a new shape that is 1-unit wide, 1-unit high, and 64-units long; the surface area would be much more than 16 times that of the small cube (Fig. 1–10*c*). The same principle works for animals: Surface area is increased relative to volume by changing compact shapes to thin or folded ones.

Because body size is related both to structure (proportions) and to function (metabolism), it is a critical determinant of many aspects of the natural history of a species. Size has important implications for temperature and water regulation (Chapter 5), locomotion (Chapter 7), feeding ecology (Chapter 9), social organization (Chapter 11), and certain aspects of reproduction (Chapter 14), to mention only some of the direct connections (see also Clutton-Brock and Harvey, 1979).

Allometry and Evolution

Large changes in size may bring with them large changes in shape. Shape changes may alter the functional properties of the animal, so selection on animal shape may restrict changes in size. If elephants got much larger, necessary increases in the thickness of the legs might drastically alter their function. If that functioning prevented the elephant from making a living, clearly selection would restrict such size increases; thus, the correlated changes in size and shape must be fundamentally advantageous if the allometric relationship is preserved. Allometry may be evident both during development of an individual and in phylogenetic series of related species.

Allometric relationships are usually described by equations of the form $Y = aX^b$ (where X is body size; Y is some dimension that indexes shape; a is a constant, as before; and b is the slope of the line). When $b = 1$, X and Y change at the same rate, and animals of different size are similar in form. If Y changes greatly for each small change of X ($b > 1$), changes in shape may very quickly limit changes in size. Gould (1966) predicted that animals possessing high b values during the growth of individuals should not grow very big—or else should show a change in b values at a certain body size.

Allometric changes in shape may sometimes necessitate still further modification of morphology to compensate for changes related to size. Gould noted, for example, that selection favored development of the elephant's trunk when allometric development of the second incisor teeth into tusks make direct acquisition of food by the lips impossible.

A phylogenetic or developmental change of size without a corresponding change of shape would probably necessitate radical changes in how the animal functions. Such changes in function are likely to be even more

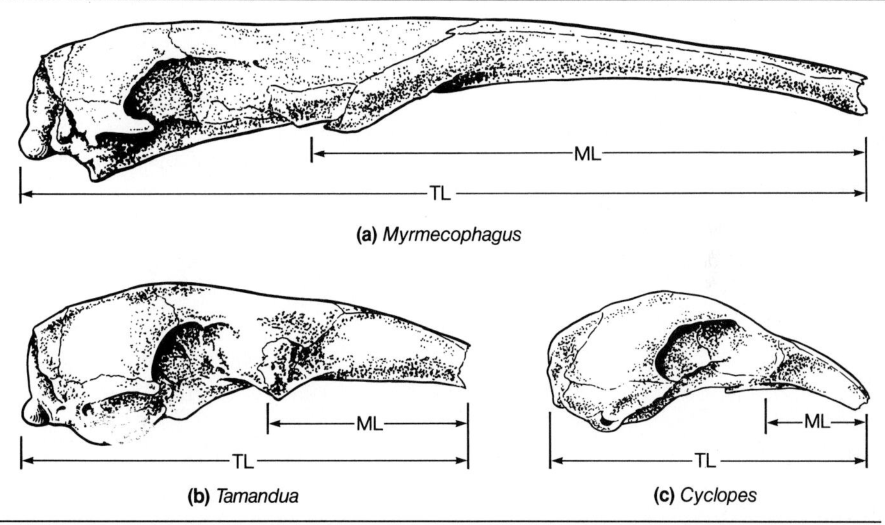

Figure 1–11 Side views of adult skulls of three species of anteaters. Note the relative length of the snout. (*a*) The giant anteater (*Myrmecophaga tridactyla*) is a ground-dweller of forest and savanna that weighs as much as 25 kg; an adult skull typically measures about 37 cm long. (*b*) The more arboreal tamanduas (two species of *Tamandua*) live in forests and are much smaller (about 6 kg); an adult skull is usually about 14 cm long. (*c*) The pygmy anteater (*Cyclopes didactylus*) is the smallest species—about 350 g, with a skull about 5 cm long—and the most arboreal and nocturnal of the three genera. (From Reeve, E.C.R., 1940. Relative growth in the snout of anteaters. *Proc. Zool. Soc. Lond.*, *100*:47–80. Reproduced by permission of The Zoological Society of London.)

fundamental than those entrained as a result of size change. A mouse and an elephant, although different in size and proportion, are functionally more similar than a real mouse and an imaginary animal of mouselike shape but elephantine size that is incapable of four-legged locomotion.

One message to be derived from the study of allometry is that seemingly enormous changes in proportion may result originally from selection for changes in size. Hence, animals of quite different geometry may, in fact, be not too distantly related. For example, anteaters of the New World tropics were once divided into two subfamilies largely on the basis of relative proportions of the snout. The large-bodied genera, *Tamandua* and *Myrmecophaga*, have a pronounced snout, as well as relatively small brains and eyes, but the smaller *Cyclopes* has a more "normal" face and relatively larger brain and eyes (Fig. 1–11). In all these genera, there is a strong positive allometric relationship between snout size and body size and a negative one for brain size and eye size. Therefore, these differences among genera are fundamentally a result of changes in size only. (The ecological factors that selected for size differences, although undoubtedly important, are not at issue here.) As a result, a subfamilial distinction hardly seemed justified.

SELECTED REFERENCES

Alexander, R.D., 1975. The search for a general theory of behavior. *Behav. Sci. 20*:77–100.

Alexander, R.M., 1971. *Size and Shape.* Arnold, London.

Amadon, D., 1950. The Hawaiian honeycreepers (Aves: Drepaniidae). *Bull. Am. Mus. Nat. Hist., 95*:151–262.

Bartholomew, G.A., and V.A. Tucker, 1964. Size, body temperature, thermal conductance, oxygen consumption, or heart rate in Australian varanid lizards. *Physiol. Zool., 37*:341–354.

Bumpus, H.C., 1898. The elimination of the unfit as illustrated by the introduced sparrow, *Passer domesticus. Biology Lectures* (11). pp. 209–226. Marine Biology Laboratory, Woods Hole, MA.

Camin, J.H., and P.R. Ehrlich, 1958. Natural selection in water snakes (*Natrix sipedon* L.) on islands in Lake Erie. *Evol., 12*:504–511.

Clutton-Brock, T.H., and P.H. Harvey, 1979. Comparison and adaptation. *Proc. R. Soc. Lond., B205*:547–565.

Cody, M.L., 1974. Optimization in ecology. *Science, 183*:1155–1164.

Cook, L.M., 1971. *Coefficients of Natural Selection.* Hutchinson, London.

Dawkins, R., 1976. *The Selfish Gene.* Oxford University Press, Oxford.

Ehrlich, P.R., and R.W. Holm, 1963. *The Process of Evolution.* McGraw-Hill Book Co., New York.

Ferris, S.D., and G.S. Whitt, 1977. Duplicate gene expression in diploid and tetraploid loaches (Cypriniformes, Cobitidae). *Biochem. Genet., 15*:1097–1112.

Frazzetta, T.H., 1975. *Complex Adaptations in Evolving Populations.* Sinauer Associates, Sunderland, MA.

Futuyma, D.J., 1979. *Evolutionary Biology.* Sinauer Associates, Sunderland, MA.

Gordon, M.S., 1977. *Animal Physiology: Principles and Adaptations.* 3rd ed. Macmillan Publishing Co., New York.

Gould, S.J., 1966. Allometry and size in ontogeny and phylogeny. *Biol. Rev., 41*:587–640.

Gould, S.J., 1978. The panda's peculiar thumb. *Nat. Hist., 87*(9):20–30.

Grant, P.R., 1972. Centripetal selection and the house sparrow. *Syst. Zool., 21*:23–30.

Grant, P.R., 1981. Speciation and the adaptive radiation of Darwin's finches. *Am. Sci.*, *69*:653–663.

Hagen, D.W., and G.E.E. Moodie, 1979. Polymorphism for breeding colors in *Gasterosteus aculeatus*. I. Their genetics and geographic distribution. *Evol.*, *33*:641–648.

Hagen, D.W., G.E.E. Moodie, and P.F. Moodie, 1980. Polymorphism for breeding colors in *Gasterosteus aculeatus*. II. Reproductive success as a result of convergence for threat display. *Evol.*, *34*:1050–1059.

Janis, C., 1976. The evolutionary strategy of the Equidae and the origins of rumen and cecal digestion. *Evol.*, *30*:757–774.

Johnson, C., 1976. *Introduction to Natural Selection*. University Park Press, Baltimore.

Johnston, R.F., D.M. Niles, and S.A. Rohwer, 1972. Hermon Bumpus and natural selection in the house sparrow *Passer domesticus*. *Evol.*, *26*:20–31.

Kleiber, M., 1932. Body size and metabolism. *Hilgardia*, *6*:315–353.

Kurland, J.A., 1980. Kin selection theory: A review and selective bibliography. *Ethol. Sociobiol.*, *1*:255–274.

Liem, K.F., 1973. Evolutionary strategies and morphological innovations: Cichlid pharyngeal jaws. *Syst. Zool.*, *22*:425–441.

Mayr, E., 1963. *Animal Species and Evolution*. Belknap Press, Cambridge, MA.

O'Donald, P., 1973. A further analysis of Bumpus' data: The intensity of natural selection. *Evol.*, *27*:398–404.

Ohno, S., 1970. *Evolution by Gene Duplication*. Springer-Verlag, New York.

Olson, S.L., and H.F. James, 1982. Fossil birds from the Hawaiian Islands: evidence for wholesale extinction by man before Western contact. *Science*, *217*:633–635.

Owen, R., 1886–1888. *On the Anatomy of Vertebrates*. 3 vols. Longmans, Green, London.

Raikow, R.J., 1974. Species-specific foraging behavior in some Hawaiian honeycreepers. *Wils. Bull.*, *86*:471–474.

Raikow, R.J., 1976. The origin and evolution of the Hawaiian honeycreepers (Drepanididae). *Living Bird*, *15*:95–117.

Reeve, E.C.R., 1940. Relative growth in the snout of anteaters. A study in the application of quantitative methods to systematics. *Proc. Zool. Soc. Lond.*, *110*:47–80.

Richards, L.P., and W.J. Bock, 1973. Function, anatomy, and adaptive evolution of the feeding apparatus in the Hawaiian honeycreeper genus *Loxops* (Drepanididae). *Ornith. Monogr.* *15*:1–173.

Ricklefs, R.E., 1973. *Ecology*. Chiron Press, Newton, MA.

Schartz, R.L., and J.C. Zimmerman, 1971. The time and energy budget of the male dickcissel (*Spiza americana*). *Condor*, *73*:65–76.

Schmidt-Nielsen, K., 1979. *Animal Physiology*. 2nd ed. Cambridge University Press, Cambridge, England.

Semler, D.E., 1971. Some aspects of adaptation in a polymorphism for breeding colors in the three-spine stickleback (*Gasterosteus aculeatus*). *J. Zool.* (Lond.), *165*:291–302.

Stebbins, G.L., and F.J. Ayala, 1981. Is a new evolutionary synthesis necessary? *Science*, *213*:967–971.

Vermeij, G.J., 1973. Adaptation, versatility, and evolution. *Syst. Zool.*, *22*:466–477.

Williams, G.C., 1966. *Adaptation and Natural Selection*. Princeton University Press, Princeton, NJ.

Wilson, E.O., and W.H. Bossert, 1971. *A Primer of Population Biology*. Sinauer Associates, Sunderland, MA.

2 *Classification, Characteristics, and Relationships*

THE PHYLUM CHORDATA

Vertebrates form one subphylum of the phylum Chordata. The phylum receives its name from the presence of a notochord ("back cord"), a supportive rod running along the dorsal side of the organism. The notochord of certain chordates is present only as a transient structure during development, but vestiges of the notochord persist in the adult of many species.

Chordates also possess a dorsal, hollow nerve cord, present at all states of development beyond the early embryo. Another characteristic is the sometimes transitory presence of gill slits (opening from the outside into a pharynx) or of dead-end pharyngeal pouches that are the phylogenetic vestiges of the gill slits. Gill slits are typically retained through adulthood by the fishes, but, in other vertebrates, they have been transformed phylogenetically into other adult structures, such as the thymus and parathyroid glands. A ventral heart or heartlike organ in chordates pumps blood through a "closed" circulatory system; that is, the blood is largely contained in blood vessels and does not slosh freely about in tissue spaces.

Chordates are segmented animals, as are many invertebrates, although the segmentation is not always readily apparent in the adult. In all segmented animals, many organs or systems, such as axial (trunk) musculature and associated nerves, are repeated along the length of the body. With several other phyla, the chordates share the characteristics of a complete digestive tract with two openings to the outside, bilateral symmetry (right and left, dorsal and ventral sides), and three embryonic germ layers (ectoderm, mesoderm, endoderm).

Two subphyla of nonvertebrate chordates include the Cephalochordata, represented by amphioxus (*Branchiostoma*), and the Urochordata. Amphioxus exhibits many of the general chordate characteristics (Fig. 2–1). The ancestors of the Urochordata or tunicates may well have been the ancestors of the vertebrates as well; the body plan of larval tunicates suggests a link between the groups. Another group called "Hemichordata," or acorn worms, are probably best omitted from the Chordata, because the structure once thought to be a notochord is probably not.

THE SUBPHYLUM VERTEBRATA

All members of the subphylum Vertebrata exhibit some degree of cephalization: they possess a head, in which the major portions of the central nervous system and a number of special sense organs are concentrated and

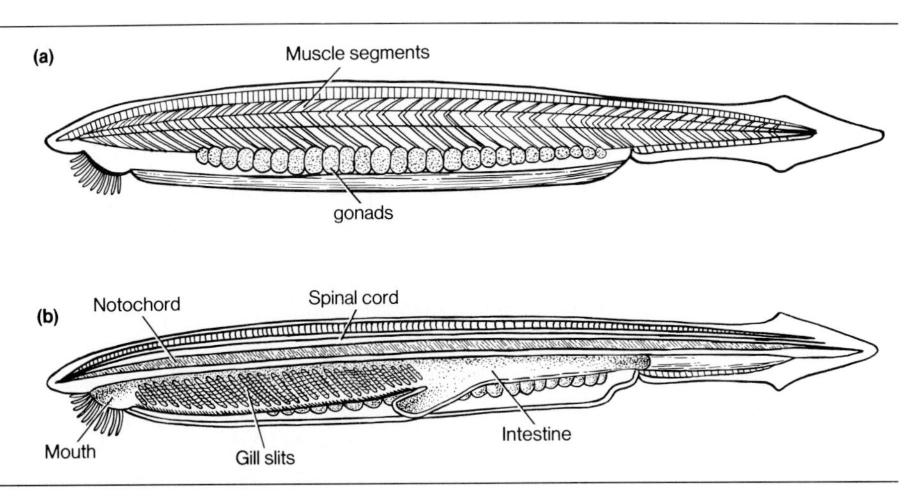

Figure 2–1 Amphioxus (or lancelet). There are about 23 species of lancelets, all marine, shallow-water, bottom-dwellers that filter tiny organisms from the water. All are small—up to about 9 cm long. This diagram of *Branchiostoma lanceolatus* exhibits several fundamental chordate characters. (After W.K. Gregory, 1951. *Evolution Emerging.* 2 vols. Macmillan Publishing Co., New York. Courtesy of The American Museum of Natural History.)

protected by a usually bony cranium. The spinal cord is paralleled and often surrounded by a chain of vertebrae that are the diagnostic feature of the group. The vertebral column is developed to different degrees in different groups of vertebrates; in some, such as the hagfishes, it is hardly distinguishable at all. Vertebrates are fundamentally sexual reproducers, although a few species are secondarily parthenogenetic (of virgin birth) and only females have been observed.

This group of chordates is very diversified and has over 40,000 species. Compared to the several hundred thousand species of insects, this is a pretty poor showing, but our biological understanding of vertebrates exceeds that of any other major taxon. Many groups of vertebrates have become extinct in past ages and their natural history is little known. Although fascinating tales can be told regarding the extinct types (anatomical specializations, phylogenetic relationships to other groups, and ecological surmises, for example), any exploration of that past world in depth must be another enterprise. In this chapter, a synopsis of vertebrate phylogeny follows the survey of present-day vertebrate groups. The rest of this section presents each of the modern vertebrate classes and introduces some basic features of their structure, life history, and classification.

Vertebrates are classified in seven classes: Agnatha, Chondrichthyes, Osteichthyes, Amphibia, Reptilia, Aves, and Mammalia. The classes of vertebrates can be grouped, somewhat informally, in several ways (Table 2–1). Sometimes the Agnatha ("without jaws") are segregated from all the rest, which are termed "Gnathostomata" ("jawed mouth"). The extant "fishy" forms—the Agnatha, Chondrichthyes, and Osteichthyes—may be clustered as a superclass Pisces, and the remaining classes—Amphibia, Reptilia, Aves, Mammalia—assembled as a superclass Tetrapoda ("four footed"). The Reptilia, Aves, and Mammalia can be distinguished from the rest by the presence of a particular membrane, the amnion, surrounding the developing embryo. These three classes together are then referred to as the Amniota and the remainder are known as the Anamniota ("without amnion"). There is nothing precedential about any of these schemes; each can be convenient for various purposes.

CLASS AGNATHA The group of vertebrates most closely linked to the invertebrate chordates is the jawless class Agnatha. Of these, the only extant subclass, sometimes called an order, is the Cyclostomata ("round mouth"), containing two distantly related groups: the lampreys (order Petromyzontiformes) and hagfishes (order Myxiniformes). There are only about 60 species of cyclostomes. They have an elongate body form, up to 90 cm in length, median but no paired fins, and no scales (Fig. 2–2). The vertebral column is represented by some cartilaginous elements next to the notochord. The rounded mouth of lampreys is a sucking organ with which the predatory species attach to other fishes. A rasping tongue scrapes at the

Table 2–1 Major groupings of vertebrate classes.

	Class of vertebrates						
Grouping	Agnatha	Chondrichthyes	Osteichthyes	Amphibia	Reptilia	Aves	Mammalia
Agnatha	+						
Gnathostomata		+	+	+	+	+	+
Pisces	+	+	+				
Tetrapoda				+	+	+	+
Anamniota	+	+	+	+			
Amniota					+	+	+

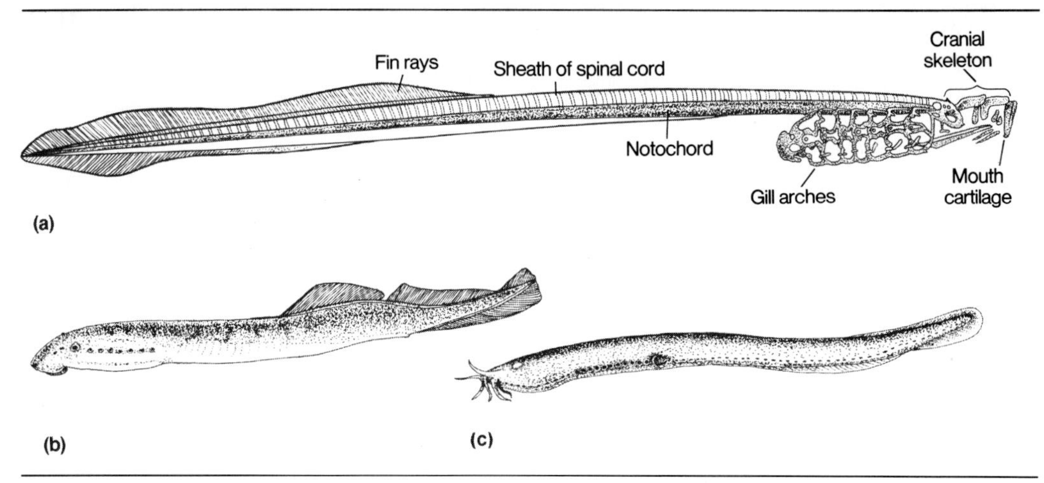

Figure 2–2 (*a*) The much-reduced skeleton of a lamprey and an exterior view of (*b*) a lamprey (*Lampetra planeri*) and (*c*) a hagfish (*Eptatretus stouti*). (From B. Dean, 1895. *Fishes Living and Fossil.* Macmillan Publishing Co., New York.)

flesh of the prey, and the lamprey sucks out the body fluids of the living prey. Some small-bodied species are not predaceous; the adults do not feed at all. Hagfishes both scavenge and burrow into dead or dying fishes, and their mouths are ringed with small tentacles instead of a sucker.

Hagfishes are entirely marine, but lampreys occur in both fresh and salt water. Adult marine lampreys migrate to freshwater rivers to breed and then die. Fertilization of the eggs is external. Eggs hatch to produce larvae that mature in fresh water, live in the bottom mud, and feed on small algae and bacteria. After several years, they undergo metamorphosis (change in form) to adults and migrate out to sea.

CLASS CHONDRICHTHYES The Chondrichthyes are the so-called cartilaginous fishes, their name referring to the absence of bone in their skeleton (Fig. 2–3). The subclass Elasmobranchii ("plate gills"), sometimes redundantly called the Selachii or cartilaginous fishes, contains the sharks, skates, and rays. Estimates of number of species range up to 800 or so. They have five to seven gill slits that open independently to the outside. The tail fin is typically asymmetrical, the spine continuing into the upper lobe, which is then larger than the lower lobe. This condition is called "heterocercal." The dorsal fins, when present, are rigid, and small scales are embedded in the skin. Most sharks are predaceous, attacking other vertebrates with a battery of effective teeth, but some, including the biggest of all, are plankton strainers. The largest is the whale shark (*Rhincodon typus*), which may attain a length of over 15 m and weigh well over 9000 kg, but the smallest, *Squaliolus laticaudus*, reaches an adult size of only about 18 cm. Skates and rays are mostly bottom-living feeders on mollusks or other inactive prey. Most breathe by inhaling water through a reduced gill slit called a spiracle,

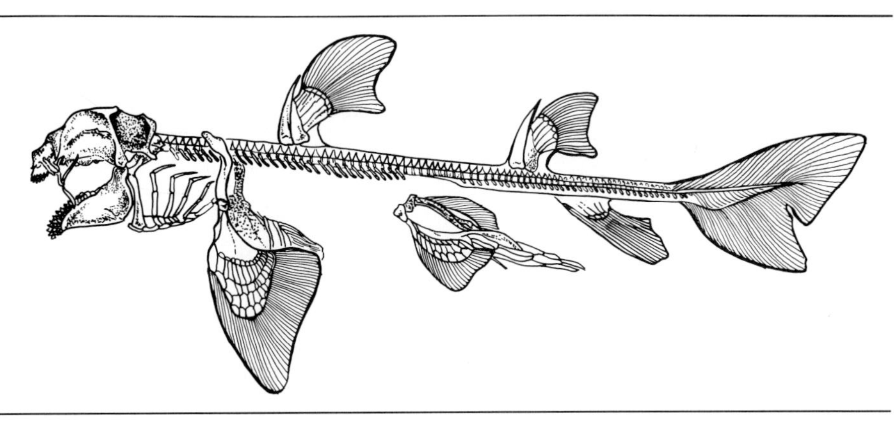

Figure 2–3 The cartilaginous skeleton of a shark (*Heterodontus* sp.). (From B. Dean, 1895. *Fishes Living and Fossil.* Macmillan Publishing Co., New York.)

in contrast to other fishes, most of which are mouth-breathers. Almost all elasmobranchs are marine, but a few live in fresh water. Modern elasmobranchs are often divided into two orders: the Pleurotremata ("lateral gills"), or Selachimorpha, contains the sharks; the Hypotremata ("ventral gills"), or Batoidimorpha, contains the rays and skates.

The subclass Holocephali ("entire head") is also known as Bradyodonti ("dull tooth") and includes the ratfishes or chimaeras, all belonging to the order Chimaeriformes. Holocephalians are exclusively marine and primarily mollusk-eaters. They have an operculum, or gill cover, over the opening of the gill slits, and a long skinny tail; the adult has no scales.

Some classificatory systems use the elasmobranch orders given above as informal subdivisions, thus raising many suborders to full ordinal status (Fig. 2–4). By one such system, there are four orders of sharklike forms. The Hexanchiformes are the frilled and cow sharks; they have six or seven gill openings and a single dorsal fin. Galeiformes, or Lamniformes, contains the majority of the sharks, such as the sand sharks, porbeagles, threshers, basking shark, bull sharks, smooth dogfishes, carpet and nurse sharks, and hammerheads. They have five gill openings and usually two dorsal fins. Some of the bull sharks live in fresh water (in the Ganges River in India and Pakistan, the Zambezi River in Africa, and Lake Nicaragua in Central America). The Squaliformes are the monk fishes and spiny dogfishes and their relations; they usually have five gill slits, two dorsal fins, each usually with an anterior spine in contrast to the Galeiformes, and no anal fins, again in contrast to the Galeiformes. The Heterodontiformes are small mollusk-eating species of the Pacific and Indian Oceans, such as the bullhead and Port Jackson shark, that have five gill slits. In contrast to the other orders, which are primarily deep-water species, this group inhabits inshore waters.

The hypotremes are commonly placed in one order, alternatively called

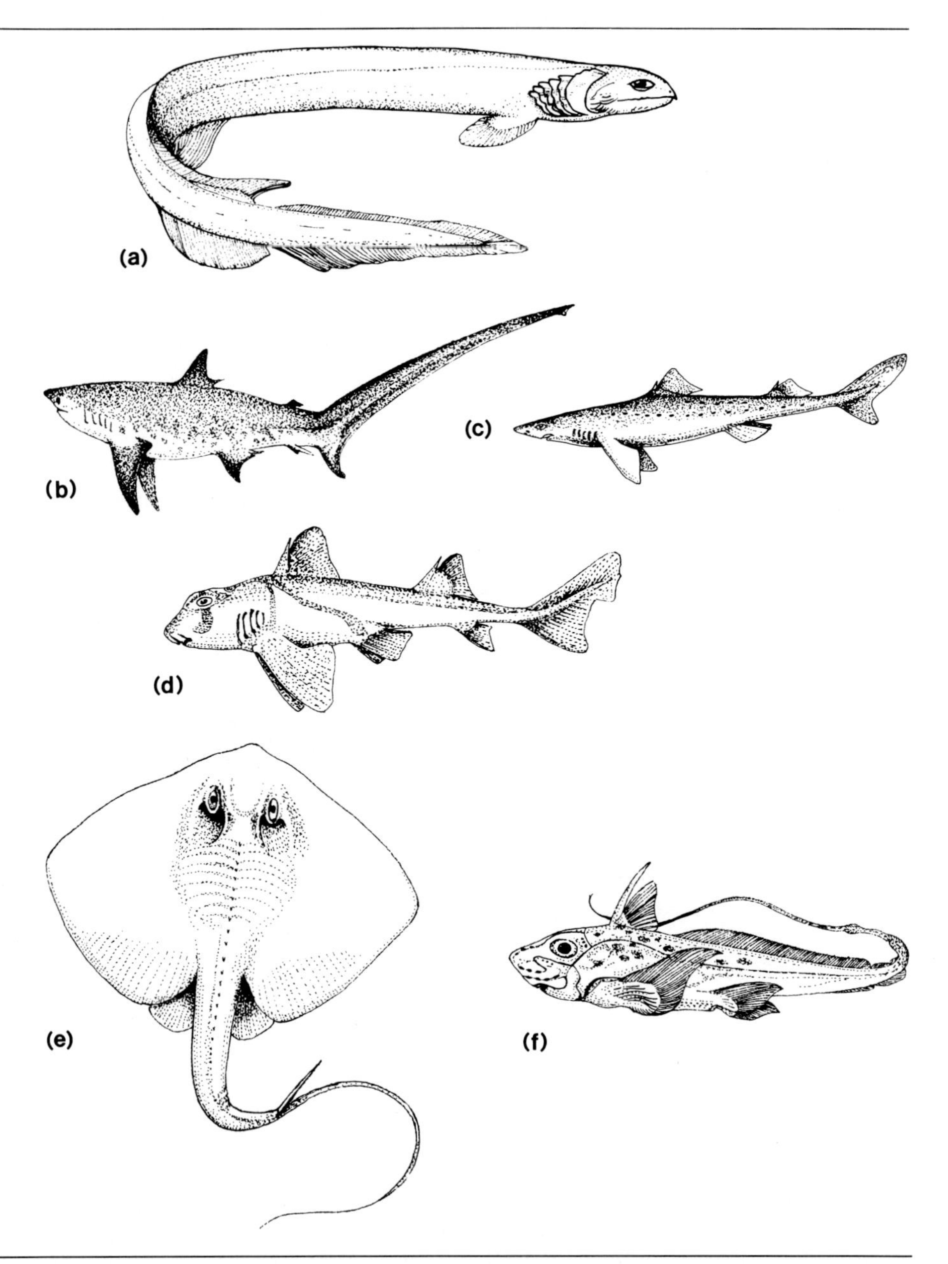

Figure 2–4 Exteriors of representatives of six orders of Chondrichthyes: (*a*) Hexanchiformes: a frilled shark, *Chlamydoselachus aguineus;* (*b*) Galeiformes: a thresher shark, *Alopias vulpina;* (*c*) Squaliformes: a spiny dogfish, *Squalus acanthius;* (*d*) Heterodontiformes: the Port Jackson shark, *Heterodontus portusjacksoni;* (*e*) Rajiformes: a stingray, *Dasyatis sabina;* and (*f*) Chimaeriformes: a ratfish, *Chimaera.*

the Rajiformes. The rays (some with electric organs in the tail), skates, stingrays, sawfishes, and their relatives form one subdivision; the torpedo rays (with electric organs in the pectoral region) form another. One family of rays (Potamotrygonidae) and some members of other families live in fresh water.

The purpose of presenting both classifications for the elasmobranchs is to illustrate the flexibility of taxonomic hierarchical arrangements. Moreover, the use of four orders for sharks emphasizes the considerable, but often neglected, diversity of this group of fishes.

Members of the Chondrichthyes possess internal fertilization, and parts of the male's pelvic fins are modified into an intromittent organ that guides sperm inside the body of the female. In most species, eggs are encased in a horny shell and are shed externally. The females of some sharks and rays, however, retain eggs in the reproductive tract, and the entire embryonic development occurs there; the young are then born alive.

Cartilaginous fishes and the bony fishes discussed later are the first living groups to exhibit paired true appendages: the pelvic and pectoral fins. Some fossil fishes called "ostracoderms" had flaps or folds on either side of the body that presumably acted as stabilizers. The advent of paired appendages supported by skeletal elements occurred in the ancestors of the cartilaginous and bony fishes and was retained in most representatives of all the later vertebrate classes. Elaboration of pectoral and pelvic appendages may have been associated with the development of active predatory habits that could be more efficiently and accurately executed if forward movement were carefully controlled by paired limbs.

CLASS OSTEICHTHYES The Osteichthyes, or bony fishes, are the largest and most diversified group of vertebrates; there are over 20,000 species. Although the degree of ossification of the skeleton varies greatly among the taxa, virtually all bony fishes possess both median and paired fins. Gill slits are characteristically covered by opercula, and almost all species are covered with scales. Most species are fully aquatic, but some are able to breathe air and live out of water for considerable periods of time. External fertilization is typical of the group, but some species have internal fertilization and give birth to living young. The remarkable variety of body forms in this class is related to the diversity of feeding habits and life-styles that bony fishes have been able to exploit. Bony fishes are found in watery habitats of all sorts and at almost all depths. Some are active predators, some are parasitic, others are filter feeders or grazers. Particular problems and adaptations associated with an aquatic existence are discussed in later chapters.

Classification of bony fishes is subject to many different rearrangements. Some systematists use over 40 orders, others perhaps as few as 25. Systematists and anatomists do not even agree altogether on the types of fishes to be included in this class. The hierarchy used here is convenient, but it will undoubtedly be subject to rearrangement—there seem to be almost as many ways to organize the fishes as there are fish systematists. What follows is based principally on Nelson (1976).

The subclass Sarcopterygii ("fleshy finned") includes lungfishes and crossopterygians, which are sometimes placed in separate subclasses (Fig. 2–5). Lungfishes (order Dipnoi) are represented by three extant genera in the tropics: one in Australia (*Neoceratodus forsteri*) and two more closely related genera in South America (*Lepidosiren paradoxa*) and Africa (*Protopterus*). These two are so distantly related to the first that Nelson (1976) puts them in two different orders. They eat invertebrates and plant material. The skeleton is little ossified and consists largely of cartilage. All three forms have lungs and nostrils that open into the mouth; they can breathe air, and two species can burrow into mud in the dry season when water disappears completely and live there for weeks until the wet season. The African lungfish are totally dependent on air breathing and will drown if held underwater.

The crossopterygians (order Crossopterygii or Coelacanthiformes) were a common type of fish millions of years ago in the Devonian period. They were thought to be extinct until the late 1930s, when a single poorly preserved specimen, called *Latimeria chalumnae,* reached the hands of biologists. Since then a number of other specimens have been obtained from the

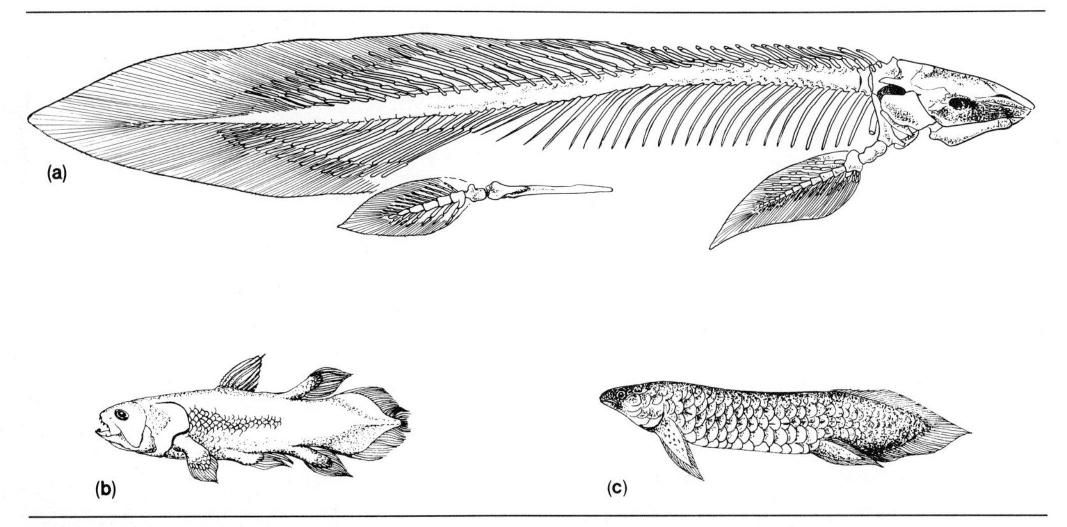

Figure 2–5 (*a*) Skeleton of a crossopterygian for comparison with that of bony fishes and amphibians. (*b*) Body form of the modern crossopt, *Latimeria.* (*c*) Body form of a modern lungfish, *Neoceratodus forsteri.* (From B. Dean, 1895. *Fishes Living and Fossil.* Macmillan Publishing Co., New York.)

Indian Ocean off the east coast of Africa. *Latimeria* belongs to a suborder called the coelacanths; it is a predator of other fishes, measures about 1.5-m long, and weighs up to 82 kg. The skeleton is mainly cartilage, and a lunglike structure contains fat rather than gases. They are not very closely related to dipnoans and may deserve to be separated from them. If this is done, the formal rubric of sarcopterygian is replaced by two subclasses: Dipneusti (lungfishes) and Crossopterygii.

The bichirs (order Polypteriformes) of central Africa, are sometimes placed with the sarcopterygians, sometimes with the actinopterygians (see following), and sometimes in a group by themselves (Brachiopterygii). They have lungs and slightly fleshy fins, but, in other respects, are like more typical ray-finned fishes.

The subclass Actinopterygii, or ray-finned fishes, today comprises the majority and greatest diversity of bony fishes. They have no internal nostrils, few have fleshy fins, and fins supported by horny rays are characteristic of the group. The names of the major divisions of this subclass are holdovers from an earlier classification scheme and may not be especially meaningful, but they nevertheless provide a convenient system for ordering the species .The superorder (or infraclass) Chondrostei ("cartilage bone") is presently represented by the sturgeons and paddlefishes (order Acipenseriformes), which have largely cartilaginous skeletons and heterocercal tails; these bottom-dwellers scavenge food or strain it from the water (Fig. 2–6).

Holostei ("complete bone") are represented only by the modern freshwater garpikes (order Lepisosteiformes or Semionotiformes) and bowfin (order Amiiformes) of North America, which may not be very closely related. The skeleton is well ossified, the tail is somewhat heterocercal, and all species are predaceous.

The third and last extant superorder, the Teleostei ("perfect bone"), are the putative ultimate in piscine evolution and the most diversified of all vertebrate groups. The tail fin is usually symmetrical or homocercal in appearance, in contrast to most other fishes (Fig. 2–7). Ossification of the skeleton is extensive. Among the largest bony fishes is the ocean sunfish, *Mola mola,* which reaches a weight of 900 kg. On the other hand, the smallest known adult teleost is *Pandaka pygmaea,* a goby of maximum length just over 1 cm. Many adult teleosts are small in body size (Miller, 1979), and the average length is much less than that of sharks or other fishes.

In the following classification scheme, the teleostean fishes are presented in a simplified fashion. Some systematists list the orders in groups supposedly representing phylogenetic lineages; others arrange them on gradients from "primitive" to "advanced" (but argue about which taxa are primitive). We will ignore these controversies here. Details of internal anatomy are usually used to distinguish the orders: jaw structure, ear-bone arrangement, vertebral modifications, and so on. These features are not

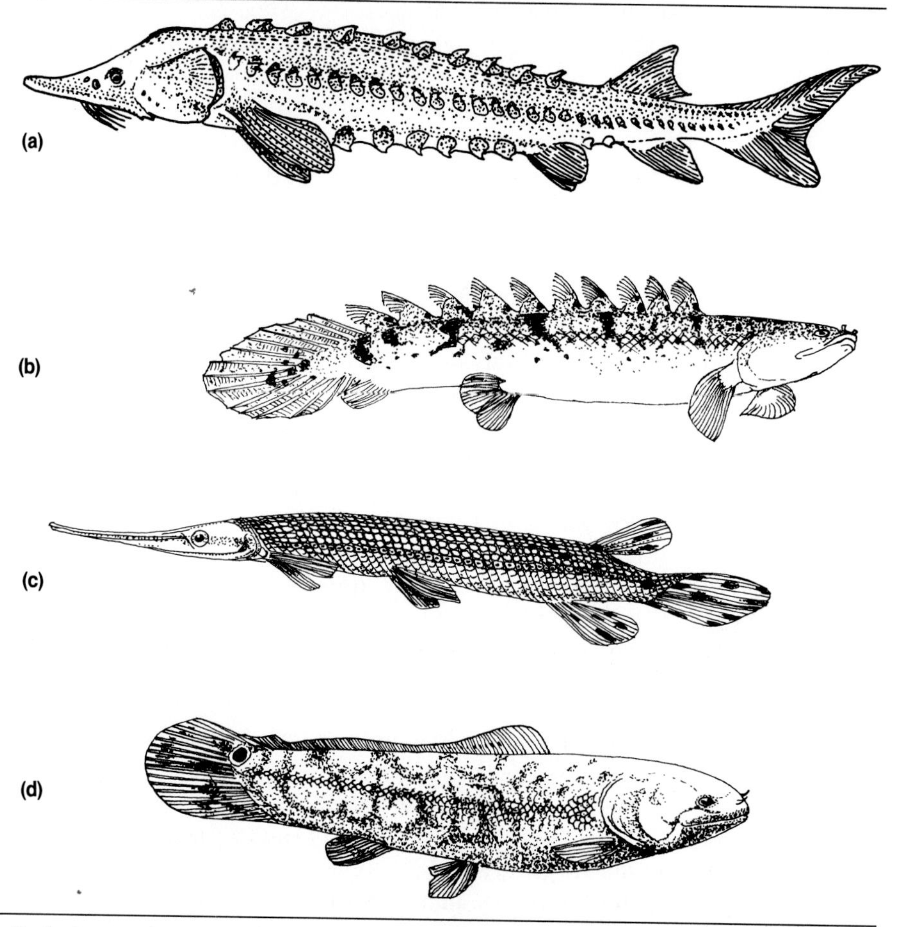

Figure 2–6 Body forms of some modern bony fishes. Chondrostei: (*a*) Acipenseriformes, a sturgeon, *Acipenser fulvescens;* (*b*) Polypteriformes, a bichir, *Polypterus delhezi;* Holostei: (*c*) Lepisosteiformes, a gar, *Lepisosteus osseus;* and (*d*) Amiiformes, the bowfin, *Amia calva.*

generally discussed in this chapter; instead, occasional mention is made of anatomical or ecological features that relate to discussions in later chapters. Examples of representatives not pictured elsewhere are found in Figures 2–8 and 2–9. The following paragraphs demarcate clusters of possibly related groups (Nelson, 1976).

The small order Osteoglossiformes includes freshwater mooneyes and bony tongues. The related elephant fishes and single species of gymnarchid constitute a separate order, Mormyriformes, many species of which are nocturnal freshwater bottom-feeders with electric organs.

The order Clupeiformes is composed of the herrings, anchovies, wolf

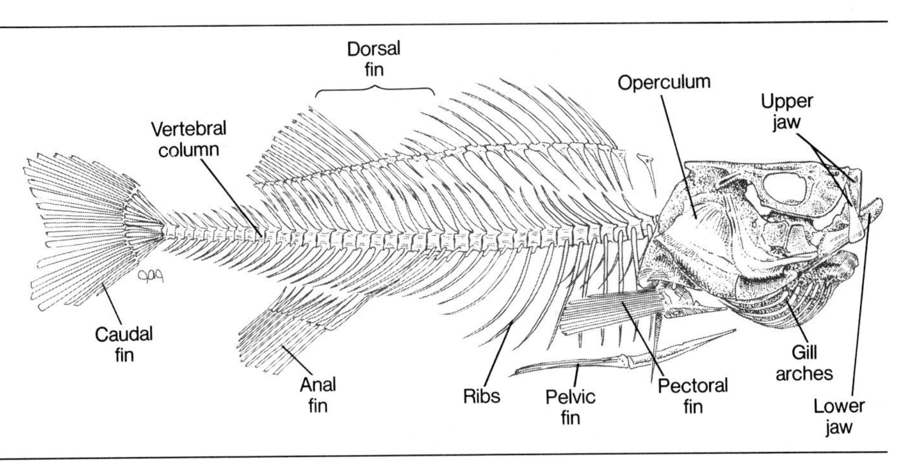

Figure 2–7 The skeleton of a teleost fish, *Perca flavescens.* (From B. Dean, 1895. *Fishes Living and Fossil.* Macmillan Publishing Co., New York.)

herrings, and their relatives. They are chiefly marine plankton-feeders, but the wolf herrings are predaceous. The tarpons and the bone fishes compose the order Elopiformes. They are typically predaceous fishes, mainly marine, and dwell in tropical and subtropical waters. The order Anguilliformes includes many kinds of true eels, such as the conger eels, freshwater eels, and, in some classifications, gulper eels. They have elongate bodies, no pelvic fins, and the dorsal and anal fins are connected to the caudal fin. The gulper eels are deep-sea fishes with huge mouths; the rest are mainly marine dwellers in rocky or coral crevices, although one group uses fresh water. The spiny eels and halosaurid eels belong to the Notacanthiformes; they are marine bottom-dwellers with dorsal fins that are not connected to the caudal fins. Tarpons, true eels, and notacanthiform eels share an unusual larval form. The leptocephalus larva is ribbonlike and transparent and goes through a complete metamorphosis in acquiring adult shape. The similarity of the larvae of these three groups suggests a common ancestry.

All of the following orders are thought—at least by some ichthyologists—to represent the major teleost radiation.

Members of the Salmoniformes are the trout and salmon, argentines, smelts, pikes, mudminnows, hatchetfishes, dragonfishes, and associated taxa. Most are cold-water inhabitants of both fresh and salt water.

The principally tropical Gonorhynchiformes (milkfishes) may be related to the herring group, but Nelson (1976) links them with the following groups. Cypriniformes are the minnows, suckers, electric eels, characins, knifefishes, and loaches. This is primarily a freshwater group, with more than 3000 species, and comprises the second largest order of fishes. A

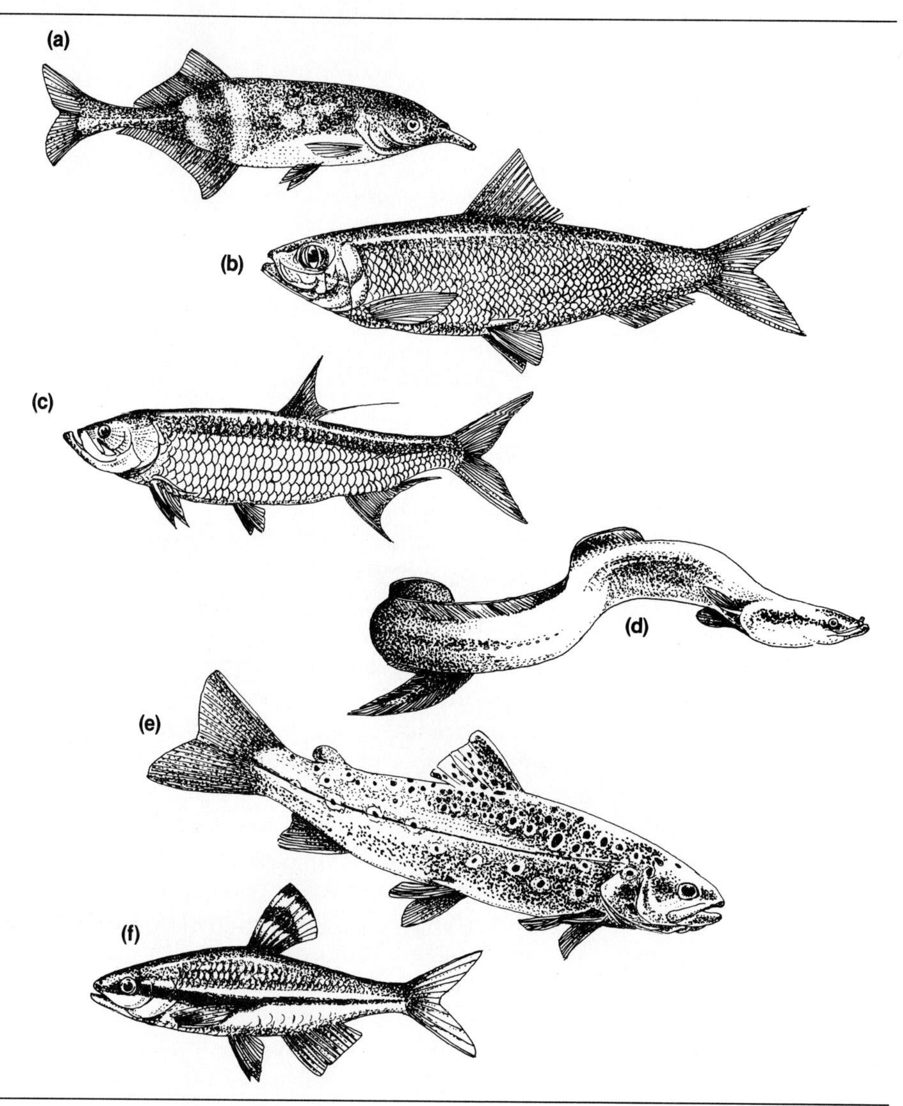

Figure 2–8 An array of teleost fishes from different orders. (*a*) Mormyriformes: an elephant fish, *Gnathonemus petersi;* (*b*) Clupeiformes: a herring, *Clupea harengus;* (*c*) Elopiformes: a tarpon, *Megalops atlanticus;* (*d*) Anguilliformes: an eel, *Anguilla anguilla;* (*e*) Salmoniformes: a trout, *Salmo trutta;* and (*f*) Cypriniformes: a shiner, *Notropis hypselopterus.*

chain of small bones called weberian ossicles characteristically connects the swim bladder and the ear and functions in hearing. The 2000 or more species of catfishes are sometimes included with the Cypriniformes and are sometimes placed in a separate order—Siluriformes. The Siluriformes are basically nocturnal and primarily freshwater forms. None have scales, and

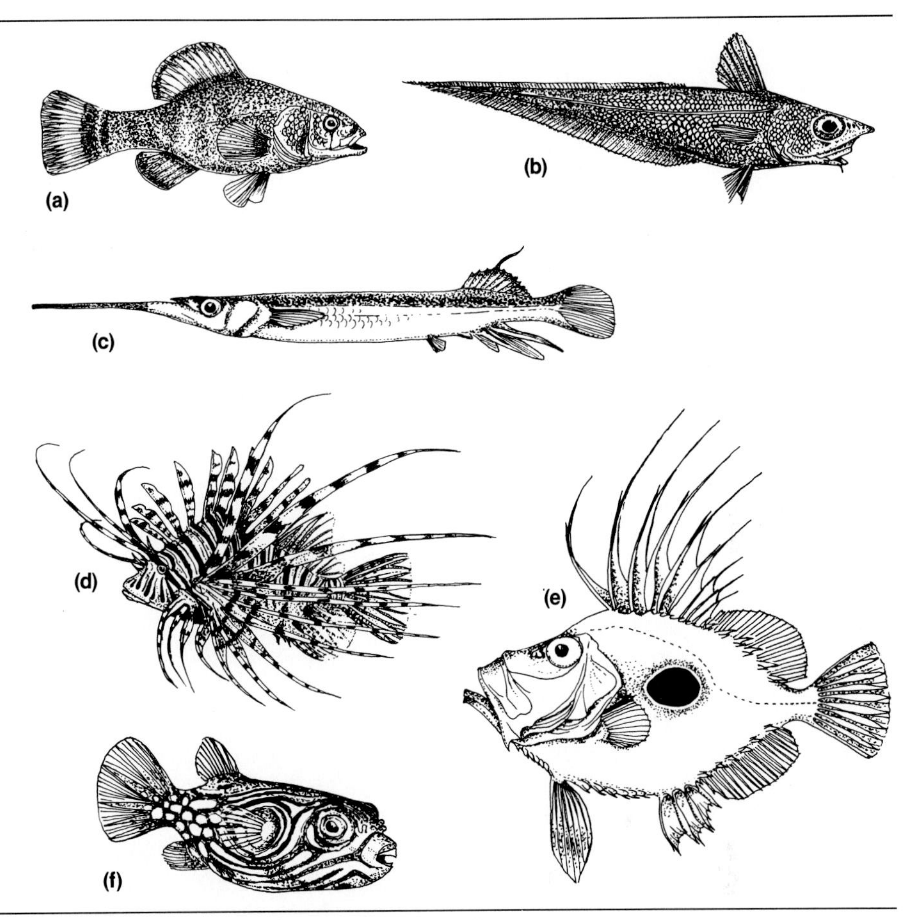

Figure 2–9 More teleost fishes. The twelve examples in Figures 2–8 and 2–9 were selected to illustrate a variety of body form in fishes not illustrated elsewhere in the book. Not all orders are represented here. (*a*) Percopsiformes; a pirate perch, *Aphreododerus sayanus;* (*b*) Gadiformes: a grenadier, *Coelorhynchus carminatus;* (*c*) Atheriniformes: a halfbeak, *Zenarchopterus dispar;* (*d*) Scorpaeniformes: a scorpionfish, *Pterois volitans;* (*e*) Zeiformes: a john dory, *Zeus japonicus;* and (*f*) Tetraodontiformes: a pufferfish, *Arothron reticularis.*

some, such as the North American madtoms (*Noturus*), have poisonous spines.

Lanternfishes, lancetfishes, and lizardfishes compose the Myctophiformes. These are mainly oceanic, mid-water or deep-water species, whose systematic treatment is variable.

Beardfishes (order Polymyxiiformes) are a small group of marine species and have been classified in several ways. Trout perch, pirate perch, and cavefishes are inhabitants of North American freshwaters and constitute the order Percopsiformes. Pirate perch are curious because the anogenital opening migrates forward during development to a position between the gills; the young are brooded in the mouth. The Gadiformes are codfishes and grenadiers, mainly marine, with pelvic fins anterior to the pectorals. The pearlfishes are a family of gadiform fish specialized for living in the body cavities of sea cucumbers. Toadfishes and midshipmen, in the order Batrachoidiformes, are principally marine; some are notable for the ability to produce sounds with the swim bladder.

The order Lophiiformes includes the anglerfishes, batfishes, and goosefishes. They are mainly marine, deep-water, predatory forms. The swim bladder is generally closed in the adult, pectoral fins may be modified for walking on the bottom, and the first ray of the dorsal fin is often modified as a lure for other fishes that become prey. Male anglerfishes of several families are parasitic on the females.

Halfbeaks, flyingfishes, and needlefishes are Beloniformes; these fishes are sometimes included with the topminnows, four-eyed fishes, silversides, and live-bearers of the order Cyprinodontiformes in a composite order—Atheriniformes. Most species are freshwater; many species are surface-feeders in ephemeral, transitory habitats. The Lampridiformes contains an assortment of very dissimilar species with peculiar mouths: The opahs, crestfishes, oarfishes, and their allies are all oceanic. Zeiformes includes the dories and boarfishes, a deep and mid-water oceanic group with laterally compressed, deep bodies and large eyes. The order Beryciformes includes the deep-sea pricklefishes, pinecone fishes, lantern-eyes, and squirrel fishes, a rather diverse marine group. The Syngnathiformes comprise the seahorses, pipefishes, and trumpetfishes, characterized by a tubular snout. Most fishes in this order are marine, and male parental care is well-developed in many species. A somewhat similar group, possibly related, is the order Gasterosteiformes, the sticklebacks, found in both marine and fresh waters. The Synbranchiformes are "swamp eels," but they are not true eels. They have elongate bodies with almost no fins. They live primarily in tropical fresh water and all can breathe air. Scorpionfishes, velvetfishes, flatheads, and sculpins belong to the Scorpaeniformes. They are marine fishes primarily, although the sculpins have also invaded fresh water. Poisonous spines are found in several species of scorpionfishes. Flying gurnards are placed in an order Dactylopteriformes, a benthic group; they are reputed to walk about on the seafloor and are capable of producing sounds. The order Pegasiformes, or seamoths, are small, long-snouted, armor-plated fishes of the Indo-Pacific tropics. The order Perciformes is the largest and most diverse teleost group, with over 7000 species (over 40 percent of all living fish species and about 35 percent of all living families of fishes). Its members inhabit virtually all available habitats, but about 75 percent of them are coastal marine fishes. These species generally have a closed swim bladder but are difficult to characterize morphologically. Many body shapes and life-styles are represented. This group includes gobies, blennies, groupers, perches, basses, darters, mackerel, swordfishes, cichlids, wrasses, leaffishes, stargazers, surgeonfishes, parrotfishes, drums, most icefishes, mullets, gouramis, and remoras. Remoras are especially interesting because they have a dorsal sucker used to attach themselves to sharks (and sometimes tuna or whales). Gobiesociformes are clingfishes, dragonets, and their relatives; all are bottom-living species with reduced scalation. In one family, the pelvic fins are located anterior to the pectorals and have been modified into an adhesive disk used for maintaining position on the bottom. These clingfishes inhabit intertidal zones. The order Pleuronectiformes comprises the flounders and soles, bilaterally asymmetrical fishes with flattened bodies and both eyes on the same side of the head. These predators usually live on the ocean floor, lying on one side with the eyes up; some species are more active, and some live in fresh water. The ocean sunfishes, puffers, triggerfishes, and their allies are in the order Tetraodontiformes. They lack a lateral line, are relatively slow-moving but predatory, and most are protected from predators by armor or poison. Many of them can produce sounds by grinding their teeth or vibrating the swim bladder.

CLASS AMPHIBIA The salamanders, frogs and toads, and caecilians belong to this class. All living amphibia belong to the subclass Lissamphibia ("smooth amphibia"): the salamanders compose the order Urodela or Caudata ("tailed"); the frogs and toads are the Anura ("tailless") or Salienta ("jumpers"); and caecilians belong to the order Apoda ("footless") or Gymnophiona ("naked snakes"). There are now almost 3000 species, many fewer than in the past. Most modern species are froglike. The name "amphibian" derives from the two-stage life cycle typical of the group. Fertilization of the eggs in anurans is usually external and occurs as the eggs are laid, but, in most salamanders, the male deposits a packet of sperm (a spermatophore) on the bottom of the pond or stream and the female picks up this packet with her cloaca (the chamber into which the reproductive, excretory, and digestive systems all empty). The eggs are usually laid in the water, and the larvae are usually aquatic and breathe by means of gills. As adult size is approached, the larva undergoes metamorphosis into adult form, losing the gills, sometimes the tail, and acquiring legs and lungs. Frog tadpoles are generally herbivorous (vegetarian), but salamander larvae are carnivorous (meat-eating). A few species are not dependent upon standing water because their larvae develop in the confines of the egg. A few, such as the mudpuppy (*Necturus*) and some populations of the tiger salamander (*Ambystoma tigrinum*), have lost the normal adult phase and retain juvenile morphology even when the sex organs mature. Despite the general requirement of water for reproduction, anurans have been successful in occupying many tropical and temperate habitats, including certain deserts. Amphibian skin is generally soft, moist, and glandular; in many species, it is used in respiration.

In body form, the amphibians are a diverse group. Salamanders have elongate, strong bodies with a stout tail; they swim in much the same manner as typical fishes, using the axial body musculature. The limbs of modern salamanders are usually fairly small and relatively weak and play little role in swimming; they are, of course, used in locomotion on land (Fig. 2–10*a*, *b*). In contrast to salamanders, the conformation of frogs and toads is clearly related to jumping: they possess long, strong, hind legs and a short compact trunk (Fig. 2–10*c*). Terrestrial locomotion on legs means that the head is no longer so intimately tied to the mechanics of locomotion. Most legged vertebrates, with the notable exception of the anurans, have a neck, permitting a greater freedom of head movement that facilitates feeding and seeing.

The third group, the caecilians, are wormlike burrowers without legs or good eyes (Fig. 2–10*d*). They live in moist, tropical regions, prey on invertebrates, and are seldom seen. They range in size from 6 to 15 cm. Some lay eggs, but others reportedly give birth to living young.

Amphibians and reptiles together are informally called "herptiles," or "herps" ("creeping things"), and are studied by herpetologists. This informal rubric does not imply unusually close relationships between these two classes; it is a convenience more than anything else.

CLASS REPTILIA The reptiles have dry, scaly skin and are basically terrestrial, although some have secondarily returned to the water. The skeleton of a lizard (Fig. 2–11) can be compared with that of an amphibian (Fig. 2–10). Young developing within the egg are protected by a third extraembryonic membrane, the amnion. Fertilization is internal, and both live-bearers and egg-layers occur. In contrast to most fishes, eggs are supplied with large quantities of nutritive yolk, and young reptiles emerge into the world at a rather advanced stage of development. Reptiles were far more common in past geologic times but now number about 6000 species. Most are carnivorous, although there are some herbivores and omnivores, the latter eating both plant and animal tissues. Some of the diversity of form is illustrated in Figure 2–12.

Modern reptiles fall into three subclasses. The Anapsida are presently composed solely of the order Chelonia (from the Greek word for tortoise), or Testudinata: the turtles and tortoises. The term "anapsid" ("without arches") refers to the solid contour of the turtle skull, which is unbroken by fenestrae or "windows" for muscle insertion. Turtles and tortoises are characterized by their armor, an array of bony plates, both dorsal and ventral; most modern turtles can withdraw their appendages into this protective plating. Most turtles live in ponds and marshes, and the sea turtles

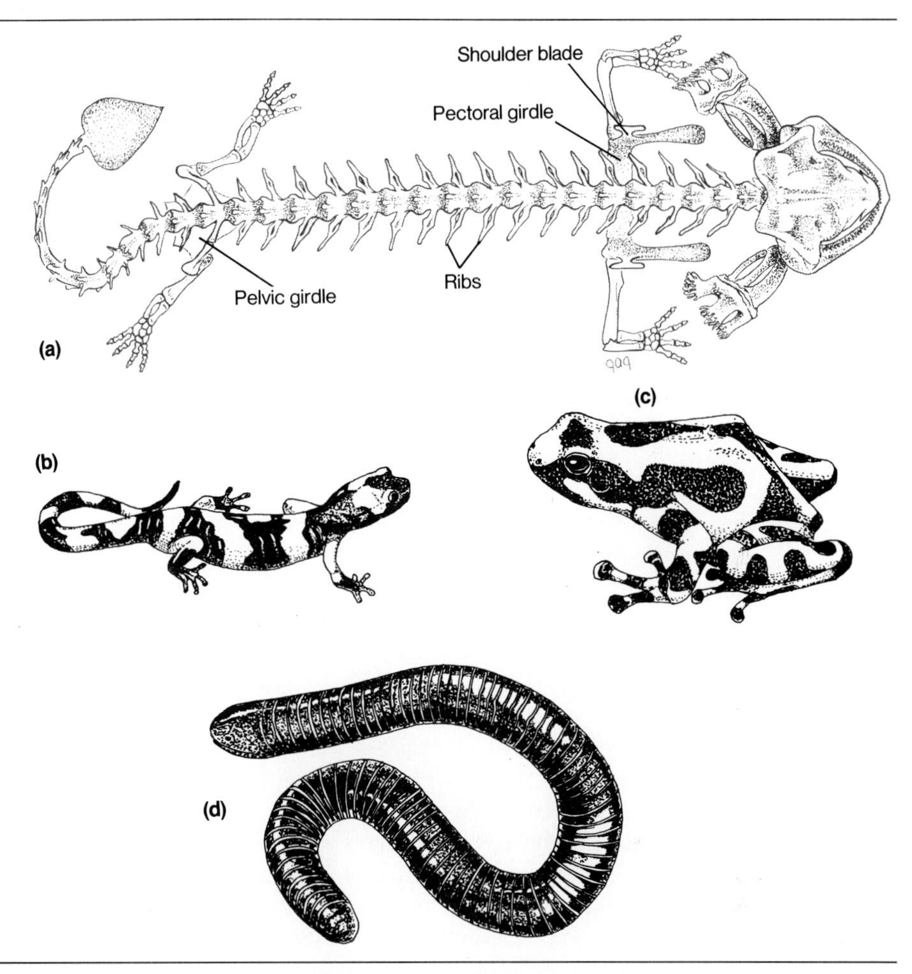

Figure 2–10 (*a*) The skeleton of a salamander, dorsal view. (Reprinted from *Vertebrate Paleontology* by A.S. Romer by permission of The University of Chicago Press.) (*b*) A salamander, *Ensatina eschscholtzi;* (*c*) a frog, *Dendrobates aliratus;* and (*d*) a caecilian, *Siphonops annulata.*

spend almost their entire lives at sea, but tortoises are completely terrestrial. All chelonians lay their eggs on land, commonly in nests dug in the soil.

The subclass Lepidosauria ("scaled reptiles") includes two orders: the Rhynchocephalia ("snout-headed") and the Squamata ("scaled"). The first of these is now represented by a single species (the tuatara, *Sphenodon punctatus*) that is now found only on a few small islands off the coast of New Zealand. This marvelous reminder of things past looks superficially like a lizard but is distinguished by certain vertebral and other anatomical characteristics.

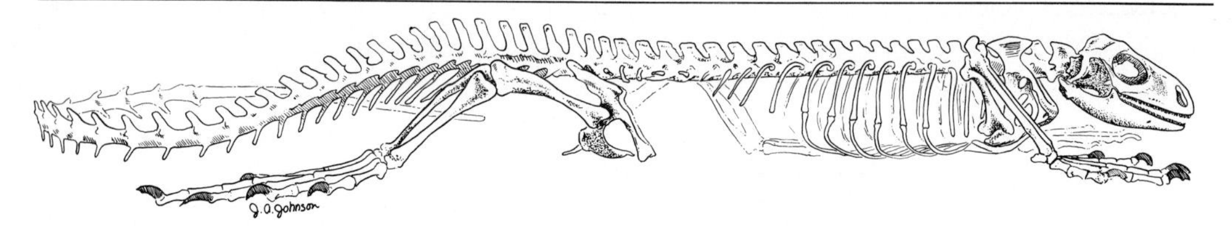

Figure 2–11 The skeleton of a lizard (*Iquana*).

The squamates are the numerous lizards (suborder Lacertilia or Sauria), snakes (suborder Ophidia or Serpentes), and amphisbaenians (suborder Amphisbaenia, whose place in the taxonomic hierarchy is still rather unsettled). Lizards make their living in a variety of ways—from active carnivory of other vertebrates to many styles of insectivory to flower or algae munching. They range in size from the 4-m predatory monitors (*Varanus*) of the East Indies to tiny insect-eating species only a few centimeters long. Most run on four legs, although some can also run bipedally, some regularly swim, and many climb. A few species, such as *Ophisaurus,* the glass lizards, have no legs.

An additional group of squamates that is often separated from the lizards is the amphisbaenians, most of which are limbless and all of which are tropical or subtropical burrowers that prey mainly on arthropods. Their name comes from their wormlike shape, suggesting (to some at least) that they can move both fore and aft equally well, although recent work has demonstrated that they cannot.

Snakes differ from most lizards in the absence of legs and, hence, have a totally different style of locomotion. All snakes are carnivorous, although the African genus *Dasypeltis* specializes on eggs. Snakes can open their mouths very wide, due to a loosely arranged jaw suspension, and can swallow their prey whole; in fact, many snakes can swallow objects much larger than their own diameter. Most species of snakes are ground-surface dwellers, but some burrow, and others climb; the sea snakes have become fully aquatic. Three families have specialized in the use of venom injected by anterior fangs: the cobras and coral snakes (Elapidae), the sea snakes (Hydrophiidae), and the vipers and pit vipers (Viperidae). Venomous species are found in the Colubridae as well, but these are generally rear-

Figure 2–12 Various reptiles: (*a*) a lizard, *Sceloporus poinsetti;* (*b*) the tuatara, *Sphenodon punctatus;* (*c*) a snake, *Lampropeltis getulus;* (*d*) an amphisbaenian, *Rhineura floridana;* (*e*) a crocodilian, *Alligator mississippiensis;* and (*f*) a turtle, *Pseudemys rubriventris.*

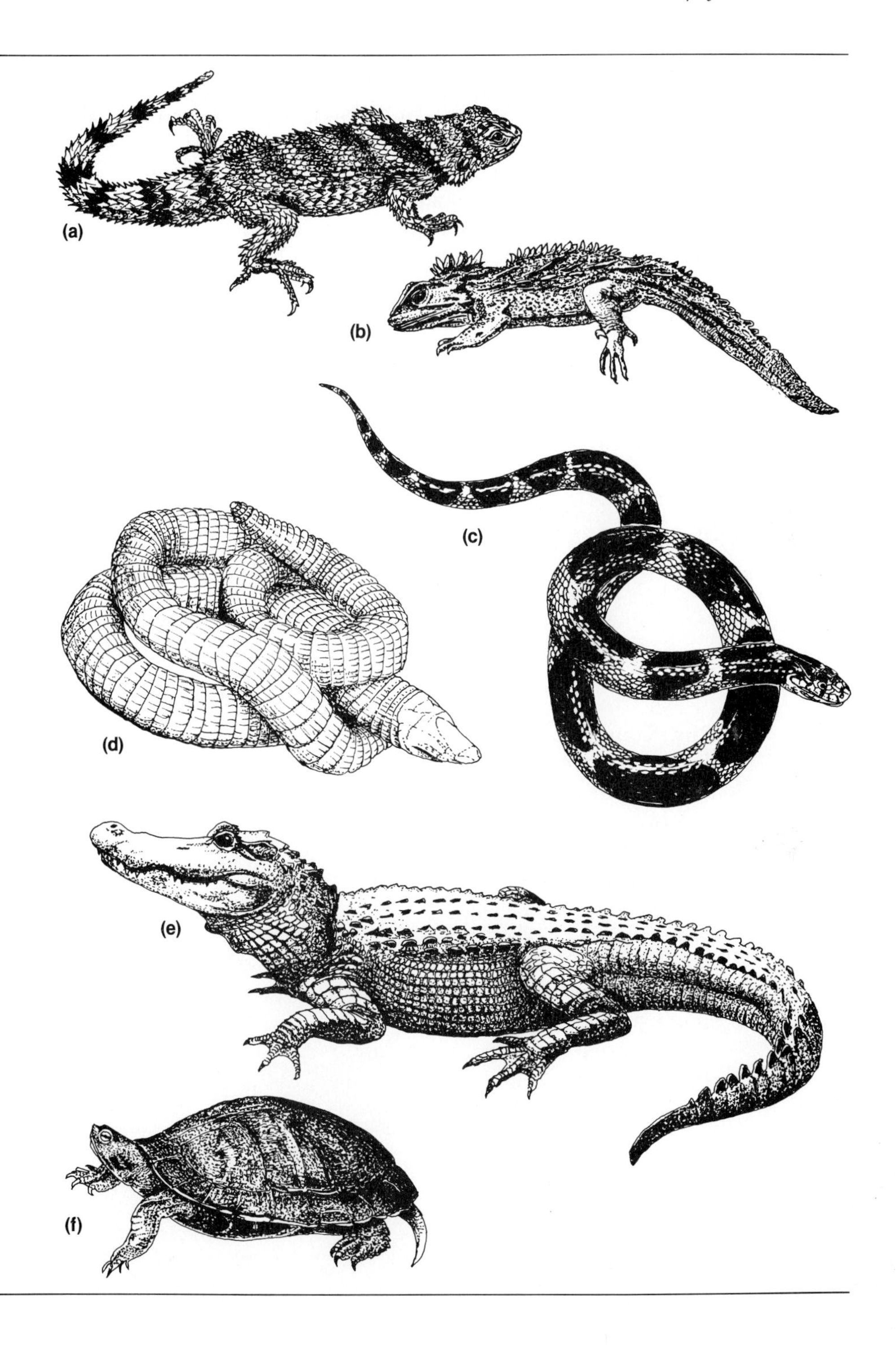
(a)
(b)
(c)
(d)
(e)
(f)

fanged and less dangerous to large mammals. Snakes vary greatly in size—from inconspicuous earth-dwellers to pythons over 10-m long.

The only living members of the subclass Archosauria ("ruling reptiles") are the crocodiles, alligators, caimans, and gharials in the order Crocodilia. They are amphibious, spending a great deal of time in the water but nesting on land or piles of floating debris. All crocodilians are predatory. Archosaurs and lepidosaurs are sometimes put together into the subclass Diapsida ("two-arched").

CLASS AVES All birds have wings and feathers and are usually placed in a class by themselves. In many respects, however, their anatomical features (Fig. 2–13) betray a close relationship to reptiles, especially the archosaurs, and many researchers now consider them little more than glorified reptiles. No modern birds have teeth, although some extinct kinds did; the jaws are modified to form a horn-covered beak or bill. Like most mammals and some reptiles, birds usually possess an ability to maintain a high and relatively constant body temperature. Also like the mammals, the circulatory system is effectively dual, with separate pulmonary (lung) and body circuits. These last two characteristics are associated with a high metabolic

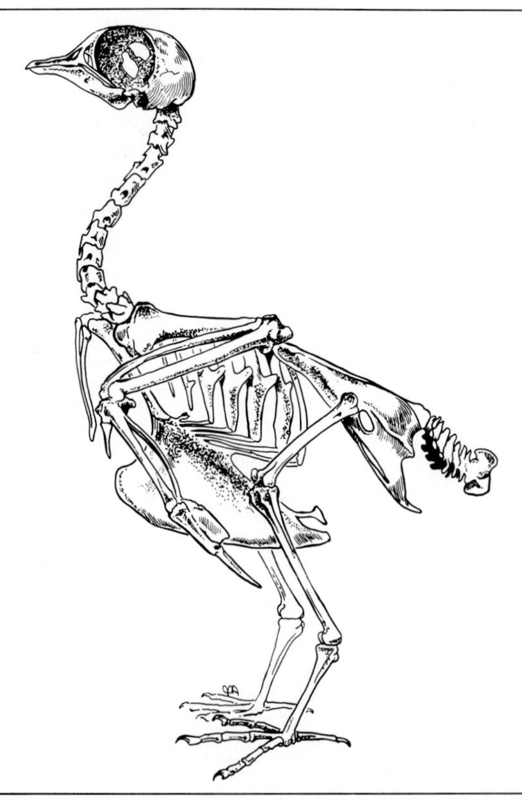

Figure 2–13
The skeleton of a pigeon. Note the large keel on the sternum for the attachment of flight muscles, the reduced tail, the compact trunk, the overlapping reinforcing projections of the ribs, and the reduction of bones in the distal portion of the wing. Birds stand on their toes. The foot and ankle bones have been incorporated into the leg, making a three-segmented leg. (From G. Heilmann, 1926. *The Origin of Birds.* Witherby, London.)

rate and an active existence. Birds have large brains that are especially well developed in the optic region and sometimes in other sensory regions. The skeletons of flying birds tend to be hollow and very light, and, in many species, air sacs located at various places in the body contribute to lightness. The feet, and sometimes the legs, are scaled. Most birds can fly; the major exceptions in extant taxa are the penguins and the terrestrial runners, such as ostriches and emus. Some birds can fly or swim better than they can walk (loons, penguins, swifts). All birds lay eggs, and most build some kind of nest in which the eggs are placed. Avian diets may be composed of carrion, vertebrates, invertebrates, nectar, fruit, seeds, vegetation, and various combinations of these. Adults range in size from the 2-g bee hummingbird (*Mellisuga helenae*) of the West Indies to the 150-kg ostriches (*Struthio*) and cassowaries (*Casuarina*); some extinct birds were even larger.

There are over 8600 extant species of birds, but Romer (1966), for one, considered that the diversity of structure within the group is less than in most other vertebrate classes, even those with fewer species. Systematists sometimes create more than 25 orders, but most of the differences among them are of the magnitude used to distinguish families in other classes of vertebrates. The anatomical characters used by taxonomists to distinguish different taxa are many and varied, but they are often too subtle or too obscure to be of use to most students of natural history. A variety of body forms are displayed in Figures 2–14 and 2–15. Phylogenetic relationships among the avian orders are much debated and difficult to ascertain (Olson, 1982), so the following summary does not represent evolutionary patterns among the orders of birds.

Struthioniformes (ostriches), Rheiformes (rheas), and Casuariiformes (cassowaries and emus) are all large birds; they cannot fly, do not have a keel on the sternum (breastbone), and have only two or three toes. The ostrich is found in Africa and Asia, the rheas in South America, and the cassowaries and emu in the Australian region. Another flightless order, Apterygiiformes, includes the kiwi of New Zealand. Members of these four orders have soft, often hairlike, plumage.

The penguins, Sphenisciformes, are also flightless and highly adapted to an aquatic mode of life. In contrast to other birds, in which the body feathers grow only from certain parts of the skin, penguin feathers grow all over the body. They inhabit mainly antarctic and subantarctic regions, but one species, *Spheniscus mendiculus,* lives on the equatorial Galapagos Islands, which are served by the cold Humboldt current flowing northward along the west coast of South America.

Tinamiformes are superficially chicken-like birds of Latin America that are primarily terrestrial and seldom fly. Males do much of the parental care of the young. The eggs are typically colorful, with an unusually high gloss.

Gaviiformes (loons) and Podicipediformes (grebes) are diving birds with webbed toes or lobed toes, respectively. Both kinds of birds have the ability

Figure 2–14 A sampling of birds in various orders. (*a*) Casuariiformes: a cassowary, *Casuarius casuarius;* (*b*) Apterygiiformes: the kiwi, *Apteryx australis;* (*c*) Podicipediformes: a grebe, *Podiceps nigricollis;* (*d*) Gaviiformes: a loon, *Gavia immer;* (*e*) Procellariiformes: an albatross, *Diomedea exulans;* and (*f*) Pelecaniformes: a pelican, *Pelecanus occidentalis.*

to expel the air from their bodies and plumage and to sink slowly beneath the surface of the water.

The order Procellariiformes includes the petrels, albatrosses, and shearwaters, marine birds with tube-shaped nostrils, webbed toes, and usually long, narrow wings. All are highly pelagic, lay only single eggs in each nesting attempt, and feed the newly hatched young on oil regurgitated from the stomach.

Pelicaniformes are the pelicans, tropicbirds, gannets, cormorants, anhingas, and frigate birds. Almost all have gular (throat) pouches, and the nostrils are reduced or absent in most species.

Herons, storks, ibises, flamingos, and their relatives constitute the order Ciconiiformes (Ardeiformes). They are primarily wading birds and have

long bills, necks, and legs, but some species feed on land.

The order Anseriformes includes the ducks, geese, swans, and screamers, aquatic birds with broad bills and at least partially webbed feet. They occupy all kinds of aquatic habitat, from torrential mountain streams to ponds and swamps to ocean.

Vultures, hawks, eagles, falcons, and the secretary bird belong to the order Falconiformes (Accipitriformes). They are predatory or scavenging birds, have hooked bills, and are usually excellent fliers.

The Galliformes are the megapodes, curassows, grouse, quail and pheasants, turkeys, and the curious hoatzin. All are vegetarian ground birds that are seldom capable of long-distance flight.

Gruiformes (Ralliformes) are a varied group that includes the cranes, trumpeters, rails, coots, finfoots, and bustards. Most species are aquatic; many others live in grassland.

Figure 2–15 More birds. (*a*) Ciconiiformes: a heron, *Ardea herodias;* (*b*) Anseriformes: a duck *Lophodytes cucullatus;* (*c*) Falconiformes: a hawk, *Buteo jamaicensis;* (*d*) Lariformes: a puffin, *Fratercula arctica;* (*e*) Psittaciformes: a parrot, *Amazona ochrocephala;* and (*f*) Piciformes: a hornbill, *Buceros bicornis.*

The order Charadriiformes contains the many species of shorebirds: the jacanas, plovers, sandpipers, phalaropes, coursers, skuas, gulls, terns, auks, and puffins. The last five kinds of shorebirds are sometimes put into a separate order—Lariformes. Most members of this group are aquatic, have webbed toes, and are strong fliers. Most species eat primarily fish or aquatic invertebrates, but the skuas and their closest relatives eat birds and mammals.

Pigeons and doves constitute the order Columbiformes; they are characterized by the presence of a crop that produces "pigeon's milk" to feed the young. Sand grouse are sometimes put with the Columbiformes and sometimes with the Charadriiformes.

Psittaciformes are the parrots, macaws, and lories. They are primarily seed-, fruit-, and nectar-eaters, and inhabit tropical and subtropical regions around the world. Fast fliers and good climbers, they have a hooked bill and two forward and two rear toes on each foot.

The cuckoos, roadrunners, and touracos belong to the Cuculiformes. Many cuckoos of the Old World are brood parasites, laying their eggs in the nests of other birds. They have two front toes, one hind toe, and another hind toe that can be moved forward.

The owls are members of the order Strigiformes. They are nocturnal predators with excellent vision and hearing, soft plumage, quiet flight, hooked bills and talons.

Caprimulgiformes includes the oilbird, frogmouths, potoos, and nightjars. These birds, generally characterized by huge gapes fringed with bristles, hunt aerial insects on the wing. The South American oilbird is an exception because it feeds largely on palm nuts. Most species are nocturnal or crepuscular (active at twilight).

The Apodiformes are distinguished by small legs and feet and excellent aerial capacities. The swifts are aerial insectivores; the hummingbirds are insect- and nectar-feeders. They are probably not closely related.

Mousebirds or colies, found only in central and southern Africa, constitute the Coliiformes. They are dull-colored cavity-nesters that have two reversible toes and uncertain taxonomic affinities.

The trogons and the spectacular quetzal of the New World tropics are Trogoniformes. They are cavity-nesters, fruit- and insect-eaters, and have weak feet and colorful plumage.

The Coraciiformes comprise the kingfishers, motmots, bee-eaters, rollers, hoopoe, and hornbills. They are cavity-nesting carnivores and insectivores with strong bills.

The Piciformes, like the Coraciiformes, are cavity-nesters with sharp, strong bills. The jacamars, puffbirds, barbets, honeyguides, toucans, and woodpeckers belong to this group. Many species are primarily insectivorous but may also eat fruit. Some species (such as toucans) eat lizards as well as fruit.

The final avian order is by far the largest. About 60 percent of all bird species belong to the Passeriformes, the perching birds. Among the many varieties of passerines (or passeriforms) are the woodcreepers, ovenbirds, antbirds, cotingas, manakins, tyrant flycatchers, and lyrebirds, which are collectively called "suboscines" because their singing capacity is less than that of the remaining kinds of passerines. All other passerines are called "oscines"; they have a more highly developed syrinx (voice box) and more varied and complex vocal capacities. Included in this group are larks, swallows, crows and jays, birds of paradise, titmice, babblers, bulbuls, wrens, thrashers, Old World warblers, New World wood warblers, Old World flycatchers, shrikes, honeyeaters, sunbirds, white-eyes, blackbirds, tanagers, buntings and sparrows, goldfinches, waxbills, and weaver finches. The habits of passerines defy adequate summary. Feeding, nesting, habitat selection, size, and morphology are highly varied, reflecting the extensive adaptive radiation of this group.

CLASS MAMMALIA Mammals are characterized by mammary glands used for nursing the young and a more or less hairy skin. Like the birds, they have an efficient circulatory system and usually maintain a high metabolic rate. The brain, especially the cerebral area, is well developed. One of the chief distinguishing features is the formation of the lower jaw from a single pair of bones, the dentaries, rather than the several bones in the jaws of reptiles. Other diagnostic characteristics of modern mammals include three small bones in the middle ear (rather than only one or two), a double set of bony knobs (condyles) at the base of the skull, a tympanic bone supporting the eardrum, and multiple roots on the cheek teeth. The skeleton (Fig. 2–16), in contrast to the skeleton of typical reptiles, is highly ossified. Unlike many fishes, amphibians, and reptiles, mammals have teeth that are borne only at the margins of the jaws and not on the palatal area. Teeth have become highly differentiated in different mammalian orders. The dietary

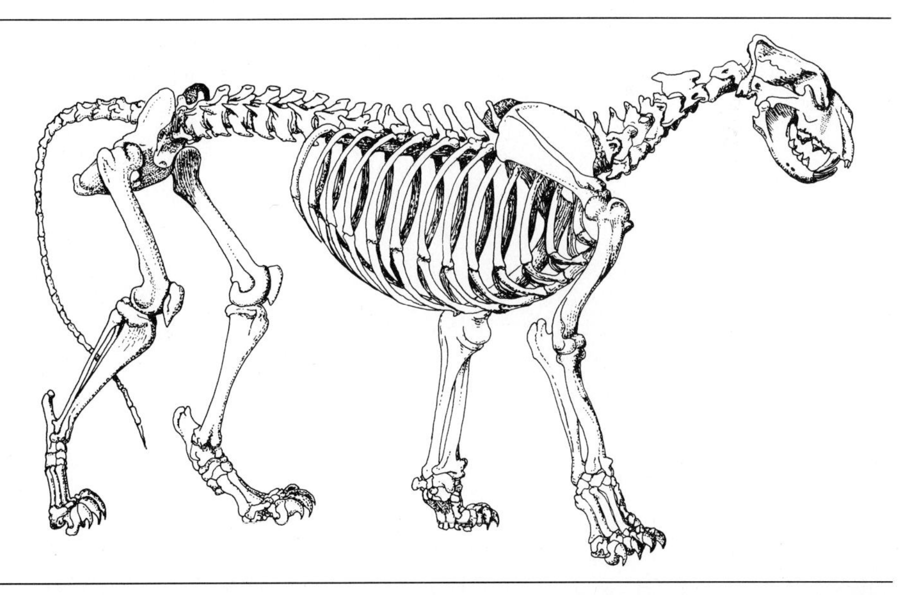

Figure 2–16 The skeleton of a mammal: the lion (*Panthera leo*). (From R. Owen, F.R.S., 1886–1888. *On the Anatomy of Vertebrates.* Vol. 2. *Birds and Mammals.* Longmans, Green, London.)

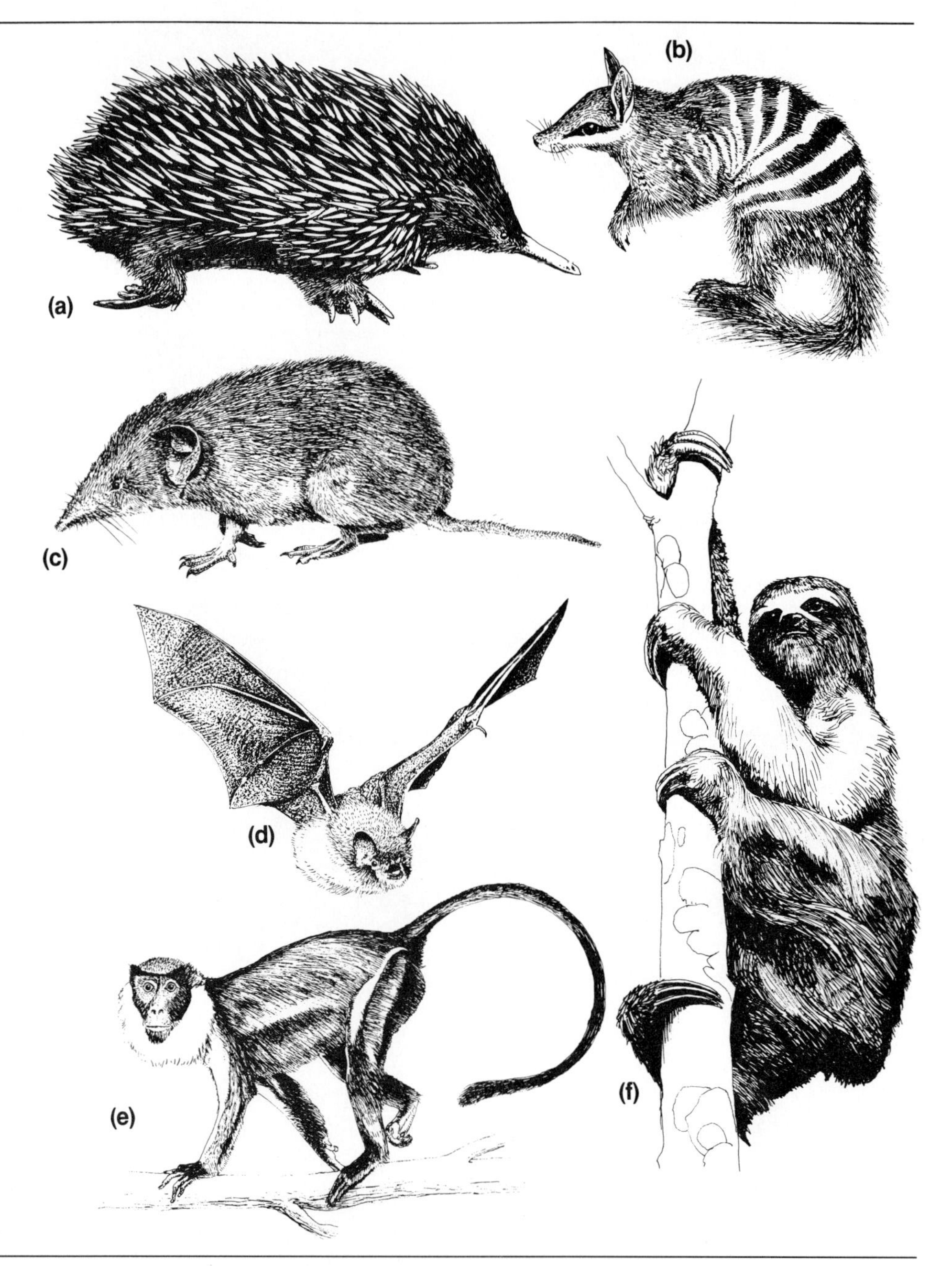

Figure 2–17 A diversity of mammals. (*a*) Monotremata: an echidna, *Tachyglossus aculeatus;* (*b*) Marsupialia: the numbat, *Myrmecobius fasciatus;* (*c*) Insectivora: a shrew, *Crocidura suaveolens;* (*d*) Chiroptera: a bat, *Myotis nattereri;* (*e*) Primates: a monkey, *Cercopithecus diana;* and (*f*) Edentata: a sloth, *Bradypus tridactylus.*

Figure 2–18 More mammals. (*a*) Rodentia: an agouti, *Dasyprocta aguti;* (*b*) Cetacea: a whale, *Balaenoptera musculus;* (*c*) Carnivora: a bear, *Ursus arctos;* (*d*) Tubulidentata: the aardvark, *Orycteropus afer;* (*e*) Hyracoidea: a hyrax, *Procavia capensis;* and (*f*) Sirenia: a manatee, *Trichechus manatus.*

and locomotor habits of mammals are as varied as those of birds. Mammals range in size from tiny shrews (*Microsorex hoyi, Sorex minutus, Suncus etruscus*) weighing only about 2 g to the giant blue whale (*Balaenoptera musculus*) measuring as much as 31 m and weighing up to 160,000 kg. The approximately 4000 species of mammals fall into three major groups (Figs. 2–17, 2–18).

The subclass Prototheria ("first beasts") contains a single modern order—Monotremata ("one orifice"; that is, a cloaca)—and is represented today only by two families in Australia, Tasmania, and New Guinea: the duck-billed platypus (one species) and the spiny anteaters or echidnas (two species). Monotremes are distinguished from other mammals not only by the cloaca but by numerous skeletal features, including the presence of cervical ribs, certain reptilian features of the limb girdles, and the lack of a bony housing for the ear. All lay eggs in underground nests; echidnas usually produce one egg, but the platypus normally lays two. Female echidnas develop a ventral pouch in which the young are incubated; the platypus has no pouch. The young obtain milk by sucking or licking the belly fur in the platypus or sucking from two specialized areas in the pouch of the echidnas; nipples are not present on the mammaries. Echidnas eat primarily ants, termites, and other invertebrates; duckbills eat mainly aquatic invertebrates. The adults are toothless, but young duckbills have teeth.

The two remaining groups are in the subclass Theria. The first is the infraclass Metatheria ("middle beasts") with one order, the Marsupialia ("pouched"), which contains about 240 species. Marsupials are distinguished from the following group by certain features of the jaw and the presence of marsupial bones anterior to the rest of the pelvis. Most female marsupials have a ventral pouch (marsupium) or at least marsupial folds in which the young are carried and fed. Some members of the families Didelphidae (opossums) and Dasyuridae (mainly carnivorous marsupials somewhat parallel to the true cats and shrews) lack a marsupium altogether; the young apparently just cling to the female's belly. In the water opossum (*Chironectes minimus*), the male has a pouch too, but his marsupium holds his testes, not young opossums. The gestation period, or time of pregnancy, is short. Young marsupials are born at a very early stage of development but are able to crawl to the female's nipples, where they cling for a period of time longer than the gestation period. At one time, this order was well represented around the world, but it is now concentrated in the southwestern Pacific region, particularly Australia. Kangaroos, wombats, bandicoots, and phalangers are examples of this group. The opossums of the Americas and a group of small shrewlike species in South America are remaining distributional outposts for the order. The largest marsupial is perhaps the great grey kangaroo (*Macropus giganteus*), which stands as much as 2 m tall and weighs up to 90 kg.

Members of the infraclass Eutheria, or placentals, are all viviparous (bearing living young), but their offspring emerge at a much more advanced stage than marsupials and no marsupium is present to house the young. Young in the uterus are nourished from the mother's circulation through a placenta. This characteristic is not diagnostic of the group, however, because one family of marsupials, the Peramolidae (or bandicoots), has a similar kind of placenta, and simpler placentas are found in other marsupials. Placentals have no pouch, no marsupial bones, a different shape of lower jaw, and other distinguishing features. This group is now the dominant group of mammals throughout most of the world. Depending on the systematist, there are 15 to 17 orders. The first six orders summarized in the following paragraphs are thought to be more closely related to each other than to the remaining groups; the next eight orders apparently form another broad group. The remaining two orders (Edentata, Pholidota) are of uncertain phylogenetic relationship (Vaughn, 1978).

The order Insectivora includes the moles, shrews, tenrecs, hedgehogs, golden moles, and the solenodon of Haiti. All feed primarily on invertebrates; some are aquatic, and some of the terrestrial species burrow.

The order Dermoptera contains a single genus of gliding mammals, the "flying lemurs," found only in southeastern Asia and outlying islands.

The order Chiroptera includes the bats, the only true flying mammals. The forelimbs are modified into membranous wings. In terms of number of species, this is the second most numerous order of mammals, with over 850 species. Some are insectivorous or carnivorous; others eat fruit, nectar, or blood.

The cosmopolitan order Primates includes the tree shrews, lemurs, tarsiers, marmosets, all of the monkeys, and the large anthropoid apes and humans. Dietary habits range from exclusively vegetarian to principally meat-eating.

Pikas, hares, and rabbits belong to the order Lagomorpha. They have negligible tails and, in contrast to rodents, two pairs of upper incisor teeth (one pair very small). None are very large and all are herbivorous.

The order Rodentia contains almost 1700 species divided into three major suborders. Canine teeth are absent. Most are small to moderate in size and mainly herbivorous or seed-eating. The suborder Sciuromorpha includes the "mountain beaver," squirrels, chipmunks, pocket gophers, kangaroo rats, pocket mice, beaver, and spring hares. Myomorpha contains rats, mice, dormice, jerboas, jumping mice, and mole rats. The suborder Hystricomorpha, or Caviamorpha, includes Old and New World porcupines, cavies, capybaras, agoutis, chincillas, spiny rats, and tuco-tucos. The phylogenetic relationships within this largest of mammalian orders are not well understood.

The largest mammals of all are in the aquatic order Cetacea: the worldwide whales, dolphins, and porpoises. Some of these mammals reach a length of more than 30 m. Some (the largest) have no teeth and strain small invertebrates from the water by means of a whalebone (baleen) strainer. These are sometimes placed in a separate order—Mysticeti. The more numerous toothed whales are carnivorous and are sometimes put in the order Odontoceti. Whales are essentially hairless and strongly modified structurally in ways suitable for living continually in water.

Carnivora are small- to middle-sized mammals found all over the world. As the name suggests, they are usually meat-eaters (especially other vertebrates) with well-developed cutting teeth. Several species, however, also include insects, fruits, and other plant materials in their diets. Wolves, foxes, bears, cats, raccoons, weasels, hyenas, and civets belong to this order. The Pinnipedia are sometimes considered to be a suborder of carnivores but may be given ordinal rank themselves. The Pinnipedia are carnivorous seals, sea lions, and walruses, all highly adapted to an aquatic existence with limbs modified into flippers.

The order Tubulidentata contains only the aardvark of South Africa, a medium-sized creature specialized for digging up ant and termite nests.

The order Proboscidea comprises the elephants of Asia and Africa. They are the largest land mammals, weighing as much as 5900 kg. The nose is elongated into a prehensile trunk or proboscis, and the upper incisor teeth develop into tusks.

The order Hyracoidea are the hyraxes, or Old World conies of Africa and southwestern Asia. They are small, hoofed, climbing vegetarians.

Manatees and dugongs belong to the order Sirenia, a herbivorous group that is totally aquatic. Extant species are restricted to tropical waters.

The order Perissodactyla contains the odd-toed hoofed mammals in which the digits typically have been reduced to one or three functional toes. They presently range over Africa, Asia, and Central and South America, but formerly they were found elsewhere as well. Some of these mammals have become feral in parts of their former range—wild horses and burros of the western United States, for example. This group includes horses and asses, tapirs, and rhinoceroses. The largest, the rhinos, weigh up to 2800 kg. All are herbivorous but they dwell in a variety of habitats. Together with the order Artiodactyla they comprise the ungulates, the name referring to their tiptoe stance, in which the nail generally is modified to form a hoof. The artiodactyls are the even-toed ungulates (two or four toes), including pigs and peccaries, hippopotamuses, camels and llamas, deer, giraffe, pronghorn, sheep, goats, and antelopes. They are essentially cosmopolitan but did not occur naturally in Australia until introduced by man. Some ungulates achieve rather large sizes; hippos may weigh 3600 kg. The nonpiglike artiodactyls are called ruminants after their compartmented stomach (one chamber is called the rumen). The ruminants are cud-chewers, regurgitating swallowed vegetation to be chewed a second time.

Members of the order Edentata have enlarged front claws used for digging or hanging from trees. Some species (the anteaters) are toothless, as the ordinal name implies; others have some molar teeth and canines. This order includes the anteaters of tropical America, the American tree sloths, and armadillos.

One genus of pangolins, or scaly anteaters, constitutes the entire order Pholidota. They live in Africa and Asia on a diet primarily of insects; teeth are absent.

This summary of who's who among the vertebrates is intended to serve as a brief introduction to nomenclature and major taxonomic groups. In the chapters that follow, particular survival problems and their solutions by the vertebrate creatures, with many examples from vertebrates around the world, are discussed.

VERTEBRATE ORIGINS AND PHYLOGENY

Vertebrate phylogeny (history of descent) has been reconstructed primarily by paleontologists, who study the remains of early forms preserved as fossils in the rocks. The fossil record is very good in later periods but quite fragmentary for the earliest stages. The older rocks have, of course, been exposed to a greater accumulation of geological wear and tear, so early fossils have a higher probability of destruction by natural processes such as erosion. Furthermore, the earliest vertebrates and some of the important links between early major groups may well have been small, rather delicate, and, thus, poor material for long-term preservation. It is not surprising, therefore, that the early record for vertebrates is spotty and subject to several different scientific interpretations, as we will see. Later phylogenetic trends and steps are quite well documented and less controversial.

Vertebrate history goes back at least 500 million years (Fig. 2–19) and probably much longer. The earliest fossil records are just fragments, bits of bony armor or scales, found in Ordovician rocks of both the U.S.S.R. and the western United States. The first well-defined group of known vertebrates were ostracoderms, small fishes seldom more than 30-cm long and sometimes as small as 6 cm. The name "ostracoderm" (shell-skinned) derives from the bony armor that most of them bore, probably as protection against invertebrate predators. Some ostracoderms also had bony internal skeletons as well. Ostracoderms were very abundant and diversified in Silurian and Devonian times. All were jawless. They are classified in the Agnatha, together with the modern cyclostomes that seem to be their closest living relatives but that may, instead, have diverged from very primitive jawed fishes.

In retrospect, it seems clear that the early vertebrates possessed some characteristic that effectively opened up a new major adaptive zone. Perhaps that door-opening invention was the backbone itself; however, in the virtual absence of information about physiology, soft anatomy, and behavior, this is only speculation. An internal, jointed skeleton may permit a greater range or precision of movement, heavier musculature, and, ultimately, a more active life-style and a larger body size than allowed by the external skeleton of arthropods and many other invertebrates.

The next major breakthrough was apparently the invention of jaws; some ancient ostracoderm-like fish perhaps gave rise to the first jawed vertebrates (gnathostomes). At about the same time, true paired fins became evident and probably contributed to the entering of the new adaptive zone. Jaws were derived from anterior gill arches, and paired fins came perhaps from lateral, stabilizing folds of the body wall. Paired fins made

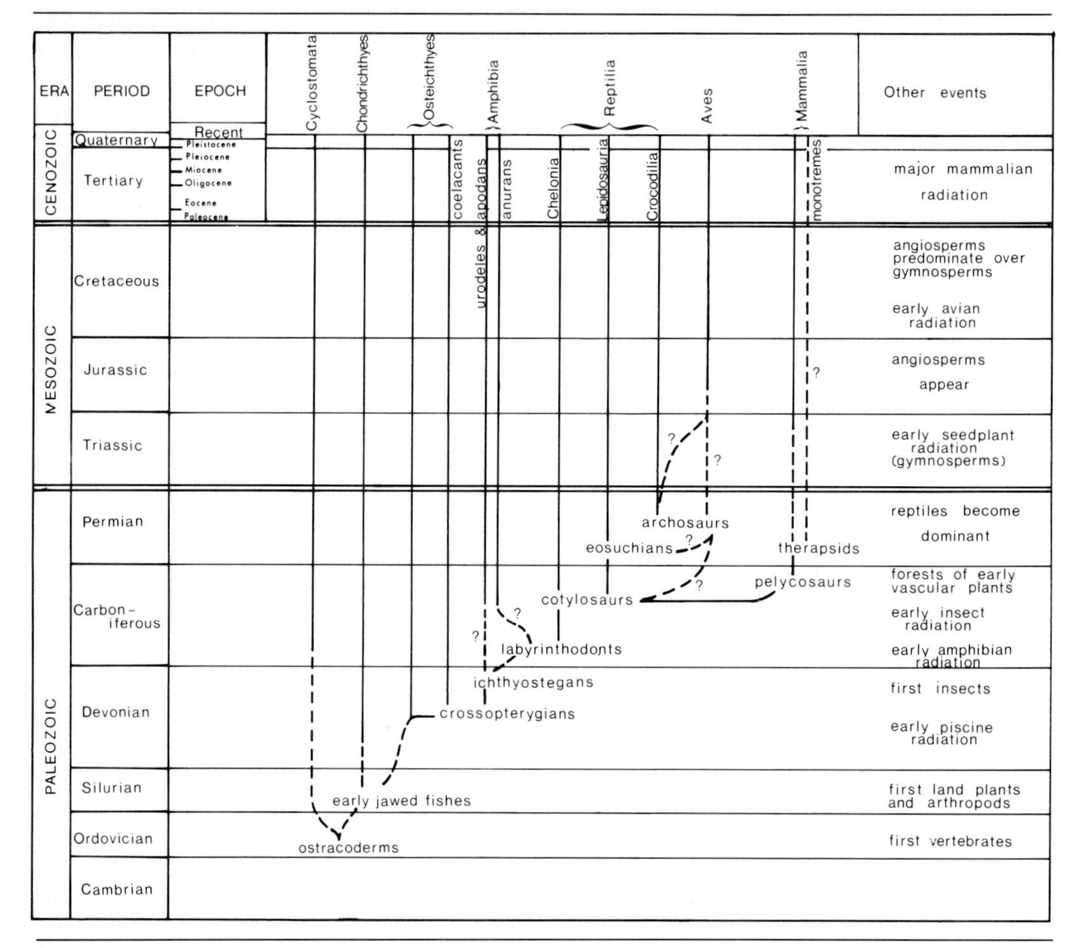

Figure 2–19 A simplified family tree of the vertebrates, on a geological time scale.

possible a greater precision in maneuvering and, with the advent of jaws, permitted an actively predaceous mode of foraging.

At least two early groups of gnathostomes are known. Quite a diversity of placoderms ("plate-skinned") existed, especially in the Devonian period; some were tiny, but others reached a length of about 9 m. Many lived in fresh water, but some were marine. Another group of fishes with jaws and paired fins, the acanthodians (the so-called spiny "sharks"), actually preceded the first placoderms proper. Acanthodians lived primarily in fresh water and were noted for the spines at the leading edges of many of their fins. They were sometimes classified with the placoderms, but many paleontologists maintain acanthodians and placoderms as two separate groups and suggest that the acanthodians are the more likely ancestors of both Chondrichthyes and Osteichthyes. Others maintain that neither group is directly ancestral to the modern jawed fishes. In any event, the latter two classes appeared in the fossil record in the Devonian period and probably originated during the preceding Silurian period.

Most paleontologists place the Chondrichthyes and Osteichthyes as independent lines of descent from the early jawed fishes. Zangerl and Case (1976), however, have suggested that the cartilaginous fishes may actually have preceded the placoderms and acanthodians and simply do not appear in the fossil record (cartilage does not fossilize well). Their study of the jaw and gill structure of a fossil shark (albeit a much later species from the Carboniferous period) has shown that certain sharks had a jaw that is a highly plausible predecessor to the jaws of other gnathostomes. Apparently, we cannot tell from the very spotty fossil record of these early fishes whether or not any such plausibly built sharks actually lived early enough to give rise to all the later gnathostomes. Nevertheless, this hypothesis is interesting and useful, because it necessitates a reassessment of early vertebrate phylogeny.

Both modern classes of fishes have, of course, persisted to the present day, but very early in the radiation of bony fishes there arose the crossopterygians. One branch of the crossopts is represented by the living coelacanth, *Latimeria;* the other branch, known as rhipidistians, lived mostly in fresh water and were the ancestors of the amphibians. The next major evolutionary "invention" in vertebrate history was the conversion of paired, fleshy fins into legs. This development may have been favored because it allowed the fleshy-finned fishes to hunt in novel ways, but it also permitted the protoamphibians to move on land as well as in water. What caused these explorers to leave water, even temporarily, is a matter of speculation. It may have been the hunt for new food resources on the shores, escape from aquatic predators, or perhaps a search for new ponds and streams as their old homes dried up. Because the ancestral forms already had lungs as well as stout limbs and may have emerged on land sometimes, amphibians were, in a sense, preadapted for the move onto land. However it happened, by the end of the Devonian period, creatures like *Ichthyostega* were present. *Ichthyostega* possessed several crossopterygian features, including a fishlike tail fin, a few cranial bones shared with fishes, and a two-parted braincase, as in the crossopts. In other respects, it was an amphibian and, thus, provides a good link between these groups.

The Carboniferous and Permian periods witnessed a burgeoning of amphibian types. Among the descendants of the ichthyostegans in the Carboniferous period were a diverse assemblage called the labyrinthodonts (named for the labyrinthine folds in the tooth enamel found only in crossopts and early amphibians). *Ichthyostega* itself is often included in the labyrinthodonts. The question of whether or not modern amphibians descended from the labyrinthodonts seems to be debated. Some paleontologists believe amphibians to be derived from a small-bodied group called "lepospondyls" that may have come from the ichthyostegans independently of the labyrinthodonts.

In any case, the labyrinthodonts are clearly the ancestors of the reptiles. Among the Permian fossils of Texas is a creature called *Seymouria* that is almost equally amphibian and reptile. Although *Seymouria* itself lived too late to be an actual ancestor of the reptiles, it exemplifies the type of animal that made this transition. Reptiles perfected the move to land begun by the amphibians. This development was made possible primarily by the evolution of an egg protected from desiccation by an additional embryonic membrane, the amnion. A shell around the entire egg provided some protection and support; thus, reproduction became independent of standing water, although Romer (1966) noted that the adults may have remained tied to water for a little longer. Because we do not know just which of the transitional forms began the new kind of egg, it is hard to draw sharp lines between ancient amphibians and ancient reptiles, even though the modern types are quite distinct. Even the skeletal features, with which paleontologists usually work, are intermediate.

Sometime in the Carboniferous, however, a group of distinctly reptilian beasts arose. These are called "cotylosaurs" ("cuplizards," named for the cuplike shape of one of the vertebral elements), but they are better known as "stem reptiles" because they seem to be the stock or stem from which all remaining forms arose. Reptiles and amphibians were numerous in the early Permian, and, throughout the Mesozoic era, the reptiles dominated the land.

The turtles are the direct descendants of the cotylosaur type. Early in their evolutionary history, the cotylosaurs also gave rise, by still-debated routes, to several extinct and three surviving lineages. One group, called "eosuchians" ("dawn crocodiles"), emerged from the basic stem in Permian times and eventually gave rise to the modern reptiles. Another group, the pelycosaurs ("bowl lizards"), diverged even earlier from the basic stem and led to the therapsids that produced the mammals. The third group, which may have arisen from the early eosuchians or directly from the stem reptiles, were the archosaurs ("ruling lizards"), which produced both modern crocodilians and perhaps the birds.

The modern lepidosaurs are descendants of the eosuchians. The rhynchocephalians split off from the early eosuchians, however, flourished moderately in the Triassic and Jurassic periods, and then dwindled until they were represented solely by the remnant *Sphenodon*. The squamates blossomed later than the rhynchocephalians and are still doing fairly well today. Nevertheless, reptiles as a group are far less diversified now than they were during the Mesozoic era.

The archosaurs included an incredible array of dinosaurs of two types, the pterosaurs (all extinct at the end of the Cretaceous period), a diversity of alligator-like creatures now reduced in variety, and perhaps a line that may have led to birds. Each of these five lines radiated from a primitive group of archosaurs called "thecodonts" ("socket tooth"). Archosaurs typically shared with the lepidosaurs the characteristic of two openings in the temporal region of the skull, but they differed in other skeletal features. The most flourishing modern descendants of the thecodont archosaurs, according to the classic view, are the birds, first recorded in the Jurassic period in Germany. The earliest known avian fossil, *Archaeopteryx*, bears many resemblances to reptiles: teeth on the jaws, a long tail, and other reptilian skeletal features. Most conspicuous, however, were the feathers covering the body, which clearly distinguish this animal from more typical reptiles. Some paleontologists have argued that birds probably descended from birdlike dinosaurs called "ornithischians." Had it not been for the feathers, *Archaeopteryx* would have been classified as an ornithischian dinosaur. This kind of evidence has resulted in the suggestion that the birds be classified with the dinosaurs rather than separately. Birds did not greatly diversify until the Tertiary era, although they lost the most reptilian of their skeletal features during the Cretaceous period. At some unknown point, they became "warm-blooded." Most of the modern orders of birds were represented in the Eocene period of the Tertiary era.

The branch of cotylosaurs that produced the mammalian lineage was characterized by a single temporal opening in the skull. The first pelycosaurs are known from the late Carboniferous and were succeeded by the therapsids during the Permian. There were many different kinds of therapsids, large and small, meat-eating and vegetarian, but only one small-bodied kind persisted into the Jurassic. Many kinds of therapsids gradually acquired a suite of mammal-like skeletal features, and somewhere along the line, they also became quite "warm-blooded." Mammals are presumed to have originated toward the end of the Triassic. They remained rather rare in the fossil record until the Cenozoic era, when they (and the birds) radiated extensively. Most of the present-day orders of mammals were present at least by the Eocene period and probably originated in the Paleocene.

The next major breakthrough in vertebrate evolution was achieved by both birds and mammals (or their predecessors, still classified as reptiles). The condition of warm-bloodedness implies a high metabolic rate and the ability to regulate body temperature internally, rather than depending on the environment for heat. We will see later that this ability opened new vistas for many vertebrates.

This has been a rapid-fire survey of the major events in vertebrate phylogeny with an emphasis on the broad picture pieced together from the past. The reconstructions of the vertebrate family tree are, by necessity, inferential; the action occurred long ago and left only fragmentary traces of the whole story. Many of the uncertainties will undoubtedly remain unresolved. Even if every single fossil were discovered and interpreted correctly, many past vertebrate types never became fossilized or their remains were subsequently ruined by geological events that changed the rock formations. Despite the many missing links, paleontologists have been extraordinarily successful in creating a plausible account of a highly successful dynasty.

SELECTED REFERENCES

Alexander, R.M., 1981. *The Chordates.* 2nd ed. Cambridge University Press, Cambridge, England.

Bakker, R.T., 1971. Dinosaur physiology and the origin of mammals. *Evol., 25*:636–658. (See also rebuttals and counterrebuttals by several authors in *Evol., 27* and *28.*)

Bakker, R.T., 1975. Dinosaur renaissance. *Sci. Am., 232*(4):58–78.

Barrington, E.J.W., and R.P.S. Jeffries (Eds.), 1975. Protochordates. *Symp. Zool. Soc. Lond., 36*:1–361.

Bellairs, A., 1970. *The Life of Reptiles.* 2 vols. Universe Books, New York.

Bellairs, A., and C.B. Cox, 1976. *Morphology and Biology of Reptiles.* Academic Press, New York.

Brodal, A., and R. Fänge (Eds.), 1963. *The Biology of Myxine.* Universitetsforlaget, Oslo, Norway.

Budker, P., 1971. *The Life of Sharks.* Columbia University Press, New York.

Cochran, D.M., 1961. *Living Amphibians of the World.* Doubleday, Garden City, NY.

Colbert, E.H., 1969. *Evolution of the Vertebrates.* 2nd ed. John Wiley & Sons, New York.

Dean, B., 1895. *Fishes, Living and Fossil.* Macmillan Publishing Co., New York.

Dorst, J., 1974. *The Life of Birds.* 2 vols. Columbia University Press, New York.

DuBois, A.B., G.A. Cavagna, and R.S. Fox, 1975. The forces resisting locomotion in bluefish. *In* T.Y.T. Wu, C.J. Brokaw, and C. Brennen (Eds.), *Swimming and Flying in Nature.* Vol. 2. Plenum Publishing, New York.

Dunson, W.A., 1975. *The Biology of Sea Snakes.* University Park Press, Baltimore.

Ewer, R.F., 1973. *The Carnivores.* Cornell University Press, Ithaca, NY.

Gans, C., 1975. *Reptiles of the World.* Bantam Books, New York.

Gans, C. (Ed.), 1969–1981. *Biology of the Reptilia.* 11 vols. Academic Press, New York.

Gilbert, P.W., R.F. Mathewson, and D.P. Rall, 1967. *Sharks, Skates and Rays.* Johns Hopkins University Press, Baltimore.

Gilliard, E.T., 1958. *Living Birds of the World.* Doubleday, Garden City, NY.

Gosline, W.A., 1971. *Functional Morphology and Classification of Teleostean Fishes.* University Press of Hawaii, Honolulu.

Greenwood, P.H., R.S. Miles, and C. Patterson, 1973. Interrelationships of fishes. *Zool. J. Linn. Soc., 53* (Suppl. 1). Academic Press, New York.

Greenwood, P.H., D.E. Rosen, S.A. Weitzman, and G.S. Myers, 1966. Phyletic studies of teleostean fishes, with a provisional classification of living forms. *Bull. Am. Mus. Nat. Hist., 131*(4):339–456.

Gregory, W.K., 1951. *Evolution Emerging.* 2 vols. Macmillan Publishing Co., New York.

Guggisberg, C.A.W., 1972. *Crocodiles.* Stackpole Books, Harrisburg, PA.

Hardisty, M.W., and J.C. Potter, 1971, 1972. *The Biology of Lampreys.* 2 vols. Academic Press, New York.

Harless, M., and H. Morlock, 1979. *Turtles.* Wiley-Interscience, New York.

Harrison, M.L., 1971. *The Life of Mammals.* Vols. 1 and 2. Universe Books, New York.

Heilmann, G., 1926. *The Origin of Birds.*

Witherby, London.
Herald, E.S., 1961. *Living Fishes of the World.* Doubleday, Garden City, NY.
Hildebrand, M., 1974. *Analysis of Vertebrate Structure.* John Wiley & Sons, New York.
Hopson, J.A., 1977. Relative brain size and behavior in archosaurian reptiles. *Annu. Rev. Ecol. Syst., 8:*429–448.
Lagler, K.F., J.E. Bardach, R.R. Miller, and R.R.M. Passino, 1977. *Ichthyology.* 2nd ed. John Wiley & Sons, New York.
Marshall, N.B., 1966. *The Life of Fishes.* World Publishing Co., Cleveland.
McFarland, W.N., F.H. Pough, T.J. Cade, and J.B. Heiser, 1979. *Vertebrate Life.* Macmillan Publishing Co., New York.
Miller, P.J., 1979. Adaptiveness and implications of small size in teleosts. *Symp. Zool. Soc. Lond., 44:*263–306.
Nelson, J.S., 1976. *Fishes of the World.* John Wiley & Sons, New York.
Norman, J.R., and P.H. Greenwood, 1963. *A History of Fishes.* Hill & Wang, New York.
Oliver, J.A., 1955. *The Natural History of North American Amphibians and Reptiles.* Van Nostrand Reinhold, New York.
Olson, E.C., 1971. *Vertebrate Paleozoology.* Wiley-Interscience, New York.
Olson, S.L., 1982. A critique of Cracraft's classification of birds. *Auk, 99:*733–739.
Orr, R.T., 1982. *Vertebrate Biology.* 5th ed. Saunders College Publishing, Philadelphia.
Owen, R., 1886–1888. *On the Anatomy of Vertebrates.* Vol. 2. Longmans, Green, London.
Porter, K.R., 1972. *Herpetology.* Saunders College Publishing, Philadelphia.
Romer, A.S., 1966. *Vertebrate Paleontology.* 3rd ed. University of Chicago Press, Chicago.
Romer, A.S., 1970. *The Vertebrate Body.* 5th ed. Saunders College Publishing, Philadelphia.
Sanderson, I.T., 1967. *Living Mammals of the World.* Doubleday, Garden City, NY.
Schmidt, K.P., and R.F. Inger, 1957. *Living Reptiles of the World.* Doubleday, Garden City, NY.
Scott, W.B., and E.J. Crossman, 1973. *Freshwater Fishes of Canada.* Fisheries Research Board of Canada, Ottawa.
Sterba, G., 1966. *Freshwater Fishes of the World.* Rev. ed. Pet Library, New York.
Stonehouse, B., and D. Gilmore (Eds.), 1977. *The Biology of Marsupials.* University Park Press, Baltimore.
Taylor, E.H., 1968. *The Caecilians of the World: A Taxonomic Review.* University of Kansas Press, Lawrence.
Tracy, C.R., 1976. Tyrannosaurus: evidence for endothermy? *Am. Nat., 110:*1105–1106.
Van Tyne, J., and A.J. Berger, 1976. *Fundamentals of Ornithology.* 2nd ed. John Wiley & Sons, New York.
Vaughan, T.A., 1978. *Mammalogy.* 2nd ed. Saunders College Publishing, Philadelphia.
Vial, J.L. (Ed.), 1973. *Evolutionary Biology of the Anurans.* University of Missouri Press, Columbia.
Walker, E.P., 1975. *Mammals of the World.* 3rd ed. 2 vols. Johns Hopkins University Press, Baltimore.
Webster, D., and M. Webster, 1974. *Comparative Vertebrate Morphology.* Academic Press, New York.
Welty, J.C., 1975. *The Life of Birds.* 2nd ed. Saunders College Publishing, Philadelphia.
Yapp, W.B., 1965. *Vertebrates: Their Structure and Life.* Oxford University Press, Oxford.
Young, J.Z., 1981. *The Life of Vertebrates.* 3rd ed. Oxford University Press, Oxford.
Zangerl, R., and G.R. Case, 1976. *Cobelodus aculeatus* (Cope), an anacanthous shark from Pennsylvanian black shales of North America. *Paleontographica, 154:*107–157.

Species Accounts of Vertebrates

The following books offer species accounts of many vertebrates, especially those of North America. Not listed are the many state bird and mammal books, the excellent field guides, or the taxonomic keys.

Bigelow, H.B., and W.C. Schroeder, 1953. *Fishes of the Western North Atlantic.* 6 vols. Sears Foundation of Marine Research, Yale University, New Haven.
Bishop, S.C., 1943. *Handbook of Salamanders: The*

Salamanders of the United States, of Canada, and of Lower California. Comstock, Ithaca, NY.

Carr, A., 1952. *Handbook of Turtles: The Turtles of the United States, Canada, and Baja California.* Comstock, Ithaca, NY.

Ernst, C.H., and R.W. Barbour, 1972. *Turtles of the United States.* University Press of Kentucky, Lexington.

Guggisberg, C.A.W., 1972. *Crocodiles.* Stackpole Books, Harrisburg, PA.

Ingles, L.G., 1965. *Mammals of the Pacific States.* Stanford University Press, Stanford, CA.

Klauber, L.M., 1956. *Rattlesnakes.* University of California Press, Berkeley.

Palmer, R.S. (Ed.), 1962. *Handbook of North American Birds.* 3 vols. Yale University Press, New Haven.

Peterson, R.L., 1966. *The Mammals of Eastern Canada.* Oxford University Press, Toronto.

Pflieger, W.L., 1975. *The Fishes of Missouri.* Missouri Department of Conservation.

Pope, C.H., 1955. *The Reptile World.* Alfred A. Knopf, New York.

Smith, H.M., 1946. *Handbook of Lizards: Lizards of the United States and Canada.* Comstock, Ithaca, NY.

Stebbins, R.C., 1951. *Amphibians of Western North America.* University of California Press, Berkeley.

Stebbins, R.C., 1954. *Amphibians and Reptiles of Western North America.* McGraw-Hill Book Co., New York.

Wright, A.H., and A.A. Wright, 1949. *Handbook of Frogs and Toads of the United States and Canada.* Comstock, Ithaca, NY.

Wright, A.H., and A.A. Wright, 1957. *Handbook of Snakes of the United States and Canada.* Comstock, Ithaca, NY.

3 *Distributions: Ecology and Geography*

Zoogeography is the study of the distribution of animals on the earth; it concerns not only who lives where, but also why they live where they do. Much zoogeography deals with distribution on a global and continental scale, but since many of the patterns and processes occur on smaller scales as well, we can legitimately include at least some aspects of distribution on the scale of the individual organism and community. Whittaker (1975) defined a community as an assemblage of populations of organisms living together and linked together by their effects on each other and their responses to the environment. Only seldom are the borders of a community sharply demarcated; usually, one grades into another. For our purposes, the fishes in a pond, the birds of a woodlot, or all the animals of a mountain meadow can constitute a community, the limits of which depend on the purpose of the study.

Why does an animal live where it does? The geographic range of a species is limited. No species is completely cosmopolitan; some have extremely broad ranges, and others have severely limited ones. What factors limit species' ranges? Sometimes a species does not occur naturally in every suitable environment. The ring-necked pheasant (*Phasianus colchicus*) of Asia, for example, is well adapted to certain parts of North America where it has been introduced, but it did not occur there naturally (Fig. 3–1). The mosquitofish, *Gambusia affinis,* a native of southeastern North America, has been successfully introduced around the world. Historical factors may determine range limits, if not all environments are accessible to

Figure 3–1 The ring-necked pheasant, *Phasianus colchicus.* A native of central and southern Asia, it has been successfully introduced into Europe, North America, New Zealand, Hawaii, and other islands. Its common name is deceptive because males of some forms have no white neck ring. The scientific name stems from the legend that the bird first became known to Europeans when brought home from Asia Minor (near the river Phasis in the old province of Colchis) by returning Greek Argonauts. Some historians believe that the skin and plumage of this bird may have been Jason's golden fleece. A black-and-white photograph hardly does justice to the resplendent plumage. (Photo by C. Beamer, *Champaign-Urbana News Gazette.*)

potential colonizers at the right time, and, obviously, both physical and biological factors of the environment itself may contribute to range limits of a species. No organisms have universal adaptations that permit survival and reproduction under all possible conditions. The factor(s) limiting the range of a species differ among species and sometimes even among the populations that compose a species. Commonly, several factors combine to restrict a species' range.

Zoogeographically it is of little interest to make a catalog of limiting factors for a species or any other taxon. Far more interesting is the search for patterns. How many species live in a certain place? To what other species are they related? What factors might control the number of species in a given place? What are the temporal patterns of distributional changes? In what circumstances do different range-limiting factors operate?

We will begin with a brief account of continental history and its effect on animal distribution, followed by a survey of present zoogeographic units. We will then investigate the climatic and ecological factors that help determine animal distributions on a smaller scale. The chapter concludes with a summary of many of these factors as seen in island microcosms.

CONTINENTAL DRIFT AND ANIMAL DISTRIBUTION

The continents have not always been in their present locations, nor have they always had the same size, shape, or connections among themselves. Changes in climate and topography associated with the spatial relationship of the continents have created both restrictions and opportunities for invasion, adaptive radiation, and faunal differentiation.

Movement of the continents is called "continental drift." Overlying a hot, molten interior, the surface of the earth is a crust composed of several rigid, moving plates that are up to 100-km thick. At some plate boundaries, massive convection currents in the interior bring material up to the surface between the plates, a process that laterally displaces older surface material. One area of upwelling forced the separation of South America and Africa and the consequent expansion of the Atlantic Ocean. The present mid-Atlantic Ridge marks the region of upwelling, which continues today. Separation of continents is accompanied by the downward movement of the ocean floor in other regions as the edge of one plate folds under the edge of another, forming a trench. Downward-moving material may pass beneath the edge of a continent, and this movement may cause earthquakes and mountain formation (for example, the Andes region of South America). Collision of continental masses propelled by these forces and moving at the estimated rate of 5 to 10 cm/year also results in the crumpling and upthrusting of new mountain ranges where none had previously existed; the Himalayas between India and northern Asia are an example.

These grand-scale events may have been going on as long as animals have lived on earth and probably much longer. Because the vertebrate fossil record is very spotty before the Silurian period of the Paleozoic era, we can begin with intercontinental relations at that time. Only the more recent of these events (usually) have had detectable effects on the distribution of present-day vertebrate groups, but more ancient events undoubtedly created conditions that have had an impact on the patterns of adaptive radiation. Therefore, a brief summary of our present understanding of early continental events is appropriate here, even if their influence on modern patterns must remain largely hypothetical. Only the movements of major continental pieces are traced here; smaller bits were added to some of the continental masses at various times.

In Silurian and Devonian times, a supercontinent called Gondwanaland, which was composed of the landmasses we know as South America, central and south Africa, Australia, Antarctica, and India, lay in the southern hemisphere; the South Pole was somewhere under present-day Africa. (China and southeastern Asia are omitted from this discussion because their early whereabouts are uncertain.) Europe and North America (previously separate) were united on the equator, and Asia lay to the north (Fig. 3–2). During these periods, the early evolution of fishes and amphibians left fossil records, especially in the then-tropical supercontinent of Euramerica. A similar state of continental affairs persisted through the Carboniferous and lower Permian periods, and the early reptiles are also best known from Euramerica.

During these periods, the supercontinents were moving, and, during the Permian period, all three apparently met and merged to form the mighty landmass called Pangaea (Fig. 3–3). Gondwanaland moved north until Australia-Antarctica was near the South Pole, and Asia migrated southward. Extensive faunal similarities among all the formerly separate continents became apparent in the upper Permian and Triassic. A major radiation of reptiles occupied the upper Permian and spilled new groups into virtually all parts of Pangaea.

Throughout this early time, radical changes in sea level further modified the sizes of continents and the distances between them. These changes continued through later times and their effects are more readily detected because the fossil record is better. Figure 3–4 shows potential changes in modern continental shapes that would result from a 180-m raising or lowering of present-day sea levels. Uncertainties still exist concerning the temporal patterns of sea-level changes, however, and the paleogeography of post-Triassic times is often confusing and controversial. We know, at least, that continents as well as seas rose and fell at various times.

During the Jurassic, Gondwanaland became increasingly separate from the rest (an incipient supercontinent called Laurasia) as the Atlantic Ocean gradually opened the gap (Fig. 3–5). Distributional patterns of dinosaurs suggest the presence of remaining land routes between Europe and Africa and between Alaska and Siberia. Evidence indicates that mild climates

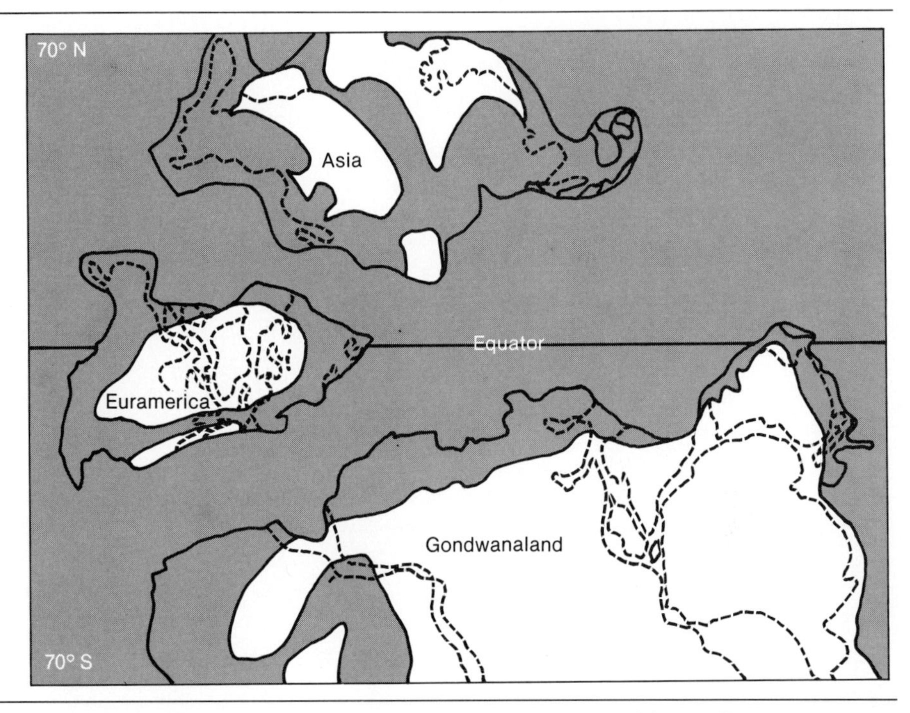

Figure 3–2 A reconstruction of the positions of the continents in Silurian-Devonian times. Oceans are stippled. Edges of the continental shelf, a zone of shallow water around each coastline, are shown by solid lines. Present-day coastlines are indicated by dashed lines when they are far from the coast or continental shelf. The indicated latitudinal positions are better known and more accurate than the longitudes. (Modified from C.B. Cox, 1974. Vertebrate paleodistributional patterns and continental drift. *J. Biogeog.*, *1*:75–94.)

obtained during the Mesozoic and high-latitude land bridges between the supercontinents may have provided migration routes for many kinds of vertebrates.

The tendencies of the Jurassic continued in the Cretaceous period (Figs. 3-6, 3-7). Widening of the Atlantic finally separated Euramerica from Africa, and, by the end of the period, South America had been isolated from Africa. Australia and Antarctica remained together but separated from Africa (as did India), although a connection with South America probably existed until the late Cretaceous. Wherever the marsupial mammals may have originated (a controversial matter), they may have reached Australia from South America across Antarctica. The subsequent separation of these three landmasses (perhaps late in the Cretaceous) may have prevented the arrival in Australia of the early placental mammals, which might have originated in Asia. A partial barrier to vertebrate migration developed be-

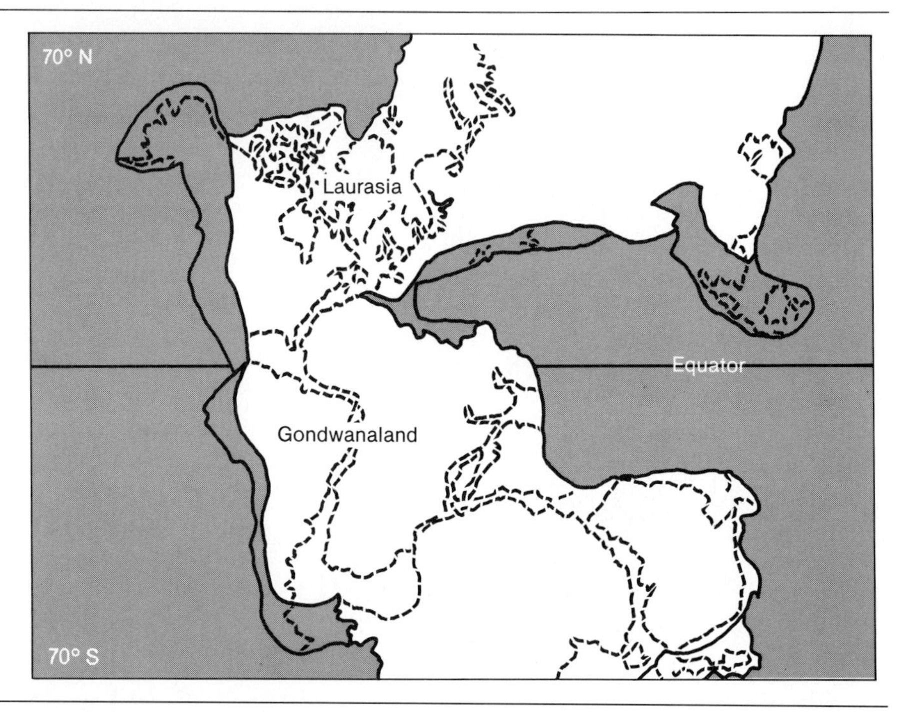

Figure 3–3 A reconstruction of continental positions in the late Permian and Triassic periods. All the continents here have merged into Pangaea. (Modified from C.B. Cox, 1974. Vertebrate paleodistributional patterns and continental drift. *J. Biogeog.*, 1:75–94.)

tween North America and Asia, and although some groups migrated eastward, westward movement apparently was restricted.

North America was separated from Europe quite early in the Tertiary (Eocene epoch), and a lowered sea level permitted a land connection between Europe and Asia. India finally moved north to join Asia. Still later, during the Pliocene, a land route was reestablished between North and South America. The emergence of the Central American bridge between North and South America permitted a diversity of placental mammals to invade the southern continent. At roughly the same time, many native South American groups, notably marsupial carnivores and (placental) ungulates went extinct. Endemic South American rodents and edentates were much reduced in diversity. During the same period, of course, some South American mammals moved north into Central America and a number of Central American endemics disappeared, but only a few (including opossums and armadillos) reached and survived in temperate North America (Marshall et al., 1982).

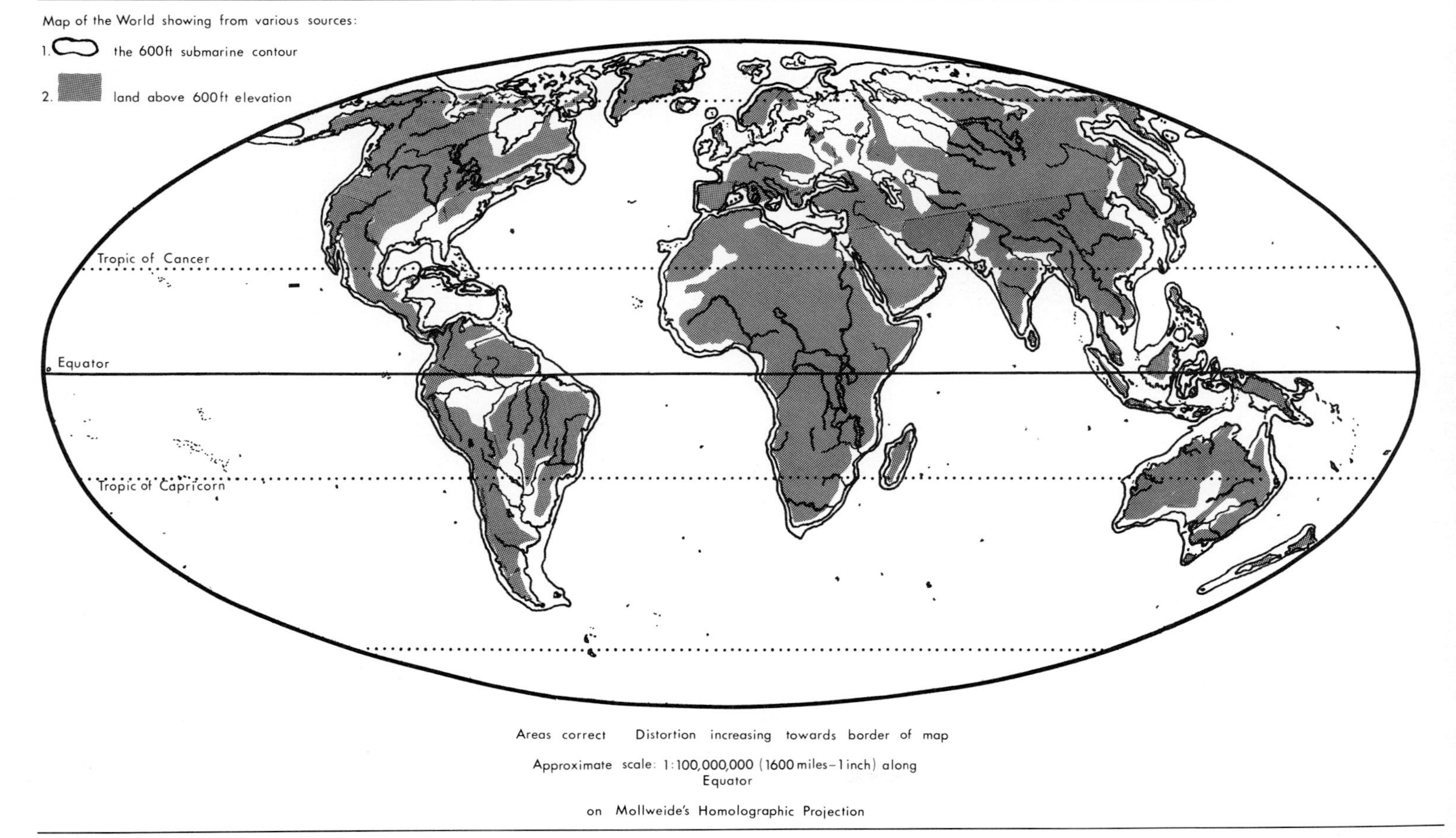

Figure 3–4 A map of present-day continents showing the potential effects of sea-level changes on continental shapes and interconnections. This example exhibits the contours created by a drop or rise of 180 m from present levels. (Modified from R. Good, 1974. *The Geography of Flowering Plants.* Longman Group Limited, London.)

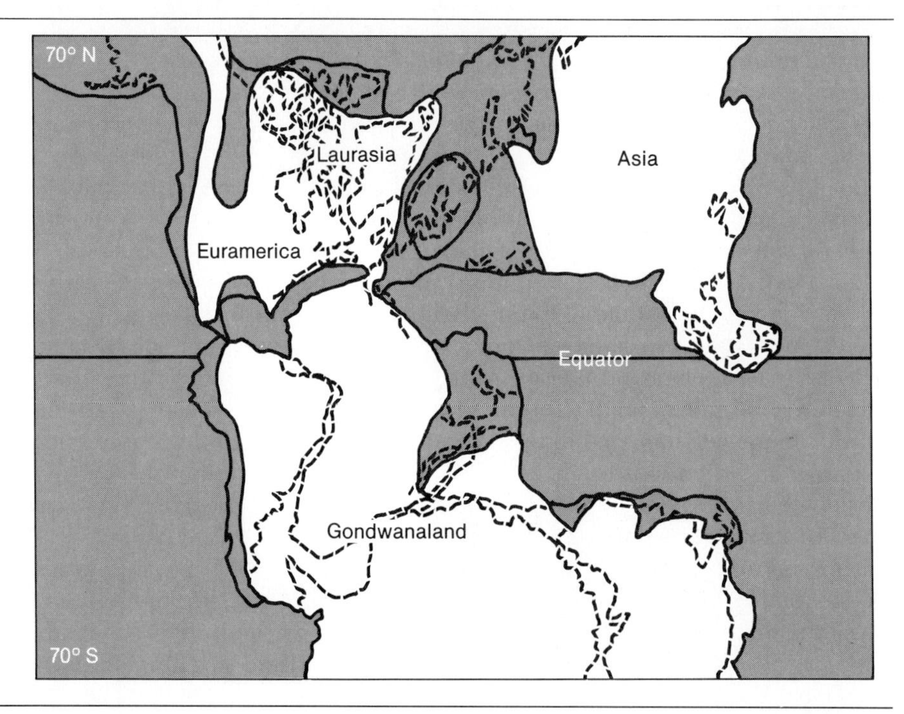

Figure 3–5 A reconstruction of continental positions in the mid-Jurassic period. (Modified from C.B. Cox, 1974. Vertebrate paleodistributional patterns and continental drift. *J. Biogeog.*, 1:75–04.)

As the landmasses became increasingly isolated, worldwide climatic changes compressed the tropical zone toward the equator and produced greater climatic zonation latitudinally. Climatic zonation tended to increase as the Tertiary and Quaternary progressed. Both large-scale climatic changes and invasions by new species changed the distribution of previous inhabitants. As we have seen, the invasion of South America by placental mammals and of Central America by marsupials was associated with range shifts and extinctions of certain earlier residents. Climatic alterations have produced dramatic changes in distribution. A good example is the major shift of distribution of both the giraffe and the elephant in Africa: Both species occurred in the Saharan region thousands of years ago but retreated southward as the climate became drier (McIntosh and McIntosh, 1981).

In some cases, ranges were so greatly reduced that formerly widespread forms were now found in relatively small areas or, sometimes, in several small areas isolated from each other by vast stretches of land; these are relict distributions. A classic example of a relict distribution is provided by

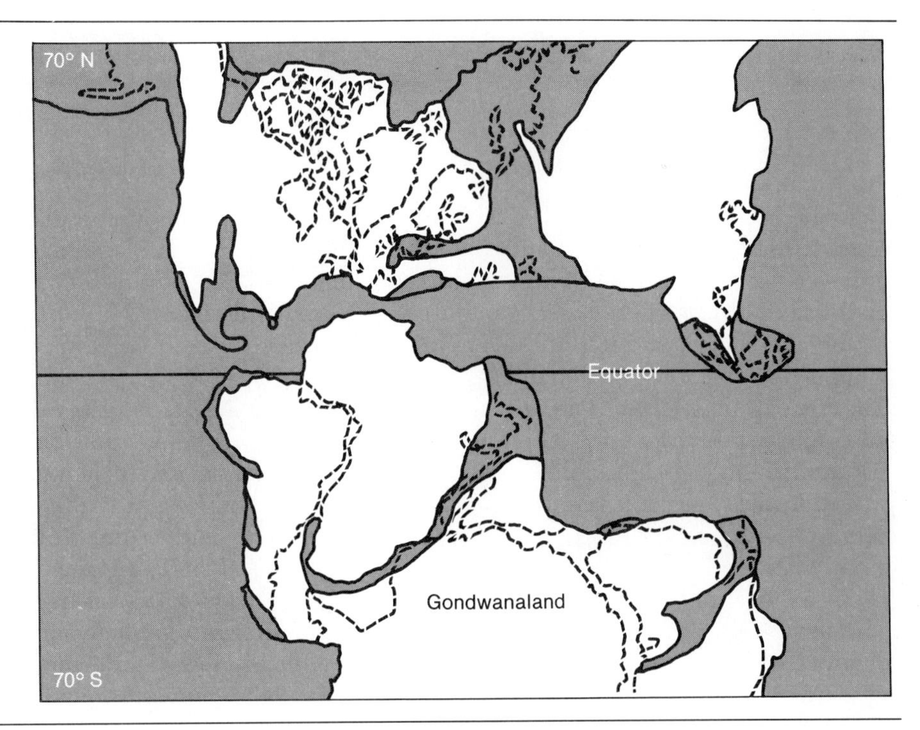

Figure 3–6 Continental positions in the early Cretaceous period. (Modified from C.B. Cox, 1974. Vertebrate paleodistributional patterns and continental drift. *J. Biogeog.*, *1*:75–94.)

the African crocodile (*Crocodylus niloticus*). Formerly distributed over much of Africa, including what is now the Sahara Desert, this crocodile's range was constricted by climatic changes. A relict population, however, could be found in the freshwater pools of a mountain range in the middle of the otherwise unsuitable Sahara about 1300 km from the nearest remaining populations along the major African rivers. (This relict may now have been hunted to extinction.) Restriction of former geographic range has sometimes resulted in extremely disjunct family distributions—the peculiar chondrostean bony fishes, the paddlefishes, occur in China and the eastern United States, and the tapirs are found in Latin America and southeastern Asia, for example.

Distribution patterns of many modern vertebrates provide evidence of earlier events on a continental scale. Freshwater fishes seem to have radiated primarily in the two large landmasses of Gondwanaland and Laurasia, with later interchange among faunas of these primary sources. The freshwater fish faunas of South America and Africa have many elements in common. About five families (including lepidosirenid lungfishes, cichlids,

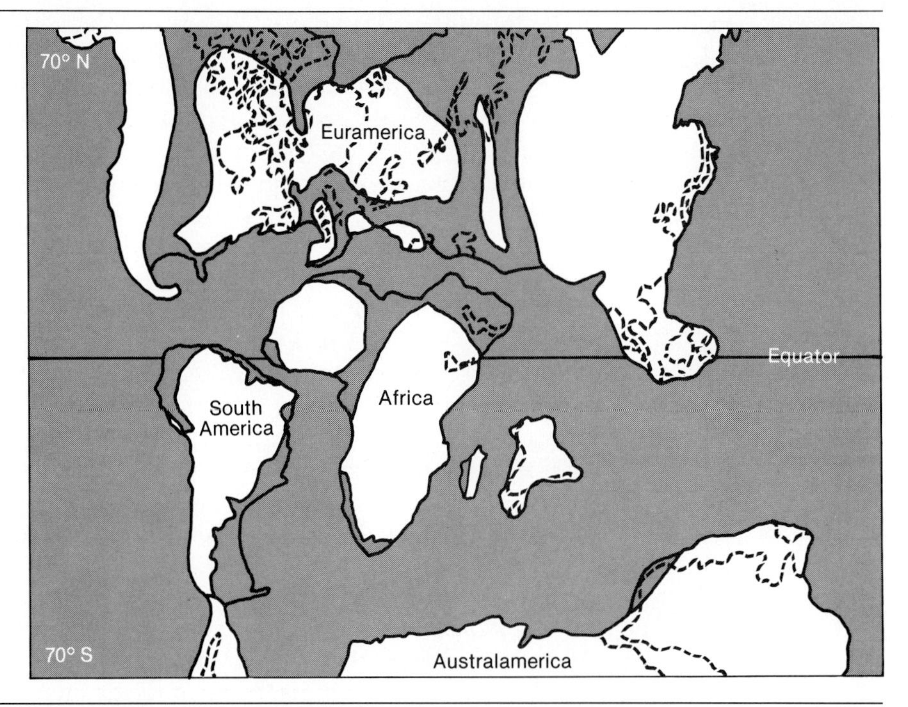

Figure 3–7 Continental positions in the late Cretaceous period. (Modified from C.B. Cox, 1974. Vertebrate paleodistributional patterns and continental drift. *J. Biogeog.*, *1*:75–94.)

and characids) are shared by these continents. Catfishes (Siluriformes) underwent separate but quite parallel radiations in South America and Africa, and some forms later invaded the northern continents. The Ceratodontidae (including the modern Australian lungfish, *Neoceratodus*) are now confined to Australia, but this distribution is a relict one, much reduced from a broader distribution during the Mesozoic. Ancient members of this family probably reached Asia when India separated from Africa and moved north. Several different groups had centers of distribution in the northern supercontinent of Laurasia. The families to which sturgeons and bowfins belong are among those with Laurasian records dating from the Cretaceous. The minnow-sucker group (four families) of Cypriniformes has an entirely Laurasian distribution, except for a relatively recent invasion of Africa. All Percidae and Percopsiformes (three families), as well as a number of Salmoniformes (pike, mudminnows, trout), have a northern distribution.

Distributional histories of modern amphibia are more difficult. Present evidence supports a Gondwanaland origin for caecilians and a Laurasian one for salamanders (which are represented in North Africa by a few salamandrids, the newts, and in South America only by some plethodontids, the lungless salamanders). Most frog families may have originated in Gondwanaland, but some, including the rhinophrynids (one species of burrowing toad) of Mexico and Central America and pelobatids (spadefoot toads) of North America and Eurasia, seem to be clearly Laurasian. Reptilian distributions are also somewhat confusing. Many groups of apparent Laurasian or Gondwanaland origin have since invaded other areas and are sometimes now absent from their original centers of distribution; a few clear examples follow. Chelyid turtles now reside in South America, Australia, and New Guinea, which indicates a Gondwanaland origin. Iguanid lizards are now found primarily in Latin America, and South American fossils of this family are known from the Cretaceous. These lizards later dispersed to and radiated in North America. Iguanids also occur in Madagascar, however, which is strong evidence of their former existence in Africa and, hence, of their origin in old Gondwanaland. On the other hand, the helodermatid lizards (gila monsters and relatives) now have a relict distribution in southwestern North America, but early Tertiary fossils show they occurred in Europe as well.

Intercontinental connections, especially during the prolonged Cretaceous breakup of Gondwanaland, influenced the geographic distribution of many modern avian groups: the "ratites" (ostriches, emus, cassowaries, rheas, kiwis, and tinamous), penguins, some of the galliforms, the suboscine passerines, and perhaps also the pigeons, parrots, and cuckoo-like birds. All these groups are widely distributed today, especially in the southern hemisphere. The oscine passerines, on the other hand, probably originated in Eurasia and reached the New World mostly in the early Tertiary.

Most present-day mammalian families seem to be no older than the Tertiary; hence, continental drift had little effect on their present distribution patterns. Most major groups appear to be derived from taxa that originated in Laurasia and subsequently invaded Africa and South America. Thus, the proboscideans, primates, carnivores, and rodents are among those with a northern origin, although the first two groups are no longer represented in North America and northern Eurasia. Certain rodents (caviamorphs) and primates in South America seem to be related to those of Africa, however, and may have reached South America in the early Tertiary before these two southern continents were so greatly separated.

PRESENT ZOOGEOGRAPHIC UNITS

Dividing lines between major faunal regions are arbitrary divisions established for convenience in studying distributional relationships. Divisions are usually established where major differences in the fauna occur. Sometimes larger taxa such as families are confined to one portion of the globe, and sometimes the range limits of several species may coincide. In either case, one could observe a more or less abrupt change in the composition of the fauna in the vicinity of the dividing line.

Oceanic

Marine zoogeography is less well documented than terrestrial zoogeography, perhaps largely because of the difficulties in obtaining good distributional information on animals that are relatively inaccessible to human study. This is especially true of deep-sea organisms, about which little is known. We can, however, at least begin to outline the major zoogeographic regions now perceived by oceanic zoogeographers (Fig. 3–8).

Continental shelves of the tropical oceans have been divided into four huge faunal regions, one on each side of the Atlantic and the Pacific: *Indo-West Pacific, Eastern Pacific, Western Atlantic,* and *Eastern Atlantic.* Each is separated from the others by an intervening continental landmass or deep watermasses, and each is characterized by numerous endemic taxa (groups restricted to a given area), of which only the fishes are mentioned here. Only about 60 species of Pacific shore fishes are common to both eastern and western Pacific continental shelves, but the western Pacific shelf has almost 400 species of its own, and the eastern Pacific shelf has about 600 of its own. The Indo-West Pacific has over 30 families of deep-sea and coastal fishes found nowhere else. Similarly, the New World landmass separating the tropical Pacific from the tropical Atlantic seems to have been a highly effective barrier; only about 1 percent of the shore-fish species (excluding the circumtropical ones) can be found on both sides of the Isthmus of Panama.

Five zoogeographic regions can be distinguished in the shallow, warm-temperature water of the southern hemisphere: *Southern Australia, Northern New Zealand, Western South America, Eastern South America,* and *Southern Africa.* Some fish families, such as the Latridae (trumpeters) and Aplodactylidae, are widespread in such waters (but not typically found elsewhere). Other families or genera may be found only in one of these faunal regions. The Southern Australia region is the most distinctive, with about six endemic fish families: Alabidae (singleslit eels), Siphonognathidae, Peronedysidae, Dinolestidae (sometimes pooled with the family Apogonidae, which is not restricted to this region), and Gnathanacanthidae (red-velvet fish), and Pataecidae. Three others are shared only with the New Zealand area. Southern Africa is distinguished by one endemic subfamily or family: the Halidesmidae (sometimes a subfamily of the Congrogadidae of the Indo-West Pacific area). The other warm-temperate regions of the southern hemisphere are apparently distinctive only at the generic level.

Four comparable regions are in the northern hemisphere: *Mediterranean-Atlantic, Carolina* (Gulf of Mexico and southeastern Atlantic coast of North America), *California* (southern California coast and the Gulf of California in

Mexico), and *Japan* (South Japan, southern Korea, and the straits between Taiwan and China). Fish families are generally shared with the regions to the south (especially) and north, and regional distinctions seem to be made primarily at the species level.

Cold-temperate faunal regions in the southern hemisphere include *Southern South America, Tasmania* (including southeastern Australia), *Southern New Zealand,* and the *Subantarctic* (small oceanic islands around Antarctica, a circumpolar region because currents flow uninterrupted around the southernmost continent). Five related families are confined to the southern cold-temperate regions; Bovichthyidae, Nototheniidae (cod icefishes), Harpagiferidae (may be included in the preceding family), Chaenichthyidae (crocodile icefishes), and Bathydraconidae. Distinctions among the four faunal regions in this category appear to be found in the distributions of species. Likewise, there are four faunal regions in corresponding zones in the northern hemisphere: *Western* and *Eastern Atlantic Boreal* and *Western* and *Eastern Pacific Boreal* regions. Faunal distinctions are based mainly on the coincidence of species' limits, although each of the Pacific Boreal regions has one unshared family. Each polar zone—the Arctic and the Antarctic—constitutes a single faunal region that is distinguished from neighboring zones primarily by species' distributions.

Regional divisions of fauna in the open ocean and in the deep sea are not at all clear. It is customary to divide these faunas vertically (zones are based on such factors as depth and light penetration) rather than horizontally; therefore, zoogeographic regions are not demarcated.

Terrestrial and Freshwater

For many years, zoogeographers have divided the earth's land surfaces into several major faunal regions (Fig. 3–9). Each region has inhabitants that are, on average, more similar taxonomically to each other than to inhabitants of other major regions. These major regions are clearest in terms of the distributions of mammals and birds but also apply to other nonmarine vertebrates. Distinctions between the regions often depend on differences at the family level, in addition to genus and species distributions; hence, the terrestrial regions are probably less finely divided than the oceanic ones. Darlington (1957) provides a detailed survey of modern distributions of freshwater and terrestrial vertebrates, but his book was written before continental drift was well accepted, so the mechanisms that led to present distributions need to be reinterpreted.

The *Australian* region is most distinct. It is the major remaining center of marsupial mammal distribution; six marsupial families live only there. Monotreme mammals and about 14 avian families, including the Cracticidae (bellmagpies), Ptilonorhynchidae (bowerbirds), and Paradiseidae (birds of paradise) are found nowhere else. About five families of fresh-

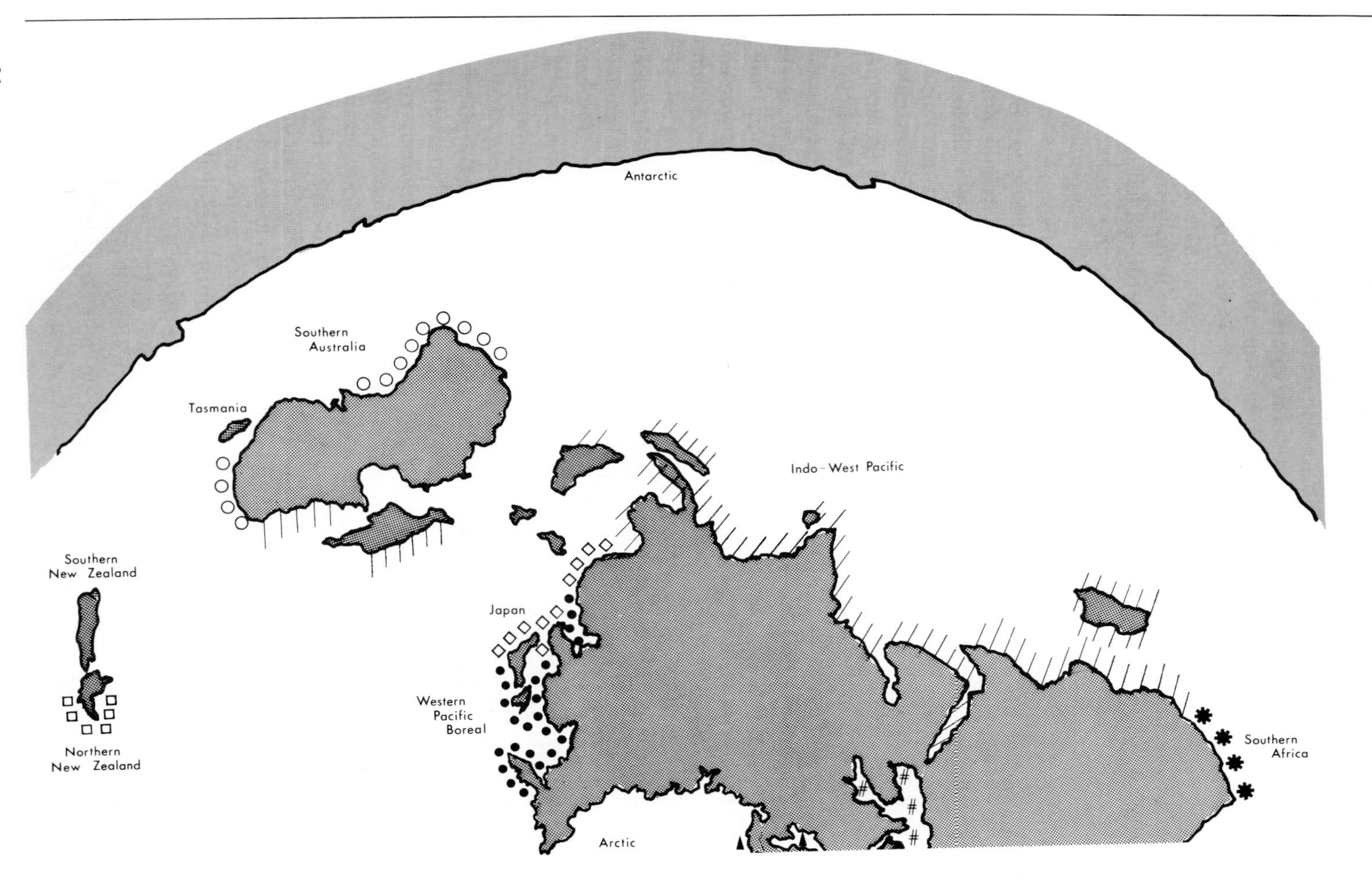

Antarctic
Southern Australia
Tasmania
Indo-West Pacific
Southern New Zealand
Japan
Western Pacific Boreal
Northern New Zealand
Southern Africa
Arctic

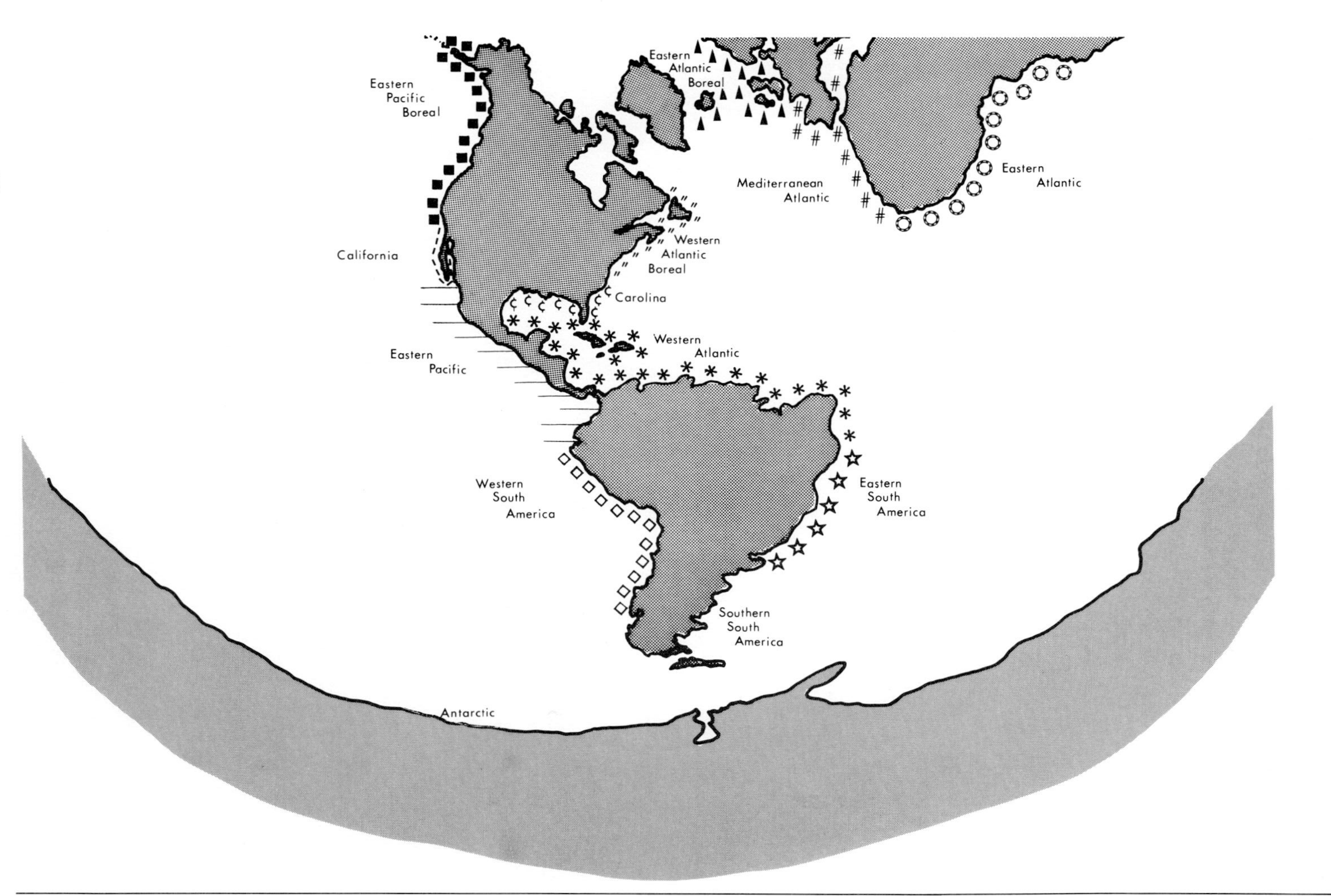

Figure 3–8 Zoogeographic regions of the oceans. The approximate extent of each region is indicated by the patterns of symbols along the coasts. The Subantarctic region consists of a series of small subpolar islands (very difficult to represent on a map of this scale) and is not indicated. (From *Marine Zoogeography* by J.C. Briggs. Copyright © 1974 McGraw-Hill Book Company. Used with permission of McGraw-Hill Book Company.)

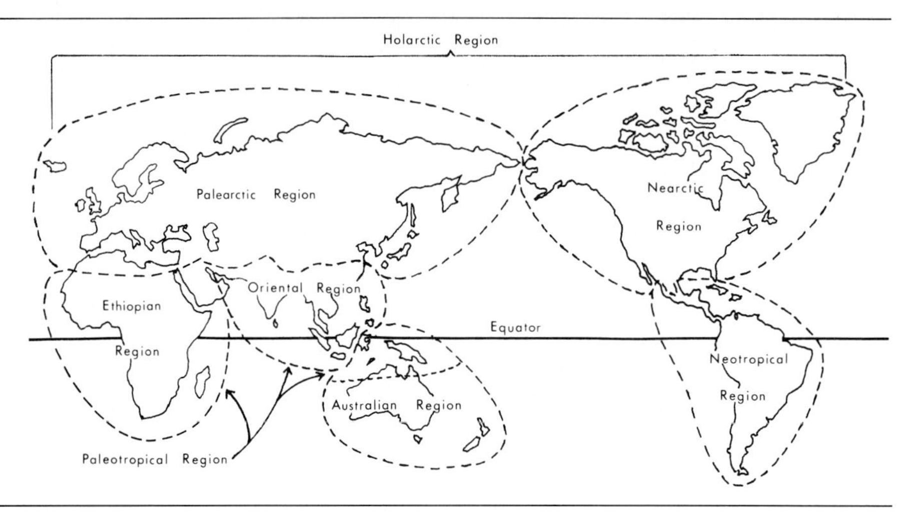

Figure 3–9 Terrestrial zoogeographic regions.

water fishes are characteristic of this region, which has a very poor freshwater fish fauna.

The *Neotropical* region is almost as distinct as the Australian. It is the center for caviamorph rodents (about 10 families occur only there) and most of the edentates; moreover, it has a very diversified bat fauna, with perhaps four endemic families. This region has even more (about 31) endemic bird families—including the Formicariidae (antbirds), Ramphastidae (toucans), and Pipridae (manakins)—than Australia. The neotropics have about 30 families of freshwater fishes found nowhere else; about 13 of them are cypriniform, and another 13 or so are catfishes. Dendrobatid, centrolenid, rhinodermatid, and almost all leptodactylid frogs are found only in the neotropics, as are the caimans of the crocodilian line.

The major part of the northern hemisphere is sometimes included in a giant *Holarctic* region. This zone is characterized, among the mammals, by wolves, hares, true beavers, and moose (European "elk") and by a number of bird families, including the Alcidae (auks), Gaviidae (loons), and Tetraonidae (grouse). The Holarctic can also be split into the *Nearctic* (North America) and *Palearctic* (Europe, North Africa, and northern Asia regions); this is perhaps the division favored by many north–temperate-region biologists who are more familiar with their own faunas. Faunal differences between Nearctic and Palearctic are usually found below the family level, although each has a small number of endemic families—the Spalacidae (mole rats) of the Palearctic and the Antilocapridae (pronghorn) and Aplodontidae ("mountain beaver") of the Nearctic, for example. Turkeys (Meleagriidae) are found in the Nearctic and hedge sparrows (Prunellidae) in the Palearctic. Helodermatid lizards occur only in the Nearctic, as do perhaps seven families of freshwater fishes, including the sunfishes and basses (Centrarchidae) and gars and bowfins (suborder Holostei).

The *Ethiopian* region comprises Africa (south and central), Arabia, and Madagascar. This area is characterized by at least seven distinctive mammal families, including the Giraffidae (giraffes), a number of distinctive antelope species and genera, cordylid (girdle-tailed) lizards, and several freshwater fish families, such as the mormyrids or elephantfishes. It also contains a small number of endemic bird families, such as the Coliidae (colies), Musophagidae (touracos), and Numididae (guinea fowl).

The *Oriental* region includes India, southeastern Asia, and the western end of the East Indies. It is famous for gibbons, tigers, crocodilian gharials, and one endemic bird family (Irenidae, leafbirds). There are about 15 endemic families of freshwater fishes and perhaps 4 endemic families of mammals. Together, the Ethiopian, Oriental, and northern Australian regions are called the *Paleotropical* region.

Despite the marked taxonomic distinctiveness of the regional faunas, sometimes remarkable parallels can be observed in the adaptations and ecological roles of certain species. Mammals that specialize on eating ants and termites are found in the neotropics, Africa, and Australia but they belong to very divergent taxa. This adaptive zone is occupied in Africa and southern Asia by the pangolins (order Pholidota), the aardvark (order Tubulidenta), and the aardwolf *Proteles cristatus* (order Carnivora). In the neotropics it is occupied by the family Myrmecophagidae (order Edentata) and in Australia by the echidnas (order Monotremata) and the marsupial numbats, *Myrmecobius.* Specialized nectar-feeding birds in the New World belong mainly to the family Trochilidae, or the hummingbirds (order Apodiformes). In tropical Africa and Asia this zone is occupied mainly by sunbirds (family Nectariniidae, Passeriformes) and in the Australian region by honeyeaters (family Meliphagidae, Passeriformes) and a few parrots (family Psittacidae, Psittaciformes). Other birds, not to mention insects and bats, also eat nectar, and many hummingbirds, sunbirds, and honeyeaters often eat insects. The point is that, in each case, a different taxon has exploited and radiated in an adaptive zone (here defined by feeding habits) represented in each major tropical landmass. Similarly, the agamid lizards of the Old World are often cited as "ecological equivalents" of the New World iguanids; moreover, the Australian "mountain devil" (*Moloch horridus*) may be seen as a counterpart of the North American horned lizards (*Phrynosoma* spp.), for both are slow and spiny and eat large quantities of ants (Fig. 3–10). The Australian death adder (*Acanthophis antarcticus*), an elapid snake, is convergent in many respects with the viperid snakes; its foraging behavior, jaw morphology, and body shape are more similar to the unrelated vipers than to other elapids (Shine, 1980). However, one must be careful when drawing these ecological parallels. Because the phylogeny and detailed ecology (in terms of potentially competing species in the same habitat, available food resources, habitat structure, and so on) of

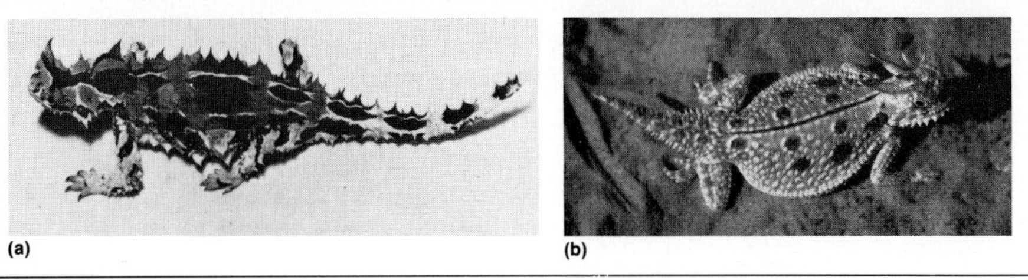

Figure 3–10 Ecological parallels on two continents: (*a*) *Moloch horridus* of Australia and (*b*) *Phrynosoma* of North America. Both have specialized to some extent on an armored defense and a diet of ants or termites. (*a*, Photo by E. R. Pianka; *b*, photo by S. Prchak.)

each of the "equivalent" species or groups is different, no exact correspondence can be expected (see Eisenberg, 1982).

In contrast to the ecological parallels or convergences just described, whole ecological categories may be missing from some zoogeographic regions. Thus the neotropics has no burrowing, insect-eating, molelike mammals (although some rodents tend in this direction), but molelike mammals belonging to different families are found on every other continent. Australia has no subterranean mammals that eat primarily roots and tubers, but large parrots called cockatoos sometimes dig for shallow roots. Africa has the molerats (Bathyergidae), and South America has some rodents that burrow for roots part of the time. Africa has a terrestrial ungulate (the giraffe) that feeds on treetop foliage; Africa and the neotropics have arboreal primates that do so. Only the neotropics has blood-eating bats; fish-catching bats are known only in the New World. Similarly, the arid grasslands of North America have no long-legged, running birds corresponding to the ostriches of Africa or the rheas of South America. The reasons for such "gaps" in the fauna of any region may be historical and geographical (related to past history of both taxon and landmass). They may also be ecological (related to recent conditions of habitat and other resources). Some of the factors and processes that determine animal distributions are surveyed in the later parts of this chapter.

CLIMATIC PATTERNS

Distributions of animals cannot be fully understood without at least a summary of world patterns of climate, which influence animal distributions both directly and indirectly through effects on vegetation. Climate is usually described in terms of temperature and rainfall. The main factors are the amount and temporal scheduling of heat and precipitation, their variability, and their predictability. Many of the geographic differences in these variables result from the shape of the earth, its orientation with respect to its major heat source (the sun), and its rotation. The size, shape, and location of land and water masses on the earth's surface also contribute to the determination of climate.

A fundamental climatic pattern underlying all others is one of zonation with respect to distance from the equator. Because the earth's axis is tipped away from the perpendicular with respect to its plane of movement around the sun (Fig. 3–11), incident radiation from the sun is not always perpendicular at the equator. Instead, the most northerly and southerly latitudes at which the sun is directly "overhead" alternate seasonally as a function of the position of the earth relative to the sun. These latitudes are 23°27′ north and south of the equator and define the limits of the so-called tropical zone.

Although the north and south angle of the sun's rays varies seasonally within the tropical zone, the variation is not great compared to the rest of the earth. Because the sun's rays are always close to perpendicular (in the north and south plane) within this zone, heat from solar irradiation is concentrated: each ray strikes a smaller area when it falls perpendicularly than when it strikes at an angle and thus yields more heat per unit area. Furthermore, angular rays outside the tropics must penetrate a thicker layer of air, so that more of the solar energy is reflected back into space. As a result, daily and especially seasonal changes in temperature within the tropics tend to be small and the average temperature is high. Closer to the poles, incident radiation is more and more angular and increasingly seasonal. Near the poles, the sun never sets for a number of days during the summer and never rises for a number of days during the winter. At the poles themselves, six months of continual sun alternate with six months of continual darkness. Seasonal extremes of temperature generally become greater and greater with decreasing distance from the poles.

Because bodies of water can absorb great quantities of heat, they can act as heat storage areas and exert significant modifying effects on temperatures over nearby land, especially those downwind. "Continental" climates, with extremes of heat and cold, are usually found far from or upwind of the effects of large bodies of water. Large expanses of water may modify the fundamental latitudinal zonation pattern of temperature. Examples of global temperature patterns are shown in Figure 3–12.

Wind patterns depend on atmospheric pressure, which varies with latitude and forces set up by the earth's rotation. Zones of enduring high pressure are found just outside the tropics in both hemispheres and at the poles. Winds blow away from high-pressure zones into low-pressure zones (Fig. 3–13). Furthermore, high temperatures in the tropics warm the air. As the warm air rises, it cools and can hold less water, which falls as rain. The cooled air descends again at about 30°N and 30°S latitude, increasing in temperature and taking up (and not releasing) atmospheric

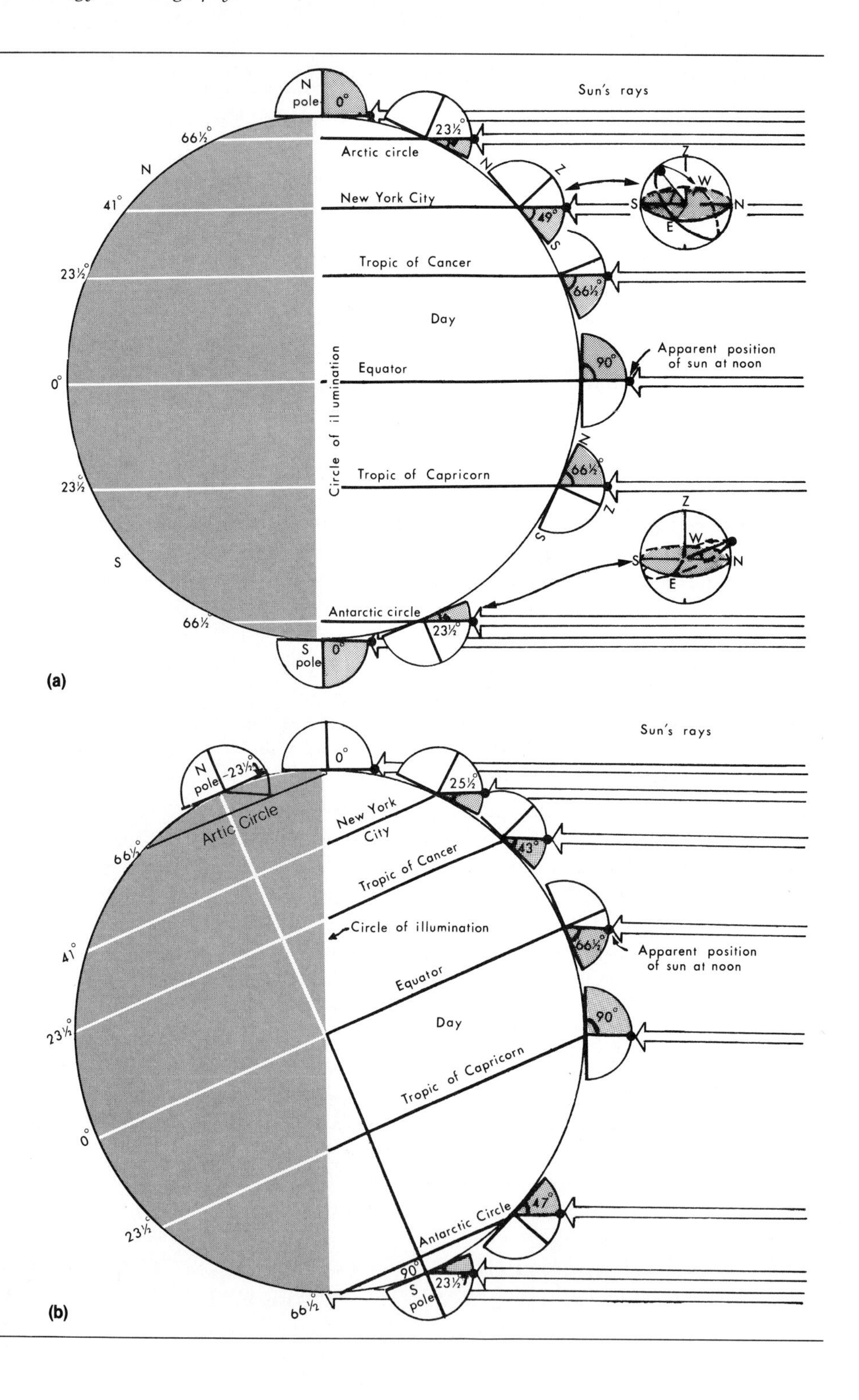
Sun's rays
N pole
0°
23½°
66½°
Arctic circle
N
Z
W
S
E
New York City
41°
49°
Tropic of Cancer
23½°
66½°
Day
Apparent position of sun at noon
Equator
90°
0°
Circle of illumination
Tropic of Capricorn
Antarctic circle
S pole
(a)
Sun's rays
N pole
-23½°
0°
25½°
Artic Circle
New York City
Tropic of Cancer
43°
Circle of illumination
Equator
66½°
Apparent position of sun at noon
Day
90°
Tropic of Capricorn
Antarctic Circle
47°
S pole
23½°
(b)

Figure 3–11 Seasonal changes in illumination of the earth. The angle of incidence of solar rays (*a*) at the equinox and (*b*) at the northern winter solstice. (From A.N. Strahler, 1975. *Physical Geography*. 4th ed. John Wiley & Sons, New York.)

water. Rising air masses in the tropics draw in air masses from just outside the tropics, thus creating the trade winds. Because of this general pattern of air movements, many of the great deserts of the world are located near 30° latitude.

Rotation of the earth imparts directional changes to the movement of air (and water) masses and thus has an enormous impact on climatic patterns. The surface of the earth rotates from west to east, and the closer to the poles, the slower the speed of rotation of a point on the earth's surface. An object at the equator is moving eastward at the same speed as the earth's surface in that area, of course, but if that object moves northward into regions that are rotating more slowly, it retains its original momentum; hence, it is deflected eastward relative to the surface. An object moving southward from the equator will also tend to be deflected to the east. Air experiences this eastward deflection, which increases with increasing distance from the equator; conversely, air is deflected westward as it approaches the equator. Thus, the easterly trade winds in the northern hemisphere come from the northeast and those in the southern hemisphere come from the southeast. (Winds are named by the direction from which they blow.)

Air descending at 30°N and 30°S flows not only toward the equator as it descends but also toward the poles. The earth's rotation coupled with the momentum of the air masses imparts an eastward direction to the movement of air in the so-called temperate regions. Winds in these latitudes, therefore, are prevailing westerlies.

These same tendencies affect the movement of water, but winds have an additional and important effect on water currents (Fig. 3–14). Winds push water in the direction the wind is headed until the water bumps into continental landmasses and is deflected; thus, the prevailing westerly winds push water against the west side of South America. Part of this water is turned northward, forming the cold Humboldt current, whose influence is felt along the west coast of South America. Similarly, the Gulf Stream begins as the northeast trade winds force water against the coast of Central America, where it deflects northward, carrying warm water along the eastern seaboard of North America. The momentum of the Gulf Stream, influenced by the earth's rotation, carries warm water across the Atlantic, then southward along the western side of Europe, and back toward Central America. Winds over these currents carry to the downwind landmasses marked temperature effects that establish climatic regimes un-

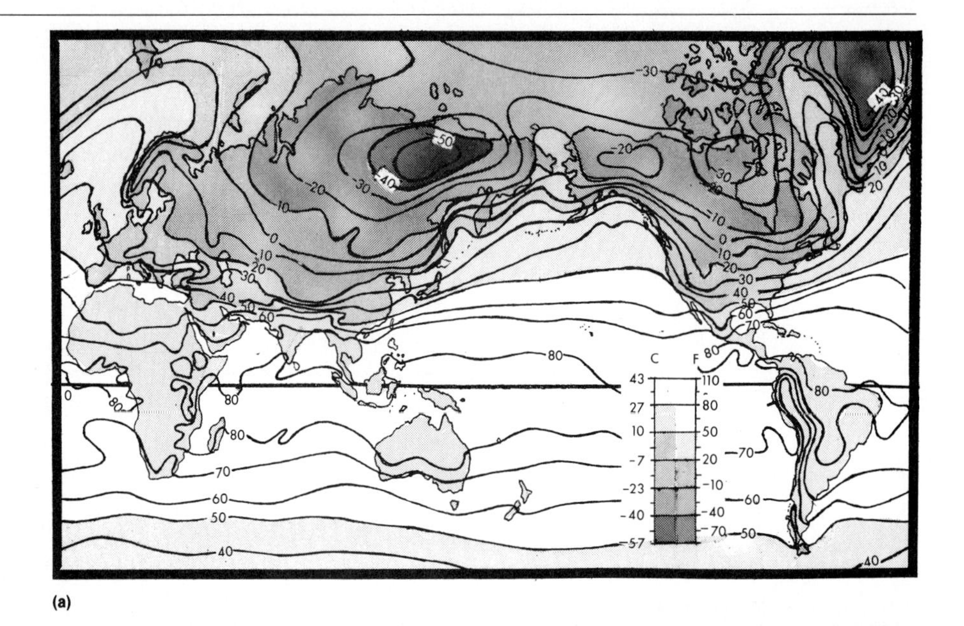

(a)

(b)

like those without the presence of the currents. Thus Britain's climate, modified by a branch from the warm Gulf Stream, is significantly warmer than that of eastern North America at the same latitude.

These patterns of air and water movement influence rainfall, but to

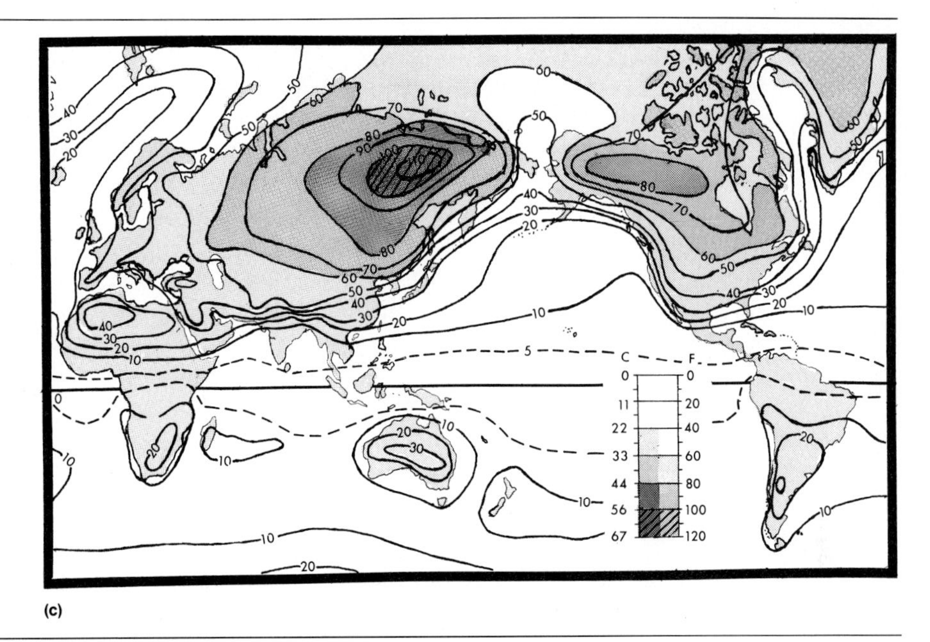

Figure 3–12 (*a*) World temperature pattern in the northern hemisphere winter (January). The zones are defined by the average temperature at sea level. (*b*) The pattern in summer (July). Note that the warmest zones are much farther north in July than in January, owing to the more northerly "position" of the sun in July (see Fig. 3–11). (*c*) Annual range of air temperatures computed as the difference between the January and July means. (*a* to *c* from A.N. Strahler, 1975. *Physical Geography*. 4th ed. John Wiley & Sons, New York.)

understand their operation, we must also look at the spatial relations and temperature differentials of oceans and continents and the effects of altitude. Air masses moving over warm water pick up moisture by evaporation. When these winds reach land that is cooler than the water, the air is cooled and releases some of its moisture as precipitation. The western sides of continents at high north-temperate latitudes tend to be very wet and support temperate rain-forest vegetation of the type found on the west coast of Canada and the northwestern United States. Eastward moving air is forced to rise when it meets the mountains of western North America; as it cools, it releases still more water, contributing to the moisture on the western side of the ranges and to the dry "rain shadow" on the eastern, leeward, side. Closer to the equator, but still in temperate zones, the landmasses are warmer—often warmer than the ocean. Here the westerly winds pick up heat and moisture as they pass over land, so these areas tend to be very dry; this contributes to the formation of deserts at 30°N latitude. In the trade-wind zones, the same principles operate; thus, rain

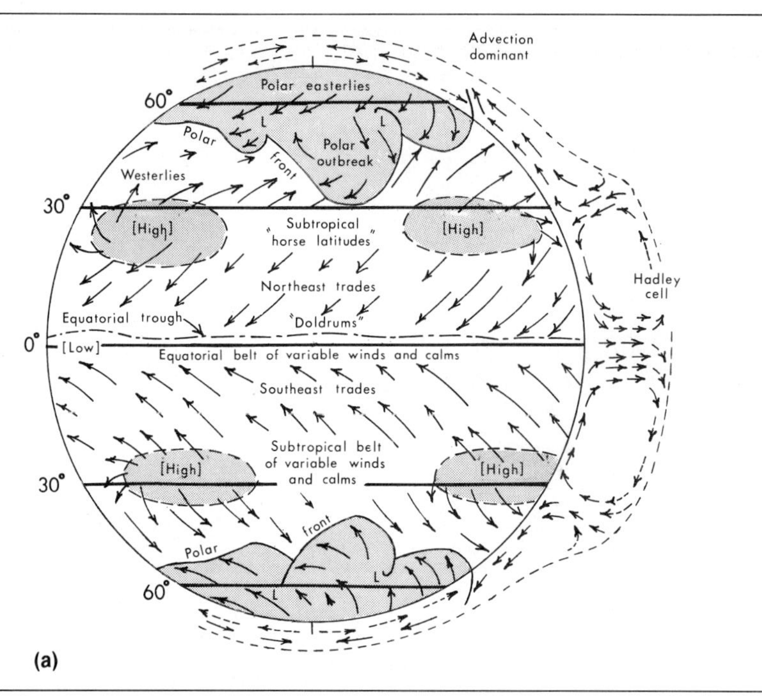

Figure 3–13 (*a*) Wind patterns over the globe. (From A.N. Strahler, 1975. *Physical Geography*. 4th ed. John Wiley & Sons, New York.) (*b*) The wind pattern in January. (*c*) The wind pattern in July. (*b* and *c* from *An Introduction to Climate* by G.T. Trewartha. Copyright © 1968 McGraw-Hill Book Company. Used with permission of McGraw-Hill Book Company.)

often falls as air moves over the warm oceans to the cooler land, and the windward sides of landmasses are often wetter than the leeward sides. If a landmass is seasonally warmed much faster than the oceans, rising warm air over the land continually pulls in air laden with moisture from over the oceans. The rising air expands, cools, and releases its water in tropical deluges called monsoons. The global pattern of average annual rainfall and its variability are shown in Figure 3–15.

The temperate regions are really misnamed. In fact, they are only sometimes more moderate than the tropics and may be less predictable than polar regions. Temperate regions are frequently characterized by extreme and rapid changes of temperature, great variation in the sequences of sunny and rainy or cloudy days, long periods of stormy rainy weather, and precipitation in many forms, including hail and sleet. Short-term climatic variability and predictability of patterns have seldom been used in bioclimatic studies. Averages and extremes are common means of summarizing

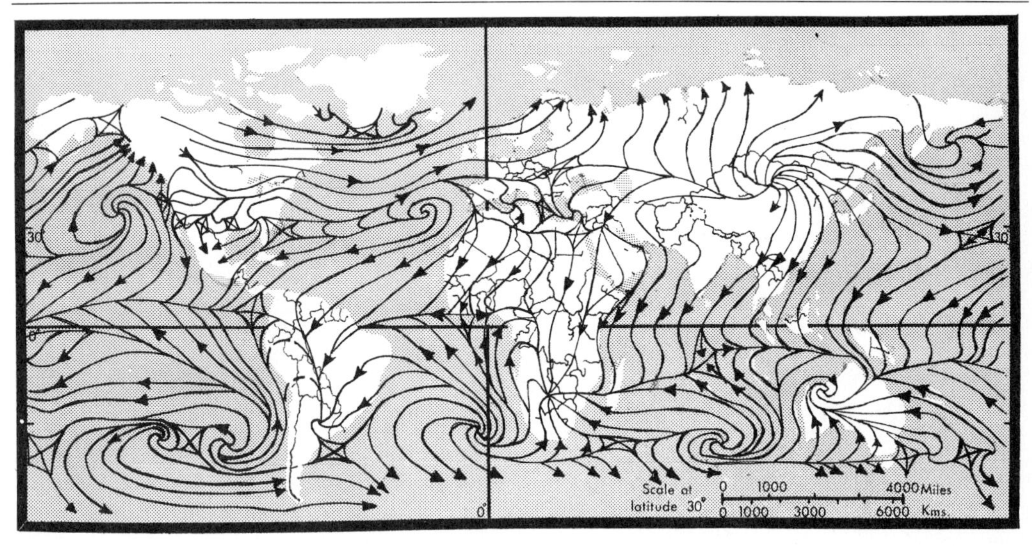

(b)

(c)

climate, and only recently have seasonal and daily variability and predictability begun to be emphasized.

Variability and predictability are not the same. Some climatic factors may be variable in an unpredictable way, but others vary quite predictably. The time of sunrise and sunset (and thus day length) are variable but highly predictable. Rainfall in deserts is predictably low, but there is a great

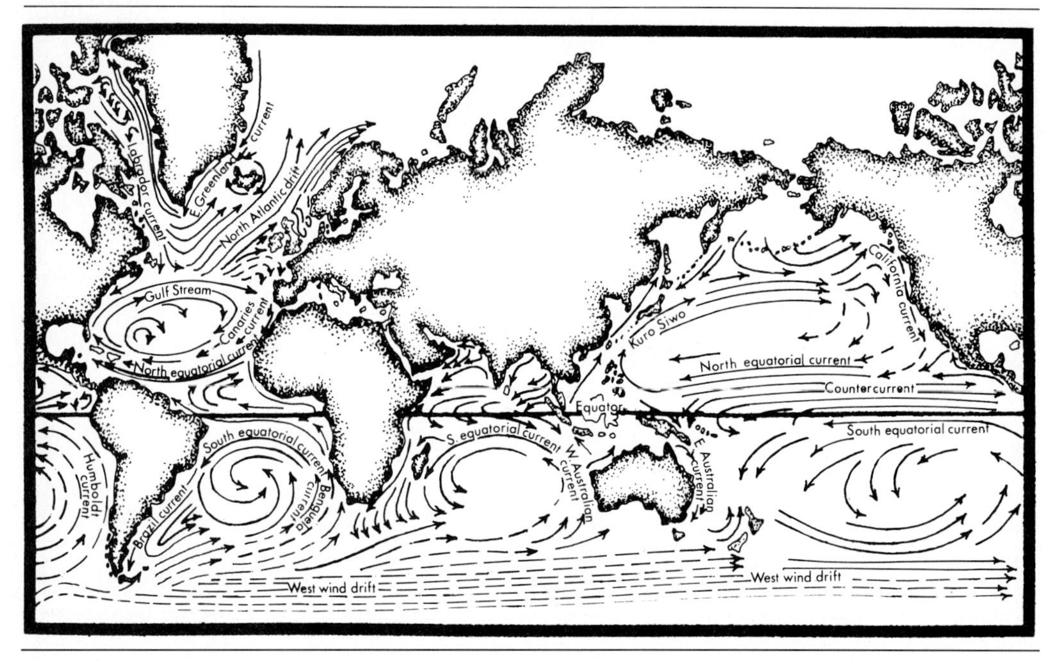

Figure 3–14 Major oceanic currents. Cold currents are indicated by dashed lines, warm ones by solid lines. (Modified from R.H. MacArthur and J.H. Connell, 1966. *The Biology of Populations.* John Wiley & Sons, New York, and from *An Introduction to Climate* by G.T. Trewartha. Copyright © 1968 McGraw-Hill Book Company. Used with permission of McGraw-Hill Book Company.)

variability around the average (see Fig. 3–15*c*). The onset of the rainy season, the timing of snow melt, or the length of Indian summer in the autumn are often unpredictably variable. The distinction is not at all trivial. The ways that animals can adapt to variability may be very different from their adaptations to unpredictability, and the effects on community composition and species distributions may be considerable (Karr, 1976a; Wiens, 1974b).

Climate and Vegetation

Climatic patterns obviously affect the distribution of vegetation types across the surface of the earth. Wet areas usually support dense, rather tall vegetation, often a rain forest. The very driest areas support no vegetation except ephemeral plants found only after sporadic rainfall. Desert vegetation is generally short, well-spaced, and often seasonally or permanently leafless. Grasslands are found mainly in areas of intermediate-to-low annual rainfall (25–75 cm/year) that is distributed irregularly throughout the

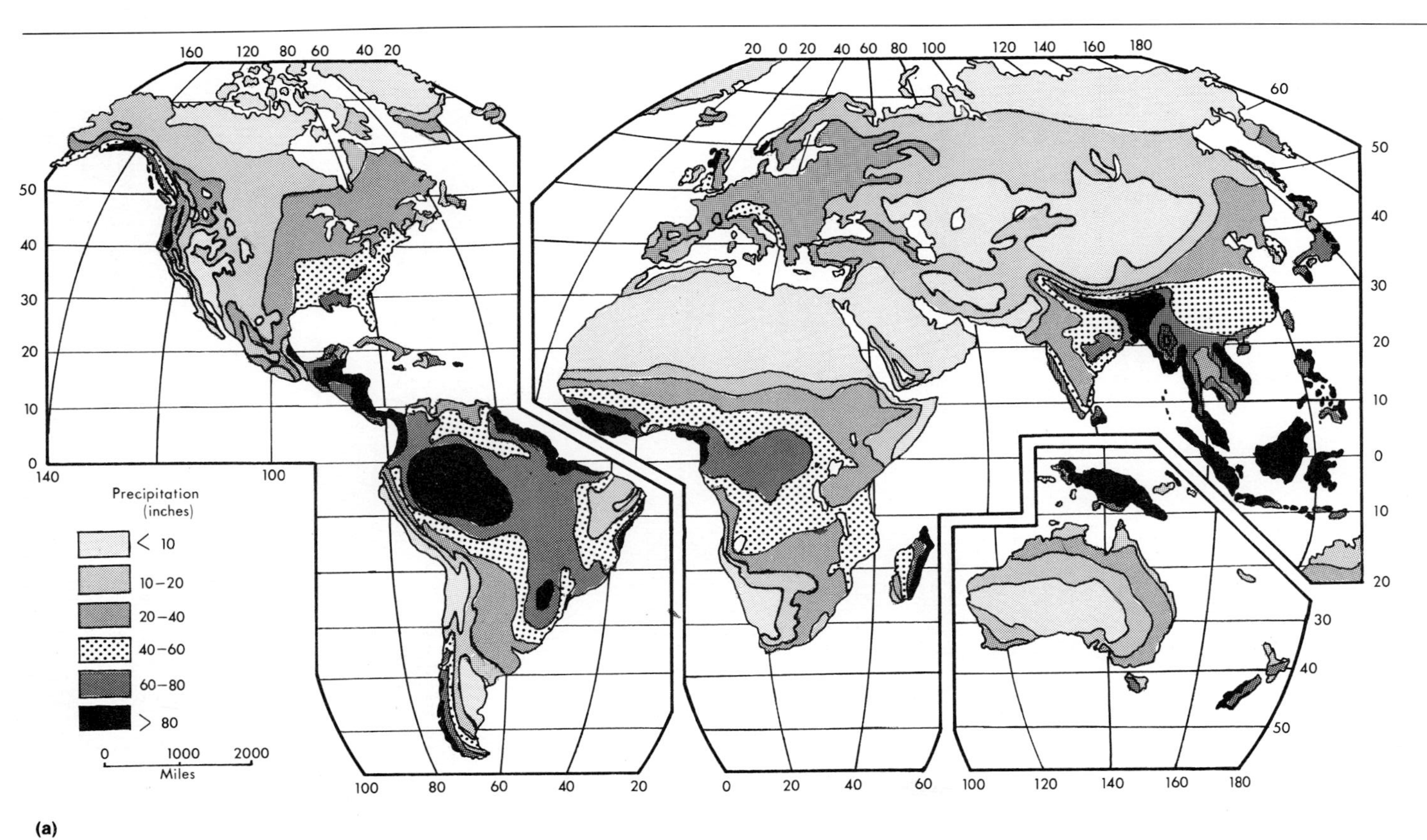

Figure 3–15a See page 101 for legend.

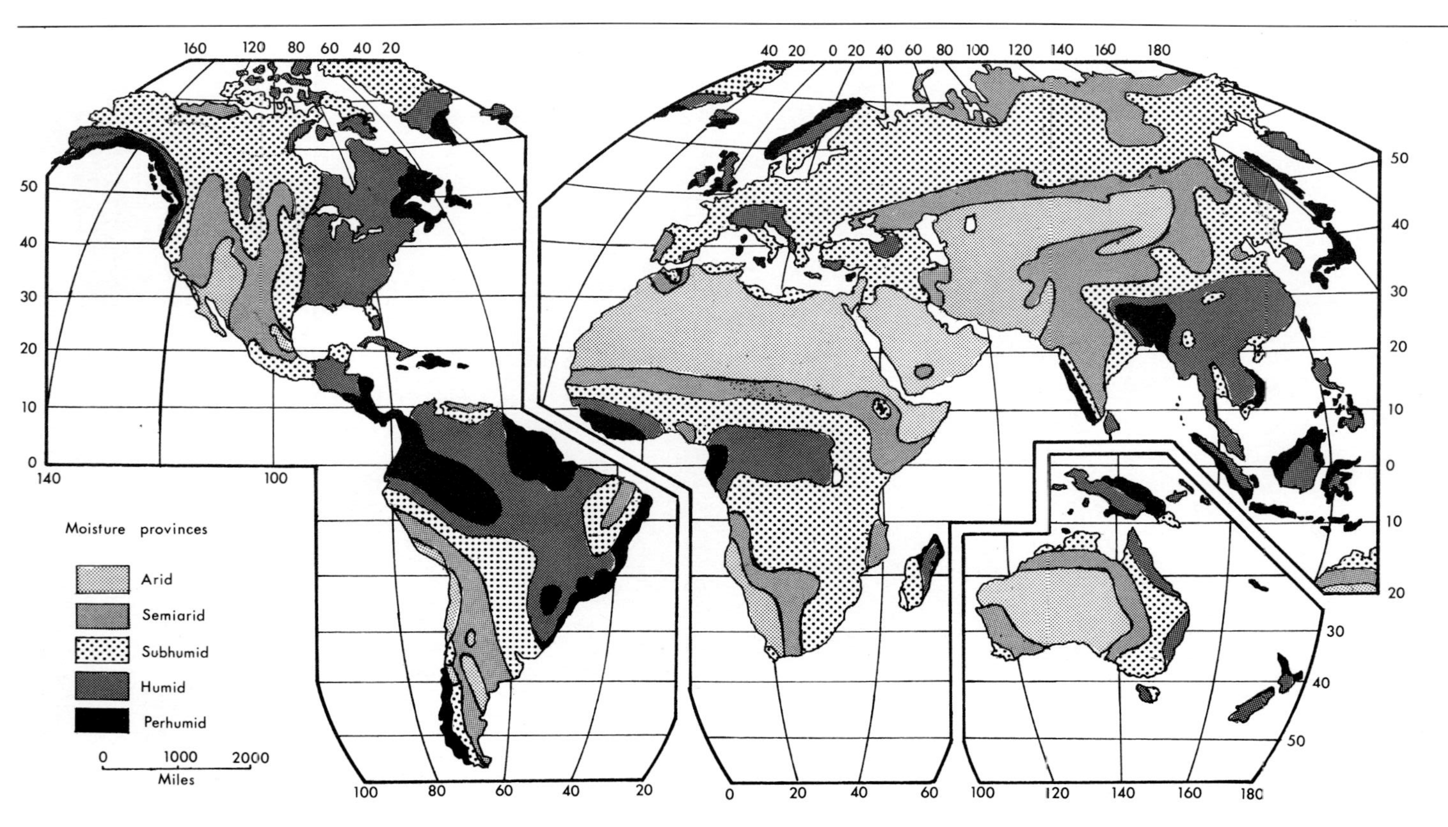
Moisture provinces
Arid
Semiarid
Subhumid
Humid
Perhumid
0 1000 2000
Miles

(b)

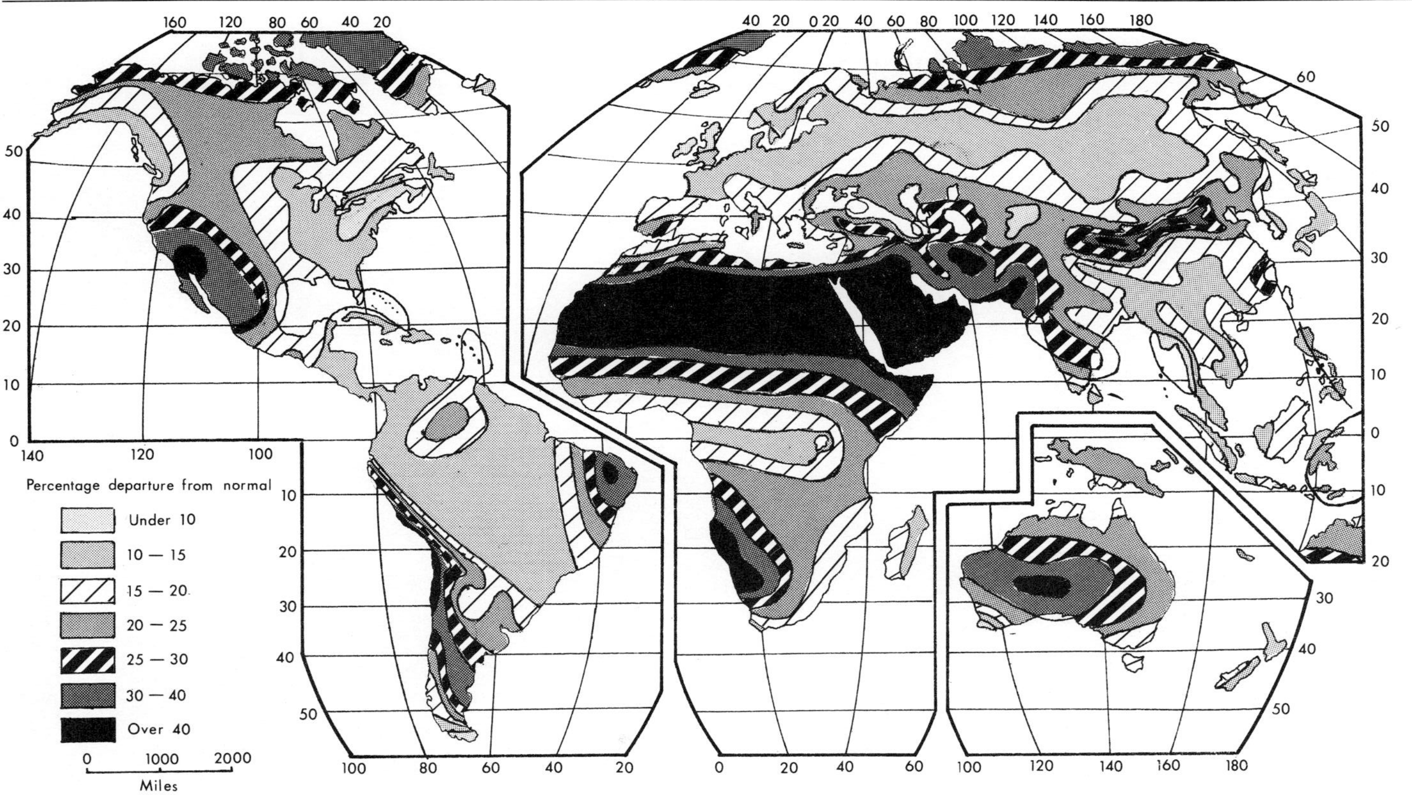

(c)

Figure 3–15 (*a*) Global patterns of rainfall. (Modified from J.E. Oliver, 1973. *Climate and Man's Environment*. John Wiley & Sons, New York.) (*b*) World distribution of effective precipitation as indexed by humidity and calculated from a ratio between annual precipitation and annual average temperature. Note the differences between this map and the previous one. (From *An Introduction to Climate* by G.T. Trewartha. Copyright © 1968 McGraw-Hill Book Company. Used with permission of McGraw-Hill Book Company.) (*c*) Global patterns of variability in precipitation. (From A.N. Strahler, 1975. *Physical Geography*. 4th ed. John Wiley & Sons, New York.)

year or in a short rainy season. Seasonal and deciduous forests are commonly intermediate in moisture, between rain forest and grassland. Coniferous forests in the northern hemisphere characterize high-latitude, cool, wet regions and certain warmer and drier areas as well. The highest latitudes are carpeted with low-growing plants or perpetual ice and snow. Precipitation here is very low, similar to deserts, but temperatures are so low that evaporation is reduced. A comparison of the rainfall and temperature maps (Figs. 3–12, 3–15) with global vegetation will illustrate the association of general climatic and vegetational features (Figs. 3–16, 3–17).

Each of the major vegetation formations, with its associated animals, is sometimes called a biome. South America, for instance, has a large area of tropical rain forest along its northeastern and northwestern coasts, the southeastern coast of Brazil, and in the Amazon basin. Temperate rain forest is characteristic of the cool, moist strip on the southwest extremity of the continent, just south of a mild-climate ("Mediterranean") scrub forest. Desert-like conditions are found along the central west coast, in northeastern Brazil, and on the inland side of the south-central Andes. Cool-temperate vegetation analogous to the northern coniferous (boreal forest or taiga) forest of the Holarctic is found on the slopes of the Andes, which are crowned in some areas by low-growing alpine vegetation called páramo. The rest of South America is typified by grassland. Each major habitat type or biome is usually characterized by concentrations of certain categories of animals: monkeys and parrots in the forests, seed-eating ground birds and small mammals in deserts and grasslands, and so on. Each biome represents a special set of conditions for life, and its vertebrate inhabitants often possess particular adaptations that permit successful existence under those conditions.

The different biomes can be subdivided of course. The grasslands may be divided into short-grass and tall-grass units, for example, and the tropical rain forest into units that differ somewhat in precipitation schedules and altitude.

Historically, changes in world climate patterns have wrought changes in the distribution of biomes across the continents, with concomitant effects on the opportunities for speciation and reinvasion. South America, for example, was subjected to extreme climatic changes during Pleistocene times and endured major alternations of wet and dry periods that corresponded to glacial advances and retreats at more polar latitudes. As a result of these climatic changes, the tropical lowland forest may have undergone several cycles of expansion followed by contraction and fragmentation (Fig. 3–18). Many birds of the tropical lowland forest are extremely sedentary and very reluctant to cross ecological barriers such as large rivers, mountain ranges, or different vegetation types. South American forest birds, therefore, may have had ample opportunity for speciation in geographic isolation during the periods when the lowland forest was broken up into noncontiguous small blocks. Similar opportunities for African forest birds may have been considerably less because the area occupied by expansion and contraction of forest may have been less, and fewer major topographic features disrupted the landscape. In addition, humans have occupied Africa far longer than South America, so Africa has a longer history of habitat modification and destruction. These historical-geographic differences in two large tropical areas may account partially for observed differences in the diversity of the avian fauna in the two regions (Karr, 1976b).

Microclimate

Climatic patterns that prevail on a regional scale may be much modified on a local scale. Small-scale climate is referred to as microclimate. Daytime temperatures may be high at the ground or water surface, but they decrease markedly with increasing distance from the surface. Springtime air temperatures are often well above soil temperatures, and the reverse is true in the autumn, because the soil warms and cools more slowly than the air. Small differences in surface topography may cause large differences in temperature. A north-facing slope in the northern hemisphere may never receive direct solar radiation, so it is generally cooler than a south-facing slope. Furthermore, a north slope is typically moister, because evaporation is less. Even small irregularities on the surface can cause sharp heterogeneity in temperature and moisture conditions. Reflections of the sun's rays from small rocks and ridges may create local hot spots, and the same rocks and ridges may deflect precipitation and runoff, creating moisture differentials.

If vegetation is present, microclimatic conditions may be additionally modified. Vegetation usually results in decreased wind velocities, decreased rates of heating and cooling (and hence a smaller range of temperature extremes), and changes in the movement patterns of moisture. Wind speeds of 24 km/h above the canopy of a forest may be reduced to only 4 km/h inside the forest. On a hot day in the temperate zone, temperature excursions at the surface of bare soil might be as much as 45°C, but as little as 15°C beneath a dense canopy of vegetation. The average temperature for a summer month in the temperate zone may be 4 or 5°C cooler in a forest than in an adjacent open field. When rain falls on plants, some of the water is intercepted and does not reach the ground, but some runs down the plant stems, creating variation in soil moisture. Local variations in microclimate are often reflected by differences in the plant species that grow there (certain plant species in warm, dry spots, and others in wet, cool ones, for example). Different species of plants may affect nearby microclimatic conditions in different ways. Some may collect moisture better than others, some may let more sunlight penetrate to the ground, and so on. Microcli-

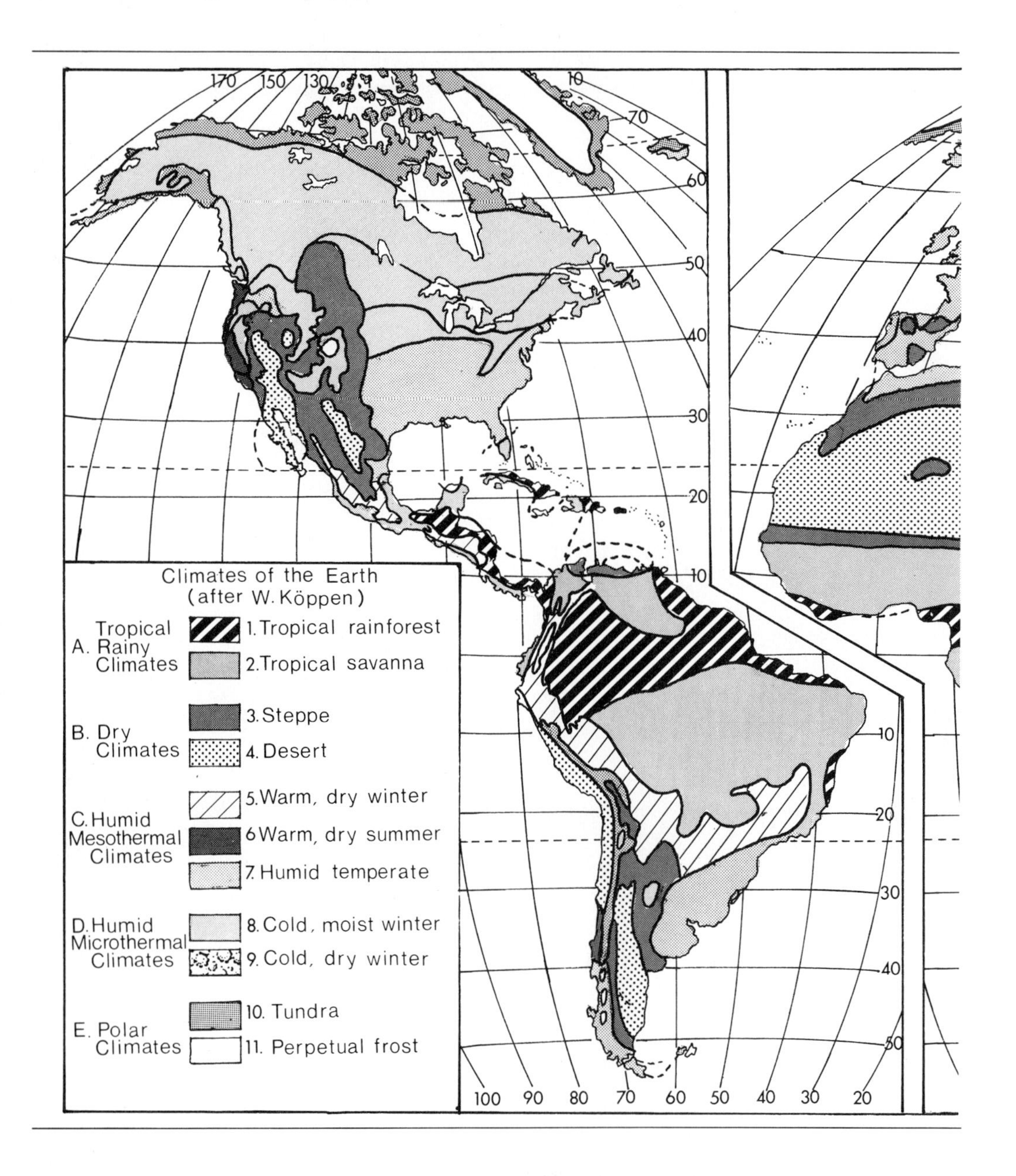
Climates of the Earth
(after W. Köppen)
A. Tropical Rainy Climates
1. Tropical rainforest
2. Tropical savanna
B. Dry Climates
3. Steppe
4. Desert
C. Humid Mesothermal Climates
5. Warm, dry winter
6 Warm, dry summer
7. Humid temperate
D. Humid Microthermal Climates
8. Cold, moist winter
9. Cold, dry winter
E. Polar Climates
10. Tundra
11. Perpetual frost
170
150
130
10
70
60
50
40
30
20
10
10
20
30
40
50
100
90
80
70
60
50
40
30
20

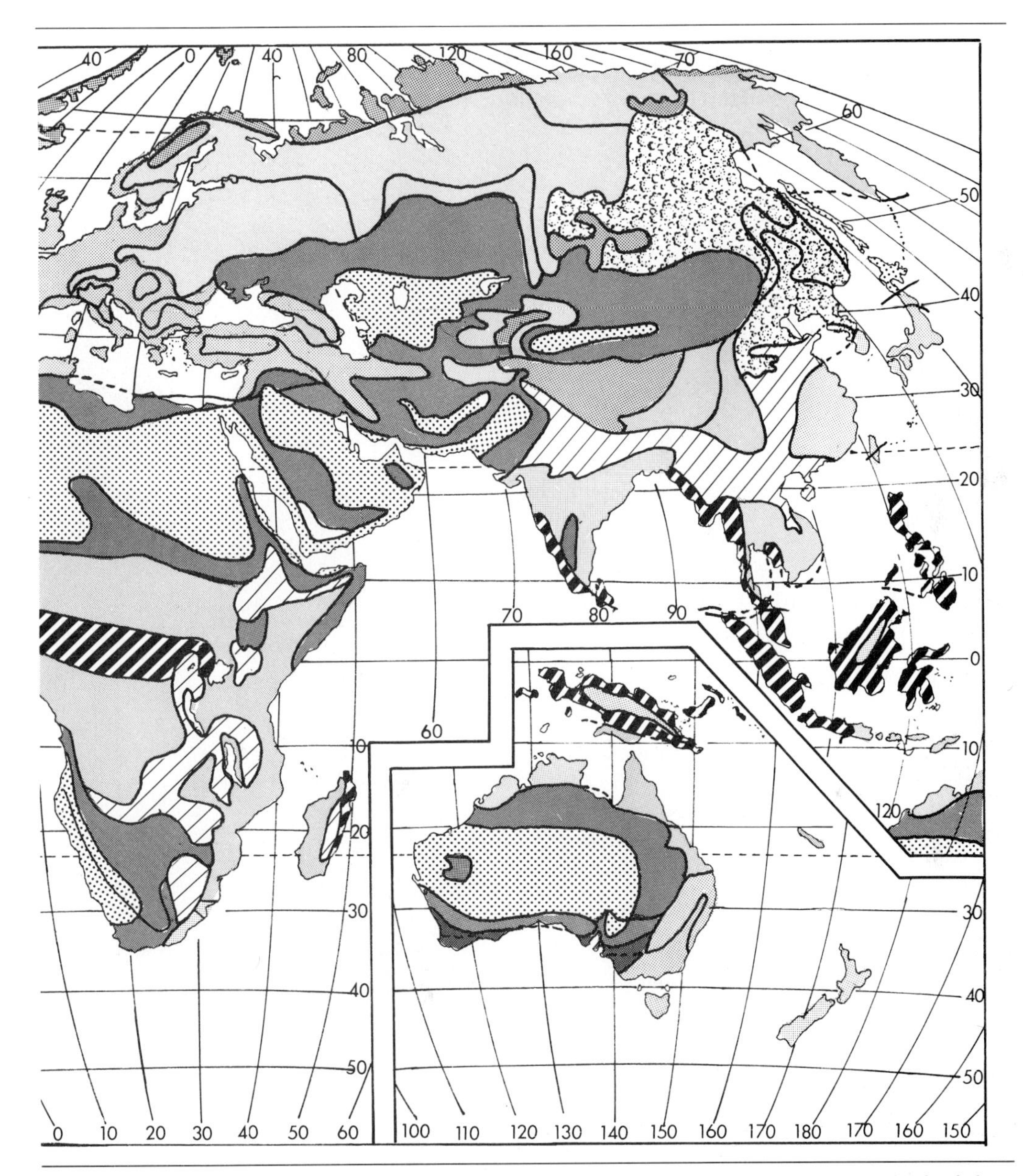

Figure 3–16 A classification of world climates that includes some aspects of seasonal scheduling of precipitation with mean precipitation and temperatures. (From *An Introduction to Climate* by G.T. Trewartha. Copyright © 1968 McGraw-Hill Book Company. Used with permission of McGraw-Hill Book Company.)

NATURAL VEGETATION REGIONS
OF THE WORLD
Simplified and modified from a world vegetation map by H. Brockman Jerosch, 1951, showing vegetation-classes according to the classification of Edward Rübel.
Map classes
Equivalent formation classes
A. Equitorial & Tropical Rain forest
1. Equitorial rain forest
2. Tropical rain forest
B. Temperate Rain forest (Laurel forest)
4. Temperate rain forest (laurel forest)
C. Raingreen forest, woodland, scrub, & savanna
3. Monsoon forest (tropical deciduous)
8. Savanna-woodland
9. Thornbush & tropical scrub
10. Savanna
D. Evergreen-Hardwood Forest
7 Evergreen-hardwood (sclerophyll forest)
E Summer-Green Deciduous Forest
5 Summer-green deciduous forest
F Needleleaf Forest
6 Needleleaf forest
G Steppe & Prairie Grasslands
14 Prairie
15 Steppe
H Dry Desert & Semidesert
17 Dry desert
11 Semidesert
I Tundra (Arctic & Alpine)
16 Grassy tundra
17 Cold woodland
18 Arctic fell-field
J Icecaps & Glaciers
Highland areas with alpine tundra
Tropic of Cancer
0
1,000
2,000
3,000
Miles

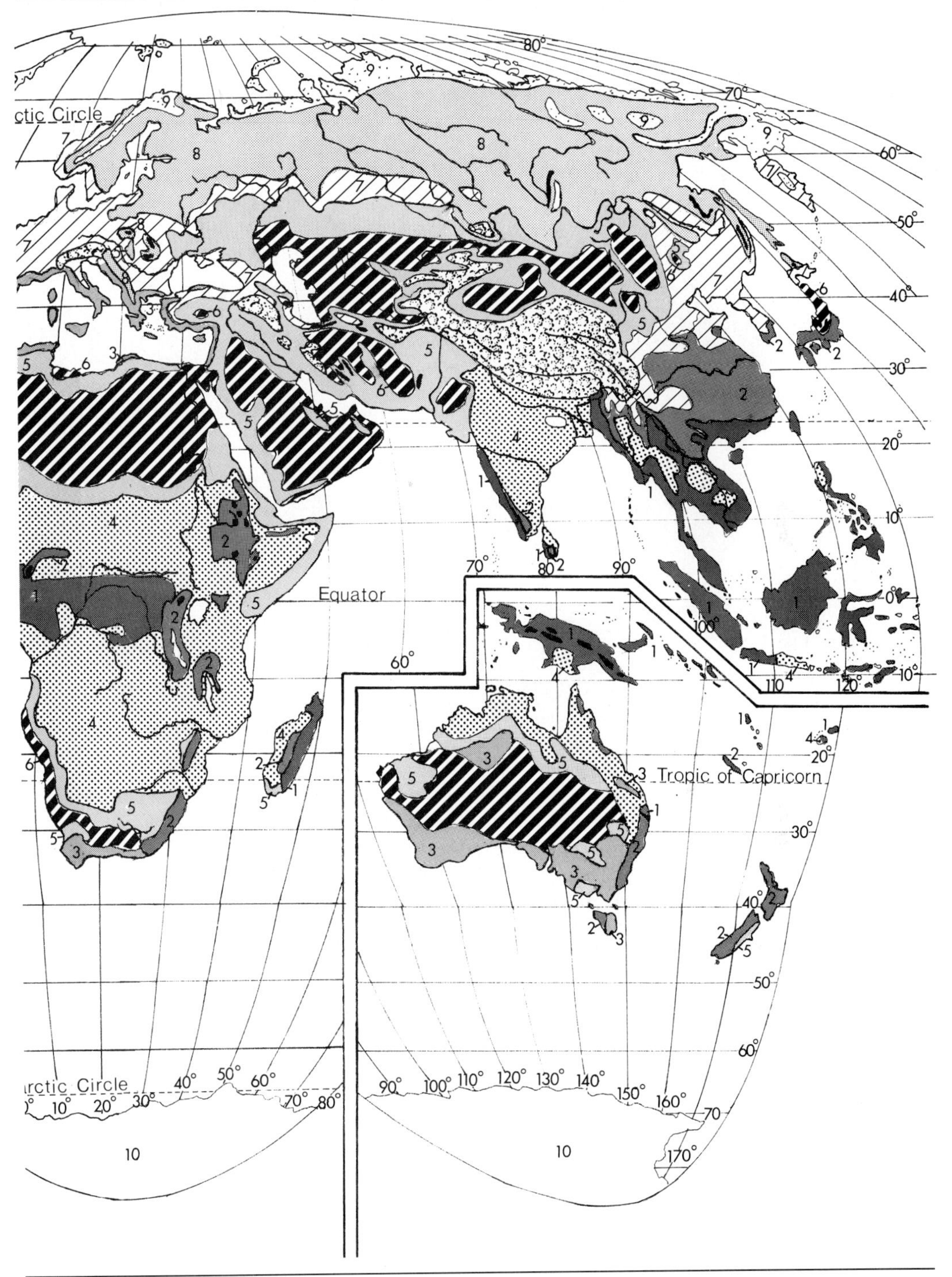

Figure 3–17 Map of the major vegetation types of the world. (From A.N. Strahler, 1975. *Physical Geography*. 4th ed. John Wiley & Sons, New York.)

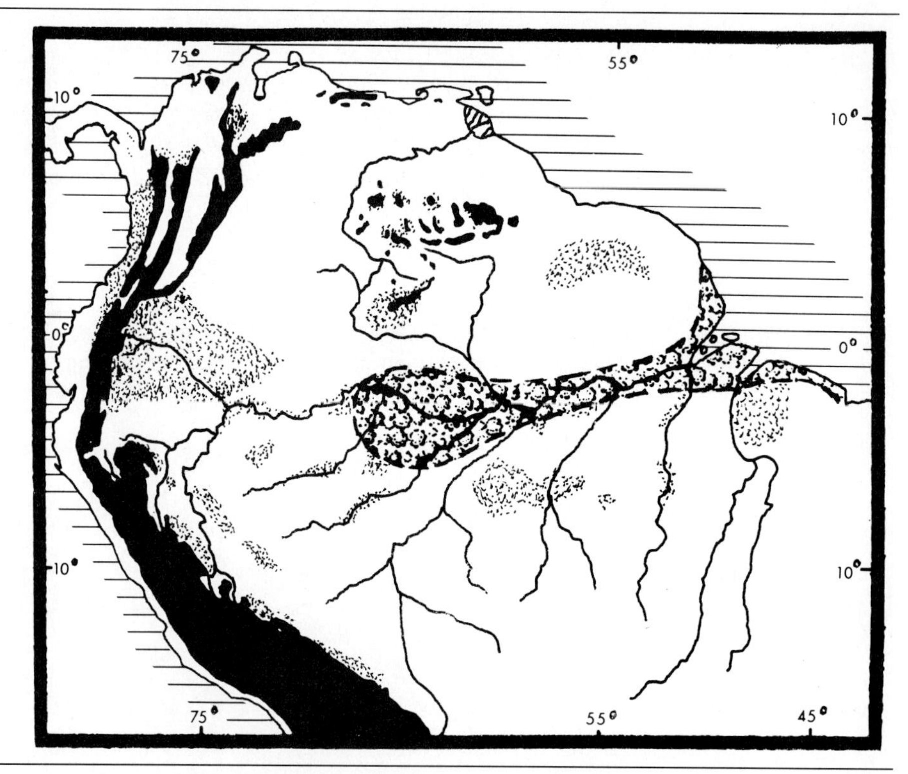

Figure 3–18 Hypothesized fragmentation of the lowland tropical forest of South America during dry climatic periods of the Pleistocene. Much of the Amazon valley was inundated by seawater during interglacial periods. High mountain areas (more than 1000 m) are indicated in black. (From J. Haffer, 1969. Speciation in Amazonia forest birds. *Science, 165*:131–137. Copyright © 1969 by the American Association for the Advancement of Science.)

matic differences may also contribute to local differences in the density of vegetation. Thus, the general tendency is for vegetation to moderate existing climatic extremes; within these moderated limits, plants may create small-scale heterogeneity in microclimate.

Vertebrates often respond to microclimatic differentials. Desert animals frequently seek shade or an underground retreat on a hot day. On a windy winter day, birds often forage in protected sites—on the downwind sides of tree trunks or other windbreaks, for example. Terrestrial salamanders, such as those of the genus *Plethodon,* are typically found under logs on the forest floor, where humidity is higher and temperatures cooler than in the open. Moreover, the care invested by a parental mammal or bird often involves provision of a special environment for the young; the warm microclimate of a nest or den is important to their survival. The distribution of some vertebrates, including many terrestrial amphibians, may be closely related to microclimatic conditions, but other vertebrates, especially birds and mammals, seek special microclimates primarily in extreme environments or under special conditions. Adaptations related to climate on both a small and a large scale are discussed primarily in Chapter 5. The general problem of habitat selection is presented briefly later in this chapter.

ECOLOGICAL DETERMINANTS OF DISTRIBUTION

This section begins with a primarily abstract discussion of ecological niches and habitat selection. Then we will survey the processes and conditions that determine niche size and shape and habitat occupancy and, thus, whether or not a species is found in a certain place. For this discussion, let us assume that the species can reach that certain place—that the place is within the geographic range of the species. It is manifestly absurd to talk about habitat selection by kangaroos in Greenland, for instance, when those fortunate creatures have never been—and presumably never will be—there.

Ecological Niches

Niches can be understood in terms of all the conditions that permit an organism to survive and reproduce. How an animal behaves, where and what it eats, its nesting requirements, its temperature and humidity tolerances—all help to define the niche of a species or population. A kingfisher, for example, usually requires an earthen bank in which to nest, a supply of small fishes in surface waters, a perch from which to spot them, and, presumably, various other things. Not all dirt banks are adequate, however; some fishes are too big or too small, some temperatures are too hot or too cold, and probably no animal can use the entire array of conditions present.

Ecologists often think of each kind of requirement as an axis or dimension in a multidimensional graph. Possible nest sites for kingfishers may lie on one axis (one extreme might be soft soil and the other one hard-packed clay) and food-particle size on another axis. Because every animal has many requirements, there are many dimensions, only three of which can be easily represented on paper (Fig. 3–19). Whatever the axis, an animal uses only part of the available variety of conditions. The utilized range of all axes taken together defines an imaginary, multidimensional space that represents the species' niche. Niche "size" is determined by the width of the range along any axis or set of axes.

So far, we have neglected the impact of other species, although no real species exists in isolation from all others. Other species—competitors,

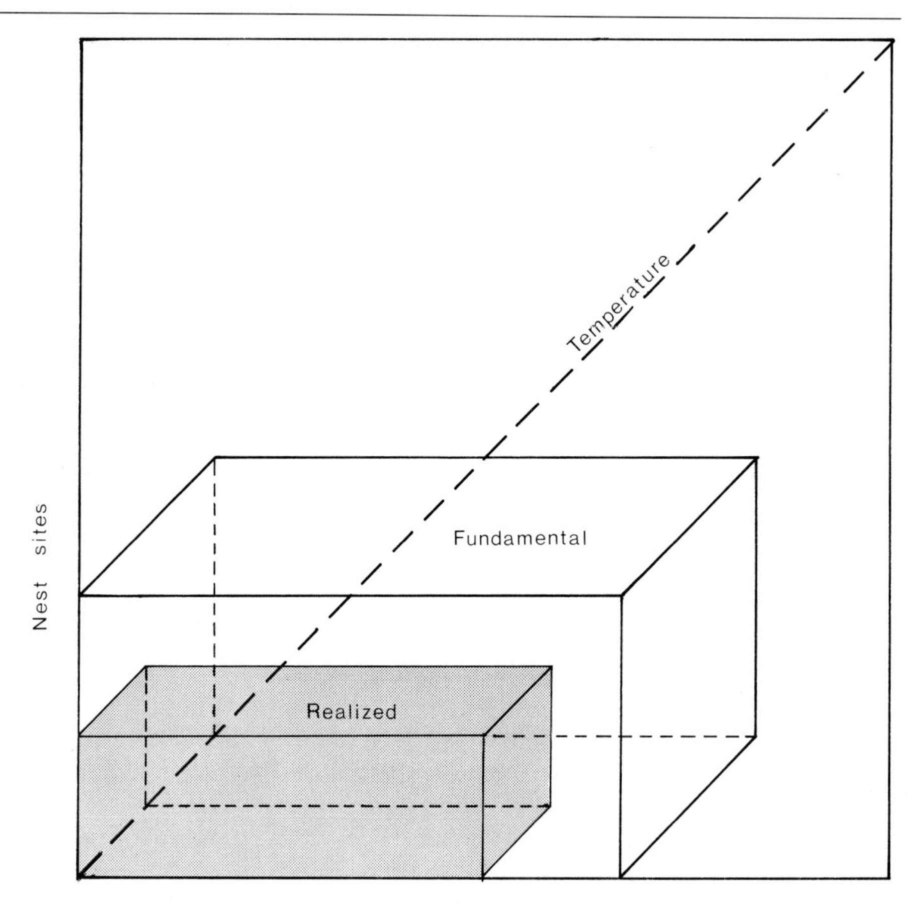

Figure 3–19 Three-dimensional representation of an ecological niche. Niches are really multidimensional, but that is impossible to present in a diagram. In this simplified example, three dimensions are shown. The length of each axis represents the potential array of conditions, and the indicated portion of each axis represents the part of that array actually usable by the species in question. By considering all three axes simultaneously, a polygonal space is defined that contains all the conditions (from three axes) suitable for this species. This imaginary space is the species' niche. In this diagram, the polygon has square corners, indicating that the three dimensions do not interact; choice of nest sites, for example, does not vary with temperature. In actuality, dimensions are likely to interact. Competition or predation may restrict the potential or fundamental niche space of the species, and the realized niche is smaller.

predators, parasites, even prey—can have a profound influence on how much of the potential niche space a species may actually occupy. Usually, only part of the potential space is occupied. Restriction of niche space by other species may mean that behavior is not so varied, that a small range of food sizes is taken, or that fewer of the possible nest sites are used. In some cases, such restriction may have no perceptible effect on where the species lives, but, frequently, a reduction of choice of food items or nest sites must mean that certain places are no longer habitable. Consequently, the influence of other species often affects distribution, at least on a small scale.

Habitat Selection

A habitat can be rather loosely defined as a place in which to live (within the animal's geographic range). Habitat selection implies that animals are somehow capable of choosing their living place, and, indeed, vertebrates generally are. How do they do this? The answer can be approached by investigating sensory systems, the effects of experience, physiological and psychological states, and environmental cues that trigger a settling response. For instance, we might find that certain vegetation configurations will induce a bird in breeding condition to settle in a particular place or that specific substrates and water depths will induce a fish to cease wandering and remain in that place. These are proximate factors that represent significant indicators of appropriate places and can therefore explain *how*—never *why*—an animal responds to a certain cue.

The fundamental explanation of why animals live in certain habitats involves ultimate factors that give selective advantage to certain behavior patterns. For every species, there is some habitat or set of habitats (the optimum) that is very suitable to its characteristics; there are other habitats that are less suitable; and there are still others that are totally unsuitable. If the population is very small, all the individuals may be able to find room in the optimum habitat, but as population size and intraspecific competition increase, the optimum habitat will be filled and some individuals will spill over into suboptimal habitats. If the population increases still more, some individuals will be forced to move into unsuitable habitats where the probability of survival or successful reproduction are very low.

The great tit (*Parus major*) in Europe characteristically breeds in oak woods. Krebs (1971), however, has shown that, in England, some birds, mostly young males, establish territories and breed in hedgerows. Nesting success in hedgerows was significantly lower than in oak woods, due largely to predation, and hedgerow birds often moved into the woods when space became available. Such domination of the best habitat by some individuals and the resulting occupation of less suitable space by others has been observed in many species.

What makes one habitat suitable and another one less so? Every existing species has a complement of physiological, morphological, and behavioral characteristics that permit it to do some things well, in some places, at some times. A seal is equipped to swim, a mole to dig, and a bat to fly. None of them would have the remotest chance of living like the others. Similarly, the bill of a kiwi (*Apteryx australis*) in New Zealand is well-suited for probing in soft earth. A kiwi, which is flightless, would quickly starve if prevented from foraging in its usual places and forced to try to glean insects from leaves or extract nectar from flowers. One can see that existing physiological, morphological, and behavioral limitations determine an animal's living space (seals in the water, moles underground, bats in the air, kiwis on the ground, and so on). In the short-term sense, this must often be true. On the other hand, it can be argued that, in the long run (and sometimes in the short run too), many other factors may be equally important in determining where an animal lives. The physiological, morphological, and behavioral features have evolved because they conferred adaptive value on the individuals when other factors compelled the species to live in certain places. These other factors involve interactions with other species: competitors, predators, parasites, and prey. Obviously, then, habitat suitability is a product of an animal's history, present environmental constraints, and past and present selection pressures.

All of this, however, still does not permit us to say, specifically, why a wood thrush (*Hylocichla mustelina*) lives in forest understory and a red-eyed vireo (*Vireo olivaceus*) in the canopy; nor does it explain why a prairie kingsnake (*Lampropeltis calligaster*) is seldom found in the woods. Knowledge of specific cases depends on detailed study of each situation. Even when we can predict that a certain place would be suitable for a certain animal species, it is often hard to assess the specific factors that make it appropriate. Limiting factors are seldom single entities. For example, fishes often have specific temperature-tolerance limits, beyond which they die, but many other factors interact with temperature effects—the activity of the fish, the time of year, the age of the fish, nutritional state of the fish, and so on. Furthermore, the lethal limits do not adequately define the range over which normal activity may take place. If the fish becomes lethargic at higher (but still nonlethal) temperatures, it may be unusually susceptible to predation or be unable to forage successfully. Furthermore, extreme conditions may exert great effects on vertebrate populations only in some years, so that habitat occupation and possibly the intensity of natural selection fluctuate. For instance, severe winters may impose unusual food shortages, enforced population movements, or high mortality on bird populations, with the possible results of lowered breeding populations in subsequent years (see Graber and Graber, 1979). Similarly, the effects of one species on another often vary in time and space; thus, one species may oust another in some conditions but not others. A variety of ecological limitations on animal distributions are explored in the next section.

Ecological Limitations

Competition depends on some limited resource for which the demand exceeds the supply. As a result, some individuals obtain reduced amounts of a resource and, consequently, have reduced fitness. Overt aggression is only sometimes involved. More often, certain individuals exploit the resource more effectively than others. Competition among members of the same species (intraspecific competition) tends to encourage divergence among members of the species. Any individual that loses in the competition for the original resource can maintain its fitness if it can exploit additional resources not used by the other members of the population. The entire range of potential resources is called a resource spectrum. Intraspecific competition tends to increase the part of a resource spectrum used by a species. On the other hand, competition among members of different species (interspecific competition) tends to restrict such intraspecific divergence along a resource spectrum. In this case, the "additional" resources may already be used effectively by members of another species, who might then outcompete the expanding members of the first species; thus, interspecific competition tends to decrease the portion of the resource spectrum used by a species. Expansion along the resource spectrum produced by intraspecific competition will usually be balanced at some point by compression as a result of interspecific competition (Fig. 3–20). Competition between species tends to confine each species to a part of the available resource spectrum.

Ecologically similar species occur in a series along a resource spectrum. At least four species of kingfishers can be found in lowland, forested regions of Costa Rica and Panama (Fig. 3–21). They differ in body size and bill size and probably capture prey of different average sizes. Because each species can forage efficiently on different prey, they often occur in somewhat different habitats: the smallest kingfisher may occur along smaller streams, which support smaller fishes; the largest is characteristic of large streams and coastlines. All four may live in close proximity to each other if the streams of the area contain a sufficient variety of prey sizes. The kingfishers can be arranged in a series of increasing or decreasing body or bill size. The difference between the average prey sizes of adjacent members of the series would be one measure of the ecological distance between the kingfisher species. With respect to prey size, the predators are a certain average distance apart on this dimension, which is divided among all four niches. If this resource axis is fully utilized by these four species, no new species that also requires this axis can move in. Note, however, that, in this

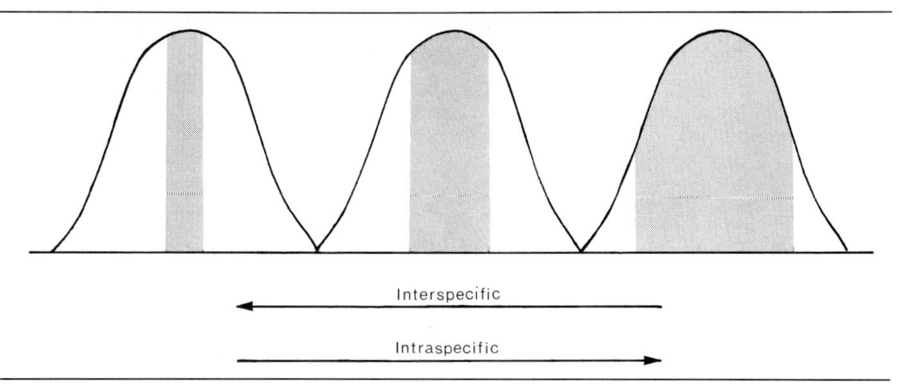

Figure 3–20 Effects of intraspecific and interspecific competition. When intraspecific competition dominates, the species spreads out into marginal areas. When interspecific competition is intense, the species tends to stay in a narrow range of optimum conditions (shaded area). (From Svardson, G., 1949. Competition and habitat selection in birds. *Oikos, 1*:157–174.)

example, interspecific competition for food is only assumed; it has not been demonstrated.

In fact, it is quite possible for body sizes and foraging styles to evolve without regard to competition. The species may now coexist in some areas because selection has favored increased foraging efficiency in times of food scarcity, irrespective of interactions with other kingfishers. Certain species

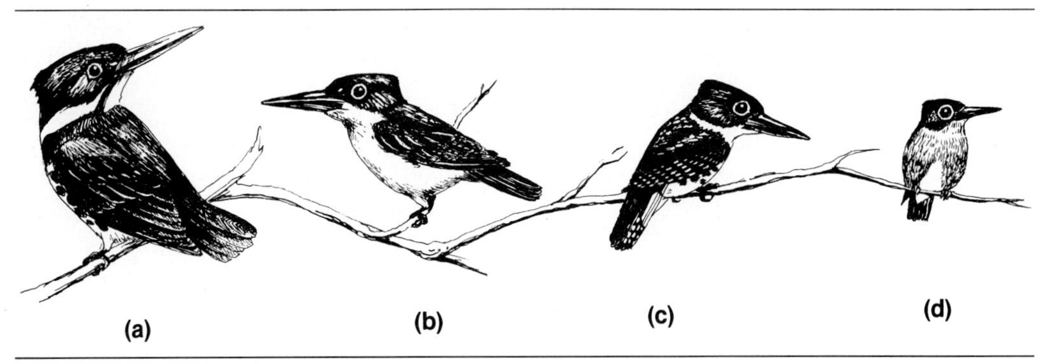

Figure 3–21 Four congeneric species of kingfishers found along streams in Costa Rica and Panama: (*a*) Amazon kingfisher, *Chloroceryle amazona,* 28 to 30 cm long; (*b*) green and rufous kingfisher, *C. inda,* 20 to 22 cm long; (*c*) green kingfisher, *C. americana,* 18 to 20 cm long; and (*d*) pygmy kingfisher, *C. aenea,* 14 cm long. Note that the two middle-sized species are much closer in size to each other than to the remaining species. Two other, larger, species of the genus *Ceryle* are most common on coastlines and lagoons.

may now fit together because they were already sufficiently different, not because competition forced divergence (Recher and Recher, 1980). It is not easy to determine the actual importance of competition to the evolution of species' differences and ecological distances in specific cases.

Interspecific competition obviously can prevent species from moving into new habitats or finding enough food or nest sites to support a population. The process of competition takes time, however; food (or other) resources usually are not depleted instantly. Therefore, environmental fluctuations, if rapid enough, may prevent one species from excluding another by shifting the competitive balance from one to the other and back again, thus maintaining both species in the community. Alternatively, if habitat patches are continually destroyed and reestablished, some species may maintain themselves by flitting among the ephemeral patches and simply avoiding the impact of their competitors. Again, if predators or climatic catastrophes reduce the potentially competing populations below the limiting level of resource demand, competition does not occur and both species continue. The outcome of competition in terms of animal distribution thus clearly depends on the distribution of resources in space and time relative to the levels of use by potential competitors. Competition has been assumed more often than it is demonstrated and must not be invoked uncritically.

Experimental manipulations are useful for revealing the competitive effects on both distribution and niche size because of the relative ease of controlling variables extraneous to the point of the experiment. One such experiment was performed by Werner and Hall (1976) with three species of sunfishes (Centrarchidae): *Lepomis macrochirus* (bluegill), *L. gibbosus* (pumpkinseed), and *L. cyanellus* (green sunfish). These three North American species (Fig. 3–22) are about the same body size and often coexist, with some segregation of habitats, in small lakes. The green sunfish is usually found in shallow water close to shore; pumpkinseeds predominate near the bottom; and bluegills live in upper water levels. Habitats do overlap, however. Populations of each of the three species were grown separately in experimental ponds, and one other pond held equal populations of all three species. All three species foraged primarily from the leaves of submerged vegetation when they were the only species in the pond. But when all three were together, the green sunfish occupied this foraging area, the pumpkinseeds switched to bottom-living prey, and the bluegills switched to open-water prey. Both switches entailed a change to prey of smaller average size. Presence of other species (the green sunfish, in this case) resulted in a niche shift and a habitat shift for the remaining species.

Another study, this one performed in natural habitats, showed that the range of habitat occupied by marsh-nesting red-winged blackbirds (*Agelaius phoeniceus*) was restricted as a result of invasion of the marsh by yellow-headed blackbirds (Orians and Willson, 1964). The larger male yel-

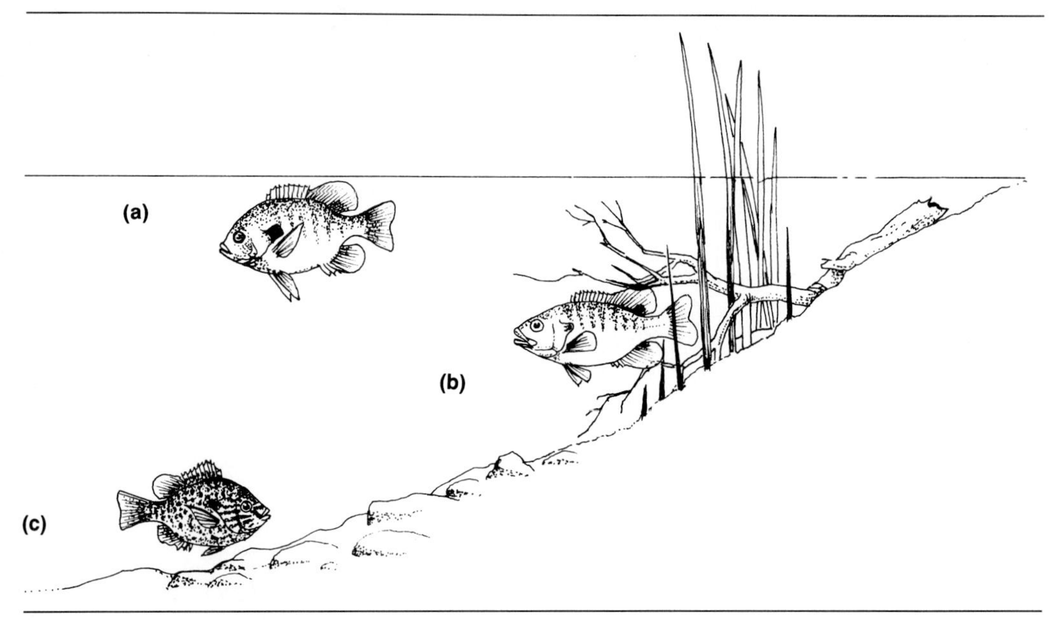

Figure 3–22 Three species of *Lepomis* (Centrarchidae): (*a*) *Lepomis macrochirus;* (*b*) *L. gibbosus;* and (*c*) *L. cyanellus.* When *L. cyanellus* is present, the others are forced into less favored parts of the pond.

lowheads were dominant over the smaller male redwings that had, at first, occupied all the vegetated parts of the marsh. Eventually, the redwing territories were confined to the vegetation near the shore as the yellowhead males established territories over deeper water (Fig. 3–23). Interspecific competition for space clearly influenced the habitat occupied by the redwings.

Competition is not always so direct and easily observed. Species with only slight similarities of ecology may exert a cumulative competitive effect on other species in the form of diffuse competition. Terborgh and Weske (1975) studied the birds along an altitudinal transect up a mountain in Peru and found that, as expected, the composition of the avifauna changed with elevation. Some of the additions or deletions of species could be accounted for by changes in habitat, and some by direct competition and replacement of one species by a similar one. About a third of the elevational changes in avifauna were ascribed to diffuse competition between sets of species with partly overlapping requirements. Similarly, Keast (1969a) showed that, in the absence of certain species on the island of Tasmania, other, sometimes unrelated, species expanded their foraging niches by hunting in the vacated feeding sites as well as in the sites they customarily used in mainland Australia.

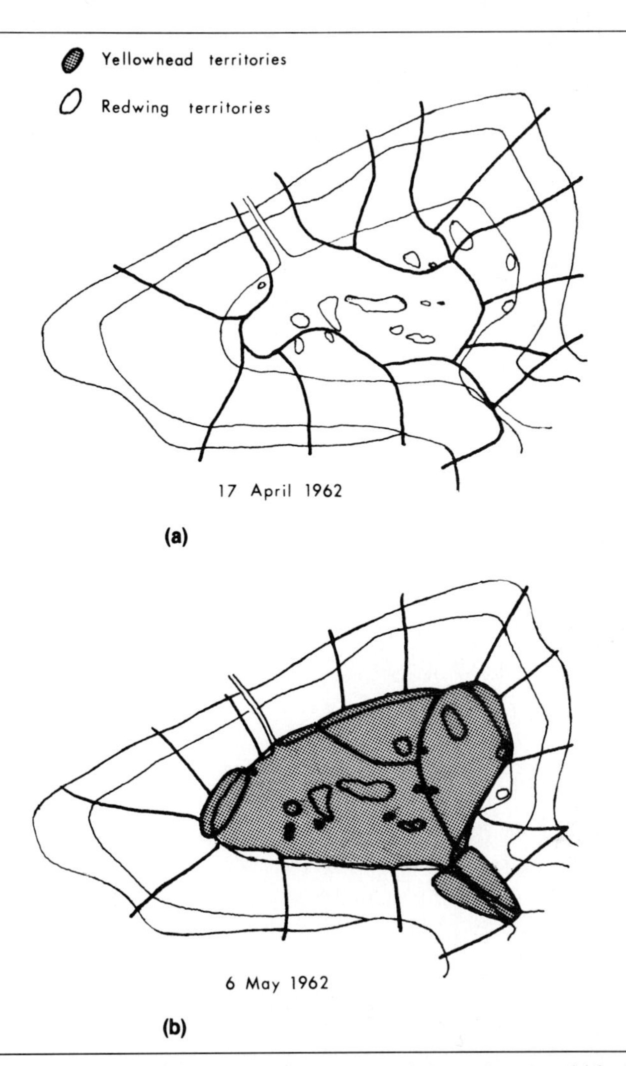

Figure 3–23 Distribution of breeding territories of the red-winged blackbird in a marsh in eastern Washington before and after arrival of yellow-headed blackbirds. Data were collected at the Turnbull National Wildlife Refuge in 1962. (Modified from G.H. Orians and Mary F. Willson, 1964. Interspecific territoriality in birds. *Ecol.*, *45*:736–745. Copyright © 1964, the Ecological Society of America.)

Predation may also limit a species' distribution. This has been demonstrated time and again by the impact of humans on many wild species. Unfortunately, the effects of human predation are usually complicated and exacerbated by simultaneous habitat destruction or competition with domestic animals. Thus, it is hard to discriminate the results of predation alone. It seems likely, however, that the decimation of the American bison (*Bison bison*) was due primarily to extraordinarily intensive hunting in the middle of the nineteenth century. It is equally likely, at least in principle, that predation may act as a limit to distribution in circumstances not so dominated by that rather unusual animal, the human. One can easily imagine a situation in which one kind of animal is faced with such heavy predation that it is simply unable to occupy certain habitats, although it is difficult to provide a clear-cut example of this with nonhuman vertebrates. Many examples can be found in studies in invertebrates, however: for instance, Connell (1971) showed that a barnacle, *Balanus glandula,* could survive only at the upper limit of the intertidal zone because predators—mainly snails—quickly ate all the young barnacles that settled in the lower intertidal zones. Hairston (1980) has suggested that predation may be more likely than competition in determining the distributions of *Desmognathus* salamanders in the Appalachian Mountains; he emphasizes the need for suitable experiments to document the possibility.

Parasitism is similar to predation in many respects, but the parasite is usually smaller than its animal host and seldom kills its host immediately (and, in fact, may never kill it). One of the best examples of limitation of a host population by parasites is drawn from a seminatural experiment that is now a classic. European rabbits (*Oryctolagus cuniculus*) were introduced into Australia in 1859. By 1900 they had multiplied so greatly that millions of hectares of vegetation were drastically altered by their feeding; this created a severe problem for the Australian ranchers, who depended on natural vegetation to feed their stock. None of the early efforts at rabbit control worked, but finally a mosquito-borne parasite (a myxoma virus that causes a disease called myxomatosis in rabbits) was introduced from South America. The effect of this parasite was sweeping: population sizes of rabbits were drastically reduced (more than 99 percent of the rabbits died), and the vegetation began to recover. This case involved two separate introductions of nonnative animals, but the principle can easily be applied to native vertebrates anywhere. It is, of course, less easy to demonstrate in a totally natural situation, in which the host has not increased so monstrously precisely because it has not been freed of its natural parasites.

Another kind of interspecific interaction that is often neglected in considerations of species' distributions is now gaining recognition. It concerns the interaction of predator and prey—but from the predator's point of view rather than the prey's. Animals and plants that are eaten by animals are not just helpless victims in an evolutionary sense. Predation establishes selection for prey or host characteristics that improve the probability of escape; therefore, every improved technique of predation may eventually be matched by an improved technique of escape. This relationship produces what has been called an "evolutionary race" between predator (to eat) and prey (not to be eaten). Escape techniques for prey are many and varied. Obviously, the predators are not always eating all of their prey, so the prey probably often possess successful escape techniques. Some prey may even be ahead in the evolutionary race, in that they have somehow found an escape so successful that they have become unsatisfactory or inadequate food for their (former) predators. As a result, the food resources available to the predator are reduced and the predator may be forced to change the habitats in which it forages. I know of no good examples of this effect among the vertebrates, but current studies, especially of insects and plants, provide abundant evidence that such a process has been very common and important.

Interaction of parasite and host is similar in many respects; however, it may be to the parasite's advantage not to kill its host too quickly, before the parasite has reproduced. In the case of the rabbits in Australia, after the first epidemics of myxomatosis the impact of the virus diminished as more and more rabbits had resistant genotypes; the virus also decreased in virulence. Selection had favored rabbit genotypes that were less susceptible and, thus, had lower mortality and virus genotypes that did not kill the host so quickly, thus increasing the reproduction and dispersal of virus offspring.

In a uniform environment, competition and predation are likely to result in the extinction of one or several interacting species. The losers have no retreat, no portion of the environment in which they can elude the others; therefore, spatial heterogeneity is an important factor in determining the outcome of many species interactions. Up to a point, where there is more spatial heterogeneity, there are more species. Obviously, this heterogeneity must exist at a scale suitable for the organisms in question. Mice and elephants will see their environments from different points of view: a hectare of grassland may be an empire for a mouse but little more than a turning-around place for an elephant.

The effects of spatial heterogeneity can easily be seen in a comparison of bird communities in North America. Many studies have demonstrated that the number of bird species increases in a series of habitats from grassland through shrubland to forest (Fig. 3–24). Obviously, the vertical dimension of the vegetation increases along this sequence, opening up the possibility of vertical stratification; some birds specialize in using the canopy, others the midlevel, and still others the layers near the ground. More than that, the addition of different forms of vegetation (shrubs and then trees) also creates horizontal heterogeneity (shrub thickets interspersed with open, grassy areas, small clusters of trees separated by shrubs and grassy patches, and so on). Horizontal patchiness makes possible the addition of bird species that can use these different patches (the yellow-breasted chats (*Icteria virens*) in the thickets, and field sparrows (*Spizella pusilla*) in the intervening grassland with scattered shrubs, for instance) and different combinations of patches. Together, vertical and horizontal heterogeneity open up new combinations of patches. The addition of trees to an area presents new substrates (bark) for foraging and new nest sites for cavity-nesting birds such as woodpeckers and nuthatches. Trees also create open spaces between canopy and ground layers that can be used by other kinds of foragers, including pewees (*Contopus virens*) and redstarts (*Setophaga ruticilla*), which forage by sallying forth from a perch to an open space to snatch an insect from the air.

Different kinds of forests, however, contain different numbers of species, which cannot be accounted for by simple differences in spatial heterogeneity. Consider one example: Neotropical lowland forests support a much higher number of bird species per unit area than north temperate forests. A Panamanian rain forest may contain 56 bird species on a 2-hectare study plot, but a forest study plot of the same size in Illinois will contain only about 32 species of breeding birds. Unlike most of the breeding birds of the Panamanian forest, many of the Illinois species are not year-round residents. Some of them, such as the wood thrush and Kentucky warbler (*Oporornis formosus*), spend at least part of their nonbreeding season in the same forest with all the neotropical residents.

The greater species number of neotropical lowland forest may be accounted for primarily by the presence of additional food resources that are not abundant or persistent in temperate forests. Particularly important are fruits, large insects, and other vertebrates, including lizards and snakes; these added reliable food resources permit additional dietary specializations as well as additional foraging styles. (Tropical forests also support more bat species than temperate forests, partly because of the greater fruit and flower resources.) Furthermore, persistence of the canopy throughout the year in a tropical rain forest, in contrast to a deciduous forest, main-

Figure 3–24 Schematic representation of a vegetation gradient from open fields to deciduous forest with some typical bird species in east-central Illinois. The increasing vertical dimension contributes also to increased horizontal heterogeneity. (Based on data in M.F. Willson, 1974. Avian community organization and habitat structure. *Ecol.*, *55*:1017–1029, and various University of Illinois graduate theses.)

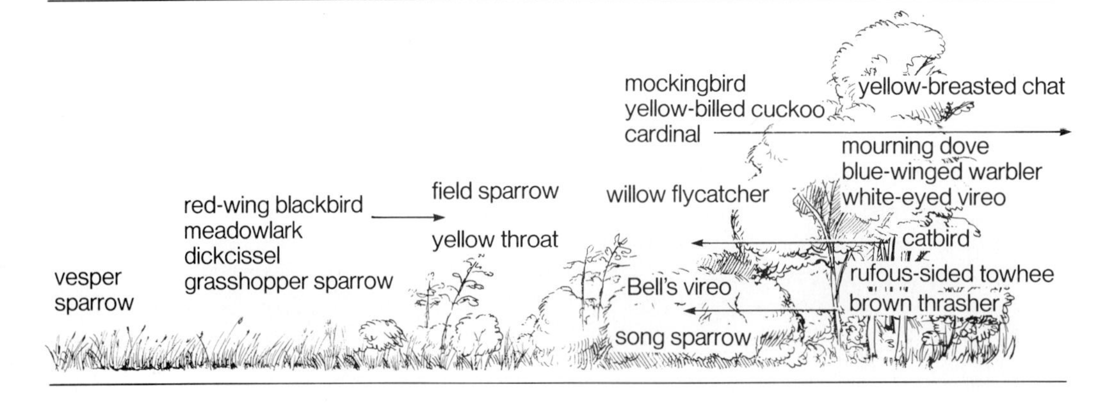

tains a more constant microclimate for bird species of the lower levels of the forest. Consequently, resource availability is seasonally more constant, and narrow foraging specializations may be more successful (Karr, 1976a). Spatial heterogeneity and resource diversity and reliability clearly increase the number of available retreats in which species may persist in the face of pressure of other species. Further study will probably show still other factors that contribute to regional differences in faunal diversity.

Not all tropical lowland forests support an avifauna as rich as that of the neotropics, however, even when habitat structure is similar. Forests in West Africa, for example, have fewer bird species than Panamanian forests, probably for several reasons. There may be a lower availability of fruits and a higher number of fruit-eating, arboreal mammals such as squirrels, and furthermore, historical factors may have restrained the adaptive radiation of birds in Africa (Karr, 1976b, 1980).

The dynamic nature of interspecific interactions, the possibility of evolution of new tolerance limits and new capacities for resource use, and changes in environmental conditions—all emphasize that animal distributions are not static, but are, in fact, quite the contrary. Some species' distributions have changed drastically or are still changing rapidly. The American opossum (*Didelphis virginiana*) and the raccoon (*Procyon lotor*) expanded their ranges northward by 320 km or more between 1930 and

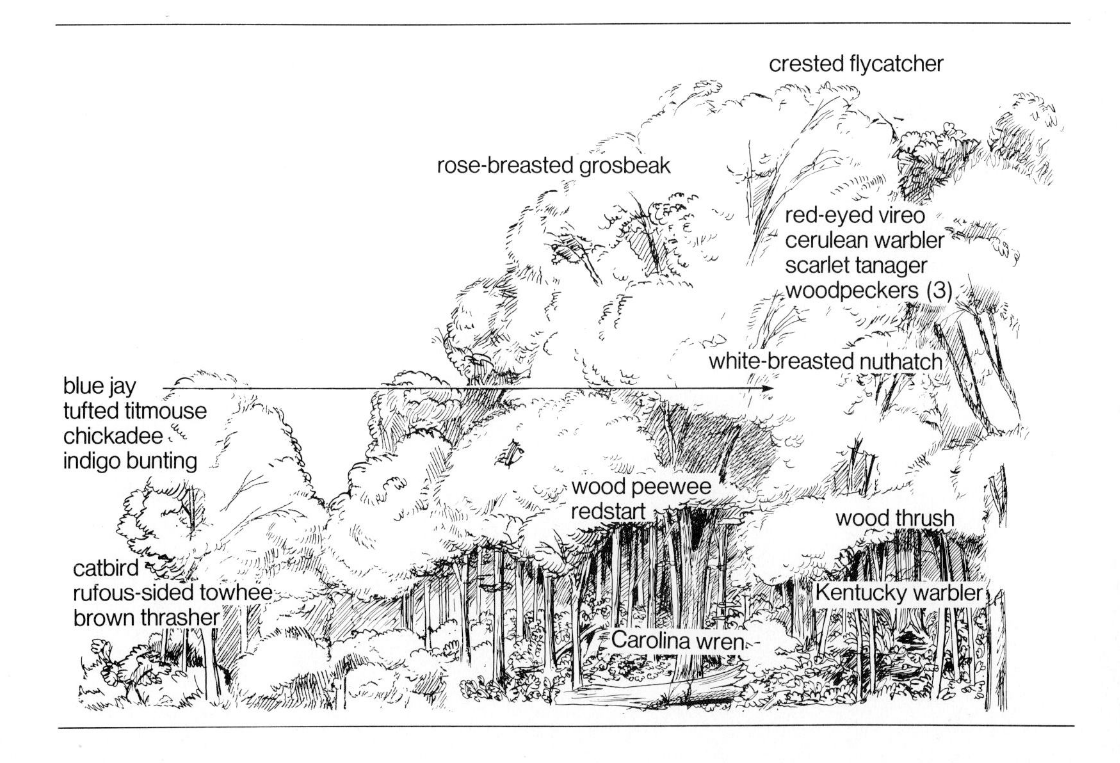

1960. The coyote (*Canis latrans*), after an initial range contraction due to human influence, has again expanded greatly into eastern North America, even becoming semiurbanized in places. Tapirs, camels, primates, and wild horses were native in North America in the Tertiary period, but have since gone extinct in that area; they are survived by natural populations of modern species only in other parts of the world.

THE ISLAND BIOGEOGRAPHIC MODEL

Because they are relatively small and isolated and have smaller assemblages of species than the larger continental landmasses, islands can serve as microcosms for the study of processes that control community structure and animal distributions. Much of our understanding of such processes has come from studies of oceanic islands, but "islands" can readily be found within continents as well. Woodlots, mountaintops, ponds, bogs, marshes, caves, canyons, and beaches are frequently viewed as habitat islands in a surrounding sea of different habitats. Therefore, principles derived from a study of oceanic islands can be directly applied to a number of mainland situations and indirectly applied to virtually all mainland faunal assemblages.

Zoogeographers and ecologists once thought of island faunas as depauperate because few animals were able to reach them. This view implies that island faunal diversity is not in equilibrium because more species would live there if only they could get there. The model of MacArthur and Wilson (1967) and subsequent research by many others, however, suggests that island faunas may be at equilibrium between the arrival (often by immigration, sometimes by speciation) and the extinction of species. Although various disruptions (climatic changes, volcanic eruptions, storms, and floods) may disturb such an equilibrium and an island fauna may take some time to return to a balance point, a fundamental equilibrium point can usually be predicted, and deviations from it add to our insight of the processes involved. Most island models to date have emphasized primarily numbers of species, but the kinds and identities of island species are of obvious importance to an eventual complete understanding of island community patterns and processes and their consequences for vertebrate distributions. In the following discussion, we will consider numbers of species and kinds of species, and try to exemplify the major points of biological significance.

Not unexpectedly, species number tends to increase with island size. Jamaica has 6 or 7 species of *Anolis* lizards, Puerto Rico has 11, Hispaniola has 20 or 21, and Cuba has 22 (Williams, 1969). As is often the case, more extensive data are available for birds (Fig. 3–25 shows data for the birds of islands around New Guinea). Large islands obviously provide larger targets for dispersing animals and often support more varied habitats. If other

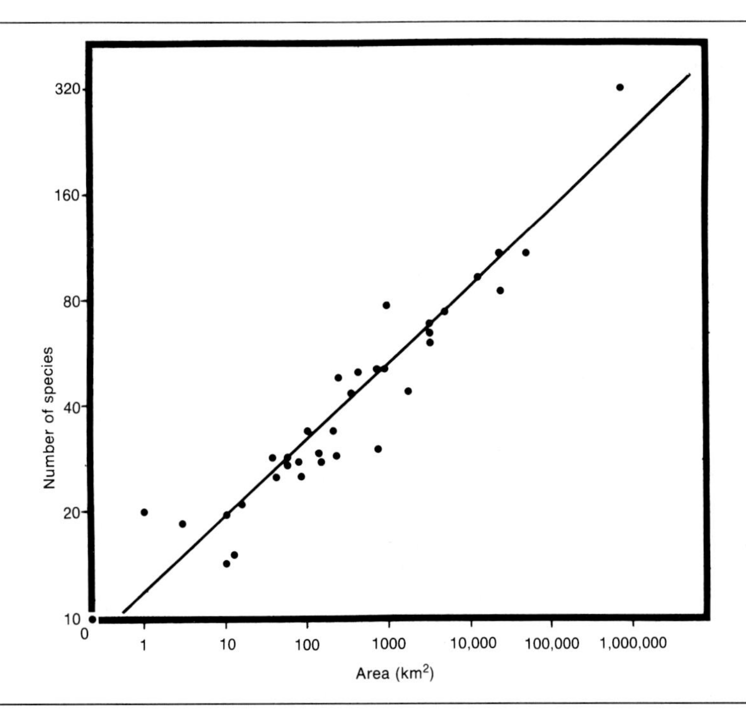

Figure 3–25 Number of resident land and freshwater bird species on some islands off the coast of New Guinea. Note that the number of species increases with area. (From J.M. Diamond, 1973. Distributional ecology of New Guinea birds. *Science, 179*:759–769. Copyright © 1973 by the American Association for the Advancement of Science.)

factors were similar, we would also expect that the extinction rate of colonists on large islands would be less than that on small islands. On large islands, each species can maintain a larger population size, and a larger population is less subject to random fluctuations that would reduce the population to zero. Furthermore, we would predict that an island close to a source of immigrants (a mainland or another island) would have a higher immigration rate than one that is far from such a source. As the number of species already present on an island increases, the probability of successful immigration decreases because more and more of the available species are already there. The probability of local extinction increases because competition or other interactions tend to reduce population size and, thus, increase the risk of severe random fluctuations. These effects of area and distance on the balance of inflow and outgo of island species are diagrammed in Figure 3–26.

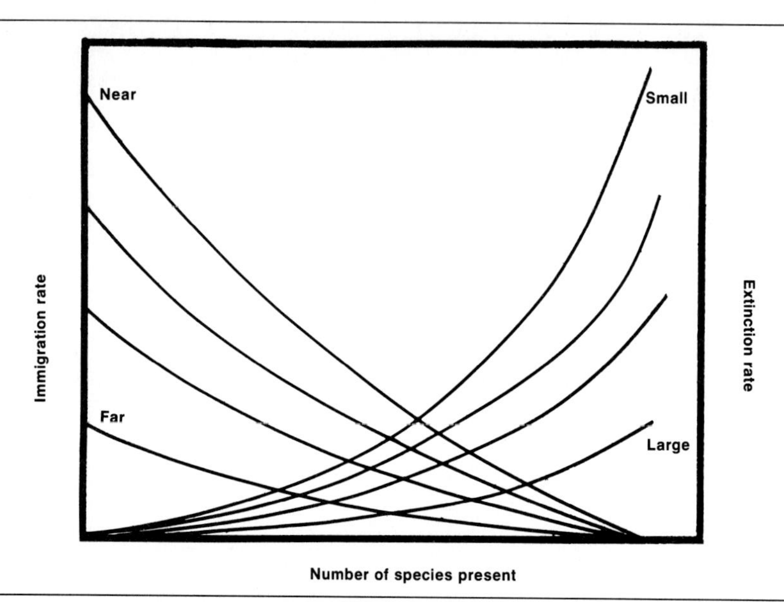

Figure 3–26 Equilibrium models of biotas on several islands varying in size and in distance from the main source area. An increase in distance (near to far) lowers the immigration curve. An increase in island size (small to large) lowers the extinction curve. The equilibrium number of species for any island is found where the relevant immigration and extinction curves cross. (Modified from R.H. MacArthur and E.O. Wilson, 1967. *The Theory of Island Biogeography.* Copyright © 1967 by Princeton University Press. Reprinted by permission of Princeton University Press.)

Because the processes of immigration and extinction go on continuously, the equilibrium is a dynamic one. Although species number may hover about a constant, the identity of the species present may change from year to year. Studies of island birds in the West Indies and near New Guinea have estimated that sometimes over 60 percent of the species may go extinct on an island or reinvade that island in as little as 50 years (Diamond, 1973). Actual rates of change of species composition are much higher: a species may colonize, go extinct, and recolonize many times during a long period of time, but such changes often go undetected in the censuses of the island fauna. Furthermore, similar and adjacent islands with the same sources of immigrants often support quite different collections of species. Despite the shifts in species composition from island to island and from time to time, assemblages of species on islands are not entirely random. We can discern several characteristics of colonizing species and determine roughly which kinds of colonists tend to persist together in a community.

Perhaps the fundamental characteristic of a good colonist is an ability to travel. Wide-ranging individuals have a higher probability of finding a new habitat than sedentary ones. In addition, good colonists tend to be fairly general in habitat or other resource requirements; often they occupy variable habitats in the source area. In a sense, then, they are preadapted to be successful colonists. Some, but not all, good colonists are able to produce large numbers of offspring in a short period of time by means of frequent reproduction, reproduction early in life, or perhaps large litters. Therefore, not all species in a source fauna on the mainland are likely to be found as island immigrants.

Even if specific identities may vary in time or space, certain consistencies may appear in species assemblages. An example is provided by Diamond's (1973, 1975) data on fruit pigeons of the New Guinea region. Eighteen species can be found on New Guinea: small pigeons of the genus *Ptilinopus* and large ones of the genus *Ducula*. All live in tree canopies and eat soft fruit, but no more than eight can be found coexisting in the same habitat (lowland rain forest), and among those eight species each is roughly 1½ times the size of the next smaller species. Larger pigeons eat larger fruits, on the average, and can only perch on larger branches. Other pigeon species live in other habitats on New Guinea. In the islands around New Guinea, these species occur with differing frequency and in different combinations; if lowland rain-forest species happen to be absent, species of other habitats often expand to occupy the rain forest.

Although the particular species may vary from place to place, size relations among the pigeons on different islands show some patterns. In Figure 3–27, the eight lowland rain forest species on New Guinea are categorized by weight into eight levels. Proceeding down the list of islands in decreasing order of number of pigeon species in lowland rain forest, we see that certain size categories drop out. By the time species number has decreased to eight species for the whole island, six of these occupy lowland rain forest and size levels 1 (the smallest *Ptilinopus*) and 5 (the largest *Ptilinopus*) are no longer represented. When only two fruit pigeons occur on an island, they are always a middle-sized (level 3 or 4) *Ptilinopus* and a middle-sized (level 7) *Ducula* (not always the same species, however). Thus, it appears that only certain size combinations are likely. Other pigeons may immigrate to the depauperate islands but are unable to establish populations. Diamond inferred that competition from resident species is too severe and the ecological distance between species cannot be reduced by the insertion of an additional category. Great care is needed in making such inferences (Connor and Simberloff, 1979; Diamond and Gilpin, 1982; Gilpin and Diamond, 1982), and factors in addition to competition may be involved.

Rules suggested by the study of islands can be applied to continental patterns as well. Speciation, immigration, and extinction rates can be compared on landmasses of different area and with access to different source faunas. Habitat fragmentation and differentiation vary with conditions and

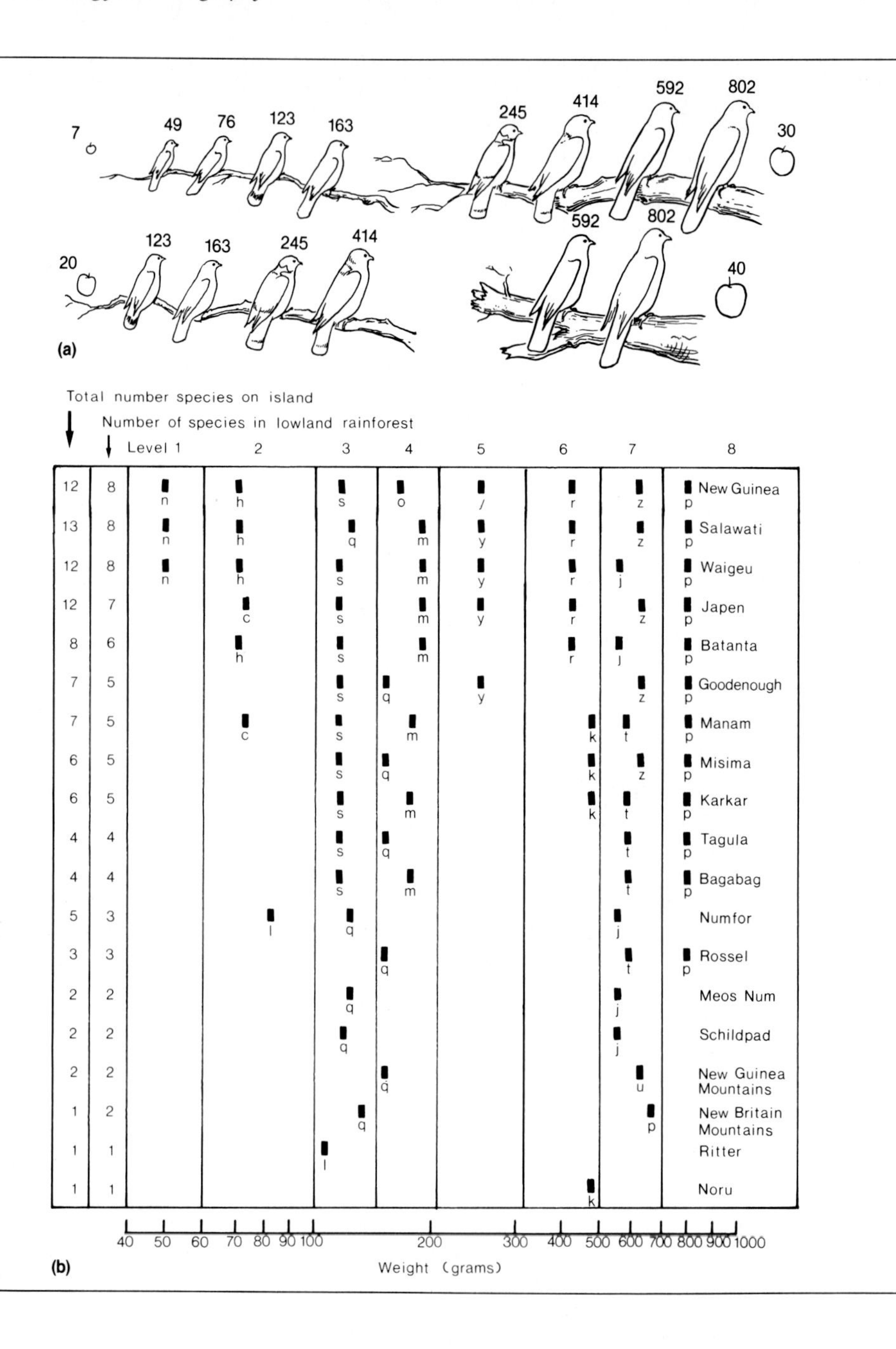

are reflected in opportunities for speciation. Climatic regimes help determine resource predictability and the opportunity for ecological specialization, as well as what interspecific ecological distances are feasible. The

Figure 3–27 (*a*) Fruit pigeons of New Guinea lowland rain forests. Birds are arrayed by average weights (g). Up to four species may feed in a tree bearing fruits of a certain size (mm), and large pigeons can feed on larger fruits than can small pigeons. Pigeons, indicated by their weights, are *Ptilinopus nanus,* 49 g; *P. pulchellus,* 76 g; *P. superbus,* 123 g; *P. ornatus,* 163 g; *P. perlatus,* 245 g; *Ducula rufigaster,* 414 g; *D. zoeae,* 592 g; *D. pinon,* 802 g. (From J.M. Diamond, 1973. Distributional ecology of New Guinea birds. *Science, 179*:759–769. Copyright © 1973 by the American Association for the Advancement of Science.) (*b*) Size distributions of fruit pigeons on islands around New Guinea that support different numbers of species. Note that certain size classes appear to drop out as fewer species are present. Each species is represented by a letter; those in level 6 and above are *Ducula,* the remainder are *Ptilinopus.* Note that different species may occupy certain size levels and that some species change levels. (From J.M. Diamond, 1975. Assembly of species communities. *In* M.J. Cody and J.M. Diamond [Eds.], *Ecology and Evolution of Communities,* pp. 342–444. Belknap Press, Cambridge, MA.)

rules about all such processes lead eventually to construction of models of continental dynamic equilibria that are analogous to those created for islands (Rosenzweig 1975; Cody 1975).

SELECTED REFERENCES

Briggs, J.C., 1974. *Marine Zoogeography.* McGraw-Hill Book Co., New York.

Brown, J.L., 1969. Territorial behavior and population regulation in birds. *Wils. Bull., 81*:293–329.

CLIMAP Project Members, 1976. The surface of the ice-age earth. *Science, 191*:1131–1144.

Cody, M.L., 1975. Towards a theory of continental species diversity. *In* M.L. Cody and J.M. Diamond (Eds.), *Ecology and Evolution of Communities.* Belknap Press, Cambridge, MA.

Cody, M.L., and J.M. Diamond (Eds.), 1975. *Ecology and Evolution of Communities.* Belknap Press, Cambridge, MA.

Connell, J.H., 1971. On the role of natural enemies in preventing competitive exclusion in some marine animals and in rain forest trees. *In* P.J. denBoer and G.R. Gradwell (Eds.), *Dynamics of Population,* pp. 298–310. Centre for Agricultural Publications and Documentation, Wageningen, Netherlands.

Connor, E.F., and D. Simberloff, 1979. The assembly of species communities: chance or competition? *Ecol., 60*:1132–1140.

Cox, C.B., 1974. Vertebrate palaeodistributional patterns and continental drift. *J. Biogeog. 1*:75–94.

Cracraft, J., 1973. Continental drift, paleoclimatology, and the evolution of biogeography of birds. *J. Zool.* (Lond.), *169*:455–545.

Cracraft, J., 1974. Continental drift and vertebrate distribution. *Annu. Rev. Ecol. Syst., 5*:215–261.

Cushing, D.H., and J.J. Walsh, 1976. *The Ecology of the Seas.* Blackwell, Oxford, England.

Darlington, P.J., 1957. *Zoogeography.* John Wiley & Sons, New York.

deVos, A., 1964. Range changes of mammals of the Great Lakes Region. *Am. Midl. Nat., 71*:210–231.

Diamond, J.M., 1973. Distributional ecology of New Guinea birds. *Science, 179*:755–769.

Diamond, J.M., 1975. Assembly of species communities. *In* M.J. Cody and J.M. Diamond (Eds.), *Ecology and Evolution of Communities.* pp. 342–444. Belknap Press, Cambridge, MA.

Diamond, J.M., and M.E. Gilpin, 1982. Examination of the "null" model of Connor and

Simberloff for species co-existence on islands. *Oecologia, 52*:64–74.

Eisenberg, J.F., 1982. *The Mammalian Radiations.* University of Chicago Press, Chicago.

Elton, C.S., 1958. *The Ecology of Invasions by Animals and Plants.* Methuen, London.

Fretwell, S.D., and H.L. Lucas, 1970. On territorial behavior and other factors influencing habitat distribution in birds. I. Theoretical development. *Acta Biotheor. 19*:16–36.

Gilpin, M.E., and J.M. Diamond, 1982. Factors contributing to non-randomness in species co-occurrences on islands. *Oecologia, 52*:75–84.

Good, R., 1974. *The Geography of the Flowering Plants.* Longmans, Green, London.

Graber, J.W., and R.R. Graber, 1979. Severe winter weather and bird populations in southern Illinois. *Wils. Bull., 91*:88–103.

Haffer, J., 1969. Speciation in Amazonian forest birds. *Science, 165*:131–137.

Hairston, N.G., 1980. Species packing in the salamander genus *Desmognathus:* What are the interspecific interactions involved? *Am. Natur., 115*:354–366.

Hays, J.D., J. Imbrie, and N.J. Shackleton, 1976. Variations in the earth's orbit: pacemaker of the ice ages. *Science, 194*:1121–1132.

Hershkovitz, P., 1969. The evolution of mammals on southern continents. VI. The recent mammals of the neotropical region: a zoogeographic and ecological review. *Q. Rev. Biol., 44*:1–70.

Janzen, D.H., 1967. Why mountain passes are higher in the tropics. *Am. Natur., 101*:233–249.

Karr, J.R., 1976a. Seasonality, resource availability, and community diversity in tropical bird communities. *Am. Natur., 110*:973–994.

Karr, J.R., 1976b. Within- and between-habitat avian diversity in African and neotropical lowland habitats. *Ecol. Monogr., 46*:457–481.

Karr, J.R., 1980. Geographical variation in the avifaunas of tropical forest undergrowth. *Auk, 97*:283–298.

Keast, A.J., 1969a. Adaptive evolution and shifts in niche occupation in island birds. *Biotropica, 2*:61–75.

Keast, A.J., 1969b. Evolution of mammals in southern continents. VII. Comparisons of the contemporary mammalian faunas of the southern continents. *Q. Rev. Biol., 44*:121–167.

Krebs, J.R., 1971. Territory and breeding density in the great tit, *Parus major* L. *Ecol., 52*:2–22.

Lansberg, H.E. (Ed.), 1969–1976. World Survey of Climatology. 15 vols. Elsevier, Amsterdam.

Lister, B., 1976. The nature of niche expansion in West Indian *Anolis* lizards. I. Ecological consequences of reduced competition. *Evol., 30*:659–676. II. Evolutionary components. *Evol., 30*:677–692.

Longwell, C.R., et al., 1969. *Physical Geography.* John Wiley & Sons, New York.

Lowe-McConnell, R.H., 1975. *Fish Communities in Tropical Freshwaters.* E. Arnold, London.

MacArthur, R.H., 1972. *Geographical Ecology.* Harper & Row, New York.

MacArthur, R.H., and J.H. Connell, 1966. *The Biology of Populations.* John Wiley & Sons, New York.

MacArthur, R.H., and E.O. Wilson, 1967. *Island Biogeography.* Princeton University Press, Princeton, NJ.

Marshall, L.G., et al., 1982. Mammalian evolution and the great American interchange. *Science, 215*:1351–1357.

McIntosh, S.K., and R.J. McIntosh, 1981. West African prehistory. *Am. Sci., 69*:601–613.

Oliver, J.E., 1973. *Climate and Man's Environment.* John Wiley & Sons, New York.

Orians, G.H., and M.F. Willson, 1964. Interspecific territories of birds. *Ecol., 45*:736–745.

Partridge, L., 1978. Habitat selection. *In* J.R. Krebs and N.B. Davies (Eds.), *Behavioral Ecology: An Evolutionary Approach,* pp. 351–376. Sinauer Associates, Sunderland, MA.

Pianka, E.R., 1974. *Evolutionary Ecology.* Harper & Row, New York.

Pianka, E.R., and H.D. Pianka, 1970. The ecology of *Moloch horridus* (Lacertilia: Agamidae) in western Australia. *Copeia, 1970*:90–103.

Recher, H.F., and J.A. Recher, 1980. Why are there different kinds of herons? *Trans. Linn. Soc. N.Y.*, *9*:135–158.

Rosenzweig, M.L., 1975. On continental steady states of species diversity. *In* M.L. Cody and J.M. Diamond (Eds.), *Ecology and Evolution of Communities,* pp. 121–140. Belknap Press, Cambridge, MA.

Roth, R.R., 1976. Spatial heterogeneity and bird species diversity. *Ecol.*, *57*:773–782.

Savage, J.M., 1973. The geographic distribution of frogs: patterns and predictions. *In* J.L. Vial (Ed.), *Evolutionary Biology of the Anurans,* pp. 351–445. University of Missouri Press, Columbia.

Shine, R., 1980. Ecology of the Australian death adder *Acanthophis antarcticus* (Elapidae): evidence for convergence with the Viperidae. *Herpetologica, 36*:281–289.

Strahler, A., 1963. *The Earth Sciences.* Harper & Row, New York.

Strahler, A., 1975. *Physical Geography.* 4th ed. John Wiley & Sons, New York.

Svardson, G., 1949. Competition and habitat selection in birds. *Oikos, 1*:157–174.

Terborgh, J., and J.S. Weske, 1975. The role of competition in the distribution of Andean birds. *Ecol., 56*:562–576.

Trewartha, G.T., 1968. *An Introduction to Climate.* McGraw-Hill Book Co., New York.

Udvardy, M.D.F., 1969. *Dynamic Zoogeography.* Van Nostrand, Reinhold, New York.

Vuilleumier, F., 1975. Zoogeography. *In* D.S. Farner and J.R. King (Eds.), *Avian Biology.* Vol. 5, pp. 421–496. Academic Press, New York.

Werner, E.E., and D.J. Hall, 1976. Niche shifts in sunfishes: Experimental evidence and significance. *Science, 191*:404–406.

Whittaker, R.H., 1975. *Communities and Ecosystems.* 2nd ed. Macmillan Publishing Co., New York.

Wiens, J.A., 1974a. Habitat heterogeneity and avian community structure in North American grasslands. *Am. Midl. Nat., 91*:195–213.

Wiens, J.A., 1974b. Climatic instability and the "ecological saturation" of bird communities in North American grasslands. *Condor, 76*:385–400.

Williams, E.E., 1969. The ecology of colonization as seen in the zoogeography of anoline lizards on small islands. *Q. Rev. Biol., 44*:345–389.

Willson, M.F., 1974. Avian community organization and habitat structure. *Ecol., 55*:1017–1029.

Wilson, J.W., 1974. Analytical zoogeography of North American mammals. *Evol., 28*:124–140.

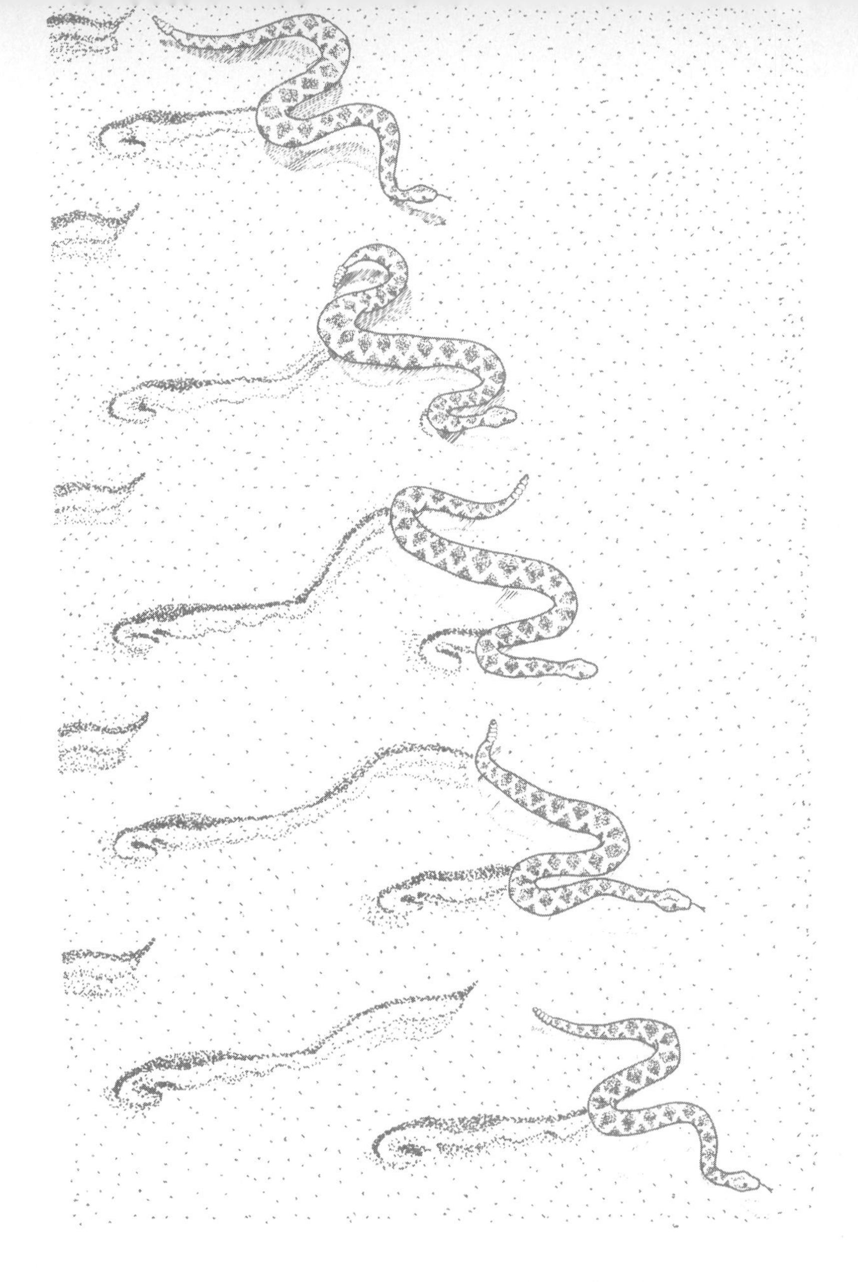

PART 2

RELATIONS WITH THE PHYSICAL ENVIRONMENT

4 Perception

Most vertebrates are active animals that live in variable environments; such factors as light, heat, pressure, and the proximity of other animals are seldom constant in time and space. The more variable the external environment, the more advantageous are sense organs capable of perceiving those variations. An animal living in perpetual darkness would find light-sensitive organs such as eyes of no use and perhaps even a liability, because eyes might be an extra site of injury and infection. Dwellers in the dark often have very small eyes compared to their relatives that encounter light; sometimes they have no eyes at all. Since all blind vertebrate species have descended from ancestors that had functional eyes, the inference is that selection in such conditions has favored eyeless genotypes at the expense of those with eyes. Blind or nearly blind vertebrates include a few fishes from the deepest ocean waters and over 40 species of cave-dwelling fishes such as the Congo barb (*Caecobarbus geertsi*).

External sensory systems not only enable animals to monitor their surroundings but also permit the evolution of complex behavior patterns by which active animals regulate their environments. Sense organs and the accompanying behavior patterns allow animals to avoid many dangers and also to seek and find shelter, food, mates, offspring, suitable habitats, and microhabitats. Some fishes, for example, can tolerate only a narrow range of temperatures; if accidentally displaced to either warmer or cooler waters, they will move in search of suitable temperatures. The shore fish, the California opaleye (*Girella nigricans*), showed a clear temperature preference for water about 26°C (Fig. 4–1), but the impact of such preferences on fitness are often unknown. Obviously, sensory mechanisms and associated behavior require interpretation and coordination by the nervous system, but because that exercise would take us too far into the realm of physiology, we will concentrate here on interactions between vertebrates and their surroundings.

From the point of view of an individual animal, senses define and limit the known world. The behavior of each individual is, in part, constrained

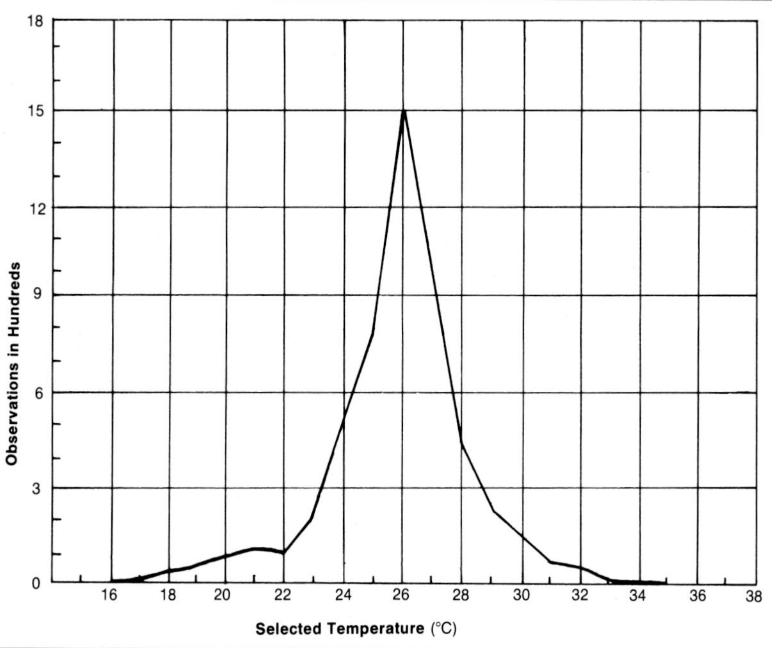

Figure 4–1 Temperature selection of 39 juvenile *Girella nigricans* in the laboratory. Individuals of this age class are typically found in tidal pools at low tide. Adults live in ocean waters of varied temperatures and breeding occurs at sea, but the young move to the intertidal "nursery" area, where most of the growth and achievement of adult morphology take place. Norris (1963) speculated that narrow temperature preferences of the juveniles may direct them to suitable nursery locations. (Modified from K.S. Norris, 1963. The functions of temperature in the ecology of the percoid fish *Girella nigricans* (Ayres). *Ecol. Monogr.*, 33:23–62. Copyright © 1963, the Ecological Society of America.)

by the limits of its perception. Techniques of food-finding, courtship display, social communication, and predator detection must reflect the animal's perceptual capacities (along with its system of information-processing). The sensory information received by different kinds of vertebrates may differ greatly from species to species; different environments have favored the refinement of different sensory capacities.

This chapter is organized around a list of stimuli important to vertebrates—light, heat, mechanical vibration (in gases, liquid, and solids), chemicals, and electricity. Sometimes the animals themselves produce stimuli that are received by other animals or stimuli that are reflected by the environment and received by the individual that produced them. For each category of stimuli, we will survey the structure and function of the associated sense organs and examine their diversity. Whenever possible, species differences are related to differences in environment and activity.

LIGHT

The phenomenon we call light is but a small part of the range of electromagnetic radiation from the sun that impinges on the earth. Radiant energy occurs in packages called quanta and travels very much like waves. The longer the wavelength or the lower the frequency of waves per unit time, the lower the energy content of each quantum. Molecules absorb such radiant energy; however, the effect of quanta of different wavelengths varies greatly. Short wavelengths (less than about 300 nanometers) contain so much energy that they break up the absorbing molecule; those longer than about 1000 nm are too puny to induce significant molecular changes. Consequently, all living creatures utilize only intermediate wavelengths within this range for photobiological reactions. Light-activated reactions are mediated by carotenoids, pigments that absorb wavelengths between 300 and 800 nm especially well. Plants use these pigments in photosynthesis. Animals, which derive their carotenoids from plants, use them to activate sensory nerves. Actually, the earth's atmosphere markedly restricts the spectrum of solar radiation that reaches the surface of the earth, and at sea level, 90 percent of the solar energy is composed of wavelengths between 300 and 1100 nm. Thus, it is not surprising that photobiological systems are particularly adapted to this portion of the total spectrum.

Within the biologically useful part of the spectrum, however, differences in photosensitivity occur among different groups of organisms. Humans and most other vertebrates can see wavelengths somewhere between about 400 and 750 nm. We call this portion of the spectrum "visible," and many vertebrates perceive the various wavelengths within the visible spectrum as colors. The shorter wavelengths are seen by humans as violet and the longer visible wavelengths as red, with all the colors of the rainbow in between (see Fig. 4–2). Some insects, such as bees, can see still shorter —ultraviolet—wavelengths; hummingbirds and a few other birds can see in the near-ultraviolet range, down to about 370 nm (Goldsmith, 1980). Some bacteria and green plants are photosensitive to wavelengths longer than those in the visible spectrum.

Animals use various characteristics of light to obtain different kinds of information. Color, a function of wavelength, aids in discerning shapes. Differences in intensity produce bright-dark contrast, which also functions in the perception of shape. Some fishes, amphibians, and birds, as well as invertebrates, can use another characteristic of light: polarization. Ordinary light is unpolarized and vibrates in all planes perpendicular to the axis of the light ray, but polarized light, reflected by air molecules, vibrates mostly in one plane. The degree of plane polarization of light from the sky depends on the angle between the point in the sky and the sun's rays; therefore, animals able to perceive polarized light can use such information to determine the location of the sun even when the sun itself is obscured. This ability aids a number of vertebrates in direction-finding.

Thus far we have dealt primarily with the features and uses of light transmitted through air. Light is also transmitted through water, of course, but the characteristics of light that has passed through water are quite different from those of air-transmitted light. At a depth of 25 m, for instance, 90 percent of the available solar energy falls between wavelengths of about 350 and 600 nm (violet to orange), and increasing depths shift the spectrum still further into the blues. Material dissolved and suspended in the water not only reduces light penetration but shifts the available spectrum toward longer wavelengths (yellow and orange). Aquatic animals, therefore, have access to different wavelength spectra, depending on the depth and turbidity of the water; often they have maximum sensitivities that correspond to the available wavelengths. Seasonal changes in spectral sensitivity have been demonstrated in the rudd (*Scardinius erythrophthalmus*) and several other species, but the adaptive function of these changes is not well studied.

At a depth of about 100 m, enough light penetrates clear ocean water to allow green plants to perform photosynthesis. Between 100 and about 800 to 1000 m is a twilight zone of progressively dimmer light. Below 1000 m is a perpetually sunless zone, at least in tropical and subtropical waters. The threshold of light penetration in temperate and polar waters is reported to be deeper. Suspended materials are found primarily in the upper 100 m; below that point, most water is quite clear.

Photosensitivity

Some fishes and amphibians and many invertebrates have light receptors scattered widely in the skin and can respond, by avoidance, to sudden

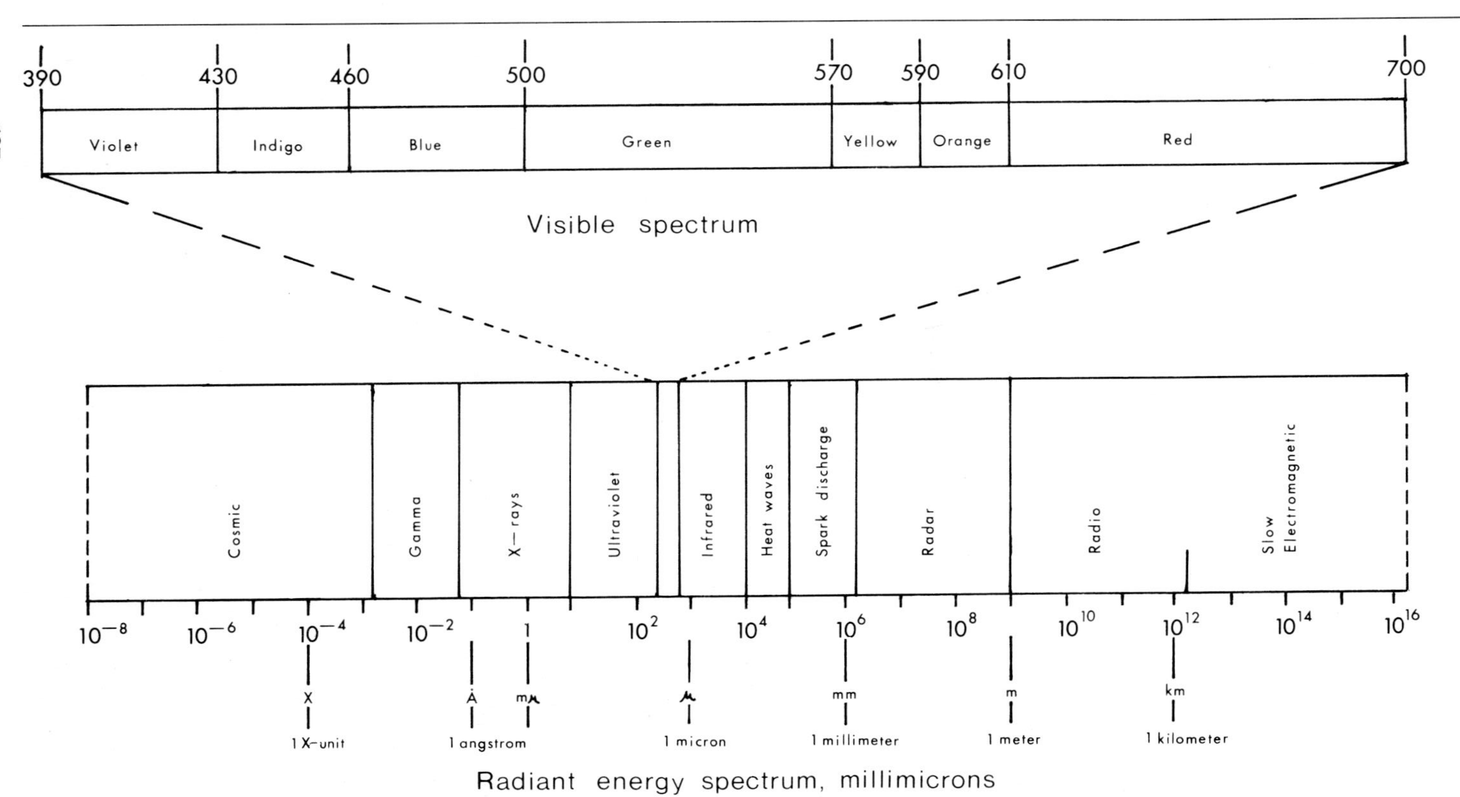

Figure 4–2 The spectrum of radiant energy available to animals and the "visible" spectrum. (Modified from P. Marler and W.J. Hamilton, 1966. *Mechanisms of Animal Behavior*. John Wiley & Sons, New York.)

changes of illumination. Almost all vertebrates, however, have more specialized and localized photoreceptors that permit directional sensitivity.

Many fishes and some birds possess a median photoreceptive organ on top of the head. The so-called median eye of the tuatara and many lizards, lampreys, and frogs is a light-sensitive organ with a retina. In some lizards and in the tuatara, this organ has a lens, but it appears that the function of this organ is not the formation of visual images. One function may be the reception of light stimuli that trigger secretion of substances governing such physiological activities as day-night cycles of activity and color change related to the regulation of body temperature.

The median eye of lizards is found in most genera of four major families—the Agamidae, Iguanidae, Lacertidae, and Scincidae—but it is missing in the Gekkonidae, Teidae, Helodermatidae, and some other small families (Gundy et al., 1975). Genera with median eyes are not ecologically or behaviorally similar to each other, except that they usually live outside the tropics, where one of the main functions of this organ may be not only thermoregulation but also the monitoring of day length, which is important in regulating the reproductive cycle.

Almost all vertebrates have functional image-forming eyes. The fundamental process of vision is absorption of visible wavelengths by certain large molecules called the visual pigments, located in the photosensitive cells of the retina at the back of the eye. There are two basic types of photoreceptive cells: rods and cones. Rods are very sensitive and can respond even to dim light, but they are relatively poor for acuity (resolution of detail) because large numbers of them transmit information through one nerve fiber. Cones, on the other hand, are less sensitive and, thus, function only in bright light. Different cones contain different visual pigments and respond to different wavelengths; cones, therefore, function in color discrimination. Small numbers of cones, sometimes only one, are served by each nerve fiber; hence, cones permit good resolution. The photochemical reaction in the rods and cones is transformed into an electrical impulse that passes along the sensory nerves to the brain.

Basic structural features of the eyes of all vertebrates are rather similar (Fig. 4–3). Some vertebrates have one or more retinal areas called a fovea, a small spot in which cones are concentrated and join nerves at nearly a 1:1 ratio. Because of the low ratio of sensory cells to nerves, visual acuity is enhanced. The iris is a muscular diaphragm that controls the size of the pupil, the opening through which light enters the inner part of the eye. The falciform process of fishes, the conus papillaris of squamates, and the pecten of birds may function in retinal nutrition or in the detection of movement by casting a shadow in the retina, but little is known of their operation.

As light rays enter the eye, they are bent or refracted so that they are focused on the retina, permitting formation of an image. Refraction of light

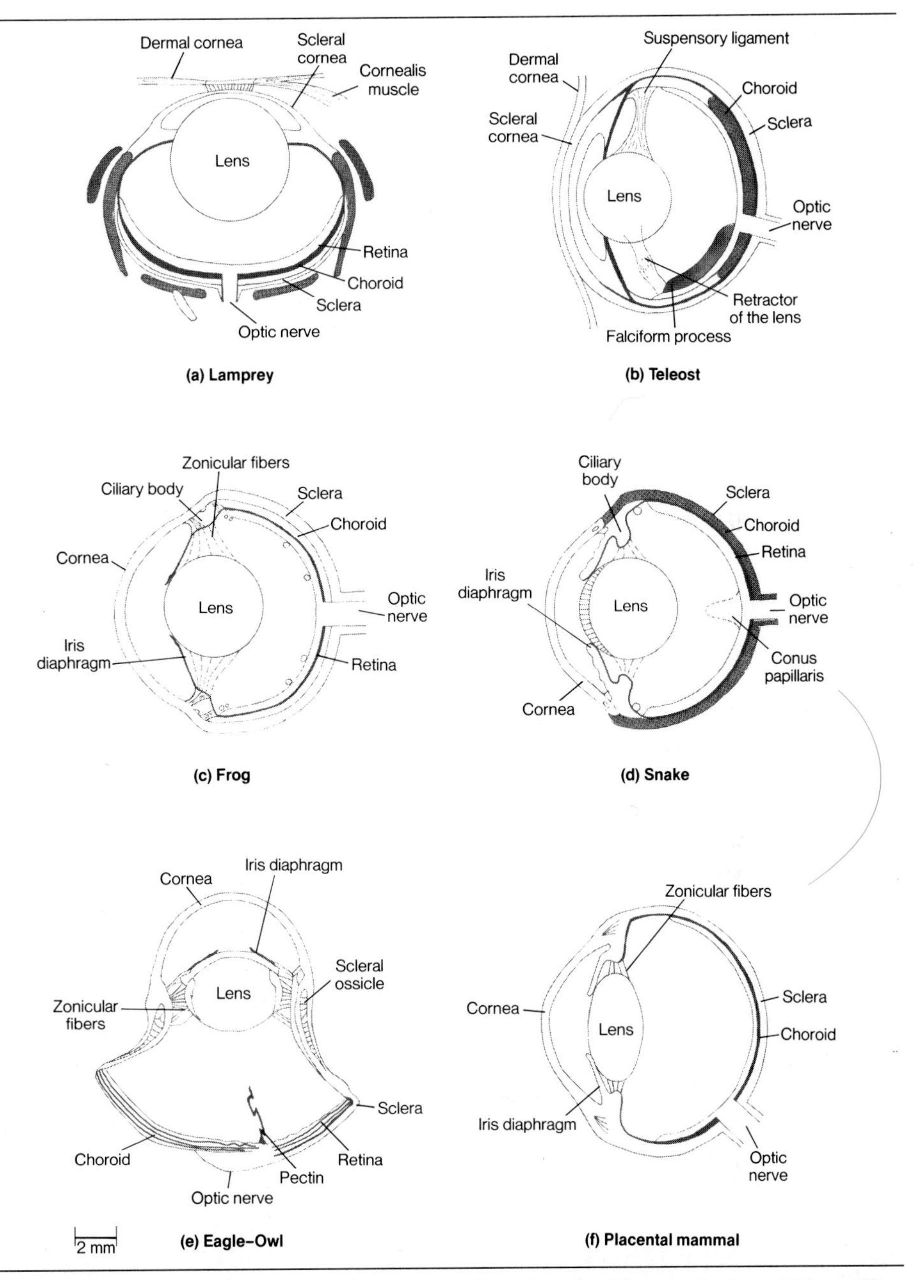

Figure 4–3 Diagrams of a variety of vertebrate eyes. (Reprinted with permission of Macmillan Publishing Co., Inc. from *Chordate Structure and Function* by Allyn T. Waterman. Copyright © 1971 by Allyn T. Waterman.)

rays may be achieved by the cornea (the transparent covering of the front of the eye) or the lens. The cornea does most of the refraction of light that passes through air, but the lens is the main refractor for animals that receive light rays through water. Focusing at different distances is called accommodation; it is accomplished chiefly by moving the lens backward for distance focus (in lampreys and most bony fishes), by moving the lens forward for near focus (in elasmobranchs, amphibians, and snakes), or by changing the shape of the lens (in lampreys, birds, and mammals). The problem of accommodation is handled in other ways in some species. In some bats, the retina is folded so that rods are in focus at different distances from the lens. In rays and horses, the retina is sloping so that visual cells lie at graduated distances from the lens. These unadjustable arrangements are not well adapted for visual acuity, but they function adequately when sensitivity is more important.

Short wavelengths are refracted more sharply than longer ones, tending to produce chromatic aberration. This means that the short wavelengths focus nearer the lens, usually off the retina, creating a blurred outline of the image. In animals with color vision, a narrow rainbow-colored outline is produced. This problem is reduced by filtering out shorter wavelengths. In many diurnal (day-active) vertebrates, the lens or the cornea is yellow and filters out ultraviolet and violet wavelengths. Some reptiles and birds have colored oil droplets in the sensory cells that may serve the same purpose. Great refraction of shorter wavelengths, especially by the lens, is probably one reason why most vertebrate eyes have become adapted not to use ultraviolet wavelengths.

This section has reviewed the general features of vertebrate eyes. Specific adaptations—and some curiosities—are presented in the next sections: a diversity of structural arrangements, color vision, and adaptations related to an amphibious life.

Structural Diversity

The visual field is the size of the area scanned by the eyes and it differs greatly among vertebrates. Most terrestrial vertebrates have laterally placed, movable eyeballs that can be rotated to increase the area viewed; the visual field is often almost 360°. In addition, most land vertebrates have necks and can turn their heads, thereby increasing the size of the area scanned. Owls have large, relatively immobile, forward-facing eyeballs with a visual field of only about 60 to 70°. They compensate for this feature by an extraordinary ability to turn their heads (an owl can look almost directly behind itself by twisting its head around). Fishes, in contrast, typically have no neck and cannot turn their heads much, but the lens commonly protrudes through the pupil and, thus, increases the visual field. *Bathylychnops,* a deep-sea teleost, has a most unusual means of increasing its field of vision (Fig. 4–4): On the dorsal side of its flattened snout, it has a pair of eyes that look laterodorsally, but a small, secondary eye, complete with lens and retina, bulges from each eyeball, hangs out over the jaw, and looks ventrally and somewhat backwards. The secondary eyes are developmental derivatives of the main eyes; their particular selective value is not known.

Many vertebrates with laterally positioned eyes have completely or mostly nonoverlapping visual fields for each eye (Fig. 4–5). This monocular vision is usually the case for fishes and a number of mammals, such as some whales and rabbits, and many birds. Other vertebrates have well-developed binocular vision, with frontally positioned eyes that have broadly overlapping visual fields. Binocular vision is one means of enhancing depth perception and fine discrimination of distances. It is particularly useful in accurate localization of objects; therefore, it is well-developed in predatory vertebrates that depend on an accurate strike for capturing prey

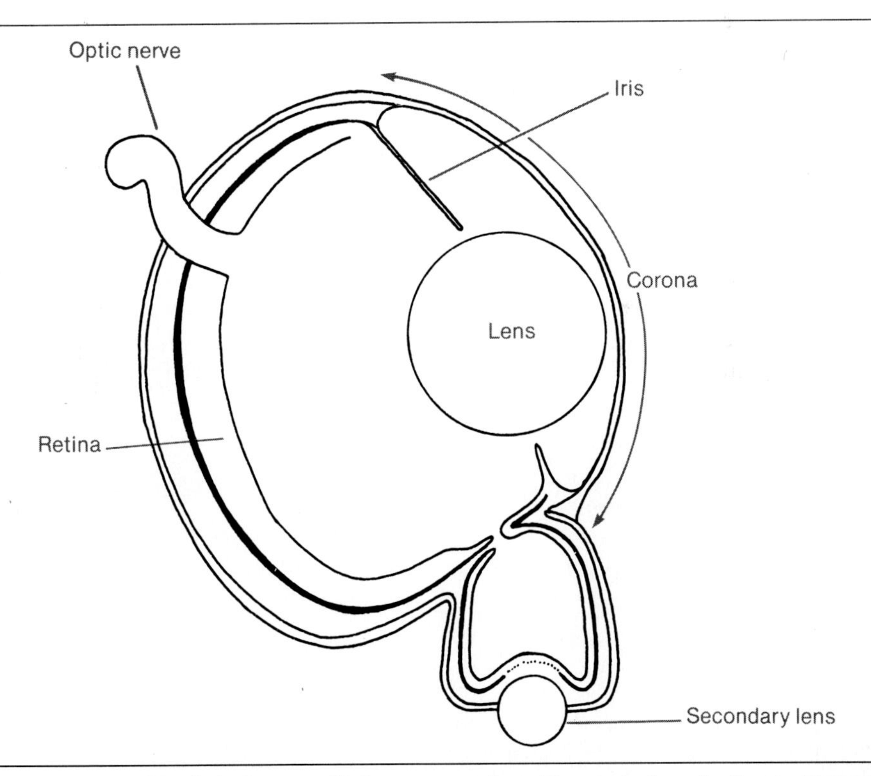

Figure 4–4 The eye of *Bathylychnops exilis* in cross-section. The main lens is directed laterally from the fish's body; the secondary one is directed downward. (Reprinted by permission from *Nature,* Vol. *207,* pp. 1260–1262. Copyright © 1965 Macmillan Journals Limited.)

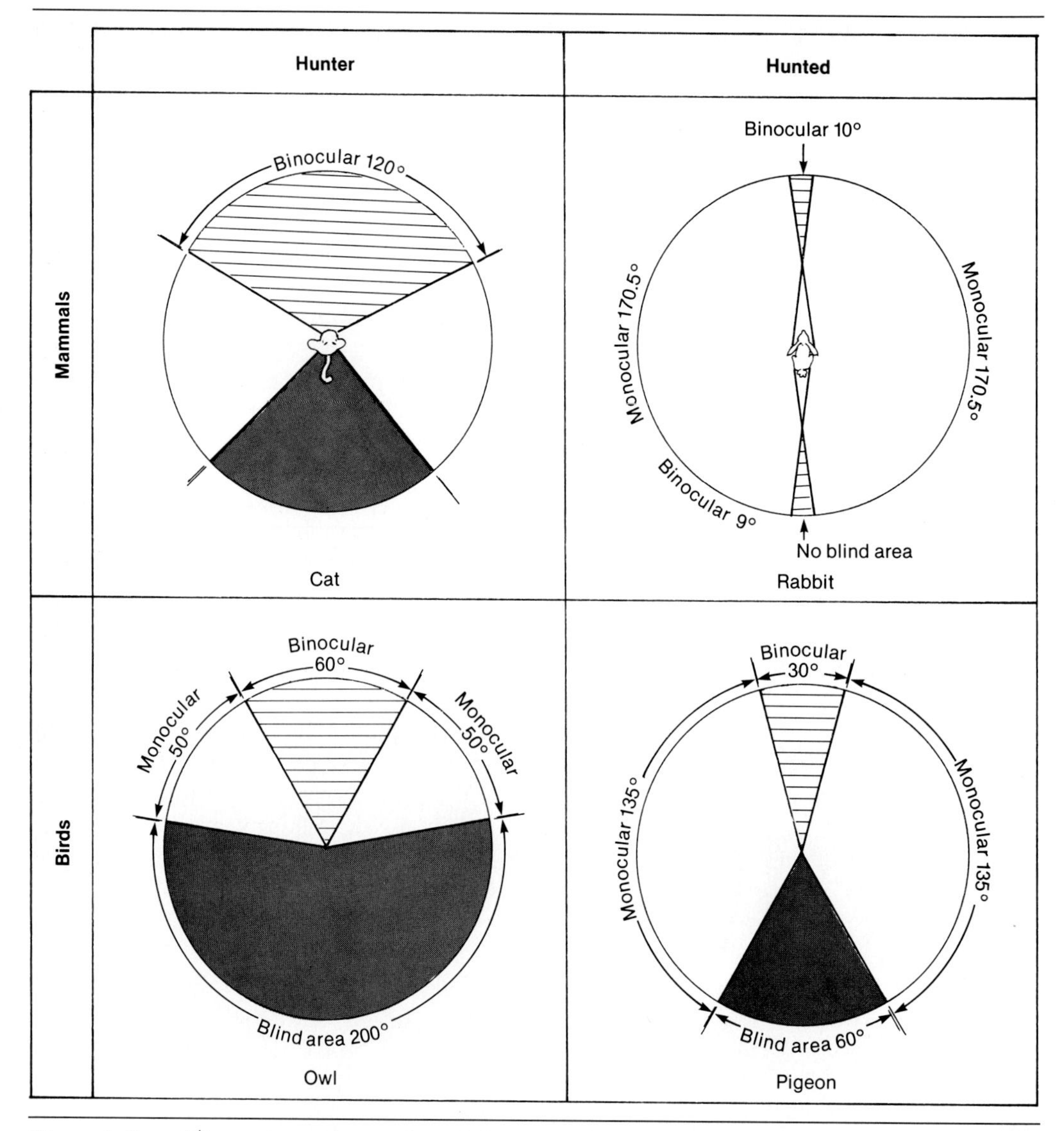

Figure 4–5 Schematic representation of the visual field of two predatory and two prey species of mammal and bird. Note the larger binocular field and larger blind areas of the predators. (Modified from P. Marler and W.J. Hamilton, 1966. *Mechanisms of Animal Behavior.* John Wiley & Sons, New York.)

and sometimes in species that move about in highly three-dimensional environments such as treetops, where a misstep may have dire consequences.

Certain vertebrates have both monocular and binocular vision. The laterally placed eyes have somewhat overlapping visual fields. A number of diurnal avian hunters, such as hawks and swallows, have both types of vision and so do some snakes that prey on birds (such as the long-nosed tree snake, *Dryophis mycterizans,* of the East Indies). Binocular vision in these species is forward, but in the woodcock (*Philohela minor*), binocular vision may be better to the rear. This ground bird feeds by probing its long bill into soft soil, so it is very vulnerable to attack from behind. Binocular vision is also found in a few fishes, such as sea horses, and in true chameleons, which can move each eye independently of the other.

Foveas contribute to visual resolving power and are often found in animals that have binocular vision—another adaptation to precision in discriminating and localizing objects such as prey. Some species, including hawks and swallows, have two foveas in each eye: one for lateral, monocular vision and the sighting of prey and one for forward, binocular vision and accurate attack on prey.

Birds' eyes have extremely high acuity; a soaring hawk, for example, can spot a mouse from an incredible height. The usual explanation for such great resolving power is the density of visual cells in the retina coupled with a low ratio of visual cells to the nerve fibers serving them; this arrangement permits fine discrimination (Fox et al., 1976). Furthermore, the eyes of birds are very large, relative to their body size, so they can contain more rods and cones than the smaller eye of a similar-sized mammal. Some evidence, however, suggests that at least a part of the hawk's ability to sight a mouse far below may be its great ability to track the moving prey, not merely its visual acuity (Sillman, 1973).

Many species normally active in dim light, including many midwater fishes, often have large eyes with protruding corneas that can gather light from a great area and large retinas containing mostly, or exclusively, rods. Benthic (bottom-dwelling) and abyssal (from the deepest oceans) fishes often have reduced eyes; however, a weird abyssal fish, *Ipnops,* has huge and unusual eyes. This fish has retinas that lie under large, translucent bony plates atop its flattened head (Fig. 4–6). *Ipnops* lives below the light zone, even below the twilight zone, where the only light to be perceived is that from luminescent organs. The significance of this light sensitivity to *Ipnops* is not known.

Many nocturnal and some diurnal vertebrates have "eye shine," usually due to a tapetum lucidum, a layer of reflective material behind the retina. By reflecting light back through the retina, the tapetum lucidum effectively increases the contrast between an object in the visual field and its background.

The shape and size of the pupils vary enormously among vertebrates (Walls, 1942). A sampling of that variability is presented in Figure 4–7. Vertebrates that roam about only at night or only during the day typically have round pupils. Some mainly nocturnal species may also be abroad in daylight—reptiles may emerge to bask in the sun, for instance, and many

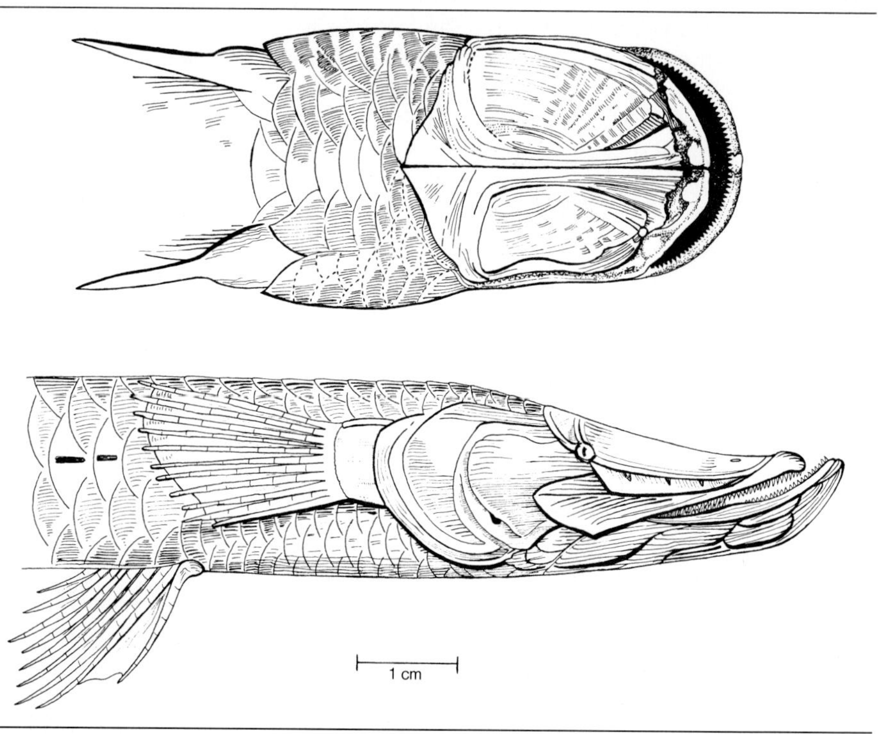

Figure 4–6 *Ipnops murrayi:* dorsal and lateral view of the head. (From O. Monk, 1959. The eye of *Ipnops murrayi. Galathea Rept.*, 3:81.)

cats move about during the day. These species have vertical pupils, especially when the pupils are contracted. Vertical pupils can close more completely than round ones; they probably function as protection from excessive sunlight for those animals whose primary activity occurs at night and whose retinas are very sensitive. The pupils of gekkonid lizards can contract down to a vertical row of pinhole openings, each of which permits light to fall on the retina. Apparently, the several images thus formed are superimposed on the retina with no loss of acuity and permit a gain in depth of field (the distance over which objects are in focus) with sufficient illumination to permit good vision. Horizontal pupils in many ungulates and whales permit scanning of wide areas in open habitats.

Eyes of terrestrial vertebrates are frequently protected from scratches, excessive light, and desiccation by movable eyelids and special secretions. A third eyelid, the transparent nictitating membrane, is present in many terrestrial creatures but tends to be small in arboreal and nocturnal species. Nictitating membranes are well developed in polar bears (*Thalarctos maritimus*) and reindeer (*Rangifer tarandus*), which use them as protection against snow blindness. Birds in flight close these membranes; this permits continued vision while protecting the cornea from drying out. Diving birds also

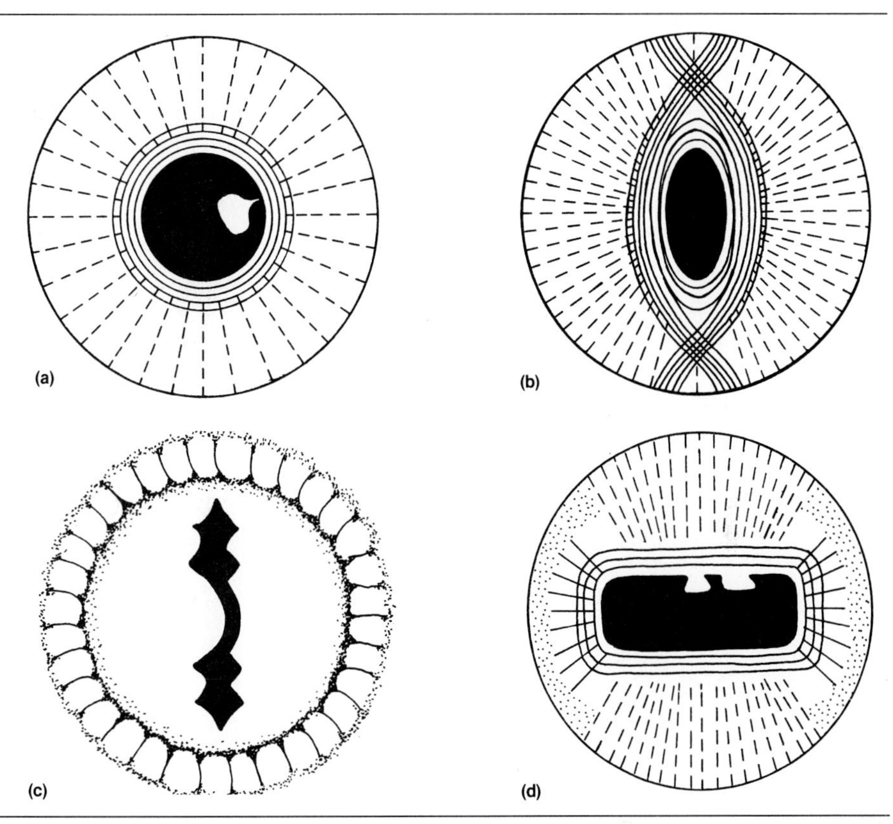

Figure 4–7 Four pupil shapes common in vertebrate eyes: (*a*) A round pupil in wholly diurnal and wholly nocturnal mammals; (*b*) a vertical-slit pupil of nocturnal mammals that also are active during the day; (*c*) the multiple pin-hole vertical slit of a gecko; and (*d*) a horizontal pupil such as is found in many ungulates and whales. (Modified from G.L. Walls, 1942. *The Vertebrate Eye.* Cranbrook Institute of Science, Bloomfield Hills, MI, and from *The Life of Reptiles* by A.D'A. Bellairs. Published by Universe Books, New York, 1970.)

draw the nictitating membrane across the eye. Snakes, some lizards and amphibians, and lampreys have a sort of dermal (skin) spectacle permanently drawn over each eye as a protective device. Aquatic mammals protect the eyeball from salt water with an oily, mucous secretion.

Color Vision

Sensitivity to different wavelengths and the range of color perceived by species with color vision vary significantly among the vertebrates, in part as a function of habitat. The filtering property of water shifts the spectrum of light that is available for aquatic organisms and emphasizes shorter wavelengths than the aerial spectrum. Fishes usually have many rods in the retina, perhaps because underwater light is significantly dimmer than light in air. Deep-water fishes are most sensitive to wavelengths in the blue range, which are transmitted best in deep water; pinnipeds and whales also exhibit this shifted sensitivity. On the other hand, fishes of turbid and fresh water are especially sensitive to midspectrum yellow-green wavelengths, which provide the maximum available light in these habitats. In contrast, terrestrial vertebrates frequently see best in the longer wavelengths.

Good color vision, and sometimes all-cone retinas, are found in many fishes (including some sharks) and frogs, most reptiles and birds, some primate mammals, and a few other species. Color vision may be weakly developed in other groups, but evaluation of the quality of color vision of some animals remains controversial because experimental studies of color vision are difficult to execute properly (Jacobs, 1981). Obviously, color vision enhances the ability to discern contrasts and favors discrimination of objects from their surroundings. Color vision might increase rates of successful prey capture, improve predator detection, and enhance powers of communication with conspecifics.

Amphibious Adaptations

Many amphibious fishes have eyes that are especially adapted for vision in both air and water. The "four-eyed" fishes (*Anableps*) of Central and South American have two pupils in each eye: one for aerial vision and one for aquatic vision. The single lens is asymmetrical: The portion used for aerial vision, in which part of the light refraction is accomplished by the cornea, is less convex than the part used for aquatic vision, in which all light refraction is done by the lens. (The optical properties of the cornea are like those of water, so the cornea in water can accomplish no bending of light rays.) *Anableps* can swim with half of each eye out of the water and half of it submerged (Fig. 4–8); it apparently uses this dual vision to locate prey in the surface film as well as to look out for predators.

Other amphibious fishes such as the mudskippers, *Periophthalmus,* are so well adapted for semiterrestrial life that their aquatic vision is quite poor. Their eyes protrude above the head and can be raised, lowered, and turned independently of each other. Their cornea is strongly curved, and the lens flattened. In water, where the cornea is optically inactive, visual images do not fall on the retina and vision is poor. In the rockskippers (such as *Mnierpes*), the eyes are located and moved about in a way similar to the mudskipper's, but, in addition, each cornea is divided into two flattened windows, one anterior to the other. Flat corneas have no differential effect in bending light rays in either air or water, so the entire job of focusing is left to the lens.

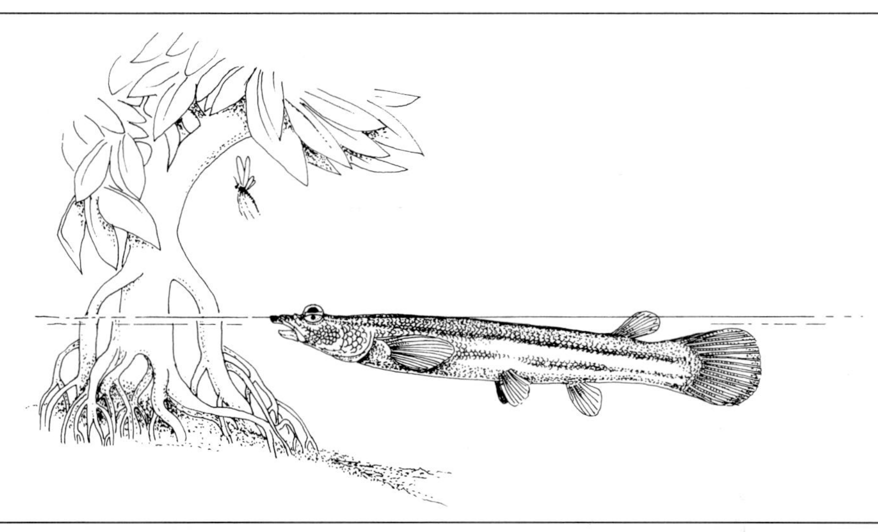

Figure 4–8 A four-eyed fish (*Anableps*) at the surface of the water, where it can use both aerial and aquatic vision.

Other amphibious vertebrates that utilize both aquatic and aerial vision also have special optical adaptations. Pond-dwelling turtles are well adapted to an amphibious existence: Their lens is very soft and readily deformed to focus both in air and in water, so the lens can compensate for the optical inactivity of the cornea in water. Marine and terrestrial turtles have a lesser range of accommodation. The European kingfisher (*Alcedo*) probably uses the central fovea and lateral vision of one eye to focus on aquatic prey when it is in the air and shifts to the temporal fovea and binocular forward vision as it enters the water to capture a fish (Fig. 4–9). As in some of the amphibious fishes, the lens of the kingfisher is asymmetrical; light passes through different parts of the lens for aerial and for aquatic vision. Diving birds, especially cormorants, have unusually deformable lenses for focusing in both air and water. Other divers, such as diving ducks, loons, and auks, also have somewhat flexible lenses, but a window in the nictitating membrane over the eye functions as an additional lens for aquatic vision. Penguins, on the other hand, are very nearsighted out of water, perhaps because they do not fly and selection for clear distant vision has not been particularly strong.

Light Production

Some animals produce their own light and respond both to their own light reflected from objects in the environment and to light generated by other individuals. Several kinds of deep-sea fishes, both elasmobranches and teleosts, as well as many invertebrates, are capable of producing light

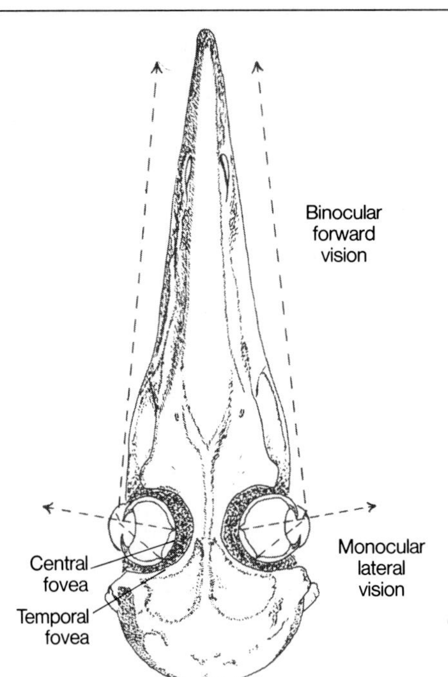

Figure 4–9 Monocular and binocular vision in a kingfisher (note the two foveas in each eye).

(bioluminescence). In fact, about two-thirds of the midwater oceanic fishes are luminescent. A few families of teleosts (the Anomalopidae, for example) have special organs inhabited by symbiotic luminescent bacteria. *Searsia* produces luminous extracellular secretions but, more commonly, intracellular mechanisms produce intermittent light flashes. Light organs are commonly found in the skin over a large part of the body but are sometimes concentrated in the head area, near the eyes, on the tongue or nose, or inside the mouth. They may even be situated around the esophagus (*Leiognathus equulus*) or associated with the intestine (*Apogon*) or the anal region (see Nicol, 1979). Light from the interior light organs is dispersed by translucent ventral muscles and, thus, is emitted from the ventral body surface. Some light organs have lenses, presumably to focus the light, and many have reflectors. The light emitted is usually blue or blue-green—the range of wavelengths to which the eyes of these fishes are most sensitive. Some species, however, produce red or yellow light and are also sensitive to those wavelengths.

Bioluminescence in fishes has several functions. In certain instances, a kind of camouflage is suggested (Chapter 10). The midshipman (*Porichthyes*) flashes its light when attacked and is usually avoided by the would-be predator. In other fishes, light may aid in foraging or in communication among conspecific individuals during courtship, aggression, or group behavior. A study of *Photoblepharon palpebratus*, a beryciform fish, both in the Red Sea and in the laboratory, suggested that light production was used in foraging, perhaps in escaping predators, and in intraspecific communication (Morin et al., 1975). Deep-sea anglerfishes (*Ceratius*) use a luminescent fintip as a lure for prey.

HEAT

Vertebrates sense environmental temperatures indirectly; that is, they sense the temperature of the skin and associated organs. Dermal heat and cold receptors are distinct from each other.

The pit vipers (Viperidae) have the most sensitive heat receptor known. They feed primarily on small mammals, mainly at night. The pit organs (one on either side) are located between eye and nostril (Fig. 4–10), are highly sensitive to temperature differentials and are also directionally responsive. The pit organ is a small unit consisting of an inner and outer chamber, separated by a thin membrane containing blood vessels and many nerve endings. Exactly how it works is not clear, but its effectiveness is well established. A pair of pit organs scans a field of about 180° in front of

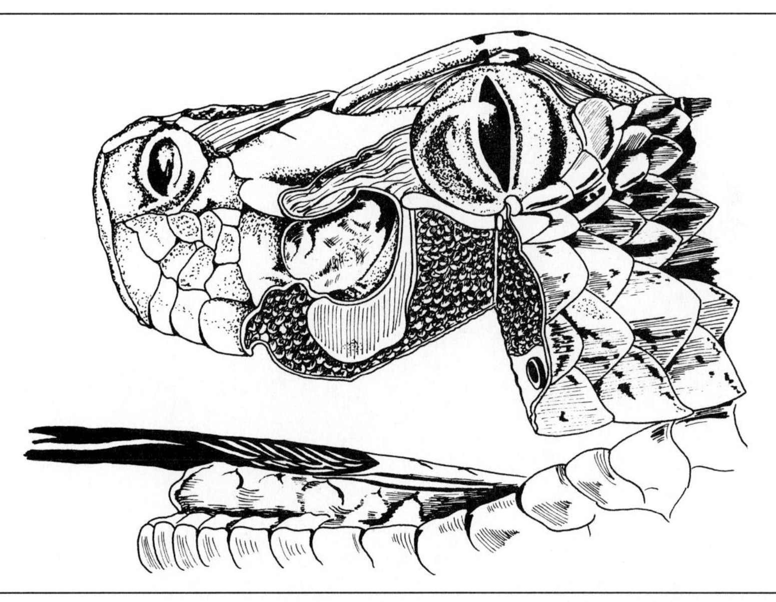

Figure 4–10 Cutaway view of a pit viper's head to show the pit organ. (Reprinted with permission of Macmillan Publishing Co., Inc., from *Animal Physiology: Principles and Function* by Malcolm S. Gordon. Copyright © 1968 by Malcolm S. Gordon. Adapted from T.H. Bullock and F.P.J. Diecke, 1956. Properties of an infra-red receptor. *J. Physiol.* [Lond.], *134*:47–87.)

the snake. A rattler, by means of its pit organs, can detect, in half a second, a living mouse 10°C warmer than the environment at a distance of almost half a meter. Furthermore, it can strike accurately when all of its other senses are occluded. Some other snakes, especially the boas, have rows of less specialized heat-sensitive organs in the lips.

Many nocturnal predators of small mammals, including certain cats and owls, have very sensitive sense organs (eyes and ears, respectively), but apparently only these snakes have heat-sensitive receptors associated with prey-finding. Clearly, such a device in a hunter that has a high body temperature (as do both cats and owls) could not function unless it were somehow completely insulated from the body heat of its owner, and total isolation of that device probably would be very difficult.

ELECTRICITY

Some fishes, including lampreys, ratfishes, certain catfishes, and elasmobranchs, can detect, but not transmit, electrical signals. This ability might be related to the detection of magnetic fields used in navigation: A water current moving across the earth's magnetic field would generate a small electrostatic field that might be detectable by such fishes. Detection of electric signals definitely works in detection of prey; a shark (*Scylliorhinus canicula*), for example, can locate a flatfish buried beneath the sand, using only its electroreceptors.

Still other fishes can both emit and receive electrical signals. Strong electrical outputs can be used for stunning prey; for example, the Pacific electric ray (*Torpedo californica*) captures fishes by folding them in the pectoral fins and shocking them into immobility. Weaker electric fields commonly establish a field of perception around the body by which any object entering or leaving the field can be detected. Electrical signals can be used in locating prey, in navigating dark or turbid waters, and in social interactions. They are also used to facilitate formation and maintenance of groups in juvenile *Marcusenius cyprinoides,* an African mormyrid fish that lives in turbid water and is active at night (Moller, 1976). A partial list of these so-called electric fishes (more than 500 species are known so far) is presented in Table 4–1. Electric receptors are morphologically similar to lateral line organs and may be derived from them. Electrical generating organs (which are discussed below) are derived from muscle (except in the gymnotid, *Electrophorus*) and are variably located on the body and tail, depending on the species, and in the pectoral fins of torpedo rays (Fig. 4–11).

Several species of electric fishes may be found in the same waters, which presents the potential problem of interference with each other's signals. One solution to the problem is differential sensitivity to different frequencies, which sometimes also reflect the varied uses of electrical fields to different species. For example, *Apteronotus* (Apteronotidae) and *Eigenmannia*

Table 4–1 Electric fishes.

Common name	*Family*	*Genus*	*Discharge*	*Distribution*
Skates	Rajidae	*Raja*	Weak	Marine, cosmopolitan
Torpedo rays	Torpedinidae	Several genera	Strong	Marine, cosmopolitan
Mormyrid	Mormyridae	Several genera	Weak	Freshwater, Africa
Gymnarchus	Gymnarchidae	One species	Weak	Freshwater, Africa
Gymnotid eels; Electric eel; Knifefish	Electrophoridae	*Electrophorus*	Strong	Freshwater, South America
	Gymnotidae	*Gymnotus* (one species)	Weak	Freshwater, South America
	Sternopygidae	Four or five genera	Weak	Freshwater, South America
	Rhamphichthyidae	Two species in two genera	Weak	Freshwater, South America
	Sternarchidae	About nine genera	Weak	Freshwater, South America
Electric catfish	Malapteruridae	*Malapterurus* (one species)	Strong	Freshwater, Africa
Stargazers	Uranoscopidae	*Astroscopus*	Strong	Marine, West Atlantic

(From Bennett, 1970.)

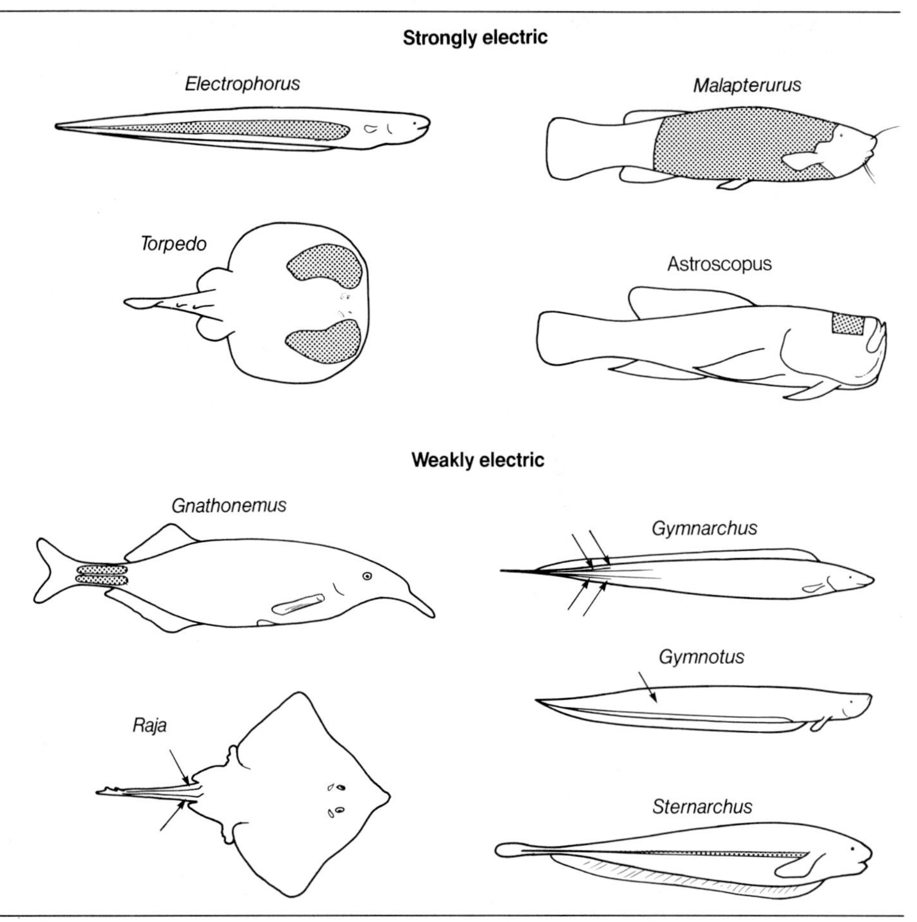

Figure 4–11 Locations of the electric organs of some electric fishes. Side views are given except for *Torpedo* and *Raja*, which are viewed dorsally. Electric organs are shaded. (Reproduced, with permission, from the *Annual Review of Physiology*, Vol. 32. Copyright © by Annual Reviews Inc.)

(Rhamphichthyidae) are both gymnotoid electric fishes whose ranges in South America overlap. *Apteronotus* is territorial and probably predatory on other fishes in midwater; *Eigenmannia* is a gregarious bottom-dweller and feeds on invertebrates. *Eigenmannia* is relatively insensitive to low frequency and responds best to frequencies close to those of its own emission (250–600 kHz or thousand cycles/sec). This suggests that communication of some sort is a primary function of this sensory system. In contrast, *Apteronotus* is sensitive to a wide range of frequencies, including low frequencies characteristic of nonelectric prey species as well as frequencies (750–12,500 kHz) used in conspecific communication. Even within a species, mutual interference or "jamming" of electrical fields is a problem. Two ways of avoiding this problem are individual frequency specializations and shifts of frequency in response to jamming. Temporal coding of electric signals is another means of conspecific recognition (Hopkins and Bass, 1981).

MAGNETISM

The use of magnetic fields for avian orientation has been controversial for many years. Recently, however, a number of experiments have shown that at least some birds and fishes clearly possess means of sensing a magnetic field and orienting their movements appropriately. The morphology and physiology of this sensory system are still totally unknown, but the ability of these animals to respond has been amply demonstrated. Several species appear to be able to use this sense in homing or in orientation associated with migration, especially when other cues are rendered useless.

MECHANICAL VIBRATION

Undoubtedly, all vertebrates can detect mechanical vibrations when in body contact with a vibrating solid object. This ability permits vertebrates to detect the approach of a large object (a source of possible danger) and may help burrowing vertebrates to discern their prey. More interesting are the varied means of detecting vibration in fluid media—air and water. Many lungfishes, elasmobranchs, teleosts, amphibian tadpoles, adult aquatic salamanders, and a few highly aquatic frogs possess a lateral-line system sensitive to water displacement (Fig. 4–12). This system consists of a series of sensory cells in pits, tunnels, or grooves over the head and body. Very active fishes usually have better developed lateral-line systems than do less active fishes, and the rougher the waters inhabited by a species, the better the protection for the sensory cells. Most experimental evidence suggests that the main function of these systems is the detection of nearby objects, either moving or stationary, that deflect water and generate currents. The vertebrate's lateral-line system permits it to locate prey, mates, or obstacles in its immediate surroundings. This sensory system is used by elasmobranchs to detect electric fields.

Fluid-borne vibrations of certain frequencies and intensities can be perceived as sound. Air and water have very different acoustical properties because of their different densities. More energy is required to initiate an underwater sound (there are more molecules per unit volume to be set in motion) than an aerial sound; but once in motion, underwater sound transmission is more rapid. Sound can travel almost five times as fast in water (about 1500 m/sec) as in air (330 m/sec). Salinity increases the speed of sound slightly, to about 1540 m/sec. Increasing water depth decreases its

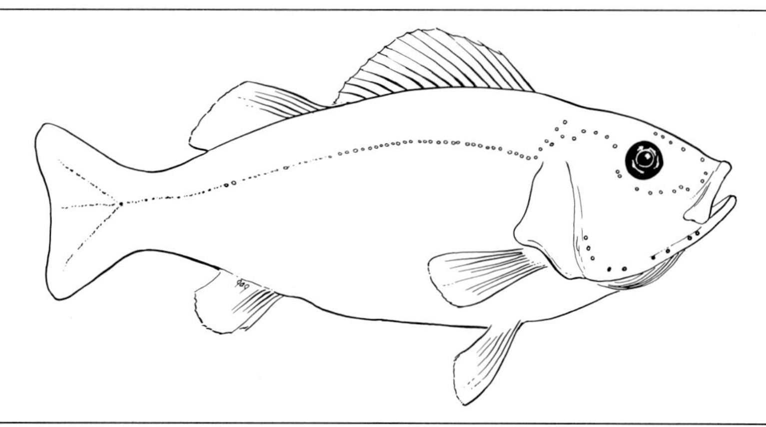

Figure 4–12 Side view of a teleost showing the location of a lateral line system and related sensory organs.

speed; at a little more than 900 m, for instance, velocity is reduced to about 1460 m/sec. A difference in velocity is associated with a difference in wavelength of sound in air and in water: wavelengths of underwater sounds are almost five times as long (at the same frequency) as those in air. Furthermore, sound waves are reflected from the surface, from the bottom, and from interfaces between layers of water of different temperatures. These reflections enhance sound transmission.

The major properties of sound are intensity, or loudness (a function of amplitude of vibration), pitch composition (a function of the frequencies generated), and patterning of the sound through time. Members of all vertebrate classes can sense all three characteristics to varying degrees; birds and mammals are best able to sense and use all three (Fig. 4–13). Ears are devices that transform vibration of air or water into vibration of internal fluids and, ultimately, into vibration of membranes containing special sensory cells. Vertebrate ears, except in fishes, typically consist of three portions. The outer ear is the opening through which sound waves pass to the interior. Most mammals have an external flap, or pinna, that aids in the collection and funneling of sounds into the opening. Pinnae are absent in cetaceans, seals of the family Phocidae, and some insectivores. Many birds have modified feathers fringing the ear opening that serve as funnels for sound waves. The middle ear is separated from the outer ear by the eardrum. The eardrum vibrates in response to incoming, fluid-borne vibrations and sets the bony elements of the middle ear in motion. These elements convey the vibrations to the inner ear, which contains the sensory cells. (The inner ear is also a balance organ.)

This basic function is common to all vertebrate ears, but the mechanism by which vibrations are translated from external to internal varies widely.

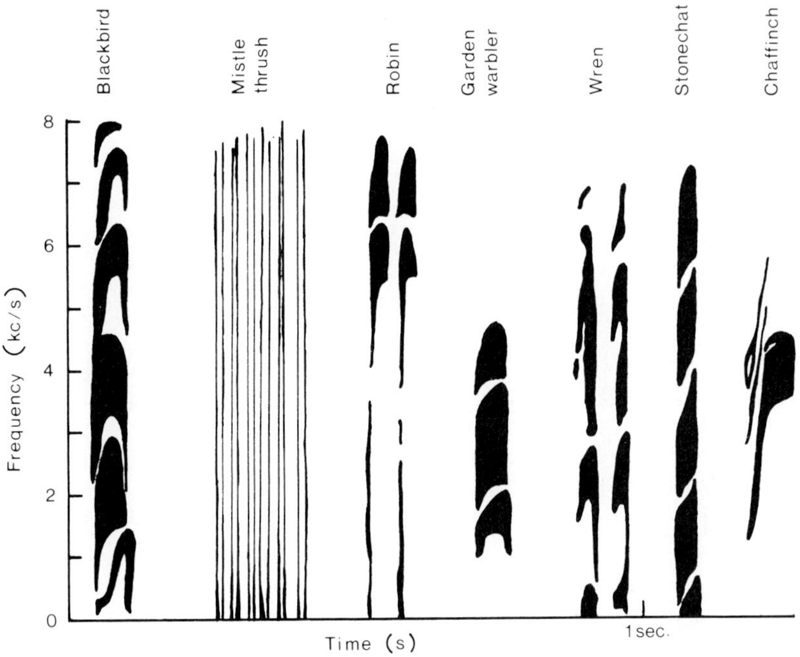

Figure 4–13 Visual representation of some sound characteristics of the calls of some British birds. Note the different frequency ranges for different species. Note also the different temporal patterns both of the entire call and of certain frequencies within the call, especially as seen in the blackbird, the wren, and the chaffinch. (Modified from P. Marler, 1959. *Darwin's Biological Work.* P.R. Bell [Ed.]. Cambridge University Press, Cambridge, England, pp. 150–206.)

Fishes do not have external ears, and most hear best at the lower frequencies (less than approximately 2 kHz). Low-frequency sounds carry farther than high-frequency sounds. Sound vibrations in the water are transmitted through the body. Otoliths ("ear stones") in the inner ear of fishes intercept sound waves and vibrate, causing nearby receptors to set off nerve impulses. In some species, including cypriniform fishes and their allies as well as the unrelated elephantfishes, the gas-filled swim bladder intercepts sound waves and directly or indirectly transmits vibrations to the otoliths. The minnows (Cypriniformes) possess a chain of small bones extending from the swim bladder to the inner ear (Fig. 4–14). These bones greatly increase auditory sensitivity and the capacity for frequency discrimination; they also increase the range of perceptible frequencies up to approximately 8 kHz.

Anurans have an eardrum that is coupled to the inner ear by a bone, the stapes (Fig. 4–15). One species of leptodactylid frog (*Eleutherodactylus co-*

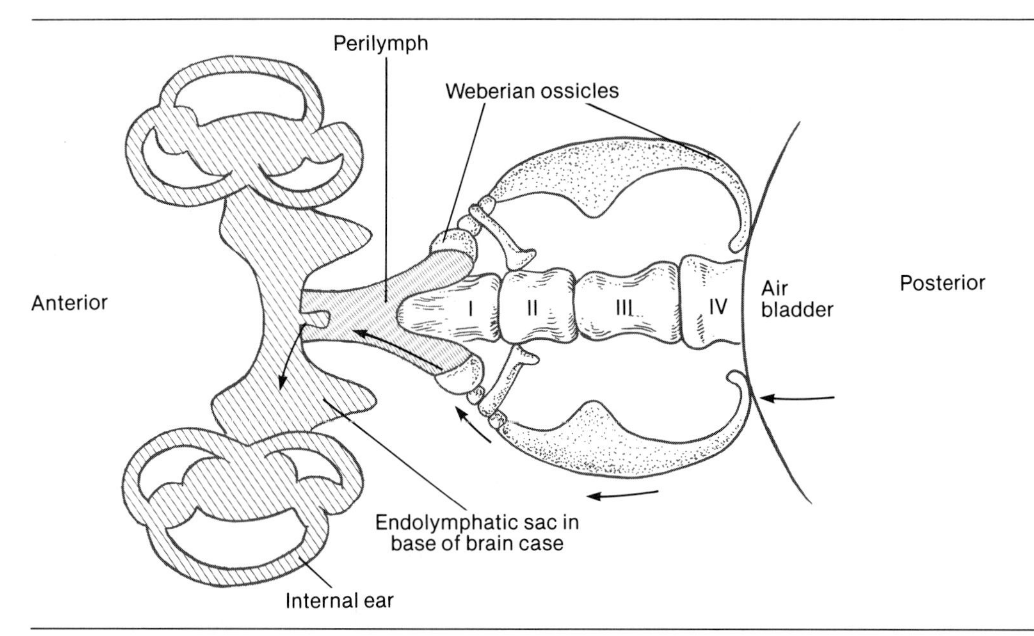

Figure 4–14 Dorsal view of the weberian ossicles of a cypriniform fish, connecting the swim bladder with the inner ear.

qui) in Puerto Rico is unusual in that males and females are acoustically sensitive to different frequencies. The male's call has two parts; males respond to one part and females to the other (Narins and Capranica, 1976). Salamanders do not have eardrums. Consequently, in aquatic species, sound is transmitted to the stapes mainly via the bones of the head; in terrestrial species, it is transmitted via the bones of the forelimb. Some amphibians are reported to hear frequencies as high as 10 kHz, but the auditory range of most amphibians is well below about 3 kHz. Reptiles also are rather limited in acoustic sensitivity. Snakes can hear only very low-frequency airborne sounds in the range of 0.1 to 0.7 kHz. They lack eardrums but can pick up these sounds through the quadrate bone, from which the jaw is suspended. They can also pick up substrate vibrations through the skull; these vibrations are then transmitted to the inner ear via the stapes. Most turtles also lack eardrums, but the skin of the head can pick up and transmit low-frequency vibrations. Crocodilians and most lizards have ears with a stapes that transmits vibrations. The hearing range of reptiles is reportedly similar to the extremes reported for amphibia.

Hearing is undoubtedly best developed in birds and mammals. Birds have eardrums that transmit vibrations to a stapes. Most mammals have eardrums that transmit vibrations to a series of three bones (the malleus, incus, and stapes), which is located in the middle ear. Whales are exceptional among mammals in that sound waves apparently pass through fat

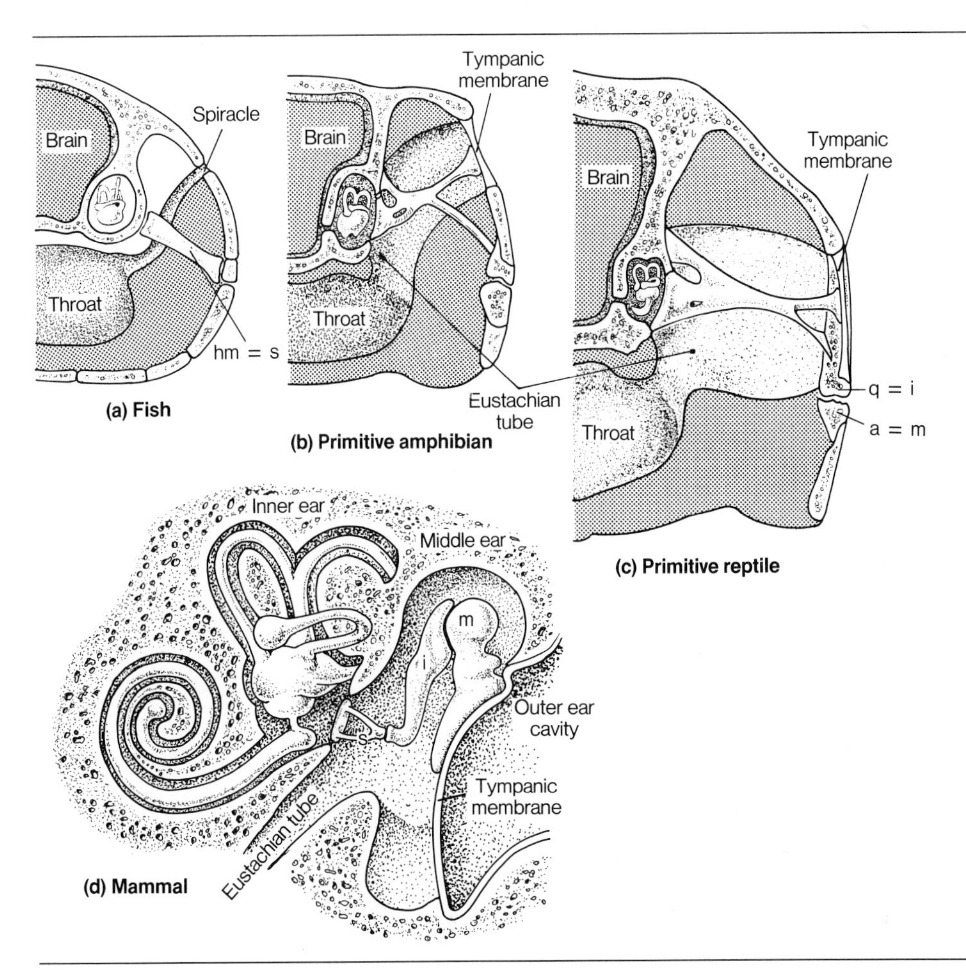

Figure 4–15 The basic arrangements of some vertebrate ears. (*a*) Cross-section of the head in the ear region of a fish. The hyomandibular bone (hm) functions here as part of the jaw suspension, but phylogenetically it becomes the stapes. (*b*) Cross-section through the ear region of the head of an amphibian. (*c*) Cross-section through the ear region of the head of a reptile. The bones labeled *q* and *a* are the quadrate and articular bones, which become the incus and malleus of mammals. (*d*) Cross-section through the ear region of the head of a mammal. The middle ear bones are labeled *m* for malleus, *i* for incus, *s* for stapes. (Modified from A.S. Romer, 1955. *The Vertebrate Body*. 2nd ed. W.B. Saunders, Philadelphia.)

and major bones of the skull, or perhaps through the lower jaw, to the bony capsule around the middle and inner ears. The external ear of a whale is permanently plugged with wax but may, nevertheless, transmit some sound. Auditory sensitivity of mammals and birds is very good. The normal range of perceptible frequencies may be as high as 20 kHz, or even 30 kHz, in rodents and shrews. Bats and cetaceans can hear frequencies as high as 180 kHz. Birds are better than most mammals in the temporal resolution of sounds; they can discriminate faster changes of temporal patterns. Some of the important elements of bird song, therefore, are undetectable to the unaided human ear.

Most vertebrates can detect sound, but only some have well-developed sound production and acoustical communication. Assorted hissing, clicking, scraping, and puffing noises are produced by many species, including a rather large array of fishes. More elaborate communication systems are known in anurans, birds, mammals, and certain fishes, such as macrourids and brotulids on the continental slopes at depths of 200 to 1000 m.

Acoustical Specializations

Three of the most interesting and unusual acoustical adaptations are seen in the enlarged middle ears of kangaroo rats and a number of other rodents; in the ear shape and ear position of owls; and in the remarkable echolocation ability of many bats, whales, pinnipeds, and insectivores.

Kangaroo rats (*Dipodomys,* family Heteromyidae) are nocturnal inhabitants of the warm deserts of North America. The bones of the middle ear are of a shape and size that permit small pressures on the eardrum to be transformed into large pressures in the inner ear. The eardrum itself is unusually large. Furthermore, the bony casing around the middle ear of the kangaroo rat is enormously expanded so that the volume of the air-filled cavity within is much larger than that of most mammals. When the eardrum moves inward during vibration, there is less resistance from compression of the air in the cavity and the vibration is damped less than it would be by a smaller middle-ear cavity. The effect is especially great in the reception of low-frequency sounds, which induce rather large movements of the eardrum.

These structural features contribute to the extraordinary sensitivity of the kangaroo rats' ears (Webster, 1962, 1966). Their ears are most sensitive to low-intensity, low-frequency sounds in the 1- to 3-kHz range. Such sounds are produced by two of the rats' natural predators, rattlesnakes and owls, just as they approach their prey. Both are extremely quiet predators, but the tiny sounds they make are enough for a normal kangaroo rat to hear and enable it to get away. Webster partially filled the middle-ear cavity with modeling clay, thus experimentally reducing the volume of air contained within the cavity. Rats with reduced middle-ear volumes could no longer escape from their predators. This is a nice demonstration of the adaptive value of these unusual ears.

Several other small mammals also have enlarged middle ears, and it is tempting to suppose that the function of their middle ears may be similar to those of kangaroo rats. Increased auditory sensitivity could also be of value in intraspecific communication, especially if it is advantageous for members of one species to be able to communicate with each other without revealing their presence to other species. Enlarged middle ears are found primarily in rodents, including other heteromyids, some of the South American hystricomorphs, Old World gerbils (family Cricetidae), jerboas (family Dipodidae), the African springhaas (Pedetidae), and the hopping mice of Australia (*Notomys* of the family Muridae). Members of one insectivore family, Macroscelididae, also have large middle ears; this family includes the 28 species of elephant shrews in Africa. They are unusual among insectivores in their ability to run rapidly and to bound. In fact, many of the mammals that have large middle ears move by hopping and bounding. Unusually sensitive ears are particularly advantageous when the animal is capable, as these jumpers are, of effective evasive action by means of a sudden change of direction.

The external ears—and sometimes the skin flaps around the ears—of many owls are strikingly asymmetrical in size and position (Fig. 4–16*a*). Asymmetry is reportedly more marked in species that hunt mainly at night and locate their prey largely by ear, such as *Asio otus* (the long-eared owl) and *Tyto* (the barn owls). Asymmetry is less marked in species that are more diurnal, such as *Athene,* the little owl of Europe.

Payne (1971), Konishi (1973), and others have shown that functional asymmetry of ears separated by a broad skull facilitates the precise localization of prey sounds because sounds arrive at each ear at slightly different times and with different patterns of intensity (Fig. 4–16*b*). An owl orients its attack by turning its head until both ears are maximally stimulated; hearing is very directional. Owls with asymmetrical ears, such as the barn owl (*Tyto alba*), are capable of capturing mice in complete darkness by using only acoustical signals. Many owls also have a facial disk of specialized, dense feathers that acts as a sound catcher and enhances directional hearing (Fig. 4–17). The degree of development of the facial disk is purportedly correlated with the degree of nocturnal activity. The barred owl (*Strix varia*), saw-whet owl (*Aegolius acadicus*), and the European tawny owl (*Strix aluco*), for instance, are very nocturnal and have well-developed facial disks. North American burrowing owls (*Speotyto cunicularia*) and ferruginous owls (*Glaucidium brasilianum*) are predominantly diurnal and have very poor facial disks. Screech owls (*Otus asio*) are intermediate in both characteristics. A facial disk is also evident in the diurnal marsh hawk or hen harrier (*Circus cyaneus*), which may forage, in part, by hearing its prey as it flies low over meadows and marshes.

A number of vertebrates emit high-frequency sounds and locate prey or other objects by listening to the echoes reflected from those objects. A sonar system such as this clearly requires special auditory capacities—not to mention the ability to produce such sounds in the first place. The sonar systems of insectivorous bats are the best known, especially those of the suborder Microchiroptera, which includes the majority of bats. Sonar calls

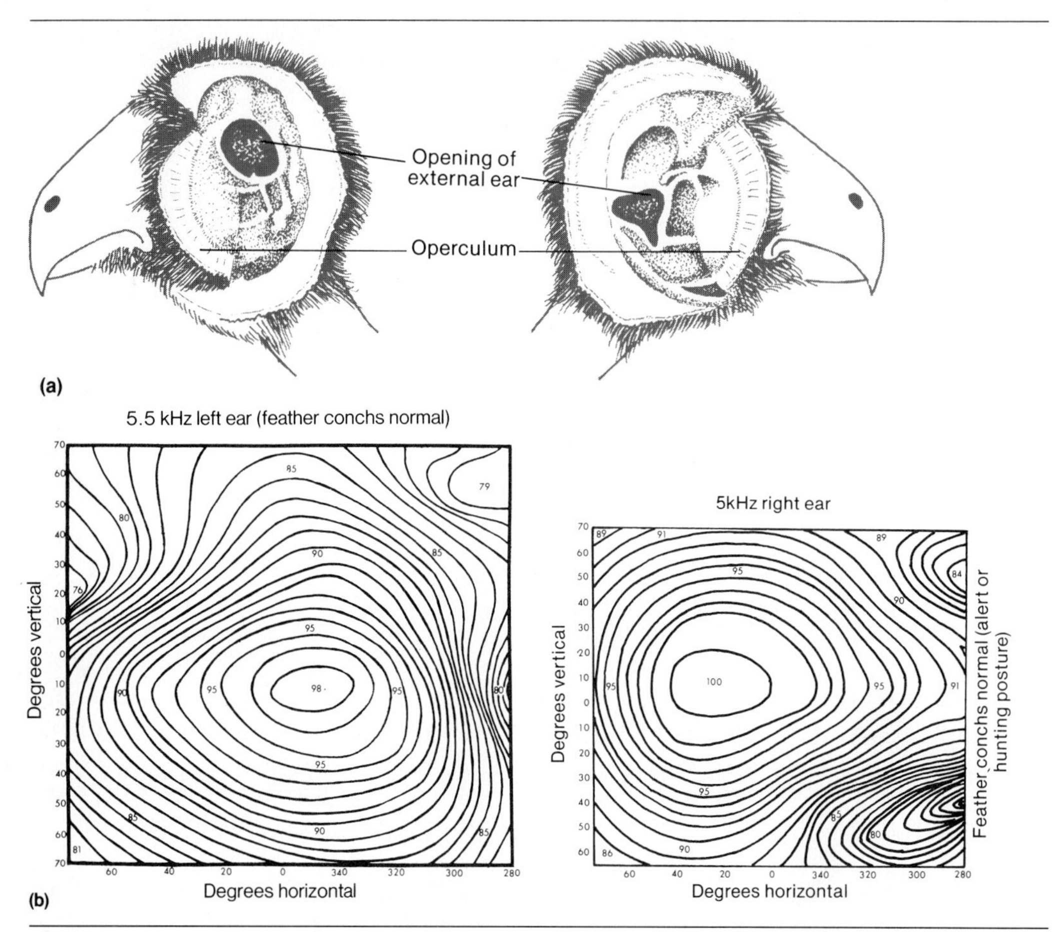

Figure 4–16 (*a*) Left and right external ear openings of an owl (*Asio*). Note the marked asymmetry of size, shape, and position. (Modified from J.C. Welty, 1982. *The Life of Birds.* 3rd ed. Saunders College Publishing, Philadelphia.) (*b*) Mercator projections of the barn owl's directional sensitivity when tones of constant intensity are directed at an ear from different angles. Sensitivity is recorded in decibels; isointensity lines are drawn every decibel. The sound intensity at the site of the owl's ear without the head in place is 100 db. Points on the plots are points at which data were taken. The owl, if it were drawn on the plot, would face the observer from 0° vertical and 0° horizontal. These diagrams can be read like topographic maps, with peaks at the high values and valleys at the low ones. (Modified from R.S. Payne, 1971. Acoustic locations of prey by barn owls [*Tyto alba*]. *J. Exp. Biol.*, 54:535–573.)

of the Microchiroptera vary from 25 to more than 100 kHz and are in the ultrasonic range for humans, whose limit is about 18 to 20 kHz. Sonar calls of many bats are of very high intensity. At maximum intensity, a bat's

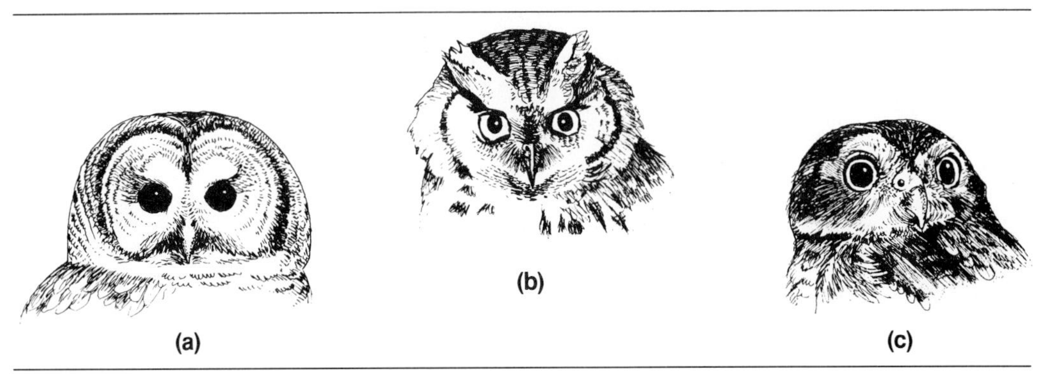

Figure 4–17 The faces of some owls, showing the contrast in development of the facial disk: (*a*) *Strix varia,* the barred owl; (*b*) *Otus asio,* the screech owl; and (*c*) *Glaucidium brasilianum,* the ferruginous owl.

ultrasonic call may have a force of as much as 20 to 70 times that produced by a jackhammer going full blast. High-intensity calls are characteristic of bats that catch flying insects; the calls of bats that feed mainly on fruit, nectar, or sitting insects are softer. The loud calls are probably adaptive in detection of prey at greater distances, which is especially important for capturing active, elusive insects. Such intensity could injure the ears of the bats themselves, were they not protected during the emission of ultrasonic pulses. The inner and middle ears are often partially isolated from the rest of the skull by fat deposits and blood-filled cavities, so that sounds produced by the vocal cords in the larynx are not transmitted internally to the ear. At the time of pulse emission, muscular contraction changes the tension of the eardrum and the position of the ear bones; these return to normal in time for pulse reception. The external ear pinna probably functions in providing directional information by selective blocking of returning sound waves. Some bats have an additional flap of skin, the tragus, in front of the opening and the pinna; the tragus may refine the selective return of sound waves (Fig. 4–18). Selective receipt of sound can be used to enhance the differences in sound quality reaching the two ears, thus facilitating localization of objects.

The sonar used in a typical hunting sequence includes three phases: search, in which the bat emits fairly long, slow pulses as it cruises about for insect prey; approach, in which the bat emits more rapid pulses as it closes in on the prey; and terminal, in which the predator emits a buzz of short, fast pulses just before the attack. This search-and-capture sequence can be very effective, even though insectivorous bats are often highly selective feeders (Buchler, 1976). *Myotis lucifugus,* the little brown bat of North America, for example, can detect prey up to about 75-cm away in an arc of about 120°. On some occasions, *Myotis* have been observed to capture two

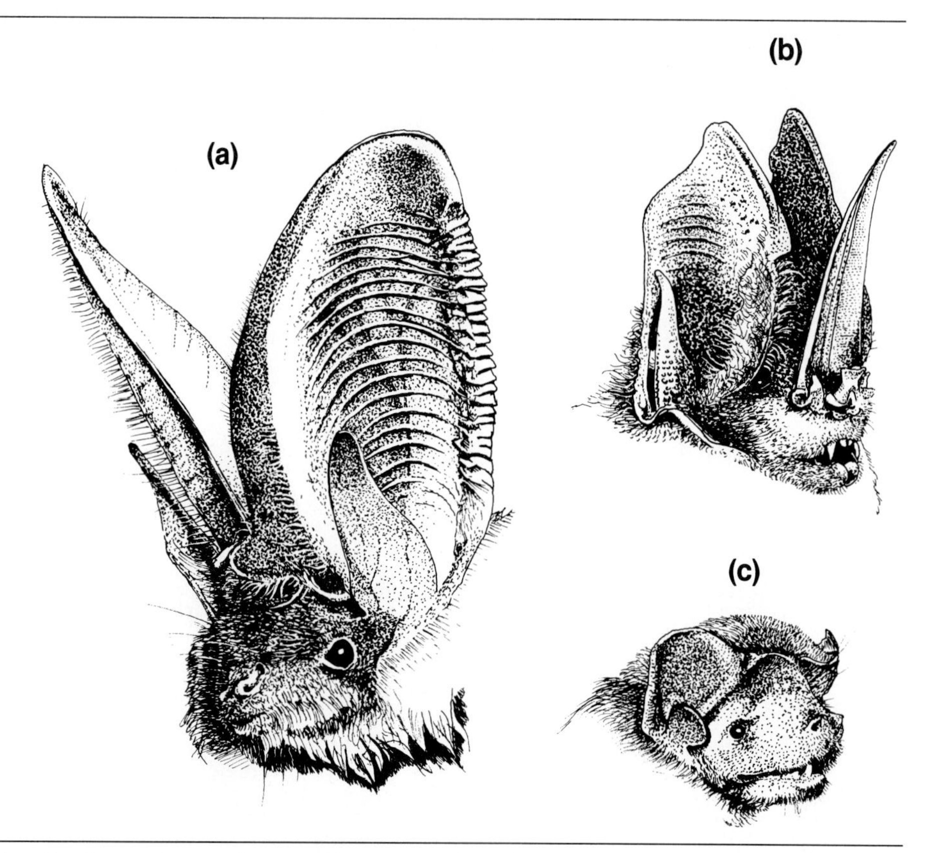

Figure 4–18 The faces of several microchiropteran bats, showing a variety of nose shapes and eye and ear sizes. (*a*) *Plecotus auritus* (Vespertilionidae), (*b*) *Lonchorhinus aurita* (Phyllostomatidae), (*c*) *Promops* sp. (Molossidae).

insects in one second, using as few as four pulses to locate and catch the second prey. A 7- to 8-g *Myotis lucifugus* has been known to catch an average of a gram of insects per hour. The eastern pipistrelle (*Pipistrellus subflavus*) may catch as much as 25 percent of its 5- to 6-g body weight in insects in about 30 minutes. One gram of insects is about 5000 gnats or 65 middle-sized moths, which is perhaps an average of 500 insects an hour or one every 7 seconds. The sonar system is even effective in capturing fish. Fish-eating bats, such as certain *Noctilio* species of Latin America, gaff their prey from the surface waters after having located the quarry by sonar.

Most bats emit ultrasounds through the open mouth, but the Old World horseshoe bats, the Rhinolophidae, beam their pulses through their noses with the aid of their peculiar nasal appendages. Rhinolophids are also unusual in that their ears are very mobile and can be turned to receive echoes directionally. The large, fruit-eating bats of the Old World suborder Megachiroptera, in contrast to most Microchiroptera, have very large eyes, small ears, and little or no echolocation. Species of the genus *Rousettus*, however, have a low-frequency sonar produced by tongue-clicking and used as an auxiliary orientation device in their roostcaves. Microchiroptera that use sonar in prey capture also use their echolocation capacity to avoid obstacles in their flight paths.

Sonar is well developed in porpoises and other toothed whales. Because light is transmitted poorly through water, an additional sensory capacity is probably adaptive in improving both prey capture and avoidance of obstacles. The bottle-nosed dolphin (*Tursiops truncatus*) can be trained to discriminate, without error, between two sizes of fish prey under conditions in which the use of vision is impossible. This species has no sense of smell and uses its sonar to locate and identify prey, as well as to avoid obstacles. The sperm whale (*Physeter catodon*) emits a series of clicks and probably uses their echoes to locate prey in deep, dark waters. A fascinating whale, the blind river dolphin (*Platanista gangetica*), inhabits the large muddy rivers of India and Pakistan. Its eyes are tiny and have no lenses, so they cannot form images. Its sonar is well developed, however, and the dolphin emits an almost continuous series of pulses, undoubtedly "seeing" with its ears. How whales emit their clicking sounds is not yet understood, but they have fatty lumps on their foreheads that beam the pulses directionally. The river dolphin emits sounds through its blowhole; the sounds are then focused by the fat deposit and incredible bony outgrowths of the skull (Fig. 4–19). The dolphin swims along the bottom on its side, apparently scanning the area ahead of it with a narrow beam of sound.

Sonar echolocation is also known to occur in certain insectivores, such as tenrecs (Tenrecidae) and shrews (Soricidae). The Weddell seal (*Leptonychotes weddelli*) of Antarctica emits ultrasonic chirps incessantly as it swims under the ice from air hole to air hole.

Among birds, the guácharo, or oilbird (*Steatornis caripensis*), of South America has a sonar of low frequency (within the human hearing range) used for orientation in the dark (Konishi and Knudsen, 1979). It also has large eyes that are sensitive in dim light. Guácharos roost and nest in caves but emerge at twilight to feed on fruits. Some species of *Collocalia*, the bird's-nest-soup swiftlets of South Asia, also have a form of sonar used in orientation.

A very different function for ultrasonic calls is found in the young of all cricetid and murid rodents (such as *Dicrostonyx* lemmings) that have been studied so far. In these rodents, females with very young offspring respond to these calls quickly, so the sounds probably function as a distress signal. Ultrasonic waves attenuate rapidly and are therefore best adapted to short-range communication. Furthermore, these sounds are difficult to localize and so are less likely to increase the risk of detection by a predator than would sound waves of lower frequency.

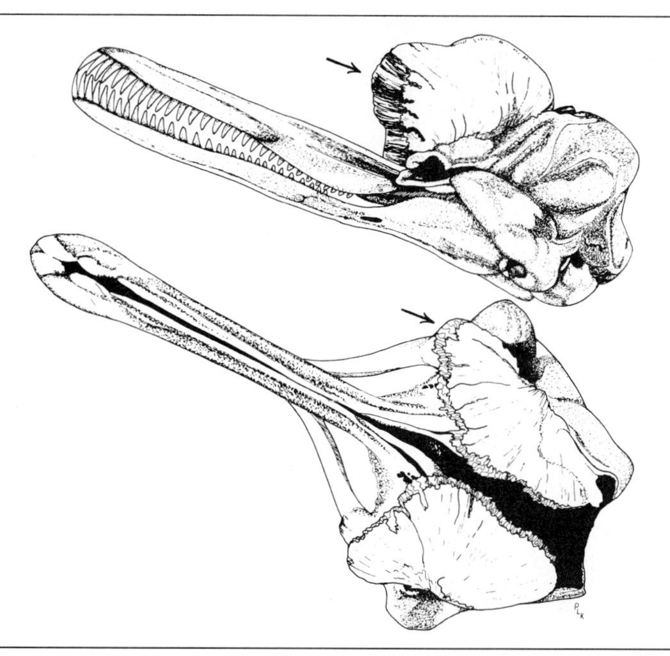

Figure 4–19 The skull of the blind river dolphin, *Platanista gangetica.* Note the grotesque bony extensions above the jaw. (From E.S. Herald, et al., 1969. Blind river dolphin: first side-swimming cetacean. *Science, 166*:1408–1410. Copyright © 1969 by the American Association for the Advancement of Science.)

CHEMICALS

The chemoreceptive senses of taste and olfaction in vertebrates are generally distinguishable, although the distinction is rather arbitrary. Taste usually functions in short-range perception. Taste receptors consist of several sensory cells located in a small depression and are usually associated with feeding structures, such as the tongue and lips. Taste receptors in fishes, however, may be located in many parts of the body, including the skin, fins, and fleshy appendages called barbels, depending on the species. For most vertebrates, taste probably functions primarily in the selection and sorting of food items. It has an added role in courtship for many species. Usually, information is derived from objects near the receiver.

Olfactory receptors are usually associated with respiratory organs and are located in the nostrils. They are sensory cells in the nasal lining and commonly transmit information derived from more distant parts of the environment. Certain fishes, including salmon (*Oncorhynchus*) and eels (*Anguilla*), have well-developed olfactory capacities that are essential in migratory homing, but others (halfbeaks, flying fishes, and relatives) have a much reduced olfactory sense. Reptiles supplement the usual olfactory receptors with Jacobson's organs, which are located in the roof of the mouth. Airborne molecules are picked up on the extruded tongue and then inserted into this organ. Controversy continues about the olfactory capacity of birds: Apparently, some birds, such as the kiwi (*Apteryx australis*), the storm petrels (*Oceanodroma*), and the turkey vulture (*Cathartes aura*), can definitely smell and use this faculty in foraging or homing. Other birds, such as parrots and passerines, may lack this ability or have little olfactory capacity. Almost all mammals, except the porpoises and dolphins, can smell.

Chemotransmission is possible at much greater distances in air than in water, and diffusion of the stimulatory molecules is much faster. Chemoreception by aquatic animals is therefore primarily short-range, because communication is most useful if the message is not delayed. Terrestrial animals have made use of long-range chemoreception: Distant but rapid detection of prey, predator, and mate is obviously more feasible for air-breathers.

Chemical stimulation functions in food-finding for many vertebrates; carnivorous mammals often track their prey by scent, for example. Salmon and some other fishes use smell to guide them to their "home" stream for spawning. Chemical signals among conspecific individuals are called pheromones and may be used in recognition of species, sex, individuals, and physiological condition. Many mammals, for example, use chemical signals in recognizing sexual partners of the appropriate kind and physiological state. Parent mammals often recognize their own offspring by means of characteristic individual scents. Chemical signals of young jewelfishes (*Hemichromus bimaculatus*) induce parental behavior by adults and aggregation by other young jewelfishes. Furthermore, scent is used by many mammals, including wolves, to mark the borders of their territories or occupied areas (Peters and Mech, 1975). Alarm substances released by injured minnows and amphibian tadpoles elicit escape or hiding behavior by conspecific associates.

SUMMARY

Vertebrate sense organs are many and varied; they monitor the changing environment in diverse ways and are important in all aspects of a vertebrate's life.

Land-living vertebrates use part of the available spectrum of light for vision; species that live underwater have a much restricted spectrum because of the light-filtering effect of water. Although many vertebrates possess a photosensitive median eye, the primary light-using sense organs are image-forming eyes. The eyes of all vertebrates are constructed in the same general way—with a cornea, lens, and retina—but considerable structural diversity is related to specific ways of life. Acrobatic and highly predatory species usually have binocular vision and high visual acuity that pinpoint objects in three dimensions. Species that are active in dim light often have large, highly sensitive eyes, but some have nonvisual, usually auditory, systems for orienting and foraging.

Directional temperature sensors are known in two families of snakes.

A number of fishes can detect electrical signals; some can emit them as well. Recently, it has become clear that some birds and fishes can sense magnetic fields and may use them in orientation.

The sound-carrying properties of water are very different from those of air, but vertebrates in both media use acoustical signals. Details of ear morphology vary greatly and limit the auditory sensory capacity. Most fishes, amphibians, and reptiles utilize primarily lower frequencies, and certain birds and mammals can both emit and receive ultrahigh frequencies. Directional sensitivity is especially well developed in species that move about in the dark.

Chemoreception is very important to most vertebrates (with some notable exceptions, especially in whales and birds) and, like the other senses, functions in foraging, sex, and social communication.

SELECTED REFERENCES

General accounts for each of the classes can be found in the references at the end of Chapter 2.

Adler, K., and D.H. Taylor, 1973. Extraocular perception of polarized light by orienting salamanders. *J. Comp. Physiol., 87*:203–212.

Albers, V.M. (Ed.), 1967. *Underwater Acoustics.* Plenum Publishing, New York.

Alcock, J., 1975. *Animal Behavior.* Sinauer Associates, Sunderland, MA.

Ali, M.A. (Ed.), 1974. *Vision in Fishes: New Approaches in Research.* Plenum Publishing, New York.

Ali, M.A. (Ed.), 1980. *Environmental Physiology of Fishes.* Plenum Publishing, New York.

Barrett, R., 1970. The pit organ of snakes. *In* C. Gans and T.S. Parsons (Eds.), *Biology of the Reptilia.* Vol. 2B, pp. 277–300. Academic Press, New York.

Bellairs, A., 1970. *The Life of Reptiles.* Universe Books, New York.

Bennett, M.V.L., 1970. Comparative physiology: electric organs. *Annu. Rev. Physiol., 32*:471–528.

Brach, V., 1977. The functional significance of the avian pecten: a review. *Condor, 79*:321–327.

Bray, R.N., and M.A. Hixon, 1978. Nightshocker: predatory behavior of the Pacific electric ray (*Torpedo californica*). *Science, 200*:333–334.

Brett, J.R., 1957. The eye. *In* M.E. Brown (Ed.), *The Physiology of Fishes.* Vol. 2, pp. 121–154. Academic Press, New York.

Brown, J.L., 1975. *The Evolution of Behavior.* W.W. Norton & Co., New York.

Buchler, E.R., 1976. Prey selection by *Myotis lucifugus* (Chiroptera: Vespertilionidae). *Am. Nat., 110*:619–628.

Bullock, T.H., 1973. Seeing the world through a new sense: electroreception in fish. *Am. Sci., 61*:316–325.

Bullock, T.H., and F.P.J. Diecke, 1956. Properties of an infra-red receptor. *J. Physiol.* (Lond.) *134*:47–87.

Cahan, P.H. (Ed.), 1967. *Lateral Line Detectors.* Indiana University Press, Bloomington.

Denton, D.A., and J.P. Coghlan, 1975. *Olfaction and Taste.* Vol. 5. Academic Press, New York.

Disler, N.N., 1960. *Lateral Line Sense Organs and Their Importance in Fish Behavior.* (1971 translation from Russian.) National Technical Information Service.

Duke-Elder, S., 1958. *System of Ophthalmology.* Vol. 1. The Eye in Evolution. Kimpton, London.

Erulkar, S.D., 1972. Comparative aspects of spatial localization of sound. *Physiol. Rev.* *52*:237–360.

Farner, D.S., and J.R. King, 1973. *Avian Biology.* Vol. 3. Academic Press, New York.

Fenton, M.B., 1974. The role of echolocation in the evolution of bats. *Am. Nat., 108*:386–388.

Fenton, M.B., 1975. Acuity of echolocation in *Collocalia hirundinacea* (Aves: Apodidae), with comments on the distributions of echolocating swiftlets and molossid bats. *Biotropica, 7*:1–7.

Fields, R.D., and G.D. Lange, 1979. Electroreception in the ratfish (*Hydrolagus colliei*). *Science, 207*:547–548.

Fite, K.V., 1976. *The Amphibian Visual System.* Academic Press, New York.

Fox, R., S.W. Lehmkuhle, and R.C. Bush, 1977. Stereopsis in the falcon. *Science, 197*:79–81.

Fox, R., S.W. Lehmkuhle, and D.H. Westendorf, 1976. Falcon visual acuity. *Science, 192*:263–265.

Goldsmith, T.H., 1980. Hummingbirds see near ultraviolet light. *Science, 207*:786–788.

Gordon, M.S., et al., 1977. *Animal Physiology: Principles and Adaptations.* 3rd ed. Macmillan Publishing Co., New York.

Gould, E., 1955. The feeding efficiency of insectivorous bats. *J. Mammal., 36*:399–407.

Graham, J.B., 1971. Aerial vision in amphibious fishes. *Fauna, 3*:14–23.

Griffin, D.R., 1958. *Listening in the Dark.* Yale University Press, New Haven.

Gundy, G.C., C.L. Ralph, and G.Z. Wurst, 1975. Parietal eyes in lizards: zoogeographic correlates. *Science, 190*:671–673.

Hairston, N.G., Li, K.T., and S.S. Easter, 1982. Fish vision and the detection of planktonic prey. *Science, 218*:1240–1242.

Heiligenberg, A., 1973. Electrolocation of objects in the electric fish *Eigenmannia* (Rhamphichthyidae: Gymnotoidei). *J. Comp. Physiol., 87*:137–164.

Herald, E.S., et al., 1969. Blind river dolphin: first side-swimming cetacean. *Science, 166*:1408–1410.

Herman, L.M., et al., 1975. Bottle-nosed dolphin: double-slit pupil yields equivalent aerial and underwater diurnal acuity. *Science, 189*:650–652.

Hoar, W.S., and D.J. Randall (Eds.), 1969, 1971. *Fish Physiology.* Vols. 3, 5. Academic Press, New York.

Hopkins, C.D., 1974. Electric communication in fish. *Am. Sci., 62*:426–437.

Hopkins, C.D., and A.H. Bass, 1981. Temporal coding of species recognition signals in an electric fish. *Science, 212*:85–87.

Jacobs, G.H., 1977. Visual sensitivity: Significant within-species variations in a non-human primate. *Science, 197*:499–500.

Jacobs, G.H., 1981. *Comparative Color Vision.* Academic Press, New York.

Kalmijn, A.J., 1982. Electric and magnetic field detection in elasmobranch fishes. *Science, 218*:916–918.

Keeton, W.T., T.S. Larkin, and D.M. Windsor, 1974. Normal fluctuations in the earth's magnetic field influence pigeon orientation. *J. Comp. Physiol., 95*:95–103.

Knudson, E.I., 1974. Behavioral thresholds to electric signals in high frequency electric fish. *J. Comp. Physiol., 91*:333–353.

Konishi, M., 1973. How the owl tracks its prey. *Am. Sci., 61*:414–424.

Konishi, M., and E.I. Knudson, 1979. The oilbird: hearing and echolocation. *Science, 204*:425–427.

Kreithen, M.L., and W.T. Keeton, 1974. Detection of polarized light by the homing pigeon, *Columba livia. J. Comp. Physiol., 89*:83–92.

Kunz, T.H., 1974. Feeding ecology of a temperate insectivorous bat (*Myotis velifer*). *Ecol., 55*:693–711.

Larkin, P.R., and P.J. Sutherland, 1977. Migrating birds respond to Project Seafarer's electromagnetic field. *Science, 195*:777–779.

Lawrence, B.D., and J.A. Simmons, 1982. Echolocation in bats: The external ear and perception of the vertical position of targets. *Science, 218*:481–483.

Manley, G.A., 1972. A review of some current concepts of the functional evolution of the ear in terrestrial vertebrates. *Evol., 26*:608–621.

Marler, P., 1959. Developments in the study of animal communication. *In* P.R. Bell (Ed.), *Darwin's Biological Work,* pp. 150–206. Cambridge University Press, Cambridge, England.

Marler, P., and W.J. Hamilton, 1966. *Mechanisms of Animal Behavior.* John Wiley & Sons, New York.

Marshall, N.B., 1971. *Explorations in the Life of Fishes.* Harvard University Press, Cambridge, MA.

Moller, P., 1976. Electric signals and schooling behavior in a weakly electric fish, *Marcusenius cyprinoides* L. (Moromyriformes). *Science, 193*:697–699.

Monk, O., 1966. The eyes of *Ipnops murrayi* Günther, 1878. *Galathea Rept., 3*:79–87.

Morin, J. G., et al., 1975. Light for all reasons: versatility in the behavioral repertoire of the flashlight fish. *Science, 190*:74–76.

Narins, P.M., and R.R. Capranica, 1976. Sexual differences in the auditory system of the tree frog *Eleutherodactylus coqui. Science, 192*:378–380.

Nicol, J.A.C., 1969. Bioluminescence. *In* W.S. Hoar and D.J. Randall (Eds.), *Fish Physiology.* Vol. 3, pp. 355–400. Academic Press, New York.

Norris, K.S., 1963. The functions of temperature in the ecology of the percoid fish *Girella nigricans* (Ayres). *Ecol. Monogr., 33*:23–62.

Novick, A., 1977. Acoustic orientation. *In* W.A. Wimsatt (Ed.), *Biology of Bats.* Vol. 3, pp. 74–287. Academic Press, New York.

Payne, R.S., 1971. Acoustic location of prey by barn owls (*Tyto alba*). *J. Exp. Biol., 54*:535–573.

Peavey, W.C., S.L. Meyer, and O. Munk, 1965. A 'four-eyed' fish from the deep-sea: *Bathylychnops exilis* Cohen 1958. *Nature, 207*:1260–1262.

Peters, R.P., and L.D. Mech, 1975. Scent-marking in wolves. *Am. Sci., 63*:628–637.

Prosser, C.L., 1973. *Comparative Animal Physiology.* 3rd ed. Saunders College Publishing, Philadelphia.

Pycraft, W.P., 1910. *A History of Birds.* Methuen, London.

Romer, A.S., 1955. *The Vertebrate Body.* 2nd ed. Saunders College Publishing, Philadelphia.

Sales, G., and D. Pye, 1974. *Ultrasonic Communication by Animals.* Chapman and Hall, London.

Sillman, A.J., 1973. Avian vision. *In* D.S. Farner and J.R. King (Eds.), *Avian Biology.* Vol. 3, pp. 349–387. Academic Press, New York.

Smith, J.C., 1975. Sound communication in rodents. *Symp. Zool. Soc. Lond., 37*:317–330.

Underwood, H., 1977. Circadian organization in lizards: the role of the pineal organ. *Science, 195*:587–589.

Walls, G.L., 1942. *The Vertebrate Eye.* Cranbrook Institute of Science, Bloomfield Hills, MI.

Waterman, A.T., et al., 1971. *Chordate Structure and Function.* Macmillan Publishing Co., New York.

Webster, D.B., 1962. A function of the enlarged middle ear cavities of the kangaroo rat, *Dipodomys. Physiol. Zool., 35*:248–255.

Webster, D.B., 1966. Ear structure and function in modern mammals. *Am. Zool., 6*:451–466.

Welty, J.C., 1982. *The Life of Birds.* 3rd ed. Saunders College Publishing, Philadelphia.

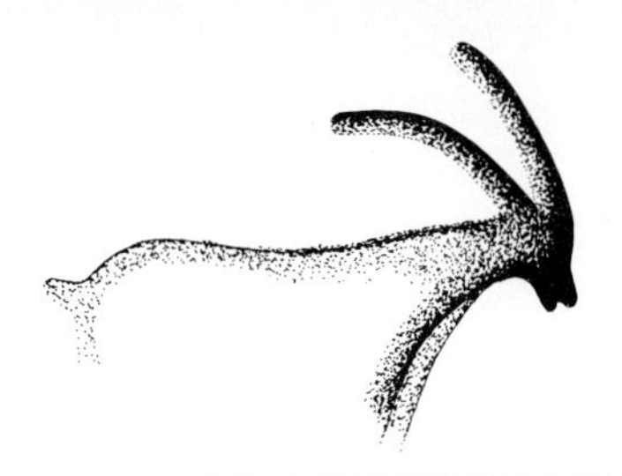

5 Thermoregulation and Osmoregulation

ENVIRONMENTAL TEMPERATURES

The temperature extremes of the universe are approximately −273°C (absolute zero) at the low end of the scale and in the millions of degrees at the upper end, but the body temperatures of animals usually fall between about −2°C and about 45°C. Although animals may tolerate environmental temperatures beyond this range for a period of time, their activity is usually much reduced. Generally, they must maintain their body temperatures well within those limits. The lower limit is just below the freezing point of water, on which all body chemistry depends; the upper limit is near the temperature at which some animal proteins begin to denature. Within these restrictions, the carbon-based chemistry of earthly animals is stable enough to permit the development of complex molecular relationships and yet variable enough to allow for the evolution of different chemically reactive systems. A complex set of genetically directed chemical interactions provides the basis for the evolution of physiological, structural, and behavioral functions. A degree of variability within reaction systems that normally operate in a predictable way is essential to the radiation of biological diversity.

Temperature, Water, and Life

Temperature is a fundamental biological variable. Most biochemical reactions are temperature-sensitive: even slight changes in body temperature may have profound effects on rates of chemical interactions in the body. An increase in temperature generally accelerates physiological processes, at least within the normal thermal limits of the animal. If a reaction doubles in rate when body temperature is raised from 0 to 10°C, a subsequent increase of 10°C often doubles it again, thus raising the reaction rate to four times that of the original. A further rise to 30°C might increase the rate to a level eight times that of the initial rate at 0°C. This exponential kind of relationship is often characteristic of many biochemical reactions. Even within the lethal limits of an organism, therefore, temperature relations are extremely critical. Rates of reaction obviously affect the entire energy budget of an animal, and, generally, there is an optimal temperature range within which animal bodies function best. Furthermore, the high body temperature or fever that often accompany infectious diseases appears to be crucial in fighting the infection; thus, it is part of the cure as much as a symptom of illness, at least in lizards and probably in other vertebrates as well (Kluger, 1976, 1979).

Water is indispensable to life on earth. It is the basic solvent for biochemical reactions and the major component by weight in most animal cells and bodies. Most vertebrates must keep their water content within rather narrow limits: too much water dilutes the body fluids excessively; too little results in overconcentration of dissolved and suspended particles and slowed transport of essential materials and wastes. Both effects are detrimental to continued life. Water is not only critical in regulating osmotic processes and transporting body materials in solution from one organ to another, it is also closely linked to the problems of temperature regulation, which are discussed in the following sections.

Temperature Variations

The problems of temperature regulation are usually less severe for aquatic vertebrates than for terrestrial vertebrates, because most aquatic species do not face the extremes of environmental temperatures to which many land animals are exposed. Water has a far greater capacity to store heat than does air and also reduces the amount of incoming solar radiation. Natural bodies of water, with the exception of geothermal hot springs or pools, usually have temperatures in the range of −2 to 40°C. Few aquatic vertebrates can survive and complete a life cycle outside this range. The desert pupfish (*Cyprinodon diabolis*) of Devil's Hole, Nevada, lives in warm springs with temperatures averaging about 34°C. The upper lethal limit for this small fish is about 43°C, the highest known for any fish species. At the other end of the aquatic thermal spectrum are species such as the Antarctic icefishes of the genus *Trematomus*. These fishes exist in an environment with a rather constant temperature of almost −2°C and are extremely sensitive to heat.

Unlike the fishes, aquatic mammals and birds, which have high body temperatures, encounter a problem produced by the high thermal conductivity of water compared to that of air. Consequently, they are faced with a potentially high rate of heat loss. In these species, special heat conservation mechanisms have evolved, which are discussed in this chapter.

Terrestrial vertebrates, on the other hand, may be faced with much greater temperature extremes. Vertebrates of the Arctic and Antarctic experience months of continual subzero temperatures during the long polar winters. Emperor penguins (*Aptenodytes forsteri*) near the South Pole even begin their breeding season during this time. At the other extreme, diurnal animals of hot deserts may encounter both high air temperatures (over 50°C) and blazing hot (over 70°C) substrates. One can see that the evolutionary transition from aquatic to terrestrial life represented a major shift in temperature adaptations and the opening of a new adaptive zone.

The adaptations discussed in this chapter are mainly those of individual physiology and morphology. The equally important and fascinating adaptations related to reproduction are considered in Chapter 14.

MODES OF THERMOREGULATION

The body temperature of an animal results from the balance or imbalance between heat gain and heat loss (Fig. 5–1). External sources of heat include radiation from the sun and radiation of longer wavelengths from the ground, the atmosphere, and objects in the environment, conduction from the air and solid objects in contact with the body, and convection by air or water currents around the body. In addition, all living animals generate heat internally during the process of metabolism. Heat may be lost in several ways: by radiation back to the environment, convection, conduction, and evaporation of water.

Whether heat is lost or gained depends primarily on temperature differences between the body surface and the environment: If the body surface is warmer than the environment, heat tends to be lost; if it is cooler, heat tends to be gained. The control of heat gain and loss, and, hence, body temperature, is known as thermoregulation. Rates of heat gain or loss can

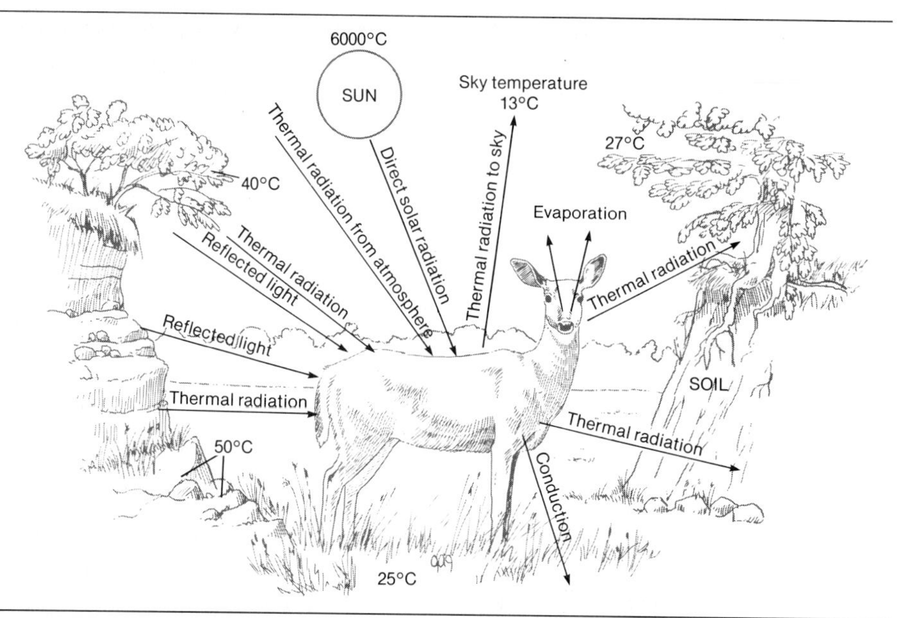

Figure 5–1 Diagram of the energy exchanges between an animal and its environment on a warm day. (Modified with permission of Macmillan Publishing Co., Inc. from *Animal Physiology: Principles and Functions* by Malcolm S. Gordon, et al. Copyright © 1977 by Malcolm S. Gordon.)

be modified by insulation provided by fat layers, fur, and feathers, and by the degree of fur or feather fluffing. Body temperature can also be regulated by the degree of dilation of surface blood vessels carrying heat from the body core to the surface and by other circulatory mechanisms that will be discussed later. Behavioral adjustments such as postural changes, shade-seeking, and basking also contribute to control of body temperature. Evaporation of water can be an effective means of heat loss for some terrestrial animals because of the large quantity of heat needed to turn water from a liquid into vapor. If a surface bears a film of water, much more heat is required to vaporize the water and remove it from the surface than would be required to heat the air next to a dry surface. By driving off water, then, more heat can be removed from the surface than would otherwise be possible. The water takes with it the heat that produced the vapor and leaves a cooler surface behind.

Rates of heat production, heat gain from the environment, and heat loss to the environment depend on the species and the circumstances. Body temperature may be largely a function of surrounding (ambient) temperatures, heat production plus controlled heat loss, or some combination of these factors.

Body temperatures in some animals are at least partly dependent on the environment; that is, the animals are ectothermic ("external heat"). Metabolic rates in most ectotherms are low, and relatively little heat is produced in ordinary circumstances. Periods of high muscular activity may increase heat production temporarily, but the major source of heat or cooling is the external environment. Ectotherms are also called poikilotherms ("variable heat") in reference to the potential lability of their body temperatures. Some ectotherms live in thermally variable environments and their body temperatures vary with that of the environment. Others live in thermally constant environments, such as the oceanic deeps, and hence have a constant body temperature similar to that of their surroundings. These latter species are customarily considered to be poikilothermic, despite their constant temperature. If their thermal environment should ever change, they would lack the ability to maintain their body temperature at the former level and would then have a variable body temperature. Other ectotherms possess specific behavior patterns that control rates of heat gain and loss in relation to the environment. Basking in the sun or on hot substrates, moving in and out of shade, changing body posture or orientation to the sun's rays—these are examples of thermoregulatory behavior with which ectotherms in variable environments can adjust their body temperatures away from ambient air temperatures.

Fishes, amphibians, and reptiles are generally classed as ectotherms; this classification, however convenient, is an oversimplification. Some reptiles are capable of physiological heat production and thermoregulation and approach the degree of internal regulation displayed by mammals. Certain fishes are able to control the temperature in particular parts of the body. On the other hand, some birds and mammals, when torpid or hibernating, have reduced abilities to generate internal heat and may approach ectothermy. Thus, the distinction between ectothermy and endothermy (internal heating) is not a sharp one. Nevertheless, the classification of ectotherm or poikilotherm is still convenient, as long as its limitations are kept in mind.

Other animals, known as endotherms, produce and retain sufficient internal, metabolic heat to warm the body. Many endotherms are homeothermic: They maintain a nearly constant body temperature higher than that of the environment. Most birds and mammals, and some reptiles, are regularly homeothermic. Avian body temperatures are commonly between 40 and 44°C, but the temperature of flightless species, such as emus and penguins, is 37 to 39°C. Monotremes maintain temperatures of about 31°C. The temperatures of placental and marsupial mammals are usually in the range of 36 to 40°C.

The capacity for heat production may vary with ecological circumstances. Desert-dwelling individuals of various deer-mouse (*Peromyscus*) populations have lower metabolic rates than individuals living in other habitats (McNab and Morrison, 1963; McNab, 1974). Individuals of high-latitude populations sometimes have higher metabolic rates than individuals from lower latitudes. The house sparrow in North America is an example: The metabolic rate of a caged sparrow in Arizona and Mexico, in January, averaged 18 kcal/day; the rate of birds in central Illinois was just over 26 kcal/day; and in northern Manitoba, the rate exceeded 34 kcal/day. Differences were smaller, but followed a similar trend, in summer (Kendeigh, 1976). Such trends do not appear among rodents at different latitudes, however (McNab, 1974).

Although metabolic rates of endotherms vary with body weight, in general, interesting modifications occur in the shape of the relationship (McNab 1974, 1979). Small mammals that weigh over 100 g exhibit the normal correlation between metabolic rate and body weight with an exponent (b) of about 0.75 in the equation $M = aW^b$. Below 100 g body weight, the slope of the line is much less (only about 0.45). Very small mammals, such as shrews, have extraordinarily high metabolic rates—higher than would be predicted on the basis of body size alone. The carnivorous weasels and their relatives show a similar shift, but the critical size in this group is about 1 kg, perhaps because weasels are characteristically elongate in shape and have a rather high surface/volume ratio. Underground-dwelling mammals that weigh more than 80 g have metabolic rates lower than expected from their body size; smaller ones (<60 g), however, have metabolic rates higher than expected. Clearly, small size often brings with it a metabolic cost that affects the animal's entire ecology (McNab, 1980). If the animal cannot pay this metabolic cost, it may carry more insulation or become dormant under stressful conditions.

A number of vertebrates are facultative endotherms, or heterotherms; that is, sometimes their body temperatures are precisely regulated and at other times they are not.

Some reptiles have the physiological capacity to control rates of heat gain and heat loss. Bartholomew and Tucker (1963, 1964) found that body warming is a faster process than cooling in the agamid lizard *Amphibolurus barbatus* of Australia and in the paleotropical monitor lizards (Varanidae); therefore, periods of warm body temperatures are extended. This differential rate is effected primarily by circulatory changes in heat transport (perhaps including the shunting of blood to the body core from the appendages, which are allowed to cool rapidly) in the agamid and by endogenous heat production in the varanids. The large lizard *Tiliqua scincoides* of Australia has such effective endogenous heat production that it is virtually endothermic. In addition, one species of python (*Python molurus*) generates enough physiological heat to maintain a body temperature more than 7°C above ambient and uses this body heat to incubate its eggs. An increasing amount of evidence indicates that the reptilian ancestors of birds and mammals were good physiological thermoregulators.

Dormant birds and mammals have a reduced metabolic rate and may approach ambient temperatures, but they do not lose all capacity to regulate internal temperature while dormant. Furthermore, very young birds and mammals are often not fully endothermic; therefore, they are at least partly dependent on parental care for temperature control. Young mammals usually possess some thermogenic ("heat-generating") abilities because of their deposits of brown fat. This fat, metabolically much more active than ordinary fat, is well vascularized and generates considerable metabolic heat. The amount of brown fat decreases as physiological temperature regulation develops in the growing young. Birds appear to lack brown fat and their young generally have a greater dependence on parental body heat than do young mammals.

The ability to maintain high temperatures in only part of the body is called regional heterothermy. Some large sharks (*Lamna, Isurus*) and tuna (*Thunnus, Katsuwonus*) are able to maintain high temperatures in the red muscles used in fast, sustained swimming by means of a countercurrent heat exchange system (Fig. 5–2). Oxygenated blood leaves the gills at water temperature, reaching the red muscles in small arteries that run closely parallel to the many small veins coming from the muscles. In this way, arterial blood that was originally at water temperature picks up heat from the warm venous blood flowing in the opposite direction. Thus, heat produced by the muscles is conserved by the incoming arterial blood, maintaining a muscle temperature higher than the ambient water temperature (Fig. 5–3) and permitting faster muscular contraction and greater power output. Heat generated by muscle contraction could also be conserved by insulating the entire body, but these fishes would obtain no advantage from this arrangement because the blood would be cooled to ambient temperatures at the respiratory surfaces of the gills in any case. Therefore, it is more efficient to conserve heat only in the specific parts of the body requiring higher temperatures. Muscles that are reliably warm may be particularly important in fishes that swim long distances (often at very high speeds) and encounter significant changes in water temperature that otherwise (in fully ectothermic fishes) would tend to produce changes in speed. Perhaps the "warm" fishes exploit their food resources in some way that has made this heat exchange mechanism particularly useful.

The brain and eyes of swordfish (*Xiphias gladius*) are warmed by a heating mechanism located in one of the eye muscles (Carey, 1982). These fishes execute large vertical migrations on a daily basis and encounter temperatures that differ by as much as 19°C. A rapid chilling of the brain could impair the functioning of the central nervous system, and Carey suggested that the brain heater is probably an adaptation that allows the swordfish to function well at a variety of water depths. Other fishes are thought to have similar mechanisms.

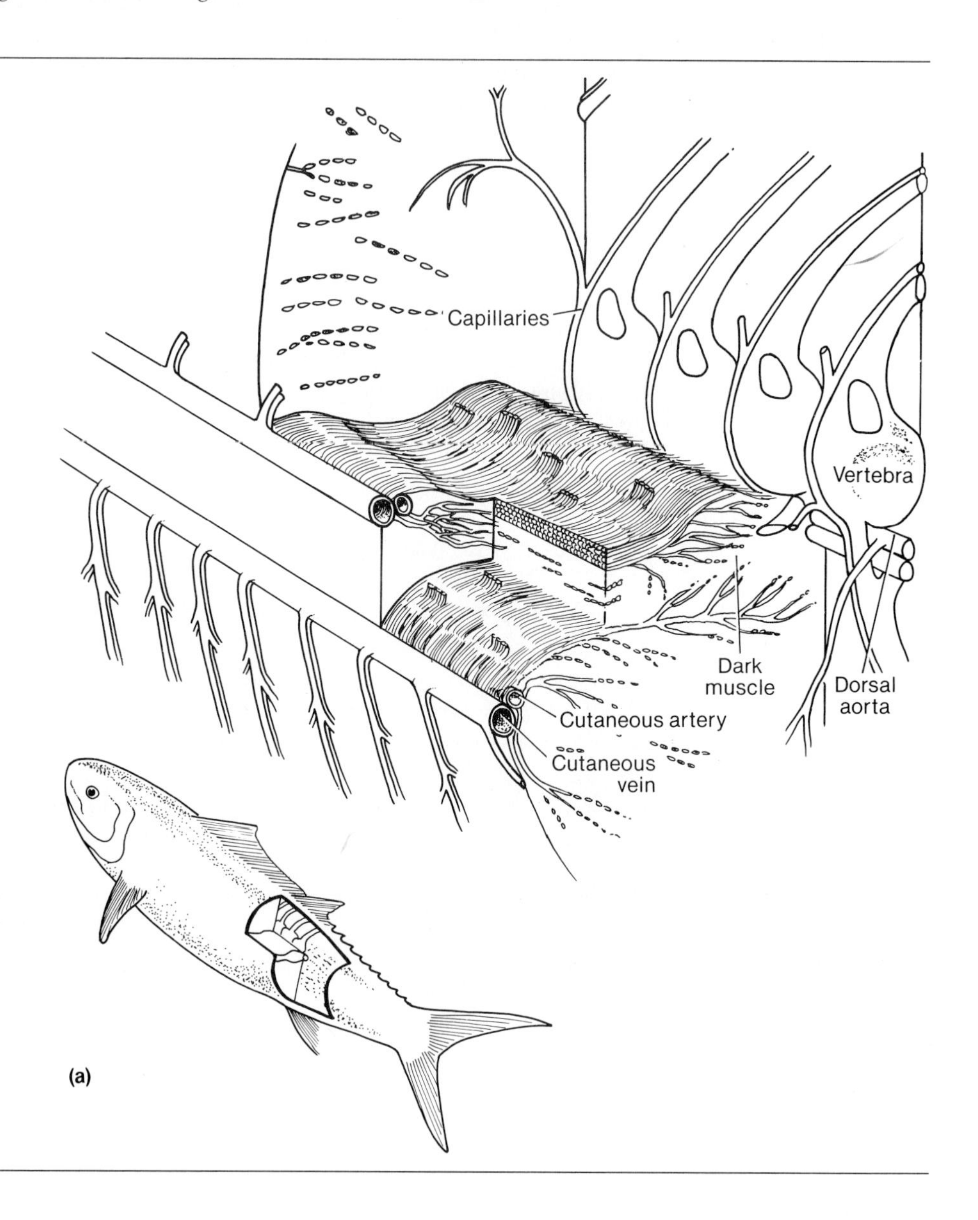

Costs of Endothermy

Endothermy is metabolically costly: As much as 80 to 90 percent of the energy from metabolism in the form of heat may be used to maintain body temperature. At some point, the amount of energy needed for thermoregulation is minimal and may remain at that level over a range of ambient temperatures called the thermoneutral zone (Fig. 5–4). This zone sometimes spans a broad range and sometimes a very narrow range of tempera-

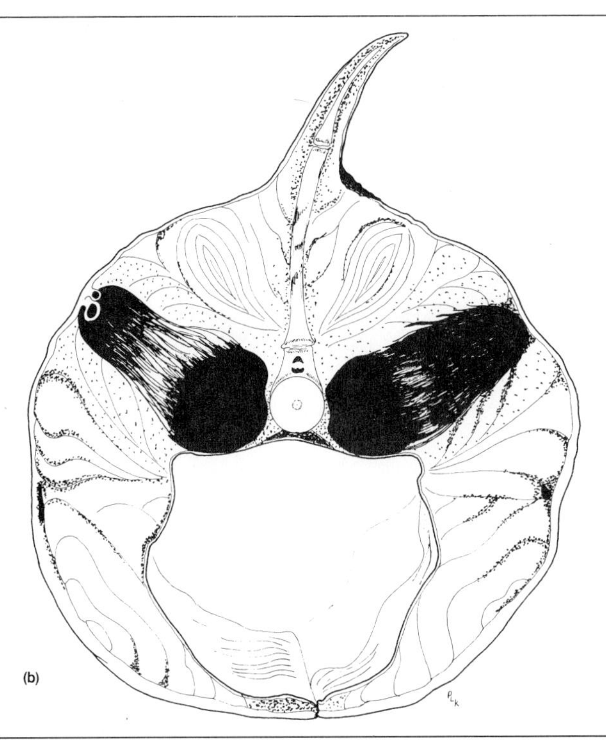

Figure 5–2 (*a*) Diagram of the blood supply of the heat exchange mechanism in a fish with regional heterothermy. Blood arrives primarily in large surface blood vessels and returns primarily via a parallel route after passing through the capillaries of the exchanger. (Modified from F.G. Carey and J.M. Teal, 1966. Regulation of body temperature by the bluefin tuna. *PNAS, 56*:1464–1469. Reprinted with permission of the authors.) (*b*) Cross section through a porbeagle shark (*Lamna nasus*) at the level of the first dorsal fin. Note the dark (red) muscle area served by the heat exchanger. (Reprinted with permission from *Comp. Biochem. Physiol.*, Vol. 28, Francis G. Carey and John M. Teal, Mako and porbeagle: warm-bodied sharks. Copyright © 1969, Pergamon Press, Ltd.)

tures, depending on the species, season, and time of day. At temperatures below the thermoneutral zone, increased heat production is required to maintain body temperatures. Within and above the zone, body temperature is regulated, at least in part, by changes in rates of heat loss to the environment. Above the thermoneutral zone, metabolism increases and heat production occurs due to the expenditure of energy needed for increased heat loss, especially in evaporation of water.

The evolutionary "choice" of endothermy and homeothermy necessitates the expenditure of energy and water; therefore, conservation of energy and water may be crucial. The differences in the relationships of the metabolic rate (involving energy expenditure) and the evaporation rate

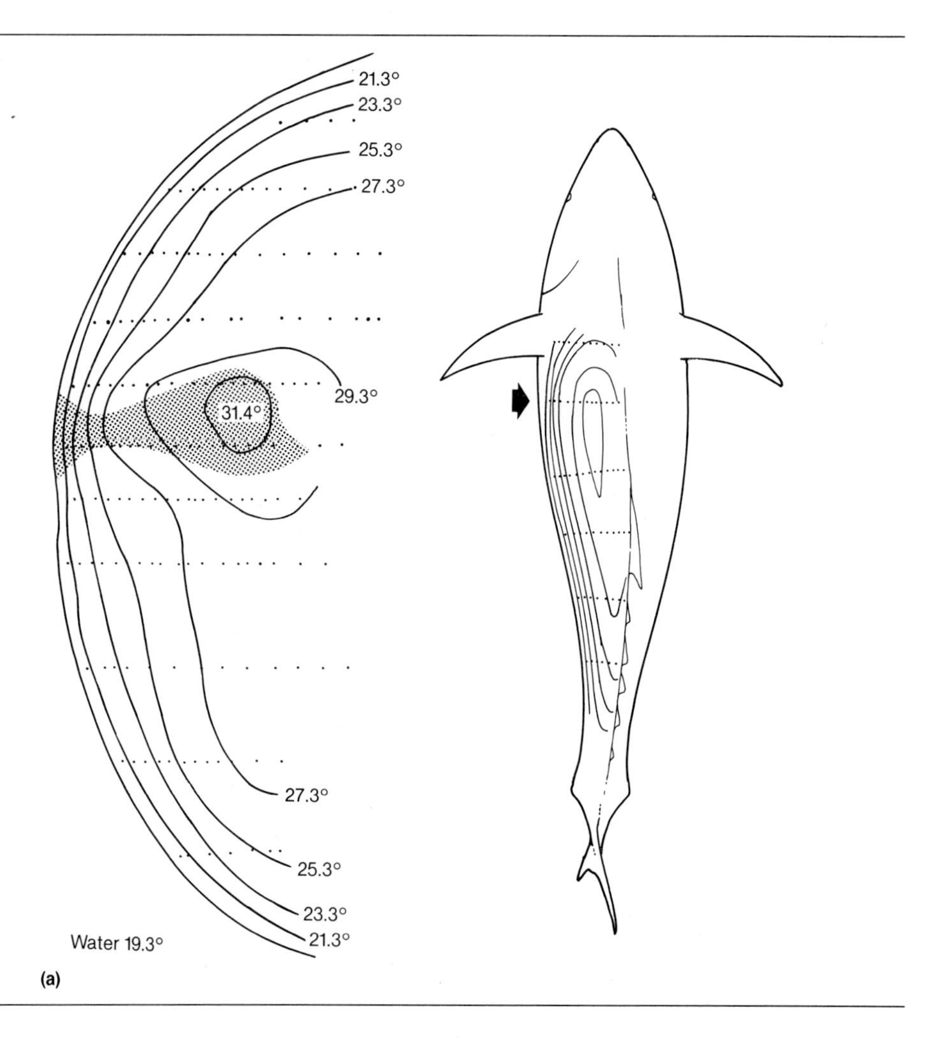

(involving water expenditure) in a hypothetical bird in relation to ambient temperature are shown in Figure 5–5. The curves have very different shapes, indicating that simultaneous optimization of both water and energy economy is impossible. Maintenance of a constant body temperature necessitates equal rates of heat gain and heat loss. High metabolic rates, and consequent high heat production, require heat dissipation at an equal rate. Lower metabolic rates would be more economical of energy, but lower body temperatures would necessitate greater evaporation rates at high ambient temperatures, because the gradient for heat loss from the animal to the environment is small. This tactic, therefore, would be uneconomical of water supplies. On the other hand, a metabolically costly high body temperature creates a steep gradient for heat dissipation by non-evaporative means; thus, it reduces the expenditure of water in evaporative heat loss.

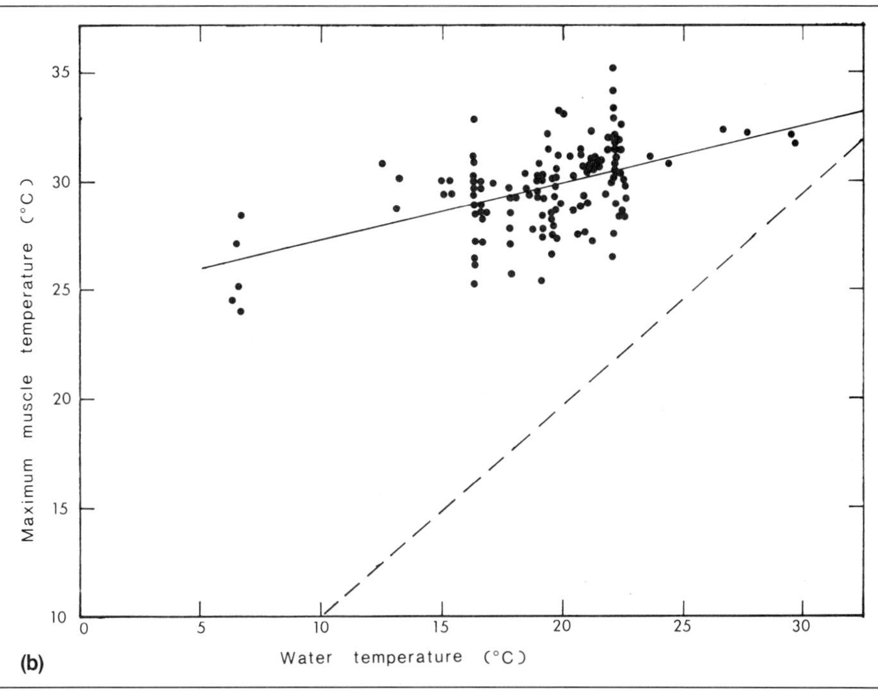

Figure 5–3 (*a*) Temperature distribution in the muscles of a bluefin tuna (*Thunnus thynnus*). (*b*) Muscle temperature of bluefin tuna over a range of different water temperatures. The muscle temperature varies only 5°C, although the water temperature varies 20°C. (*a* and *b* reprinted with permission from *Comp. Biochem. Physiol.*, Vol. 28, Francis G. Carey and John M. Teal, Regulation of body temperature by the bluefin tuna. Copyright © 1969, Pergamon Press, Ltd.)

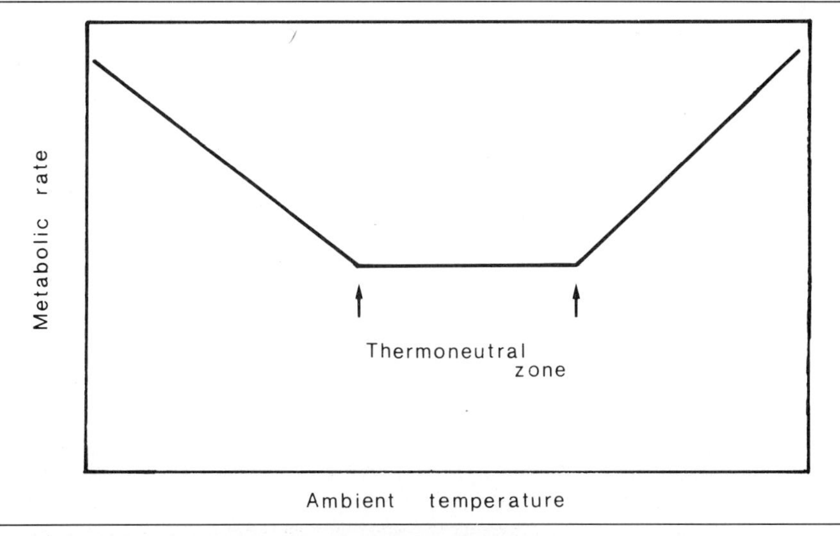

Figure 5–4 Graph of a hypothetical thermoneutral zone.

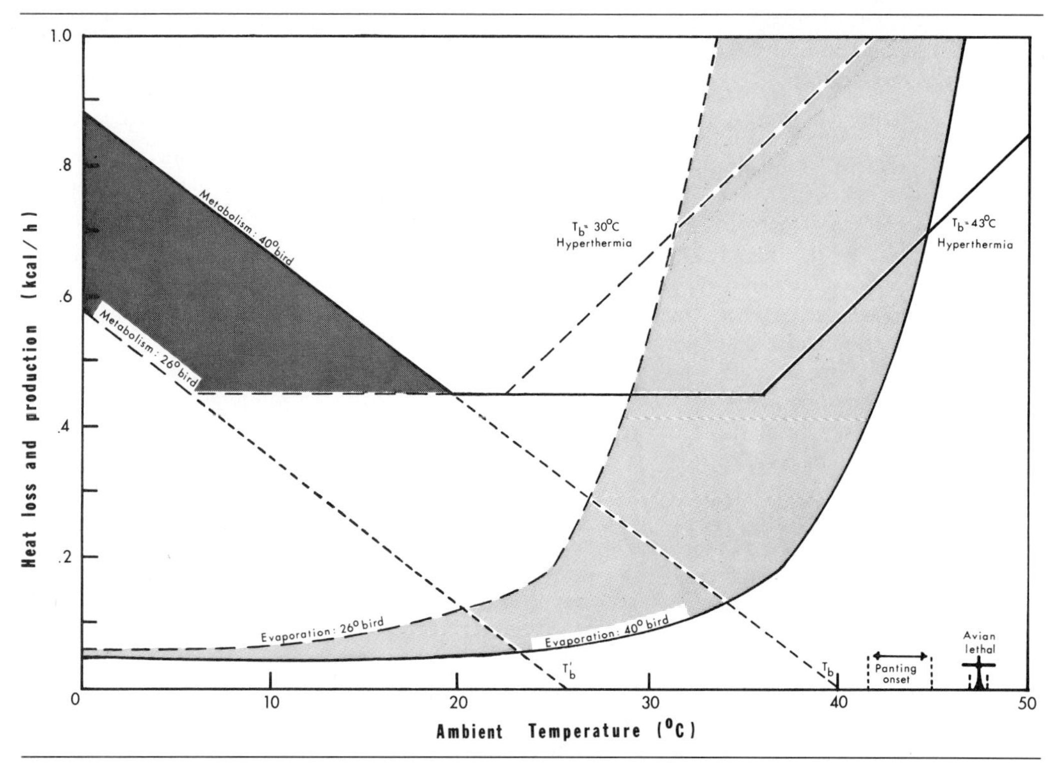

Figure 5–5 Comparison of metabolic (energy) and evaporative (water) costs of maintaining a high body temperature in a hypothetical bird. Water costs are measured here in the amount of heat removed by evaporation; the conversion factor is 0.58 kcal/g water. A bird with a body temperature of 40°C is compared to a bird similar in all respects except that its body temperature is regulated at 26°C. The savings in energy achieved by the lower metabolic rate and lower body temperature is indicated for a given environmental temperature by the vertical distance between the curves for metabolic rates in the shaded area. The water savings of the higher metabolic rate and higher body temperature are indicated by the vertical distance between the evaporation curves, divided by 0.58 kcal/g water. Note that the 26°C bird becomes overheated (hyperthermic) at a lower temperature than the 40°C bird and is spending much more water at a lower ambient temperature. At no place on the graph is there a point that minimizes *both* energy and water costs. (Modified from W.A. Calder and J.R. King, 1974. Thermal caloric relations of birds. *In* D.S. Farner and J.R. King [Eds.], *Avian Biology*, pp. 259–413. Academic Press, New York.)

The evolution of high metabolic rates in homeotherms demanded high rates of breathing to supply oxygen for metabolism. These high rates were accompanied by the risk of desiccation. Calder and King (1974) suggest that natural selection may have favored the evolution of high body temperatures in homeotherms as a means of economizing on water but at the cost of energy expenditure. High levels of energy expenditure necessitate the immediate availability of oxygen for metabolism. Oxygen is present in higher concentrations in air than in water, so endothermy and homeothermy are primarily found in air-breathing species.

At the cost of relatively great energy intake and expenditure, endotherms have achieved a significant degree of independence from environmental temperatures. Hence, they are able to be active at various times of the day or year when ectotherms are incapacitated. A constant body temperature permits maintenance of steady levels of metabolism and activity over a wide range of environmental temperatures. Endotherms, therefore, have been able to exploit a number of adaptive zones (such as year-round activity in nonaquatic polar regions) and ecological niches that are not available to ectotherms.

On the other hand, we cannot properly view the terrestrial ectotherms as merely primitive or evolutionary deadends. Their mode of thermoregulation is generally associated with rather small body size, which may allow them to occupy ecological niches that endotherms cannot use (Pough, 1980). The body forms of ectotherms are more variable than those of endotherms, because they are less constrained by heat-conservation problems. Metabolic rates are low—about 10 to 20 percent of the rates for endotherms of similar size. Furthermore, because so little energy is expended on heat production, more of the animal's energy budget is devoted to weight gain or reproduction. Although they cannot maintain high activity levels for very long, they can exploit highly seasonal or unpredictable environments and occupy low-oxygen habitats that are severely stressful for endotherms.

ADAPTATIONS TO ENVIRONMENTAL EXTREMES

This section presents some general physiological and morphological means of adapting to the environmental extremes of heat and cold.

Dormancy

A dormant animal is "sleeping," or torpid, in a state of inactivity and suspended animation. Dormancy is a means of avoiding many kinds of environmental unsuitability, such as thermal extremes, desiccation, and food shortage. Physiologists do not agree on the terminology for dormancy. For endotherms, perhaps "adaptive hypothermia" (lowered temperature) is a less controversial term, but the general term "dormancy" will be used here. Dormancy involves a range of responses that must have evolved independently a number of times, because it occurs in many unrelated species. Some endotherms have a daily cycle of torpor, some enter into dormancy on a seasonal cycle, and some do so in response to a sudden change in environmental suitability. Deep dormancy on a seasonal basis is called hibernation, if it occurs in winter, and estivation, if it occurs in summer. A condition sometimes called partial hibernation (or partial estivation) involves a partial torpor with just a few degrees of hypothermia. Physiologically, all these states are rather similar; they differ primarily in the rate and extent of decline in body temperature and the rate and frequency of arousal. They may also differ in the stimuli that trigger the entry into and arousal from hypothermia, but they do have many features in common.

The physiology of dormancy involves a slowing down of many body processes: the higher levels of the central nervous system, endocrine activity, respiration, and circulation. The lowered respiration reduces evaporative water loss from the lungs; this is particularly important in estivators. The clotting ability of the blood is decreased (an adaptation to slowly moving blood), and the concentration of red blood cells is increased, enlarging the oxygen-carrying capacity of the blood. Body temperature drops and approaches ambient temperature, but, in mammals and birds at least, temperature regulation is not totally abandoned. Animals do not become completely unresponsive to external stimuli during dormancy and can be aroused. Actually, they rouse themselves from time to time to move around a bit and void wastes. Black bears (*Ursus americanus*) go into torpor in winter, but they rouse easily and frequently and the females customarily give birth during this time. Regular seasonal hibernators and estivators often prepare an energy supply for dormancy by accumulating fat (as much as 80 percent of lean body weight in the golden-mantled ground squirrel, *Citellus lateralis*). Others store food or use both fat buildup and food storage.

Dormancy saves considerable amounts of energy. The oxygen consumption of active bats increases with decreasing air temperature, but the rate of oxygen consumption in torpid bats decreases. A 20°C drop in ambient temperature from 35 to 15°C results in a fortyfold difference in oxygen consumption, so a torpid bat at 15°C burns its fuel at one fortieth the rate of an active bat at the same temperature. Differences in oxygen consumption in torpid and active bats as a function of ambient temperature are shown in Figure 5–6. The two species of bats tested were *Eptesicus fuscus*, the North American big brown bat, which weighs about 16 g, and *Tadarida mexicana*, the Mexican free-tailed bat, which weighs about 10 g. The big brown bat normally hibernates, at least in the northern portion of its range; the free-tailed bat is migratory and apparently does not hibernate. Despite these differences in biology, both species have very similar metabolic responses to decreasing environmental temperatures.

Figure 5–6 (*a*) Body temperatures of two species of bats when the individuals were active and maintained a body temperature (T_0) greater than ambient (except above about 30°C) and when they were torpid and allowed T_b to track T_a. (*b*) Oxygen consumption of active and torpid bats. (Modified from C.F. Herried and K. Schmidt-Nielsen, 1966. Oxygen consumption, temperature, and water loss in bats from different environments. *Am. J. Physiol.*, *211*:1108–1112.)

The energy cost of hypothermia is a function of body size (Fig. 5–7). Small endotherms use dormancy as an adaptive response to unfavorable

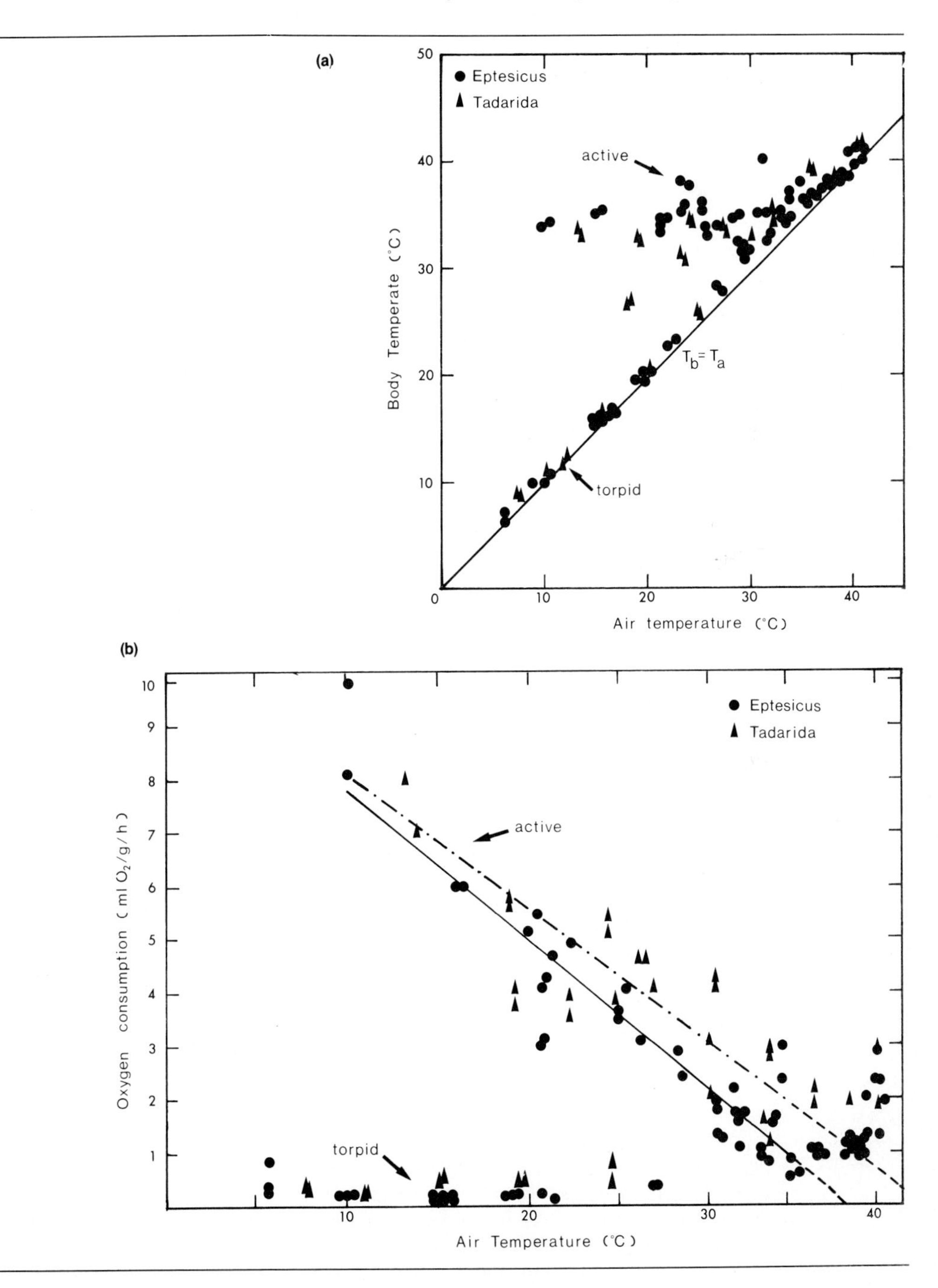

conditions more frequently than do large species because their costs of homeothermy are greater: The ratio of surface area to body volume is larger and the metabolic rate per gram is higher. Furthermore, the total mass to be cooled or warmed is less in small individuals, so small-bodied homeotherms can cool and warm more rapidly than large species. A small animal like a hummingbird might require a small fraction (about one eighty-fifth) of its daily energy budget to rewarm its body from 10 to 37°C after a period of torpor. A 200-kg bear, however, has been estimated to need its entire daily energy budget to warm its body about the same amount. Such considerations clearly place size limits on the occurrence of frequent and marked hypothermia among vertebrates. Torpor is more likely to be of selective advantage in dealing with temperature extremes in small-bodied species than in large ones.

Tucker (1965) calculated the total energy costs of torpor in the California pocket mouse, *Perognathus californicus*, by including the cost of rewarming. Pocket mice go into torpor at ambient temperatures between 15 and 32°C. At 15°C the heat production necessary to resume normal body functioning

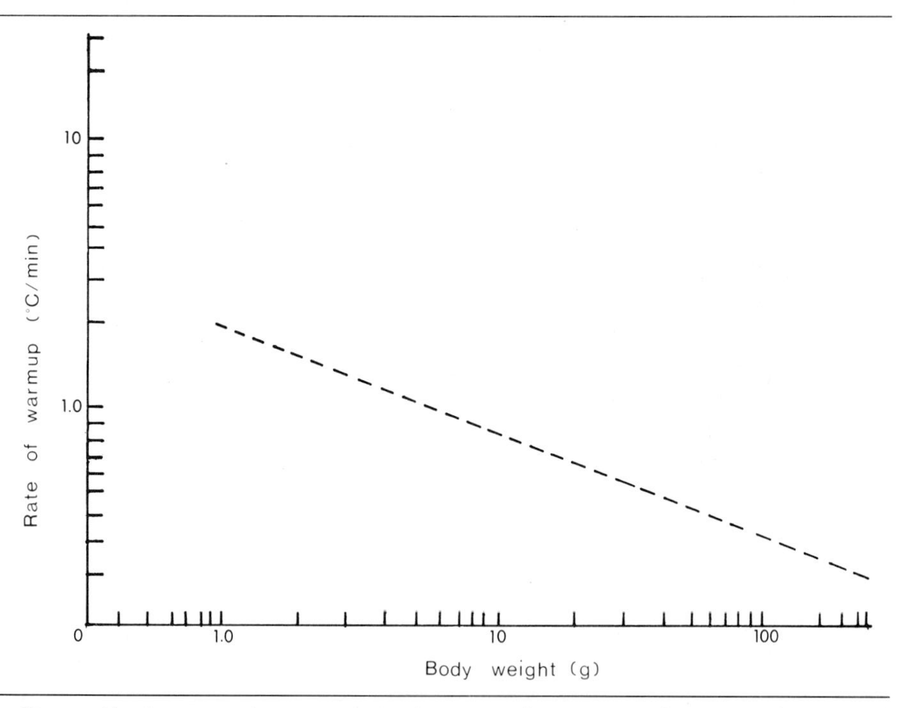

Figure 5–7 Rates of body rewarming as an animal emerges from torpor shown as a function of body weight for mammals and birds. (Modified from B. Heinrich and G.A. Bartholomew, 1971. An analysis of pre-flight warm-up in the sphinx moth, *Manduca sexta. J. Exp. Biol.*, 55:223–239.)

is 10 to 15 times the minimum level of oxygen consumption. Despite this cost of arousal, even if a pocket mouse goes into torpor and immediately begins to arouse (the whole process taking about 3 hr), oxygen consumption is only about 55 percent of the cost of homeothermic metabolism. If the mouse remained torpid for 10 hr, just under 20 percent of the normal energy expenditure would be required. Three quarters of this expenditure is the cost of arousal.

Larger species may combine limited hypothermia at night with basking behavior during the day as a means of obtaining the advantages of reduced metabolism without incurring the metabolic costs of rewarming. This combination of features has been reported for several birds, including anis (*Crotophaga*), roadrunners (*Geococcyx californicus*), and turkey vultures (*Cathartes aura*).

Seasonal hypothermia of endotherms is found in one species of North American caprimulgid bird, the poorwill (*Phalaenoptilus nuttalli*), a desert bird that hibernates in rock crevices (Fig. 5–8). The phenomenon is much more common among mammals. Members of several mammalian orders are hibernators or estivators to some degree: Some members of Monotremata, Marsupialia, Insectivora, Rodentia, Carnivora, Chiroptera (Microchiroptera), and Primates (prosimians), for example. Facultative and daily hypothermia in the avian world is known among hummingbirds (Trochilidae), swifts (Apodidae), and nightjars (Caprimulgidae). Irregular occurrences of hypothermia, or slight declines of body temperature, are reported for colies (Coliidae), some doves (Columbidae), anis (Cuculidae), a vulture (Cathartidae), and swallows (Hirundinidae), among others.

In bats and birds, a capacity for dormancy can be correlated with the nature of the food supply. In general, insectivorous birds and bats that feed on the wing and nectarivorous (nectar-eating) birds more readily enter into torpor than do those with other diets. In contrast, nectar-, fruit-, and meat-eating bats of similar sizes have higher basal metabolic rates and are usually good thermoregulators. Blood-eating bats such as the vampire bats (*Desmodus*) are intermediate. In a general way, these differences can be linked to the temporal stability of the food sources: Insects and possibly nectar, at least in some regions, may be less reliable resources than fruit (in the tropics) or vertebrate prey. Nectar-eating bats obtain amino acids from eating pollen, but nectar-eating birds usually get nitrogenous compounds from insects in their diet. Therefore, perhaps the difference in thermoregulation between nectarivorous birds and bats may be correlated with their dependence on insects. Sporadic food resources are probably a particularly severe problem for species whose method of foraging requires high levels of energy expenditure. Constant flight while foraging is costly and may be impossible to maintain when food is scarce.

Ectotherms may become dormant as an escape from unsuitable conditions. Amphibians and reptiles of temperate and boreal regions are usually seasonally inactive. Many retreat to crevices or burrows, and some am-

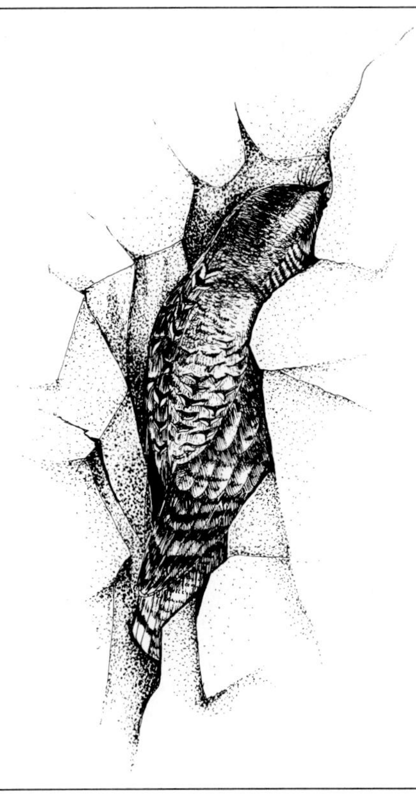

Figure 5–8 The poorwill (*Phalaenoptilus nuttalli*) hibernates in rock crevices.

phibians enter dormancy in underwater retreats. Changing temperatures, or sometimes aridity, are often the stimulus for becoming torpid. One example of seasonal dormancy in reptiles is provided by *Dipsosaurus dorsalis* (Moberly, 1963). This desert iguana in southern California (Fig. 5–9) is inactive from October to March, or about 42 percent of the year. Entry into and emergence from torpor are correlated with soil temperatures. The hibernation period is spent in a burrow, usually 5 to 15 cm under the surface. Metabolic rate is reduced and oxygen consumption is lowered as much as 50 percent when temperatures are in the range of 30 to 40°C. The change in metabolic rate is less at lower temperatures. The lizard lives on stored fat that constitutes about 5 percent of its body weight—enough to last about 150 days at prevailing temperatures. Chuckwallas (*Sauromalus obesus*) in the Mohave Desert begin to lose weight in May as their plant food becomes increasingly dry; they begin to be inactive and cease feeding in June or July; and in October they begin hibernation that continues until April. Green sea turtles (*Chelonia mydas*) become torpid in winter in the Gulf of California and can be found lying on the bottom in an inactive condition.

Certain fish species hibernate in the mud of lake bottoms. *Carassius*

Figure 5–9 The desert iguana (*Dipsosaurus dorsalis*) is inactive from October to March. (Photo by S. Prchak.)

auratus, the crucian carp, and *Dallia pectoralis*, called a "blackfish," live in the arctic part of the Soviet Union and reach very low metabolic levels in dormancy. They reportedly can survive even though frozen in a chunk of ice. Lungfishes of Africa and South America (*Protopterus* and *Lepidosiren*) estivate in mud burrows when drought and seasonal drying render their habitats unsuitable.

Geographic and Seasonal Changes

Many species change their temperature tolerances with the seasons. This physiological acclimatization permits them to tolerate colder temperatures in winter than in summer and, conversely, warmer temperatures in summer than in winter. The brown bullhead (*Ictalurus nebulosus*) of eastern North America has an upper lethal limit of about 36°C in summer but only 28°C in winter. For the willow ptarmigan (*Lagopus lagopus*) in Alaska, the lower end of the thermoneutral zone, at which maintenance of body temperature necessitates an increase of metabolic rate, is at 7.7°C in summer and −6.3°C in winter. Similar seasonal differences in both upper and lower temperature tolerances and extent of the thermoneutral zone have been reported for many vertebrates.

Geographic variation in temperature tolerances is common not only among species but within species. House sparrows were introduced to North America in the 1800s and have spread across the continent. Now individuals from different populations show adaptations to local conditions, not only in metabolic rates, as discussed earlier, but also in temperature limits. In winter, house sparrows from Manitoba have a low-temperature limit of about −31°C, but individuals from Florida have a corresponding limit of about −23°C.

The range of temperature tolerances in anuran amphibians is closely correlated with the variability of environmental temperatures: High-altitude frogs and toads are exposed to greater temperature variability and have broader temperature tolerances. The leopard frog, *Rana pipiens*, in central Mexico demonstrates altitudinal differences in temperature responses. Embryos of a high-altitude (3000 m) population develop more slowly than those of a low-altitude (350 m) population; embryos of an intermediate population are intermediate in development rate (Fig. 5–10). The rate was measured during the early embryonic stages until gill circulation began. Differences among populations were evident even in laboratory culture. Thus, these differences apparently had a genetic basis, which indicates that natural selection could account for the evolution of these different temperature responses (Ruibal, 1955). Great care must be exer-

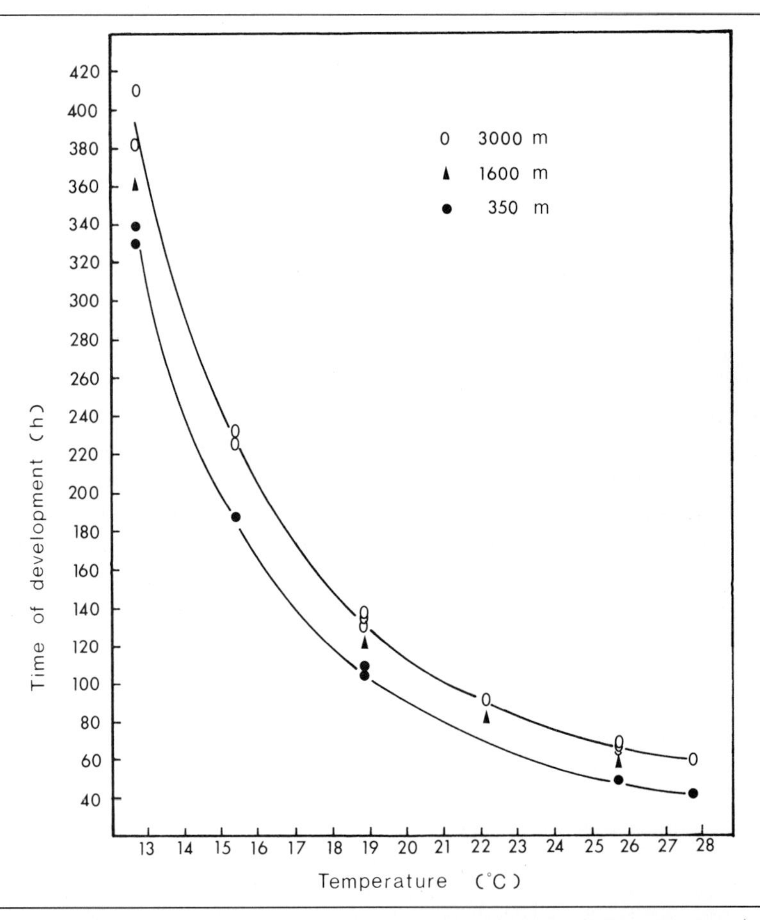

Figure 5–10 The rate of larval development of leopard frogs at different altitudes in Mexico. (From R. Ruibal, 1955. A study of altitudinal races in *Rana pipiens*. *Evolution*, 9:322–338.

cised in concluding that all such differences are entirely genetically based, however. Altitudinal variation in larval development times of the wood frog (*Rana sylvatica*) is partly controlled genetically, but strong environmental effects were evident (Berven, 1982). Control of developmental rate in the green frog (*R. clamitans*) was chiefly environmental. In fact, underlying genetic differences in the populations at different elevations would have produced slower development at low elevations (that is, the *opposite* of the observed trend) if they were not overridden by environmental differences (see Chapter 14).

Geographic variation in temperature tolerances that limit the temperature range over which activity takes place may sometimes be restricted by the presence of ecologically similar and presumably competing species. For example, *Anolis oculatus,* an iguanid lizard, is the only anole on the island of Dominica in the West Indies. It occupies a wide range of habitats and is active at many temperatures. In contrast, Cuba is inhabited by many species of *Anolis,* most of which tend to have narrower temperature tolerances than *A. oculatus* on Dominica. One hypothesis to explain this difference is that interspecific competition imposes habitat and microhabitat restrictions on each species, with the result that each species is exposed to less variation of environmental temperature.

Geographic variation in the need for drinking water and in salt tolerance is demonstrated by different populations of the North American savanna sparrow, *Passercula sandwichensis.* Individuals from populations living in salt marshes are able to drink sea water with no ill effects. In fact, members of one of these populations were able to go without drinking for long periods of time. In contrast, members of populations in other habitats lost weight when water similar to sea water was the only fluid available for drinking and were unable to go without drinking.

It is essential to remember, however, that such relationships are not as simple as they may appear. Temperature tolerance is also modified by the individual's physical and nutritional condition, as well as by interactions of temperature with other environmental variables, such as humidity. Furthermore, gene flow among local population segments may prevent fine adjustments of temperature adaptations to the local environment of each population segment. Sometimes the lack of appropriate genetic changes results in lack of correlation between individual tolerances and environmental conditions.

Bergmann's Rule

Endotherms at high latitudes are sometimes larger than their low-latitude relatives; this relationship is known as Bergmann's rule. Although this so-called rule was originally conceived for interspecific comparisons of closely related species, it has been extended to intraspecific (James, 1970) and altitudinal comparisons. The wood rats (*Neotoma*) in the western United States, for example, are almost three times as large in cool environments as in warm ones. To explain this correlation, ecological physiologists noted that large-bodied endotherms have a lower rate of heat loss relative to body weight than small-bodied ones. Large sizes may also permit an extension of the lower limit of the zone of thermal neutrality (when one is present) or lower the extreme limits of cold tolerance.

Large endotherms require more total food intake than small ones, however, so the maintenance costs increase with body size despite the savings in heat loss. If natural selection favors large size for some other reason (perhaps an advantage in competition or reproduction), the decreased surface/volume ratio reduces the energetic disadvantage but does not result in an absolute reduction of energy expenditure. Other explanations of Bergmann's rule would be useful, therefore.

The effects of temperature are not always direct, for humidity, direct insolation, and many other climatic variables interact with temperature. Associations of body size with climatic variables are frequently closer when more than one variable is considered (James, 1970); nevertheless, the physiological advantages of these correlations remain unconfirmed.

Another possible adaptive advantage of large size with respect to cold-temperature physiology is that large animals may have a greater ability than small ones to endure food shortages and fasting. Data on birds of various sizes suggest that survival times without food increase with body size. This effect is apparently greater at low environmental temperatures (−18 to −13°C) than at moderately cool temperatures (+2 to +6°C).

Actually, the commonness of Bergmann's "rule" is open to discussion: It is not a good generalization. McNab's (1971) study of 47 species of North American mammals, for example, showed that body size of many species was independent of latitude. Only 32 percent of the species studied showed a significant positive correlation of body size and latitude, and negative correlations were almost as common. Any explanations of latitudinal trends in body size, therefore, must account for the occurrence of such trends in some taxa but not in others. Metabolic savings potentially apply to all cold-climate endotherms, so additional factors must be invoked.

For mammalian carnivores and seed-eaters, McNab suggested that latitudinal gradients in body size might be accounted for by a decrease in average food size with increasing latitude. Assuming that large predators of a set of similar species generally eat large prey, such a decrease would often result in the omission of the largest members of a set. Moreover, McNab argued, there would be an increase in the size of the remaining smaller members of the set in the absence of their supposed competitors (Fig. 5–11). If the largest members of the set are not absent, this trend is not observed. Grazers and browsers, in contrast, do not seem to fit this hypothesis, perhaps because their food is generally packaged in relatively common but less discrete units than the food of carnivores and granivores. Also, the food of plant-eaters may not decrease in size with increasing latitude. McNab's interesting concept does not explain or document the hypothesized decrease in average prey size with increasing latitude; nor does it explain why the larger members of some sets (*Martes*) drop out at high latitudes and others (*Canis*) do not. In addition, the reality of competition among the members of each set requires documentation.

Another possible ecological basis for occasional latitudinal gradients in body size involves the ability of larger individuals to produce larger litters of young than small individuals of the same species. A correlation of size and reproductive output within a species is common in ectotherms. Litter size of birds and mammals often increases with latitude as well and may be associated with larger body size (in addition to other factors). This hypothesis is not well documented, but it may pertain to some species.

Ectotherms—a number of fishes (including members of the family Cyprinidae) and snakes—sometimes follow Bergmann's rule. The heat conservation argument is certainly less germane for them than for endotherms. In fact, smaller body sizes in environments of low temperature and variable weather may be adaptive to ectotherms by permitting rapid warming. A 108-g alpine lizard (*Liolaemus multiformis*), for instance, was able to raise its body temperature from near freezing to about 30°C above ambient (0°C) during one hour of basking in the sun. (This lizard happens to live in the Andes of Peru, but it demonstrates the possible rates of warming achieved by medium-sized lizards.) In contrast, *Amphibolurus ornatus,* an Australian species, is considerably smaller, weighing 15 to 20 g. This lizard is capable of warming rates as great as 1°C/min in the field. Its body temperature at emergence in the morning averaged about 25°C, but the mean body temperature of an active individual was about 37°C (Bradshaw and Main, 1968). This 12°C increase could be achieved in about 12 min; thus, the basis for ectotherms conforming to Bergmann's rule is likely to be found outside of simple maxims of thermoregulation.

We do not yet know if a single explanation for Bergmann's rule (when it holds true) is to be found. Perhaps many different factors have established a latitudinal trend in body size of different groups. Indeed, it seems to be a rather feeble generalization that offers little explanation.

Allen's Rule

The size of extremities, such as ears and limbs, in endotherms sometimes decreases with increasing latitude. This relationship is known as Allen's rule. An early explanation of this occasional trend suggests that small ears in cold regions reduced heat loss and large ears in warm areas were adaptive as heat radiators. Certainly, the enormous ears of jackrabbits (*Lepus*)

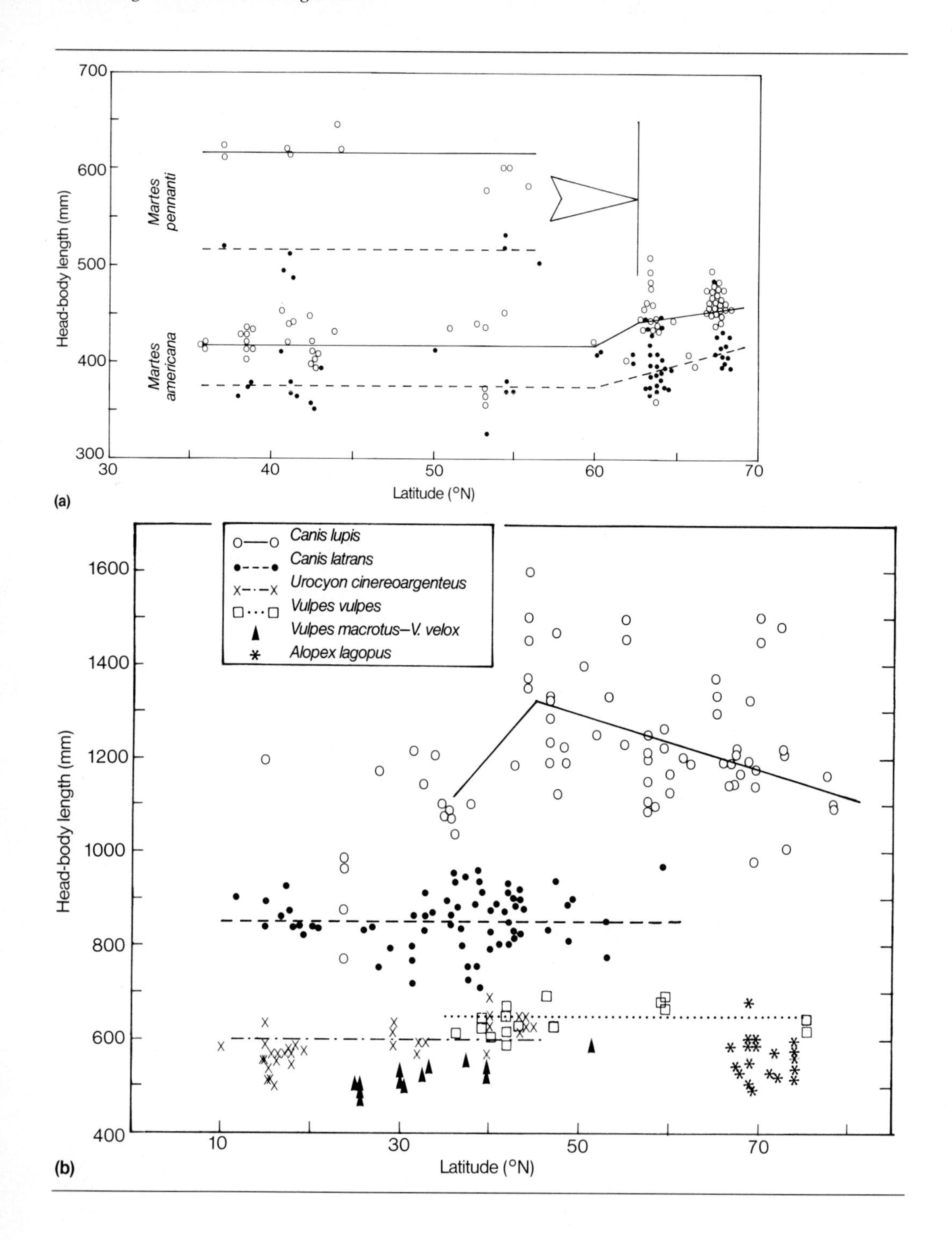
Head-body length (mm)
Latitude (°N)
Martes pennanti
Martes americana
(a)
(b)
Canis lupis
Canis latrans
Urocyon cinereoargenteus
Vulpes vulpes
Vulpes macrotus–V. velox
Alopex lagopus

Figure 5–11 (*a*) Body size of the fisher (*Martes pennanti*) and the marten (*M. americana*) as a function of latitude. The northern limit of the fisher's range in western Canada is about 62°N. At this latitude, the body size of the marten increases. (*b*) For comparison, another set of predatory mammals, of the family Canidae, is shown. The largest predator in this set is the wolf (*Canis lupus*), which does not drop out at high latitudes. The smaller species—coyote and foxes—do not increase in size. (Modified from B.K. McNab, 1971. On the ecological significance of Bergmann's Rule. *Ecol.*, *52*:845–854. Copyright © 1971, the Ecological Society of America.)

(Fig. 5–12) do radiate heat and may well be important in thermoregulation, at least at some temperature ranges (Hill and Veghte, 1976). However, another kind of selective pressure may have contributed to latitudinal gradients in ear size. Sound travels best in cool, moist air and least well in warm, dry air; thus, the large ears of some low-latitude animals may also be explained as an adaptation for improved hearing.

Reduction of relative wing or leg length in cold regions may also conserve heat. Other factors impinge upon lengths of appendages, however. Wing length can be associated with habitat or distance of migration. Relative leg length in birds is related to foraging techniques and sites. One can see that many selective factors may be involved with Allen's rule.

Figure 5–12 The ears of a jackrabbit, *Lepus californicus*.

ADAPTATIONS TO LOW TEMPERATURES

Terrestrial

Many arctic or high-altitude endotherms can endure a lowered temperature in certain parts of the body without apparent discomfort. Reindeer (*Rangifer tarandus*) from Scandinavia maintain an interior body temperature of 38°C at an air temperature of −31°C, but the temperature of the nose declines to 20°C and the feet to 9°C. A countercurrent exchange system has evolved in which heat from arteries carrying blood to the extremities is transferred to blood in veins returning to the body. In this way, peripheral tissues receive a normal oxygen supply, heat is conserved, and energy is not wasted by heating the extremities.

Color may contribute to the thermoregulation at cool, as well as warm, temperatures. In cool weather, the roadrunner orients itself with its back to the sun, drops its wings from their normal resting position over the back, and elevates its dorsal feathers, exposing black skin. By doing this, the roadrunner absorbs much solar radiation, which contributes to body warmth. The dark colors of Arctic animals such as musk oxen (*Ovibos moschatus*) and ravens (*Corvus corax*) seem anomalous, if black colors enhance heat absorption. These animals might overheat in the constant sun of summer, and during the sunless winter, their dark colors provide no thermoregulating advantages. The absorptive properties of dark colors and reflective properties of light colors can be adjusted by changes in posture. Furthermore, the importance of color in thermoregulation diminishes with increasing wind speeds (Lustick et al., 1980). Perhaps Arctic animals can combine such variables to avoid serious thermoregulatory problems.

Some birds and mammals—the gentoo penguin (*Pygoscelis papua*), the pygmy nuthatch (*Sitta pygmaea*), and the tree creeper (*Certhia brachydactyla*), for example—conserve energy by huddling together. A number of vespertilionid bats frequently cluster together when roosting. A tropical phyllostomid, *Phyllostomus discolor*, also conserves heat by forming tight clusters.

Large endothermic vertebrates living in cold environments generally have heavier fur or feathers or thicker fat layers than similar species in warm environments. Small species usually are not especially well insulated, however, probably because heavy insulation would too greatly impede locomotion.

Small arctic birds known as redpolls (*Acanthis*) are capable of increasing their resting metabolic rate over five times the minimum daytime rate. When environmental temperatures are well below 0°C, these birds regularly metabolize at rates two or three times the minimum. The metabolic cost of keeping warm is less for some arctic-adapted species than for those of warmer regions, however (Fig. 5–13). In contrast to the redpolls, the young of some sandpipers (*Calidris*) that breed in the tundra save energy by allowing their body temperatures to drop as low as 30°C, yet they

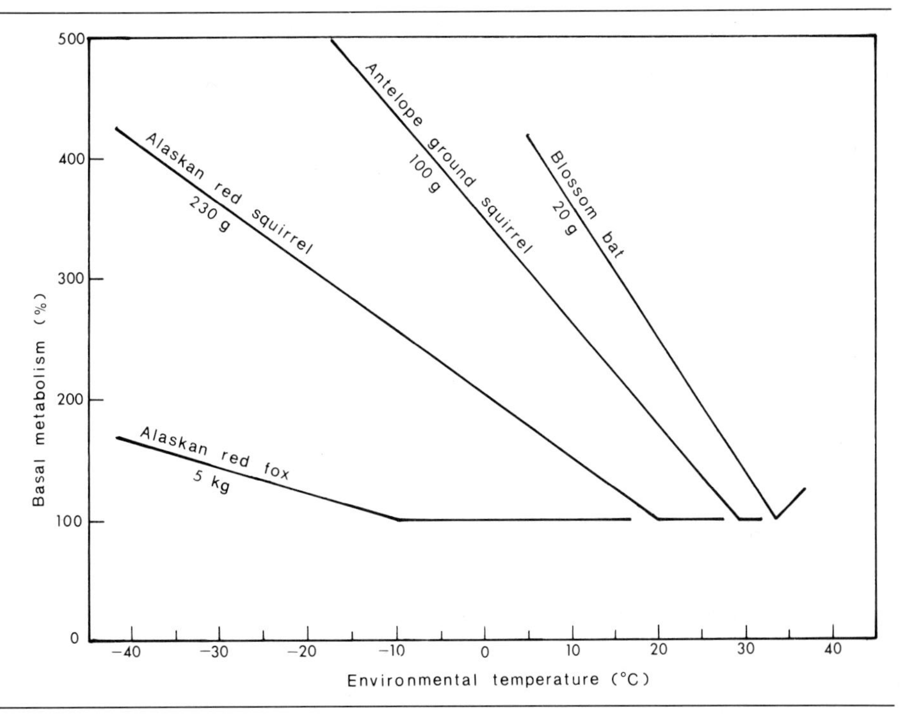

Figure 5–13 Relative changes of metabolic rate versus ambient temperature in some animals adapted to different conditions. (Reprinted with permission of Macmillan Publishing Co., Inc. from *Animal Physiology: Principles and Functions* by Malcolm S. Gordon, et al. Copyright © 1977 by Malcolm S. Gordon.)

somehow remain active and grow very rapidly. Small body size creates problems for thermoregulation in homeothermic animals: The smaller the body, the higher the cost of warming each gram of body weight and the greater the relative surface area for heat loss. This probably places a limit on the minimum size feasible for a cold-climate endotherm.

Some vertebrates escape the most severe cold by dwelling in burrows. All terrestrial ectotherms in cold climates do so regularly. Small mammals such as mice and voles create extensive systems of burrows under an insulating layer of snow. Temperatures beneath a layer of snow can be significantly warmer than those on the surface. A blanket of snow that is 60-cm deep can make the difference between a temperature of −46°C in the open air and −9°C at the surface of the soil. Ptarmigan (*Lagopus*) and redpolls habitually burrow in the snow at night. A good snow cover is apparently important to the winter survival of red-backed voles (*Clethrionomys*) and deer mice (*Peromyscus*) of boreal, subarctic regions. Snow cover is less protective in high-arctic regions, for strong winds blow and pack the snow layers.

A number of vertebrate species migrate seasonally, thereby escaping severely low temperatures. Some travel very long distances, almost from pole to pole; others migrate shorter distances, altitudinally, or from tundra to tree line.

Aquatic

An increased ability to maintain high activity despite low body temperature seems to be a primary adaptation to endure cold in many ectothermic species. Deep-water benthic fishes (such as flounder, *Pseudopleuronectes*) in polar and subpolar regions, for instance, live at below-freezing temperatures throughout the year; the pressure of the water at such depths allows them to remain functional in a supercooled state. In contrast, shallow-water or surface species (such as the icefishes, *Trematomus*) in the same regions may come in contact with ice crystals that would induce lethal ice formation in the body. Ice formation is prevented, however, by the presence in the blood of unusually high concentrations of glycoprotein or peptide molecules that serve as antifreeze by impeding the growth of internal ice crystals (DeVries, 1982).

Aquatic endotherms are especially threatened by potential heat loss because water conducts heat more rapidly than air. Diving king penguins (*Aptenodytes longirostris*) increase their metabolic rate over two times the standard rate (Kooyman et al., 1982). Aquatic mammals of frigid arctic and antarctic waters may have a somewhat higher metabolic rate than would be predicted from their body size alone, but this does not seem to be true for the harp seal (*Phoca groenlandica*). Nevertheless, body temperatures are similar to those of other mammals. A higher metabolic rate tends to compensate for heat loss, but an effective insulation seems to be the primary means of coping with life in icy waters. Thick deposits of subcutaneous fat permit skin temperatures to approach the ambient temperature as interior temperatures remain at normal levels. The blubber layer does not extend to the flukes and flippers, which are well vascularized and potential sites of substantial heat loss. There is a countercurrent heat exchange system, however: Arteries to the appendages are surrounded by veins, and arterial heat is transferred to venous blood, preventing excessive heat loss. Certain aquatic vertebrates are insulated to some degree by air bubbles trapped in their fur or feathers (for example, muskrats [*Ondatra*], river otters [*Lutra*], fur seals [*Callorhinus*], and loons [*Gavia*]. The fur of polar bears (*Thalarctos*) and beavers (*Castor*), however, reportedly loses most of its insulative value when submerged.

Some cold-water vertebrates occasionally swim in warmer waters or emerge on land, where their adaptations for heat conservation may become a distinct disadvantage. Seals solve the problem of overheating in two ways. At times of heat stress, the flow of blood through the blubber layer to the skin is increased and thus permits greater heat dissipation.

Also, the heat exchange apparatus is reportedly arranged so that an increased blood flow to the appendages causes the arteries to swell, collapsing the surrounding veins. Venous blood then returns to the central part of the body through alternate veins located nearer the surface, bypasses the heat exchanger, and returns to the body at a relatively cool temperature.

ADAPTATIONS TO HIGH TEMPERATURES

Terrestrial

The problems of temperature extremes and desiccation presented by life out of water are epitomized by a warm desert environment. Just as daily temperature fluctuations are greater on land than in water, those in the desert are often greater than in other terrestrial habitats. Nights can be bitterly cold, although days may be very warm. Technically, desert areas are characterized by low rainfall and high evaporation rates and may be either hot or cold, but most people are familiar with warm deserts, so they are discussed in this section.

Most adaptations to a desert existence are features that are already necessary for terrestrial living but that are present in a highly developed form. The inhabitants of warm deserts face not only the problem of great temperature fluctuations but also the related challenges of high temperature and aridity. Furthermore, reflection of heat from surface irregularities on the desert floor can produce great variations in temperature within short distances and create microsites of extreme heat. The amount of external heat encountered by diurnal desert animals is potentially much greater than that encountered by inhabitants of wetter areas, where dark soil, moisture, and vegetation reduce incident solar radiation and absorb much of the radiation that does penetrate the atmosphere. Heat stress in deserts can be so great that tiny hatchling side-blotched lizards (*Uta stansburiana*) may die from heat if forced to remain in the full sun of midsummer even for the length of time needed to move 1 m.

Aridity results from low precipitation and high rates of evaporation. Standing water is seldom encountered in deserts in most seasons, so dew condensed on surfaces near the ground becomes a major source of water to many organisms. The known range of the sand-diving lizard (*Aporosaura anchietae*) in the Namib Desert of southwest Africa is, in fact, limited to the fog belt near the coast. Relative humidities in warm deserts tend to be low. (Relative humidity is the actual amount of water in the air compared [in ratio form] to the amount the air could hold, if saturated, at the same temperature. Warm air can hold more water vapor than cold air.) Heat dissipation by means of evaporation is therefore less difficult in arid environments because there is relatively little water vapor in the air already. This form of heat loss necessitates water loss, which in turn creates a serious difficulty since, by definition, deserts do not have much available water. Therefore, many adaptations to desert life involve obtaining and conserving water as well as dealing even more directly with temperature-related problems.

COPING WITH HIGH TEMPERATURES Some desert representatives of all four terrestrial or semiterrestrial classes of vertebrates escape excessive heat by retreating to burrows during the hottest, driest times. The air in underground dwellings is both cooler and moister than that on the surface. The moisture of respired air is partially retained within the confines of the burrow, and protection from sun and wind reduces water loss and incoming heat from solar radiation. If the surface temperature is 65°C, the temperature within a burrow 45 cm below the surface may be as cool as 27°C. Even a shallow burrow only 10 cm below ground level may provide temperatures as much as 17°C cooler than those at the surface.

Nonbreeding anuran adults are often dormant, hidden in burrows until the rainy season. *Cyclorana* frogs in Australia burrow more than a meter underground, where water from only the heaviest rains can seep. North American desert spadefoot toads (*Scaphiopus*) may remain buried for as long as 3 years if adequate rains do not fall. The North American gopher tortoise (*Gopherus agassizi*) usually avoids temperature extremes by hiding in holes. An Australian gecko, *Rhynchoedura ornata*, hides in spider burrows during the day. Most diurnal lizards and snakes avoid extreme heat by emerging for activity mainly at intermediate temperatures—that is, midday in cooler months and morning and evening in summer. A few species, such as night lizards (*Xantusia*) in North America and *Heteronota binoei* in Australia, have escaped the problem by becoming nocturnal. Such species often are intolerant of high temperatures. The Australian species has a preferred temperature of about 30 to 34°C, which is several degrees below those of more diurnal species (Main, 1976). On the other hand, certain squamates, including the desert iguana, are unusually tolerant of high temperatures and can remain active later in the day than most other lizards of the same region.

Most desert birds move into the shade and become less active during the hottest part of a summer day. Some, such as the North American rock wren (*Salpinctes obsoletus*), retreat to crevices and nooks in rock. The elf owl (*Micrathene whitneyi*) retires to a cavity in a tree or saguaro cactus.

Small mammals are commonly nocturnal and rest in burrows during the day. Some of the smaller mammals are tolerant of increases in body temperature and are diurnal (the white-tailed antelope ground squirrel [*Ammospermophilus leucurus*] of North American deserts, for instance). This ground squirrel will frequently go into its burrows to cool off during the day and may avoid activity at the very highest temperatures.

The round-tailed ground squirrel (*Spermophilus tereticaudus*) and the Mohave ground squirrel (*S. mohavensis*) in warm deserts of western America also are diurnal. In contrast to the antelope ground squirrel, both have a lower basal metabolic rate than would be predicted on the basis of body size; therefore, they have reduced heat production. Many desert mammals (kangaroos, other marsupials, and ground squirrels, for example) and birds have a lower metabolic rate than expected for their body size. This feature may be adaptive in lowering the water requirement for thermoregulation: Less heat is produced, so there is less to be dissipated (Brown and Main, 1967; Dawson, 1972; Trost, 1972). This suggestion seems contradictory to the theoretical considerations of Calder and King, which were discussed earlier. A real contradiction may not exist, however. Calder and King dealt with the origin of high body temperature and high metabolic rates as a means of conserving water. The first homeotherms presumably were not faced with the extremes of high temperature and aridity that characterize warm deserts. Nonevaporative heat loss necessitates maintenance of a body temperature that is greater than the ambient temperature, but at very high ambient temperatures, this means an impossibly high body temperature. The alternative seems to be a reduction of metabolic rate and heat production, less dependence on heat loss via a temperature gradient from body to environment, a tolerance of slight rises in body temperature, and shade-seeking behavior.

REGULATING HEAT GAIN AND LOSS Diurnal squamates achieve a considerable degree of thermoregulation by moving in and out of shade, orienting to the sun's rays, expanding or contracting the body to change the surface area exposed to thermal radiation, and lifting the appendages off the hot substrate. Desert reptiles make frequent use of these tactics, but even nondesert species use them extensively. For lizards of semiopen habitats, behavioral thermoregulation may be relatively easy. The availability of sun and shade patches permits quick access to differing ambient temperatures. Species living in the shade of warm forests, however, may find behavioral thermoregulation very costly if suitable sunny patches are few and far between. One solution to this difficulty in forests is the evolution of wide temperature tolerances for activity. *Anolis polylepis* in the forests of Costa Rica may have wider temperature tolerances for activity than nonforest species of the genus.

Snakes and lizards in the desert have two temperature problems: They must lose excess heat gained from exposure to the sun's rays; however, in the morning and in cool weather they must acquire enough heat to be active. Species that use the sun to increase body temperature are called heliotherms. Color is probably a major regulator of temperature in some lizards and functions both in the increase of the warming rate and in the prevention of overheating. These lizards are dark when they first emerge from their burrows; therefore, they are able to absorb solar radiation and warm up quickly. They become pale and less absorptive as they become warm enough to be active. On cool days, lizards may stay dark all day.

To some extent, the effectiveness of color change in regulating heat flux varies with body size. Moreover, not all families utilize color change to the same degree. Color change is well developed in small iguanid and agamid lizards but less well developed in teids, lacertids, and skinks (these lizards are primarily thigmothermic; that is, they adjust temperature by contact with the substrate). The thigmothermic lizards are generally more mobile hunters than the small iguanids and agamids, which often use sit-and-wait tactics for getting food. In large iguanids such as the chuckwalla of the North American warm deserts, the young show more color changeability than the adults. The body size of the adult is so large (sometimes over 120 g) that color changes are relatively ineffective in adjusting heat transfer (Fig. 5–14). However, even some small iguanids (*Uma,* for example) have adopted a different tactic. Species of *Uma* live in western North America on barren sand dunes of relatively uniform color. Color control of heat exchange in *Uma* has apparently been lost in response to selective pressure to match the background for protective coloration. Heat control, therefore, is achieved primarily by contact with the substrate. Furthermore, individuals of *Uta stansburiana* that live on lava flows are often permanently dark and have little color-changing ability. Once again, selection for protection from predators has possibly outweighed selection for color changes in thermoregulation.

Despite the apparent advantages of white or light colors in reducing heat gain, white is not a common color in desert animals. Individual lizards

Figure 5–14 The chuckwalla. In the adult, color changes have little effect on heat transfer. (Photo by S. Prchak.)

in North American deserts may be active at both warmer and cooler times of day and many can change their colors. In areas inhabited by greater numbers of species, however, there may be greater temporal segregation of the species; some species will be active at different times than others, thus maintaining both light and dark types. Lizard species active at high (but sublethal) temperatures are often pale, but species active only in the morning and evening are likely to be dark, which aids in heat gain.

In mammals, the methods of adapting to the desert vary with body size and diet. Carnivorous mammals such as bobcats (*Lynx rufus*) often live in many habitats and, in some respects, are preadapted for life in the desert. Animal prey contains relatively more water than the food of herbivores, and most mammalian carnivores are able to travel fairly long distances in search of food and water; therefore, they can afford to expend water in thermoregulation. Large herbivores such as camels or deer cannot escape the heat as well as small ones, since they are not adapted to burrowing, but they have some advantages due to their large size. Because the body mass is relatively great and the surface/volume ratio is small, body temperature increases slowly. The large mass of these herbivores permits heat storage without much rise in body temperature. In addition, large desert herbivores are quite mobile and can travel to seek shade and water. Access to water means that they can afford evaporative heat loss better than small forms.

Comparative study of several African antelope exemplifies both similarities and differences of thermal adaptations adopted by related animals living in desert or near-desert habitats (Taylor, 1970, 1972) (Fig. 5–15). Body temperature of the large eland (*Taurotragus oryx*) rises during the day, but because the animal is large (weighing up to 1000 kg), the rise is usually small. Under conditions of dehydration, smaller antelopes, such as Grant's gazelle (*Gazella granti,* which weighs up to 50 kg) and the oryx (*Oryx beisa,* which weighs up to 300 kg) allow their body temperatures to rise sharply—as high as 42°C. In so doing, they increase the thermal gradient from body to environment (or at least reduce the gradient from environment to body) and, hence, reduce evaporation. Eland, oryx, and gazelles have metabolic rates that are somewhat higher than might be predicted from their body weights. When the animals are hydrated, water loss is correspondingly high, but when they are in a dehydrated condition, metabolic rate is decreased in proportion to the extent of dehydration. All three antelope can survive without drinking. Eland seek shade at midday and eat large quantities of *Acacia* leaves that contain sufficient water to meet their normal needs. Oryx and gazelles are not shade-seekers and eat (at some seasons, anyway) mainly dry grass. How do they obtain sufficient water? Dry grass can absorb enough moisture from the night air to increase its moisture content as much as fortyfold. By foraging at night (as they apparently often do), these two species could probably meet their water requirements.

Figure 5–15 (*a*) Eland, (*b*) oryx, and (*c*) Grant's gazelle. All three can survive without drinking.

Heat dissipation can be facilitated by circulatory changes in exposed regions of the body. Dilation of blood vessels in the unfeathered legs of birds or the ears of mammals, for example, permits more heat to reach the surface, where it can be lost.

Insulation can be an important regulator of the rate of heat gain and loss in mammals and birds. The fur of large desert mammals is often quite reflective and thicker dorsally than ventrally. This characteristic helps to reduce heat gain from above while permitting heat loss from the shaded venter. The fur of camels (*Camelus dromedarius*) is effective in reducing heat gain: A camel experimentally shaved so that it was no longer insulated against incoming radiation used 50 percent more water in evaporative heat loss than an intact camel exposed to the same conditions (Schmidt-Nielsen et al., 1957). In several species (camels and desert sheep, for example), fat deposits are localized in a dorsal hump or in the tail, permitting the rest of the body to radiate excess heat more rapidly than if the fat were evenly distributed and acting as general insulation. Birds and mammals may flatten their plumage or fur to increase the rate of heat loss or ruffle their feathers or fur to reduce the rate of heat gain.

EVAPORATIVE WATER LOSS Contrary to what is sometimes thought, the skin of squamates is not impervious to water. Many species can both absorb and lose water through this organ. Cutaneous water loss may account for as much as 75 percent of the total evaporation of desert reptiles. Nevertheless, one adaptation of desert reptiles to their arid environment is the reduction of evaporative water loss. At 23°C reptiles from aquatic and moist habitats can lose water to the surrounding air more than 10 times faster via evaporation than can desert reptiles. Specifically, a chuckwalla may evaporate about 3 μg of water per square centimeter of body surface per hour, whereas a water snake (*Nerodia*) or a caiman (*Caiman*) may lose water at a rate of 41 to 65 μg at the same relative humidity.

Birds have no sweat glands, and evaporation from respiratory surfaces is one principal means of heat loss. Cutaneous water loss from the permeable skin may be even higher than from the respiratory surfaces. Some birds have supplementary means of increasing evaporative heat loss in hot weather. The wood stork (*Mycteria americana*), and other storks as well, directs a fluid urine onto its bare legs, thus increasing evaporation from these surfaces. These are not birds of the deserts and probably can afford to expend water for this form of heat control.

Large mammals typically have sweat glands, but small placentals and marsupials often lack them entirely. Evaporative heat loss is very costly for small mammals: Theoretically, a 50-g kangaroo rat (*Dipodomys*) might lose as much as 13 percent of its body weight per hour if the animal were exposed to the desert sun (Fig. 5–16). Sweat glands are plainly disadvantageous to small mammals because of the great rate of water loss. A number of mammals (including kangaroo rats), and some birds, further reduce evaporative water loss by a marked cooling of exhaled air as it passes out through unusually constricted nasal passages. Cooled air holds less water vapor than warm air, so water condenses on the internal surfaces of the nose, thus diminishing the amount of water lost by this route. Ground squirrels (*Spermophilus*) and large kangaroos spread saliva over the fur under some conditions of thermal stress, and squirrel monkeys (*Saimiri*), although they are forest animals rather than desert dwellers, may utilize urine in this way. This means of heat loss is rather inefficient, however.

Water may be evaporated by panting or, more efficiently, by fluttering the gular region (the floor of the mouth and the upper end of the esophagus). Gular flutter, found in many unrelated species of birds (pelicans, quail, owls) and in some lizards (such as *Dipsosaurus dorsalis*), is a very economical means of heat transfer. In contrast to panting, in which thoracic and abdominal muscles are used, gular flutter involves only the muscles of the throat region. Pigeons and doves also have an inflatable esophagus with a vascular network that serves as an important heat exchanger.

A potential disadvantage of fluttering and especially panting is that it requires muscular effort, producing heat. In fact, rapid panting reportedly could produce more heat than is lost by evaporation. This difficulty is minimized by using the elasticity of the respiratory system. Panting and gular flutter occur at the resonant frequency (the natural frequency of oscillation of the elastic tissues) of the respiratory and oral chambers; thus,

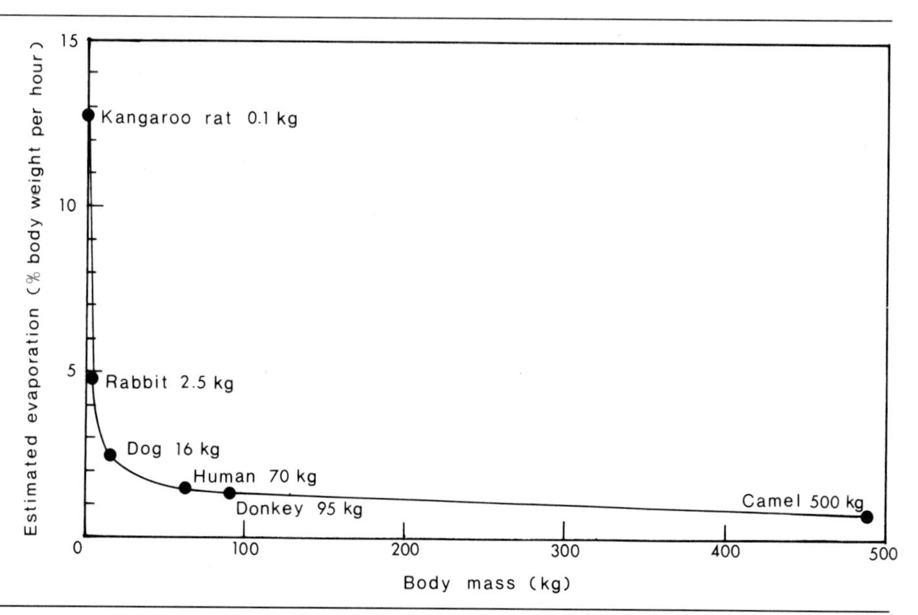

Figure 5–16 Relative amount of evaporative water loss required to maintain constant body temperature, calculated as a function of body size and surface area. (Modified from K. Schmidt-Nielsen, 1964. *Desert Animals.* Oxford University Press, Oxford, England.)

they require minimal energy and produce minimal heat.

Panting and gular flutter have at least three potential advantages over sweating as a means of evaporative heat loss. The air of the respiratory passages is moving anyway, facilitating evaporation, but sweat may or may not be exposed to wind. Furthermore, although water loss from the skin would be impeded by a fur covering (in mammals), an insulating coat may be of value in retarding heat gain. Evaporative heat loss by panting and fluttering avoids these contradictory functions (Bligh, 1972). Moreover, salt is not usually lost from the body in panting or gular fluttering, so that ionic balance is not disturbed.

A comparative study of two large kangaroos in Australia illustrates the value of panting and interspecific differences in its functioning. The red kangaroo (*Megaleia rufa*) is widely distributed in dry, open plains and seeks the feeble shade of small trees on hot summer days. The euro (*Macropus robustus*) lives in rocky hills and retreats to caves and crevices in the heat of the day. The fur of the red kangaroo is more reflective and insulative than that of the euro, but the euro's refuge protects against almost all incoming radiant heat. Body temperatures of both species at thermoneutrality are about 35.5°C, but at very high ambient temperatures, they sometimes rise slightly to 37.7°C. For both species, the main route of evaporative heat loss at high ambient temperatures is panting. Panting is so effective that at 45°C all metabolic heat and an almost equal amount of environmental heat could be dissipated by this means. However, the panting rate (and presumably the metabolic cost) for the red kangaroo was considerably lower (roughly 70 percent at 40°C) than the rate for the euro. Perhaps a larger proportion of the air in the red kangaroo's lungs is exchanged at each breath, thus permitting greater heat (and water) loss at each exhalation (Dawson, 1972).

An additional advantage of panting may be that blood to the brain can be selectively cooled in animals subjected to sudden heat stress. The brain is very sensitive to heat, but in some mammals and birds, it is protected from temperature increases by an ingenious arrangement of blood vessels in the head (Fig. 5–17). The main artery to the brain divides, at the base of the skull, into many small arteries, which merge again before continuing to the brain. The network of little arteries is surrounded by a large venous sinus filled with blood that has just been cooled in the nasal passage (or in the well-vascularized horns of species such as sheep, *Ovis*, and goats, *Capra*). The arterial blood is cooled in this way and brain temperatures may remain several degrees cooler than body temperature. A running Thompson's gazelle (*Gazella thompsonii*) from eastern Africa may experience an increase in body temperature of 5°C but a brain temperature increase of only 2°C. This structural arrangement is reported from other bovid ungulates and pigs (species that frequently depend on running to escape their

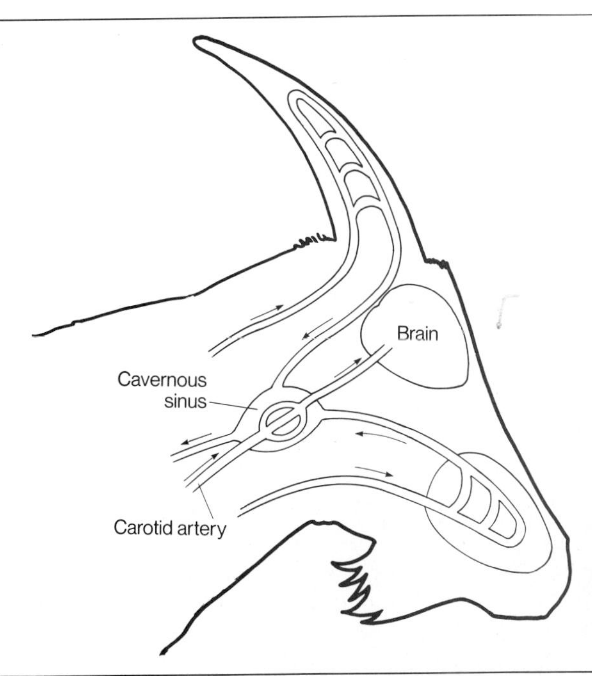

Figure 5–17 Diagram of a goat's cranial sinus and arterial network that cool the blood going to the brain. (Modified from R. McN. Alexander, 1975. *The Chordates*. Cambridge University Press, Cambridge, England.)

predators), cats, and various species of birds. Similarly, in the chuckwalla, selective cooling of blood to the head is achieved by the passage of cranial arteries that are immediately next to the pharyngeal surface, which is cooled by evaporation in panting. The effectiveness of this arrangement is demonstrated in Figure 5–18.

ACQUISITION OF WATER Whatever tactics desert vertebrates have evolved to deal with high temperatures, one problem is common to all of them: obtaining water. Adult amphibians of arid regions can capture water quickly when it becomes available. Their principle sources of water intake are puddles or damp ground, and water is taken up through the thin ventral skin. Among *Neobatrachus* frogs in Australia, the rate of water absorption is fastest among species from the driest areas. The same species also have the lowest rate of evaporative water loss, although this characteristic may be a function of large size rather than metabolic physiology. Some anurans may also store water under the skin or in the peritoneal cavity. Desert anurans

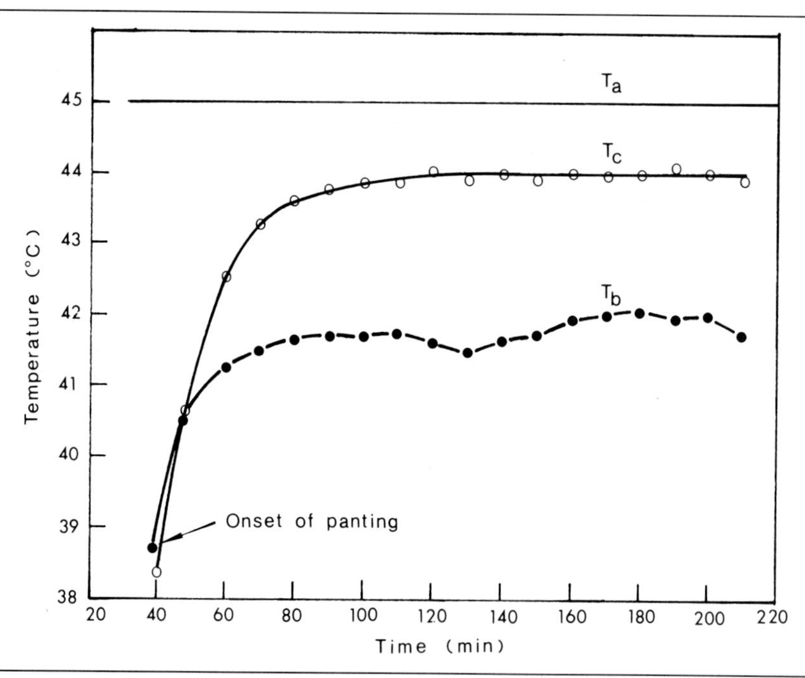

Figure 5–18 When a chuckwalla is moved from an ambient temperature of 15°C to 45°C, brain temperature rises with the temperature of the rest of the body (here measured in the cloaca). At about 42°C, however, brain temperature increases no more: It remains about 2°C below body temperature and about 3°C below ambient. (From E.C. Crawford, 1972. Brain and body temperatures in a panting lizard. *Science, 177*:431–433. Copyright © 1972 by the American Association for the Advancement of Science.)

such as *Neobatrachus* and *Cyclorana* (and *Phyllomedusa nebulosa* of South America) store water in the bladder and, thus, may increase their body weight as much as 50 percent. The stored water is returned to the circulatory system as needed to maintain the normal concentrations of solutes and cells in the blood and lymph. This living storage tank is exploited by the Australian aborigines as a mobile supply of drinking water.

Reptiles need standing water (at least dewdrops) for drinking in addition to water produced from metabolism of food. One Australian lizard, *Moloch horridus*, has a very specialized means of obtaining water. Special capillary channels on the body surface pick up dew or rain and carry it over the skin to the mouth, where it is taken up by mucus and swallowed. If only salt water is available, some lizards (as well as sea snakes and sea turtles) will drink the salty water and void excess salt by nasal secretion. Gopher tortoises can store water and also tolerate water loss; they may also dig basins to collect rainwater for drinking.

The Australian *Amphibolurus ornatus*, which seasonally feeds mainly on ants rich in salt, retains the excess salt, and voids it when the occasional summer storms provide enough water. The extra salt aids water retention in the body between storms. Individuals with the capacity for salt retention grow more slowly but survive drought more successfully than faster-growing individuals that have lesser capacity for salt retention. The fast growers lose weight (mostly water) and, at 20 percent weight loss, switch their diet to a different kind of ant. At 30 percent weight loss, they stop eating altogether and often die. Although the fast growers fail to retain sufficient body water during a drought, they are more resistant to cold in winter. Apparently this advantage is sufficient to maintain the genotype in the population (Bradshaw, 1970, 1971).

Most desert birds need to drink in addition to obtaining metabolic water from oxidation of food. However, some bird species from dry regions around the world have evolved the capacity to survive at a constant weight without drinking. All their water comes from metabolic water and free water in their food. Most of these birds are small passerines. They are usually larks or finchlike birds, such as the horned lark (*Eremophila alpestris*) and black-throated sparrow (*Amphispiza bilineata*) of North America, and members of the genera *Amadina, Lonchura, Fringillaria,* and *Spizocorys,* among others, of the Old World. One small nonpasserine, the budgerigar (*Melopsittacus undulatus*) from Australia, is also known to have this ability. Some desert (and marine) birds, including roadrunners, can drink salt water and excrete the excess salt through their nasal glands.

Desert birds commonly provide water to their nestlings in insect food and by regurgitation. Sand grouse (*Pterocles*), which live in the deserts of Africa, bring water to their precocial, seed-eating offspring in their feathers (Cade and MacLean, 1967). Sand grouse (Fig. 5–19) fly many kilometers from the nesting area to reach water holes. The male rumples his breast feathers against the ground and water then is easily taken up when he

Figure 5–19 A male *Pterocles* sand grouse. Its modified belly feathers enable it to transport water to its young.

wades into the shallows to drink. The abdominal feathers of *Pterocles* males are especially modified for increased capacity to carry water. The water-holding capacity of a male is 15 to 20 mg of water per milligram of dry feather; that of a female is 11 to 13 mg water. Compare this to a capacity of only 5 or 6 mg of water per milligram of dry feather in other birds. The total carrying capacity of an adult male has been estimated at 25 to 40 ml of water, 10 to 18 ml of which remain after evaporation on the long flight from water hole to nest area. When the male returns to the nest, the young draw the feathers through their bills to obtain water. They apparently do not drink in the usual avian fashion, which is by filling the mouth with standing water and tipping the head to let it run down the throat. Why only the males have such a highly developed water-carrying capacity is still a matter for speculation.

Desert mammals obtain water in several ways. As mentioned earlier, carnivores and some herbivores can often obtain sufficient water from their food and are able to travel to water holes. Kangaroo rats and many pocket mice (family Heteromyidae) and the Australian murid hopping mice (*Notomys*) can exist solely on metabolic water obtained from seeds, so long as the air is not completely dry. One cricetid rodent, *Peromyscus crinitus*, can do so too, at least in the laboratory. In contrast, camels require large amounts of water, although they can tolerate long periods (perhaps a week) of dehydration and of concomitant increases in body temperature. They have the ability to replace lost water in the tissues very quickly and can tolerate the resulting dilution of tissue fluids; in less than 10 min, a camel can take in over 100 L of water, which is more than one third of its body weight.

CONSERVING WATER Water can be conserved not only by reduction of evaporative water loss but also by adaptations in the excretory system. Nitrogenous wastes from the digestion of protein may be excreted as ammonia, urea, or uric acid and its salts. Uric acid can be stored and excreted in crystalline form; thus, it requires much less water for its elimination than does urea. A mammal would typically require about 60 ml of water to excrete 1 g of urea, but a bird needs only about 2 ml to void the same amount of uric acid. Even the efficient kidneys of desert rodents cannot concentrate urea as much as a bird can concentrate uric acid. The urine of most terrestrial adult amphibians, reptiles, and birds has a high concentration of uric acid and is often excreted as a semisolid or a solid. Terrestrial reptiles tend to excrete more uric acid than aquatic species, which often void nitrogenous wastes as urea and ammonia. The proportion of uric acid in the urine of the gopher tortoise increases with the animal's degree of dehydration. Some amphibians also shift their excretory products according to the availability of water. Many species of birds also reduce excretory water by water resorption in the kidney when drinking water is scarce.

Adult mammals excrete nitrogenous wastes as urea, which usually requires a good supply of water for its elimination. Why mammals do not excrete uric acid, in view of its apparent advantages, is not entirely clear. Packard (1966) suggested that perhaps the early mammals were not as water-limited as reptiles; therefore, the water was more available for excretion. Furthermore, the embryos of egg-laying vertebrates could not excrete ammonia (which is toxic) or urea (which would upset the osmotic balance within the egg); thus, uric acid, which is insoluble, solved the problem of waste disposal. Embryos of mammals can rely on their mother's circulation, via the placenta, to remove wastes, which are then excreted by the mother's kidney (Bligh et al., 1976). In any event, some desert mammals are able to excrete urea without losing much water (Fig. 5–20). The kidneys of many desert mammals such as *Dipodomys*, the African jerboas (*Jaculus jaculus*), and the hopping mice of Australia are famous for their ability to concentrate nitrogenous wastes (Table 5–1). A high degree of water resorption by the kidneys permits excretion of a concentrated urine, often in very tiny quantities. Water is also resorbed by the large intestine so that the feces, too, are very dry. The Saharan sandrat (*Psammomys obesus*) eats vegetation that contains a great deal of salt and has the most efficient kidneys known. Much less is known about other small mammals that also survive successfully in the desert. Perhaps a combination of daily torpor and the modest effectiveness of their kidneys in conserving water have facilitated survival in these species (MacMillen, 1972).

Ruminant ungulates and at least some kangaroos, such as the euro, have the ability to recycle urea formed by the digestion of nitrogenous foods. Urea, with saliva, reenters the digestive tract directly through the stomach wall and can be digested (Brown and Main, 1967). This feature increases the efficiency of nitrogen extraction, which is probably of particular importance to animals whose food items contain little protein. It also results in unusually low urea concentrations in the urine, which surely contribute to water conservation in these mammals.

Figure 5–20 (*a*) A kangaroo rat (*Dipodomys*) and (*b*) an Australian hopping mouse (*Notomys*). Both desert mammals can excrete urea without losing much water. Notice also the similarity in body conformation.

Aquatic

Penguins are typically considered to be well adapted to the cold waters in which they feed; many species breed in cold habitats at high southern latitudes. Some species nest at relatively low latitudes, however. One example is the jackass penguin, *Spheniscus demersus,* which breeds on subtropical islands off the west coast of Africa (Fig. 5–21). The water currents in this area are cold and rich in food for the penguins, but the birds nest on land, where ambient temperature may reach 40°C and the sun's rays warm all exposed objects. These penguins, when on land, are faced with a problem of high temperatures. One of their solutions to this problem is the use of burrows for nests; another is that many adults spend the day at sea. Those that remain on land either sit in the burrows or, if they are outside, expose their well-insulated backs to the sun, allowing convection and radiation of heat from the shaded belly, wings, and feet. Furthermore, exposed individuals often allow their body temperatures to rise slightly, from about 38.7 to about 40°C. Like the pinnipeds discussed earlier, these fundamentally aquatic penguins have adaptations that permit them at least the limited use of warm terrestrial habitats.

Some kinds of fishes are very well adapted to warm springs or seasonal pools. Many of these species are highly resistant to temperature shock from variable temperatures; some can tolerate changes as great as 30°C. High activity levels, however, may render certain individuals particularly susceptible to heat death. In a warm pool in Coahuila, Mexico, pupfish

Table 5–1 Comparative urine concentrations of selected mammals.

Species	*Maximum urea concentration (mmol/L)*	*Maximum electrolyte (salt) concentration (meq/L)*
Sciuridae		
Ammospermophilus leucurus (North America)	2860	1200
Heteromyidae		
Dipodomys merriami (North America)	3840	1200
D. spectabilis (North America)	2710	—
Cricetidae		
Neotoma albigula (North America)	1980	714
Gerbillus gerbillus (Egypt)	3410	1600
Psammomys obesus (Africa)	2850	1920
Dipodidae		
Jaculus jaculus (northern Africa)	4320	1530
Muridae		
Rattus norvegicus (albino) (laboratory)	2410	760
Notomys alexis (Australia)	5430	675
N. cervinus (Australia)	3430	1085
Leggadina hermansburgensis (Australia)	3920	—
Hominidae		
Homo sapiens (cosmopolitan)	792	460
Felidae		
Felis domesticus (almost cosmopolitan)	2330	600

Most of these mammals are desert species, but humans, the domestic cat, and the laboratory rat are shown for comparison.
(Data from MacMillen and Lee, 1967, 1969, and Schmidt-Nielsen, 1964.)

males that were actively defending territories were the only individuals that died. Less active individuals survived exposure to the same environmental temperatures.

Bodies of water in very hot climates often have high salt and low oxygen concentrations. These factors create additional problems for their inhabitants.

OSMOTIC BALANCE

All animals require moderate amounts of water. Salts and other solutes in body fluids are important regulators of water content and, in turn, may be partially regulated by their solvent. This mutual regulation is accomplished by the process of osmosis, which is the differential diffusion of molecules

Figure 5–21 A jackass penguin (*Spheniscus demersus*). It is adapted for warm land as well as for cold water.

through a membrane (a cell membrane in this case) that is permeable only to some of the molecules. Specifically, osmosis is defined as the movement of water between two solutions containing different solute concentrations and separated by a membrane permeable only to water. Few biological membranes actually have such restricted permeability, and some solutes tend to move across them, but the principles are nevertheless the same. Basically, water tends to diffuse across the membrane until it achieves equal concentrations on both sides, and no net movement of water molecules occurs thereafter. If the solutes are the same on both sides of the membrane, then they too become equally concentrated on both sides. If a certain kind of solute molecule is found on only one side of the membrane and it cannot pass through, then water continues to diffuse toward the side containing these molecules.

Because living organisms always contain some molecules not found outside their bodies (at least not in the same concentrations) the movement of water and solutes is clearly important to the balance of the chemical conditions in the body. Those chemical conditions provide the proper environment for biochemical reactions of all sorts. Furthermore, osmotic regulation can control the amount of water that enters the body, and, thus, control changes in body or cell volume that could disrupt physiological processes. Solute movement can be biologically controlled not only by water movement and passive transport of solutes but also by active transport of ions or molecules by membranes. An equilibrated state in which no net movement of molecules occurs is isotonic. Many animals are capable of high degrees of osmotic regulation. Interestingly, the molecules used in osmoregulation by different organisms are not very diverse, perhaps because only a few types are both economical to build and compatible with other cell functions (Yancey et al., 1982).

Aquatic environments inhabited by vertebrates vary enormously in their relative concentrations of water and solutes. Sea water usually contains about 3½ percent salt (35 g of salt per 1000 ml of sea water), mostly sodium and chloride ions, and seldom varies greatly in its chemical composition. Inland water, on the other hand, varies widely according to the kind of rock with which it has been associated. Not only does the amount of salt vary greatly, but the proportion of various ions also differs. (The Great Salt Lake of Utah and the Dead Sea in Israel are both saturated with salts, but sodium chloride predominates in the former and magnesium chloride in the latter. Neither body of water supports any vertebrate life.) In ordinary fresh water, ionic concentrations can vary by a factor of 30 times or even more; for example, very "hard" fresh water may contain more than 30 times the sodium ions, 20 times the calcium ions, and more than 400 times the chloride ions of "soft" fresh water. Between fresh water and sea water lies the brackish water of estuaries, where rivers enter the seas. The salinity (salt concentration) of estuaries lies between the salinities of fresh water and salt water (approximately between 0.5 and 3.0 percent); it often varies seasonally with tides and the outflow from rivers. Of course, terrestrial environments provide an almost waterless medium in which to live. In these areas, both water and solute loss can create potentially serious problems, some of which have already been discussed.

Vertebrates have radiated into most of these environments; therefore, they have encountered widely differing and sometimes variable osmotic environments. For every aquatic species, osmotic balance is a problem that is intimately related to the medium in which the animal lives.

Osmotic balance is maintained primarily by the presence of urea in the blood of most cartilaginous fishes and in the crossopterygian *Latimeria*. Urea, a metabolic waste product, keeps the osmotic concentration of elasmobranch blood higher than that of the surrounding medium. As a result, water tends to move into the body; excess water is voided in a dilute urine. Sharks and their relatives do not drink. Even freshwater sharks—although the concentration of urea in their blood is less than in marine forms—have enough urea in the blood to cause an influx of water; they too have a dilute urine. An excess of certain salts in the food of marine elasmobranchs can be eliminated by rectal glands. In contrast, freshwater rays living in the Amazon and Orinoco Rivers of South America osmoregulate much like freshwater teleosts.

Hagfishes are unusual in that their blood has high concentrations of inorganic salts and is similar in osmotic concentration to the surrounding sea water. However, ionic balances are regulated by these fishes and do not passively reflect the composition of the medium.

Salt concentration in the blood of marine teleosts is less than that in seawater, so water tends to be lost from the body. Marine bony fishes drink water and ingest it with their food; they excrete excess salt primarily from the gills. Nitrogenous wastes are also commonly excreted through the gills in the form of ammonia, which requires no water for its excretion. In contrast, freshwater teleosts seldom drink and have a blood-salt concentration that is much higher than that of their medium. They excrete copious dilute urine and can take up salt through the gills as well as in their food.

Some fishes migrate between fresh and salt water during the life cycle and encounter quite different osmotic environments. Adult Pacific salmon and sea-going trout (*Onchorhynchus* and *Salmo*) depart from their oceanic hunting grounds and ascend freshwater streams to breed. Eels (*Anguilla*) leave fresh water to breed in the sea. Both changes of habitat obviously necessitate major changes in osmotic balance. In fact, the osmotic concentrations in the blood of these fishes may change as much as 25 or 30 percent—and may do so even before the animal actually changes habitats.

Amphibia take up water mainly through the skin and usually need free-standing, fresh water to survive. Most species have little salt tolerance. Two exceptions are the green toad, *Bufo viridis,* of brackish waters in Eurasia, and the crab-eating frog, *Rana cancrivora,* which inhabits coastal mangrove swamps of southeastern Asia. These two species do not lose water to the environment because they maintain a high osmotic concentration in their blood. The frog does this mainly by means of urea in the blood, and the toad chiefly by means of inorganic salts. A few salamanders (*Batrachoseps*) can also tolerate salt water to some degree. Some desert amphibians, such as a spadefoot toad, *Scaphiopus couchii* of North America, may also develop high urea levels when water is not available. They may be able to extract water osmotically from the soil.

Aquatic reptiles, birds, and mammals have basically terrestrial adaptations, although a number of species have secondarily adapted to an aquatic existence. In general, the osmotic concentration in the body fluids in marine reptiles, birds, and mammals is less than that of the medium. These species acquire water from their food and from drinking. As a rule, reptiles excrete excess ingested salt from glands in the optic, nasal, or palatal region; this mechanism is found in sea turtles (such as *Chelonia*), sea snakes, and Galapagos lizards (*Amblyrhynchus*). Whales, which have a very high salt intake with their food, have a salt concentration of the blood that is somewhat higher than that of terrestrial mammals, although it is lower than the osmotic concentration of the ocean. Whales may also be able to excrete excess salt. Other marine mammals, such as the manatees, seals, and sea lions, apparently maintain osmotic concentrations in their body fluids that are similar to those in whales.

SUMMARY

Animals usually live within a relatively narrow temperature range—approximately −2 to +50°C. Temperature tolerances vary among species and within species both seasonally and geographically. Aquatic vertebrates generally face much less variable environmental temperatures than do terrestrial species. Accordingly, they have fewer adaptations for dealing with extreme temperatures.

Birds and mammals and some reptiles are usually considered to be endothermic; fishes, amphibians, and other reptiles are usually called ectothermic. This dichotomy is blurred by many intermediate conditions. Many species are heterothermic (sometimes endothermic and sometimes ectothermic), often as a function of age, season, or activity.

The thermoregulatory effectiveness of latitudinal gradients in body size (Bergmann's rule) and appendage size (Allen's rule) is debatable. Several alternative explanations for these gradients may be suggested: competition, selection for high reproductive rates, improvement of hearing when sound travels poorly, and the effects of habitat, foraging habits, and migratory behavior.

Adaptations to low temperatures include cold tolerance, insulation (with special devices for heat dissipation when needed), and avoidance by migration, burrow-dwelling, and dormancy.

Terrestrial adaptations to high temperatures include avoidance of temperature extremes by migration, living in burrows, nocturnality, or dormancy, and temporary rises in body temperature. Some species have special methods of water intake and storage. Evaporative water loss by panting, gular flutter, or sweating must be achieved as economically as possible—for example, by movement of respiratory air at resonant frequencies of body structures, by heat exchange in respiratory passages, and by the lack of sweat glands (especially in small endotherms). Water conservation in excretion is also common.

Aquatic adaptations to high temperatures include resistance to temperature shock and mechanisms related to osmotic balance.

Aquatic environments vary greatly in solute content. Vertebrates of saline water face the potential problem of great losses of water. Elasmobranchs meet this problem by having a high internal osmotic concentration, which is due to the presence of urea (causing an influx of water) and by excreting both salts and dilute urine. Marine teleosts, on the other hand, drink sea water and excrete excess salts. Freshwater teleosts actively take up salt and void great quantities of dilute urine. Few amphibians are salt-tolerant. Those that are salt-tolerant maintain osmotic balance by means of urea or salts in the blood. Marine reptiles, birds, and mammals drink sea water and excrete excess salt.

SELECTED REFERENCES

Alexander, R.McN., 1975. *The Chordates.* Cambridge University Press, Cambridge, England.

Ali, M.A. (Ed.), 1980. *Environmental Physiology of Fishes.* Plenum Publishing, New York.

Barnett, R. J., 1977. Bergmann's rule and variation in structures related to feeding in the gray squirrel. *Evol., 31*:538–545.

Bartholomew, G.A., and V.A. Tucker, 1963. Control of changes in body temperature, metabolism and circulation by the agamid lizard, *Amphibolurus barbatus. Physiol. Zool.,*

36:199–218.

Bartholomew, G.A., and V.A. Tucker, 1964. Size, body temperatures, thermal conductance, oxygen consumption, and heart rate in Australian varanid lizards. *Physiol. Zool., 37*:341–354.

Bartholomew, G.A., V.A. Tucker, and A.K. Lee, 1965. Oxygen consumption, thermal conductance, and heart rate in the Australian skink *Tiliqua scincoides. Copeia, 1965*:169–173.

Bennett, A.F., and J.A. Ruben, 1979. Endothermy and activity in vertebrates. *Science, 206*:649–654.

Berven, K.A., 1982. The genetic basis of altitudinal variation in the wood frog *Rana sylvatica*. II. An experimental analysis of larval development. *Oecologia, 52*:360–369.

Bligh, J., 1972. Evaporative heat loss in hot arid environments. *In* G.M.O. Maloiy (Ed.), Comparative physiology of desert animals. *Symp. Zool. Soc. Lond., 31*:357–369.

Bligh, J., J.L. Cloudsley-Thompson, and A.G. Macdonald (Eds.), 1976. *Environmental Physiology of Animals.* John Wiley & Sons, New York.

Bradshaw, S.D., 1970. Seasonal changes in the water and electrolyte metabolism of *Amphibolurus* lizards in the field. *Comp. Biochem. Physiol., 36*:689–719.

Bradshaw, S.D., 1971. Growth and mortality in a field population of *Amphibolurus* lizards exposed to seasonal cold and aridity. *J. Zool.* (Lond.), *165*:1–25.

Bradshaw, S.D., and A.R. Main, 1968. Behavioural attitudes and regulation of temperature in *Amphibolurus* lizards. *J. Zool.* (Lond.), *154*:193–221.

Brown, G.D., and A.R. Main, 1967. Studies on marsupial nutrition. V. The nitrogen requirements of the euro, *Macropus robustus. Aust. J. Zool., 15*:7–27.

Brown, G.W., 1968, 1974. *Desert Biology.* Vols. 1 and 2. Academic Press, New York.

Brown, J.H., and C.R. Feldmeth, 1971. Evolution in constant and fluctuating environments: thermal tolerances of desert pupfish (*Cyprinodon*). *Evol., 25*:390–398.

Cade, T.J., and G.L. Maclean, 1967. Transport of water by adult sand grouse to their young. *Condor, 69*:323–343.

Calder, W.A., 1974. Consequences of body size for avian energetics. *Publ. Nutt. Ornith. Club, 15*:86–144.

Calder, W.A., and J.R. King, 1974. Thermal and caloric relations of birds. *In* D.S. Farner and J.R. King (Eds.), *Avian Biology,* pp. 259–413. Academic Press, New York.

Carey, F.G., 1982. A brain heater in the swordfish. *Science, 216*:1327–1329.

Carey, F.G., and J.M. Teal, 1966. Heat conservation in tuna fish muscle. *PNAS, 56*:1464–1469.

Carey, F.G., and J.M. Teal, 1969. Regulation of body temperature by the bluefin tuna. *Comp. Biochem. Physiol., 28*:205–213.

Cloudsley-Thompson, J.L., 1971. *The Temperature and Water Relations of Reptiles.* Merrow, Watford, England.

Cloudsley-Thompson, J.L., and M.J. Chadwick, 1964. *Life in Deserts.* Dufour, Philadelphia.

Crawford, E.C., 1972. Brain and body temperatures in a panting lizard. *Science, 177*:431–433.

Dawson, T.J., 1972. Thermoregulation in Australian desert kangaroos. *In* G.M.O. Maloiy (Ed.), Comparative physiology of desert animals. *Symp. Zool. Soc. Lond., 31*:133–146.

DeVries, A.L., 1982. Biological antifreeze agents in coldwater fishes. *Comp. Biochem. Physiol., 73A*:627–640.

Dizon, A.E., and R.W. Brill, 1979. Thermoregulation in tunas. *Am. Zool., 19*:249–265.

Felger, R.S., K. Cliffton, and P.J. Regal, 1976. Winter dormancy in sea turtles: independent discovery and exploitation in the Gulf of California by two local cultures. *Science, 191*:283–284.

Frith, M.H., 1959. Ecology of wild ducks in inland Australia. *In* A. Keast, R.L. Crocker, and C.S. Christian (Eds.), *Biogeography and Ecology in Australia,* pp. 383–395. W. Junk, The Hague.

Frost, P.G.H., W.R. Siegfried, and A.E. Burger, 1976. Behavioral adaptations of the jackass penguin, *Spheniscus demersus,* to a hot, arid environment. *J. Zool.* (Lond.), *179*:165–187.

Fuentes, E.R., and F.M. Jaksic, 1979. Latitudinal size variation of Chilean foxes: tests of alternative hypotheses. *Ecol., 60*:43–47.

Fuller, W.A., L.L. Stebbins, and G.R. Dyke, 1969. Overwintering of small mammals near Great Slave Lake, northern Canada. *Arctic, 22*:34–55.

Gans, C., and W.R. Dawson, 1976. *Biology of the Reptilia.* Vol. 5. Academic Press, New York. (Contains several relevant survey chapters.)

Gaunt, S.L.L., 1980. Thermoregulation in doves (Columbidae): a novel esophogeal heat exchanger. *Science, 210*:445–447.

Goodall, D.W. (Ed.), 1976. *Evolution of Desert Biota.* University of Texas Press, Austin.

Gordon, M.S., et al., 1977. *Animal Physiology: Principles and Adaptations.* 3rd ed. Macmillan Publishing Co., New York.

Graham, J.B., 1975. Heat exchange in the yellowfin tuna, *Thunnus albacares,* and skipjack tuna, *Katsuwonus pelamis,* and the adaptive significance of elevated body temperatures in scombroid fishes. *Fisheries Bull, 73*:219–229.

Hadley, N.F. (Ed.), 1975. *Environmental Physiology of Desert Organisms.* Dowden, Hutchinson & Ross, Stroudsburg, PA.

Hamilton, W.J., 1973. *Life's Color Code.* McGraw-Hill Book Co., New York.

Hannon, J.P., and E. Viereck (Eds.), 1962. *Comparative Physiology of Temperature Regulation.* Arctic Aeromedical Lab., Fort Wainwright, AK.

Heinrich, B., 1977. Why have some animals evolved to regulate a high body temperature? *Am. Nat., 111*:623–640.

Heinrich, B., and G.A. Bartholomew, 1971. An analysis of preflight warm-up in the sphinx moth, *Manduca sexta. J. Exp. Biol., 55*:223–239.

Herreid, C.F., and K. Schmidt-Nielsen, 1966. Oxygen consumption, temperature, and water loss in bats from different environments. *Am. J. Physiol., 211*:1108–1112.

Hill, R.W., 1976. *Comparative Physiology of Animals.* Harper and Row, New York.

Hill, R.W., and J.W. Veghte, 1976. Jackrabbit ears: surface temperatures and vascular responses. *Science, 194*:436–438.

Huey, R.B., 1974. Behavioral thermoregulation in lizards: Importance of associated costs. *Science, 184*:1001–1003.

Huey, R.B., and E.R. Pianka, 1977. Seasonal variation in thermoregulatory behavior and body temperature of diurnal Kalahari lizards. *Ecol., 58*:1066–1075.

Huey, R.B., and M. Slatkin, 1976. Costs and benefits of lizard thermoregulation. *Q. Rev. Biol., 51*:363–384.

Huey, R.B., and T.P. Webster, 1976. Thermal biology of *Anolis* lizards in a complex fauna: the *cristatellis* group in Puerto Rico. *Ecol., 57*:985–994.

Irving, L., 1972. *Arctic Life of Birds and Mammals.* Springer-Verlag, New York.

James, F.C., 1970. Geographic size variation in birds and its relation to climate. *Ecol., 51*:365–390.

Keast, A., 1959. Australian birds: their zoogeograph and adaptations to an arid climate. *In* A. Keast, R.L. Crocker, and C.S. Christian (Eds.), *Biogeography and Ecology in Australia,* pp. 89–114. W. Junk, The Hague.

Kendeigh, S.C., 1976. Latitudinal trends in the metabolic adjustments of the house sparrow. *Ecol., 57*:508–519.

Kendeigh, S.C., and C.R. Blem, 1974. Metabolic adaptations to local climate in birds. *Comp. Biochem. Physiol., 48A*:175–187.

King, J.R., 1974. Seasonal allocation of time and energy resources in birds. *Publ. Nutt. Ornith. Club, 15*:4–70.

Kluger, M.J., 1976. The importance of being feverish. *Nat. Hist.* (Jan.):71–75.

Kluger, M.J., 1979. Fever in ectotherms: evolutionary implications. *Am. Zool., 19*:295–304.

Kooyman, G.L., et al., 1982. Diving depths and energy requirements of king penguins. *Science, 217*:726–727.

Lustick, S., M. Adam, and A. Hinko, 1980. Interaction between posture, color, and the radiative heat load in birds. *Science, 208*:1052–1053.

MacMillen, R.E., 1972. Water economy of nocturnal desert rodents. *In* G.M.O. Maloiy (Ed.), Comparative physiology of desert animals. *Symp. Zool. Soc. Lond., 31*:147–174.

MacMillen, R.E., and A.K. Lee, 1967. Australian desert mice: independence of exogenous water. *Science, 158*:383–385.

MacMillen, R.E., and A.K. Lee, 1969. Water metabolism of Australian hopping mice. *Comp. Biochem. Physiol., 28*:493–514.

Main, A.R., 1968. Ecology, systematics, and evolution of Australian frogs. *Adv. Ecol. Res., 5*:37–86.

Main, A.R., 1976. Adaptation of Australian vertebrates to desert conditions. *In* D.W. Goodall (Ed.), *Evolution of Desert Biota,* pp. 101–131. University of Texas Press, Austin.

Maloiy, G.M.O. (Ed.), 1972. Comparative physiology of desert animals. *Symp. Zool. Soc. Lond., 31*:1–413.

McNab, B.K., 1969. The economics of temperature regulation in neotropical bats. *Comp. Biochem. Physiol., 31*:227–268.

McNab, B.K., 1971. On the ecological significance of Bergmann's rule. *Ecol., 52*:845–854.

McNab, B.K., 1974. The energetics of endotherms. *Ohio J. Sci., 74*:370–380.

McNab, B.K., 1978. The evolution of endothermy in the phylogeny of mammals. *Am. Nat., 112*:1–21.

McNab, B.K., 1979. The influence of body size on the energetics and distribution of fossorial and burrowing mammals. *Ecol., 60*:1010–1021.

McNab, B.K., 1980. Food habits, energetics, and the population biology of mammals. *Am. Nat., 116*:106–124.

McNab, B.K., and P. Morrison, 1963. Body temperature and metabolism in subspecies of *Peromyscus* from arid and mesic environments. *Ecol. Monogr., 33*:63–82.

Moberly, W.R., 1963. Hibernation in the desert iguana, *Dipsosaurus dorsalis. Physiol. Zool., 36*:152–160.

Norris, K., 1967. Color adaptation in desert reptiles. *In* W.W. Milstead (Ed.), *Lizard Ecology: A Symposium.* University of Missouri Press, Columbia.

Packard, G.C., 1966. The influence of ambient temperatures and aridity on molds of reproduction and excretion of amniote vertebrates. *Am. Nat., 100*:667–682.

Pearson, O.P., and D.F. Bradford, 1976. Thermoregulation of lizards and toads at high altitudes in Peru. *Copeia, 1976*:155–170.

Pond, C.M., 1978. Morphological aspects and the ecological and mechanical consequences of fat deposition in wild vertebrates. *Annu. Rev. Ecol. Syst., 9*:519–570.

Pough, F.H., 1980. The advantages of ectothermy for tetrapods. *Am. Nat., 115*:92–112.

Precht, H., et al., 1973. *Temperature and Life.* Springer-Verlag, New York.

Prosser, C.L., 1973. *Comparative Animal Physiology.* 3rd ed. Saunders College Publishing, Philadelphia.

Pruitt, W.O., 1970. Some ecological aspects of snow, pp. 83–99. *In Ecology of Subarctic Regions.* UNESCO, Paris.

Ronald, K., and J.L. Dougan, 1982. The ice lover: biology of the harp seal (*Phoca groenlandica*). *Science, 215*:928–933.

Ruibal, R., 1955. A study of altitudinal races in *Rana pipiens. Evol., 9*:322–338.

Schmidt-Nielsen, K., 1964. *Desert Animals.* Oxford University Press, Oxford.

Schmidt-Nielsen, K., 1972. *How Animals Work.* Cambridge University Press, Cambridge, England.

Schmidt-Nielsen, K., 1979. *Animal Physiology.* 2nd ed. Cambridge University Press, Cambridge, England.

Schmidt-Nielsen, K., B. Schmidt-Nielsen, S.A. Jarnum, and T.R. Houpt, 1957. Body temperature of the camel and its relation to water economy. *Am. J. Physiol., 188*:103–112.

Searcy, W.A., 1980. Optimal body sizes at different ambient temperatures: an energetics explanation of Bergmann's rule. *J. Theor. Biol., 83*:579–593.

Skadhauge, E., 1975. Renal and cloacal transport of salt and water. *Symp. Zool. Soc. Lond., 35*:97–106.

Snyder, G.K., and W.W. Weathers, 1975. Temperature adaptations in amphibians. *Am.*

Nat., *109*:93–101.

Swan, H., 1974. *Thermoregulation and Bioenergetics.* Elsevier, New York.

Taplin, L.E., and G.C. Grigg, 1981. Salt glands in the tongue of the estuarine crocodile *Crocodylus porosus. Science, 212*:1045–1047.

Taylor, C.R., 1970. Strategies of temperature regulation: effect of evaporation in East African ungulates. *Am. J. Physiol., 219*:1131–1135.

Taylor, C.R., 1972. The desert gazelle: a paradox resolved. *In* G.M.O. Maloiy (Ed.), Comparative physiology of desert animals. *Symp. Zool. Soc. Lond., 31*:215–227.

Tracy, C.R., 1977. Minimum size of mammalian homeotherms: role of the thermal environment. *Science, 198*:1034–1035.

Trost, C.H., 1972. Adaptations of horned larks (*Eremophila alpestris*) to hot environments. *Auk, 89*:506–527.

Tucker, V.A., 1965. The relation between the torpor cycle and heat exchange in the pocket mouse, *Perognathus californicus. J. Cell Comp. Physiol., 65*:405–414.

Vernberg, F.J. (Ed.), 1975. *Physiological Adaptations to the Environment.* Intext, New York. [A 1973 AIBS Symposium.]

Walsberg, G.E., G.S. Campbell, and J.R. King, 1978. Animal coat color and radiative heat gain: a re-evaluation. *J. Comp. Physiol., 126*:211–222.

Whittow, G.C. (Ed.), 1970–1973. *Comparative Physiology of Thermoregulation.* Vols. 1–3. Academic Press, New York.

Yancey, P.H., M.E. Clark, S.C. Hand, R.D. Bowlus, and G.N. Somero, 1982. Living with water stress: evolution of osmolyte systems. *Science, 217*:1214–1222.

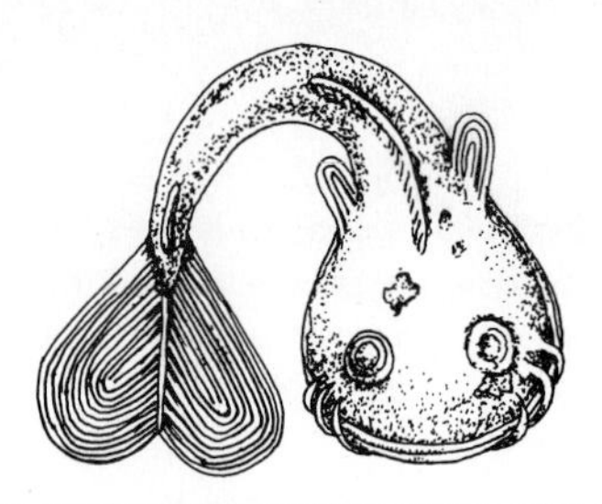

6 Respiration

All vertebrates obtain energy from eating food and render their foods into biochemically usable forms by the process of oxidation, a chemical combination with oxygen (O_2). Two of the major components of food, namely fats and carbohydrates, are composed of carbon and oxygen and are oxidized to form carbon dioxide (CO_2) and water. The third major component, protein, also oxidizes to CO_2 and water, but it results in nitrogenous wastes as well. Excess water can be reused or eliminated in many ways. Nitrogen-containing waste materials are generally removed by the kidneys; CO_2 must also be removed. Obtaining O_2 and disposing of CO_2 are the two primary functions of respiration. Gas exchange is accomplished by respiratory surfaces in a variety of organs, including lungs, gills, and skin.

The process of oxidation takes place in cells. As large molecules are gradually broken down, energy is released at several points in the sequential process and stored in high-energy phosphate bonds. The first part of the sequence can occur in the absence of an O_2 supply and results in the production of pyruvic acid. With the addition of hydrogen, pyruvic acid can be converted to lactic acid for storage until O_2 becomes available. The second part of the sequence requires a constant availability of O_2. It results in final breakdown to CO_2 and water and yields more energy than the first part of the sequence. The first series of intracellular reactions is called anaerobic (without air) metabolism; the second is called aerobic metabolism.

As might be expected, O_2 consumption varies greatly with taxon, body size, and activity. Ectothermic, large-bodied, or sluggish animals generally have lower metabolic demands than endothermic, small-bodied, or active ones. Thus, we find that a typical O_2 consumption rate for a relatively inactive fish such as a carp (*Cyprinus carpio*) is about 100 ml/kg/hr while that for an active pike (*Esox*) of about the same size is about 350 ml/kg/hr. A house mouse (a laboratory strain of *Mus musculus*) consumes 2500 ml/kg/hr at rest; a running mouse uses 20,000 ml/kg/hr.

In addition to variations in O_2 needs, there are great variations in the availability of O_2 as a function of environmental conditions. Air-breathing animals and water-breathing animals face different problems because of the properties of the medium in which the respiratory gases are found. Air is composed of about 20 percent O_2, about 20 times the O_2 content of well-aerated water. Not only is there more O_2 in air than in water, but O_2 molecules can move more readily in air. Because air is much less dense and viscous than water, aerial respiratory organs may need to do less work to move the medium to and from the respiratory surfaces. Not all water is well-aerated, however; increasing temperature and salinity reduce O_2 availability. Moreover, many bodies of water have low-O_2 zones. In swamps, at deeper levels of some very deep lakes, and in polluted waters, a limited O_2 supply is rapidly depleted by the metabolism of microorganisms. Sometimes the deep ocean basins also are low in available O_2. Vertebrates living in such aquatic habitats face rather special problems in obtaining sufficient O_2. On the other hand, elimination of CO_2 is generally not difficult for aquatic vertebrates. Carbon dioxide is highly soluble and diffuses readily in water; most waters have low concentrations of gaseous CO_2.

GAS TRANSPORT IN THE BODY

Carbon dioxide is carried in temporary chemical combination with constituents of the plasma and in cells in the blood. In contrast, only a tiny amount of O_2 is carried in the plasma in most vertebrates. Almost all vertebrates have red blood cells containing hemoglobin molecules, protein molecules containing iron, that bind and transport O_2 molecules from the respiratory surfaces to the rest of the body. Blood with hemoglobin can carry about 100 times as much O_2 as blood with O_2 in solution. The only vertebrates known not to have red blood cells and hemoglobin are eel larvae and icefishes (Chaenichthyidae), although relatives of these fishes have reduced hemoglobin levels.

A major advantage of red blood cells is that the chemical environment within the cells can differ from that in the plasma, facilitating control of the chemical environment of hemoglobin. The interaction of hemoglobin with O_2 is affected by many substances, so the regulation of such substances is important to the ability of hemoglobin to transport O_2. Furthermore, Snyder (1977) noted that red blood cells usually occur in animals that have hemoglobin of low molecular weight. These small hemoglobins would be removed by the kidneys if they were not in cellular packages. (The exceptions seem to be invertebrates that have small hemoglobins but no excretory filtering mechanisms through which they could be lost.) Hemoglobin in the plasma also would increase the plasma osmotic pressure, which affects normal distribution and movement of fluids. Hemoglobin sequestered into red blood cells does not have this effect because the cell membrane is impermeable to many osmotically important ions. Thus, presence of red blood cells also may prevent osmotic imbalances that would be especially great in animals containing small hemoglobin molecules in large numbers.

Why have small hemoglobins in the first place? Snyder suggested that, at most concentrations of hemoglobin, small hemoglobins have a greater capacity for O_2 transport than large molecules. Although hemoglobins of large molecular weight can bind as much O_2 per unit weight as smaller hemoglobins, large molecules have a greater impact on blood viscosity and on the rate of transport of the O_2-carrying molecules. The evolution of red blood cells may have permitted the further reduction of the molecular weight of hemoglobin and, hence, the evolution of higher concentrations of hemoglobin.

The binding of O_2 molecules to hemoglobin is reversible, and O_2 is unloaded in the tissues. The affinity of hemoglobin for O_2 is sensitive to many factors, including O_2 concentration, CO_2 concentration, and blood acidity (pH). Because each iron atom in hemoglobin can hold one O_2 molecule, there is clearly a limit to the amount of O_2 that can be carried. At any given O_2 concentration (or pressure), a certain proportion of hemoglobin will be oxygenated; that is, at any given concentration, the blood has a certain capacity to take on more O_2 and, conversely, a certain tendency to unload the O_2 it already carries. This relationship can be graphed in what is called an O_2 dissociation curve (Fig. 6–1). At intermediate O_2 concentrations, small changes in concentration effect major changes in the amount of O_2 bound to hemoglobin. At higher O_2 concentrations, changes in concentration have decreasing effects on O_2 binding, so the curve flattens out.

Oxygen dissociation curves can be used to illustrate the effects of CO_2 concentration and acidity on O_2 binding (Fig. 6–1*b*). At increased concentration of CO_2 and increased acidity, the ability of hemoglobin to bind O_2 decreases, so that at a given O_2 concentration less oxygen is held; conse-

Figure 6–1 (*a*) An oxygen dissociation curve: the degree of saturation of hemoglobin with oxygen as a function of increasing concentration of pressure of oxygen. (*b*) Shifts in oxygen dissociation with increasing acidity (decreasing pH). (Reprinted with permission of Macmillan Publishing Co., Inc. from *Animal Physiology: Principles and Functions* by Malcolm S. Gordon et al. Copyright © 1977 by Malcolm S. Gordon.)

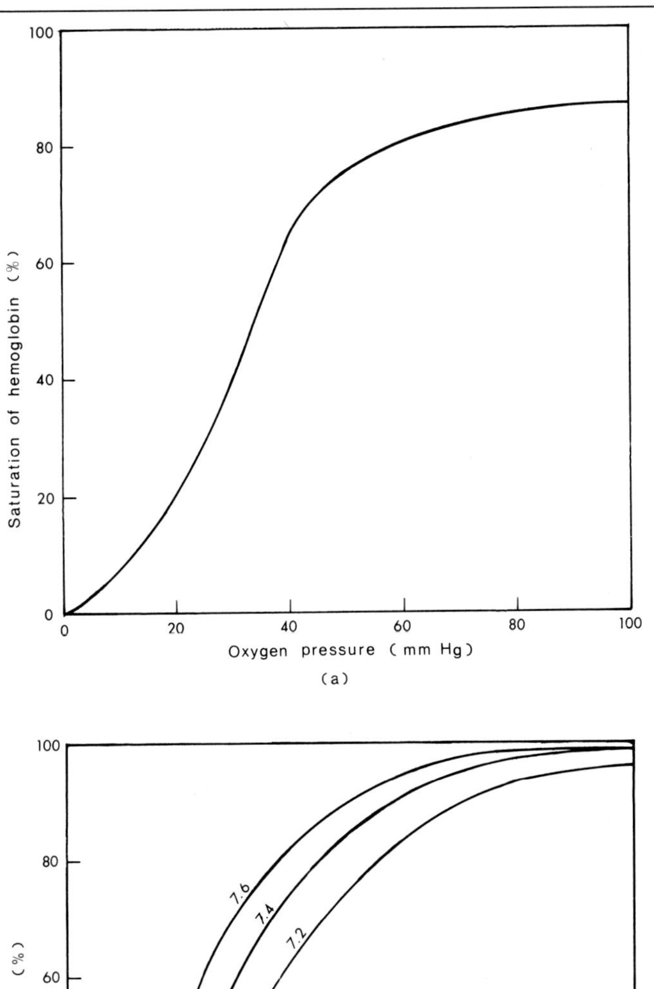
100
80
60
40
20
0
Saturation of hemoglobin (%)
0
20
40
60
80
100
Oxygen pressure (mm Hg)
(a)

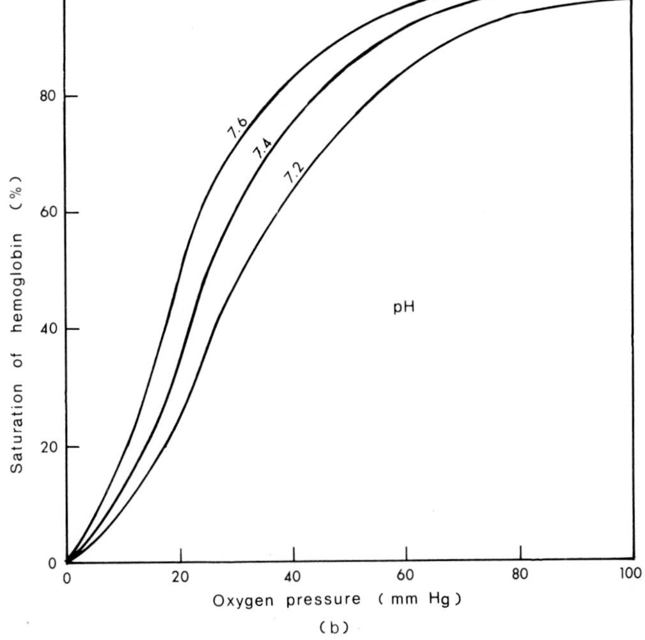
100
80
60
40
20
0
Saturation of hemoglobin (%)
7.6
7.4
7.2
pH
0
20
40
60
80
100
Oxygen pressure (mm Hg)
(b)

quently, the dissociation curve shifts to the right. Hemoglobin therefore tends to unload O_2 in the tissues, where both CO_2 concentration and acidity are higher.

The hemoglobin of frog tadpoles is much less sensitive to pH than that of adult frogs. This characteristic seems to be adaptive in that tadpoles are more likely to encounter high CO_2 concentrations in stagnant pools where O_2 concentration is already low. If the maintenance of growth rate is important to tadpole success, modifications that avoid lowered metabolism might be of great importance. The pH sensitivity of small mammals is considerably greater than that of large ones (Fig. 6–2). Thus, O_2 can be unloaded at a faster rate in small mammals, which have a higher metabolic rate and O_2 demand per unit weight.

Temperature can have profound effects on the affinity of hemoglobin for O_2. Because increased temperature lowers the dissociation curve, less O_2 can be taken up and O_2 is unloaded more readily. These effects are well known in fishes, as illustrated in Figure 6–3. Note that different species respond differently to changing temperatures. Eels (*Anguilla japonica*), for instance, are much less sensitive to temperature change than either speck-

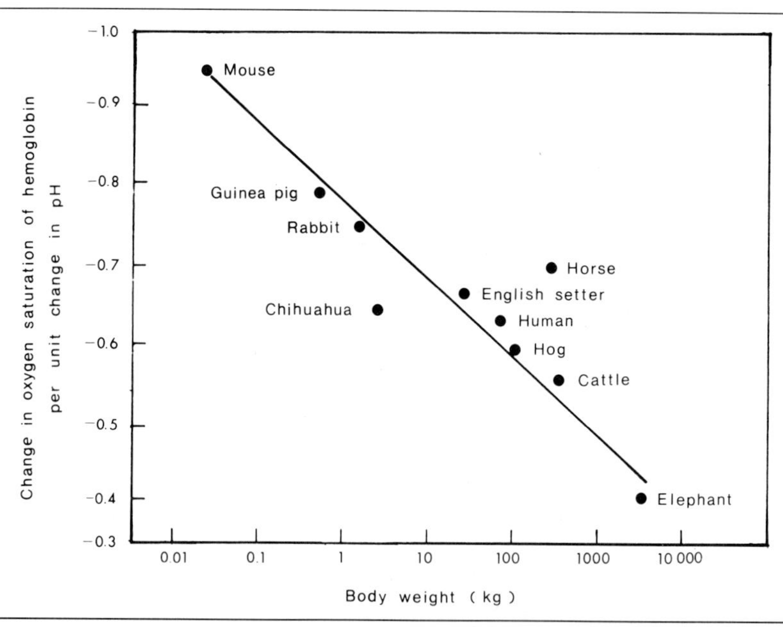

Figure 6–2 The rate of change of oxygen saturation of hemoglobin per unit change in pH versus body size in mammals. The hemoglobin of small mammals is more sensitive to changes in acidity than is the hemoglobin of large mammals. (Modified from A. Riggs, 1959. The nature and significance of the Bohr effect in mammalian hemoglobins. *J. Genet. Physiol.*, 43:737–752. Reprinted with permission of the author.)

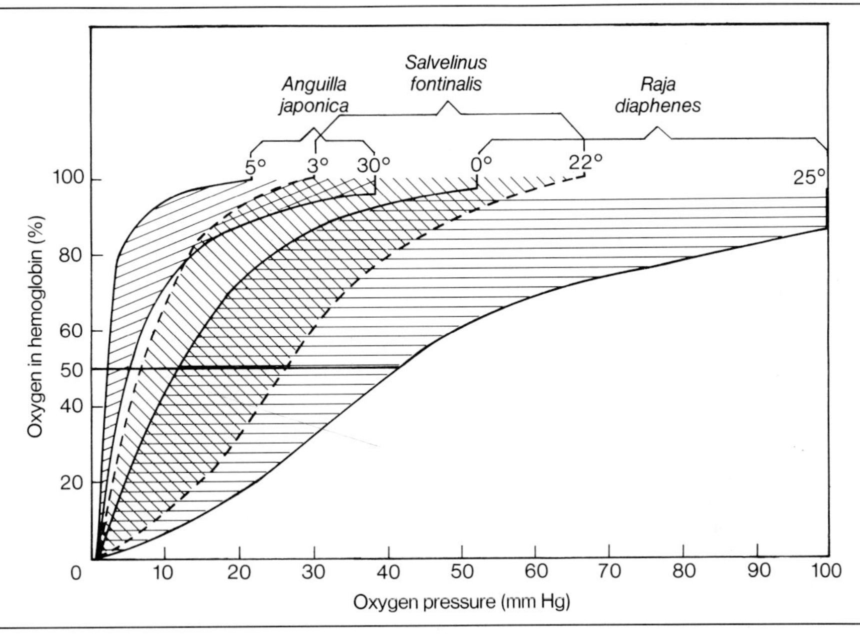

Figure 6–3 Temperature sensitivity of the hemoglobin of three fish species. (Modified from F.E.J. Fry, 1957. The aquatic respiration of fish. *In* M.E. Brown [Ed.], *The Physiology of Fishes.* Vol. 1, pp. 1–63. Academic Press, New York.)

led trout (*Salvelinus fontinalis*) or skates (*Raja diaphenes*). More comparative data would be useful here to suggest ecological associations. Probably neither *Salvelinus* nor *Raja* (this species has never been captured in waters more than 19°C) occupies widely varying habitats, but *Anguilla* is found, at various times, in both salt and fresh water and even moving overland. In such circumstances, relative insensitivity to temperature may well be advantageous.

Temperature effects on O_2 affinity are also known in homeotherms. Higher temperatures result primarily in increased rates of O_2 release in the tissues. A concomitant decrease in O_2 uptake is probably not serious, given the relatively high concentration of O_2 in the air. Schmidt-Nielsen (1979) suggested that this effect may be adaptive, because increased body temperatures occur during exercise and fever when O_2 consumption also increases.

Species differences in O_2 dissociation curves can sometimes be correlated with habitat and activity level. Mackerel (*Scomber*) are fast-moving fishes of well-aerated waters and have a lower O_2 dissociation curve than the toadfishes (*Opsanus tau*), which are sluggish bottom-dwellers often found in less well-oxygenated waters (Fig. 6–4). Toadfish blood has a high

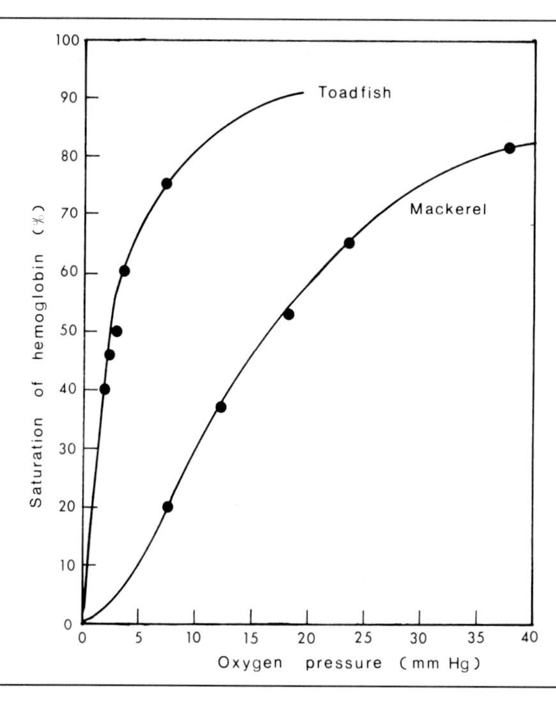

Figure 6–4 Oxygen dissociation curves for the active mackerel and the sluggish toadfish. (Modified from F.G. Hall and F.H. McCutcheon, 1938. The affinity of hemoglobin for oxygen in marine fishes. *J. Comp. Cell Physiol.*, *11*:205–212.)

affinity for O_2 and picks it up well from the environment; since the metabolic rate is relatively low, the reduced release of O_2 in the tissues is seemingly less critical. In contrast, because of the high concentration of O_2 in the normal habitat, mackerel blood will pick up a full load of O_2 despite the lowered binding affinity of mackerel hemoglobin. The lower dissociation curve means that O_2 can be delivered to tissues at a high rate.

The hemoglobin of at least some high-altitude terrestrial mammals shows a greater affinity for O_2 than that of low-altitude species. Both the llama and the vicuña (*Lama* spp.), which are native to alpine habitats in South America, show steeper O_2 dissociation curves than mammals that customarily live near sea level. There is, of course, less O_2 available in the rarefied atmosphere at high elevations.

In general, the hemoglobins of terrestrial vertebrates have lower O_2 dissociation curves than the hemoglobins of aquatic forms. Furthermore, air-breathing fishes of the Amazon have blood with a lower O_2 affinity than water-breathers. The higher O_2 affinity of hemoglobin in fully aquatic species improves the extraction of O_2 from a medium that contains much less O_2 than does air.

The environment of embryos of non–egg-laying vertebrates is rather special. Hemoglobin of the fetus is always different from that of the mother and has a high O_2 affinity, which permits uptake of O_2 from maternal blood. In addition, avian embryos (in chickens, at least) and frog tadpoles have higher dissociation curves than those of adults.

Interspecific differences in body size are associated with differences in metabolic rate; hence, it is not surprising to find that O_2 dissociation curves also vary with body size. These comparisons have been demonstrated for mammals ranging in size from elephant to mouse (Fig. 6–5). The normal pressure of O_2 in mammalian lungs is sufficient to load all the hemoglobin with O_2, so differences in the ability to pick up O_2 molecules do not appear to be crucial. The rate of unloading O_2 in the tissues may be very important, however, because small mammals have a much higher metabolic rate per gram of tissue than large ones and require higher rates of O_2 delivery. Part of this demand is met by increasing the delivery of blood to the tissues—the density of capillaries in the tissues of small mammals is greater, and therefore O_2 diffuses from the vessels to the tissues over a shorter distance. Furthermore, by reducing the O_2 affinity of hemoglobin, O_2 is given up more readily to the tissues of small mammals.

THE ANATOMY OF TRANSPORT SYSTEMS

Blood obviously transports many things—nutrients, wastes, hormones, heat, and gases. However, many of the major changes of vertebrate circu-

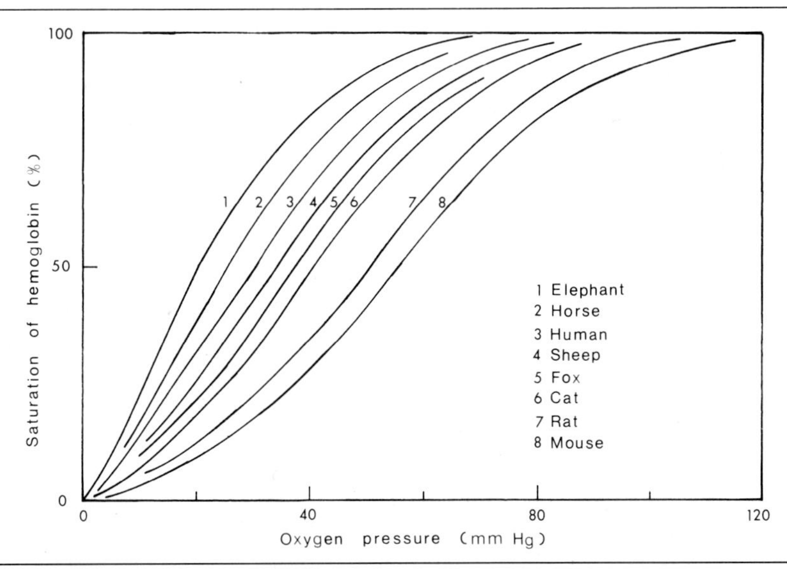

Figure 6–5 Oxygen dissociation curves of hemoglobin of various mammals. Smaller mammals have lower oxygen affinity. (Modified from K. Schmidt-Nielsen, 1970. Energy metabolism, body size, and problems of scaling. *Fed. Proc.*, *29*:1524–1532, and from K. Schmidt-Nielsen, 1975. Recent advances in avian respiration. *Symp. Zool. Soc. Lond.*, *35*:33–46.)

lation systems during the course of evolution can be associated with respiratory differences, so a cursory description of some fundamental differences in circulation patterns is relevant here.

Vertebrate circulatory systems are basically closed; that is, blood moves from the heart through the body in a continuous series of arteries, capillaries, and veins. The only exceptions are among the cyclostomes: Hagfishes have a partially open circulation with large blood sinuses in certain body regions. The pump that moves blood through the blood vessels is, of course, the heart, aided by the elastic recoil of blood vessel walls. Again hagfishes are unusual, for they have several hearts, in various regions of the body, located to augment the pumping action.

Beyond these general characteristics, the basic trends found among vertebrates involve increasing separation of oxygenated and unoxygenated blood and increasing blood pressure and flow rates. Both these characteristics are associated with the transitions from aquatic to terrestrial living and from ectothermy to endothermy.

Typical fish circulatory patterns are termed "single" because blood is transported to the body tissues and to the gills in a single loop (Fig. 6–6). Unoxygenated blood leaves the heart, passes through capillary beds in the gills, where it receives oxygen, and thence to the rest of the body and back to the heart. A marked drop in blood pressure occurs during passage of blood through the gills. In modern anurans, however, the lungs and body have partially separate circulatory channels that are the beginnings of a dual circulation system. Blood leaving the heart is routed, without a large drop in pressure caused by a capillary bed, to either the body or the lungs. Blood returning to the heart from these places is partially separated in the heart, although some mixing of oxygenated and unoxygenated blood may occur. Salamanders have a less well-developed division of circulatory pathways.

A dual circulatory system is clearly advantageous to an animal that uses lungs. Blood is moved at higher pressure, and blood that reaches body tissues carries a good supply of O_2 for metabolism; unoxygenated blood returning from the tissues is sent to the lungs. Mixture of the two types of blood obviously reduces the amount of O_2 that can be carried to the tissues. For many amphibians, an imperfect separation of pulmonary and body circulation may not pose a serious problem because cutaneous (skin) respiration can provide an additional source of O_2 to blood passing to the body. Reptiles can achieve a more functional segregation of the two circuits, although anatomically the separation is not quite complete. Many reptiles can shunt blood from one side of the central circulation to the other, according to whether the animal is diving or not. Lungfish circulation is functionally similar to that of reptiles. Separation of lung and body circulation is complete in both birds and mammals.

We can observe a tendency toward a double circulatory system in air-breathing vertebrates. Dual systems appear to be best developed in the two classes in which endothermy is most prevalent. Why should such an association exist? Metabolic rates are generally highest in endotherms. A flying bird, for example, can use O_2 about 30 times as fast as a lizard of similar weight at maximum metabolism. High metabolic rates must have evolved in conjunction with a circulatory system that could deliver well-oxygenated blood at high pressure and, thus, achieve high rates of O_2 delivery to the tissues. This problem was solved anatomically in several ways, as we have seen, but the functional solution was to avoid the pressure drop created by friction of the blood against the walls of the capillary networks in the respiratory organs. A fully double circulation also allows separate regulation of pressure in the two circuits, and pressure in vessels to the lungs is generally lower than in those to the rest of the body. High pressure in lung capillaries has deleterious effects on gas exchange that are negligible or absent in gills. Excessive blood pressure tends to force fluid from the vessels into air spaces of the lungs, obviously interfering with gas transfers. But gills are bathed in water anyway, so fluid loss is both lower and less important. Alternatively, capillaries in the lungs might have thicker walls to prevent bursting, but thick walls would slow the diffusion of O_2.

We can only guess at the selection pressures that originally led to the evolution of air-breathing in ancient fishes. Researchers generally seem to agree that conditions creating low O_2 availability in water gave adaptive value to alternative modes of respiration. One solution to the problem involved the ability (facultative at first) to breathe air. Just what factors may have been important in reducing O_2 availability in certain habitats is the subject of some debate. The classic view is that high temperatures in shallow, stagnant bodies of fresh water provided the likely conditions that favored air-breathing. Packard (1974) has suggested that high salt concentrations, which reduce O_2 availability while they increase the amount of work required to move water over the gills, could also have made air-breathing adaptive. However, more recent arguments suggest that prevailing salinities were insufficient to favor a switch to air-breathing and that the classic view is more probable (Graham et al., 1978). In any case, the ready availability of O_2 in air made possible the achievement of high metabolic rates and phylogenetically new ways of exploiting environments.

AQUATIC RESPIRATION

Cyclostomes, cartilaginous fishes, most bony fishes, and some amphibia and reptiles obtain O_2 from water. Simple diffusion (movement of molecules from a region of high concentration to an area of low concentration) of O_2 across the skin and mucous membranes of the digestive tract may suffice for existence in some cases, especially at low temperatures. A Pacific sea snake, *Pelamis platurus*, can acquire 33 percent of its O_2 from the water by means of its skin. The highly aquatic soft-shelled turtles (*Trionyx*) perform as much as 70 percent of their respiration through the skin; their throat lining accounts for another significant amount of gas exchange. But most fully aquatic vertebrates breathe by means of gills.

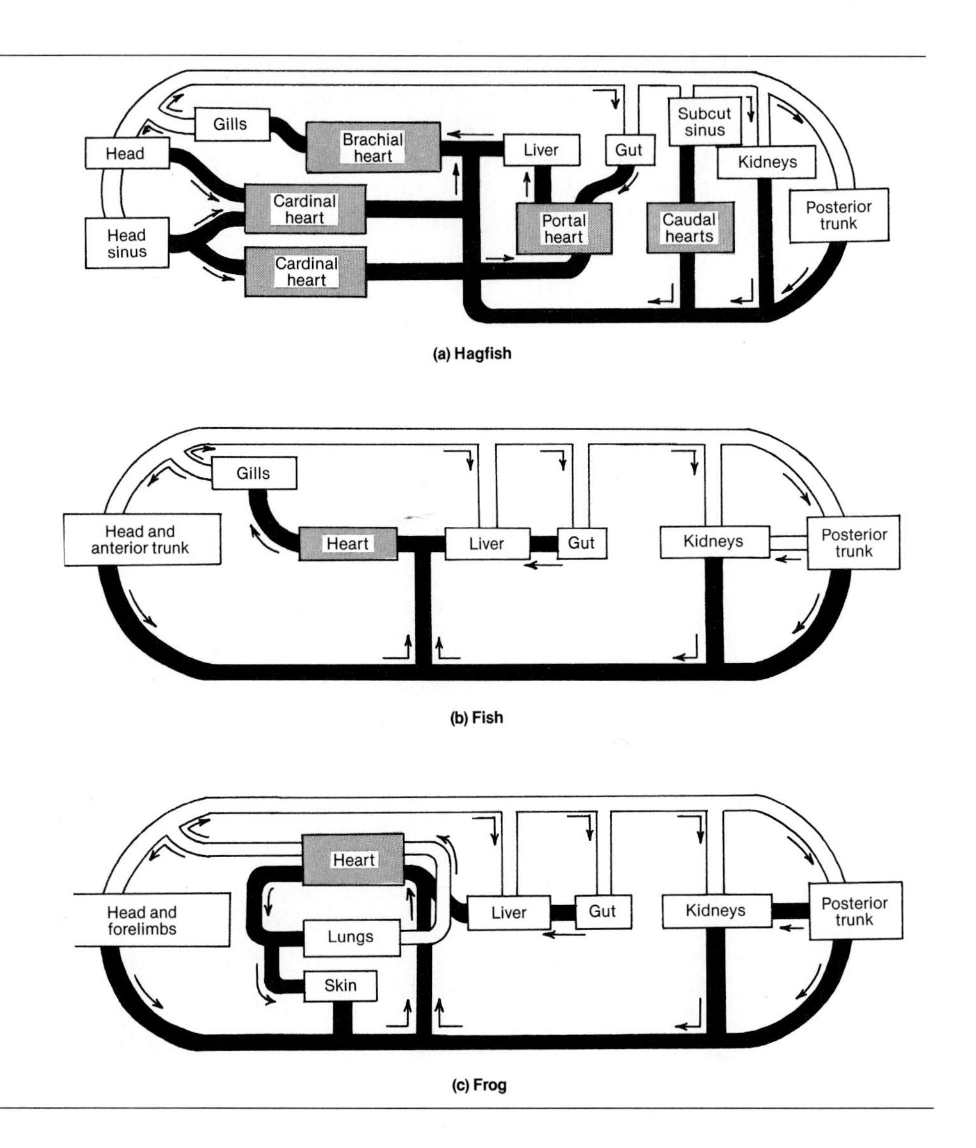

Although structural details vary among taxa, basically gills are highly vascularized surfaces derived from an outpocketing of the embryonic gut (Fig. 6–7). Fish gills are internal and enclosed in a gill chamber, but aquatic salamanders often have external many-branched gills that protrude into the water. The thin surface membranes of gills permit ready inward diffusion of O_2 and outward diffusion of CO_2. Moreover, the many blood vessels and large surface area of the gill provide a great interface for gas exchange. In very active species of fishes, such as mackerel, the surface area of the gills may be many times that of more sluggish species like the toadfish. Total gill surface area tends to increase with body size, although

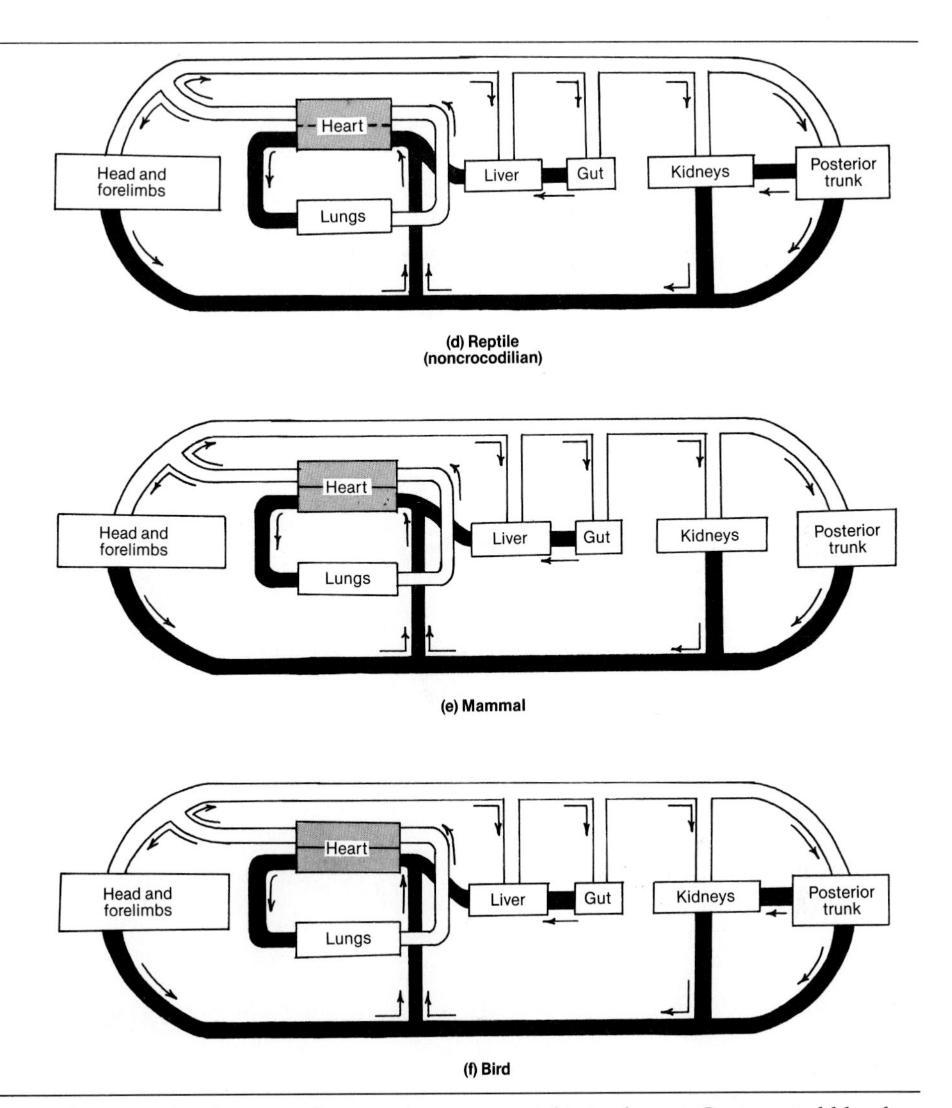

Figure 6–6 The main circulatory pathways of various vertebrate classes. Oxygenated blood or mainly oxygenated blood is shown in white; less well-oxygenated blood is depicted in black. The noncrocodilian reptile heart is incompletely divided. In crocodiles, this division is complete, but a small, more anterior opening remains between two major blood vessels. (Reprinted with permission of Macmillan Publishing Co., Inc. from *Chordate Structures and Function* by Allyn T. Waterman. Copyright © 1971 by Allyn T. Waterman.)

large fishes usually possess a smaller gill surface relative to body size than small fishes do. That is, the increase in gill area does not occur at the same rate as the increase in body size. Total respiratory surfaces of fishes are

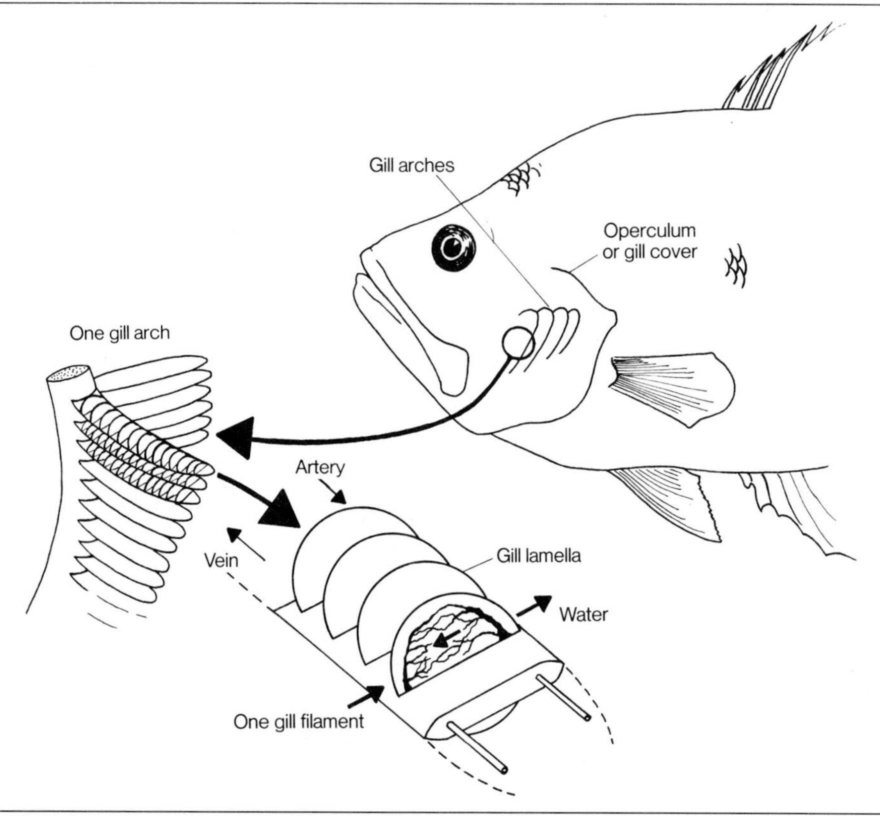

Figure 6–7 Diagrammatic representation of the gills of fishes. Fishes typically have several gill arches on each side, each arch bearing two rows of gill filaments. Each filament has a series of thin plates (lamellae) in which gas exchange is effected. (Modified from D.J. Randall, 1968. Functional morphology of the heart in fishes. *Am. Zool.*, *8*:179–189.)

generally less than those of similarly sized mammals, except in tuna, which are unusually active fishes (Fig. 6–8). External gill size of some aquatic amphibian larvae is inversely correlated with the degree of aeration of their normal habitat.

A major evolutionary solution to the respiratory problems of an aquatic existence is the countercurrent flow of water over the gill and blood flowing in the gill (Fig. 6–9). Blood and water flow in opposite directions, maximizing the effectiveness of picking up O_2. As blood leaves the exchange surface in the gill, it meets (across a membrane) fully oxygenated water, and O_2 diffuses across that membrane into the blood. As water continues to flow across the surface, it meets blood of lower and lower O_2 content, and O_2 continues to pass from water to blood. Blood entering the

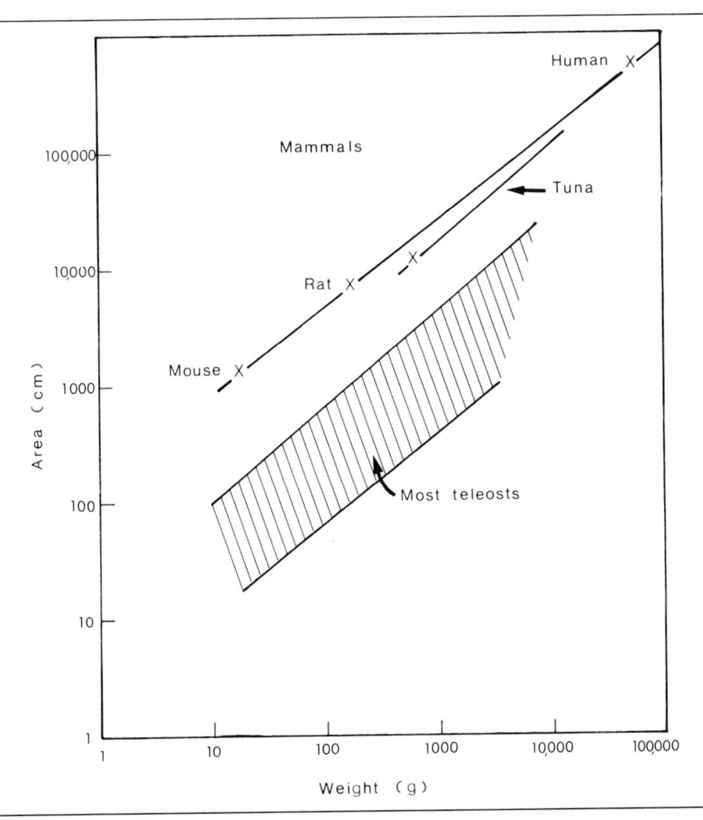

Figure 6–8 Correlation of respiratory surface area and body size in mammals and in fishes. (From D.J. Randall, 1970. Gas exchange in fish. *In* W.S. Hoar and D.J. Randall [Eds.], *Fish Physiology*, pp. 253–286. Academic Press, New York. Adapted from B.S. Muir, 1969. Gill dimensions as a function of fish size. J. Fisheries Research Board of Canada, *26*:165–170.)

gill encounters water of higher and higher O_2 content and continues to pick up O_2, so that blood leaving the gill contains almost as much O_2 as the water entering the gill. This would be totally impossible if both fluids flowed in the same direction. The exchange system is so effective that 80 to 90 percent of the available O_2 can be extracted from the water, in amazing contrast to the 20 to 30 percent effectiveness of mammalian lungs in extracting O_2 from air. The effectiveness of the countercurrent system has been demonstrated by experimentally forcing water to flow in the same direction as the blood. For the tench (*Tinca tinca*), this reversal resulted in an O_2 uptake of less than one fifth the normal amount.

Many fishes move water over the gills by pumping it from the mouth to the gill chamber; however, fast swimmers generally swim with open mouths and, in that way, achieve a high rate of water flow over the gills.

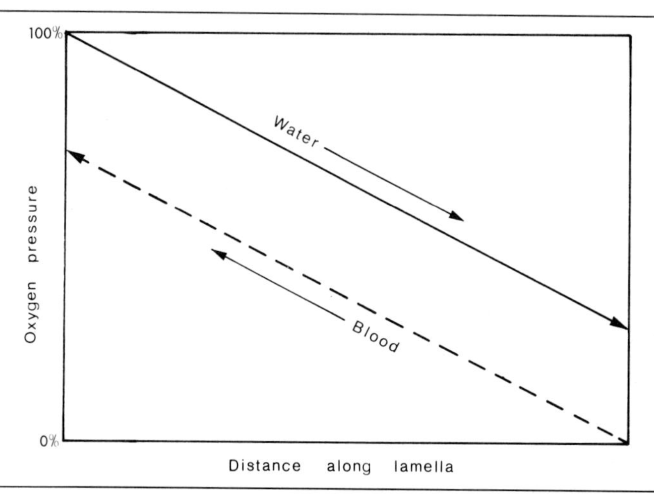

Figure 6–9 Diagram illustrating the effectiveness of the countercurrent exchange mechanism in a fish gill. Blood leaves the gill with an oxygen pressure almost as high as that of the incoming water; water leaving the gill is almost depleted of oxygen. (Modified from K. Schmidt-Nielsen, 1979. *Animal Physiology,* 2nd ed. Cambridge University Press.)

Some species, including mackerel and tuna, apparently have no pumping mechanism at all: They rely on constant motion for their O_2 supply. The flow of water over the gills in all fishes is in one direction and virtually continuous; the same appears to be true of hagfishes, which also breathe through their skin. Lampreys, however, both inhale and exhale through the gill openings (the mouth is frequently attached to prey for long periods of time), but valves in the gill chambers cause water to pass over the gas-exchange surfaces only during exhalation. Young individuals of *Monopterus albus,* which breathes air by the time it is 2 weeks old, obtain oxygen by sweeping well-oxygenated surface water back over the body, in which the surface blood supply flows mainly foward. The entire fishlet functions as a countercurrent exchanger and can survive when its ponds become stagnant (Liem, 1981).

The Antarctic icefishes, such as *Pagetopsis macropterus* (Fig. 6–10) and other chaenichthyids, lack red blood cells and hemoglobin. They live in cold, well-aerated waters, and many, but perhaps not all, of them are relatively inactive. Even so, it is not altogether clear how they obtain enough O_2 to survive and carry on the essential activities of foraging and reproduction. No doubt several factors are involved, although their relative importance is uncertain. The skin is highly vascularized, and much gas exchange appears to be cutaneous, although this arrangement is not peculiar to these fishes. Other factors are the relatively large blood volume and rapid circulation. In some species, but apparently not all, the metabolic rate

Figure 6–10 An icefish (*Pagetopsis macroptera*). This fish has no red blood cells and no hemoglobin but nevertheless obtains sufficient oxygen. (Photo by A.L. DeVries.)

is lower than in other, unrelated, Antarctic fishes. Interestingly, gill surface areas are often small, relative to the size of the fish; this could be related to a dependence on cutaneous gas transfer, although some red-blooded fishes have both large gills and well-vascularized skin. These fascinating icefishes are still a bit of a puzzle in environmental physiology.

If fishes encounter a deficient O_2 supply, they may respond in several ways. They may breathe rapidly, but this response is inefficient because the percentage of O_2 utilization decreases. Deeper breathing is more effective. Some species swim until aerated water is found. Other species (including the goldfish, *Carassius auratus*) can switch to anaerobic metabolism for long periods; these species are inhabitants of relatively unaerated water. Anaerobic metabolism also characterizes fishes of low-O_2 marine waters whose O_2 dissociation curves are remarkably low. Quantities of O_2 are probably available to these fishes only during extensive diurnal vertical migrations (Douglas et al., 1976).

The minimum O_2 requirements of a fish are closely linked to its ability to survive through the winter in bodies of water that are covered by ice and its relative adaptability to variations in water temperature and flow. At about 20°C, speckled trout require a minimum O_2 level of 2.5 parts per million (ppm); blunt-nosed minnows (*Pimephales*) need 2.25 ppm; goldfish need 0.6 ppm (Erichsen Jones, 1964). Trout typically inhabit cold, fast-running water; blunt-nosed minnows are common in small riffles of intermediate temperature and speed; and goldfish can survive an entire winter under the ice in a stagnant pond. These limits were determined in the laboratory and refer to the fish at rest; no allowance was made for normal activities. Minimum O_2 tolerances also affect the ability of the fish to survive pollution: One effect of pollution is an increase in populations of microorganisms that tend to use up the dissolved O_2.

AERIAL RESPIRATION

All birds and mammals, most reptiles, and some fishes and amphibians depend exclusively or almost exclusively on air-breathing. Air-breathers encounter readily available O_2 in all normal circumstances except at very high altitudes. Because the density of air is so much less than that of water, less work is needed to move a given volume of air. This difference permitted the evolution of respiratory mechanisms that move the O_2-bearing medium back and forth rather than solely unidirectionally, as in water-breathers. The high concentration of O_2 in air also means that energy is expended in the movement of relatively little "excess" material, compared to the large proportion of material other than O_2 in water. Back-and-forth respiration is generally less efficient than unidirectional respiration, however, because simple countercurrent mechanisms for extraction of O_2 become ineffective. As we will see, birds seem to have overcome this potential disadvantage.

Breathing air presents the danger of desiccation of the respiratory surfaces. Desiccation not only means excessive water loss from the body but also the end of O_2 intake, since O_2 enters the body in dissolved (not gaseous) form. Air-breathing vertebrates, therefore, live in moist habitats or protect the respiratory surfaces by enclosing them in the body cavity.

Many organs may be involved in aerial respiration, especially in fishes. Air-breathing fishes use lungs and vascularized air chambers associated with the gills, the skin, the swim bladder, the mouth and pharynx, the intestine, or the stomach (Table 6–1). Some fishes, although they retain an ability to breathe in water, require air and will drown if denied access to it. Some lungfishes (*Lepidosiren* and *Protopterus*) and the electric eel (*Electrophorus*) are examples of fishes that require air. Air-breathing fishes frequently inhabit warm, swampy, fresh water that contains litte O_2; the North American bowfin (*Amia calva*) is a good example of such a fish. Some, such as the mudskipper *Periophthalmus,* spend considerable time on shore at the water's edge, and *Anguilla* eels are known to make overland movements. Air-breathing fishes also live in marine habitats, primarily in intertidal zones of rocky shores, mud flats, or mangrove thickets; one species, the tarpon (*Megalops atlantica*), lives in open water.

Graham (1976) compared adaptations for air-breathing in marine and freshwater fishes. Most marine air-breathing fishes are amphibious, spending considerable periods of time both in and out of water. Their respiratory organs are relatively simple. Typically these organs involve the gills, mouth and pharynx, or skin; members of two genera (*Megalops, Gillichthyes*) use the swim bladder. Most of these species inhabit well-oxygenated water and, if amphibious, switch totally to air-breathing when they are on land. Most freshwater species are fully aquatic in all respects except breathing, although at least five genera are amphibious. In water with little O_2, freshwater air-breathers use the gills and skin for release of

Text continued on page 240.

Table 6–1 Survey of air-breathing teleost fishes. Many of the listed genera also contain species that do not breathe air. Further research will undoubtedly disclose more air-breathers.

Order	Family	Genus	Air-breathing organ	Habitat and range
Freshwater				
Osteoglossiformes	Notopteridae	*Notopterus*	Swim bladder	Flood plains and back waters: Asia
Mormyriformes	Gymnarchidae	*Gymnarchus*	Swim bladder	Swamps and rivers: Africa
Clupeiformes	Phractolaemidae	*Phractolaemus*	Swim bladder	Tropical: West Africa
	Arapaimidae	*Arapaima*	Swim bladder	Swamps and Amazon River: South America
Salmoniformes	Umbridae	*Umbra*	Swim bladder	Stagnant water: Europe and North America
Anguilliformes	Anguillidae	*Anguilla**	Skin	Rivers: Europe, Asia, Africa, North America
Symbranchiformes	Amphipnoidae	*Amphipnous*	Pharyngeal lung	Rivers, swamps, ponds: southern Asia
	Synbranchidae	*Synbranchus**	Opercular chamber	Swamps: South America
		*Monopterus**	Opercular chamber	Swamps, ponds: southern Asia
Cypriniformes	Electrophoridae	*Electrophorus*	Mouth and pharynx	Rivers and swamps: South America
	Sternarchidae	*Hypopomus*	Opercular chamber	Swamps: South America
	Saccobranchidae	*Heteropneustes*	Opercular lung	Ponds and swamps: Southeast Asia
	Clariidae	*Clarias**	Opercular lung	Ponds and swamps: Africa and Southwest Asia
	Loricariidae	*Plecostomus*	Stomach	Swamps: South America
		Ancistrus	Stomach	Swamps: South America
	Cobitidae	*Misgurnus*	Intestine	Slow rivers and pools: Europe and Asia

Table continued on following pages.

Table 6–1 Survey of air-breathing teleost fishes. Many of the listed genera also contain species that do not breathe air. Further research will undoubtedly disclose more air-breathers.

Order	*Family*	*Genus*	*Air-breathing organ*	*Habitat and range*
Cypriniformes (cont.)		*Lepidocephalichthyes*	Intestine	Slow rivers and pools: Asia
	Doradidae	*Doras*	Intestine	Rivers and swamps: South America
	Characinidae	*Erythrinus*	Swim bladder	Swamps: South America
	Callichthyidae	*Hoplosternum*	Intestine	Swamps: South America
Perciformes	Gobiidae	*Pseudapocryptes*	Opercular chamber	Pools and swamps: southern Asia
	Anabantidae	*Macropodus*	Opercular lung	Tropical ponds and ditches: Asia
		Colisa	Opercular lung	Tropical ponds and ditches: southern Asia
		Betta	Opercular lung	Tropical ponds and ditches: southern Asia
		Osphronemus	Opercular lung	Tropical ponds and ditches: southern Asia
		*Anabas**	Opercular lung	Tropical ponds and ditches: Southeast Asia, tropical and southern Africa
	Channidae	*Ophicephalus*	Pharyngeal lung	Rivers, swamps, ponds: southern Asia
Marine				
Clupeiformes	Megalopidae	*Megalops*	Swim bladder	Pelagic shallow water, estuaries: tropical and subtropical
Perciformes	Blennidae	*Anadamia**	Gills, skin	Rock shores: family is mostly tropical and subtropical
		*Alticus**	Gills, skin	Surf zone on coral reefs
		*Istiblennius**	Gills, skin	High on the shore
		*Entomacrodus**	Pharyngeal area, gills, skin	Waves, zones of coral reefs, and rocky shores

Table 6–1 Survey of air-breathing teleost fishes. Many of the listed genera also contain species that do not breathe air. Further research will undoubtedly disclose more air-breathers.

Order	***Family***	***Genus***	***Air-breathing organ***	***Habitat and range***
Perciformes (cont.)		*Blennius**	Gills, skin	High on the shore
	Gobiidae	*Periophthalmus**	Gills, skin, pharyngeal area	Mangrove swamps, rocky intertidal and mud flats: family is mostly tropical and subtropical
		*Boleophthalmus**	Gills, skin, pharyngeal area	Mud, mangrove channels
		*Periophthal-madon**	Gills, skin, pharyngeal area	Mangroves
		*Scartelaos**	Gills, skin, pharyngeal area	Mangroves
		Pseudapocryptes	Gills, skin, pharyngeal area	Estuaries
		Taenioides	Gills, skin, pharyngeal area	Estuaries
		Apocryptes	Gills, skin, pharyngeal area	Estuaries
		Pisoodonophis	Gills, skin, pharyngeal area	Estuaries
		Gobionellus	Gills, skin, pharyngeal area	Mud flats
		*Gillichthyes**	Swim bladder, pharyngeal area	Tidal lagoons, estuaries
		Quietula	Pharyngeal area	Tidal flats
	Eleotridae	*Dormitator*	Skin	Brackish water: family is tropical and subtropical
	Clinidae	*Mnierpes**	Gills, skin	Intertidal: family is mostly tropical
		*Dialommus**	Gills, skin	Intertidal: family is mostly tropical

Table continued on following page.

Table 6–1 Survey of air-breathing teleost fishes. Many of the listed genera also contain species that do not breathe the air. Further research will undoubtedly disclose more air-breathers.

Order	*Family*	*Genus*	*Air-breathing organ*	*Habitat and range*
Gobiesociformes	Gobiesocidae	*Sicyases**	Gills, skin, pharyngeal area	Rocky intertidal: many oceans
		*Phallerodiscus**	Gills, skin	Rocky intertidal: many oceans
		*Gobiesox**	Gills, skin	Rocky intertidal: many oceans
		*Tomicodon**	Gills, skin	Rocky intertidal: many oceans

(From Dehadrai and Tripathi, 1976, and Graham, 1976.)
* Amphibious genera.

CO_2 and have air-breathing organs that are located internally to prevent loss of O_2 to the water. Two freshwater fishes that are exceptions (*Synbranchus, Hypopomus*) breathe air with their gills. Marine air-breathers maintain normal metabolic rates in both water and air; freshwater species frequently have reduced metabolic rates when out of water. Air-breathing species in both marine and freshwater habitats have evolved physiological capacities that permit them to exploit environments little used by other vertebrates.

Amphibians use the mouth lining, skin, and lungs to varying degrees. Even among highly aquatic species, different respiratory organs take precedence and the ability to breathe air is varied. For example, a salamander called the hellbender (*Cryptobranchus alleganiensis*) has no gills, small and poorly vascularized lungs, but a large surface area for cutaneous respiration. Almost all of its respiration is by means of the skin, although in air it is less effective because the body rests in the substrate and thus cuts off some of the skin surface, thereby reducing the hellbender's O_2 consumption. This salamander normally lives in cool, shallow, fast streams. Another highly aquatic salamander, the mudpuppy (*Necturus maculosus*), has large external gills and small lungs. It can extract adequate O_2 from well-aerated water by means of the skin and uses its very efficient gills when less O_2 is available or when activity raises O_2 demands. Mudpuppies can live in a wide variety of freshwater habitats. In air the gills are not very functional and O_2 consumption drops.

The greater siren (*Siren lacertina*) has small external gills, and the skin can account for all necessary gas exchange at low temperatures if the water is well-aerated. At higher temperatures (25°C), the large lungs become increasingly important, however, and, in very stagnant water, air-breathing is essential. Large individuals are obligate air-breathers. Sirens are often found in unaerated and warm waters, sometimes at temperatures as high as 29°C. The strange beast known as the congo eel (*Amphiuma means*) is really a salamander and has no gills. It is an obligate air-breather at 25°C but can manage with cutaneous aquatic respiration at cool temperatures. Although *Amphiuma* is frequently found in warm, swampy water, it also is found on land, burrowed into the damp ground. Both *Amphiuma* and *Siren* can undergo long periods of dormancy and fasting and are the least dependent of these four species on bodies of water for physiological survival (Guimond and Hutchinson, 1976; Ultsch, 1976).

Gas exchange through the skin tends to be less important in the more terrestrial amphibians, except in the lungless plethodontid salamanders, which are small and live in moist places. Temperature affects the relative roles of skin and lungs in both urodeles and anurans. A more or less typical example is found in the southern toad (*Bufo terrestris*): High ambient temperatures result in relatively more O_2 being taken up by the lungs (Figure 6–11). Thus, lungs provide additional O_2 when metabolic rate increases.

Reptiles, birds, and mammals respire mainly by means of lungs, but some gas exchange may take place cutaneously as well. Cutaneous O_2 uptake appears to be minimal in most species, but loss of CO_2 is known to occur cutaneously. Indeed, the large surface area in the membranous

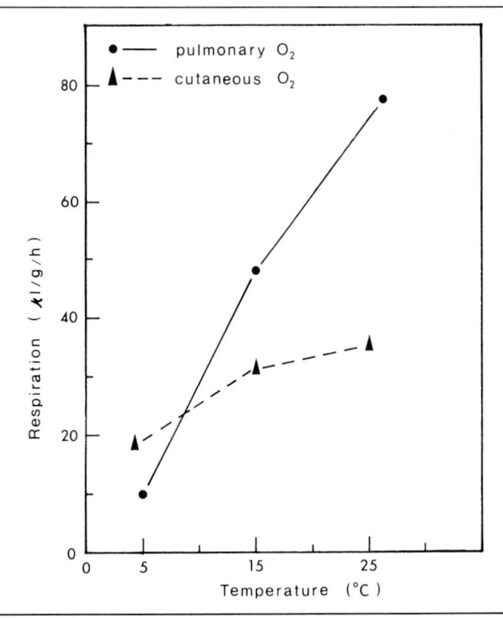

Figure 6–11 The effects of temperature on the relative contribution of skin and lungs to respiration in the European toad, *Bufo terrestris.* (Modified from V.H. Hutchinson, W. Witford, and M. Kohl, 1968. Relation of body size and surface area to gas exchange in anurans. *Physiol. Zool.*, *41*:65–85.)

wings of bats may be responsible for the elimination of fairly large amounts of CO_2 at high environmental temperatures.

Lungs of fishes and amphibians are rather simple affairs: Usually they are not greatly subdivided and, therefore, expose much less surface area than the highly subdivided lungs of birds and mammals. Reptile lungs are generally intermediate in the degree of subdivision. Lungfishes and amphibians commonly fill their lungs by "swallowing" air, forcing it to pass from the mouth into the lungs. In contrast, the lungs of most reptiles and those of birds and mammals are filled by a suction-pump mechanism. The thoracic cavity of these animals is expanded in various ways so that air pressure is greater outside than inside, causing air to flow into the lungs. The rib cage is widened (except in turtles, in which the ribs are fused to the shell), and, in mammals, the muscular diaphragm between the thoracic and abdominal cavities contracts downward.

Birds have a complex system of air passages and air sacs associated with the small (compared to mammals) lungs. The air sacs (Fig. 6–12) do not seem to be associated directly with gas exchange, which takes place in the lungs. Instead, the sacs let air move over the gas-exchange surface of the lungs in one direction (Fig. 6–13), rather than back and forth, as in a typical

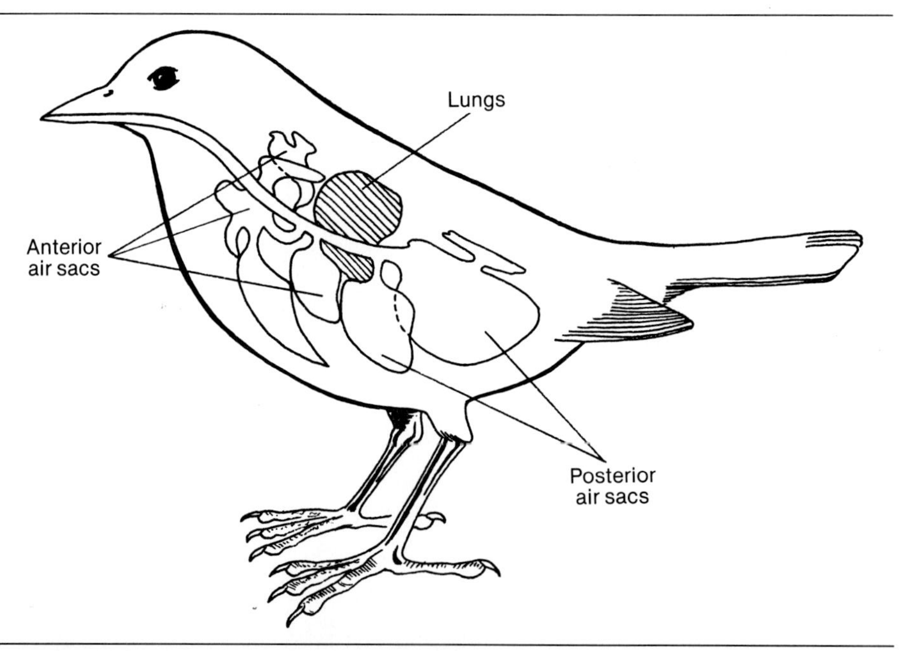

Figure 6–12 Diagram of avian air sacs and lungs. (Modified from K. Schmidt-Nielsen, 1979. *Animal Physiology: Adaptation and Environment.* Cambridge University Press, Cambridge, England.)

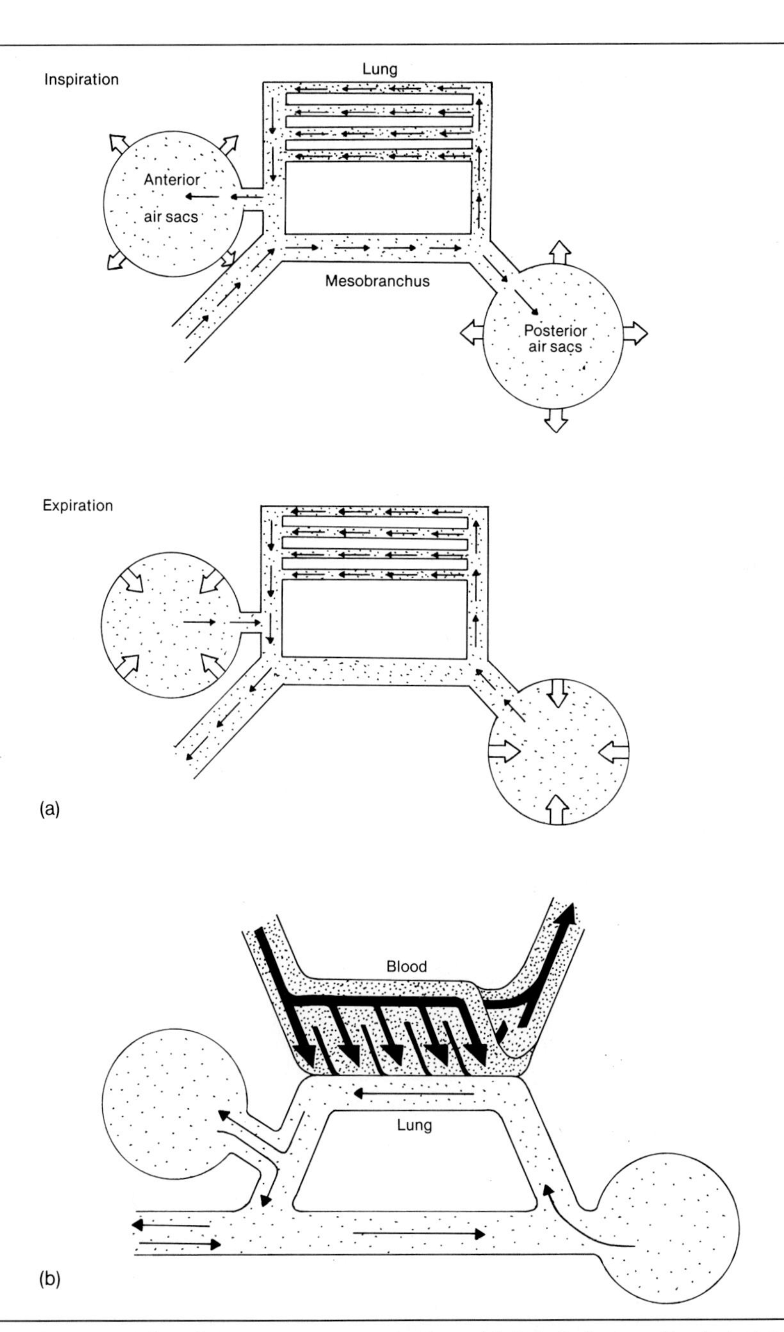

Figure 6–13 Patterns of airflow in avian respiration. (*a*) Inhalation and exhalation. (From K. Schmidt-Nielsen, 1975. Recent advances in avian respiration. *Symp. Zool. Soc. Lond.*, *35*:33–46. Reproduced with permission of the author and the Zoological Society of London.) (*b*) The crosscurrent flow in the avian lung. (Modified from P. Scheid and J. Piiper, 1972. Cross-current gas exchange in avian lung: effects of reversed parabracheal air flow in ducks. *Respir. Physiol.*, *16*:304–312.

mammalian lung. When air is inhaled, it passes to the posterior air sacs, then to the lung surfaces, the anterior sacs, and, finally, outward. The single direction of flow is advantageous in that it permits continual flow over the respiratory surfaces during both inhalation and exhalation. Furthermore, the flow pattern of air relative to the flow patterns of blood is unusual; it is called crosscurrent flow because the blood flows at approximately right angles to the air. Structurally different from a countercurrent exchanger, this arrangement nevertheless produces a similar effect: The transfer of O_2 to the blood is enhanced.

Although a good deal remains to be learned about the functioning of the avian system, it does appear to be effective. Tucker (1968) showed that at experimentally lowered O_2 pressures house sparrows were able to continue flying when house mice of the same size (with the same ability to carry O_2 in the blood and similar metabolic rates) could barely crawl. Birds can function actively at higher altitudes than mammals can. Not only is their respiratory system apparently more efficient, but also blood flow to the brain can be maintained at normal levels. At the low O_2 pressure found at high altitudes, animals breathe fast and expel large amounts of CO_2. As a result, the blood becomes too alkaline and, in mammals, blood vessels constrict, reducing the flow of blood to the brain. Blood-vessel constriction doesn't occur in ducks (or presumably in other birds), however, and the brain continues to receive a normal blood supply. Many migratory birds can clearly profit from such physiological adaptations on high-altitude migratory flights.

Mammals and birds that dive regularly have special problems of respiration because they are dependent on O_2 in air, but some aquatic mammals can remain submerged for amazing periods of time—even an hour or more. How do they manage this? We might suspect that they have some means of increasing their O_2 supply. Indeed they do, but not by increasing the size of the lungs, which might seem logical. Lung size of diving mammals is just what would be expected for animals of their body sizes; the lungs of some whales are even smaller than expected. Furthermore, many diving mammals are known to exhale before a dive and allow their lungs to collapse in response to increasing water pressure during a dive. Instead of large lungs, avian and mammalian divers often have greater blood volumes than nondivers, and considerable quantities of O_2 can be carried. In addition, higher quantities of hemoglobin are present in the muscles of divers than in the muscles of nondivers.

The storage of O_2 in the blood and muscles rather than in the lungs is important in preventing the disorder known as the bends. Increased pressure during a deep dive tends to force gases into the blood. When the diver surfaces, a too-sudden decrease of pressure allows the gases in the blood to form bubbles—a phenomenon similar to that of popping the top of a bottle of soda pop. Such bubbles get stuck in small blood vessels and can result in death. Human divers solve the problem by decompressing slowly. Vertebrates adapted for diving solve it by diving with little air. Excess dissolved gases (especially nitrogen) in the blood of whales may be dissolved in fat droplets and expelled in foam during the "blow."

Accomplished divers typically undergo pronounced changes in circulation during a dive. The heart rate often slows markedly, but this is not evident in certain free-diving birds (Kanwisher et al., 1981). Although normal circulation to heart and brain are maintained, blood is shunted away from many internal organs, skin, and skeletal muscles. These muscles then switch to anaerobic metabolism and, as a result, lactic acid builds up. After a dive, divers are able to quickly restore normal metabolism. They also can exchange the air used during a dive for a fresh supply at the completion of a dive more quickly than can nondivers.

The extent to which metabolic rates may be reduced during a dive is uncertain. Submerged, resting animals in the laboratory often exhibit low metabolic rates, but active, swimming animals in more natural circumstances may not.

SUMMARY

Oxygen is necessary for the oxidation of foods to obtain energy, and one end product of metabolism—namely, CO_2—must be eliminated by respiration. Oxygen requirements vary with metabolic rate, and, hence, with body size, activity, and thermoregulatory capacity. Oxygen is more readily available in air than in water, but many fishes are capable of very efficient extraction of O_2 from water. One major means of accomplishing such efficiency is the countercurrent flow of water and blood through the gills. Air-breathing vertebrates use many different organs to obtain O_2: lungs, which predominate in birds, mammals, most reptiles, and a few fishes; skin (very important in many amphibia and some turtles and fishes); and various portions of the gastrointestinal tract. Gas exchange in birds seems to be more effective than in other lunged species because of the air sacs that permit unidirectional flow of air through the lungs.

Oxygen-carrying capacity of vertebrate blood differs considerably among species and is also varyingly sensitive to changes in such factors as temperature, acidity, and CO_2 concentrations. Many such differences can be closely correlated with characteristics of different species, such as habitat, activity, body size, and metabolic rate. The single circulatory system of typical fishes means that oxygenated blood from the gills is delivered to tissues at fairly low pressures. Most air-breathing vertebrates have a tendency toward a double circulatory system: Oxygenated blood can usually be delivered to tissues at relatively high pressures, but the lung circulation can operate at lower pressures. The most effective separation of blood flow to lungs and body is found in birds and mammals, which achieve very high metabolic rates and endothermy.

Diving vertebrates often have a larger blood volume and O_2-storage capacity than nondivers, but their major adaptation to this habit involves change in circulation during a dive.

SELECTED REFERENCES

Alexander, R.M., 1975. *The Chordates.* Cambridge University Press, Cambridge, England.

Andersen, H.T. (Ed.), 1969. *The Biology of Marine Mammals.* Academic Press, New York.

Bellairs, A., 1970. *The Life of Reptiles.* Vol. 1. Universe Books, New York.

Butler, P.J., 1976. Gas exchange. *In* J. Bligh, J.L. Cloudsley-Thompson, and A.G. MacDonald (Eds.), *Environmental Physiology of Animals,* pp. 164–195. John Wiley & Sons, New York.

Dehadrai, P.V., and S.D. Tripathi, 1976. Environment and ecology of freshwater air-breathing teleosts. *In* G.M. Hughes (Ed.), *Respiration of Amphibious Vertebrates,* pp. 39–72. Academic Press, New York.

Douglas, E.L., W.A. Friedl, and G.V. Dickwell, 1976. Fishes in oxygen-minimum zones: blood oxygenation characteristics. *Science, 191*:957–959.

Erichsen Jones, J.R., 1964. *Fish and River Pollution.* Butterworth, London.

Fry, F.E.J., 1957. The aquatic respiration of fish. *In* M.E. Brown (Ed.), *The Physiology of Fishes.* Vol. 1, pp. 1–63. Academic Press, New York.

Gans, C., and W.R. Dawson, 1976. *Biology of the Reptilia.* Vol. 5. Academic Press, New York. (Contains several relevant survey chapters.)

Gordon, M.S., et al., 1977. *Animal Function: Principles and Adaptations.* 3rd ed. Macmillan Publishing Co., New York.

Graham, J.B., 1976. Respiratory adaptations of marine air-breathing fishes. *In* G.M. Hughes (Ed.), *Respiration of Amphibious Vertebrates.* Academic Press, New York.

Graham, J.B., R.H. Rosenblatt, and C. Gans, 1978. Vertebrate air breathing arose in freshwaters and not in the oceans. *Evol., 32*:459–463.

Guimond, R.W., and V.H. Hutchinson, 1976. Gas exchange of the giant salamanders of North America. *In* G.M. Hughes (Ed.), *Respiration of Amphibious Vertebrates,* pp. 313–338. Academic Press, New York.

Hall, F.G., and F.H. McCutcheon, 1938. The affinity of hemoglobin for oxygen in marine fishes. *J. Cell Comp. Physiol., 11*:205–212.

Hemmingsen, E.A., and E.L. Douglas, 1977. Respiratory and circulatory adaptations to the absence of hemoglobin in chaenichthyid fishes. *Proc. SCAR Symp. Antarctic Biol., 3*:479–487.

Hill, R.W., 1976. *Comparative Physiology of Animals.* Harper and Row, New York.

Hochachka, P.W., 1981. Brain, lung, and heart functions during diving and recovery. *Science, 212*:509–514.

Hochachka, P.W., and K.B. Storey, 1975. Metabolic consequences of diving in animals and man. *Science, 187*:613–621.

Hochachka, P.W., G.C. Liggins, J. Ovist, R. Schneider, M.Y. Snider, T.R. Wonders, and W.M. Zapol, 1977. Pulmonary metabolism during diving: conditioning blood for the brain. *Science, 198*:831–834.

Hughes, G.M., 1973. Comparative vertebrate ventilation and heterogeneity. *In* L. Bolis, K. Schmidt-Nielsen, and S.H.P. Maddrell (Eds.), *Comparative Physiology,* pp. 187–220. North Holland, Amsterdam.

Hughes, G.M., 1974. *Comparative Physiology of Vertebrate Respiration.* 2nd ed. Heinemann, London.

Hughes, G.M. (Ed.), 1976. *Respiration of Amphibious Vertebrates.* Academic Press, New York.

Hutchinson, V.H., W.G. Whitford, and M. Kohl, 1968. Relation of body size and surface area to gas exchange in anurans. *Physiol.*

Zool., 41:65–85.

Jakubowski, M., and J.M. Rembiszewski, 1974. Vascularization and size of respiratory surfaces of gills and skin in the Antarctic fish *Gymnodraco acuticeps* Boul. (Bathydraconidae). *Bull. Acad. Polon. Sci.* (Biol.), *22*:305–313.

Johansen, K., 1970. Air breathing in fishes. *In* W.S. Hoar and D.J. Randall (Eds.), *Fish Physiology*. Vol. 4, pp. 361–411. Academic Press, New York.

Johansen, K., C.P. Magnum, and G. Lykkeboe, 1978. Respiratory properties of the blood of Amazon fishes. *Can. J. Zool.*, *56*:898–906.

Kanwisher, J.W., G. Gabrielsen, and N. Kanwisher, 1981. Free and forced diving in birds. *Science*, *211*:717–719.

Kooyman, G.L., 1972. Deep diving behaviour and effects of pressure in reptiles, birds, and mammals. *Symp. Soc. Exp. Biol.*, *26*:285–311.

Kramer, D.L., et al., 1978. The fishes and the aquatic environment of the central Amazon basin, with particular reference to respiratory patterns. *Can. J. Zool.*, *56*:717–729.

Lasiewski, R.C., 1972. Respiratory function in birds. *In* D.S. Farner and J.R. King (Eds.), *Avian Biology*. Vol. 2, pp. 288–342. Academic Press, New York.

Liem, K., 1981. Larvae of air-breathing fishes as countercurrent flow devices in hypoxic environments. *Science*, *211*:1177–1179.

Muir, B.S., 1969. Gill dimensions as a function of fish size. *J. Fish. Res. Bd. Can.*, *26*:165–170.

Norris, K.S. (Ed.), 1966. *Whales, Porpoises, and Dolphins*. University of California Press, Berkeley.

Packard, G.C., 1974. The evolution of air-breathing in paleozoic gnathostome fishes. *Evol.*, *28*:320–325.

Piiper, J., 1978. *Respiratory Function in Birds, Adult and Embryonic*. Springer-Verlag, New York.

Piiper, J., and P. Scheid, 1973. Gas exchange in avian lungs: models and experimental evidence. *In* L. Bolis, K. Schmidt-Nielsen, and S.H.P. Maddress (Eds.), *Comparative Physiology*, pp. 161–185. North Holland, Amsterdam.

Randall, D.J., 1968. Functional morphology of the heart in fishes. *Am. Zool.*, *8*:179–189.

Randall, D.J., 1970. Gas exchange in fish. *In* W.S. Hoar and D.J. Randall (Eds.), *Fish Physiology*. Vol. 4, pp. 253–292. Academic Press, New York.

Riggs, A., 1960. The nature and significance of the Bohr effect in mammalian hemoglobins. *J. Gen. Physiol.*, *43*:737–752.

Roberts, J.L., 1975. Respiratory adaptations of aquatic animals. *In* F.J. Vernberg (Ed.), *Physiological Adaptation to Environment*, pp. 395–414. Intext, New York.

Robilliard, G.A., and P.K. Dayton. 1969. Notes on the biology of the chaenichthyid fish *Pagetopsis macropterus* from McMurdo Sound, Antarctica. Antarctic J. U.S., 4:304–306.

Scheid, P., 1980. Ventilation and gas exchange in the lung. *Proc. Int. Ornith. Cong.*, *17*:355–359.

Scheid, P., and J. Piiper, 1972. Crosscurrent gas exchange in avian lungs: effects of reversed parabronchial air flow in ducks. *Respir. Physiol.*, *16*:304–312.

Schmidt-Nielsen, K., 1970. Energy metabolism, body size, and problems of scaling. *Fed. Proc.*, *29*:1524–1532.

Schmidt-Nielsen, K., 1975. Recent advances in avian respiration. *Symp. Zool. Soc. Lond.*, *35*:33–46.

Schmidt-Nielsen, K., 1979. *Animal Physiology*. 2nd ed. Cambridge University Press, Cambridge, England.

Seymour, R.S., and H.B. Lillywhite, 1976. Blood pressure in snakes from different habitats. *Nature*, *264*:664–666.

Snyder, G.K., 1977. Blood corpuscles and blood hemoglobin: a possible example of coevolution. *Science*, *195*:412–413.

Tucker, V.A., 1968. Respiratory physiology of house sparrows in relation to high-altitude flight. *J. Exp. Biol.*, *48*:55–66.

Ultsch, G.R., 1976. Eco-physiological studies of some metabolic and respiratory adaptations of sirenid salamanders. *In* G.M. Hughes (Ed.), *Respiration of Amphibious Vertebrates*, pp. 287–312. Academic Press, New York.

Waterman, A.T., et al., 1971. *Chordate Structure and Function*. Macmillan Publishing Co., New York.

7 Locomotion

Virtually all vertebrates are actively mobile, in contrast to the passive mobility of wind-dispersed seeds or the water-borne eggs and larvae of some invertebrates. Actively mobile animals expend energy in order to move around. In return for this expenditure, they are able to exploit food and habitat resources that are not available to other organisms. With the ability to move comes the necessity of more or less complex sense organs and behavior patterns that permit the animal to distinguish objects in the environment and react to them appropriately. Mobility, along with the accompanying behavioral and sensory capacities, also makes possible the elaboration of diverse social systems that are impossible for passive creatures.

Vertebrates have evolved an impressive array of locomotor abilities. Fishes, as a group, are basically swimmers, but some species burrow, walk, leap, or soar. Among the amphibians are walkers, jumpers, burrowers, and gliders. Reptiles may run or undulate over the ground, climb, burrow, swim, or glide. Mammals are basically walkers and runners, but some characteristically glide, fly, jump, climb, brachiate (swing by the forelimbs), burrow, or swim. Birds may be the least varied of the vertebrate classes in their spectrum of locomotor styles. Most modern birds fly and walk or hop, but some swim or climb and a few burrow. Some birds have lost the ability to fly, although none are completely without wings.

Vertebrates have several ways of running, swimming, or flying, each of which may entail distinct advantages and costs. The purpose of this chapter is to outline this locomotor variety and to indicate, whenever possible, the problems and possibilities associated with the different styles.

The basic feature of all locomotion is the application of a force upon the animal's center of gravity. When animals walk, they push against the earth with their feet, but the earth resists their push because it is so much heavier than they are (Fig. 7–1). Muscular power within the body is exerted against the system of skeletal levers (and in sharks, against the hydrostatically stiffened skin). These levers apply the forces to the body's center of gravity and to the earth or surrounding medium. Friction helps prevent the feet of a walking animal from slipping on the surface, and the force of the push

Figure 7–1 An animal propels its body forward by pushing backward against the ground with its foot. As it steps off the spring-mounted platform, the propulsive force can be measured from the compression of the spring *S*. If the platform did not resist the thrust of the foot, the force applied by the foot would simply push the platform back without producing forward movement of the animal. (Modified from J. Gray, 1953. *How Animals Move.* Cambridge University Press, Cambridge, England.)

results in forward movement. Much of the force is translated into motion of the body because little energy is spent moving the earth or slipping.

Muscular effort sometimes can be exerted quite efficiently. In a vertical jump by a bipedal animal, most of the force is applied perpendicularly to the substrate and can be translated directly into upward motion. Many forces resulting in movement are applied obliquely to the substrate or medium, however, so part of the total expenditure results in a tendency toward nonforward motion.

This problem may be analyzed by means of vectors. Forces are vector quantities in that they have both direction and magnitude; they can be represented graphically by arrows whose orientation indicates the direction of the force and whose length indicates the magnitude (Fig. 7–2). When a frog jumps, it leaves the ground at an angle that can be resolved into a forward and a vertical component. The vertical component does not contribute directly to the forward-force component. Nevertheless, it contributes to forward motion by increasing the amount of time over which the forward thrust is effective, since the foot is on the ground while the body is rising, and in this way increases the distance traveled. Another example of an oblique force is provided by a swimming fish whose tail pushes against the water. The oblique resultant force can be resolved into a lateral and a forward component. The forward component creates forward motion, but the lateral component produces movement of both the medium and the anterior end of the fish.

These two examples illustrate that much of the energy expended in

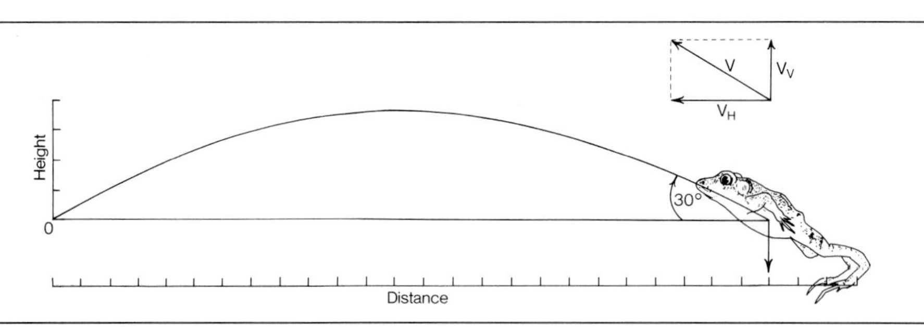

Figure 7–2 A vector diagram used in analyzing the forces involved in locomotion. In this example, a frog is jumping: Its feet push against the substrate, which resists with an equal and opposite force, labeled V in the diagram. Clearly, there is both an upward and a forward component to this force, since a jumping frog moves both up and forward. These two components are labeled V_V (vertical) and V_H (horizontal). (Modified from C. Gans, 1974. *Biomechanics: An Approach to Vertebrate Biology.* J.B. Lippincott, Philadelphia.)

locomotion may not be directly involved with propulsion. An introduction to the theory and mechanics of locomotion has been presented by Tricker and Tricker (1967) and more detailed discussions by Gray (1968), Alexander (1967, 1968), and Alexander and Goldspink (1977). A fascinating compendium of information on the effects of size on locomotion is presented by Pedley (1977).

TERRESTRIAL LOCOMOTION

Quadrupedal Walking and Running

Fundamental skeletal differences between fishes and terrestrial vertebrates indicate some basic differences between motion in water and motion on land. The pectoral girdle in fishes is attached to the skull, and the pelvic girdle is not attached to the vertebral column (see Fig. 2–7). This skeletal arrangement is clearly unsuitable for extensive, habitual movement on land. Much of the force of the hind limb pushing back on the ground would be absorbed by the compressible intervening muscles and not be transmitted to the spine to move the body forward. Furthermore, the jolt produced by the forelimb against the ground could create headaches and bad aim in prey capture. In contrast, the limb articulations of tetrapods (see Figs. 2–9, 2–11) are adapted to movement over a firm substrate. The pectoral girdle is connected to the vertebral column in a variety of taxon-specific arrangements; the pelvic girdle is firmly anchored to the spine.

Why have four limbs for locomotion? Four legs (the number defining a quadruped) are the minimum for stability in walking. At a slow walk, one foot at a time is raised off the ground and the animal's center of gravity is always over the triangle formed by the other three feet (Fig. 7–3). At a fast walk, two feet often are off the ground at the same time and some instability accompanies the increase in speed. When the animal is running, static mechanical stability is sacrificed for speed and most, or all, of the feet may be off the ground simultaneously. An animal may be more easily thrown off balance (and falling is therefore more probable) at faster gaits. Six legs (the number possessed by insects) make an animal very stable because three of the six legs could always be on the ground even at a fast walk, but an increase in the number of legs necessitates an increase in total limb weight (Alexander, 1971). Insects are generally much smaller than vertebrates and much more likely to be tossed about by air currents, so six-legged stability may be particularly advantageous for them. The effect of wind on vertebrates is much less because their surface area is much smaller relative to their volume. Vertebrates evidently have been able to manage successfully with four legs.

Salamanders, lizards, crocodilians, monotremes, and a number of therian mammals stand with their legs out to the side. When one of these animals walks or runs, the body usually undulates laterally and the limb

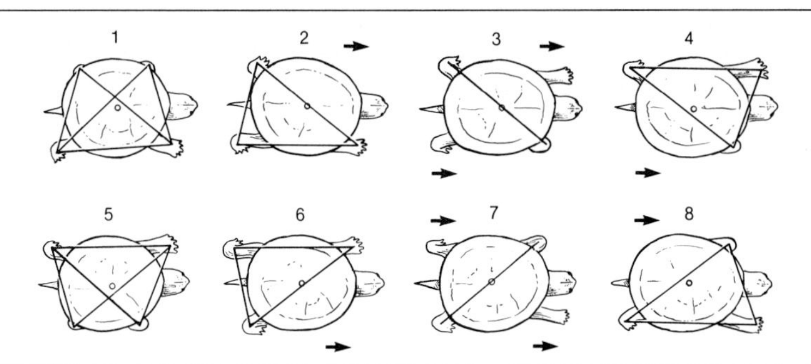

Figure 7–3 A walking turtle, *Chrysemys picta,* moving one foot at a time. Note that the center of gravity (marked by the circle on top of the shell) remains within the triangle formed by three stationary feet. (Modified from W.F. Walker, 1971. A structural and functional analysis of walking in a turtle, *Chrysemys picta marginata. J. Morphol., 134*:195–214.)

swings forward (Fig. 7–4). Bending the body increases the length of the stride; a heavy tail counterbalances the lateral body motion. Lateral limb orientation creates a posture that requires the expenditure of energy just to keep the body off the ground. Moreover, lateral undulation for terrestrial locomotion necessitates an expenditure of energy to move the whole weight of the body and tail sideways in opposite directions. This limb orientation apparently works well enough for vertebrates that move slowly, weigh little, or spend part of their time in water.

One would not expect to find legs out to the side in very active or very heavy animals; for them, it is advantageous to have the legs placed vertically under the body (Fig. 7–5). Birds, many mammals, some of the ancient dinosaurs, and some lizards and crocodilians (when running) have legs that are shifted toward the vertical position. In addition, the very slow-moving arboreal chameleons (Chamaeleontidae) have legs oriented vertically under the body, which permits them to clasp a branch effectively. This arrangement means that less muscular effort is spent holding the body off the ground while standing. Furthermore, the points of support (the feet) are below the center of gravity, permitting the animal to run and jump more effectively.

One can think of the center of gravity as a balance point for the animal. Although the center of gravity may shift with changes in posture and with the eating of a meal, it is convenient to regard it as being in one place. Gravity acts on the body as if all the mass were concentrated at that point called the center of gravity. If an animal were somehow suspended from its center of gravity, it would not tend to rotate: neither end or side would rise or sink. When support points are close to the center of gravity, the vertical thrust of the limbs is more directly applied to moving the weight upward

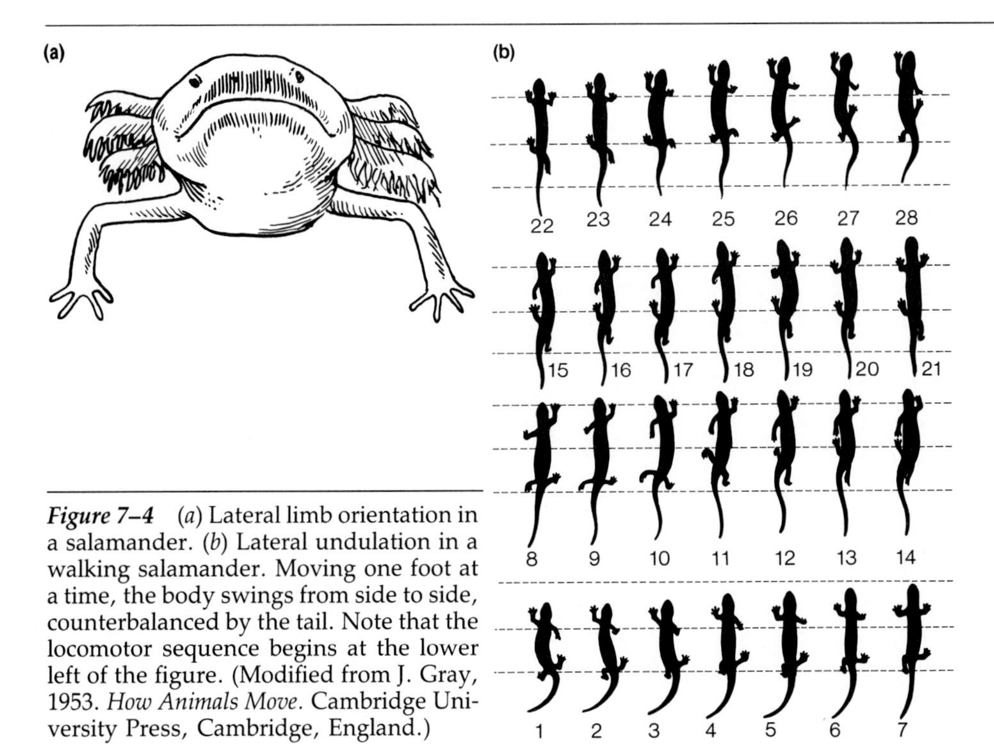

Figure 7–4 (*a*) Lateral limb orientation in a salamander. (*b*) Lateral undulation in a walking salamander. Moving one foot at a time, the body swings from side to side, counterbalanced by the tail. Note that the locomotor sequence begins at the lower left of the figure. (Modified from J. Gray, 1953. *How Animals Move*. Cambridge University Press, Cambridge, England.)

Figure 7–5 Vertical limb orientation in a scarlet ibis (*Eudocimus ruber*).

because the lateral force components are less than those generated by a laterally directed limb. Relatively more force is available for vertical motion, therefore, and there is less tendency for the body to rotate about a horizontal axis.

Mammals are characteristically quadrupedal and have ventral limb orientation, especially in the fast-moving and large-bodied species. Lateral motion of the vertebral column, a characteristic of quadrupedal lizard and salamander locomotion, is typically absent in mammals; therefore, the center of gravity tends to remain over the limbs. The heavy, counterbalancing tail that accompanies lateral undulation on land would be disadvantageous—an unnecessary and unbalancing weight—in the absence of sideways flexion. Terrestrial quadrupedal mammals usually have small, light tails.

Medium and small mammalian cursors (runners) generally have a vertical flexure of the vertebral column, especially in the lumbar region (Fig. 7–6). Like lateral flexion, this feature increases stride length but without the necessity of sideways shifts of weight. Vertical undulations are exhibited emphatically by mustelids such as weasels and mink (*Mustela*). These species commonly hunt by following their prey into burrows and crevices

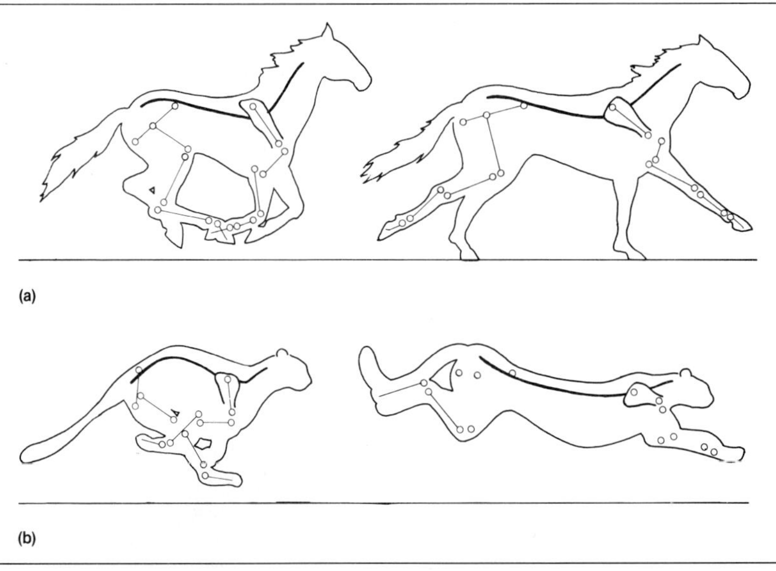

Figure 7–6 Vertical flexure of the spine in a running cheetah compared to the stiffer spine of a running horse. The cheetah's stride is longer relative to its body length than the horse's stride. (Modified from J. Gray, 1953. *How Animals Move.* Cambridge University Press, Cambridge, England. Adapted from M. Hildebrand, 1959. Motions of the running cheetah and horse. *J. Mammol.*, *40*:461–495.)

where short legs are advantageous. Weasels are also able to move rapidly in the open and over obstacles because their elongate, flexible body permits a bounding gallop. Cursorial carnivores have also increased their effective leg length by standing on their toes; thus, the metatarsals and metacarpals (foot and hand bones) contribute to the length of the limb.

In contrast, the ungulates, especially the larger, heavier species, have a stiff lumbar column. Stride length in ungulates has been increased primarily by elongation of the limbs (Fig. 7–7). Distal limb segments (those distant from where the leg is attached to the body) are commonly elongated. Moreover, the toes themselves are incorporated into the legs along with the hand or foot bones, so that most ungulates actually stand on their nails (or hooves).

The basic mammalian body plan has the heavy muscles proximal (near the point of limb attachment) in the limb; distal limb segments are lighter. This basic plan is exaggerated in the ungulates; many distal muscles are replaced by lighter tendons connected to the proximal muscle masses. The proximal concentration of limb mass reduces the arc through which the heavy muscles must be moved and, hence, the energy expenditure for moving the limb. The basic five digits have been reduced to one or two in the most cursorial ungulate species. The remaining distal limb bones are larger and stronger than any one of the multiple distal bones of the basic tetrapod design, but the total weight of the distal elements is less. Therefore, less energy may be required to swing the distal portion of the limb back and forth.

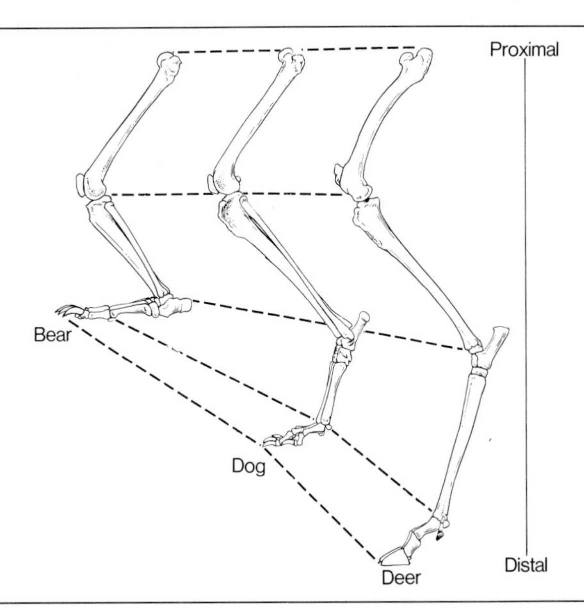

Figure 7–7 Relative lengthening of the distal limb segments in increasingly cursorial mammals. (Modified from M. Hildebrand, 1974. *The Analysis of Vertebrate Structure.* John Wiley & Sons, New York.)

Digit loss also reduces the risk of breaking the lateral digits. For a heavy animal running over uneven ground, this risk might be far from trivial. Even lightweight racing dogs, which have multiple digits, have been known to break their side toes while running on a level track. Reduction of the number of digits has been accompanied by loss or reduction of the muscles formerly used for manipulatory and twisting motions. This reduction further lightens the distal segments, of course, but it also means that the bracing effect of such muscles around the joints is lost. Instead, cursorial mammals have limb joints modified to move in only one plane, like a hinge. Joint surfaces tend to be broader and flatter or have complementary ridges and grooves oriented in the plane of motion of the joint and the animal.

The anatomy and function of the shoulder in all cursorial mammals are different from those of nonrunners. In mammals not adapted for running (and in birds, most reptiles, and amphibians), the scapula (shoulder blade) is typically held in place by the clavicle (collarbone) and perhaps other bony elements. In cursorial mammals, however, the clavicle is reduced or absent, allowing the scapula to rotate. Furthermore, the rib cage of cursors is deep and narrow, so that the scapula can rotate in the same plane as the legs are moving (Fig. 7–8). Scapular freedom effectively lengthens the legs and the stride by moving the leg pivot from the shoulder joint to a point somewhere up on the scapula. A somewhat similar arrangement is found in arboreal lizards such as *Chamaeleo dilepis* and some *Anolis;* their equivalent of the shoulder blade is not braced by a clavicle and thus is allowed to rotate on the body wall. In such species, this arrangement of the bones

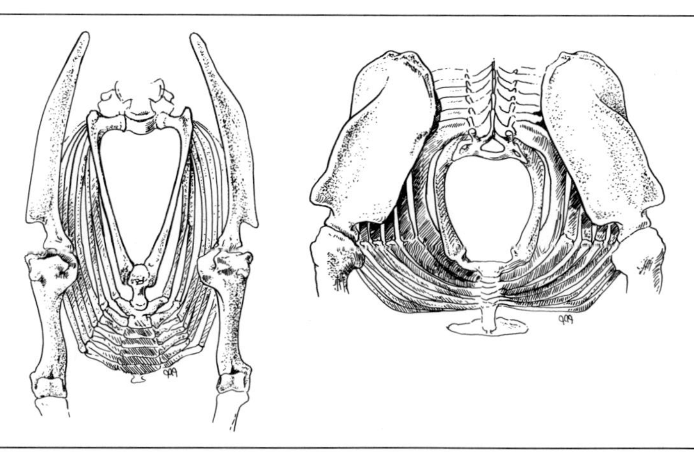

Figure 7–8 Thorax shape, position of scapula, in (*a*) a cursor (*Cervus canadensis*) and (*b*) a noncursor (*Manis javanicus*).

apparently permits a better grasp and balance on twigs and branches (Peterson, 1971).

The speeds achieved by cursorial vertebrates are quite respectable. Jackrabbits (*Lepus*), coyotes (*Canis latrans*), and red foxes (*Vulpes*) may run 65 to 70 km/hr. The cheetah (*Acinonyx jubatus*) can achieve 110 km/hr over short distances. The North American pronghorn (*Antilocapra americana*) can step out at 98 km/hr. A horse may sustain a run for 80 km at an average speed of more than 18 km/hr or sprint for a short distance at a speed of 70 km/hr. (But if the distance is short enough, perhaps the length of a basketball court, from a standing start a human can beat a horse.) Small crocodilians can gallop at an average speed of 50 km/hr (Zug, 1974).

Bipedalism and Saltation

Some vertebrates use only two limbs for locomotion; that is, they move bipedally. In these species, the hind foot is elongated and the body is held so that the center of gravity lies above the feet. Thus, a significant degree of stability is ensured as long as the feet are on the ground.

The fastest lizards are able to run bipedally. Species of *Crotaphytus* in the deserts of North America switch from quadrupedal to bipedal propulsion when great speed is required. Although the forelegs may still touch the ground, they apparently are used more for balance than for propulsion. *Crotaphytus* uses bipedal, high-speed running when attacking its prey (often other, smaller lizards) and to escape predators or unsuitable patches of habitat, such as very hot sand. The bipedal basilisk lizards (*Basiliscus*) of Central and South America are also known as "Jesus lizards" for their ability to run on water (Fig. 7–9). Basilisks (and other bipedal lizards) have elongated hind limbs that make a long stride possible. The surface area of their feet on the downstroke is very large: This feature increases support by distributing resistance of the water over a large area. The tail is an essential balance in bipedal lizard locomotion; without it, a bipedal lizard tends to fall forward or sideways.

All birds are bipedal and their legs are arranged beneath the body. Some hop, and others walk and run. Most birds are active animals with high metabolic rates and are rather agile on their feet. Avian cursors include the roadrunners (*Geococcyx*) and ostrich (*Struthio*), which can achieve speeds of 25 km/hr and 80 km/hr, respectively. Surface swimmers such as ducks employ a locomotor pattern that is similar to walking. Diving birds, including penguins and loons, have sacrificed some of their hind-limb agility for hydrodynamic advantages: Their legs are displaced posteriorly, where they operate more successfully in submerged swimming. This leg position makes walking awkward and clumsy; walking is less economical for penguins than for other birds or quadrupedal mammals (Pinshow et al., 1977).

Most bipedal mammals and amphibians move rapidly by means of saltation, or jumping. Bipedal saltators include kangaroos, kangaroo rats (*Di-*

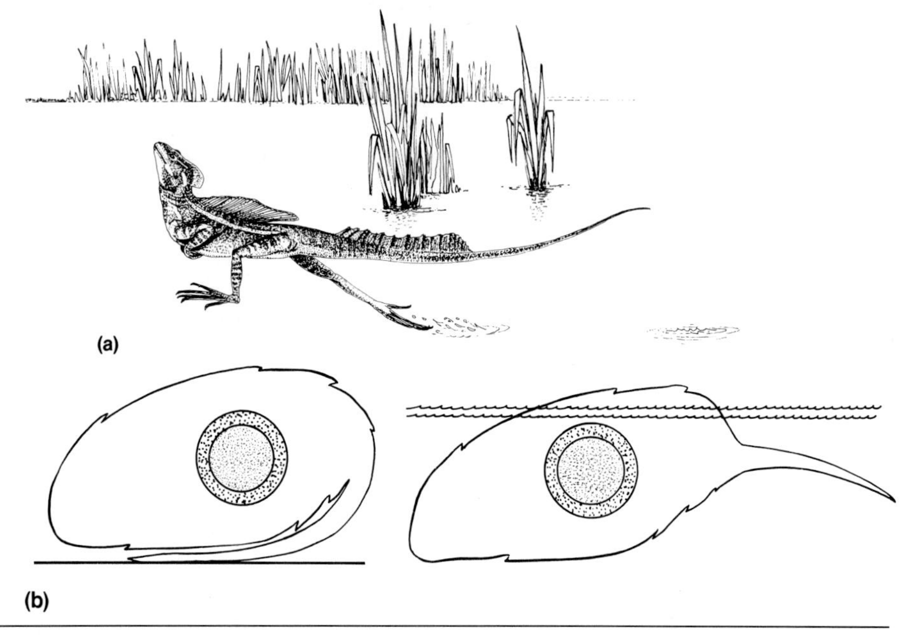

Figure 7–9 (*a*) A basilisk lizard running on the water. (*b*) The toe of a basilisk when resting on a solid substrate (left) and when pushing against the water (right). (Modified from J. Laerm, 1973. Aquatic bipedalism in the basilisk lizard: the analysis of adaptive strategy. *Am. Midl. Nat., 89*:314–333.)

podomys) and several other rodents, tarsiers (*Tarsius*) and galagos (*Galago*) among the primates, and frogs and toads. Rabbits and hares derive their main propulsive thrust from their hind limbs, but they run quadrupedally and are cursors rather than saltators.

Bipedal saltation allows more rapid acceleration from a resting position and more rapid changes of speed and direction than are possible for similar species using quadrupedal locomotion. Therefore, saltation is probably one means of reducing the risk of predation.

The body weight in most saltatorial vertebrates is concentrated over the hind limbs, and the trunk is compact. The lumbar region of the spine is particularly sturdy. These features bring the center of gravity close to the points of support and reduce excessive bending of the spine, which would shift the center of gravity. Perhaps the most obvious anatomical features associated with jumping are elongated hind legs and feet. The long hind limbs keep the feet on the ground, providing continual forward and upward thrust, while the bulk of the body rises. Furthermore, the long feet stay under the animal's center of gravity for a relatively long period of time, which contributes to a balanced jump. Saltatorial mammals typically possess conspicuous tails that act as balancers during a jump. Kangaroos have heavy caudal cantilevers. Kangaroo rats have lighter, but long and furry, tails that serve as stabilizers and orientation controls. Anurans of course characteristically do not have tails as adults. Their long hind legs are kept extended during a jump and contribute to balance, but jumping frogs and toads seem less well balanced than saltatorial mammals—on land, an anuran in a hurry often tips over when it completes a jump.

Locomotion without Limbs

Limbless locomotion among land vertebrates is found in caecilian amphibians, snakes, a few lizards, and amphisbaenians. Caecilians and amphisbaenians are fossorial forms that spend virtually their entire lives in underground burrows. The skulls of diggers in all classes of vertebrates (Fig. 7–10) tend to be very solid and compact, especially in species that use the head for digging, as limbless species must. They burrow by wedging their heads into the soil and pushing—sometimes straight ahead, sometimes from side to side, or sometimes up and down.

The amphisbaenians show a variety of head shapes and scale patterns that can be correlated, at least in part, with digging styles (Fig. 7–11). Round-headed species appear to be found in looser soil than wedge-headed ones. Species with the cranial wedge oriented in the vertical plane press alternately left and right to compress the soil and use only half the axial muscle mass (right or left) at a time. Those with the cranial wedge oriented horizontally, like a shovel, compress the soil primarily with the upper surface of their snouts and can bring to bear a large proportion of the axial muscle mass for each stroke. These species, in conjunction with their greater digging power, tend to be larger and stouter than those with a vertical cranial wedge (Gans, 1974). Forward propulsion is provided by several means, including lateral undulation, rectilinear motion, and concertina movement, which are discussed in the following paragraphs as means of locomotion in snakes.

The fundamental mode of locomotion in snakes is lateral undulation. Long and slender snakes, such as the blue racer (*Coluber constrictor*) of North America, can use this mode to better advantage than heavy-bodied snakes like the African gaboon viper (*Bitis gabonica*). In the laboratory, a snake can be made to wriggle forward across a smooth board that is studded with equally spaced rotatable pegs. The forces applied to the pegs as the snake presses against them can be monitored (Fig. 7–12). When this is done, it is seen that the body must press back against at least three pegs simultaneously for forward motion to be achieved by lateral undulation. As the muscles on one side contract, the obliquely backward thrust of the body loop creates an equal and opposite force of the peg on the snake's body. One component of that oblique force has a forward orientation and contributes to forward progress. With more lateral loops in the snake's

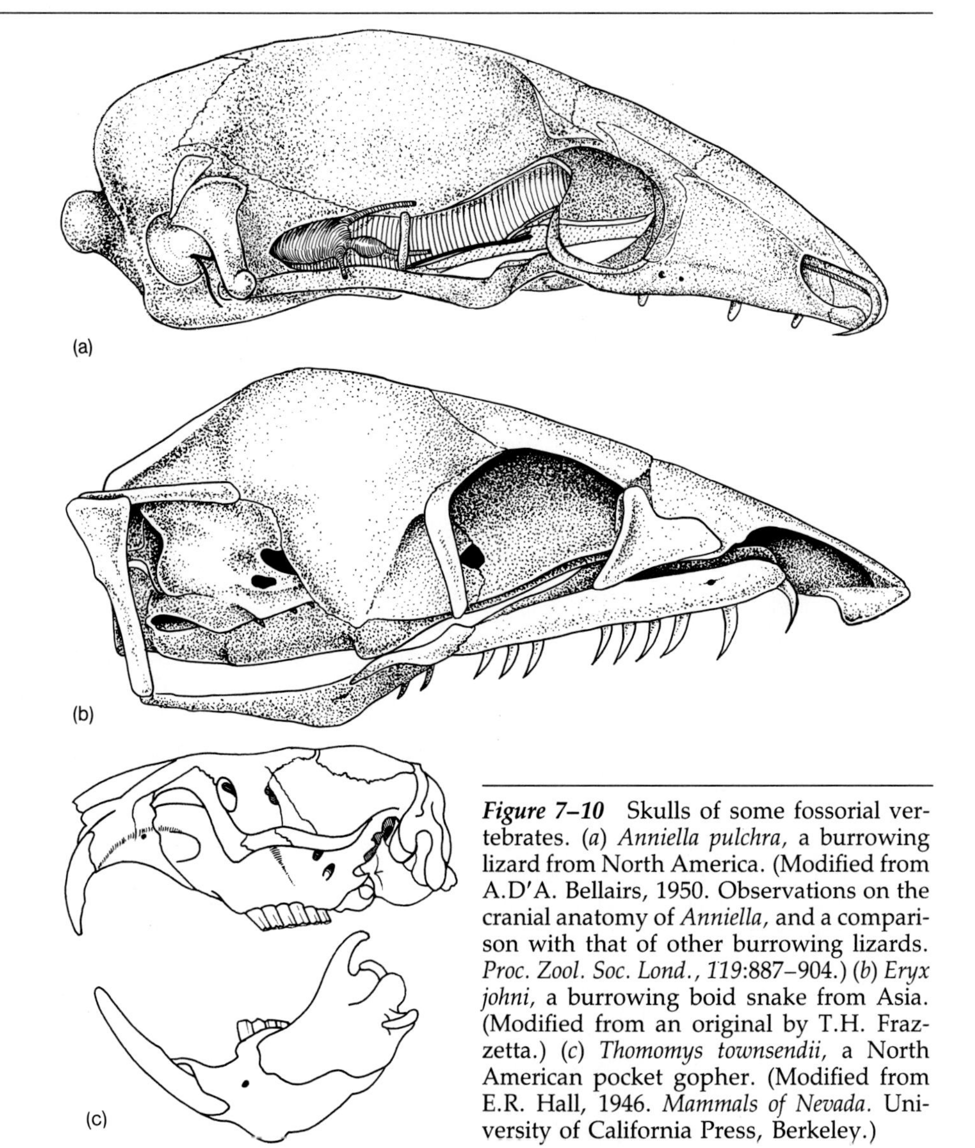

Figure 7–10 Skulls of some fossorial vertebrates. (*a*) *Anniella pulchra*, a burrowing lizard from North America. (Modified from A.D'A. Bellairs, 1950. Observations on the cranial anatomy of *Anniella*, and a comparison with that of other burrowing lizards. *Proc. Zool. Soc. Lond.*, *119*:887–904.) (*b*) *Eryx johni*, a burrowing boid snake from Asia. (Modified from an original by T.H. Frazzetta.) (*c*) *Thomomys townsendii*, a North American pocket gopher. (Modified from E.R. Hall, 1946. *Mammals of Nevada*. University of California Press, Berkeley.)

body and more pegs utilized, more lateral-force components are present and more energy is expended without forward motion. Therefore, there is a tendency for undulating snakes to use as few pegs as possible in order to increase the efficiency of forward movement. In nature, substrate irregularities such as rocks, stems, or debris serve as the "pegs." When snakes move by lateral undulation, the body does not thrash back and forth; instead, each body segment follows in the path of the one ahead of it. In

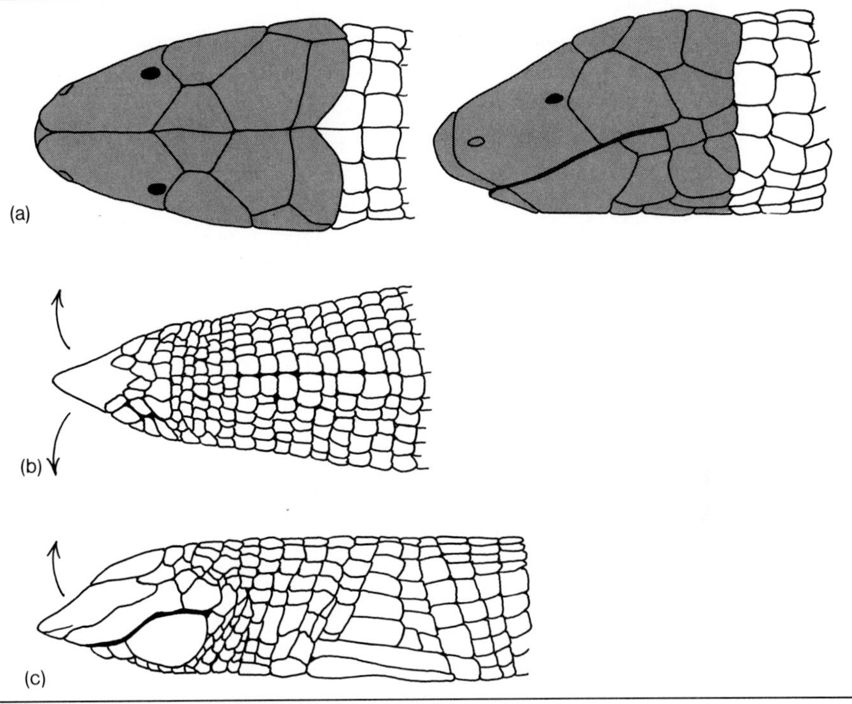

Figure 7–11 The heads of some amphisbaenians along with their digging patterns. (*a*) A round-headed form (top and side views); (*b*) a keel-headed form (top view); and (*c*) a shovel-headed form (side view). (Modified from C. Gans, 1974. *Biomechanics: An Approach to Vertebrate Biology*. J.B. Lippincott, Philadelphia.)

this way, the same "pegs" may be used continually by successive body segments. The body loops thus pass in waves along the body but remain stationary in relation to the peg.

When there are fewer than three irregularities of the substrate or when curvature of the animal's path is uniform (either straight or circular), locomotion by typical lateral undulation is impossible. There is nothing against which the snake can press backward. In these circumstances, snakes, as well as amphisbaenians, adopt a slower concertina movement. In concertina motion, one portion of the body is thrown into loops that press against the sides of the channel and act as an anchor for the rest of the body (Fig. 7–13). An anterior anchor allows the rear to be pulled forward; a posterior anchor permits the front end to be pushed ahead.

In extremely confined spaces where body curvature is not possible, many snakes, such as the boas, can still progress forward by using rectilinear movement. This rather slow form of locomotion is achieved by pulling forward the large ventral scutes of one segment of the body, planting these

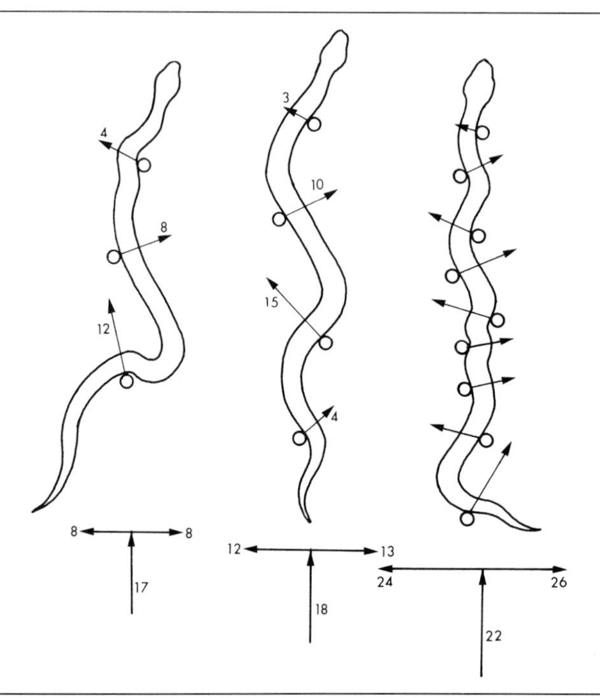

Figure 7–12 Lateral undulation in a snake. Note that the relative magnitude of forward and lateral components of forces is a function of the number of points against which the snake presses. (From Gray, J., 1968. *Animal Locomotion.* Weidenfeld and Nicholson, London; and J. Gray and H.W. Lissman, 1950. The kinetics of locomotion in the grass-snake. *J. Exp. Biol.*, *26*:354–367.)

firmly on the substrate, and then drawing the rest of the body over this segment (Fig. 7–14). Usually several portions of the body are lifted at the same time while the intervening segments are pressed to the ground. Although the ventral surface moves discontinuously forward in "steps," the vertebral column is moved smoothly forward at a nearly constant speed equal to the average speed of points on the ventral surface. It may be worth noting that this kind of movement is *not* achieved by "walking on the ribs." The ribs do not move, but muscles attached to the ribs, which are in turn attached to the spine, pull against the ventral skin. A snake in rectilinear motion may seem uncanny to an unsuspecting observer, for even though it shows little sign of movement within its body it continues to advance.

Another kind of snake locomotion—sidewinding—is exhibited by vipers and rattlesnakes of sandy deserts and is used in modified form by other species when they are forced to move across flat expanses. When a snake moves by sidewinding (Fig. 7–15), the head is held near the ground. The neck is placed on the ground and bent laterally, obtaining a hold on the ground by friction. The posterior portion of the body is progressively laid down parallel to the head but anterior to where the neck touches the ground. As segments toward the rear of the body achieve this position, the head is again lifted and moved laterally and forward. The neck is placed at

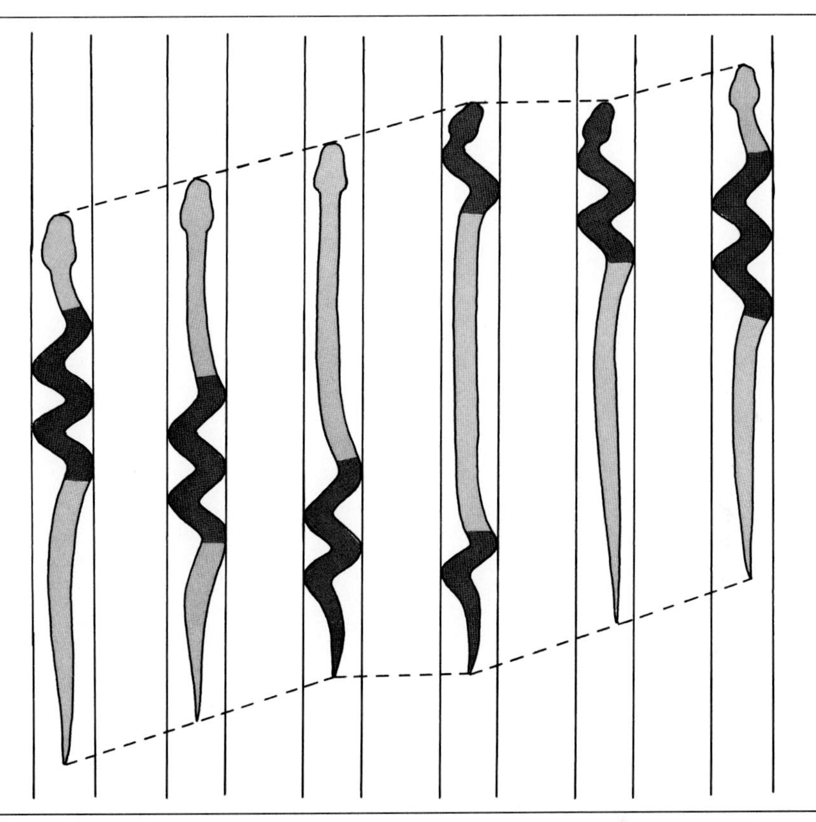

Figure 7–13 Concertina locomotion in a path of uniform curvature. Stationary portions are shown darker than moving parts. (Modified from M. Hildebrand, 1974. *The Analysis of Vertebrate Structure.* John Wiley & Sons, New York.)

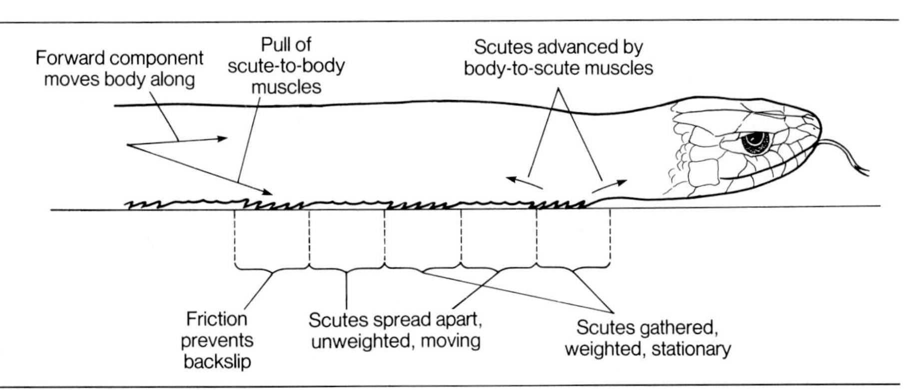

Figure 7–14 Rectilinear locomotion by a snake in a very confined space. (Modified from M. Hildebrand, 1974. *The Analysis of Vertebrate Structure.* John Wiley & Sons, New York.)

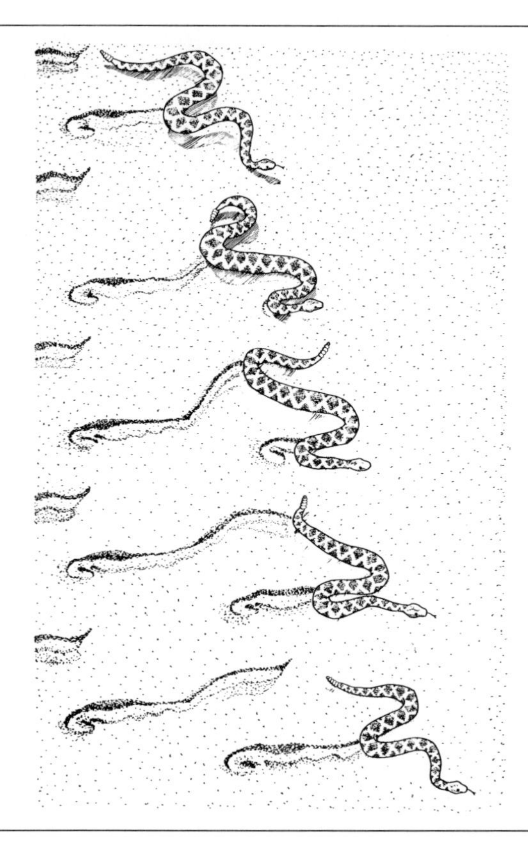

Figure 7–15 Sidewinding locomotion by a snake. The only parts of the body in contact with ground at any one time are the parts in line with the track in the sand. (Modified from C. Gans, 1974. *Biomechanics: An Approach to Vertebrate Biology*. J.B. Lippincott, Philadelphia.)

a new point of frictional purchase, and the body follows again. Several body segments may be in contact with the ground simultaneously, with the intervening loops of the body lifted and drawn to forward points of contact. Each placement of the neck begins a new track, parallel to the last and oriented obliquely to the actual path of progression.

Sidewinding is a very rapid style of locomotion, probably adaptive in speedy avoidance of undesirable situations. For desert snakes, it may also function in reducing the percentage of the body in contact with a hot substrate. It is sometimes claimed that sidewinding is mechanically the most efficient method of snake locomotion, but the necessity of lifting body loops off the ground presumably entails a significant energy cost not present in other serpent locomotor patterns.

Climbing

Members of many vertebrate taxa are excellent climbers (Fig. 7–16), and some have well-defined morphological adaptations for climbing. Scansorial (climbing) adaptations are clearly evident in several families of tree

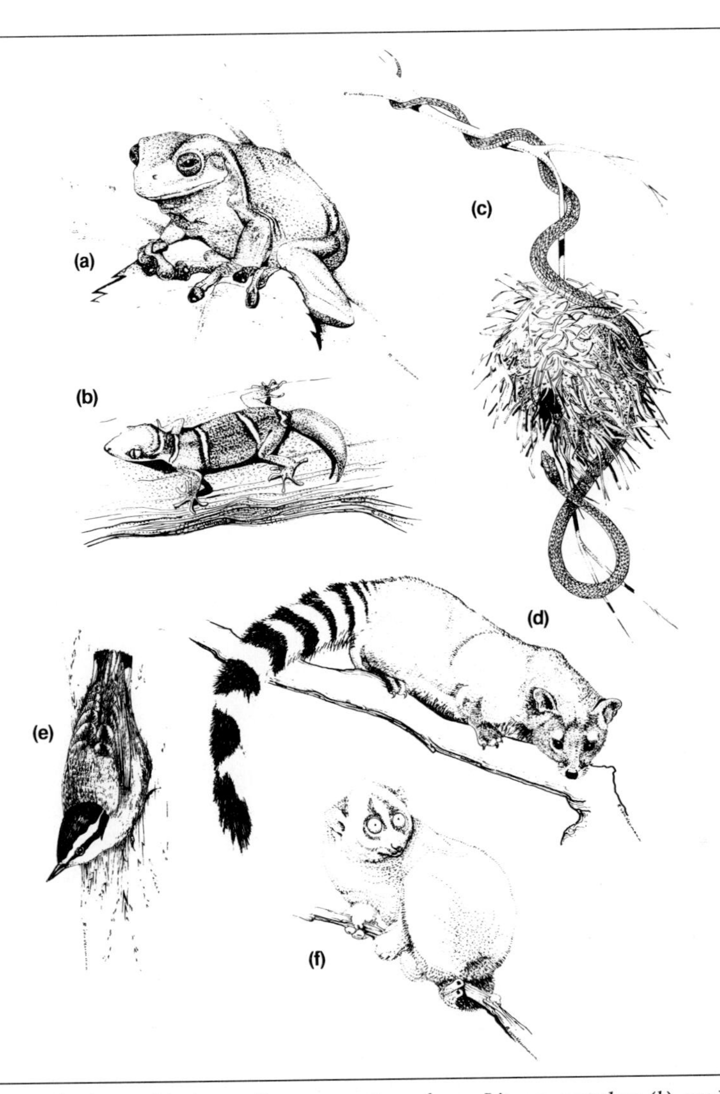

Figure 7–16 Climbers: (*a*) Australian green tree frog, *Litorea caerulea;* (*b*) gecko, *Gymnodactylis deccanensis;* (*c*) Boomslang, *Dispholidus typus;* (*d*) ringtail, *Basariscus astutus;* (*e*) nuthatch, *Sitta canadensis;* and (*f*) slow loris, *Nicticebus coucang.*

frogs, some salamanders, tree snakes, chameleons, geckos, and some other lizards. Species of a number of bird families, including nuthatches, woodpeckers, and parrots, are scansorial. Among the many climbing mammals are American opossums, Australian possums and phalangers, one genus of kangaroos (*Dendrolagus*), tree shrews, some bats, many primates, sloths, small anteaters, squirrels, mice and rats of several families, and New World porcupines. Most climbing carnivores, however, show few structural adaptations specific to this ability. The ringtail (*Bassariscus astutus*) is unusual in its ability to "chimney" or "chimney-stem" up narrow crevices by pressing its feet against opposite walls. It is also such a strong jumper that it can bound upward from wall to wall in spaces too wide to chimney.

Climbing vertebrates generally must have good vision and good coordination for negotiating their three-dimensional environment. Most scansorial vertebrates are able to move with great speed; notable exceptions include chameleons and sloths, which move with eerie slowness. Fast-moving quadrupedal climbers usually have long tails that function as balancing organs. Limbs are usually short, and the body's center of gravity is low and close to the support. However, appendages have lengthened to a remarkable degree in the arboreal primates that brachiate through the trees. In gibbons and siamangs, for instance, the forelimb is very long compared to the compact trunk and is supported by a well-developed clavicle. Joint construction allows freedom of movement in many directions. Arboreal snakes have no limbs, of course, but the vertebrae and associated muscles are modified to prevent sagging of the spine as the body spans a gap between branches.

Devices for hanging on are many and varied. Claws of climbers are usually strongly curved. Digits are frequently opposable, so that they can grasp a twig or surface irregularity. Ringtails, squirrels, and some other scansorial mammals can achieve a 180° rotation of the hind foot, permitting the claws to be used in a well-controlled, head-first descent of a cliff or a tree. Tails are used as props by woodpeckers (Picidae) and woodcreepers (Dendrocolaptidae) when they move up a vertical tree trunk. Prehensile (grasping) tails are found in chameleons, some salamanders, certain arboreal snakes, and assorted mammals, including opossums, some monkeys, kinkajous, anteaters, some mice, and a porcupine. Grasping tails often have hairless cushioned areas where they contact the substrate.

Adhesive pads are commonly present on digits or feet and hands; in tree frogs and some primates, they are conspicuous. These pads are usually rough and improve holding ability by increasing friction and catching on small irregularities. The toe disks of tree frogs are also sticky, which increases adhesion. Members of two families of bats (the neotropical Thryopteridae and the Madagascar Myzopodidae) hang in their roosts by means of elastic suction disks on the feet or at the bend of the wing.

Perhaps the most intriguing adhesion system is found in climbing geckos. The ventral surface of each toe (Fig. 7–17) has several overlapping lamellae. Each lamella has many thousand setae, which, in turn, have hundreds of bristles with expanded, cupped end plates. The toes are cushioned so that a maximum number of end plates is in contact even when the substrate is irregular. All of these minute contact points permit adhesion

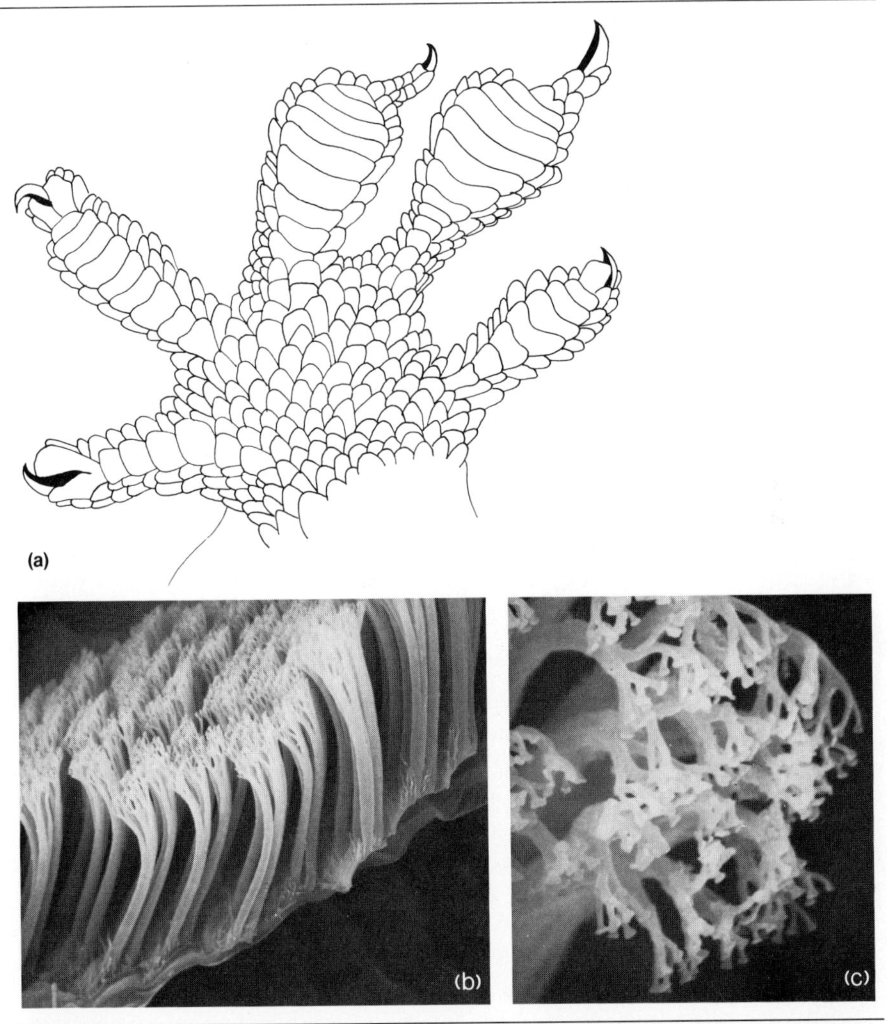

Figure 7–17 (*a*) Ventral view of a gecko foot, showing digital lamellae. (Modified from A. P. Russell, 1976. Some comments concerning interrelationships amongst gekkonine geckos. *Linn. Soc. Symp. Ser.*, 3:217–244.) The arrangement of lamellae differs among different kinds of gecko. (*b*) Scanning electron micrograph of lamellar setae, much magnified. (*c*) A still closer view of some of the thousands of tiny contact points that permit adhesion by surface tension. (*b* and *c*, SEM photos by J. Peterson, reproduced by permission of E.E. Williams and J. Peterson.)

by surface tension, as long as the setae are oriented properly and free of dirt. When a gecko moves forward, the contacts are gradually disengaged from the tip of the toe to its base. This system is so effective that even a dead gecko can stick to a wall. A similar, but less elaborate, adhesion system that employs unbranched setae is found in the unrelated American *Anolis* lizards of the family Iguanidae.

SWIMMING

Movement in water and air has attendant problems not found in locomotion on the ground. Both water and air are fluid and less resistant than the ground; they tend to move away from the push of an animal. Some energy, therefore, must be expended in moving the medium. Furthermore, water is about 800 times denser than air and, thus, has a higher resistance to forward motion. This characteristic can be countered by development of a streamlined form and smooth body surfaces that reduce both the resistance of the medium to displacement and the restraining forces of drag created along the body surface and behind the moving animal. On the other hand, the high density of water provides some support, so less energy is needed to maintain position.

Drag may be created in several ways and is an especially serious problem at high speeds. One source of drag is friction. A moving object carries with it a very thin layer of water called the boundary layer, but water layers that are increasingly farther away from the moving body move slower and slower until the distant layers are entirely motionless (Fig. 7–18). Friction between adjacent water layers creates drag, even at low speeds. As speed increases, ever smaller irregularities of the body surface interrupt the smooth flow of water over the surface, resulting in eddies and turbulence. Turbulent flow produces a thick boundary layer that increases the frictional surface and the effective size of the moving object without increasing its available power. Because turbulence increases friction, it increases drag. Frictional drag is a function of the speed and size of the object, and the density and viscosity of the medium. Larger objects have greater drag, but greater muscular power can compensate for it.

Drag is also created behind a moving object as water moves back into the space just vacated by the object. Filling in that space frequently takes a certain amount of time. As long as the backfill is incomplete, a partial vacuum exists behind the moving body and resists forward motion because the water is exerting less forward pressure on the object. Along the middle of a moving body is another area of reduced pressure where the water moves most rapidly around the object (Bernoulli's principle of physics). At higher and higher speeds, this pressure differential is another ever-increasing source of pressure drag.

Speed has a tremendous effect on total drag. Drag increases at a rate exceeding the square of the velocity, so that every increment of speed brings with it a large increase in drag. One estimate suggests that each time speed is increased by one body length per second, the energy needed to maintain it is doubled; another estimate states that a doubling of speed

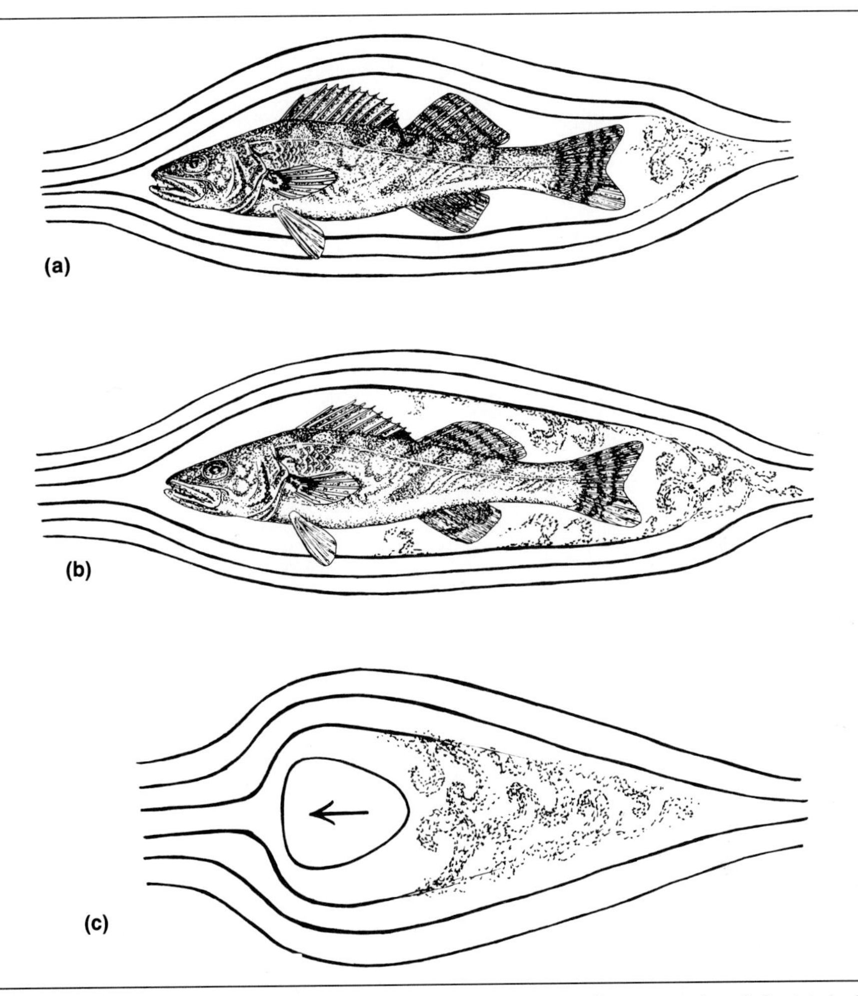

Figure 7–18 This diagram indicates the "layers" of water surrounding a moving fish. (*a*) A slow-moving fish with a small boundary layer. (*b*) A faster fish with a thicker boundary layer and greater posterior turbulence. (*c*) A symmetrical, but not streamlined, object generates greater turbulence than a streamlined one.

necessitates about a sevenfold increase in power output. Such considerations clearly place a limit on the maximum speed of aquatic locomotion for any animal.

Frictional drag is least on a short, stout body, but pressure drag is least on a long, slender body. The best compromise results in a spindle-shaped body that is thickest just anterior to the middle and has a maximum diameter of one fourth to one fifth of its length. A fast-swimming tuna closely approaches such a shape, and the bodies of all characteristically aquatic, highly mobile vertebrates are modified in this direction (Fig. 7–19). External appendages are reduced as much as possible, which minimizes turbulence; in salamanders, lizards, and crocodilians, the limbs are folded while the axial musculature provides the swimming power. When the limbs are used as paddles, they obviously must protrude into the water, at least on the power stroke. Sea turtles and penguins can rotate their forelimbs to derive power from both up strokes and down strokes. Most paddling swimmers fold their paddles on the recovery stroke, however, thus reducing drag, and extend them once again for the power stroke. Diving birds have a modified pelvic structure that places their hind limbs as a posterior extension of the basically spindle-shaped body; the leg musculature is included in the body contour.

Drag is further reduced by a smooth body covering. Fast-swimming fishes are well covered with slimy mucus and small scales or no scales at

Figure 7–19 The spindle-shaped bodies of some fishes. (*a*) A mackerel shark (*Lamna nasus*) and (*b*) a bluefin tuna (*Thunnus thynnus*). (Modified from N.B. Marshall, 1971. *The Life of Fishes.* World Publishing Co., Cleveland, OH.)

all. Whales and sea cows are essentially hairless. The feathers of penguin wings are reduced to scalelike nubbins. The furred or feathered exterior of other submerged swimmers such as otters and loons is also very smooth.

Fishes

The forward motion of a typical open-water fish is accomplished by lateral undulation. Gray (1953, 1968) studied fish locomotion by techniques similar to those used to analyze lateral undulation in snakes: A fish is placed on a horizontal pegboard (in air, not water) and the forces applied to the pegs are monitored as the fish moves forward (Fig. 7–20). A fish on a pegboard needs a minimum of three pegs to move itself forward. The pressure of the body against the peg generates an equal and opposite reaction against the animal. As shown in the vector diagram, the oblique force can be resolved into lateral and longitudinal components. As waves of lateral contraction pass along the body of a fish from head to tail, the equal and opposite forces continually change position along the fish's body and do not exactly cancel each other. Instead they create an oscillation, which is damped by the momentum of forward motion, the action of the fins, and the lateral flattening of the body form. The elongate body shape of some species results in a large number of simultaneous and opposing contractions and, thus, reduces the magnitude of the lateral vectors at each contracting position.

Flexibility of the tail fin ensures that some part of the tail is always exerting a backward thrust, and the lateral forces of the tail are reduced. The longitudinal vector represents a forward-directed force that provides propulsion. Even in rather short-bodied fishes, the tail is not simply wagged from side to side. It can be shown mathematically that tail-wagging would require more power than the waves of bending that pass along the body. Energy required for forward motion is also less when the waves increase in amplitude as they travel backwards, as is the case for typical teleosts.

The power available for locomotion varies as the weight of the fish and, therefore, as the cube of body length. Resistance, however, varies with the surface area and, hence, with the square of body length. Therefore, with increasing size, power increases faster than resistance, and large individuals should be able to move faster than small individuals of the same species. Maximum swimming speed, in absolute terms, does increase with body size, as shown in Figure 7–21, but the slope of the line is less than 1 (a 24-cm fish does not swim twice as fast as a 12-cm fish). Measured in terms of body length, therefore, large individuals may actually move more slowly than small ones. Speed is a function of tail-beat frequency (and, up to a point, amplitude), but tail-beat frequency tends to decrease with body length.

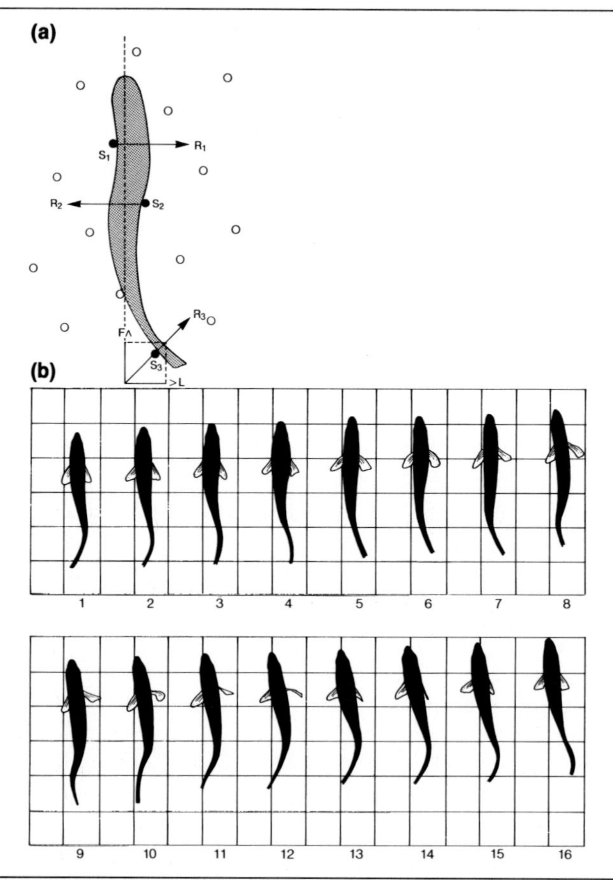

Figure 7–20 Lateral undulation by a fish swimming (in air) on a pegboard. (*a*) Three pegs are essential to forward motion (S_1, S_2, S_3) and are shaded in the diagram. The forward force F is the forward component of the pressure (R_3) exerted by peg S_3 against the tail fin. The lateral component L is equal and opposite to forces R_1 and R_2 exerted laterally by the two anterior pegs, S_1 and S_2. (*b*) Normal motions of a whiting (*Gadus merlangus*) through one complete cycle of tail motion (1-12) and the beginning of the next cycle (13-16). (Reproduced from *Animal Locomotion* by James Gray, by permission of W.W. Norton & Company, Inc. Copyright © 1968 by James Gray.)

Cruising speed, which is based on aerobic metabolism, is limited in part by the rate at which oxygen is supplied; sprint speed, which is based on anaerobic metabolism, may be limited by muscular and skeletal restrictions (Wardle, 1977). Fishes (and other vertebrates) have two kinds of locomotory muscles. Red muscles function aerobically and require a steady supply of oxygen. They contain myoglobin for oxygen transport and many blood vessels. White muscles function anaerobically and are well suited for short

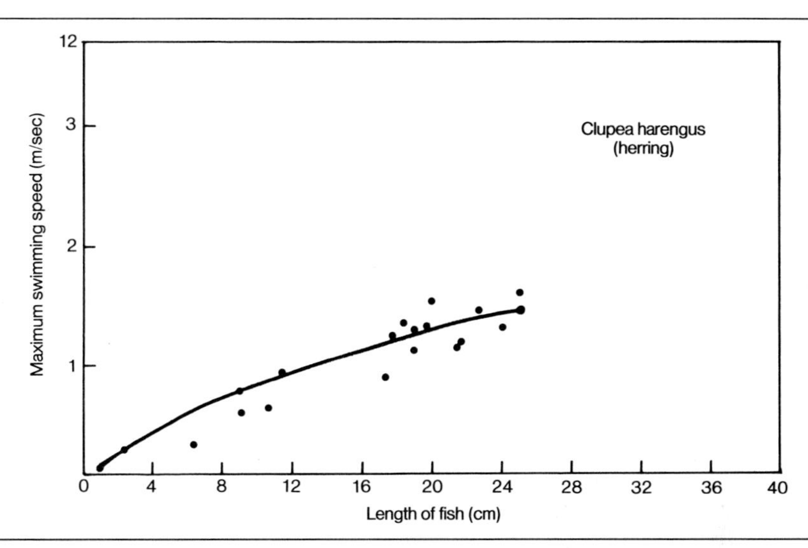

Figure 7–21 Swimming speed as a function of body size for the herring, *Clupea harengus*. (Reproduced from *Animal Locomotion* by James Gray, by permission of W.W. Norton & Company, Inc. Copyright © 1968 by James Gray. Adapted from J.H.S. Blaxter and W. Dickson, 1959. Observations on the swimming speeds of fish. *J. Cons. Int. Explor. Med.*, 24:472–479.)

sprints because the rate of fiber contraction is not limited by the rate of oxygen supply. Red muscles are used in cruising, white ones in bursts of activity. Fishes, such as tuna, that habitually travel long distances have much more red muscle than do sedentary forms.

Many teleosts (and amphibian tadpoles and aquatic salamanders) possess a peculiar means of muscular coordination particularly useful for swift, sudden movement. A system of long, large-diameter neurons with few synapses between them permits very rapid transmission of signals from brain to muscles and, hence, a quick response of the swimming muscles. Benthic, or bottom-dwelling, species and crevice-living species often lack this special device. They seldom depend on sudden, fast movement for escaping predators.

For short sprints, the wahoo (*Acanthocybium solanderi*) can attain a remarkable speed of 77 km/hr or about 20 body lengths per second. Marlin (*Makaira*) and sailfish (*Istiophorus platypterus*) may even exceed this velocity. Mackerel (*Scomber*) can reach 35 km/hr. These speeds generally represent a maximum effort and are not maintained for extended periods of time.

Many sharks and teleosts are virtually weightless in water. Certain sharks, such as the Portuguese shark (*Centroscymus coelolepis*) and the basking shark (*Cetorhinus maximus*), store huge amounts of low-density oil (up to 24 percent of body weight) in the liver and have a specific gravity very close to that of seawater. Some teleosts, such as lanternfishes (*Lampanyctus*), also store large quantities of fat and have low specific gravity. Moreover, teleost skeletons are very light, with thin areas between reinforced struts that maintain the strength of the bones. The scales are thin and bony and do not have the heavy coating found on scales of other Osteichthyes. Buoyancy in many species is controlled by the swim bladder (Fig. 7–22). Basically, this gas-filled sac serves as a hydrostatic organ, although it sometimes has auxiliary functions in hearing, sound production, and breathing. Because the specific gravity of sea water is greater than that of fresh water, freshwater fishes commonly have larger swim bladders and a lower specific gravity than marine species. Open swim bladders, found mainly in freshwater species, open into the foregut and are filled with air swallowed by the fish at the surface or from a gas gland that secretes oxygen and nitrogen. Closed swim bladders are found in many marine forms, especially those living at such depths that surface-gathered bubbles would be

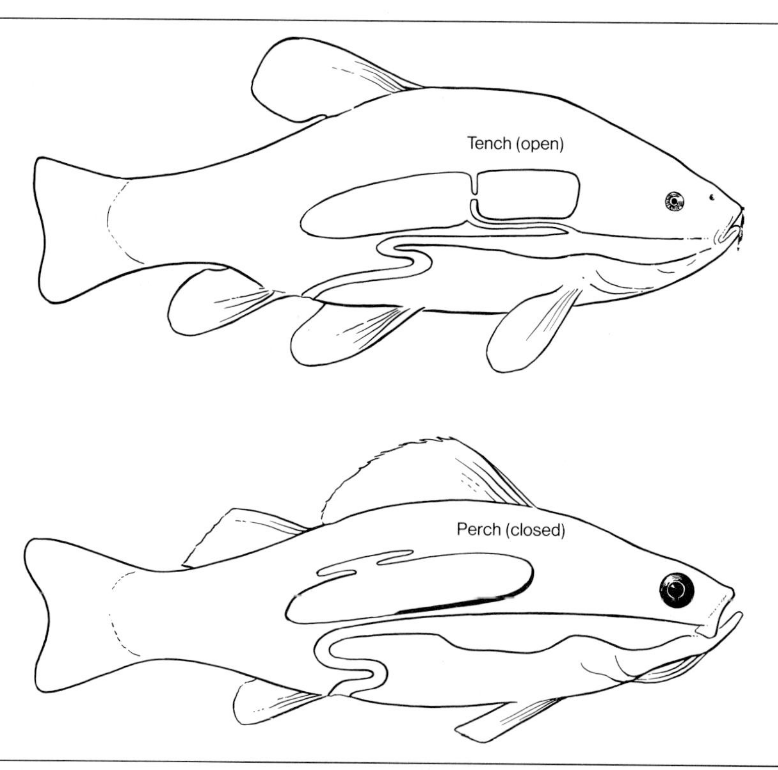

Figure 7–22 Swim bladders. (*a*) The tench's swim bladder is open; there is a duct from the swim bladder to the foregut. (*b*) The swim bladder of the perch is closed. (Modified from N.B. Marshall, 1971. *The Life of Fishes.* World Publishing Co., Cleveland, OH.)

compressed into uselessness at the animal's normal depth. Closed swim bladders are filled from a gas gland. Maintenance of weightlessness means that no energy is required to counteract gravity. Furthermore, efficiency of maneuvering and maintaining position is greater.

Possession of a swim bladder containing a certain amount of gas confers "neutral buoyancy" upon its owner at only one depth, because the gas volume varies with the water pressure. If the fish is displaced upward, the gases expand and the fish tends to rise farther. Downward displacement compresses the gases so that the fish tends to sink even more. Small displacements can be corrected rapidly by fin movement; larger ones can be altered slowly by secretion or removal of gases.

Some fishes have no swim bladder at all. It would be a disadvantage for bottom-dwellers such as flatfish (Pleuronectiformes) and sculpins (Cottidae). Certain deep-sea bony fishes such as the anglerfishes (Ceratiidae) and gulper eels (Eurypharyngidae) have diminished, if not eliminated, the problem of maintaining buoyancy by reducing the amount of heavy bone and muscle in the body; the most obvious body features are the mouth and fins.

There are at least three possible disadvantages to having a swim bladder. In the first place, these organs are sometimes located in the body cavity in such a way that the center of gravity is dorsal to the center of buoyancy. If the fish tips a bit to one side, this combination of forces (down at the center of gravity and up at the center of buoyancy) tends to turn the fish upside down. Therefore, the fins must frequently exert small forces to keep the fish upright. A second disadvantage, perhaps a critical one to some species, is that swim bladders reflect underwater sound very well. Some whales use echolocation and may be able to detect fishes with swim bladders more easily than those without. A third disadvantage is that storage of gases or fats leads to an increase in body volume and hence, surface area, thus increasing drag (Alexander, 1972). The fact that so many fishes nevertheless *have* swim bladders indicates that the advantages must be considerable.

The tail fin is asymmetrical in sharks and some bony fishes such as sturgeon (*Acipenser*) and paddlefishes (*Polyodon*). The upper part of these heterocercal tails is supported by the caudal end of the vertebral column, the lower part by fin rays. Contrary to earlier notions, the asymmetrical tail does not cause the shark to tip either up or down; the forward thrust is apparently directed through the center of gravity. The main advantage of a heterocercal tail in sharks may be its great versatility in permitting very powerful dives and climbs for oblique attacks on prey (Thomson, 1976). The function of the caudal fin rays in bony fishes with heterocercal tails is controversial. The tail fin in most modern teleosts is symmetrical (homocercal) and provides forward thrust (but little lift) during forward movement. Most teleosts have a specific gravity near that of their medium, and extra lift would interfere with direct forward motion.

The shape of the caudal fin in teleosts is related to the speed of locomotion. Fast-moving fishes usually have crescentic tails with lobes that reach beyond the turbulent wake (Fig. 7–23). The tail lobes are stiffened to deliver the powerful thrust necessary for high speed. Slower fishes and those living in nooks and crannies do not create such great turbulences. They often have rounded, flexible tail fins that can deliver thrust at any speed because some part of them is always pressed back against the water.

The pectoral fin of sharks are fairly stiff and heavy. The angle of these fins can be altered, and they can be moved back and forth. They function in steering and are probably important in providing lift. Sharks are seldom capable of sudden stops; usually they swerve around obstacles. The pectoral fins of most modern teleosts are supported by fin rays and are flexible. They are located higher on the side of the body than shark pectorals and are particularly useful in braking and turning. The pelvic fins of primitive bony fishes are located posteriorly, near the anus, but in some advanced teleosts (such as the Perciformes), they are found near the pectorals and function with the pectorals in maneuvering.

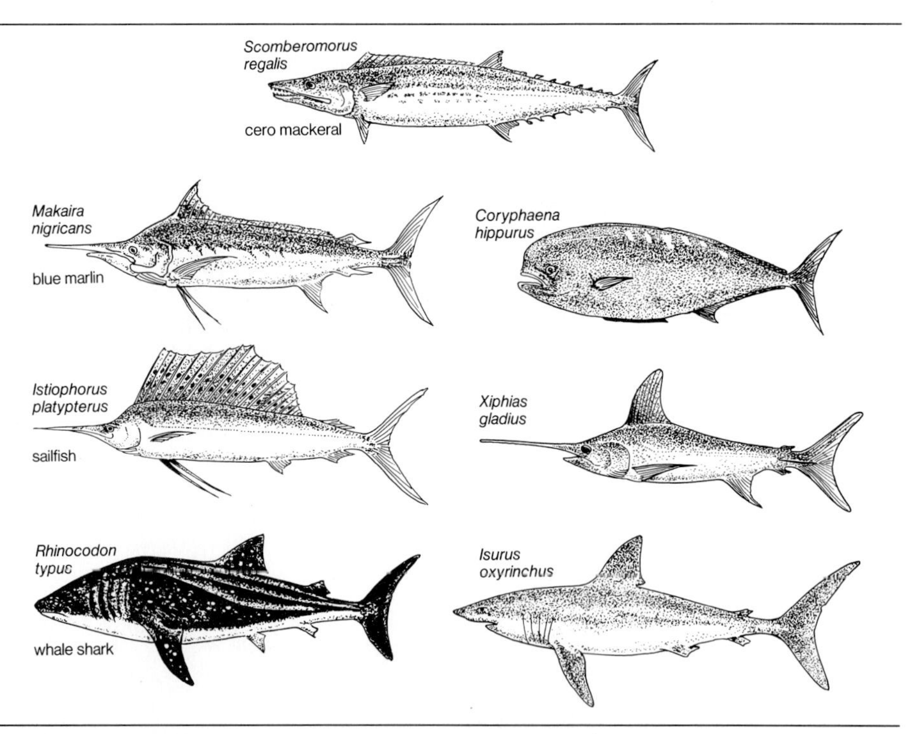

Figure 7–23 Crescentic tail fins in several fishes. (Modified from J. Lighthill, 1975. Aerodynamics of animal flight. *In* C.J. Brokaw and C. Brennen (Eds.), *Swimming and Flying in Nature.* Vol. 2, pp. 423–492. Plenum Publishing, New York.)

Not all fishes have a "typical" fish shape (Fig. 7–24). Fishes exploit their environments in many ways, and some have adapted to modes of life in which fast or efficient swimming was relinquished as the price for achieving other advantages. Body flexure and the caudal fin are used little or not at all in locomotion of many species. Among skates and most of the rays, the pectoral fins are the major source of propulsion. In species that forage for slow-moving prey or use the technique of ambush, undulations generally pass over the pectorals in much the same manner as they would pass down the body of a typical teleost. In the rays that are active hunters, however, the pectorals function essentially as flapping wings.

Many fishes of coral reefs (surgeonfishes [Acanthuridae], butterflyfishes [Chaetodontidae]) use their pectoral fins as oars. Others utilize their dorsal and anal fins in flapping, undulating, or sculling (triggerfishes [Balistidae], pufferfishes [Tetraodontidae], ocean sunfishes [Molidae]). Sea horses (*Hippocampus*) move by means of their dorsal and pectoral fins. Except for the ocean sunfish, most of these species have evolved stiff, armored bodies or globular, spiny ones, probably as a means of protection from predators. They often live in crannies provided by reefs or other reef inhabitants;

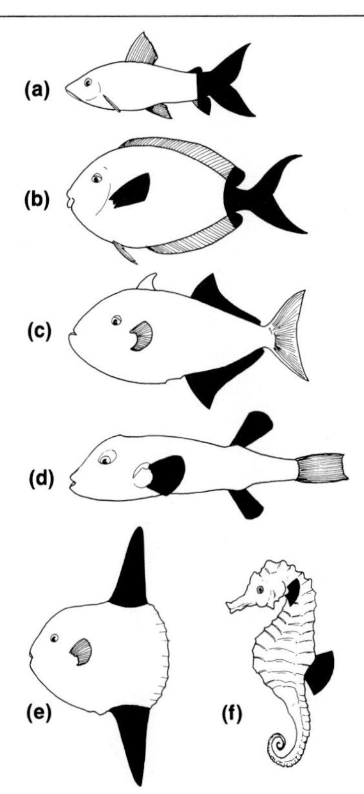

Figure 7–24 Some fishes that move without lateral undulation of the body, using only the fins. (*a*) A characin, *Leporinus*. (*b*) A surgeonfish. (*c*) A triggerfish, *Xanthichthyes*. (*d*) A pufferfish, *Sphaeroides*. (*e*) The ocean sunfish, *Mola*. (*f*) A seahorse. The principal locomotory fins are shaded black. (Modified from N.B. Marshall, 1971. *The Life of Fishes*. World Publishing Co., Cleveland, OH.)

some may use their inconspicuous but well-controlled mobility to sneak up on their prey.

Most of the bony fishes have the ossified part of the skeleton inside the main body contour; only the integumentry rays protrude. The ancient crossopterygians, in contrast, had fleshy fins with supporting bony elements that, superficially, looked like those of the later amphibian limbs. For terrestrial locomotion, however, the buoyant effect of water is absent, and limbs must support the body's weight. Thus, the articulation of limbs with body must transmit greater forces on land than in water, as discussed near the beginning of this chapter. Although some species of fishes walk and climb with their pectoral fins, their body weight is only temporarily supported by these limbs. Mudskippers (Periophthalminae) have unusually strong fins and pectoral girdles; these are the fishes best adapted to walking on land. They hunt on exposed, tidal mud flats and in mangrove swamps.

Other Vertebrates

Many nonpiscine vertebrates can swim, but this brief discussion focuses particularly on species that are highly aquatic and spend a great deal of time in the water. Surface swimmers, such as dabbling ducks (*Anas*), geese, and gulls, do not customarily swim totally immersed, and their primary aquatic adaptations are found in their webbed feet. Birds that regularly swim submerged beneath the surface of the water often use their webbed feet as rudders and propel themselves with their wings. Penguins and the alcids (puffins, murres, and so on) are among the birds that typically swim this way. Likewise, the aquatic amphibia are not sustained swimmers. Aside from lateral body compression in salamanders and webbed rear feet in some frogs, they show few specializations for aquatic locomotion.

The most intriguing modifications for swimming and diving outside the fishes are those of amniote species descended from terrestrial ancestors and secondarily adapted to life in the water. These species must solve not only the problems common to all aquatic locomotion (drag, vertical positioning, and so forth) but also the additional problems created by their terrestrial origins. All these species breathe air; therefore, water must not enter the respiratory passages and water pressure must not crush the air-filled chambers. The nostrils of many divers can be closed by special valves or erectile tissue; the lungs are usually emptied and allowed to collapse. Walls of the respiratory passages, such as the trachea and bronchioles, are reinforced by cartilage and may store small amounts of air. For homeothermic divers, thermoregulation may be a problem.

Nonpiscine swimmers do not attain the high speeds reported for the fastest fishes. Penguins may swim up to 36 km/hr; sea lions may reach a maximum of about 22 km/hr. The great blue whale (*Balaenoptera musculus*), the giant of all whales, is reported to have achieved sprints of 37 km/hr, and the finback whale (*B. physalus*) is said to sprint at up to 65 km/hr.

Loons can dive to 55 m and stay submerged for 15 min. Manatees and seals can stay underwater at least that long. The Weddell seal (*Leptonychotes weddelli*), which forages extensively under the antarctic ice, can dive to 600 m and stay underwater for an hour. Sperm whales (*Physeter catodon*) can dive more than 900 m and remain below the surface for 90 min. A bottlenose dolphin (*Tursiops truncatus*) has been recorded to submerge for as much as 2 hr.

AERIAL LOCOMOTION

Falling and Gliding

Many arboreal vertebrates possess structural and behavioral modifications that permit them to break falls, to glide, or to fly. Small creatures can sometimes launch themselves into the air and spread out their body to present a broad surface to the flow of air. Hylid tree frogs can convert a 90° drop into a 60° angle of descent. Rhacophorid frogs have large webbed feet and webs along their arms and between their trunk and thighs. When the webs are spread, the animal "parachutes" downward.

Gliders, with a customary descent angle of less than 45° from the horizontal, are found among fishes, lizards, snakes, and mammals (Fig. 7–25). Gliders climb to gain height, jump off, and drop rapidly until their broad gliding membranes catch the rapidly moving airstream, at which time the glide begins. One genus of Indo-Malayan tree snakes (*Chrysopelea*) and some geckos are able to break their falls by means of body flattening or webbed toes and body fringes. The geckos can reduce their angle of descent to almost 45° and are quite maneuverable; the snakes may do even better than this. Gliding geckos (such as *Ptychozoon lionatum* of Southeast Asia) can sometimes glide as far as 9 m. *Chrysopelea* can go at least 30 m and change direction in midglide. *Draco* lizards have lateral membranes supported by several sets of ribs that fold against the body wall when not in use; these lizards can glide as much as 24 m. Gliding mammals include the colugos (*Cynocephalus*), several species of Australian marsupial gliding possums, other phalanger-like species in several families (that can glide more than 100 m), and 15 genera of small rodents. These mammals have extensive membranes from forelimb to hind limb along the side of the body; many have long bushy tails. The colugo's gliding membrane extends from the throat to the arms and from the legs to the tail as well.

Gliding fishes build up momentum by rapid swimming near the surface of the water, then they emerge. The so-called flying fish (*Cypselurus*, family Exocoetidae) spreads its enormous pectoral fins while the lower lobe of the caudal fin continues to drive the fish forward and airspeed increases. When the airspeed reaches approximately 40 to 70 km/hr, the pelvic fins

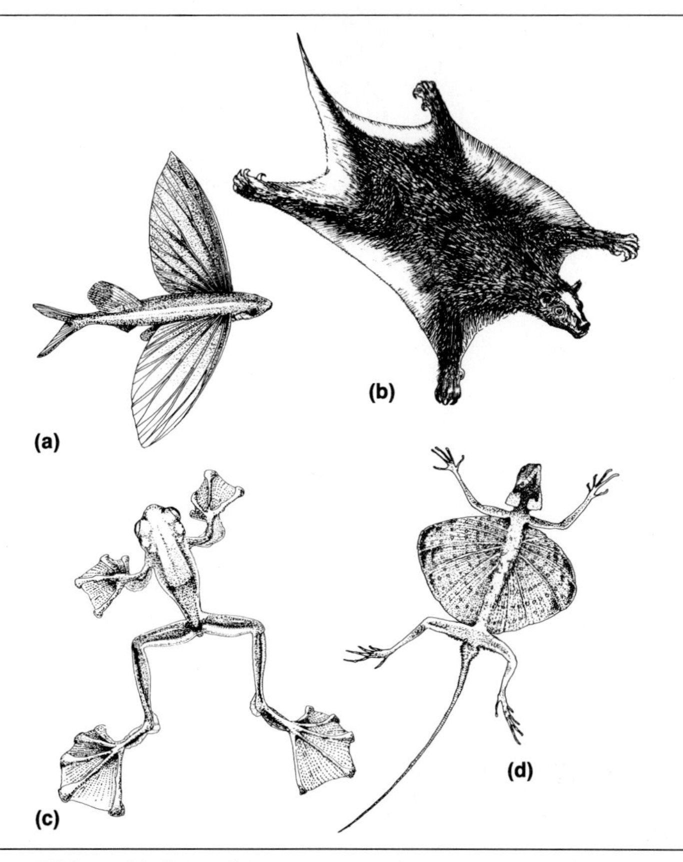

Figure 7–25 Gliders: (*a*) flying fish, *Parexocoetus mesogaster;* (*b*) colugo, *Cynocephalus;* (*c*) flying frog, *Rhacophorus nigropalmatus;* (*d*) gliding lizard, *Draco.*

are extended and the fish leaves the water completely, sometimes gliding as much as 100 m. Gliding depends on the presence of wind, and several glides can be made in succession. This unusual mode of progression may be used to elude underwater enemies and to capture fast-moving prey.

Flying

True, continuous, flapping flight in vertebrates has been achieved perhaps by the extinct pterosaurs and by two groups of modern endothermic vertebrates: bats and birds. Critical to the extensive aerodynamic success of the two extant groups of fliers is what Maynard Smith (1952) has viewed as the loss of stability in order to achieve maneuverability. In this context, stability refers to the capacity to resist perturbations without special corrective procedures. Maynard Smith makes an analogy with a transport plane versus a fighter plane. The transport plane lumbers forward through thick and thin with little adjustment; the fighter plane, which has a minimum of stability and requires readjustments if flight is perturbed, has a maximum of agility. Birds and bats are unstable but agile; they have capitalized on maneuverability.

The main flight surfaces of birds are produced by the flight feathers, supported primarily by the hand and forearm skeleton. The number of digits and metacarpal elements is reduced, so the distal portion of the wing is relatively light. Bat wings are supported by four of the five digits and the forearm (Fig. 7–26); they extend posteriorly to the hind limb and often the tail. Skeletons of flying birds are usually very light: The marrow spaces are large and often filled with air, and strength is maintained by internal struts. The flight muscles are relatively large (averaging about 15 percent of

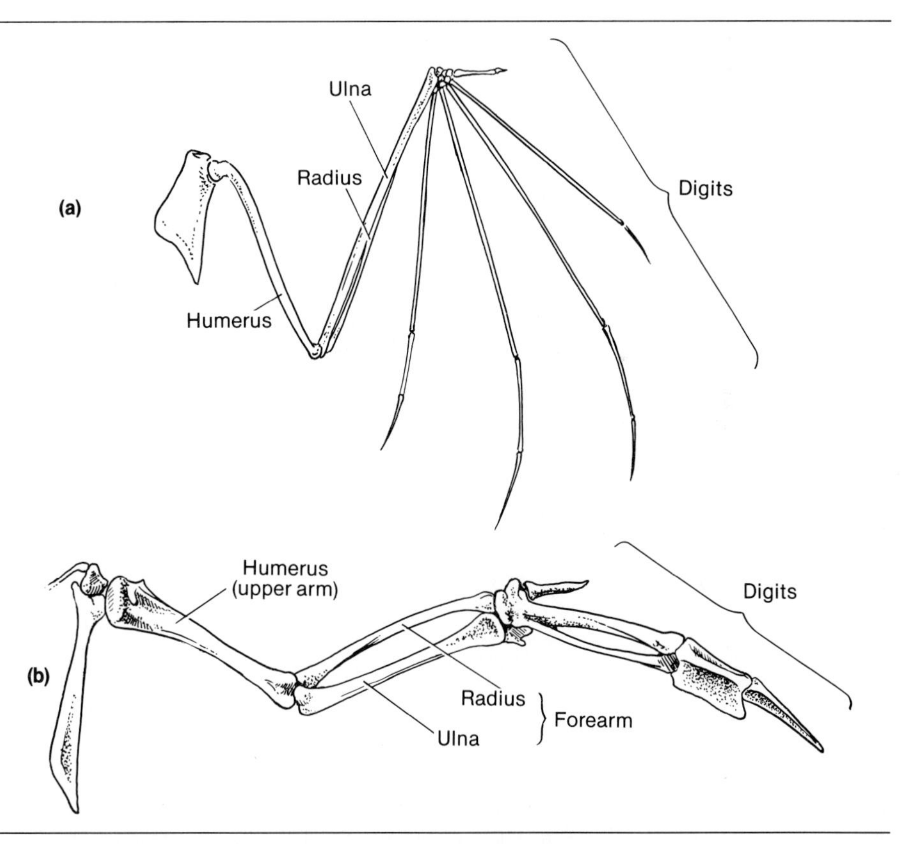

Figure 7–26 Bones of (*a*) a bird wing and (*b*) a bat wing. (Modified from J. Dorst, 1971. *The Life of Birds.* Vol. 1. Columbia University Press, New York, and from T.A. Vaughn, 1972. *Mammalogy.* W.B. Saunders, Philadelphia.)

body weight in both bats and birds), and the sternal keel, where the flight muscles attach, is greatly enlarged, especially in birds. The pectoral girdle is strong and well-braced; in birds, it is actually immobilized. The avian tail skeleton has been severely reduced to a nubbin that supports the tail feathers. Many of the trunk vertebrae of birds are restricted in mobility or even fused into a rigid column that resists deformation by the flight muscles. Bats also have compact trunks, although vertebrae are not fused. In both birds and bats, the wing joints (except the shoulder joint) are constructed so that motion is possible in only one plane.

The mechanics of flight are better known for birds than for bats, although the basic principles must be similar. The following brief account is based on studies of avian flight.

Lift is essential to counteract gravity and keep the animal airborne during level flight. During takeoff, lift must be great enough to exceed the pull of gravity and raise the animal off the ground. Lift is provided by air pressure on the underside (usually) of the wing, while the faster-moving air on the upper side creates a reduced pressure there (Fig. 7–27). Brown's (1953) work with rock doves (*Columba livia*) in slow-flapping flight has

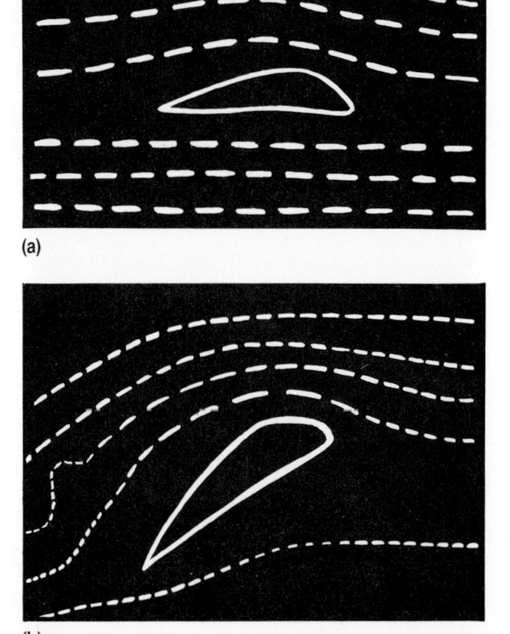

Figure 7–27 The flow of fluid around an airfoil, showing the faster flow above than below. These diagrams were made from photographs of lights moving past the object; the velocity of the moving medium is proportional to the length of the dashes. (*a*) Object in line with flow; (*b*) object has a higher angle of attack, emphasizing the differences above and below. (Modified from R.A.R. Tricker and B.J.K. Tricker, 1967. *The Science of Movement.* Published by Mills & Boon Ltd., by permission of Bell & Hyman Ltd.)

shown that, on the down stroke, lift is provided mainly by the proximal wing section. Propulsive thrust is provided primarily by the distal wing segments in two ways. The down stroke bends the flight feathers so that their ends push backward against the air. At the top of the up stroke, the distal wing segment snaps rapidly backward, so that the top of the flight feathers pushes against the air (Fig. 7–28).

Fast forward flight is seemingly quite different. The amplitude of the wing beat is less. Moreover, the forward motion of the wing on the down stroke and the thrust of the backward flick at the end of the up stroke are much reduced. Lift is provided by both proximal and distal wing segments as the moving air passes over them. Air moves faster over the upper surface, which is convex and therefore wider than the undersurface. The surface exposed to faster-moving air has less pressure than the lower surface and, consequently, produces lift.

Several species soar, using air currents instead of muscle power to remain aloft. Obstruction updrafts are found near mountains and cliffs; thermal updrafts are created by differential heating of the earth's surface, which causes air to rise (Fig. 7–29). Birds soaring in thermals commonly fly in circles around the top of the rising column of air. The size of this circle depends not only on the size of the thermal but also on wing loading

Figure 7–28 Slow-flapping flight by a rock dove. The downstroke of the wing is shown in the sequence *a* to *d*, the upstroke in *e* to *h*. The bending of the flight feathers is shown clearly at stage *c* and indicates the upward and forward thrust. Similarly, at stage 8 the backward flick of the wing tip bends the feathers against the air and produces forward thrust.

(weight per unit of wing area) of the species: Larger birds typically have greater wing loading (Fig. 7–30). So smaller soaring birds can utilize smaller, weaker thermals. Hankin's (1913) study of Indian vultures showed that the daily schedule of soaring was quite species-specific. The smallest vulture, with a wing loading of 0.27 g/cm^2, began soaring first; at approximately half-hour intervals, larger and larger birds began soaring. The fifth and largest vulture (with a wing loading of 0.75 g/cm^2) was seen only at the hottest part of the day, when thermals were strongest.

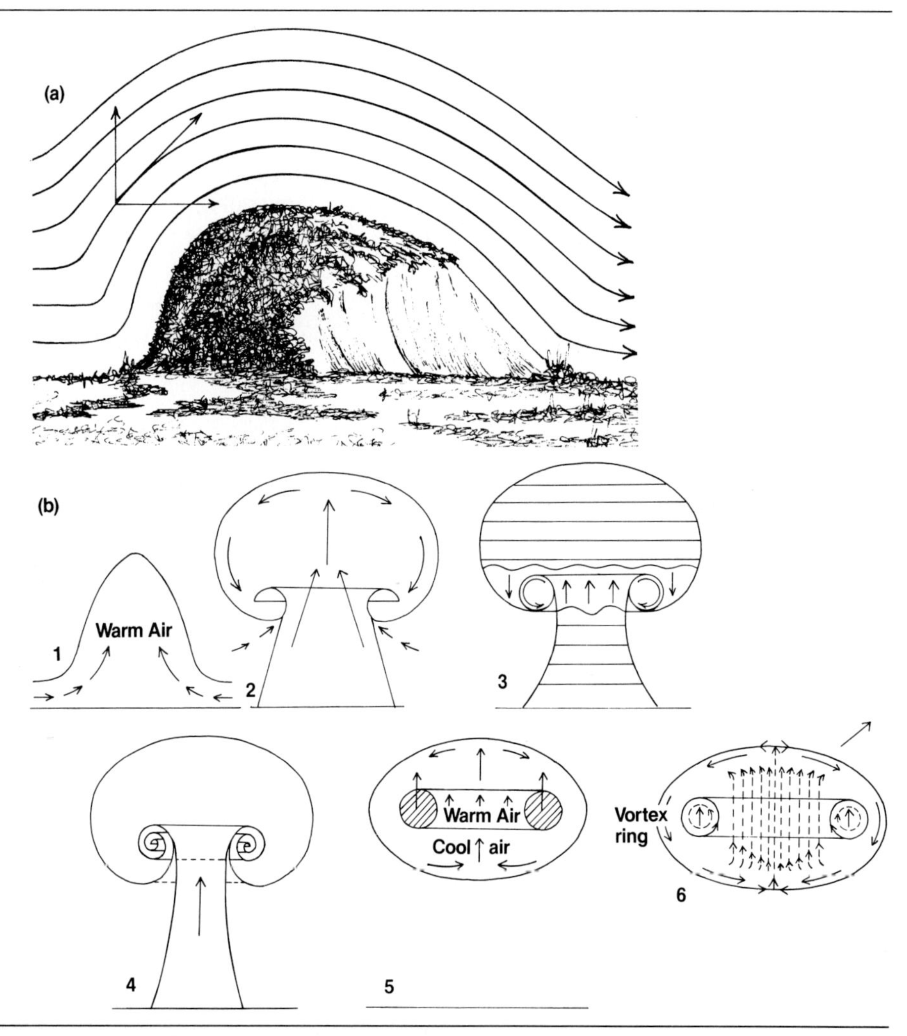

Figure 7–29 (*a*) An updraft forms when an obstruction forces the moving air to rise. (*b*) The formation of a thermal: A bird soars in a circle in the rising air in the middle of the vortex ring. (From "The soaring flight of birds" by C.D. Cone, Jr. Copyright © 1962 by Scientific American, Inc. All rights reserved.)

Once airborne, however, birds with great wing loading can travel faster and farther than species with lower wing loading. The large Rüppell's griffons (*Gyps rueppellii*) and white-backed vultures (*G. africanus*) in eastern Africa forage over a wide range, apparently depending on dead migratory ungulates for food (at least in the Serengeti plains). Several other species of vulture that have lower wing loading patrol much smaller areas and seem to utilize a greater variety of food sources.

The aerodynamic capabilities of wings vary with wing size and shape. Because drag increases with wing area, especially at high speeds, small wings are favored for fast flight. The shape of a wing is described by several specific terms. Aspect ratio relates wing area to the total wingspan (wing tip to wing tip). In a fixed wing, as on an airplane, it is calculated as wing length/wing width. Since animal wings are not fixed and have somewhat irregular shapes, aspect ratios are calculated by using the "mean aerodynamic chord" (span^2/area). Wings with a high aspect ratio are long and narrow (Fig. 7–31*a*). The narrow wingtips reduce turbulence and drag, resulting in increased lift at cruising speeds. They provide little lift, however, at slow speeds or for takeoff.

Such wings are well developed in many seabirds, which take off in open areas and do not need to achieve great rates of climb. The wandering and

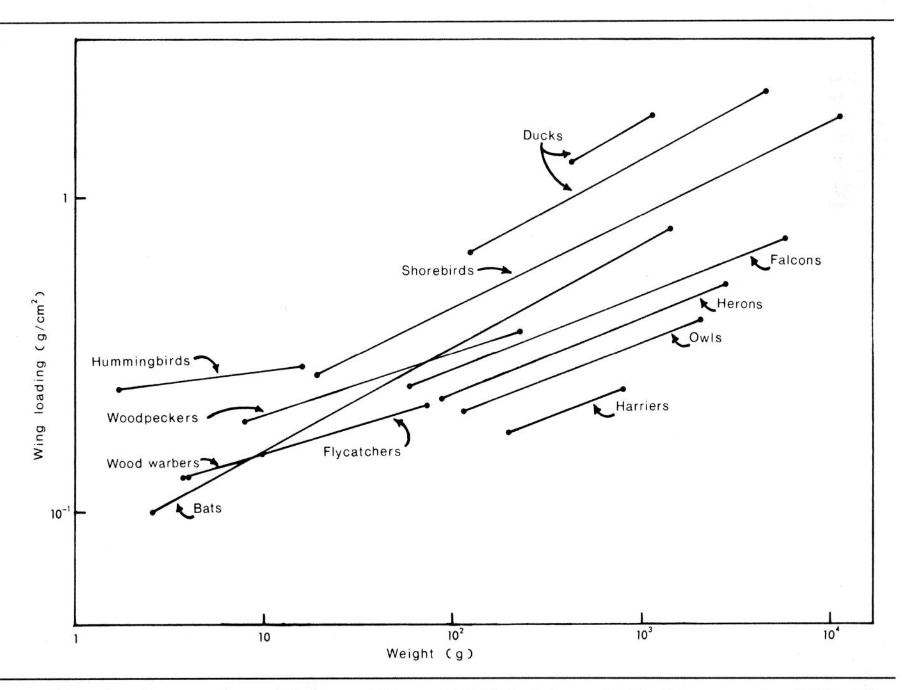

Figure 7–30 Wing-loading of some birds and bats. (Modified from C.H. Greenewalt, 1975. The flight of birds. *Trans. Am. Phil. Soc.*, *65*(4):16.)

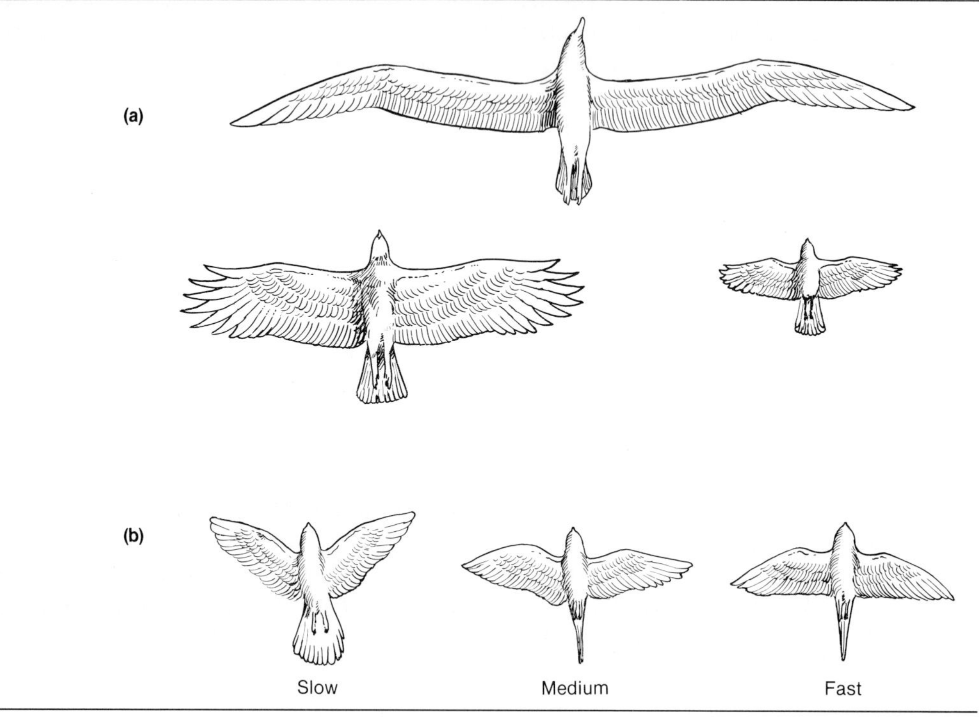

Figure 7–31 Some aspects of wing shape. (*a*) High aspect ratio in an albatross; low aspect ratio in an eagle and a crow. The eagle's wings also show slotting. (*b*) Change of wing shape with increasing speed in a parrot. (Modified from J. Lighthill, 1975. Aerodynamics of animal flight. *In* C.J. Brokaw and C. Brennen (Eds.). *Swimming and Flying in Nature.* Vol. 2, pp. 423–492. Plenum Publishing, New York.)

royal albatrosses (*Diomeda exulans* and *D. epomophora*) have wings of extremely high aspect ratios and are the largest of the albatrosses. Their wingspans are more than 3 m; their body weight, 9 to 12 kg. They are probably near the maximum size possible for this body style. A further increase of body weight would require an unreasonable expansion of wing length. These big albatrosses are found primarily in the area between 30° and 60° south latitude, where winds are strong and nearly continuous. Smaller relatives of these giants often fly in regions where winds are weaker and less reliable. Among other birds with high–aspect-ratio wings are many shorebirds (Charadriidae, Scolopacidae), swifts (Apodidae), and swallows (Hirundinidae), which characteristically fly fast or for long periods of time. The wings of swifts are stiff during fast flight, and the flight surface has a shape similar to the crescentic caudal fin of the fastest fishes (Lighthill, 1975). In fact, the wings of other birds assume this shape as flight speed increases (Fig. 7–31*b*).

Camber refers to the arching of a wing between the leading and trailing edges (Fig. 7–32*a*). High camber indicates a strongly arched wing section; low camber indicates a fairly flat wing. Wings with high camber are well adapted to high lift at low speed. Moreover, they contribute to agile flight and are found in many woodland birds and raptors. Low camber is characteristic of high-speed wings, where agility has been sacrificed for speed, and is commonly associated with high aspect ratios.

The wings of many birds are slotted (Figs. 7–31, 7–32*b,c*). Notches between the ends of the flight feathers and between the alula or "bastard wing" and the main wing reduce turbulence and, thus, increase lift. Many small birds with small, rounded wings have slotted wing tips. Short, wide wings (low aspect ratios) are an advantage in cluttered and complex environments, such as woodland, where maneuverability is critical. The slots are an adaptation that permits short wings to produce strong lift. Woodland birds that flit about actively and take off many times a day have greater slotting than less active birds. Eagles, buteo hawks, and owls also have slotted wings, but their wings have a higher aspect ratio than those of small woodland species. For them the slots create high efficiency (less drag) and greater lift at low speeds, which is advantageous for raptors that pounce on their prey and carry it off. Dabbling ducks have more wing slots than diving ducks. As a rule, dabblers cannot dive as a successful tactic of predator evasion: Slotting contributes to their ability to make a swift vertical takeoff.

The wing-beat cycle of bats is generally similar to that of birds, but bats are less diverse in flight styles. The evolution of bats has emphasized low flight speeds and great maneuverability at the expense of high-speed flight. Bat wings have high camber and can produce great lift. Bats, unlike birds, have the ability to vary the wing camber; an increase in camber increases the ability to fly very slowly. Birds can attain much higher flight speeds and longer continuous flight than bats.

Although bat wings are less diverse in shape than those of birds, many of the same adaptations are evident. Fruit-eating bats (such as *Artibeus*, family Phyllostomatidae) have higher wing loading than insect-eating vespertilionid species and can fly faster. Many vespertilionids have large wings for their size (hence, low wing loading), low aspect ratios, and high camber; they fly relatively slowly but have great agility. The pursuit of insect prey in close quarters is clearly facilitated by these wing adaptations. In contrast, the insect-eating free-tailed bats (Molossidae) have relatively long, narrow wings, low camber, and high wing loading. As another adaptation to high speeds, the bones of the forearm are streamlined in cross section (in contrast to the circular cross sections of most other bat forearms). These bats are frequently found in more open areas and are adapted to faster flight than vespertilionids. Some bats, such as many rhinolophids and members of other families as well, commonly forage in a manner

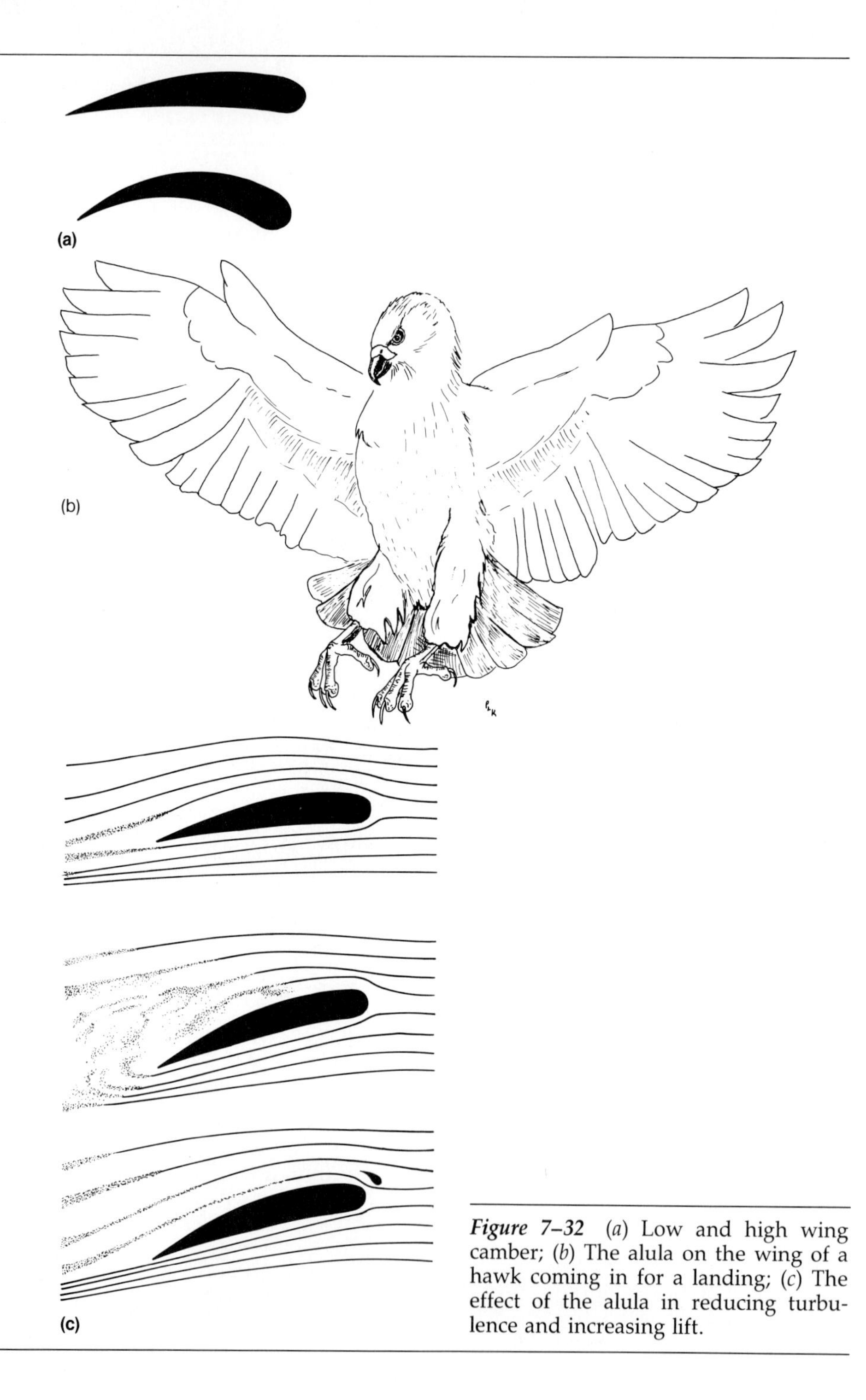

Figure 7–32 (*a*) Low and high wing camber; (*b*) The alula on the wing of a hawk coming in for a landing; (*c*) The effect of the alula in reducing turbulence and increasing lift.

similar to some tyrannid flycatchers and bluebirds (*Sialia*). Most of their hunting time is spent on a perch waiting to dart out and snatch insects from the ground or vegetation. Their wings are similar to those of the insect-eating vespertilionids, but their wing loading is higher and they are less agile.

BODY SIZE AND ENERGETICS

If several birds have the same body shape and differ only in size, we would predict that the surface area of the birds' bodies will vary approximately as the square of the linear dimensions (L^2) and the volume or weight will vary about in proportion to the cube of these dimensions (L^3). Therefore, with increasing body length, weight increases more rapidly than area of the flight surfaces and wing loading is correspondingly greater in large birds. Gray (1968) notes that, as a result, the power needed (P_N) for flight in larger birds is also great. That power is proportional to the product of the wing-loading and the drag, which is related to wing area. It can be shown algebraically that the power needed increases approximately as $L^{3.5}$. On the other hand, the power available increases with the sizes of the flight muscles. An increase in flight muscle weight is almost directly proportional to body weight increase ($W^{0.96}$) or almost proportional to L^3 (Fig. 7–33). The power needed for flapping flight, therefore, increases with body size faster than the power available ($L^{3.5}$ is greater than L^3). Eventually, a size will be reached at which a bird cannot rise off the ground under its own power. The largest flying bird known was the now-extinct vulture *Teratornis merriami;* it had an estimated weight of 23 kg and probably was not capable of prolonged flapping flight. Even present-day condors typically require good winds or ample elevation in order to become airborne, so these birds exist slightly beyond the normal limit of size for sustained flapping flight. The largest modern birds that can fly well weigh around 12 to 14 kg; these include the trumpeter swan (*Cygnus buccinator*) and the larger bustards (*Ardeotis kori, Otis tarda*).

In contrast, the largest bats weigh only about 900 g. The earlier adaptive radiation of birds may have prevented bats, which radiated later, from occupying certain ecological niches now exploited by large birds. In fact, large-bodied birds do not typically feed on the wing. There also may be functional considerations limiting the size of bats. Although bat wings can produce great lift, wing loading also tends to be high and may place a lower size limit on bats than on birds. Female bats must be able to carry their prenatal and sometimes postnatal young about while they feed. Furthermore, large bats presumably feed on large prey and need good dental armature to cope with it. That being the case, the weight of the snout complex would be difficult to reduce, so large bats might carry about a relatively greater cranial weight than birds (which have eliminated teeth

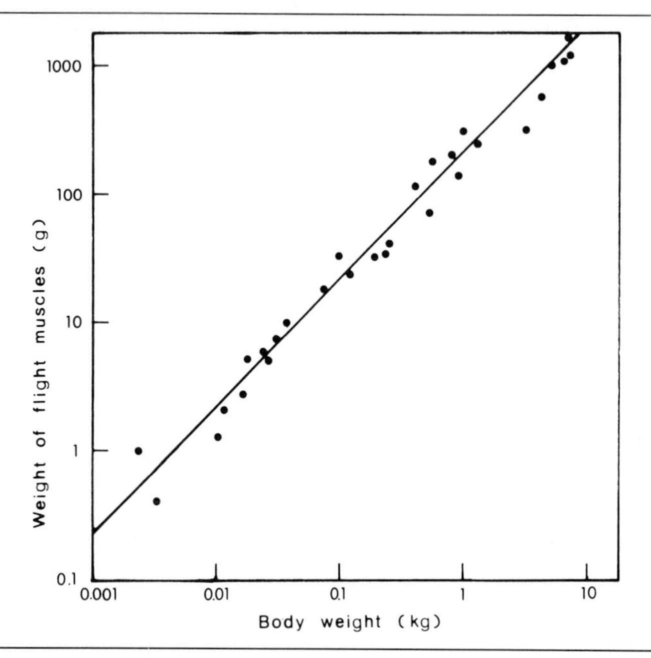

Figure 7–33 The relationship of flight muscle weight to body size. The birds range from a hummingbird to a bustard. (Modified from R. McN. Alexander, 1971. *Size and Shape.* Arnold, London.)

altogether). Factors such as these may have placed an upper limit to the size bats can achieve.

No bats seem to have evolved the soaring habit of the largest flying birds. Soaring requires much less energy than powered flight: In the case of a white-backed vulture, Pennycuick (1972b) estimated that flapping flight would cost about 30 times as much fuel as soaring does. The difference would be less for a smaller bird, so the advantages of soaring are less (but not zero) for a small bird.

Size limits for hovering birds can also be estimated. Hovering is stationary flight; that is, the bird remains suspended in one location without forward motion. Hummingbirds (Trochilidae) are classic examples of hoverers. They possess the ability to rotate their wings at the shoulder so that the wing tip performs a figure eight, and upper and lower surfaces of the wing alternately face the ground (Fig. 7–34). Since there is no forward motion, all lift must be provided by movement of the wings, and the wing beat is very rapid. The power needed for hovering is determined by body weight and wingspan. As discussed before, the power required increases with body size faster than the power available; hence, there must be some maximum size for birds that hover in a sustained fashion in still air. Alexander (1971) estimated this size at perhaps 20 g, close to the weight of the

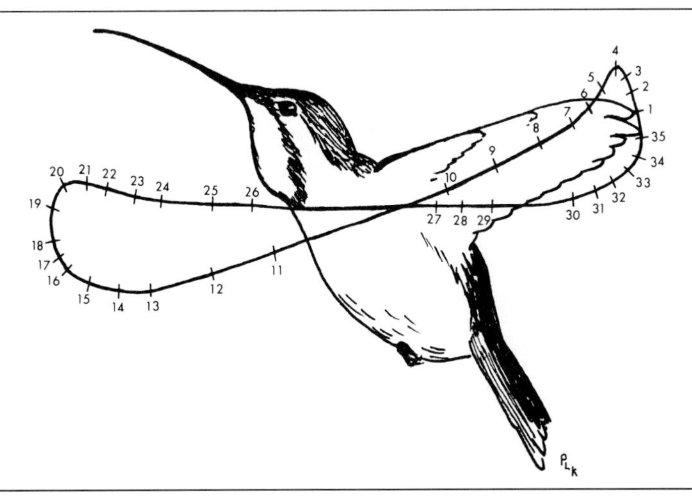

Figure 7–34 Side view of a hummingbird's hovering wing beat. The sequential position of the wing tip is indicated by the numbers. (Modified from J. Lighthill, 1975. Aerodynamics of animal flight. *In* C.J. Brokaw and C. Brennen (Eds.) *Swimming and Flying in Nature.* Vol. 2, pp. 423–492. Plenum Publishing, New York.)

largest hummingbird, *Patagona gigas.* Only hummingbirds are sustained hoverers of great control and finesse in still air. Some larger birds hover briefly (rough-legged hawk, *Buteo lagopus*) or for medium lengths of time (American kestrel, *Falco sparverius*), usually in a wind while hunting for prey below them.

What is the caloric cost of the various forms of locomotion? The total energy required for travel is a function of the animal's metabolic rate, its weight, the distance it moves, and its speed of movement. The number of legs employed in movement may not affect the energetic cost. In fact, a series of experiments with various species of apes and monkeys that can run easily on two legs or on four showed no significant difference in O_2 consumption at the two gaits over the same range of velocities (Taylor and Rowntree, 1973). When different animals of different sizes are compared, however, we find that the relative cost of locomotion for small bipeds (<1 kg) is less than that for small quadrupeds, but the relative cost for large bipeds is greater than for quadrupeds of similar size (Fedak et al., 1974; Taylor, 1977).

The metabolic cost of running increases linearly with speed (Fig. 7–35*a*) for quadrupedally running lizards and mammals, bipedally running birds, and even for undulating snakes, but, in all instances, the rate of increase is lower for larger animals. The total costs for lizards are lower than for the homeotherms because their metabolic rate is much lower, but the size-speed relationship is similar. In contrast, the metabolic cost of swimming for fishes increases exponentially with speed because of the density and viscosity of their medium (Fig. 7–35*b*). This places a relatively low limit on

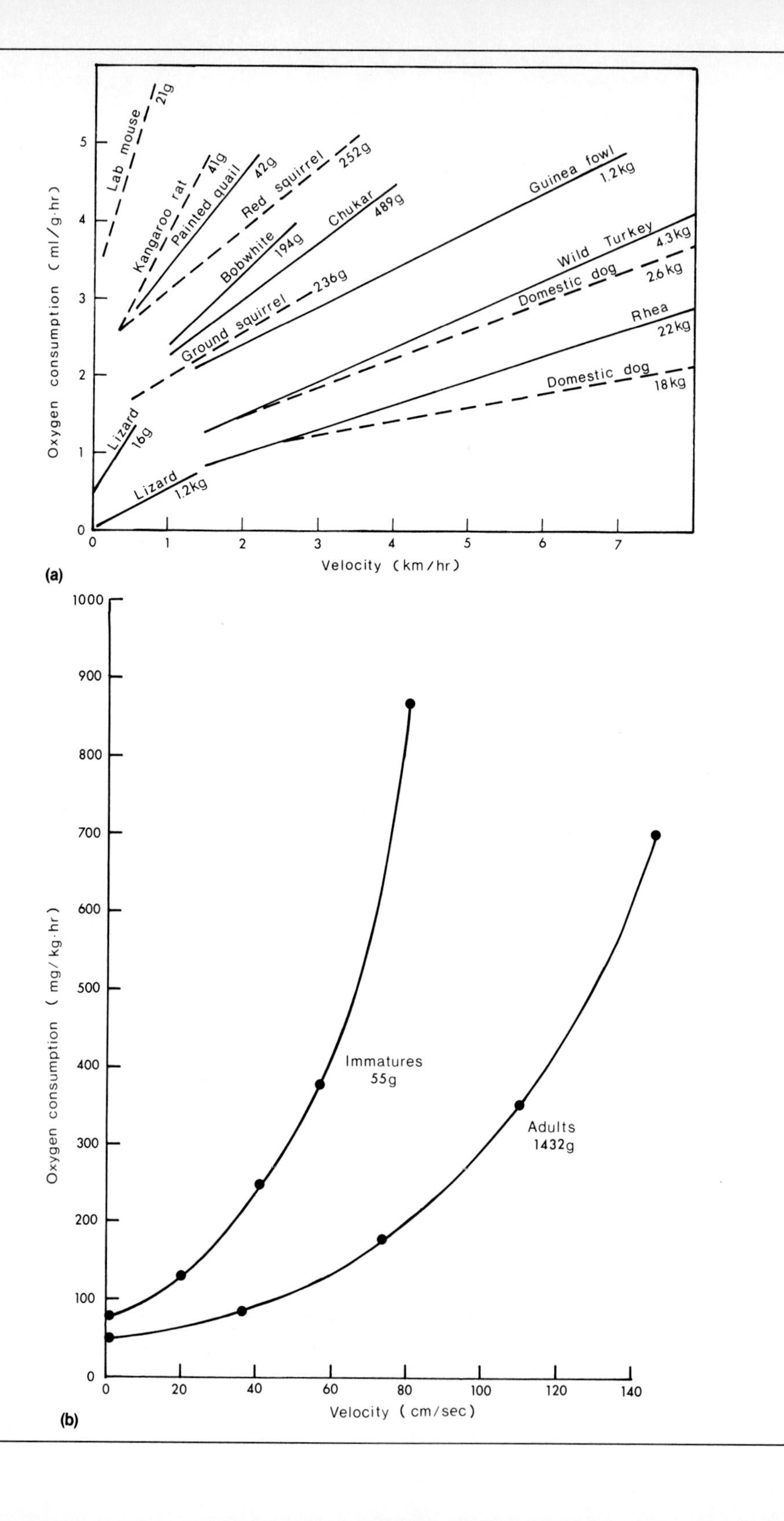
Lab mouse 21g
Kangaroo rat 41g
Painted quail 42g
Red squirrel 252g
Bobwhite 194g
Chukar 489g
Ground squirrel 236g
Guinea fowl 1.2kg
Wild Turkey 4.3kg
Domestic dog 2.6kg
Rhea 22kg
Domestic dog 18kg
Lizard 16g
Lizard 1.2kg
Oxygen consumption (ml/g·hr)
Velocity (km/hr)
(a)
Immatures 55g
Adults 1432g
Oxygen consumption (mg/kg·hr)
Velocity (cm/sec)
(b)

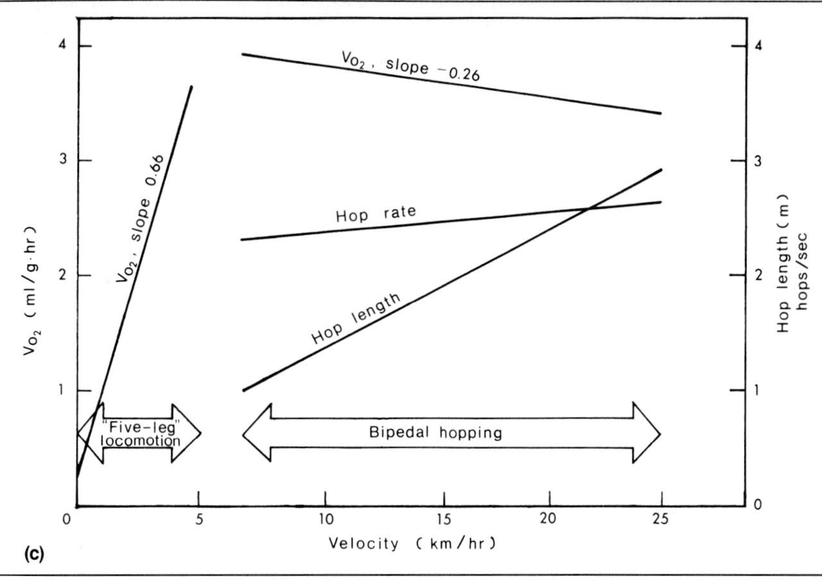

Figure 7–35 Relationship between the metabolic costs of locomotion and velocity for various vertebrates. (*a*) Quadrupedally running mammals, bipedally running birds, and quadrupedal lizards; (*b*) Large and small sockeye salmon (*Oncorhynchus nerka*); (*c*) A hopping red kangaroo (*Megaleia rufa*). (Reprinted with permission of Macmillan Publishing Co., Inc., from *Animal Physiology: Principles and Functions* by Malcolm S. Gordon et al. Copyright © 1977 by Malcolm S. Gordon.)

the speeds attainable. As for runners, however, relative metabolic costs are greater for small fishes than for large ones.

Hopping kangaroos (*Megaleia rufa*) exhibit an extraordinary relationship between cost and speed (Fig. 7–35*c*). When they are moving slowly, the tail is used as a fifth leg: The front legs and tail support the body as the hind legs are swung forward. The cost of five-legged slow hopping increases very rapidly with speed and, at about 6 km/hr, the animal shifts to bipedal hopping. As speed continues to increase, the metabolic costs actually decline slightly, chiefly because a fast-moving kangaroo makes longer hops. The elastic ligaments and tendons of the legs store energy at the end of each hop and release it at the start of the next hop. As much as 70 precent of the cost of each hop may be derived from the recoil mechanism. Quadrupeds also use elastic recoil, but the efficiency of the kangaroo's recoil system somehow increases with speed. So, as the kangaroo moves faster and faster, its elastic elements provide more of the required energy and less muscular effort is required. This relationship is usually not observed in small saltatorial mammals, however. Saltators weighing between 32 g (*Dipodomys merriami*) and 3 kg (*Pedetes capensis*) exhibit locomotor costs that are equivalent to those of quadrupedal cursors (Thompson et al., 1981).

In theory, large and small animals of similar build have the capacity for running at equal speeds. Our intuition protests this assertion, however, and Taylor's (1973) studies support our doubts. The greater energetic cost of running (per gram of body weight) for small animals, compared to the cost for large ones, means that a small runner must be able to increase its metabolic rate more than a large one in order to achieve equal speeds. Small animals have a higher per-gram metabolic rate to start with, and great increases at high speed may be limited unless the animals can utilize anaerobic metabolism. If one considered only energetics, large animals should be the fastest of all. They are not, however, perhaps because of structural limitations on heavy bodies. The greatest increase of metabolic rate at top running speed relative to resting metabolism is achieved by animals of intermediate size (Fig. 7–36). Other factors being similar, middle-sized animals should be able to reach the highest speeds.

In order to compare animals of different sizes, units must be standardized. Therefore, it is common to use energy per unit body weight per unit time (cal/g/hr) at a given speed (km/hr), which reduces to cal/g/km. This method certainly provides a first approximation, but it assumes that all speeds can be compared on an equal basis, which may not always be true. For running mammals and lizards, a linear relationship can be shown over a wide range of speeds; mice are notable exceptions (Fig. 7–37). For hopping kangaroos and flying starlings (*Sturnus vulgaris*), O_2 consumption

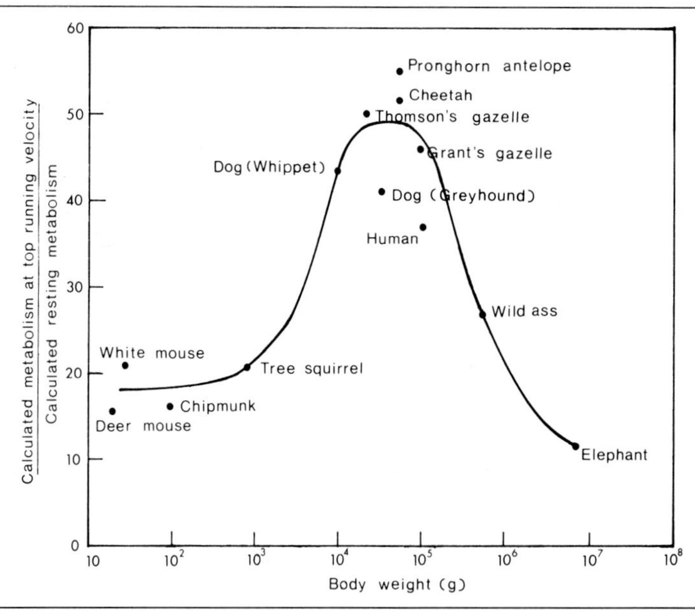

Figure 7–36 The relative increase in metabolism at top-speed running versus the logarithm of body weight in mammals. Intermediate-sized mammals can increase their metabolism more and run faster than either small or large mammals. (Modified from C.R. Taylor, 1973. Energy cost of animal locomotion. *In* L. Bodis, K. Schmidt-Nielsen, and S.H.P. Maddrell (Eds.), *Comparative Physiology*, pp. 23–42. Elsevier, Amsterdam.)

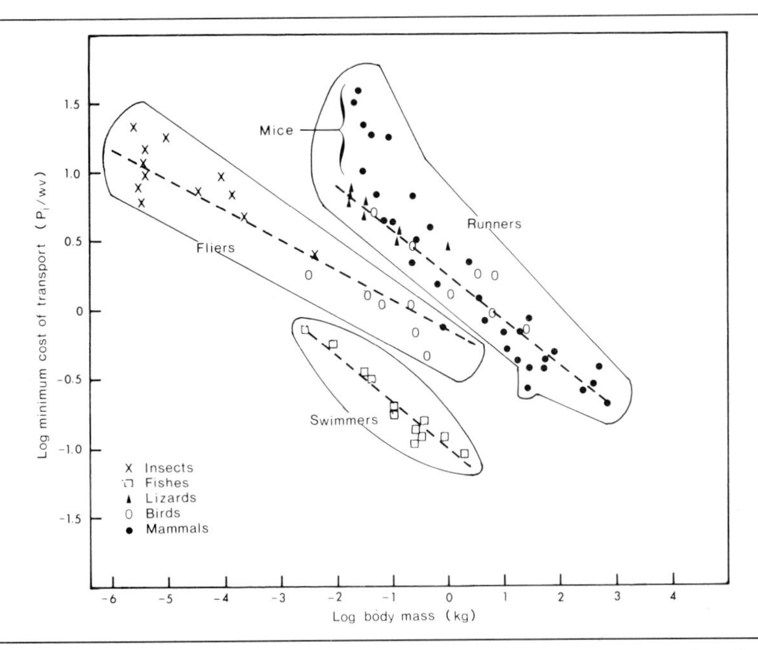

Figure 7–37 The relative cost of different modes of locomotion. (Reprinted with permission of Macmillan Publishing Co., Inc. from *Animal Physiology: Principles and Adaptations* by Malcolm S. Gordon et al. Copyright © 1977 by Malcolm S. Gordon.)

changes little with increasing speeds. Some snakes, swimming fishes, and flying budgerigars (*Melopsittacus undulatus*) exhibit an optimum speed for a given set of conditions at which the metabolic rate is lower than at other speeds. To what extent this "optimum" is actually used remains to be seen.

The estimated costs of different styles of locomotion, over a range of body sizes, can be compared. If we plot the approximate energetic cost (cal/g/km) against body weight on a log-log scale, we can see that running (and surface swimming) is more costly than flying, which is more expensive than immersed swimming. In addition, although lateral undulation on land costs about three times as much energy as lateral undulation in water, it is still less costly than quadrupedal locomotion (Chodrow and Taylor, 1973; Taylor, 1973). When flying, the large bats called "flying foxes" use more energy than birds, but less than four-footed mammals (Carpenter, 1975). Furthermore, large animals have a lower per-gram cost of locomotion and, hence, are more efficient than smaller ones.

An evolutionary increase in body size is facilitated by this efficiency relationship, but large animals still require more total food to support themselves than do small ones. Large animals can travel farther in a given amount of time and usually occupy larger hunting areas than similar, but smaller, species.

SUMMARY

The active mobility of vertebrates made possible the evolution of a diverse array of hunting and escape behaviors and social organization, all achieved by the expenditure of energy and the resultant metabolic costs. All locomotion derives from muscular forces directed against the medium or the substrate and the resulting equal and opposite forces upon the animal. Vertebrates have exploited this basic principle in a variety of ways. Some use their entire body for locomotion; others use a tail, median fins, forelimbs, hind limbs, or combinations of these.

Locomotion on four limbs at slow speed is very stable. Morphological modifications permit longer strides, and sometimes lighter limbs, for higher speeds. Stability is achieved by coordination and rapid changes of balance instead of by means of the support system. Rapid acceleration and direction changes are made possible by bipedal jumping.

Locomotion without limbs has permitted a variety of activities, from burrowing to climbing and even gliding. It is accomplished by lateral undulation, concertina motion, rectilinear locomotion, and sidewinding. Climbing adaptations include hooks, adhesive pads, and sometimes propping, grasping, or balancing tails.

In swimming, fishes have the advantage of support by the medium for much or all of the body weight; it has the disadvantages of drag and resistance to forward motion. Fishes often have swim bladders that increase buoyancy and streamlined shapes that minimize drag and resistance. Some fishes have sacrificed rapid locomotion for protection or maneuverability; in these species, the paired or median fins, in addition to serving as stabilizers, may be the means of propulsion. Nonpiscine swimmers have also become streamlined but do not reach the high speeds reported for fishes. Diving birds and mammals face potentially severe problems of temperature regulation and oxygen supply that are met at least partially, by physiological changes.

A number of vertebrates can modify their rate of falling; some can do this to such a degree as to be called gliders, but among extant vertebrates, true flight has been achieved only by birds and bats. Air resists forward motion less than water but offers less support. Lift is provided by the differential rates of air passage over and under the wing; forward propulsion is furnished by a posterior push of some portion of the wing. Wing size and shape and body size and proportion are closely related to flight abilities.

Running consumes more energy than flying, which costs more than immersed swimming. Large animals have a lower per-gram locomotion energy expenditure than small ones, so an increase in size is not accompanied by an equal increase in locomotor costs. The feasibility of evolving large body size is thus enhanced, but upper limits on body size are nevertheless subject to physical limitation.

SELECTED REFERENCES

Alexander, R.M., 1967. *Functional Design in Fishes.* Hutchinson, London.

Alexander, R.M., 1968. *Animal Mechanics.* Sidgwick & Jackson, London.

Alexander, R.M., 1971. *Size and Shape.* Arnold, London.

Alexander, R.M., 1972. The energetics of vertical migration by fishes. *Symp. Soc. Exp. Biol., 26*:273–294.

Alexander, R.M., and A. Vernon, 1975. The mechanics of hopping by kangaroos (Macropodidae). *J. Zool.* (Lond.), *177*:265–303.

Alexander, R.M., and G. Goldspink (Eds.), 1977. *Mechanical Energetics of Animal Locomotion.* Chapman and Hall, London.

Bellairs, A., 1950. Observations on the cranial anatomy of *Anniella,* and a comparison with that of other burrowing lizards. *Proc. Zool. Soc. Lond., 119*:887–904.

Bennet-Clark, H.C., 1977. Scale effects in jumping animals. *In* T.J. Pedley (Ed.), *Scale Effects in Animal Locomotion,* pp. 185–201. Academic Press, New York.

Berger, M., and J.S. Hart, 1974. Physiology and energetics of flight. *In* D.F. Farner and J.R. King (Eds.), *Avian Biology.* Vol. 4, pp. 416–477. Academic Press, New York.

Blaxter, J.H.S., and W. Dickson, 1959. Observations on the swimming speeds of fishes. *J. Cons. Int. Explor. Mer.* 24:472–479.

Bolis, L., K. Schmidt-Nielsen, and S.H.P. Maddrell (Eds.). 1973. *Comparative Physiology: Locomotion, Respiration, Transport and Blood.* North Holland, Amsterdam.

Bone, Q. 1975. Muscular and energetic aspects of fish swimming. *In* T.Y.T. Wu, C.J. Brokaw, and C. Brennen (Eds.), *Swimming and Flying in Nature.* Vol. 2, pp. 493–528. Plenum Publishing, New York.

Breder, C.M., 1926. The locomotion of fishes. *Zoologica, 4*:159–297.

Brown, R.H.J., 1948. The flight of birds. I. The flapping cycle of the pigeon. *J. Exp. Biol., 25*:322–333.

Brown, R.H.J., 1953. The flight of birds. II. Wing function in relation to flight speed. *J. Exp. Biol., 30*:90–103.

Brown, R.H.J., 1963. The flight of birds. *Biol. Rev., 38*:460–489.

Carpenter, R.E., 1975. Flight metabolism of flying foxes. *In* T.Y.T. Wu, C.J. Brokaw, and C. Brennen (Eds.), *Swimming and Flying in Nature.* Vol. 2, pp. 883–890. Plenum Publishing, New York.

Chodrow, R.E., and C.R. Taylor, 1973. Energetic cost of limbless locomotion in snakes. *Fed. Proc., 32*:422.

Cone, C.D., 1962. The soaring flight of birds. *Sci. Am., 206*(4):130–140.

Dawson, T.J., 1976. Energetic cost of locomotion in Australian hopping mice. *Nature, 259*:305–307.

Dawson, T.J., and C.R. Taylor, 1973. Energetic cost of locomotion in kangaroos. *Nature* (Lond.), *246*:313–314.

Dorst, J., 1971. *The Life of Birds.* Columbia University Press, New York.

Epting, R.J., and T.M. Casey, 1973. Power output and wing disc loading in hovering hummingbirds. *Am. Nat., 107*:761–765.

Fedak, M.A., B. Pinshaw, and K. Schmidt-Nielsen, 1974. Energy cost of bipedal running. *Am. J. Physiol., 227*:1038–1044.

Gambaryan, P.P., 1974. *How Mammals Run.* (Translation by H. Hardin). John Wiley & Sons, New York.

Gans, C., 1974. *Biomechanics.* J. B. Lippincott, Philadelphia.

Gans, C., and D. Baic, 1977. Regional specialization of reptilian scale surfaces: relation of texture and biologic role. *Science, 195*:1348–1350.

Gans, C., H.C. Dessauer, and D. Baic, 1978. Axial differences in the musculature of uropeltid snakes: the freight-train approach to burrowing. *Science, 199*:189–192.

Gordon, M.S., et al., 1977. *Animal Function: Principles and Adaptations.* 3rd ed. MacMillan Publishing Co., New York.

Gosline, W.A., 1971. *Functional Morphology and Classification of Teleostean Fishes.* University Press of Hawaii, Honolulu.

Gray, J., 1953. *How Animals Move.* Cambridge University Press, Cambridge, England.

Gray, J., 1968. *Animal Locomotion.* Weidenfeld and Nicholson, London.

Gray, J., and H.W. Lissman, 1950. The kinetics of locomotion in the grass-snake. *J. Exp. Biol., 26*:354–367.

Greenewalt, C.H., 1975. The flight of birds. *Trans. Am. Phil. Soc., 65*(4):1–67.

Hall, E.R., 1946. *Mammals of Nevada.* University of California Press, Berkeley.

Hankin, E.H., 1913. *Animal Flight.* Illife, London.

Heyer, W.R., and S. Pongsapipatana, 1970. Gliding speeds of *Ptydiozoon lionaturs* (Reptilia: Gekkonidae) and *Chrysopelea ornata* (Reptilia: Colubridae). *Herpetologica, 26*:317–319.

Hildebrand, M., 1959. Motions of the running cheetah and horse. *J. Mammal., 40*:481–495.

Hildebrand, M., 1974. *Analysis of Vertebrate Structure.* John Wiley & Sons, New York.

Jenkins, F.A., 1971. Limb posture and locomotion in the Virginia opossum (*Didelphis marsupialis*) and in other noncursorial mammals. *J. Zool.* (Lond.), *165*:303–315.

Laerm, J.R., 1973. Aquatic bipedalism in the basilisk lizard: The analysis of an adaptive strategy. *Am. Midl. Nat., 89*:314–333.

Lighthill, J., 1975. Aerodynamics of animal flight. *In* T.Y.T. Wu, C.J. Brokaw, and C. Brennen (Eds.), *Swimming and Flying in Nature.* Vol. 2, pp. 423–491. Plenum Publishing, New York.

Marcellini, D.L., and T.E. Keefer, 1976. Analysis of the gliding behavior of *Ptychozoon lionatum* (Reptilia: Gekkonidae). *Herpetologica, 32*:362–366.

Marshall, N.B., 1966. *The Life of Fishes.* World Publishing Co., Cleveland.

Marshall, N.B., 1971. *Exploration in the Life of Fishes.* Harvard University Press, Cambridge, MA.

Marshall, N.B., 1972. Swimbladder organization and depth ranges of deep-sea teleosts. *Symp. Soc. Exp. Biol., 26*:261–272.

Maynard Smith, J., 1952. The importance of the nervous system in the evolution of animal flight. *Evol., 6*:127–129.

Pedley, T. (Ed.), 1977. *Scale Effects in Animal Locomotion.* Academic Press, London.

Pennycuick, C.J., 1975. Mechanics of flight. *In* D.S. Farner and J.R. King (Eds.), *Avian Biology.* Vol. 5, pp. 1–75. Academic Press, New York.

Pennycuick, C.J., 1972a. *Animal Flight.* Arnold, London.

Pennycuick, C.J., 1972b. Soaring behavior and performance of some East African birds, observed from a motor-glider. *Ibis, 114*:178–218.

Peterson, J.A., 1971. Functional morphology of the shoulder in *Chamaeleo* and *Anolis.* (Abstract). *Am. Zool., 11*:704–705.

Pinshow, B., M.A. Fedak, and K. Schmidt-Nielsen, 1977. Terrestrial locomotion in penguins: It costs more to waddle. *Science, 195*:592–594.

Reichman, O.J., and D. Oberstein, 1977. Selection of seed distribution types by *Dipodomys merriami* and *Perognathus amplus. Ecol., 58*:636–643.

Russell, A.P., 1976. Some comments concerning interrelationships amongst gekkonine geckos. *Linn. Soc. Symp. Ser., 3*:217–244.

Savile, D.B.O., 1957. Adaptive evolution in the avian wings. *Evol., 11*:212–224.

Scholz, A., and H. Volsoe, 1959. The gliding flight of *Holaspis guentheri,* a West-African lacertid, *Copeia, 1959*:259–260.

Schmidt-Nielsen, K., 1971. Locomotion: energy cost of swimming, flying and running. *Science, 171*:222–227.

Schmidt-Nielsen, K., 1972. *How Animals Work.* Cambridge University Press, Cambridge, England.

Snyder, R.C., 1962. Adaptations for bipedal locomotion in lizards. *Am. Zool.,* 2:191–203.

Snyder, R.C., 1967. Adaptive values of bipedalism. *Am. J. Phys. Anthropol., 26*:131–134.

Stolpe, M., and K. Zimmer, 1939. *Der Vögelflug.* Akademische Verlaggesellschaft, Leipzig.

Taylor, C.R., 1973. Energy cost of animal locomotion. *In* L. Bolis, K. Schmidt-Nielsen, and S.H.P. Maddrell (Eds.), *Comparative Physiology,* pp. 23–42. North Holland, Amsterdam.

Taylor, C.R., 1977. The energetics of terrestrial locomotion and body size in vertebrates. *In* T.J. Pedley (Ed.), *Scale Effects in Animal Locomotion,* pp. 127–141. Academic Press, New

York.

Taylor, C.R., and V.J. Rowntree, 1973. Running on two or four legs: Which consumes more energy? *Science, 179*:186–187.

Thomson, K.S., 1976. On the heterocercal tail in sharks. *Paleobiol., 2*:19–38.

Thompson, S.D., R.E. MacMillen, E.M. Burke, and C.R. Taylor, 1981. The energetic cost of bipedal hopping in small mammals. *Nature, 287*:223–224.

Trapp, G.R., 1972. Some anatomical and behavioral adaptations of ringtails (*Bassariscus astutus*). *J. Mammal, 53*:549–557.

Tricker, R.A.R., and B.J.K. Tricker, 1967. *The Science of Movement.* Elsevier, New York.

Tucker, V.A., 1973. Aerial and terrestrial locomotion: a comparison of energetics. *In* L. Bolis, K. Schmidt-Nielsen, and S.H.P. Maddrell (Eds.), *Comparative Physiology,* pp. 63–76. North Holland, Amsterdam.

Tucker, V.A., 1975. Aerodynamics and energetics of vertebrate fliers. *In* T.Y.T. Wu, C.J. Brokaw, and C. Brennen (Eds.), *Swimming and Flying in Nature.* Vol. 2, pp. 845–867. Plenum Publishing, New York.

Vaughan, T.A., 1970. Flight patterns and aerodynamics. *In* W.A. Wimsatt (Ed.), *Biology of Bats.* Vol. 1, pp. 195–216. Academic Press, New York.

Vaughan, T.A., 1972. *Mammalogy.* Saunders College Publishing, Philadelphia.

Wainwright, S.A., F. Vosburgh, and J.H. Hebrank, 1978. Shark skin: function in locomotion. *Science, 202*:747–749.

Walker, W.F., 1971. A structural and functional analysis of walking in the turtle, *Chrysemys picta marginata. J. Morphol. 134*:195–214.

Wardle, C.S., 1977. Effects of size on the swimming speeds of fish. *In* T.J. Pedley (Ed.), *Scale Effects in Animal Locomotion,* pp. 299–313. Academic Press, New York.

Welty, J.C., 1982. *The Life of Birds.* 3rd ed. Saunders College Publishing, Philadelphia.

Williams, E.E., and J.A. Peterson, 1982. Convergent and alternative designs in the digital adhesive pads of scincid lizards. *Science, 215*:1509–1511.

Wu, T.Y.T., C.J. Brokaw, and C. Brennen (Eds.), 1975. *Swimming and Flying in Nature.* Vol. 2. Plenum Publishing, New York.

Zug, G.R., 1974. Crocodilian galloping: an unique gait for reptiles. *Copeia, 1974*:550–552.

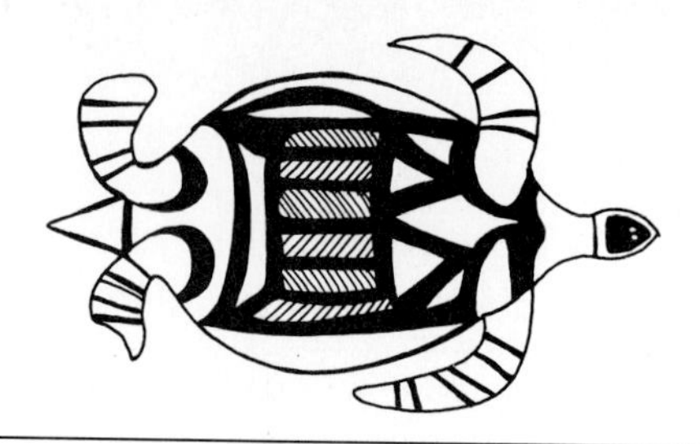

8 *Migration*

MIGRATION

Migration is a regular, usually seasonal, movement of animals from one place to another. Vertebrates are sufficiently long-lived that migrating individuals typically make at least one round trip from one region to another and back again. Sometimes the movements occur between breeding and nonbreeding grounds, and sometimes among several feeding areas. Regular migratory movements are known in all classes of vertebrates.

The migrations of birds have perhaps attracted the most notice. The winter disappearance of many bird species has been noted for hundreds of years, and the way that birds spent the winter was a subject of much debate. Some "naturalists" argued for migration and some for hibernation, and for years, no one bothered to gather any data. A prize example of early unnatural history is provided by Olaus Magnus (1490–1558), the Archbishop of Uppsala, in his *Historica de Gentibus Septentrionalibus,* published about 1555:

> Several authors who have written at length about the inestimable facts of nature have described how swallows often fly from one country to another, travelling to a warm climate for the winter months; but they have not mentioned the

denizens of northern regions which are often pulled from the water by fishermen in a large ball. They cling beak to beak, wing to wing, foot to foot, having bound themselves together in the first days of autumn in order to hide amid canes and reeds. It has been observed that when spring comes they return joyously to their old nests or build new ones, according to the dictates of nature. Occasionally young fishermen, unfamiliar with these birds, will bring up a large ball and carry it to a stove, where heat dissolves it into swallows. They fly, but only briefly, since they were separated forcibly rather than of their own volition. Old fishermen, who are wiser, put these balls back into the water whenever they find them.

Impressive long-distance migrations are known among fishes, a few reptiles, mammals, and birds. The most famous migrants among fishes are the salmon and eels. Salmon of the genus *Oncorhynchus* are native to the northern Pacific region. Adults breed in fresh water. The young migrate to the sea at species-specific ages, mature in the ocean, return as adults to their birth streams, reproduce, and then die. The total distance covered is sometimes thousands of kilometers. Other species of salmon or trout of the genus *Salmo,* such as the sea-run rainbow trout or steelhead and the Atlantic salmon, perform similar extensive migrations. They do not always die after spawning and may survive to make another round trip. Two species of eel (*Anguilla*) spawn in the Sargasso Sea and then die. The young gradually move toward the shores of either Europe or North America, depending on the species. The young mature slowly in fresh water and return to the sea as adults.

Although these cases are well known, many other fishes also make extensive migrations. A dogfish shark, *Squalus acanthius,* moves regularly between the North Atlantic and the southeast coast of the United States. Cod (*Gadus morhua*) migrate from the Barents Sea to the coast of Norway and from southern Greenland and northern Iceland to southern Iceland to breed.

The green sea turtle (*Chelonia mydas*) is a well-known migratory reptile. Some populations breed on Ascension Island in the middle of the South Atlantic and feed off the shores of South America (Fig. 8–1). Females make the 4000-km round trip about every two years (Carr, 1967, 1972; Koch et al., 1969).

The grey whale (*Eschrichtius gibbosus*) spends the summer in arctic waters but migrates for the winter, mainly to the west coast of Baja California, where the females give birth in warm lagoons. This is the longest migration known for mammals—a round trip of more than 16,000 km (Fig. 8–2). Some grey whales move south along the east coast of Siberia to Korean and Japanese waters; the grey whale population in this area is now quite small because of overexploitation by whalers in the early 1900s. Several pinnipeds also cover long distances: Female and young northern fur seals (*Callorhinus ursinus*) move from breeding areas on the islands between Alaska and Siberia to spend the winter as far south as California and Japan.

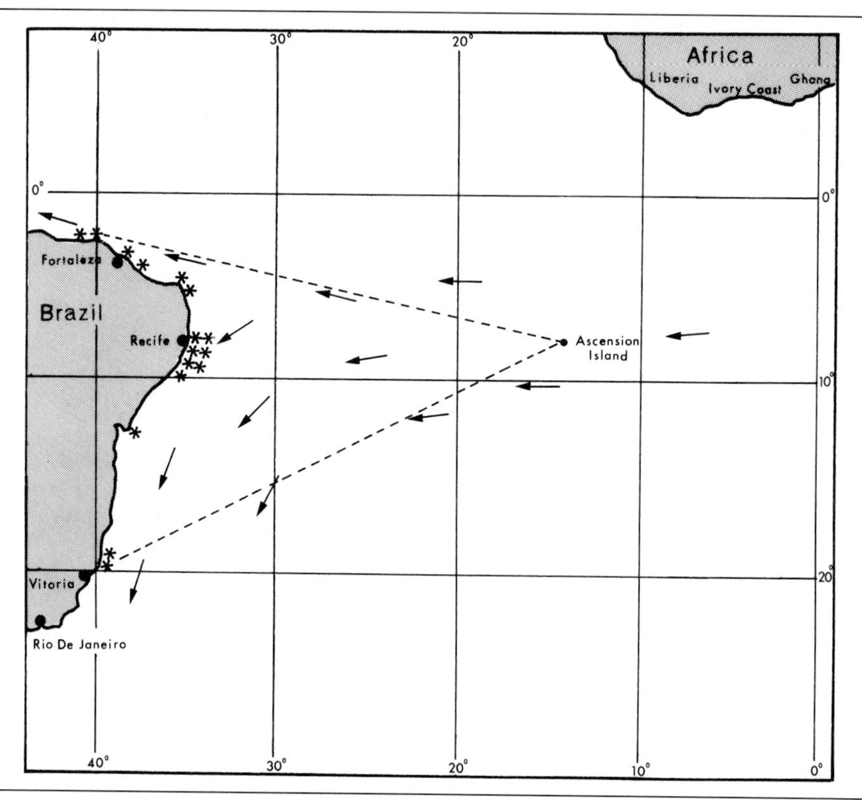

Figure 8–1 The green sea turtle's migration route between Ascension Island and Brazil. (Modified from A.F. Carr, 1972. The case for long-range chemoreceptive piloting in *Chelonia*. *In* S.R. Galler (Ed.), *Animal Orientation and Navigation*, pp. 469–483. NASA, Washington, D.C. Reprinted with permission of the National Aeronautics and Space Administration.)

Some birds perform amazing feats of seasonal migration. The bobolink (*Dolichonyx oryzivorus*) breeds in east-central North America and winters in southern Brazil and Argentina. The champion long-distance migrant is often said to be the arctic tern (*Sterna paradisea*), which annually flies from its arctic breeding grounds to the subantarctic and back again, sometimes covering 30,000 to 40,000 km a year. The short-tailed shearwater (*Puffinus tenuirostris*) breeds on the coasts of southeastern Australia and Tasmania. Each year, when the adults leave the breeding ground, they fly north past Japan, across the Bering Sea, south along the coast of western North America, and then across the Pacific back to Australian waters (Fig. 8–3). The total distance they travel may exceed 30,000 km, so perhaps they should share the title of champion with the arctic tern. Some other seabirds may

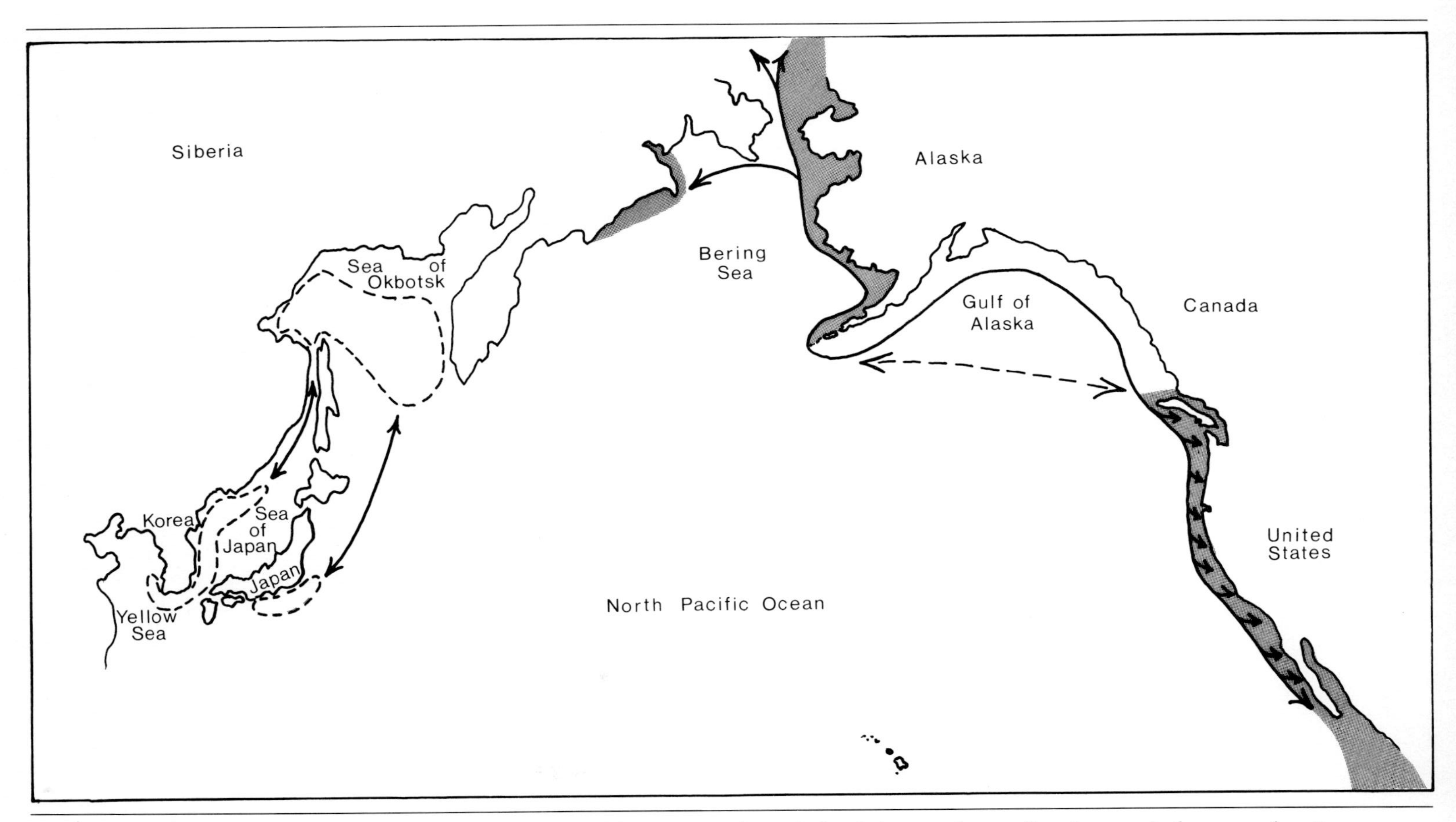

Figure 8–2 The gray whale's migration routes. Two populations are indicated; the Asian one has suffered severely from overhunting.

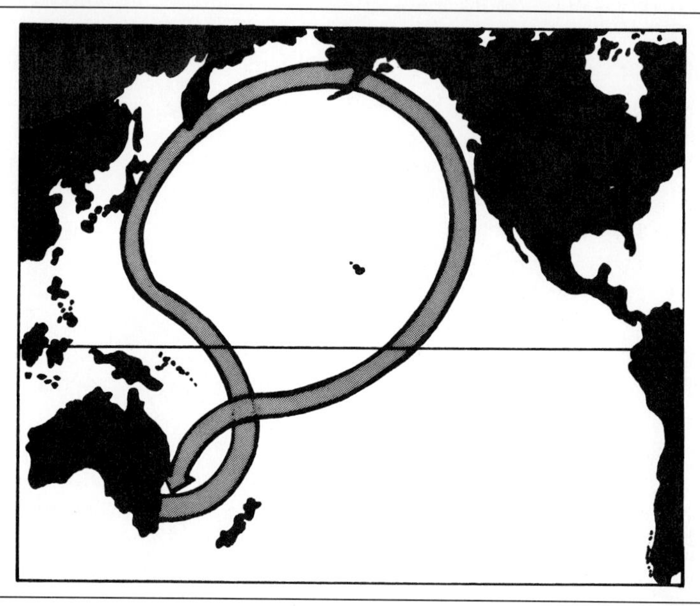

Figure 8–3 The short-tailed shearwater's annual migration route. Also known as the Tasmanian muttonbird (because of their heavy commercial exploitation for human food), these shearwaters harvest the plankton blooms during both the northern- and the southern-hemisphere summers. (From Marshall, A.J., and Serventy, D.L., 1956: The breeding of the short-tailed shearwater, *Puffinus tenuirostus* (Temminck) in relation to transequatorial migration and its environment. *Proc. Zool. Soc. Lond., 127*:489–510.)

also cover vast distances. Some migratory flights are very fast. The blue or snow goose (*Chen caerulescens*) in fall migration takes less than 60 hr to go from James Bay, Canada, to Louisiana—a distance of more than 5,000 km as the crow (or goose) flies, at an average speed of 83 km/hr.

Not all migrations are so extensive, however. Some amphibians and reptiles make regular short migrations between hideaways where they spend the winter or a dry season and their foraging or breeding grounds. The cottonmouth moccasin (*Agkistrodon piscivorus*) in the southern United States moves from its winter quarters, which are often in dry, rocky hills, to its swampy hunting and breeding area every spring and returns each autumn. The distances involved in these movements are not terribly imposing (sometimes a few hundred meters or even less), but they qualify as migrations because the movements are regular and reasonably predictable in both time and space. Numerous species of tropical birds (such as the African pennant-winged nightjar, *Cosmetornis vexillarius*) migrate back and forth across the equator over relatively short distances. Many populations of caribou (*Rangifer tarandus*) move south for the winter and north for the summer, but some Alaskan populations of caribou move from the southern slopes of the mountain ranges in summer to the wind-swept northern slopes, where the snow cover is lighter, in winter. Deer and elk (*Odocoileus* spp. and *Cervus canadensis* in North America; red deer, *Cervus elaphus,* in Europe) in mountainous regions commonly migrate altitudinally, to higher elevations in summer and to lower valleys and meadows in winter.

Although many aspects of migration are characteristic of whole species, large individual and population differences can be found. Many species are partial migrants—that is, some populations migrate but others do not. For example, the variegated flycatcher (*Empidonomus varius*) lives in South America; populations nesting in Argentina and southern Brazil winter in the Amazon River basin sympatrically with the conspecific resident population. The females of some species, such as the European chaffinch (*Fringilla coelebs*), are more migratory than the males, departing earlier and going farther. On the other hand, the males of some Arctic-breeding sandpipers leave the nesting ground long before their mates or offspring. Adult least flycatchers (*Empidonax minimus*) seem to migrate in the fall, well ahead of their offspring. The young of gannets (*Sula bassana*) nesting in Britain spend the winter in northwest Africa, but their parents can be found in France and Portugal.

Other Geographical Movements

As it is used in this text, the term "migration" refers specifically to round-trip movements and, therefore, is distinguished from both emigration and immigration. Both of these types of movement are unidirectional, either out of or into an area. They generally reflect dispersal away from a natal region into another area, where the dispersing animal may settle down. Although some authors conflate all kinds of space changes under the heading of migration, I prefer to maintain a distinction between a potentially permanent change of location (dispersal) and seasonal revisits (migration).

Dispersal occurs at some time during the life history of virtually every vertebrate species. It is commonly the younger individuals that leave the place of birth and explore other areas, eventually settling in a new home (if they survive). Adults may also disperse between successive breeding seasons. The period of dispersal is likely to be a time of considerable risk to the dispersing individual because it often is moving through unfamiliar areas where both food and hiding places may be found less quickly. Furthermore, a dispersing individual may find that all suitable habitats are already occupied by conspecifics, and, in order to establish itself in a new area, it must compete with the residents. This, too, may be costly, because residents commonly have an advantage over an intruder and because the intrusion itself requires the expenditure of energy and possibly exposure to injury. Presumably the advantages of such dispersal outweigh the possible disadvantages.

The most likely advantage of dispersal is that the animal has a greater probability of establishing itself successfully than if it stayed at home. One would expect, then, that a potential disperser would stay at home if space became available or if the chances of successful establishment elsewhere were predictably very low. At least sometimes, this may be true: For example, young male great tits (*Parus major*) in England more often claimed part of their natal territory if their father died than if he survived. Young males of several communally breeding birds (Chapter 15) commonly remain with their parents for several seasons, perhaps because their best opportunity for breeding is obtained by waiting to claim the family territory, rather than by attempting to establish in another habitat that is saturated.

A sex differential in tendency to disperse is common in both birds and mammals (Greenwood, 1980). Female birds usually disperse more than males, whereas male mammals are more usual dispersers; there are exceptions to both trends, however. One result of dispersal is that inbreeding is reduced, but because this potential disadvantage pertains to both sexes, it cannot explain the sex differentials. The basis for these patterns probably lies in the relative costs and benefits of dispersal to each sex. Greenwood suggests that males that remain in the home area obtain great benefits because the cost of establishing a territory on familiar ground may be less than that of establishing a territory in a new area. Females are less constrained because, in such species, they do not pay the costs of territoriality; they merely choose among the territorial males. These considerations would apply to many birds and to a few mammals. In contrast, many mammals do not defend resources in a territorial system; instead, females are often sought out by males and it is the females, rather than the resources, that are defended. Resident females are in a better position to benefit from staying on familiar ground than males, and dispersing males will profit from greater access to mates. These suggestions can be explored further by inspection of the exceptions to these general trends and by surveys of other taxonomic groups.

Other long-distance movements of large numbers of individuals are irregular and are called irruptions or invasions. These sporadic movements are known to occur in lemmings (*Lemmus*), squirrels (*Sciurus*), some fishes, and birds.

The snowy owl (*Nyctea scandiaca*) breeds in the Arctic and, in many areas, depends primarily on lemmings for food. Lemming population size varies greatly from year to year, in a more or less cyclic manner. Lemming population crashes are often followed by an exodus of snowy owls, which then invade north temperate regions of America and Europe. During the winter of 1974–1975, snowy owls were reported in unprecedented numbers across much of the United States, even as far south as Alabama and Texas. The previous irruption, in the winter of 1971–1972, was much smaller and did not extend so far south. Nutcrackers (*Nucifraga*) and crossbills (*Loxia*) of Europe and of North America feed heavily on pine and spruce seeds, respectively. Many species of pines and spruces are notorious for great annual variations in the size of their cone crops. When seed production is low, nutcrackers, crossbills, and other birds that depend on these seeds may irrupt in large numbers and invade regions where they are not normally found.

Some bird species are characteristically nomadic; they have no fixed pattern of movement. The rosy pastor or rose-colored starling (*Pastor roseus*) feeds chiefly on locust swarms in southeastern Europe and the Near East. Its movements are habitually irregular over a huge area, depending on the location of its prey. Desert birds in Australia are also often nomadic, wandering about a large area and breeding when they encounter a spot favored by recent rains (Fig. 8–4). About a quarter of the Australian avifauna, primarily species of the arid interior, can be called nomadic, a much larger proportion than that of the avifauna of north temperature regions.

Neither irrupting species nor nomads conform to the definition of migration. Their movements are generally irregular in both time and space.

THE ECOLOGY OF MIGRATION

Why Animals Migrate

Migration can be viewed as a means of using habitats and food resources that are available or suitable only seasonally. In the course of evolutionary history, individuals explored and successfully exploited new areas temporarily rich in resources, low in predation risks, or otherwise of great suit-

Figure 8–4 Some nomadic birds of Australia. (*a*) Flock pigeon (*Histriophaps histrionica*); (*b*) masked wood swallow (*Artamus personatus*); (*c*) pink-eared ducks (*Malacorhynchus membranaceus*).

ability and thereby increased their fitness. Implicit in the idea of successful exploitation is the ability to leave the new areas when (or before) they decline and become unsuitable. The hazards (storms, famine, getting lost), especially those of long-distance migration, may be great, so we can imagine that the benefits of seasonal use of an area must also be great. Why the individuals of some populations find seasonal movements profitable while others do not is not understood in detail. We must suppose that migrants and residents differ in ecological constraints and potentials in ways that coincide with their different seasonal habits.

Naturalists frequently assume that an animal's real "home" is the breeding ground and regard its voyages to other regions as only visits. Many temperate-zone ornithologists have held this view. Inherent in many studies of avian ecology in temperate zones is the assumption that the major selection pressures on migrants occur in the breeding grounds and that the winter season is somehow of little significance. The migrant birds, after all, have escaped the rigors of severe winters and are perhaps basking in tropical sun, or at least are conveniently out of sight! Events in the nonreproductive season are clearly of great importance as selection pressures, however, and should never be ignored—the individual must survive that season in order to breed the next. Sometimes vertebrates stay as long or even longer on the nonbreeding grounds as they do in the breeding areas. If the animals are not dormant, their interactions with other organisms may exert great selection pressures on body size and proportions, foraging behavior, and choice of foraging sites. It is quite possible that many of the ecological characteristics of migrants on the breeding grounds are partly the result of selection on the wintering grounds.

Exploitation of seasonally available food resources is clearly exemplified by the migration patterns of many whales. Migrating whales move between a summer feeding ground in polar waters and a winter mating and calving area in warmer waters. Summer is the main feeding season for such whales as the humpback (*Megaptera novaeangliae*), which feeds on small aquatic invertebrates. The animals gain weight in the summer by foraging in the highly productive polar seas but lose weight during the winter when they move to lower latitudes, where they do little feeding. The young are born with little insulating fat, perhaps as a means of minimizing body size at birth; a warm birthplace probably contributes to their survival. Females and young are the last to leave winter quarters and the first to leave the summer feeding areas. Thus they spend more of their time in warm water than do the males, but the adaptive basis of this difference remains to be explored. Migration patterns of this whale are a good example of seasonal exploitation because feeding and breeding are temporally and spatially separated. This separation makes it difficult to call just one of the regions "home" and facilitates visualization of different ecological factors. Similar principles are at work with other vertebrates, but little study has been made of the ecology of migration for any vertebrate group.

If the idea of seasonal exploitation is valid, we should be able to compare communities, tallying their proportions of migrant species, and infer that seasonal resource changes are greater in communities with larger proportions of migrants. A survey of certain bird communities in North America showed that the species of the eastern deciduous forest were less likely to be migratory than those of northern conifer forests. The chief winter foods of winter resident birds (which include woodpeckers, nuthatches, and titmice) are bark-dwelling insects and fruits or seeds. If those deciduous forests support more resident bird species than do coniferous forests, many of which are at the same latitudes as deciduous forests, it may be because deciduous forests offer the birds more ways to exploit their food resources. The variety of bark surfaces and perhaps of fruits is great in these deciduous forests. Therefore, both foraging sites and food resources may be subject to proportionately less seasonal change than they are in coniferous forests, despite the seasonal drop of deciduous leaves and the retention of needles by conifers. No one has measured seasonal changes in such resources, however, so the hypothesis is still untested.

The numbers of Australian honeyeater species (Meliphagidae) that move seasonally varies in a broad way with the reliability of rainfall (Keast, 1968). Where rains are plentiful and predictable, 70 to 85 percent of the species are year-round residents, but where rains are skimpy and erratic, fewer than 50 percent of the species are residents. Many of the seasonal movements of these honeyeaters are not proper migrations, being somewhat irregular in pattern, but Keast notes that species that are most dependent on nectar for food are the most mobile, implying both that habitats may differ in their supplies of nectar-bearing flowers and that the honeyeaters of different habitats depend on floral resources to differing degrees. This study emphasizes the likelihood that resource reliability underlies the frequency of seasonal movements in different communities.

Although seasonal changes in food supply are of considerable importance in determining migratory habits, other factors must also be involved. For instance, in Europe, the Ural owl (*Strix uralensis*) feeds on birds and mammals and nests in tree holes, which are limited in supply; a pair of these owls occupies its territory permanently. The long-eared owl (*Asio otus*) specializes on microtine rodents, breeds in readily available twig nests built and abandoned by other birds, and is largely migratory. Tengmalm's owl (*Aegolius funereus*) is a microtine specialist, like the long-eared owl, and nests in tree cavities, like the Ural owl. According to Lundberg (1979), male Tengmalm's owls are resident, defending their hole nests throughout the year. Females and young of this species are migratory, moving to areas of more abundant food in winter. Microtine populations exhibit enormous changes in abundance from year to year, and female Tengmalm's owls breed in different areas in different years, depending on where the microtines are available. This means that the resident males may not breed every year, but when they do, they mate with more than one female. Lundberg's hypothesis is that use of tree cavities for nests makes a resident habit advantageous for male Tengmalm's owls (although dietary specialization necessitates female movement) and for both male and female Ural owls, whose dietary generalization permits both sexes to stay.

Retreats from high predation may become available seasonally and provide a temporary refuge. Some Amazonian fishes move into floodplain shallows for spawning, perhaps as a means of eluding predators. Some fishes migrate between streams, for breeding, and lakes or seas, for feeding; whether they are stream fishes temporarily exploiting seasonally available food resources or lake/ocean fishes temporarily using protected spawning sites is moot. The migrations of some whales, including the grey whale and the humpback, seem to indicate a seasonal passage from a productive feeding area to a favorable environment for bearing young; in this instance, both food and refuge may be important.

Given that an animal migrates at all, the timing of migration probably has evolved in response to several ecological pressures in addition to the seasonal changes in food resources. Some animals begin to migrate long before food supplies are depleted, perhaps in response to pressures to claim nest sites or to stake out winter territories (Morton, 1976). Others may leave early because their route is long or when weather conditions for long flights are suitable. The ecological factors favoring the time of migration are undoubtedly varied—and little explored.

Ecological Problems

When a migrant animal leaves one portion of its range, it must pass through habitats that are occupied by other animals. When the animal arrives at another part of its range, that area too will have its own fauna. Furthermore, the foraging and resting sites, the microclimates, and the food items are all very likely to change as a migrant moves from place to place. Some migratory vertebrates, such as the Swainson's hawk (*Buteo swainsoni*), which migrates from North America to Argentina, may reduce the problems of finding food en route by fasting for most of the trip. Many other birds lay on large amounts of fat before embarking on their long, energetically costly, voyages. Some, like the snow goose, can fly long distances nonstop. Many migrants, however, face the problem of using an ever-changing array of resources. Imagine a bay-breasted warbler (*Dendroica castanea*) that breeds in North American spruce forests and winters in Panama and Columbia. A south-bound migrant runs out of spruce trees when it leaves the northernmost reaches of the United States, except in the Appalachian Mountains, and can be seen foraging in deciduous trees as it passes south. The configuration of vegetation in its winter range is different still, so each individual, even if it selects foraging sites as similar as possible to those provided in spruce forests, must be flexible in its habitat use. No migrant could be successful if it were too specialized to conditions found in only one portion of its range. Many ecologists have suggested that, in some ways, migrants may be more broadly adapted to a range of conditions than many nonmigratory species.

A resident fauna is bound to be encountered wherever a migrant chooses to go, which presents the possibility of intense interspecific competition. How might the migrants avoid severe competition? A few suggestions, at least for birds, have been made.

Migrants from Europe to Africa and from North America to Central and South America do not usually (there are some exceptions) winter in either desert or evergreen (rain) forests. To a lesser degree, the same tends to be true for Asian migrants. They are concentrated principally in woodland, grassland, savanna, second-growth, or high-altitude habitats whose own residents tend to be more migratory or vagrant than those of the rain forest or desert. Thus, the number of residents encountered in these habitats may be less than in habitats with a more sedentary avifauna. Use of such habitats varies from continent to continent, in part because different habitats and seasonal climatic conditions predominate in different regions. The number of wintering wood warbler (Parulidae) species decreases southward in Latin America, corresponding roughly to an increase in the number of resident birds of similar foraging habits. This may indicate that wintering migrants deploy chiefly in regions with fewer potential competitors or that they are in fact full members of these neotropical communities, returning to occupy their own ecological space in those communities, or both.

Another possibility that facilitates coexistence is reduced energy requirements. Moreau (1972) estimated that the energy demands of the migrants on the tropical wintering grounds may be about 40 percent of those on the breeding ground because the days are warmer and shorter and energy demands are not increased by breeding activities.

Diet may also play a role in reducing ecological problems. Many species in the tropics, both migrants and resident, regularly eat large quantities of fruit, thus expanding their resource base beyond that available for most temperate-zone breeding seasons. Fruits are used to a certain extent by some temperate-zone breeders, but most of these species increase their use of fruits in the nonbreeding season. North American migrants in Latin America often use abundant or sporadically available resources not fully exploited by residents. Palearctic migrants to the savannas of eastern Africa arrive during the dry season when their insect food is not very abundant except after local rainstorms. The migrants move around among these localized wet spots where insects become available. As the rainy season gets under way, they settle down and co-occur with breeding residents. In both cases, the influx of migrants seems to occur when potential competition with residents is low (Sinclair, 1978). Differences in food selection and foraging behavior may also mitigate potential competition between seasonal and permanent residents. The severity of such competition is not well-established. Some studies have indicated that migrant birds are often subordinate to resident tropical birds in direct confrontation, but other studies challenge this finding. Many studies have found that migrants and residents exhibit differences in habitat and foraging behavior, but actual competition has proved difficult to demonstrate.

In the absence of extensive published research, this section must remain inconclusive, but it does serve to emphasize that migrants must be equally well-adapted to all parts of their range. The ecological problems of life outside the breeding season (for birds anyway) are still not well understood.

MIGRATION BEHAVIOR

Stimuli for Migration

Like the proximate cues that trigger breeding activity, signals stimulate migratory behavior at the ecologically appropriate time. Many north-temperate birds respond to changing day length as one major stimulus for both preparation for migration and actual departure. The annual cycle of day lengths is invariant and broadly correlated with changing conditions; hence, it is a good general proximate cue. Its effects can be modified, however, by other factors whose importance varies greatly with the species. Some birds maintain migratory readiness in response to changing day length but do not depart in fall migration until the food supply in their northern breeding areas is diminished. The swallow (*Hirundo rustica*) and redwing (*Turdus iliaca*) of Europe reportedly behave this way. Warm temperatures may also delay departure in the fall. Many birds migrate in flocks, and a final stimulus needed for departure involves a response to a group of individuals and perhaps to some dominant, intensely stimulated members of the group. Other vertebrates can employ many of the same cues.

All such environmental influences impinge upon individual "physiological calendars." In fact, some species apparently migrate solely in response to internal endogenous rhythms, independent of environmental conditions in the region they are about to leave. Several populations of the European yellow wagtail (*Motacilla flava*) winter together in central tropical Africa, but each population leaves for the north at a different time, corresponding to the time when its breeding habitat will be suitable. Thus, the individuals that nest in southern Europe depart in February, those of central Europe in March and April, and the Scandinavian breeders in April and May. This temporal pattern suggests independent, endogenous rhythms, but it is also possible that individuals of each population respond to different cues in the same environment.

Genetic differences in migratory behavior are reported for the blackcap (*Sylvia atricapilla*). Individuals from different geographic populations exhibit very different intensities of migratory unrest (Zugunruhe) before departure and normally fly quite different distances between breeding and wintering grounds. Hybrids between such populations were intermediate between the parent stocks in their migratory behavior, a result that strongly indicates a genetic basis for the behavioral patterns (Berthold and Querner, 1981).

Preparation for migration in north-temperate birds often includes completion of a molt sequence and laying on of large fat deposits, the fuel for the long flight. Gwinner's (1975) work with *Phylloscopus* warblers in Europe exemplifies the chain of events and demonstrates some adaptive interspecific differences. The willow warbler (*P. trochilus*) is a long-distance migrant to central and southern Africa and begins migrating south in late summer. The chiffchaff (*P. collybita*) migrates only as far as the Mediterranean area and begins in late fall. Each species puts on fat and becomes restless prior to departure (Fig. 8–5). The willow warbler becomes fatter and is restless for a longer time than the chiffchaff, corresponding to the relative length of their migration routes. Furthermore, it seems that the willow warbler must depart so early that there is not enough time to finish the entire molt, which is then completed after arrival on the winter grounds.

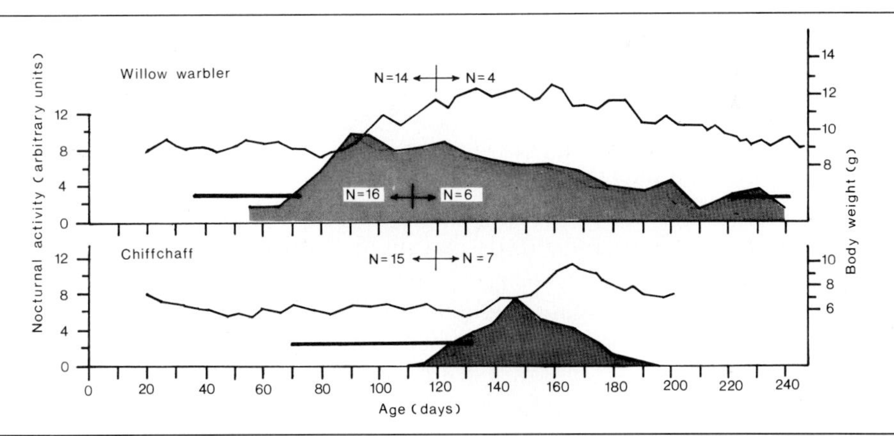

Figure 8–5 Temporal changes of body weight (line), *Zugunruhe* (shaded), and molt (bars) in the willow warbler and the chiffchaff. These differences in premigratory development are correlated with differences in migratory behavior. The willow warbler departs in late summer, the chiffchaff in late fall. (Modified from E. Gwinner, 1975. Circadian and circannual rhythms in birds. *In* D.S. Farner and J.R. King (Eds.), *Avian Biology*. Vol. 5, pp. 221–285. Academic Press, New York.)

Fat deposition is essential to successful migration for many birds. Most north-temperate migratory birds eat prodigiously before departing, as well as during the journey, and may become very fat. Blackpoll warblers (*Dendroica striata*) weigh 11 to 12 g in the nesting season, which is spent in the northeastern United States, but weigh 20 to 23 g at the time of fall migration to South America. This amount of fat is estimated to be enough fuel for a 115-hr nonstop flight—perhaps enough for the entire journey if the birds flew directly from one place to the other.

The amount of fat deposited varies significantly among species (Blem, 1980). Several long-distance migrants put on heavy fat deposits for migration, usually more than what birds migrating for shorter distances acquire. Some long-range migrants, however, feed on their way and may not carry such large deposits of fat. Both fat build-up and foraging habits may differ on spring and fall migrations of a single species. In keeping with their lower weight-specific metabolic rates, large birds may deposit smaller reserves of fat relative to their size than small birds, although the total fat reserve required to fly the same distance is, of course, greater for a larger bird.

Orientation and Navigation

Three levels of migratory orientation capability have been identified: (1) *piloting* is the ability to go to a particular location by using familiar landmarks; (2) *compass orientation* refers to the ability to go in a particular compass direction; and (3) *true navigation* is the ability to orient to a particular place from a variety of unfamiliar locations. Many vertebrates can find their way by using known landmarks that are often perceived visually. Some bats can apparently recognize landmarks with their sonar; a number of vertebrates may use olfactory cues. In general, wide-ranging animals can use landmarks better than sedentary ones can. Small salamanders, such as the red-bellied newts (*Taricha rivularis*), and painted turtles (*Chrysemys picta*), however, can return home from as far away as 3 km, apparently by using landmarks. Birds can often "home" from hundreds of kilometers away, but their use of landmarks for migratory orientation seems to be minimal. Compass orientation is known in a number of species, but true navigation has been demonstrated only in a few kinds of birds.

We are still ignorant of the orientation abilities of most vertebrates. We know virtually nothing about the abilities of most species to adjust for displacements from their course or to switch from one directional cuing system to another. We also know little about the acquisition of orientational ability by young animals or the impact of energetic costs of locomotion on the direction chosen.

Nevertheless, an extensive body of research has elucidated a good deal about the kinds of cues that animals use in orientation, even though exactly how these cues are employed is still a mystery. Because birds are by far the best-studied group, we begin with a summary of direction-finding cues in birds and conclude with a brief survey of orientation in other vertebrates. True navigation and compass orientation are emphasized.

Birds are very visual creatures, and visual orientation systems are the most thoroughly examined. The sky provides numerous visual cues that can assist direction-finding. Since about 1950, many studies have demonstrated that birds can orient with celestial cues—the sun during the day and the stars at night (see references in Emlen, 1975a).

The early work of Kramer (1950) is still a fine example of sun orientation (Fig. 8–6). Many birds exhibit migratory restlessness at the usual time of year—even when housed in a cage—as long as they are exposed to normal

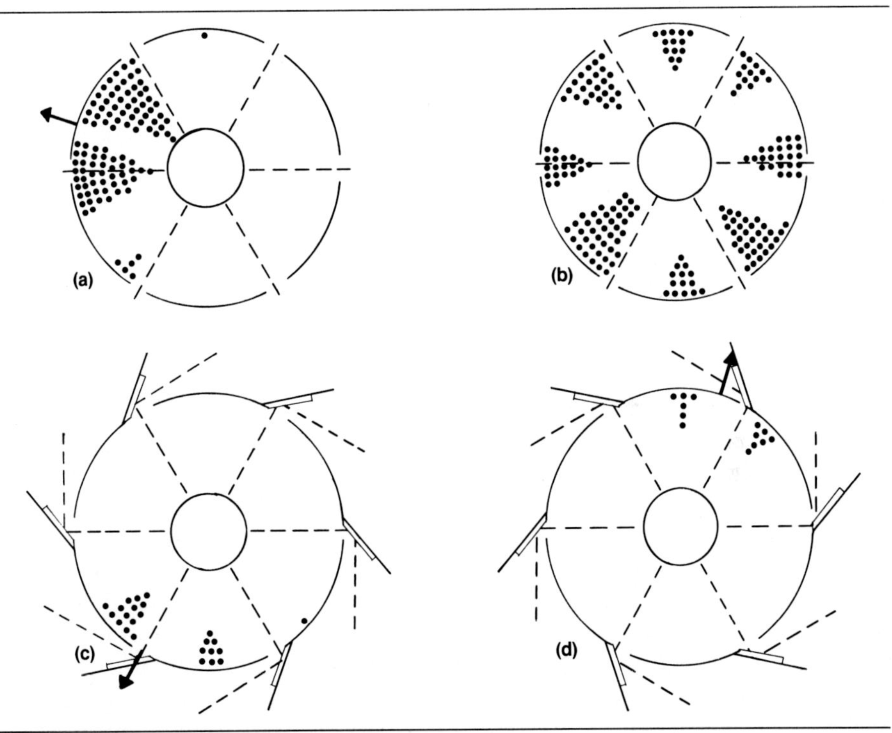

Figure 8–6 Orientation of spontaneous *Zugunruhe* in a caged European starling under various conditions of sun exposure. The bird was tested outdoors in a pavilion with six windows during the spring migration season. (*a*) Behavior under clear skies. (*b*) Behavior under total overcast. (*c*) Behavior when the image of the sun was deflected 90° clockwise by mirrors. Each dot represents 10 seconds of fluttering activity. Dashed lines show incidence of light from the sky. Solid arrow denotes mean direction of activity. (Modified from S.T. Emlen, 1975. Migration: orientation and navigation. *In* D.S. Farner and J.R. King (Eds.), *Avian Biology*. Vol. 5, pp. 129–219. Academic Press, New York.)

light-dark cycles. Kramer first observed that these fidgetings in European starlings (*Sturnus vulgaris*) were oriented in the direction appropriate to the season, as long as the sun was visible. He then obscured all landmarks and found that the starlings maintained the proper orientation under clear skies but not under overcast skies. When Kramer shifted the apparent location of the sun with mirrors, the starlings changed their directional orientation in a predictable way. Because the earth rotates, however, the sun's position in the sky changes continually during the day. Therefore, birds that orient by the sun must be able to compensate for this apparent movement, and indeed they do. Kramer's starlings, and other species as well, were trained to find food in a certain compass direction in a circular cage. Trained at one time of day and tested at another (when the sun's position was different), most birds still made the correct choice, showing that they had adjusted for the sun's apparent motion.

Such a time sense, or "internal clock," appears to be essential for birds to correct their sun orientation. A standard experimental procedure demonstrating the importance of the internal clock is the clock shift. When captive birds are subjected to a regular light-dark cycle shifted earlier or later than the natural one, many quickly reset their activity patterns and physiological rhythms in accordance with the new cycle. When a bird with a reset internal clock is then exposed to the natural daytime sky, its orientation direction is shifted in a predictable way. The sun's position changes at an approximate rate of 15° per hour; thus, a bird whose clock has been shifted six hr ahead should err by about 90° to the left (counterclockwise) of the correct orientation angle. Suppose a bird with an unmodified clock normally flies north (that is, away from the sun) at noon to go home. If its clock has been shifted six hr fast, when the bird thinks it is noon it is really 6:00 A.M. sun time. When the bird flies as usual away from the sun, it will fly due west, away from the sun in the east.

Experimental evidence to date favors the idea that birds use the sun as a compass to maintain a bearing, not as a means of determining longitude or other geographical position; that is, the sun usually does not appear to provide a "map." Sun compasses are probably important even to nocturnal migrants.

Starlings that breed near the Baltic Sea customarily migrate to Belgium, southern Britain and Ireland, and northern France for the winter (Fig. 8–7). More than 11,000 starlings were captured in the Netherlands, banded for identification, and released in Switzerland (Perdeck, 1958). Young starlings with no previous migration experience left Switzerland with the same southwestern heading they had been using and ended up in Spain, Portugal, and southern France. Adult starlings, which had made other migrations, headed northwest from Switzerland and reached their usual wintering grounds. The young birds were using directional orientation, but the adults apparently were capable of true navigation. We do not know

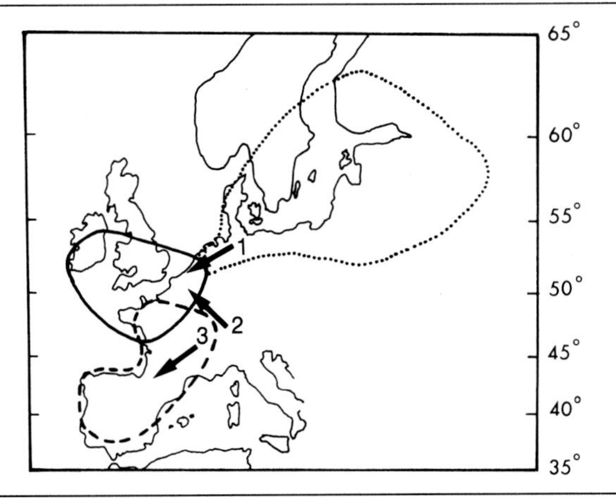

Figure 8–7 Breeding range (dotted line) and normal winter range (solid line) of starlings captured in the Netherlands and moved to Switzerland. Dashed lines show the new winter range of the displaced young birds. Arrows indicate directions of movement: (1) normal; (2) displaced adults; (3) displaced young birds. (Modified from K. Schmidt-Koenig, 1975. *Migration and Homing in Animals.* Springer-Verlag, Berlin.)

what clues the adult starlings used to determine the appropriate flight direction.

Kramer (1949) and Sauer and Sauer (1955) demonstrated that blackcaps and a few other species can use stars for orientation at night. Since their early work, stellar orientation has been demonstrated in a number of species. Several Old World migrants, including the blackcap, move long distances between temperate and tropical low latitudes and may use an internal clock to compensate for the position of the stars. Not all nocturnal, star-orienting migrants appear to use time-compensated orientation, however. Emlen (1967) devised an elegantly simple way of recording nocturnal orientation directions during Zugunruhe. His birds, indigo buntings (*Passerina cyanea*), were placed in an inverted cone of blotting paper covered by a screen top and floored with an ink pad (Fig. 8–8). Caged in this fashion, the buntings oriented their potential migratory movements under the stars or under a planetarium sky, and their inky feet left a record of their movements on the blotting-paper cone. The density of footprints in various directions showed the favored direction (Fig. 8–9). Buntings did not appear to use an internal clock in their star orientation. Emlen speculated that their intermediate-distance migrations between the eastern United States and the West Indies and Central America, always north of 10°N latitude, may not have provided selection for time-compensating abilities (Fig. 8–10).

Apparently, buntings use star patterns to give them compass bearings. Most individuals in Emlen's experiments recognized the stellar patterns in large portions of the northern sky. In fact, individual constellations or portions of sky could be blocked out without causing disorientation. Furthermore, different individual buntings often relied on different stellar cues. Emlen's buntings learned their individual orientation cues during

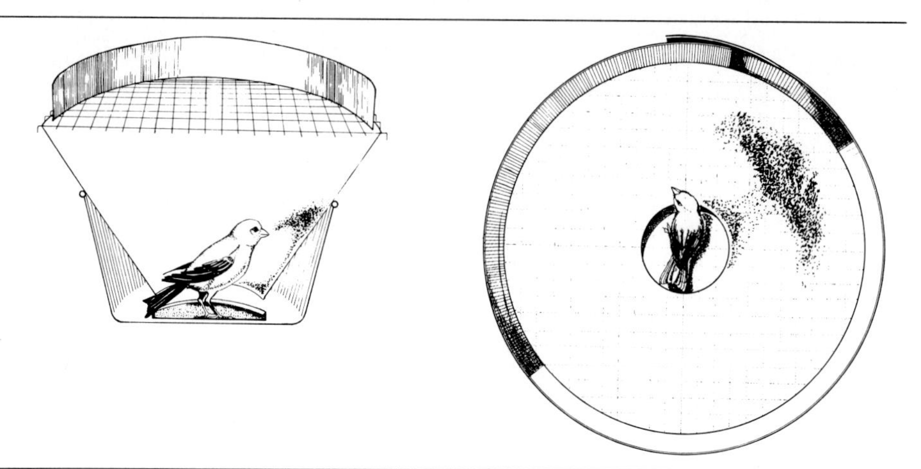

Figure 8–8 The conical experimental cage for the indigo bunting studies. The bunting stands on an ink pad. Its activity is registered on the side of the cone in the form of black footprints. (From "The stellar-orientation system of a migratory bird" by S.T. Emlen.)

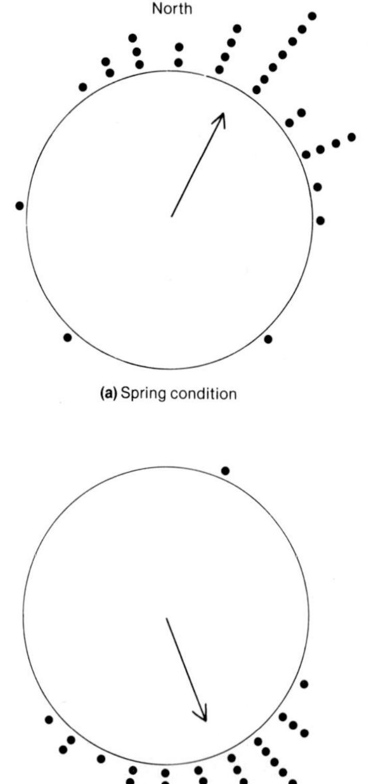

Figure 8–9 The effect of physiological conditions on orientation direction in the indigo bunting experiment. (From "The stellar-orientation system of a migratory bird" by S.T. Emlen.)

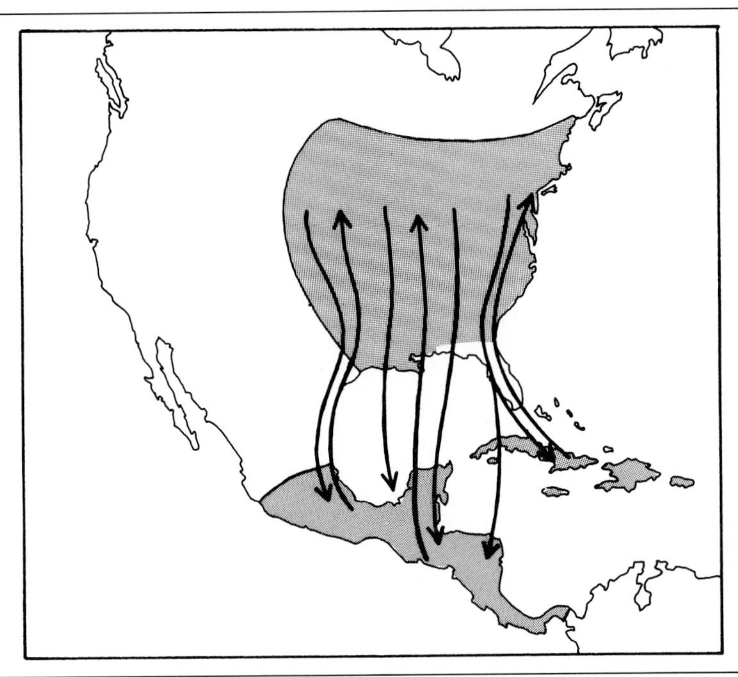

Figure 8–10 The indigo bunting's migration pattern. In late April they head north; in September and October they depart for their wintering grounds. (From "The stellar-orientation system of a migratory bird" by S.T. Emlen. Copyright © 1975 by Scientific American, Inc. All rights reserved.)

their development. This finding was in apparent contrast to Sauer's findings on European warblers and others, in which young experimental birds could orient correctly even if they had never seen the sky before the time of the experiment.

Emlen also showed that indigo buntings use the north circumpolar sky for both spring and fall migration. Furthermore, by experimentally manipulating day length, he simultaneously prepared one group of buntings for spring and another for fall migration (Emlen, 1969). Exposed to identical spring sky patterns in a planetarium, the birds ready for spring migration headed north-northeast, whereas those ready for fall migration oriented south-southeast, approximately the correct directions. Emlen then concluded that seasonal changes in hormonal condition were important in choosing a migration direction. Buntings thus use star patterns to determine direction, and something about their hormonal state influences the direction chosen. The mechanism of directional choice remains unknown.

Recent studies have suggested that a number of bird species alter their orientations in the presence of experimental magnetic fields. The orientation of the homing pigeon under overcast skies (that is, deprived of solar cues) can be changed by the application of certain magnetic fields around the bird's head. Magnetic materials have been found in pigeon heads that may be magnetic-field detectors. Ring-billed gull (*Larus delawarensis*) chicks and European robins (*Erithacus rubecula*) become disoriented in the presence of altered magnetic fields. Indigo buntings also exhibit some ability to use magnetic directional cues. At the very least, such experiments show that birds can somehow sense magnetic cues (or other cues correlated with them) and that magnetic orientation is a possibility. The importance of such potential cues is not well established for birds in natural circumstances, however. The effects of magnetic fields are usually detectable only in certain experimental setups and with certain statistical treatment of the data (Emlen, 1975a; Able, 1980). Nevertheless, there are clear hints that the system has effects on migrations of wild birds (Keeton, 1980).

Birds may be able to use other natural cues for orientation. Very low frequency sounds (<1 hz) can be heard at least by homing pigeons and may contribute to locating major landmarks at a considerable distance. Some birds may be able to use polarized light, ultraviolet light, olfaction, barometric pressure, or even gravity variations as aids in orientation (Able, 1980; Keeton, 1980).

ORIENTATION IN OTHER VERTEBRATES Sun-compass orientation has been demonstrated in several fishes, including some salmonids, several sunfishes (*Lepomis*), white bass (*Roccus chrysops*), mosquitofishes (*Gambusia affinis*), topminnows (*Fundulus notti*), and parrotfishes (*Scarus*). The parrotfishes were studied in Bermuda, where they spend the night in underwater caves but feed in the open along the shore during the day. Parrotfishes captured on the feeding grounds were displaced, released, and followed. On sunny days, almost all individuals headed for "home," but at night or on overcast days homing was much less successful (Winn et al., 1964). Some species of halfbeaks in the genera *Zenarchopterus* and *Dermogonys* are able to use polarized light for orientation (Waterman, 1972). Elasmobranchs may have the capacity to orient magnetically (Abel, 1980).

A number of amphibians have also been shown to be capable of celestial orientation, especially by the sun, but some are reportedly able to use the night sky as well. Both larval and adult bullfrogs (*Rana catesbeiana*) have a sun compass (Ferguson et al., 1968; Justis and Taylor, 1976). Adult anurans of several genera, including *Rana, Bufo, Acris, Pseudacris,* and *Ascaphus,* can orient in a direction corresponding to their home shoreline by means of celestial cues; some can also use olfactory cues. Some salamanders, such as *Taricha granulosus* and *Ambystoma tigrinum,* can also orient by means of light from the sky. At least some amphibians, including *Taricha* and the southern cricket frog (*Acris gryllus*), can perceive light and orient correctly without the use of eyes; other cranial light receptors function in this capacity. Some salamanders can orient using polarized light.

A few reptiles have been shown to be capable of using celestial cues for orientation. Two species of water snakes—the common water snake (*Nerodia sipedon*) and the queen snake (*Regina septemvittata*)—can orient to their home streamsides by using the sky. Furthermore, they have an internal clock and can compensate for the apparent movement of the sun (Newcomer et al., 1974). Western diamondback rattlesnakes (*Crotalus atrox*) are also capable of sun orientation (Landreth, 1973).

Surprisingly, orientation capacities of mammals are not very well known. The striped field mouse (*Apodemus agrarius*) and at least one vole (*Microtus*) have been shown to have some ability to use a sun compass.

The long-distance migrations of Pacific salmon are guided in fresh water by olfactory cues. Solely by means of their sense of smell, these fishes can find the "home" stream in which they hatched and to which they return to breed (Hasler et al., 1966). Individuals with experimentally occluded nostrils got lost on their upstream journey and seldom returned to the correct stream. Young coho salmon (*O. kisutch*) were exposed to water treated with traces of olfactorily distinctive chemicals before the age of seaward migration and showed that they could learn the smell of waters in which they were reared. When they matured, these individuals selected steams in which low concentrations of the same chemicals had been released. Individuals from each rearing treatment tended to select streams providing the same chemical cues (Scholz et al., 1976).

Nonavian vertebrates can probably use magnetic cues for orientation as well. Aquatic species may use currents and waves for directional information. Flying species may use the wind as a directional cue.

SUMMARY

Vertebrate migration is a regular movement of animals from one place to another and back again, usually on a seasonal basis. Migration is distinguished from irregular irruptions or invasions and from nomadism.

The ultimate factors that select for migratory behavior often involve the exploitation of temporary but seasonally predictable resources. Ecological factors and selection pressures in all parts of a species' range at all times of year are significant. Ecological problems encountered by migrants include altered habitats, microclimates, diets, and different sets of potential competitors.

The stimuli for migration vary among species. The cues are sometimes endogenous and sometimes triggered by proximate factors such as changing day length, temperatures, food supply, and gregarious behavior. Virtually all migrants store energy (as fat) for their migratory journeys.

Migrants find their way by the use of landmarks (visual, auditory, or olfactory) and by celestial and magnetic navigation.

SELECTED REFERENCES

Able, K.P., 1980. Mechanisms of orientation, navigation, and homing. *In* S.A. Gauthreaux (Ed.), *Animal Migration, Orientation, and Navigation.* pp. 283–373. Academic Press, New York.

Baker, R.R., 1978. *The Evolutionary Ecology of Animal Migration.* Holmes and Meier, New York.

Berthold, P., 1975. Migration: Control and metabolic physiology. *In* D.S. Farner and J.F. King (Eds.), *Avian Biology.* Vol. 5. pp. 77–128. Academic Press, New York.

Berthold, P., and U. Querner, 1981. Genetic basis of migratory behavior in European warblers. *Science, 212:77–79.*

Blem, C.R., 1980. The energetics of migration. *In* S.A. Gauthreaux, *Animal Migration, Orientation, and Navigation.* pp. 175–224. Academic Press, New York.

Bock, C.E., and L.W. Lepthien, 1976. Synchronous eruptions of boreal seed-eating birds. *Am. Nat., 110*:559–571.

Brown, J.L., 1975. *The Evolution of Behavior.* W.W. Norton & Co., New York.

Carr, A.F., 1967. Adaptive aspects of the scheduled travel of *Chelonia. In* R.M. Storm (Ed.), *Animal Orientation and Navigation.* pp. 35–55. Oregon State University Press, Corvallis.

Carr, A.F., 1972. The case for long-range chemoreceptive piloting in *Chelonia. In* S.R. Galler, et al. (Eds.), *Animal Orientation and Navigation.* pp. 469–483. NASA, Washington, D.C.

Chipley, R.M., 1976. The impact of wintering migrant wood warblers on resident insectivorous passerines in a subtropical Colombian oak woods. *Living Bird, 15*:119–141.

Dingle, H., 1980. Ecology and evolution of migration. *In* S.A. Gauthreaux (Ed.), *Animal Migration, Orientation, and Navigation.* pp. 1–101. Academic Press, New York.

Dorst, J., 1962. *The Migration of Birds.* Houghton-Mifflin, Boston. [Original in French, 1956]

Dorst, J., 1974. *The Life of Birds.* Vol. 2. Columbia University Press, New York.

Emlen, S.T., 1967. Migratory orientation in the indigo bunting, *Passerina cyanea.* Part 2: Mechanism of celestial orientation. *Auk, 84*:463–489.

Emlen, S.T., 1969. Bird migration: influences of physiological state upon celestial orientation. *Science, 165*:716–718.

Emlen, S.T., 1975a. Migration: orientation and navigation. *In* D.S. Farner and J.R. King (Eds.), *Avian Biology.* Vol. 5. pp. 129–219. Academic Press, New York.

Emlen, S.T., 1975b. The stellar-orientation system of a migratory bird. *Sci. Am., 233*:102–111.

Emlen, S.T., and J.T. Emlen, 1966. A technique for recording migratory orientation of captive birds. *Auk, 83*:361–367.

Emlen, S.T., W. Wiltschko, N.J. Demong, R. Wiltschko, and S. Bergman, 1976. Magnetic direction finding: evidence for its use in migratory indigo buntings. *Science, 193*:505–508.

Ernst, C.H., 1970. Homing ability in the painted turtle, *Chrysemys picta* (Schneider). *Herpetologica, 26*:399–403.

Ferguson, D.E., J.P. McKeown, D.S. Bosarge, and H.F. Landreth, 1968. Sun-compass orientation of bullfrogs. *Copeia, 1968*:230–235.

Galler, S.R., K. Schmidt-Koenig, C.J. Jacobs, and R.E. Belleville (Eds.), 1972. *Animal Orientation and Navigation.* NASA, Washington, D.C.

Greenwood, P.J., 1980. Mating systems, philopatry and dispersal in birds and mammals. *Anim. Behav., 28*:1140–1162.

Greenwood, P.J., P.H. Harvey, and C.M. Perrins, 1979. The role of dispersal in the great tit (*Parus major*): The causes, consequences and heritability of natal dispersal. *J. Anim. Ecol., 48*:123–142.

Gwinner, E., 1975. Circadian and circannual rhythms in birds. *In* D.S. Farner and J.R. King (Eds.), *Avian Biology.* Vol. 5. pp. 221–285. Academic Press, New York.

Gwinner, E., 1977. Circannual rhythms in bird migration. *Annu. Rev. Ecol. Syst., 8*:381–405.

Harden Jones, F.R., 1968. *Fish Migration.* Arnold, London.

Hasler, A.D., 1966. *Underwater Guideposts: Homing of Salmon.* University of Wisconsin Press, Madison.

Hussell, D.J.T., T. Davis, and R.D. Montgomerie, 1967. Differential fall migration of adult and immature least flycatchers. *Bird-Band, 38*:61–66.

Justis, C.S., and D.H. Taylor, 1976. Extraocular photoreception and compass orientation in larval bullfrogs, *Rana catesbeiana. Copeia, 1976*:98–105.

Karr, J.R., 1976. On the relative abundance of migrants from the north temperate zone in tropical habitats. *Wils. Bull., 88*:433–458.

Keast, A., 1959. Australian birds: Their zoogeography and adaptation to an arid continent. *In* A. Keast, A.R.L. Crocker, and C.S. Christian (Eds.), *Biogeography and Ecology in Australia.* pp. 89–114. W. Junk, The Hague.

Keast, A., 1968. Seasonal movements in the Austrailian honeyeaters (Melipagidae) and their ecological significance. *Emu, 67*:159–209.

Keast, A., and E.S. Morton (Eds.), 1980. *Migrant Birds in the Neotropics: Ecology, Behavior, Distribution, and Conservation.* Smithsonian Institution Press, Washington, D.C.

Keeton, W.T., 1980. Avian orientation and navigation: new developments in an old mystery. *Proc. Int. Ornith. Cong., 17*:137–157.

King, J.R., 1972. Adaptive periodic fat storage by birds. *Proc. Int. Ornith. Cong., 15*:200–217.

Koch, A.L., A.F. Carr, and D.W. Ehrenfeld, 1969. The problem of open-sea navigation: the migration of the green turtle to Ascension Island. *J. Theoret. Biol., 22*:163–170.

Kramer, G., 1949. Über Richtungstendenzen bei der nächtlichen Zugunruhe gekäfigter Vogel. *In* E. Mayr and E. Schüz (Eds.), *Ornithologie als Biologische Wissenschaft: Festschrift.* pp. 269–283. Springer-Verlag, Berlin.

Kramer, G., 1950. Weitere Analyse der Factoren, welche die Zugaktivität des gekäfigten Vogels orientieren. *Naturwiss., 37*:377–378.

Kramer, G., 1957. Experiments on bird orientation and their interpretation. *J. Ornith., 99*:196–227.

Kramer, G., and U. von St. Paul, 1950. Stare (*Sturnus vulgaris*) lassen sich auf Himmelsrichtungen dressieren. *Naturwiss., 37*:526.

Lack, D., and P. Lack, 1972. Wintering warblers in Jamaica. *Living Bird, 11*:129–153.

Landreth, H.F., 1973. Orientation and behavior of the rattlesnake, *Crotalus atrox. Copeia, 1973*:26–31.

Leck, C., 1972. The impact of some North American migrants at fruiting trees in Panama. *Auk, 89*:842–850.

Leggett, W.C., 1977. The ecology of fish migrations. *Annu. Rev. Ecol. Syst., 8*:285–308.

Lundberg, A., 1979. Residency, migration and a compromise: Adaptations to nest-site scarcity and food specialization in three Fennoscandian owl species. *Oecologia, 41*:273–281.

Moore, F.R., 1977. Geomagnetic disturbance and the orientation of nocturnally migrating birds. *Science, 196*:682–684.

Moreau, R.E., 1972. *The Palearctic-African Bird Migration Systems.* Academic Press, London.

Morton, E.S., 1976. The adaptive significance of dull coloration in yellow warblers. *Condor, 78*:423.

Newcomer, R.T., D.H. Taylor, and S.I. Guttman, 1974. Celestial orientation in two species of water snakes (*Natrix sipedon* and *Regina septemvittata*). *Herpetologica, 30*:194–200.

Perdeck, A.C., 1958. Two types of orientation in migrating starlings, *Sturnus vulgaris* L., and chaffinches, *Fringilla coelebs* L., as revealed by displacement experiments. *Ardea, 46*:1–37.

Raveling, D.G., 1976. Migration reversal: a regular phenomenon of Canada geese. *Science, 193*:153–154.

Rowley, I., 1974. *Bird Life.* Taplinger Publishing Co., New York.

Sauer, F., and E. Sauer, 1955. Zur Frage der nächtlichen Zugorientierung von Grasmücken. *Rev. Suisse Zool., 62*:250–259.

Schmidt-Koenig, K., 1975. *Migration and Homing in Animals.* Springer-Verlag, Berlin.

Schmidt-Koenig, K., and W.T. Keeton (Eds.), 1978. *Animal Migration, Navigation, and Homing.* Springer-Verlag, New York.

Scholz, A.T., R.M. Horrall, J.C. Cooper, and

A.K. Hasler, 1976. Imprinting to chemical cues: the basis for home stream selection in salmon. *Science, 192*:1247–1249.

Sinclair, A.R.E., 1978. Factors affecting the food supply and breeding season of resident birds and movements of Palearctic migrants in a tropical African savannah. *Ibis, 120*:480–497.

Southern, W.E., 1974. The effects of superimposed magnetic fields on gull orientation. *Wils. Bull., 86*:256–271.

Storm, R.M. (Ed.), 1967. *Animal Orientation and Navigation.* Oregon State University Press, Corvallis.

Taylor, D.H., and K. Adler, 1973. Spatial orientation by salamanders using plane-polarized light. *Science, 181*:285–287.

Taylor, D.H., and D.E. Ferguson, 1970. Extraoptic celestial orientation in the southern cricket frog *Acris gryllus. Science, 168*:390–392.

Twitty, V.C., D. Grant, and O. Anderson, 1964. Long distance homing in the newt, *Taricha rivularis. PNAS, 51*:51–58.

vander Wall, S., and R.P. Balda, 1977. Coadaptations of the Clark's nutcracker and the piñon pine for efficient seed harvest and dispersal. *Ecol. Monogr., 47*:89–111.

Walcott, C., and R.P. Green, 1974. Orientation of homing pigeons altered by a change in the direction of an applied magnetic field. *Science, 184*:180–182.

Waterman, T.H., 1972. Visual direction finding by fishes. *In* S.R. Galler, et al. (Eds.), *Animal Orientation and Navigation.* pp. 437–456. NASA, Washington, D.C.

Willson, M.F., 1976. The breeding distribution of North American migrant birds: a critique of MacArthur (1959). *Wils. Bull., 88*:582–587.

Wiltschko, R., D. Nohr, and W. Wiltschko, 1981. Pigeons with a deficient sun compass use the magnetic compass. *Science, 214*:343–345.

Winn, H.E., M. Salmon, and N. Roberts, 1964. Sun-compass orientation by parrotfishes. *Z. Tierpsychol., 21*:798–812.

PART 3

RELATIONS WITH OTHER SPECIES

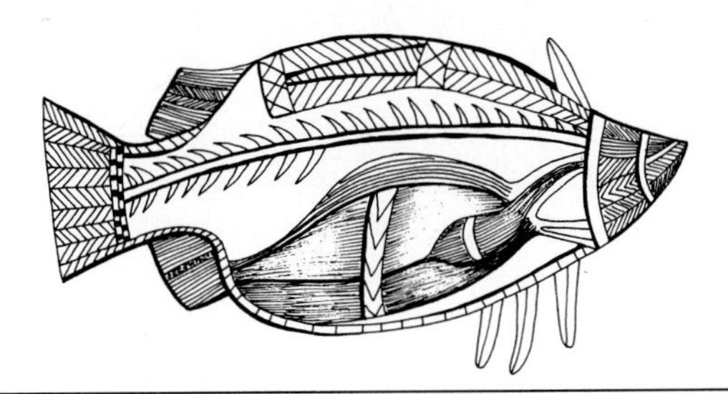

9 Food and Foraging

All vertebrates eat. They do so to provide the energy necessary for growth, maintenance, and reproduction and to acquire the nutrients, such as vitamins and minerals, needed for the proper functioning of their body chemistry. Some vertebrates eat prodigiously, but for diverse reasons. Elephants eat great volumes because they are large and their vegetable food has low concentrations of usable materials. Shrews eat large quantities for their body size because they are small and have a high metabolic rate. Sometimes an enormous intake of food is periodic in preparation for reproduction, migration, or dormancy.

Vertebrates eat almost anything organic: invertebrates, plants and plant parts, dead animals, blood, excrement, and each other. Even so, as a group they are less versatile than insects and other invertebrates, many of which can, for example, suck plant juices from particular plant species. This difference in dietary range may result largely from the significantly greater body size of most vertebrates. Most vertebrates, of necessity, are generalized feeders and eat more than one or a few species of food items. Because they are large, they require so much food that narrow specialization to a tiny array of foods is economically impossible in most cases. Invertebrates, in contrast, are often quite small; some may even pass their entire life cycle within a plant or animal host. Extreme food specialists such as these have been successful in adapting to each new defense the host has evolved, which evolutionarily reinforces their specialization. Food generalists perhaps have found it impossible to counteradapt to host or prey defenses in all their hosts or prey, thus reinforcing their inability to specialize on any one prey.

Nevertheless, within the subphylum Vertebrata a few cases of food specialization are known. Particular specializations of ecology are probably attendant upon such narrow diets, but this matter remains virtually unstudied. Perhaps the best-known food specialist among the vertebrates is the Australian koala (*Phascolarctos cinereus*), which eats only the leaves of certain kinds of *Eucalyptus* trees. Vampire bats perhaps equal the koala in specialization: They specialize on mammalian or avian blood. The African egg-eating snakes of the genus *Dasypeltis* also seem quite specialized in diet. The European badger (*Meles meles*), long thought to be a very unspecific predator, is now known to consume chiefly earthworms. However, even species we might consider to be narrow specialists—the ant- and termite-eating pangolins (*Manis*), aardvarks (*Orycteropus afer*), and anteaters (Myrmecophagidae) for example—also eat other things. Hummingbirds (New World Trochilidae), sunbirds (Afro-Asian Nectariniidae), and Australian honeyeaters (Meliphagidae) are famous for drinking nectar from flowers, but they also eat large quantities of insects.

Almost all vertebrates travel to find food and therefore require a certain amount of hunting space. Just how much depends heavily on the diet, the foraging behavior, and the body size of the hunter. The amount of space needed to provide adequate food varies with environmental conditions that influence food abundance, distribution, and diversity.

This chapter begins with some ecological considerations of foraging space and food choice and proceeds to a survey of the morphological and behavioral aspects of capturing and eating the selected food items.

FORAGING

The Use of Space

The environments of all vertebrates are heterogeneous and made up of patches. Perhaps no species occupies all of its potential environment. Instead, each species is characteristically found in certain patches, chiefly those in which the individuals can forage effectively, as well as meet their other living requirements. The environmental mosaic of patches can be seen at many levels within the geographic range of a species. On a large scale, each individual somehow chooses the habitat it will occupy. Thus, we observe that some fishes, for instance, are found in riffles in small streams, while others in the same streams are typically found in pools. The white-footed mouse, *Peromyscus leucopus*, is characteristic of wooded areas in Illinois whereas the deer mouse, *P. maniculatus*, is typical of Illinois grasslands. Male and female mice may be found in somewhat different microhabitats. On a smaller scale, not all portions of the selected habitat are exploited evenly or in proportion to their availability. Although the white-footed mouse can and does climb trees, it seems to be found more frequently on the ground.

Each forest species of American wood warbler (Parulidae) has its own typical pattern of foraging through the vegetation. For example, MacArthur (1958) studied several species of wood warblers nesting in conifer forests in Maine (Fig. 9–1). The Cape May warbler (*Dendroica tigrina*) for-

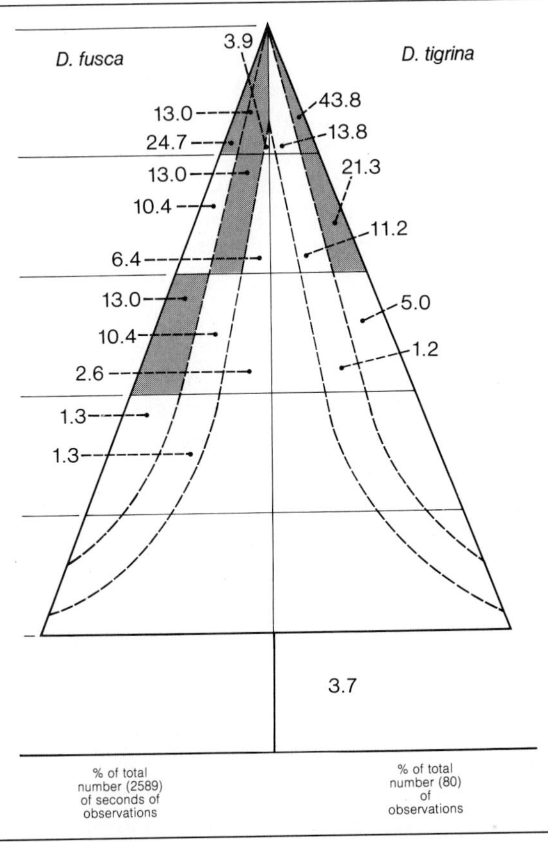

Figure 9–1 Feeding positions of Cape May warbler (*Dendroica tigrina;* N = 80 observations) and blackburnian warbler (*D. fusca,* N = 77) in a diagrammatic spruce tree. At least 50 percent of the activity is concentrated in the shaded zones. Double-headed arrows show the most common direction of movement while gleaning insects from the tree. (Modified from R.H. MacArthur, 1958. Population ecology of some warblers of northeastern coniferous forest. *Ecol.*, 39:599–619.)

aged mainly in the tops of the trees, usually moving vertically at the ends of the branches and frequently capturing insects in flight. Another species that fed in the treetops, the blackburnian warbler (*D. fusca*), tended to move along the branches and, thus, was found more frequently toward the center of the tree near the trunk. *Dendroica fusca* also made fewer foraging flights. An alternative food source for blackburnian warblers was nectar of flowers of low-growing shrubs.

What determines which patches are exploited by an organism? MacArthur and Pianka (1966) created a simple model that helps answer that question conceptually. Imagine a mosaic of environmental patches (at any scale) that contain different assortments of food items. Suppose that the foraging animal can discriminate among these patches and rank them in order of decreasing profitability. In other words, the patch from which the forager can obtain the most calories and nutrients per unit expenditure of time and energy is ranked first and the patch yielding the lowest return is ranked last. Each forager spends time foraging within a patch (foraging time) and time traveling between patches (traveling time) for each food item captured. The more kinds of patches visited, the less the traveling time per item because fewer and fewer patch types would be passed over as the animal moves through the environment. Also, as more kinds of patches are visited, foraging time per item increases as poorer and poorer patches are included in the itinerary. The best way to exploit a patchy environment thus results from the relative rates of change of traveling time and foraging time; as patches are added to the itinerary, there comes a point of diminishing returns at which the yield per expenditure for each items decreases. Consequently an "optimal forager" will, at least in theory, visit only as many kinds of patches as are profitable, and it should stay in each patch only as long as hunting there is more profitable than it is elsewhere. Since a forager probably cannot know precisely the exact profitability of all patches in its environment, we must expect that perfectly optimal use of patches is not likely.

Patch size influences the number of patch types included in the itinerary. Imagine two environments with exactly the same array of patch types but different patch sizes. Foraging time per food item within any given patch type in the two environments is identical because the patches are identical in all but size. However, traveling time per item decreases as patch size increases because fewer patches need to be visited to fill the needs of the forager. Traveling time in the environment with larger patches is less, relative to foraging time, than in other environments. As a result, the forager is more specialized to patch type in the large-patch environment (Fig. 9–2). If a patch is sufficiently large, an animal may spend all its time in one patch, reducing traveling time to zero.

MacArthur and Pianka also considered the possible effects of competitors for food on patterns of patch utilization. If competitors (or any other factor, for that matter) reduce food density within a patch, foraging time

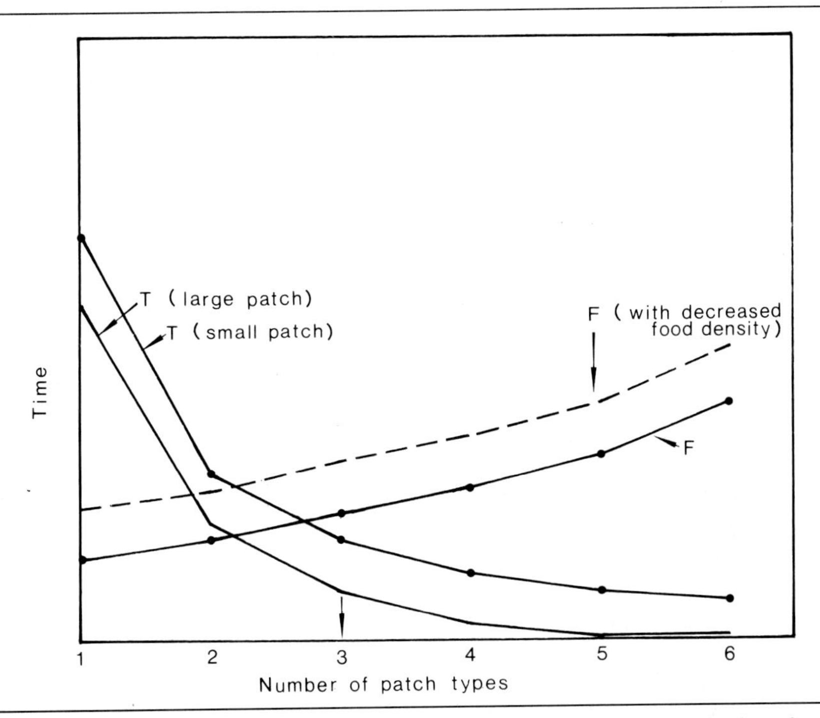

Figure 9–2 A model for the use of foraging space. The patch types are ranked in decreasing order of yield (calories per unit time). As patch types are added to an animal's itinerary, foraging time (F) per food item within a patch increases because less and less profitable patches are visited, and traveling time (T) per item (between patches) decreases. Large patches decrease the traveling time and the number of patch types visited (from A to B). A decrease in food density increases foraging time and decreases the numbers of patch types that are profitable to visit. (Reprinted from *American Naturalist* by R.H. MacArthur et al. by permission of The University of Chicago Press. Copyright © 1966 by The University of Chicago Press.)

must increase in that patch; consequently, hunting efficiency is reduced. As a result, some patches cease to be profitable. An optimal forager should reduce the number of patch types included in its itinerary, assuming the remaining patches are sufficient to sustain it.

How are patches defined? Clearly that depends on the scale being considered and on various attributes of the forager. In the case of yellow wagtails (*Motacilla flava*) in Britain, definition may be easy. Yellow wagtails feed on dung flies found on cowpats. A foraging patch for this bird could be a cowpat of an appropriate degree of freshness or, on a larger scale, a pasture with many cows. A more complex example is provided by insect-eating birds in a forest. If we suppose that no single bird is able to exploit all foods or foraging sites, then some specialization is necessary. The birds could specialize in several ways, however. If patches were defined by plant species, for example, then some species of birds would use certain plant species while other birds exploited different plant species. If usable plants were very far apart, the birds would have to fly long distances between patches. A second alternative might be to specialize on only certain types of insect prey, but this would necessitate passing by numerous potentially suitable prey; thus, this alternative seems uneconomical. Another alternative could be the use of certain parts of the plants for foraging, using many kinds of plants and prey, thus bypassing fewer food types and reducing traveling time between foraging sites.

In actuality, birds may do all of these things to varying degrees. Woodpeckers (Picidae) tend to specialize—in certain seasons, anyway—on insects found in bark crevices and under the bark of trees, although they catch flying insects and eat fruit upon occasion. When they are foraging for bark-dwelling insects, they seldom use one kind of tree to the exclusion of others. Rough-barked trees with lots of crevices for insects may be favored, however, and some seasonal change in tree-species preference may occur. Bird species that glean insects from the canopy foliage seldom specialize to tree species and take a fairly wide range of insect types. They do have species-typical patterns of hunting behavior and choice of foraging site, however, as seen in the *Dendroica* example. Furthermore, individual animals differ in their foraging efficiency and hunting styles and can be expected to prefer to forage on food items and in sites that enhance their individual success (Partridge, 1976; Grant, 1981).

Definition of a patch by food species is quite uncommon among vertebrates but can occur temporarily if some foods are unusually abundant. American robins (*Turdus migratorius*) can be seen gorging themselves on hackberries (*Celtis occidentalis*) for a few days in the fall, for example, and insect outbreaks may permit insect-eaters to zero in on one insect species for a time. Although we can seldom specify exactly what defines a patch for a given species, we can see in a general way that it must be determined largely by the economics of exploiting various patch types.

The size of the area occupied by an individual depends on the patchiness of its environment and the levels of its available resources. Size of the occupied area also changes in a predictable way with body size and hunting behavior in mammals, birds, lizards, and probably other vertebrates as well. Large vertebrates require more food than small ones, and they generally travel over a larger area to find it. The differences in total metabolic requirements of small and large vertebrates may not account for all of the differences in the sizes of their foraging ranges, however.

Figure 9–3 shows the relationships between body weight and size of foraging area as determined for mammals, birds, and lizards. Carnivores commonly have a larger foraging area than herbivores of the same size, and males often use larger areas than females use. For birds, lizards, and many mammals the slope of the line relating body size to foraging area (on a logarithmic scale) is close to 1 or even greater. In contrast, the value usually accepted for the relationship between body size and total metabolic rate is in the range of 0.60 and 0.80—a considerably lower slope than that for body size and foraging area. Some of the discrepancy may be due to the relatively greater scarcity of food items for large foragers, which tend to feed on larger food packages than do small foragers and, therefore, generally occupy particularly large areas. The differences between areas used by males and females seems to be related more closely to relative body size, at least in mammals (Harestad and Bunnell, 1979).

Data for all three vertebrate groups indicate that foraging areas are larger for meat-eating species than for leaf-eating species of the same body size. For mammals, seed- and fruit-eaters resemble meat-eaters more than leaf-eaters—perhaps because seeds and fruit, like animals, are often distributed less densely than foliage. This finding was not checked for birds, since Schoener's study dealt with breeding birds, most of which seldom feed extensively on seeds or fruits. It was also not tested for lizards, which seldom eat these foods. All three studies suggest that food density for leaf-eaters is greater than that for meat- or seed- and fruit-eaters, so that leaf-eaters need to cover less ground in order to find adequate food supplies. This should not be taken to mean, however, that leaf-eaters always feed unselectively. Howler monkeys (*Allouata palliata*) in Costa Rica are amazingly choosy, using only the leaves of certain individuals of certain tree species (Glander, 1977). Other herbivores are also quite selective.

Food Choice

The variety of foods consumed by an animal is obviously neither infinite nor completely random; each individual includes a particular spectrum of items in its diet. Dietary breadth often varies seasonally for an individual and also differs between species. The economics of foraging plays a major role in determining breadth of diet. Foraging costs not only time and energy; it often also involves the risk of attack by a predator.

For a certain dietary pattern to develop, the profits must outweigh the costs. Virtually all models of food choice assume that animals can rank food items from most to least profitable per unit of time or effort. Diets may continue to expand as long as the yield per item continues to exceed the cost of finding, capturing, and digesting it. When food is scarce, we expect animals to broaden their diets, bypassing no suitable food items, up to the point at which the cost exceeds the yield. When food is plentiful, however, the diet is expected to include only the most profitable food types. In general, this is the pattern observed in natural populations of vertebrates. White-tailed deer (*Odocoileus virginianus*) in eastern North America typically eat conifer needles only when their customary foods (twigs and leaves of many broadleaved deciduous plants) are unavailable under heavy snows. Conifer needles are unsuitable food for deer, providing essentially

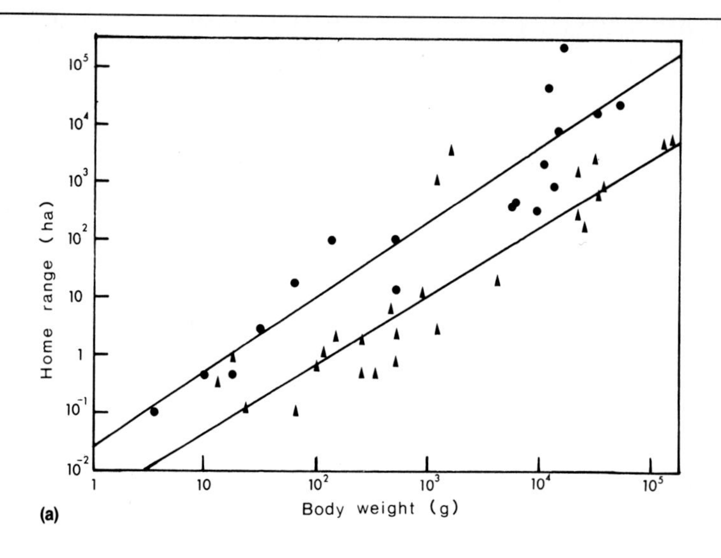
Home range (ha)
10⁻²
10⁻¹
1
10
10²
10³
10⁴
10⁵
1
10
10²
10³
10⁴
10⁵
Body weight (g)
(a)

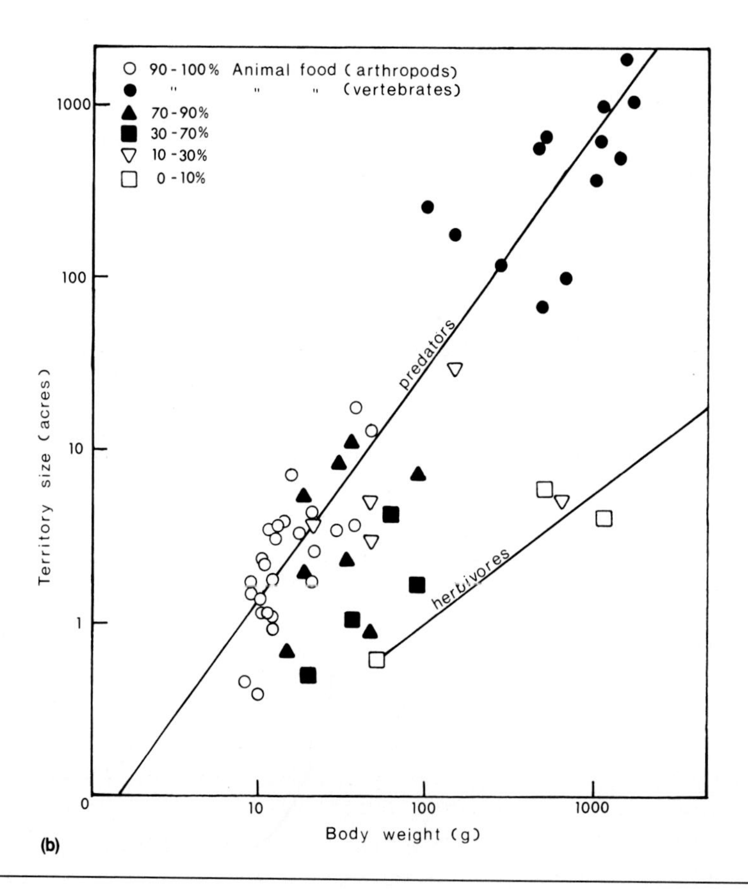
90-100% Animal food (arthropods)
" " " (vertebrates)
70-90%
30-70%
10-30%
0-10%
predators
herbivores
Territory size (acres)
0
1
10
100
1000
0
10
100
1000
Body weight (g)
(b)

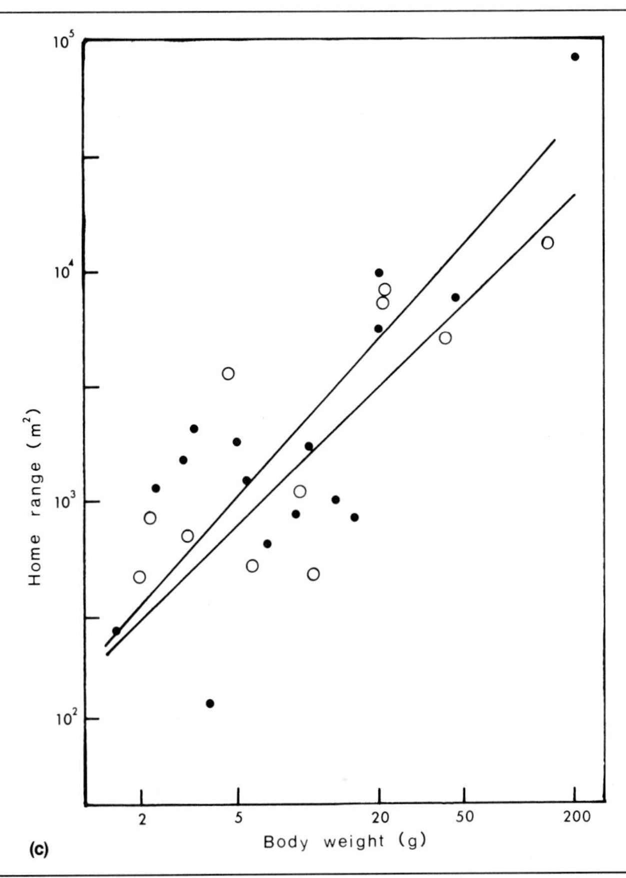

Figure 9–3 Three sets of relationships between home range size and body size. (*a*) Mammals: The upper line is the relationship for carnivores, the lower line is for herbivores. (Modified from A.S. Harestad and F.L. Bunnell, 1979. Home range and body weight—a re-evaluation. *Ecol.*, *60*:389–402. Copyright © 1979, the Ecological Society of America.) (*b*) Birds: Although the sample size for leaf-eaters is small, all the points fall well below those of meat-eaters of similar body size. (Modified from T.W. Schoener, 1968. Sizes of feeding territories of birds. *Ecol.*, *49*:123–141. Copyright © 1968, the Ecological Society of America.) (*c*) Lizards: The upper line and solid circles are for males; the lower line and open circles are for females. (Modified from F.B. Turner, R.I. Jennrich, and J.D. Weintraub, 1969. Home ranges and body sizes of lizards. *Ecol.*, *50*:1076–1081. Copyright © 1969, The Ecological Society of America.)

no nutrition, and deer can be found dead of starvation in winter with their stomachs full of conifer needles. At the other extreme, well-fed animals, often including humans and their domesticated vertebrates, are notoriously picky about what they will ingest.

More specific studies have demonstrated that predators can indeed forage so as to maximize profitability and that selectivity increases with increasing prey density. Wild 15-spined sticklebacks (*Spinachia spinachia*) of different body sizes selected prey (mysid shrimp) of a size that maximized the weight of food intake relative to the time needed to handle the prey (Fig. 9–4). Hungry individuals were less selective, taking a wider range of sizes than well-fed ones. Laboratory experiments on the foraging of great tits (*Parus major*) revealed that feeding (on large versus small pieces of mealworm) was unselective when prey density was low. When the density of large prey was increased sufficiently, the titmice became very selective, choosing mainly large, profitable items. As long as large prey were present in adequate numbers, an increase in density of small prey caused no change in selectivity: Small prey were still unprofitable (Krebs et al., 1977). In other cases, however, high densities of less rewarding prey may make them sufficiently profitable to be worth harvesting. Variability of the food reward may also influence choice.

In the real world, of course, foraging must be far more complex. Many organisms require not only energy but also specific nutrients; therefore,

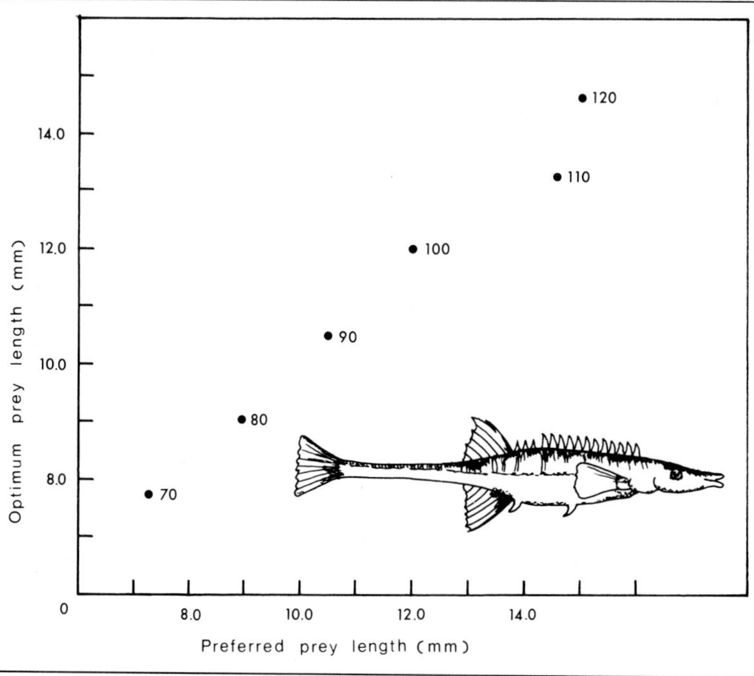

Figure 9–4 Prey selection by the 15-spined stickleback. For individuals of each size class, the actual length of prey selected corresponds closely with the optimum length, which is based on the dry weight of prey per handling time. Note the larger optima and larger prey size for larger predators (fish size is indicated by numbers next to the points). (From J.R. Krebs and N.B. Davies, 1978. *Behavioral Ecology*. Blackwell Scientific Publication Limited, New York.)

diets may not be composed simply of items of great energy yield. For example, moose (*Alces americana*) often select herbage high in sodium; brown lemmings (*Lemmus sibericus*) select a diet that increases calcium and phosphorus intake; red grouse (*Lagopus lagopus*) prefer vegetation high in nitrogen and phosphorus; and the redshank (*Tringa totanus*) rejects high-energy polychaete worms when amphipods are available. Furthermore, many potential food items are defended by morphological or chemical means; the value of including these items in a diet is reduced because of increasing handling time or energy or because of increased risk of injury or illness. Unprofitable prey may not be recognized instantly and some sampling of nonpreferred foods may occur; both will result in somewhat less selectivity of diet. When food quality is not critical, diets may change with the changing availability of different prey; the forager's diet emphasizes whatever prey is most common. Experience may modify the effectiveness of food searching and food choice. Furthermore, competition from other individuals of the same or different species, a need to perform other activities such as nest defense, the danger of predation, or the hazards of exposure to wind, rain, or temperature extremes may constrain the foraging of an individual in such a way that its diet is not "optimal" (in terms of nutritional profitability), although its overall fitness may be enhanced!

Generally, vertebrate herbivores are more likely to specialize on particular groups of species or certain parts of their food organisms than are meat-eating vertebrates (examples follow later in this chapter). Plants have evolved numerous defense mechanisms, both morphological (spines, toughness, abrasive materials such as silica) and chemical. Many of these defenses are successful against large numbers of would-be feeders, invertebrate and vertebrate, but, in most cases, at least a few animals have evolved adaptations to counter the plant defenses. Wood rats (*Neotoma*) and burros (*Equus asinus*), for example, eat cactus without injury from the formidable spines, although few other vertebrates will tackle this food source. Grass-eaters often have very hard, high-crowned or ever-growing teeth to counteract the abrasive effects of silica. Many fruit-eating birds can eat fruits poisonous to mammals, and it is likely that some birds can eat fruits that remain chemically protected from other birds. The African rock hyrax, *Procavia habessinica,* feeds extensively on the leaves and shoots of the otherwise highly toxic shrub *Phytolacca docecandra.* White-tailed deer cannot subsist on conifer needles, whereas the capercaillie (*Tetrao urogallus*) depends on them. These examples should serve to illustrate the conclusion that herbivores can be moderately specialized.

Meat-eating vertebrates, on the other hand, seldom specialize on particular body parts (vampires and Old World vultures can be considered exceptions) or on particular prey species. Once a carnivore captures its prey, it is usually capable of eating almost all of it, perhaps leaving large bones or shells for scavengers. This difference between carnivores and herbivores may be due to the mobility of most carnivore prey; once a successful attack is made it is economical to consume virtually the entire prey. Furthermore, most animal prey (with a few exceptions) are similar in nutritional value and general absence of poisonous flesh; as a result, most predators could subsist on almost any prey species captured. Food specializations for most vertebrate carnivores involve selections of certain size classes of prey, specific hunting and capturing techniques that result in certain prey types being captured, or use of different kinds of places for foraging.

Foraging behavior of predators needs to be adjusted to the behavior of their prey as well. Watchful prey can be successfully attacked only by clever or stealthy predators. Flighty prey must be approached carefully. Horned lizards (*Phrynosoma cornutum*) seem to sample their chief prey (certain ants) by taking a few here and a few there, rather than gulping down large numbers in any one place. Why do these lizards not capitalize on prey concentrations by consuming all they can catch? It seems that the ants usually cease all activity for several days if subjected to very heavy predation. Because this behavior by the prey would leave the lizard without food for a time, the sampling behavior of this predator may be an accommodation to prey behavior (Whitford and Bryant, 1979). Aggression by some ant species may also contribute to selectivity of these predators.

So far we have dealt primarily with vertebrates that eat mainly one kind of food, although those with varied diets have also been mentioned. The term "omnivore" is usually applied to animals that regularly eat both plant and animal material, and, obviously, many vertebrates fall into this category: bears (*Ursus*), various piglike mammals, humans, and many birds, to indicate just a few.

Almost no vertebrate eats exactly the same thing year-round because most vertebrate environments are seasonal; different foods are available at different times. Some species, however, undergo major shifts of diet with the season; most of the North American emberizid passerines switch from insects during the summer to seeds during the winter. Major changes in diet are often accompanied by morphological changes in the digestive tract, so that the items added to the diet are adequately digested. A seasonal change from carnivory to herbivory, for instance, would probably entail a lengthening of the intestines to accommodate the lower digestive efficiency and the slower digestive rates achieved on an herbivorous diet.

Many animals appear to try small amounts of novel foods, perhaps as a means of testing their suitability or of obtaining a required nutrient. The diets of most vertebrates, even rather specialized ones, therefore, often include "samples" of nonpreferred items. Diets are likely to change if food must be carried different distances to a storehouse. One general prediction to be made is that food selectivity will increase as the collecting area is located at greater and greater distances from the storage area, because only the more rewarding items are sufficiently profitable to compensate for the greater effort of harvesting them at greater distances. Beavers (*Castor cana-*

densis) at least sometimes conform to these expectations, as do some other harvesters, at least in some circumstances.

FOOD HANDLING

Filter-Feeding

Filter-feeding involves the straining of small aquatic organisms from their medium. This feeding mode is best developed among the fishes; examples include the freshwater gizzard shad (*Dorosoma cepedianum*), the Atlantic coast menhaden (*Brevoortia tyrannus*), and the very large basking and whale sharks (*Cetorhinus maximus, Rhincodon typus*). Most anuran tadpoles are filter-feeders on suspended algae. Filter-feeding birds include flamingos (*Phoenicopterus*); whale birds, or prions (*Pachyptila*, Procellariidae); and the shoveler duck (*Anas clypeata*). Some of the largest whales are filter-feeders: the humpback (*Megaptera novaeangliae*), the blue (*Balaenoptera musculus*), and the fin (*B. physalus*).

Filter-feeders normally must process great volumes of water to obtain sufficient quantities of their tiny food items, which are usually invertebrates or plants, but the whale shark also consumes small fishes. The menhaden, a small, herring-like fish, can reputedly take in and expel as much as 23 L of water a minute. A slowly swimming basking shark can process almost two million kg of water per hour. Water enters the mouth of a filter-feeding fish and passes posteriorly to exit from the gill openings. As it passes the gill arches, long, close-set combs called gill rakers (Fig. 9–5) strain small organisms from the water; the particles and organisms become trapped in a screen of mucus that is then swallowed. Flamingo and shoveler ducks both have beaks fringed with small projections that rake plank-

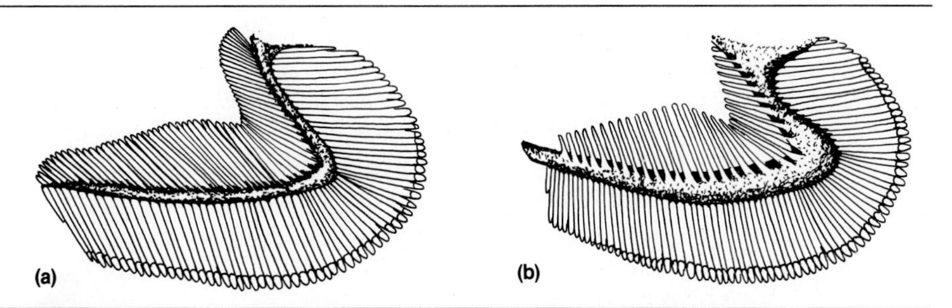

Figure 9–5 Gill rakers of (*a*) a filter-feeding fish compared to those of (*b*) a species that eats small fishes. (From J.L. Brooks and S.I. Dodson, 1965. Predation, body size, and compostotion of plankton. *Science, 150*:28–35. Copyright © 1965 by the American Association for the Advancement of Science.

tonic organisms from the water as the beak swishes from side to side. Filter-feeding whales are fitted with long, dense, keratinous plates (baleen) along the edges of the mouth (Fig. 9–6). Large volumes of water flood in when the whale opens its mouth and are squished out through the filtering baleen when the mouth closes. Most vertebrate filter-feeders are unselective of particular prey species, although the size of the openings in the filter typically determines the size class of the prey items.

A number of birds could be said to filter or "screen" insects from the air. The nightjars (Caprimulgidae) and swifts (Apodidae) hunt this way. Unlike the aquatic filterers, however, these birds simply fly around with their mouths open; they do not take in volumes of air and strain out the insects. They are probably more selective feeders than aquatic filterers.

Detritus-Feeding and Scavenging

A number of fishes are detritus-feeders, ingesting mud and ooze containing organic debris from the bottom of the bodies of water they inhabit. Examples include the sturgeons (*Acipenser*), suckers (Catostomidae), carp (*Cyprinus carpio*), and young lampreys (Petromyzontidae). Members of this category usually have fleshy lips and often have ventral mouths. They commonly move across the bottom, vacuuming up dead organic matter and mud. They are also relatively nonselective feeders but are not totally

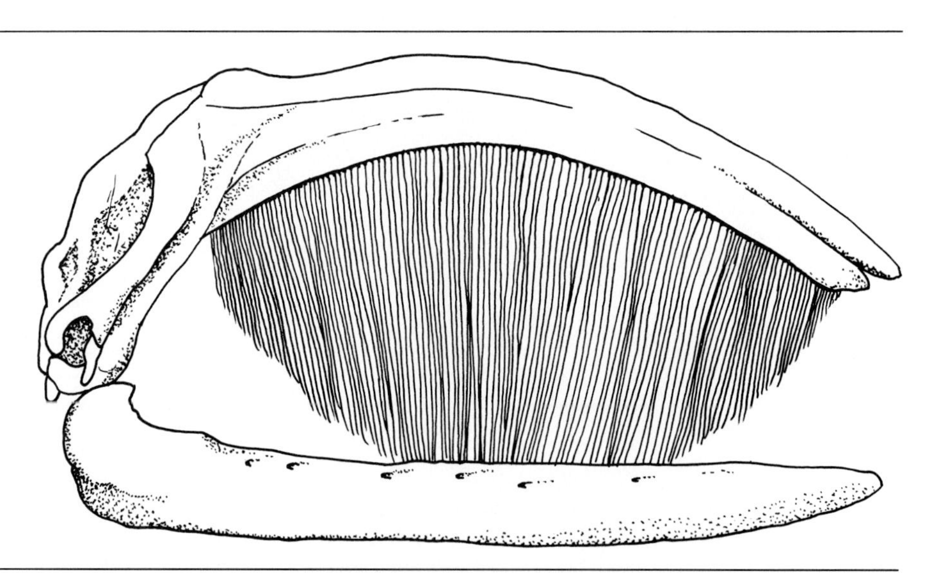

Figure 9–6 Skull of an Atlantic right whale (*Eubalaena*), total skull length about 4 m. Baleen plates are suspended from the upper jaw. (From T.A. Vaughan, 1978. *Mammalogy*. 2nd ed. Saunders College Publishing, Philadelphia.)

indiscriminate. An African fish, *Labeo velifer* (Cyprinidae), reportedly feeds on the detritus and mud that collect on the skin of the hippopotamus.

Scavenging is distinguished from detritus-feeding largely by the size of the food object. Scavengers eat dead animals (carrion) that have not yet decomposed to small bits. Carrion is a common food for vertebrates, although many may use it only on rare occasions. Ground squirrels, such as *Spermophilus richardsoni* in Alberta, sometimes may feed on carrion, venturing out on the highway to feast on their road-killed confreres. Hyenas (*Crocuta, Hyena*) are commonly considered scavengers but, in fact, often capture their own prey very effectively. On the other hand, hagfishes characteristically attach to dead or dying fishes and extract body fluids. Vultures (Cathartidae in the New World, some Acciptridae in the Old World) and the American bald eagle (*Haliaeetus leucocephalus*) are well-known avian scavengers.

A scavenging mode of life seldom brings with it very specialized morphological features, although some features may be considered specializations to certain kinds of scavenging. The tentacled oral arrangement of hagfishes, for instance, can be associated with the habit of sucking juices; the enormously strong jaws of hyenas can be linked to the advantages of cracking bones to extract additional nourishment. Vulture morphology varies with the stage at which a carcass is approached: Among the African vultures that consume small fragments or soft parts such as eyes and entrails, the bill is relatively weak and small; those tearing flesh and tendons have larger, heavier bills.

Parasitism

For vertebrates, parasitism involves attaching to living prey for a long time in order to obtain nourishment from its body. Lampreys, which attach to living fishes, rasp through the skin and extract juices; they certainly qualify for this category. So do the males of certain anglerfishes, which settle on the body of a female while still very young and become little more than a convenient bag of gametes, supported by the female.

The pearlfishes (Carapidae) frequently live inside the bodies of sea cucumbers. The Mediterranean species, *Carapus acus,* is a facultative parasite as an adult and only sometimes lives inside the body of a sea cucumber, which it enters through the anus. Pearlfishes apparently eat the gonads of the host. Young pearlfishes of a certain age, however, seem to be obligate parasites: They require the host for food and the proper conditions for maturation.

Tiny South American catfishes called candíru (*Vandellia*) normally live in the gill chambers of other fishes, nibbling gills and blood, and are effectively parasitic. Sometimes, however, they make a mistake and swim into the urethra of mammals, perhaps confusing the current produced in urination with that produced when water is expelled from gills. This mistake causes great distress to both fish and mammal, since the fish cannot get out again and is removable only by surgery.

The three genera of vampire bats (Fig. 9–7) are sometimes considered parasites; all live in Latin America. *Desmodus rotundus* finds a large mammal, such as a deer or cow, and crawls up a leg of the victim to the dorsal region, where it scrapes a hole through the skin and laps up the flowing blood. An anticoagulant produced by the bat slows the rate of clotting; thus, the same hole may be visited on successive nights. The other two vampires (*Diameus* and *Diphylla*) usually feed on bird blood. Vampires are not really proper parasites, in a narrow sense, because they fly about when not feeding, roost away from the food source, and neither attach themselves to the prey nor stay on it for extended periods of time. They might be better categorized as a special kind of meat-eater, specifically a blood-eater, or sanguinivore.

One of the Galápagos finches, *Geospiza difficilis*, is reported to peck at the wings of certain seabirds until the blood flows, whereupon the finch gets a blood meal. This habit has developed in only one population of the species, apparently. The bird eats other foods as well and is not specialized to a blood diet.

Herbivory

Herbivorous vertebrates eat plants and plant parts. Eaters of whole plants strain algae from the plankton and were considered in the discussion of the filter-feeders. It has been customary to consider the eaters of seeds and fruits as herbivores, since seeds and fruits are obviously plant parts. Some-

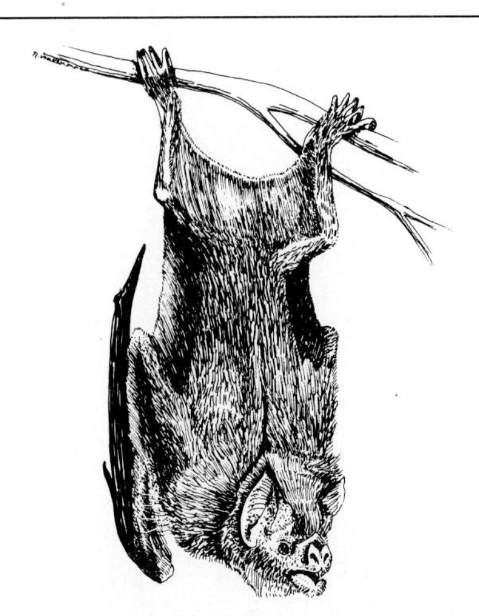

Figure 9–7 Vampire bat (*Desmodus*): It is not really a proper parasite; the vampire might better be classified as a sanguinivore. This one feeds chiefly on mammals.

times, however, the ecology and behavior of seed-eaters (granivores) and fruit-eaters (frugivores) is quite different from that of leaf-eaters (folivores), so it can be useful to consider these as a separate category. Seed- and fruit-eating is common among vertebrates, at least on a seasonal basis. Flower- and pollen-eating is another, but less well known, form of herbivory. Bark-eating is rather rare.

FRUIT-EATING Frugivory is not well developed in amphibians, reptiles, or fishes. A South American characin fish called a pacu (*Colossoma nigripennis*) eats figs and other fruits that fall into the river; a number of other tropical fishes do likewise. Fruit-eating is rather common in mammals. It is emphasized by the large Australian fruit bats or flying foxes (Megachiroptera); some of the microchiropterans, such as the tropical American *Artibeus;* numerous primates, such as the chimpanzee (*Pan troglodytes*); and some rodents, including the agoutis (*Dasyprocta*). Many other mammals eat fruits at least part of the time—black bears (*Ursus americanus*) and grizzly bears (*U. arctos*), some of the small African antelope, some tropical cats, raccoons (*Procyon*), coatis (*Nasua*), opossums (*Didelphis*), and many cricetine and echimyid rodents, for example. Even species usually thought of as meat-eaters, such as foxes (*Urocyon, Vulpes*), eat fruit. Many mammals that eat fruit probably do so opportunistically, when fruit is available, and can feed on other things if fruit is absent. Fruit-eaters in the tropics, however, may be more dependent on a year-round supply and more specialized to that diet.

Frugivorous birds are many and range from the oilbird (*Steatornis caripensis*) of northern South America, which even feeds its young largely on fruit, to some thrushes (Turdidae), some tyrant flycatchers (Tyrannidae), and many others that eat quantities of fruit only seasonally. Avian specializations for fruit-eating often include an ability to digest fruit very rapidly; the phainopepla (*Phainopepla nitens*) of the North American southwest, for example, can usually process mistletoe (*Phoradendron*) berries in less than 45 min. Thrushes often regurgitate fruit seeds in less than 5 min after swallowing the fruit, thereby decreasing their load of undigestible ballast. The bird is therefore able to ingest large quantities of food per unit time, an important ability because fruits usually contain little protein compared to other foods, and much fruit has to be eaten to garner enough protein. In some cases, fruit appears to be a valuable source of energy for migration. The gastrointestinal tract of frugivorous birds treats its contents very gently, and the seeds of the fruit are usually eliminated, either in the feces or by regurgitation, fully capable of germination. In fact, many of the plants whose fruits are eaten by vertebrates depend on these fruit-eaters to disperse their seeds in this way.

SEED-EATING Most seeds are highly nutritious food, representing the store of nutrition donated by parent plants to each of their offspring. (Although the acorns of oak trees [*Quercus*] are botanically called fruits, I choose here to include them with true seeds, in part because they have a hard outer covering, much like a seed, and in part because vertebrate consumers treat them in much the same way.) Birds, mammals, and some tropical fishes seem to have developed specializations to this kind of food, however. Granivory is often considered typical of most dipodid, cricetine, heteromyid, and sciurid rodents. Dipodid rodents include the Old World jerboa, *Salpingotus;* cricetines include the large genus *Peromyscus;* heteromyids are the pocket mice (*Perognathus*) and kangaroo rats (*Dipodomys*); and sciurids are the tree squirrels, ground squirrels, and chipmunks. Deer and pigs of various kinds may eat acorns and large seeds at times. Because herbivorous mammals usually chew their food and crushed seeds are relatively easy to digest, specializations of the gastrointestinal tract for seed-eating are not well developed. The Central American tapir (*Tapirus bairdii*) digests the seeds of certain trees, not by crushing but perhaps by retaining them in the gut until they germinate. Dental specializations are not much in evidence; the large incisors of seed-eating rodents, for example, also characterize the leaf-eaters.

Granivorous birds seasonally include many members of the bunting family (Emberizidae), the ploceid weaverbirds, and some of the Icteridae (blackbirds). Cardueline finches, including goldfinches and siskins (*Carduelis*) and crossbills (*Loxia*), are unusual among extant birds in that they feed even their young on seeds. The extinct passenger pigeon (*Ecopistes migratorius*) in North America fed heavily on acorns. The mourning dove (*Zenaidura macroura*) in North America and the wood pigeon (*Columba palumbus*) of Europe are good granivores. Lacking teeth, seed-eating birds often have strong gizzards for grinding up hard seeds, but few other general morphological features characterize this category (Fig. 9–8).

BARK-EATING Apparently known only among mammals, this dietary habit characterizes the American beaver (*Castor canadensis*), which harvests twigs and small branches of certain species and stores them underwater for winter food. The porcupine of North America (*Erethizon dorsatum*) regularly strips bark, as well as twigs and leaves, from both conifers and deciduous trees. The North American moose also eats large quantities of bark. Australian wombats (marsupials, family Phascolomidae) include bark in

Figure 9–8 Many granivorous birds have strong gizzards for grinding seeds. (*a*) The muscular stomach or gizzard of a seed-eating passerine bird (*Carduelis*), exterior view and sagittal section. (*b*) The more ordinary, glandular stomach of a heron (*Ardea*). (Modified from V. Ziswiler and D.S. Farner, 1975. Digestion and the digestive system. *In* D.S. Farner and J.R. King (Eds.), *Avian Biology*. Vol. 2, pp. 343–430. Academic Press, New York.)

their diets too (along with leaves and roots), and elephants (*Loxodonta africana*) in Africa do likewise. There seem to be no particular morphological specializations for bark-eating.

FLOWER-EATING Some vertebrates consume whole flowers. *Amazona* parrots eat *Combretum* flowers in Costa Rica, for example, and desert iguanas (*Dipsosaurus dorsalis*) may eat flowers of such species as creosote bush (*Larrea*)

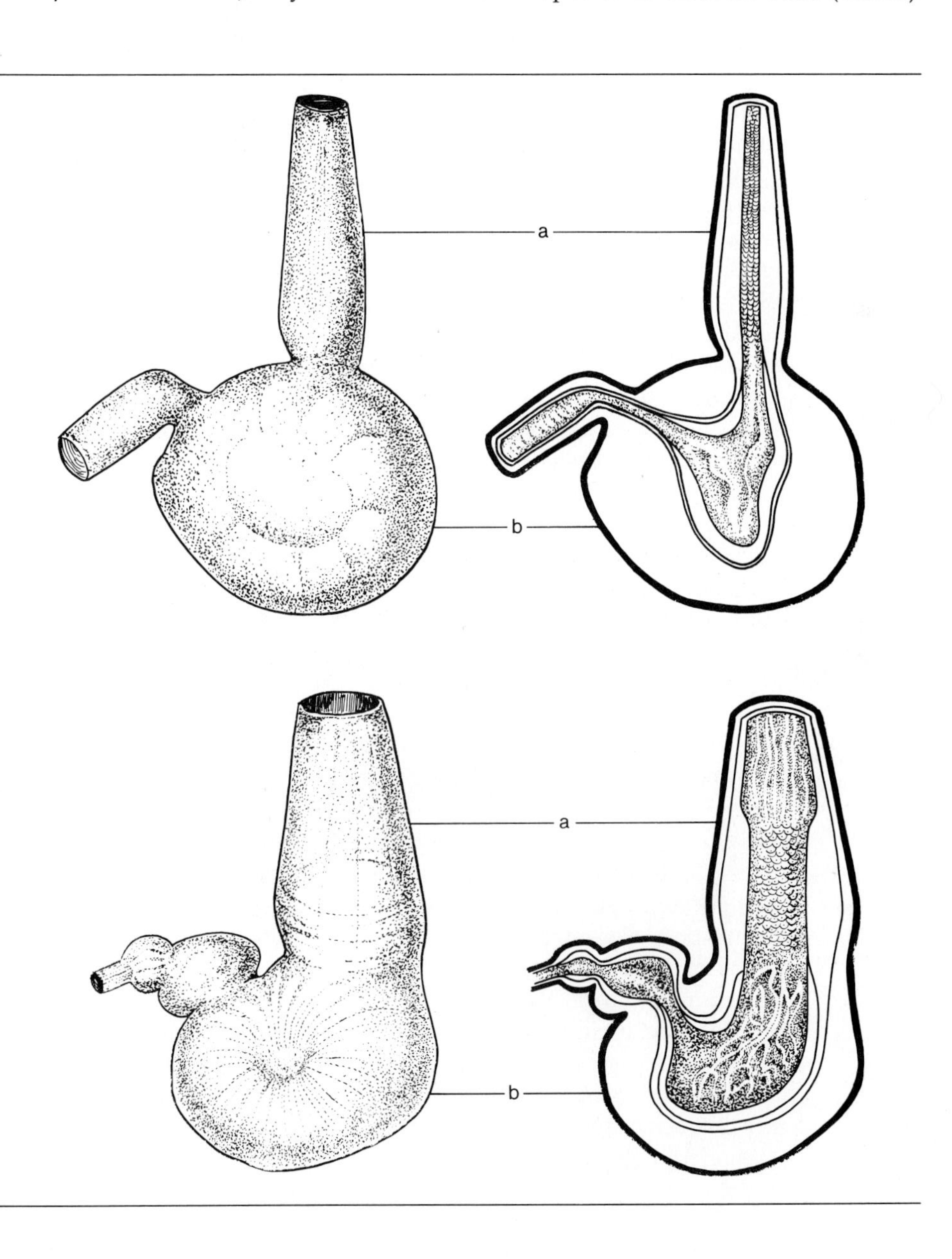

in the North American deserts. Perhaps more widespread is the habit of eating floral nectar. Many flowers produce nectar in order to attract pollinators, and numerous vertebrates capitalize on this food source. From the plant's point of view, the vertebrate exploiters are mostly thieves, although some bats and hummingbirds are good pollinators. Regular nectar-eaters include a number of bats and all hummingbirds. They also include Australian honeyeaters, sunbirds in Africa and Asia, New World orioles (*Icterus*) in some seasons, Old World flower-peckers (*Dicaeum*), New World flowerpiercers (*Diglossa*), some of the Hawaiian honeycreepers (Drepanididae), and others. About 1600 species, more than one fifth of all birds, are nectareaters at least part of the time (Welty, 1982). Nectar is the primary food of one marsupial, the Australian honey possum (*Tarsipes spenserae*).

Most nectars contain mainly sugars, but many also contain different assortments and concentrations of amino acids, the building blocks for proteins. Most vertebrates, however, require more nitrogen than can be supplied by nectar. The diet of nectarivorous vertebrates is often supplemented by insects, a fine protein source. Sometimes pollen serves as a dietary supplement yielding nitrogen. For example, nectar-feeding bats of the subfamily Glossophaginae (family Phyllostomatidae) live in tropical and subtropical America. These bats appear to enrich the protein portion of their diet by eating pollen. Plants adapted to bat pollination in Arizona are reported to produce pollen with more protein and amino acids than other plants. At least one nectarivorous bat in this region, *Leptonycterus sanborni*, eats large quantities of pollen from the plants it visits. The protein content is removed from the pollen grains by extraction in sugar solution—namely, nectar consumed at about the same time. Pollen of *Eucalyptus* flowers is also reported to be a primary food of the brush-tongued lorikeet (*Glossopsitta porphyrocephala*) of Australia.

Most regular nectar-feeders have brushy or serrated tongues with a large surface area, or grooved or tubular tongues, for sucking up nectar (Fig. 9–9). Both bats and hummingbirds usually have elongated snouts or bills designed expressly for probing into flowers.

LEAF-EATING Folivores are commonly divided into two general categories: grazers, which eat mostly grasses, and browsers, which eat mainly broadleaved plants. These categories were conceived for terrestrial folivores and do not properly encompass most aquatic vertebrates that eat plant material. The Galápagos iguana (*Amblyrhynchus cristatus*), for instance, eats macrophytic (large) algae, which technically do not have leaves. Nevertheless, I would include this iguana in the folivore category because the large blades of kelp seem to be ecologically more like leaves than the tiny algae or algal fragments consumed by some other aquatic herbivores. Some African characins (*Alestes, Distichodus*) feed on water lilies; the grass carp (*Ctenopharyngodon idellus*) is a true grazer, trimming the grasses at pond edges.

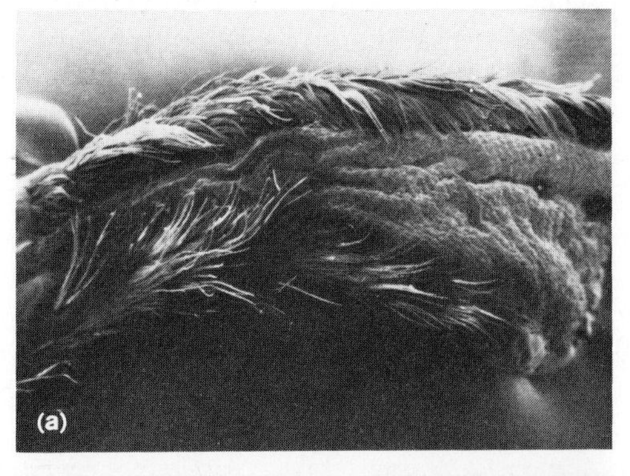

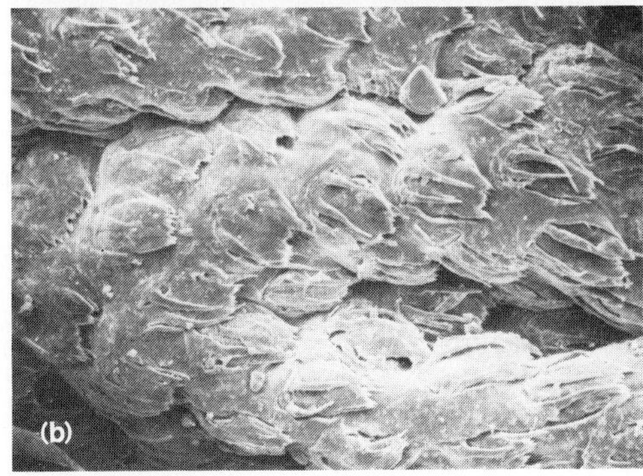

Figure 9–9 Electron micrograph of the brushy tongue of a nectar- and pollen-feeding bat, *Leptonycteris sanborni.* (Photo by D.J. Howell and N. Hodgkin.)

A few terrestrial lizards qualify as at least occasional folivores. These include the agamid *Uromastix* and several iguanas (*Ctenosaura, Iguana, Dipsosaurus*). Several bird species eat mainly leaves. The Canada goose (*Branta canadensis*) is largely a grazing bird, and many grouse (Tetraonidae) eat quantities of buds and leaves. The capercaillie of Europe feeds primarily on conifer needles in winter—in enormous quantities because these leaves are exceedingly poor in nutrients. The hoatzin (*Opisthocomus hoazin*) of South America reportedly specializes on mangrove and other tough leaves.

Mammalian folivores are many and varied. *Colobus* monkeys of the Old World and howler monkeys of the New World are largely folivorous. The microtine subfamily of cricetid rodents are folivores also: Voles (*Microtus*) consume quantities of leaves and leaf stalks, especially grasses; brown lemmings in the Arctic frequently eat mosses as well; and the muskrats (*Ondatra, Neofiber*) eat aquatic vegetation. Most of the artiodactyls (except piglike forms) and perissodactyls are either grazers or browsers. Tree sloths (Bradypodidae) are mainly folivores, as are lagomorphs, elephants, and sirenians. The koala, already mentioned, and kangaroos (Macropodidae) are examples of marsupial folivores.

Morphological adaptations for leaf-eating by vertebrates involve primarily teeth and jaws and the digestive tract. Cheek teeth in folivorous mammals usually have broad grinding surfaces that partially macerate even tough leaves before swallowing (Fig. 9–10). The jaws of folivorous mammals tend to move more parallel to one another than the shearing jaws of carnivores, which facilitates the grinding action as the lower jaw moves from side to side. Birds, lacking teeth, must fragment their leaves while plucking them or swallow them whole.

The digestive tracts of folivores have an assortment of specializations that aid in the extraction of nutrition from material that is difficult to digest. Folivorous birds usually have two large sacs, called caeca, that open near the posterior end of the small intestine. Bacteria in the caeca help break down material not digested anteriorly; a large surface area facilitates absorption of digested material. Even with enormous caeca, the capercaillie on a diet of conifer needles eats such volumes of low-quality food that 3 m of feces per day may be produced (Welty, 1982).

Mammalian folivores also frequently have symbiotic microorganisms that aid in digestion. These symbionts reside in different parts of the gastrointestinal tract in different groups of vertebrates: the intestines of elephants and horses and the chambered stomach of artiodactyls. Many large folivores regurgitate their food and chew it a second time. Kangaroos do

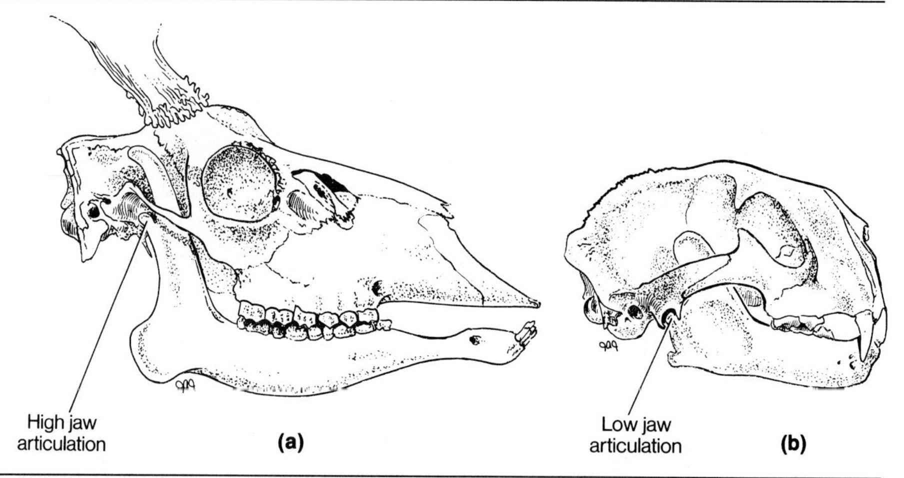

Figure 9–10 The jaws of a herbivore. (*a*) The mule-deer (*Odocoileus hemionus*) jaw articulates with the skull well above the rows of teeth. (*b*) In the cougar (*Felis concolor*), a carnivore, the jaw articulation is almost in line with the tooth rows. As a result of the high point of articulation, the cheek teeth of the herbivore meet rather evenly and grind the leafy food. However, the sharper cheek teeth of the carnivore have a more scissors-like action and slice up the prey.

so, but the artiodactyls include most famous "cud-chewers," or ruminants: the camels, deer, antelope, sheep, goats, and so on (but not pigs or hippos). Most ruminative, placental folivores have a four-chambered stomach (Fig. 9–11), but chevrotains (family Tragulidae) and camels have three-

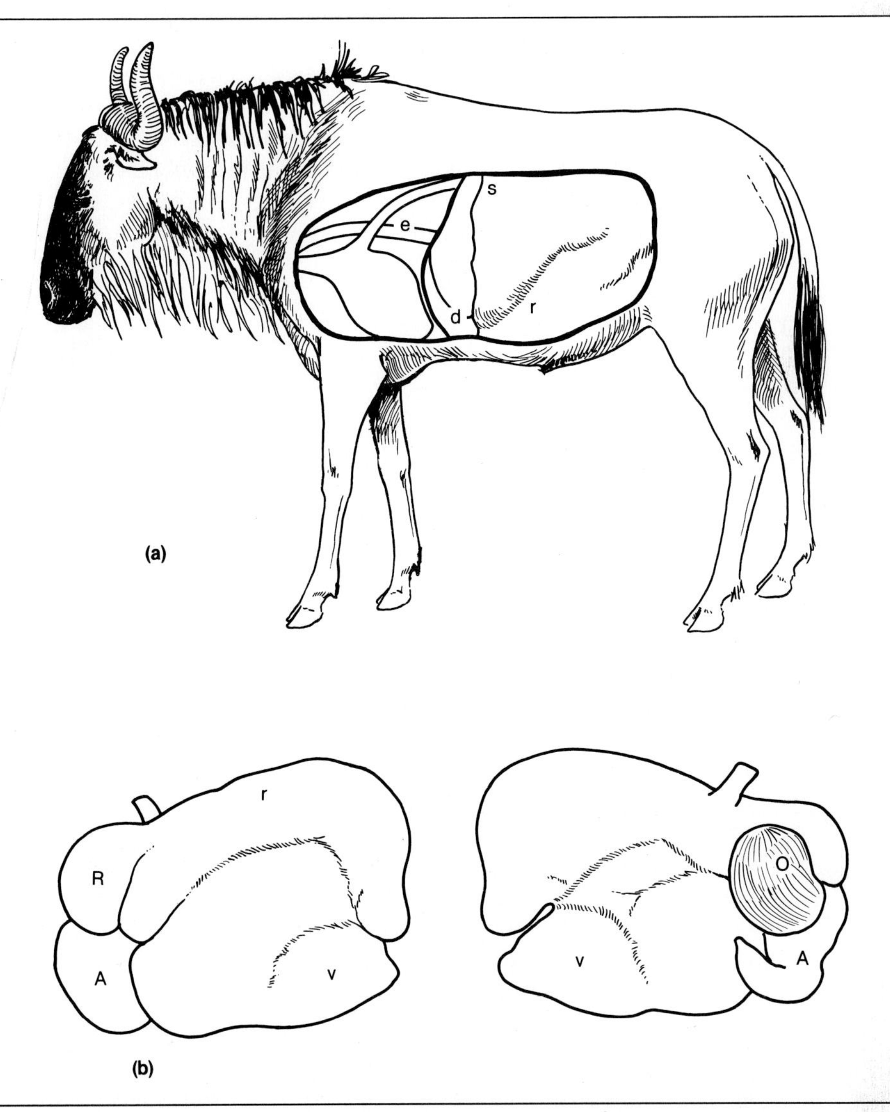

Figure 9–11 The four-chambered stomach of a ruminant—a wildebeest or gnu (*Connochaetes taurinus*)—in place and extracted from the body, viewed from the right and the left sides. e = esophagus; d = diaphragm; r = rumen; v = ventral blindsac; R = reticulum; A = abomasum; O = omasum; and r, R, A, and O are the four chambers of the stomach.

chambered stomachs. Microorganisms in the anterior part of the stomach partially break down herbage when it is first swallowed. The cud is then regurgitated for more chewing, after which the material is reswallowed and passed on to the posterior portions of the stomach and the intestine.

Some folivores have the unusual habit of reingesting their own excrement, a process called coprophagy, or refection; it is best known among lagomorphs, both the pikas (Ochotonidae) and the hares and rabbits (Leporidae), but also observed in a number of herbivorous rodents (Kenagy and Hoyt, 1980). This process not only is a means of digesting things twice and thus of extracting nutrition more completely, but may also provide certain vitamins available in quantity only after food has been processed by intestinal bacteria.

ROOT-EATING Many mammals eat roots, bulbs, tubers, or other underground parts of plants. Among these are the American pocket gophers (Geomyidae), the African springhaas (Pedetidae), the Old World porcupines (*Hystrix*), the South American tuco-tucos (*Ctenomys*, Octodonidae), and Old World mole rats of several families (such as Bathyergidae and Spalacidae). All these rodents are burrowing mammals, but some, including pocket gophers, tuco-tucos, and mole rats, are fossorial to an extreme and normally feed underground as well. Roots are often included in the varied diet of baboons (*Papio*) and the Eurasian wild boar (*Sus scrofa*), to name only two examples.

OTHER KINDS OF HERBIVORY Some miscellaneous herbivores do not fit neatly into the usual categories. A number of marine fishes, such as some parrotfishes (Scaridae) and butterflyfishes (Chaetodontidae), feed on algae by scraping it off reefs and rocks with beaklike jaws. Some members of the freshwater Cichlidae, especially in African waters, feed in a similar fashion. *Gyrinocheilus* species in southeastern Asia scrape algae off rocks and leaves; they attach to the feeding substrate by means of sucker-like lips and rasp the algae off. (Water for respiration is inhaled through special holes above the gill chambers.) The larvae of many anurans are also algae-scrapers. The pygmy parrots (*Micropsitta*) of New Guinea apparently include lichens and fungi in their diet.

A small neotropical tamarin (*Saguinus oedipus*) obtains a significant portion of its diet from tree gums or resins exuded from wounds on tree trunks. Many other primates do so (Bearder and Martin, 1980), as does the marsupial sugar glider (*Petaurus breviceps*) in some seasons (Smith, 1982). In West Africa, a small (about 350 g) prosimian, the needle-clawed bushbaby (*Euoticus elegantulus*), lives in the forest canopy and forages at night. Part of its diet is insects and fruits, but it eats quantities of gum; during the dry season, resins of certain plants become a large part of the diet. A larger sympatric prosimian, *Perodicticus potto* (about 1100 g) also eats resins along with insects and fruits, but its resin-eating is confined to the wetter seasons. When *E. elegantulus* is eating large amounts of resin in the dry season, the potto switches to fruits (Fig. 9–12). The needle-tipped claws of the bushbaby permit it to climb along tree trunks where other prosimians cannot go, and its long, protruding lower incisors are adapted for prying resin globules out of bark crevices (Charles-Dominique, 1971, 1977). Pygmy marmosets (*Cebuella pygmaea*) in South America gnaw through tree bark and eat the exuded sap. The North American sapsuckers (Picidae) drill orderly rows of holes through bark on tree trunks and large branches. They drink the sap that emerges and eat the insects attracted to it.

Carnivory

Some vertebrates eat each other; some eat invertebrates; and some eat what we can loosely call animal products. We begin with a survey of vertebrate-eaters and invertebrate-eaters, followed by a summary of morphological adaptations to meat-eating and a section on eating animal products.

Figure 9–12 Two prosimians: (*a*) *Euoticus elegantulus* eats resin during the dry season and (*b*) *Perodicticus potto* eats resin in the wet season.

EATING VERTEBRATES Predation on other vertebrates is found in some representatives of all vertebrate classes. Fishes are, of course, eaten by other fishes. Most piscivorous fishes have very pointed teeth, sometimes hinged, that bend toward the throat, facilitating capture of scaly, slippery prey (Fig. 9–13). An unusual piscivorous fish (*Rhaphiodon vulpinus*) from the Amazon River can bend its head upward about 45°. This peculiar capacity may permit the predator, with its wide gape and large fangs, to snatch its prey from below. One carnivorous fish in the Amazon ambushes

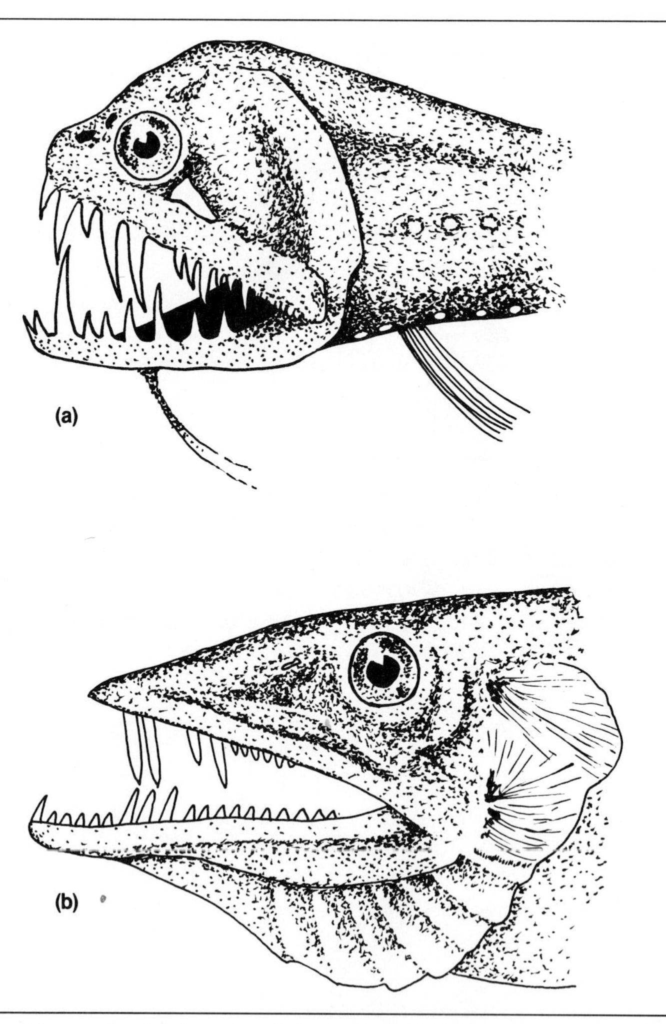

Figure 9–13 The teeth of fish-eating fishes: (*a*) *Haplostomias tentaculatus* and (*b*) *Alepisaurus ferox*. (Modified from J.R. Norman and P.H. Greenwood, 1963. *A History of Fishes*. Ernest Benn Ltd., London.)

smaller fishes in a highly specialized way. *Monocirrhus polyacanthus,* the leaf fish, has a flat body colored like a dead leaf and a chin barbel like a leaf stalk. It drifts around or lies motionless with its large fins compressed, controlling its position with tiny, transparent fins. Woe to the hapless small fish that comes too near and is engulfed by the suddenly telescopically extended gape of the "leaf." Pipefishes (Aulostomidae) and cornetfishes (Fistulariidae) sneak up on their prey by riding on the backs of harmless parrotfishes or mullet.

Some fishes, including the young of some cichlids, eat the mucus from the parents' body surface. Others eat fish skin from living prey. Some tropical and subtropical blennies (*Aspidontus rhinorhynchus,* the sabertoothed blenny of the Indian Ocean, for instance) are skin-eaters, and they capitalize on a complex situation. Fishes are subject to infestations of ectoparasites, and certain fishes, including wrasses, make their living by cleaning off the parasites and eating them. Because the "host" fishes often seek out the cleaners and offer their bodies to be cleaned, wrasses can easily approach them. By mimicking the wrasses in appearance and behavior, blennies can often get close enough to a host to slip in for a snack. Adult hosts can usually distinguish between blennies and wrasses, but young ones cannot; thus, the blennies exploit the temporary ignorance of young hosts by their "aggressive mimicry."

Anglerfishes (Antennariidae) engage in another sort of aggressive mimicry. They are benthic, sedentary hunters that resemble the algae or the sponge-encrusted rocks among which they often live. The first spine of the dorsal fin is located on the snout and resembles a worm, a crustacean, or a small fish. The immobile, hidden angler waves the lure, offering an easy meal to another fish, which then becomes dinner for the crafty angler. Some snakes, such as *Bothrops bilineatus* of South America, may also lure their prey within striking range by wiggling a colorful tail.

The crocodilian gharials and false gharials (*Gavialis, Tomistoma*) eat fishes; they have narrows snouts, at least partly for reducing water resistance to rapid closure of the mouth, that are armed with well-spaced conical teeth that can puncture scaly armor (Fig. 9–14). Alligator snapping turtles (*Macroclemys temmincki*) lure fishes into their waiting jaws with a wormlike bit of flesh on the floor of the mouth. Another turtle, the matamata (*Chelys fimbriata*) of South America, is a sluggish, bottom-dwelling, long-necked predator without the horny, beaklike jaws of most turtles (Fig. 9–15). Its strange shape and head, disguised by fleshy projections, camouflage it from unwary fishes passing by. The prey is snapped up by the turtle's circular, open mouth and swallowed whole. Sea snakes (Hydrophiidae) also eat fishes, mostly eels. When they happen to eat a spiny fish, the spines reportedly are eliminated, not by digestion or regurgitation, but by being forced out through the body walls. These snakes are highly venomous to their prey and to mammals as well. There is even a fish-eating lizard, *Neusticurus rudis,* in northern South American streams.

Figure 9–14 The snout of a gharial. The conical teeth, well spaced, are designed for puncturing scaly armor. (From J.R. Krebs and N.B. Davies, 1978. *Behavioral Ecology*. Blackwell Scientific Publications Limited, New York.)

Figure 9–15 A matamata. The head of this long-necked predator is well camouflaged when the turtle lurks in vegetation and debris.

Fish-eating birds include cormorants (Phalacrocoracidae), many alcids, mergansers (*Mergus, Lophodytes*), and loons (*Gavia*); all of these birds dive for their prey. Skimmers (*Rhynchops*) and some gulls (Laridae) take fishes as they fly over the water surface. Herons (Ardeidae) are waders that snatch fishes from the shallows with their bills. Ospreys (*Pandion*) use their feet to grab surface fishes. One Asian genus of owl (*Ketopa*) depends on fishes, and many of the antarctic penguins are piscivores. Piscivorous birds may have hooked or serrated bills (cormorants, mergansers); long, straight, spearlike bills (anhingas); forceps (herons); or roughened, spiny feet (osprey) to hold their prey (Fig. 9–16). Skimmers have an elongated lower bill and fly right at the water surface to scoop up small fishes.

Mammals that specialize on fishes include a number of toothed whales of several families, such as the arctic white or beluga whale (*Delphinapterus leucas*); the bouto (*Inia geoffrensis*) of the Amazon and Orinoco rivers; the Atlantic common or harbor porpoise (*Phocoena phocoena*); the cosmopolitan common or saddle-backed dolphin (*Delphinus delphis*); and the false killer whale (*Pseudorca crassidens*). Some bats of the genus *Noctilio* (Vespertilionidae) gaff fishes from near the water surface. Fishes are important in the diets of many pinnipeds, especially some phocids such as the antarctic Weddell seal (*Leptonychotes weddelli*), the north-temperate gray seal (*Halichoerus grypus*), and the common seal (*Phoca vitulina*). The fishing cat (*Felis viverrina*) of the East Indies and southern Asia and the river otters (*Lutra*) are piscivores, as are the brown or grizzly bear (*Ursus arctos*) and polar bear (*Thalarctos maritimus*) at times. Perhaps surprisingly, there are several genera of fish-eating cricetid rodents; they inhabit creeks and rivulets in South America.

Amphibian-eaters include frog-eating water snakes (*Nerodia*) and garter snakes (*Thamnophis*) and toad-eating hognose snakes (*Heterodon*) of North America. Among birds, the herons and storks often eat anurans. Amphibians seldom figure largely in the diet of mammals but are sometimes eaten by raccoons (*Procyon lotor*) and other mammals. Fringe-lipped bats (*Trachops cirrhosus*) commonly prey on vocalizing males of neotropical frogs.

The predators of lizards include other lizards (such as the leopard lizards, *Crotaphytus wizlizenii,* in North America); snakes (some vine snakes); birds, such as the American roadrunner (*Geococcyx*); and occasionally mammals. Snakes are eaten by other snakes: The king snakes (*Lampropeltis*) are well-known for eating rattlesnakes, and some cobras (*Ophiophagus*) are named for their snake-eating ability. Some birds, especially in the tropics, eat quantities of snakes: the African secretary bird (*Sagittarius serpentarius*), the neotropical laughing falcon (*Herpetotheres cachannius*), and *Leucopternis* hawks. The North American red-shouldered hawk (*Buteo lineatus*) also takes many snakes. Among mammals, mongooses (*Herpestes, Ichneumia*) in Africa and Asia have been immortalized by Rudyard Kipling as killers of snakes, which are a normal part of their diets.

Bird-eaters are found among fishes, some of which (pike, bass) gobble down young ducklings, but no fishes seem to specialize on birds. Young

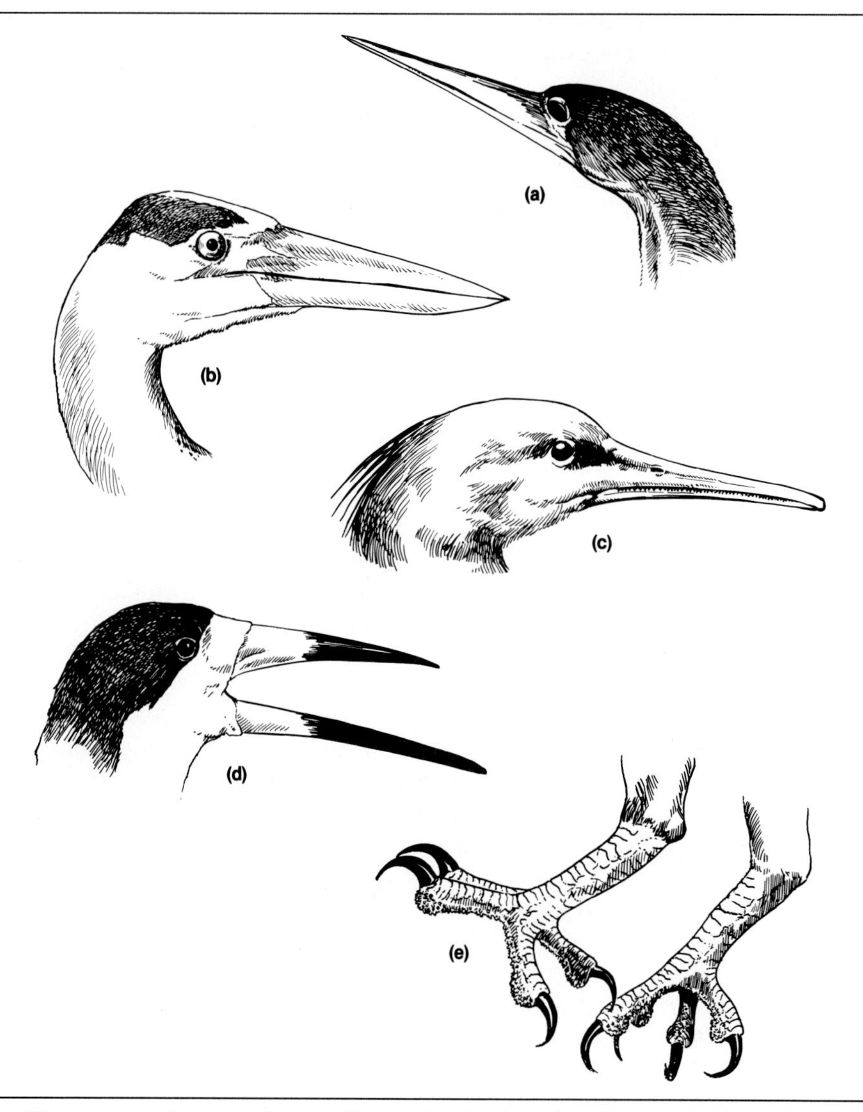

Figure 9–16 The capture devices of some fish-eating birds. (*a*) Anhinga, *Anhinga;* (*b*) heron, *Ardea;* (*c*) loon, *Gavia;* (*d*) skimmer, *Rhynchops;* (*e*) osprey, *Pandion.*

birds are included in the diets of lizards, many snakes (such as racers, *Coluber*), birds (such as blue jays, *Cyanocitta cristata*), and mammals (such as raccoons and skunks, *Mephitis* and *Spilogale*). Specialists on adult birds might include the bird-hawks (*Accipiter*) and some of the larger falcons (such as the peregrine, *Falco peregrinus*), the leopard seal (*Hydrurga leptonyx*) that attacks penguins, and several arboreal tropical snakes.

Primary eaters of mammals include few fishes (not even most sharks), amphibians (but bullfrogs, *Rana catesbeiana,* sometimes eat rats and mice), or reptiles, although the 3-m, 160-kg komodo dragon (*Varanus komodoensis*), which is native to some small islands east of Java, can eat small deer. Avian specialists on mammals should include many owls and buteonid hawks. Some larger raptors specialize on larger mammals such as monkeys and sloths: the crowned hawk-eagle (*Stephanoaetus coronatus*) in Africa, harpy eagle (*Harpia harpyja*) in South America, and monkey-eating eagle (*Pithecophaga jeffreyi*) of the Philippines (Fig. 9–17). The zone-tailed hawk, *Buteo albonotatus,* is reported to mimic turkey vultures (*Cathartes aura*) in plumage and soaring style and is often found flying with these vultures in Mexico and the southwestern United States. Perhaps it uses this mimicry as a means of concealing its presence from prey. Many canids and felids as well as a number of mustelids, such as the pine marten (*Martes americana*), and weasels (*Mustela*), are specialized mammal-eaters. Many species of the marsupial family Dasyuridae are predaceous on other marsupials, as well as on members of other vertebrate groups and insects. The short-tailed shrew, *Blarina brevicauda,* can kill mice and voles larger than itself. Killer whales (*Orcinus orca*) are the only whales that feed on homeotherms, such as porpoises, dolphins, young baleen whales, seals, and seabirds, in addition to fishes and squid.

The hunting success of vertebrates preying on other vertebrates varies greatly, as evidenced in Table 9–1. High rates of failure to capture prey are least costly when the next opportunity for attack will occur in a short time or when the size of the food reward is very great. Animals foraging off their home range—on migration, for example—are also likely to be less successful than those on familiar ground.

Intraspecific predation or cannibalism is far more prevalent than commonly believed and occurs in all major vertebrate taxa (Polis, 1981). Adults eat other adults, juveniles, and eggs, and juveniles consume both each other and eggs. Cannibalism sometimes occurs in response to very low food availability, and parents may feed some of their young to the remainder, as a means of conserving their parental expenditure in times of food shortage. Moreover, in certain sharks, one embryo eats all the other embryos and eggs in the same oviduct; *in utero* cannibalism is also reported for some salamanders and for ratfishes. This is clearly a form of maternal provisioning of the young, which permits offspring to be born at a relatively advanced stage of development; it is not induced by starvation conditions. A number of species are polymorphic for the cannibalistic trait. Spadefoot toads and certain salamanders produce some cannibalistic larvae that are larger and grow faster than their noncannibalistic siblings (Chapter 14). Cannibalistic young of several fish species are larger than the rest and voraciously gobble up the smaller individuals; presumably they grow very quickly as a result. Certain adults of herring gulls (*Larus argentatus*) and other species are far more cannibalistic than others. Given the

Figure 9–17 Eagles that eat medium-sized mammals: (*a*) African crowned hawk eagle; (*b*) neotropical harpy eagle; and (*c*) Phillipine monkey-eating eagle.

Table 9–1 Hunting success of some vertebrate predators.

Predator	***Prey***	***Successful captures (%)***
Forster's tern (*Sterna forsteri*)	Fishes	24
American kestrel (*Falco sparverius*)	Rodents (and perhaps insects)	33
Osprey (*Pandion haliaetus*)	Fishes	80–96
Various raptors on migration	Birds	4–11
Black bear (*Ursus americanus*)	Salmon	39
Coyote (*Canis latrans*)	Snowshoe hare	10
Jackals (*Canis* spp.)	Young gazelle	33
Wolf (*Canis lupus*)	Moose	8
Cape hunting dog (*Lycaon pictus*) in groups	Young gazelles	95–100
	Older gazelles	49–66
	Wildebeest	66–75
	Warthog	57
Spotted hyena (*Crocuta crocuta*) in groups	Wildebeest, gazelle, zebra	32–44
Lynx (*Felis lynx*)	Snowshoe hare	16–42
	Ruffed grouse	13
Cougar (*Felis concolor*)	Deer, elk	82 (excludes hunts aborted early)
Lion (*Panthera leo*)	Various antelope, zebra	15–32
	Warthog	47
Cheetah (*Acinonyx jubatus*)	Gazelle	70
Chimpanzee (*Pan troglodytes*)	Various mammals	40
	Young baboons	36
	Red colobus monkeys	32–56

(From Busse, 1978; Curio, 1976; and Schaller, 1972.)

apparent advantage of cannibalism, it is easy to wonder why it is not still more common. The benefits and costs of this habit have yet to be assayed analytically.

EATING INVERTEBRATES There are myriad kinds of invertebrates, and most are probably eaten by vertebrates at one time or another. Here, however, we deal with three particular kinds of invertebrate-eaters: plankton-feeders (planktivores), mollusk-feeders, and the broad category of arthropod-eaters.

Planktivores that eat zooplankton can be far more selective of individual prey than filter-feeders. The alewife (*Alosa pseudoharengus*) is a herring-like fish that invades freshwater bodies, sometimes seasonally, sometimes permanently. It is an obligate planktivore with very close gill rakers. It feeds preferentially on the largest cladocera (such as *Daphnia*) and copepods and eats small crustaceans only when the large ones are gone. Rainbow trout (*Salmo gairdneri*) and yellow perch (*Perca flavescens*) are facultative planktivores that eat *Daphnia* but switch to nonplanktonic food if *Daphnia* is absent. Such selectivity seems to be characteristic of almost all marine and freshwater planktivorous fishes, as well as salamanders and birds (such as the red phalarope, *Phalaropus fulicarius*).

Mollusk-eaters are found in all vertebrate classes. Some fishes, including the Port Jackson sharks (*Heterodontus*) and the eagle rays (*Myliobatis*), have massive plates of flat teeth for crushing the hard shells of their prey (Fig. 9–18). Other fishes, including cod, swallow mollusks whole and crunch them in the stomach. The snail- and mussel-eating caiman lizard, *Draecaena guianensis,* lives in tidal marshes of northeastern South America and munches its prey with broad, flat teeth. Slender tree snakes of the subfamily Dipsadinae feed mainly on slugs and snails; *Dipsas* extracts snails from their shells with its elongated teeth. Mollusk-eating birds include the snail, or Everglades, kite (*Rostrhamus sociabilis*), which extracts its favorite snails with a slender, hooked bill. The limpkin (*Aramus guarauna*) reportedly wedges a snail in mud with the opening facing up; it then tears off the snail's operculum (the cover of the opening) and pulls the snail from the shell with its long bill. The oystercatcher (*Haematopus ostralegus*) hammers open its mussel prey in tidal flats or snatches it from between the opened shells in shallow water. In addition to such specialists, a number of birds include mollusks in their diets: shorebirds, European thrushes, and gulls, to name a few.

Mammals in this category are the sea otter (*Enhydra lutris*), which also eats sea urchins and fishes, and the walrus (*Odobenus rosmarus*), which roots its bivalve prey from the bottom with its tusks. Giant squid are common prey for sperm whales (*Physeter catodon*), and various cephalopods are eaten by the pygmy sperm whale (*Kogia breviceps*) and the narwhal (*Monodon monoceros*).

Arthropods are important in the diets of so many vertebrates that only a brief survey can be presented here. Cleaner fishes make their living by cleaning arthropod ectoparasites, fungi, and diseased skin from other fishes. The cleaners usually establish a particular site as their home base and wait there for the arrival of the customers. Some species clean the surface, others work over the mouth and throat, and still others enter the gill chamber to clean the gills. These unusual fishes seldom get eaten by their clients, even when working in the risky region of the teeth and mouth. Most cleaners are marked by distinctive color patterns and behavior that advertise their identity to the other fishes, which usually respond to their ministrations by becoming very placid and cooperative.

Free-living insects are consumed by fishes such as bass (*Micropterus*) and trout (*Salmo*), by salamander adults and larvae, and by many adult anu-

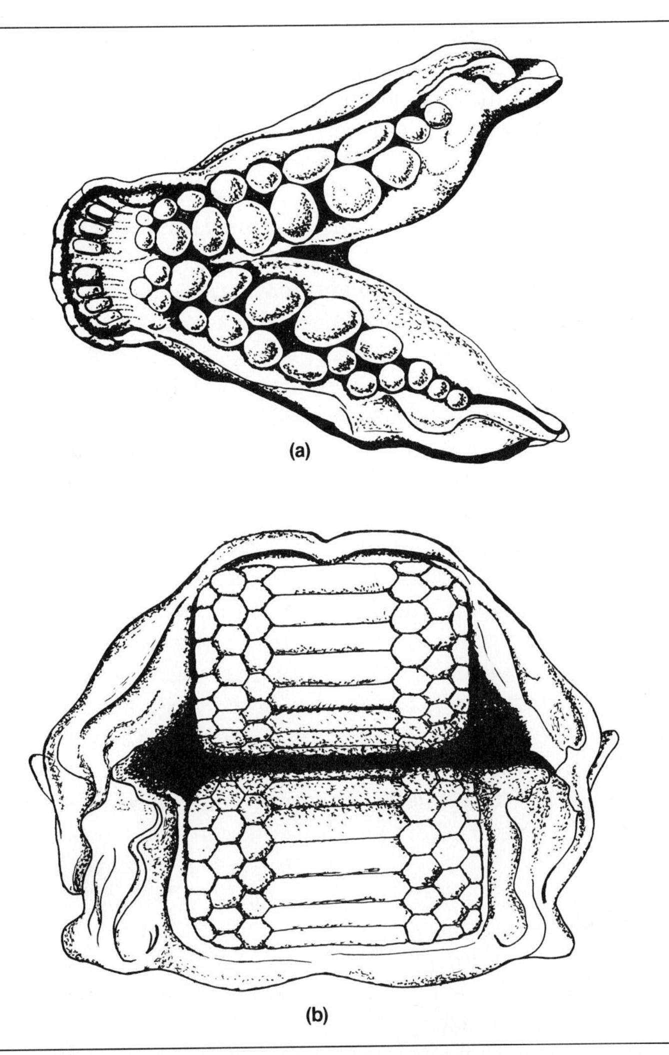

Figure 9–18 The teeth of some mollusk-eating fishes: (*a*) a porgy (*Sparidae*) and (*b*) an eagle ray (*Myliobates*). (Modified from K.F. Lagler, J.E. Bardach, and P.R. Miller, 1962. *Ichthyology*. John Wiley & Sons, New York.)

rans. Insectivorous frogs and toads can flip out a long tongue, attached at the front of the mouth, to snap up a passing insect. Many of the smaller snakes are insectivores, and the young of species that hunt vertebrates are often insectivorous when they are small. The North American queen snake (*Regina septemvittata*) feeds a great deal on crayfish; a venomous snake (*Fordonia*) on Java eats crabs, and its venom is specific to that prey alone. Most lizards are mainly insect-eating, and some have specialized on ants: *Phrynosoma* in North America, *Moloch horridus* in Australia, and the arboreal *Draco* in Asia. The true chameleons (*Chamaeleo*) of the Old World have a marvelous insect-catching device in their enormously long tongue, which is about 1½ times the body length and equipped with a sticky bulb at the end. The chameleons stalk their prey with amazing slowness, then shoot the tongue forward at high speed (Fig. 9–19). A highly specialized arthropod-eater is a burrowing snake, *Leptotyphlops phenops*, which feeds on the contents of termite abdomens, leaving the abdominal exoskeleton, thorax, and head behind.

Birds capture arthropods in many ways and places. Gleaning birds pick their prey from surfaces (ground, bark, leaves, and so on). Warblers (Parulidae) and vireos (Vireonidae) glean leaves, for instance. Titmice, chickadees (Paridae), and creepers (Certhiidae) often glean bark. Meadowlarks (Icteridae) glean the ground and low vegetation. A few avian gleaners pick arthropods from the exteriors of other vertebrates. Tickbirds or oxpeckers (*Buphagus*) in Africa ride about on rhinoceroses, buffalo, and other large mammals, grooming off their ticks and any infesting insects. In addition, one of the ground finches on the Galápagos Islands, *Geospiza fuliginosa*, removes ticks from marine iguanas and giant tortoises (*Testudo elephantopus*). Salliers or hawkers snatch insects on short flights between perches; this group includes many tyrant flycatchers (Tyrannidae) and the American redstart (*Setaphaga ruticilla*). Other birds, such as swallows (Hirundindae), are aerial foragers that snap up insect after insect on a prolonged, continuous flight. Many woodpeckers flake bits of bark or drill into the bark in search of insects.

Mammalian insectivores and arthropod-eaters include many of the aerially feeding bats, ground-living shrews, small rodents such as the North

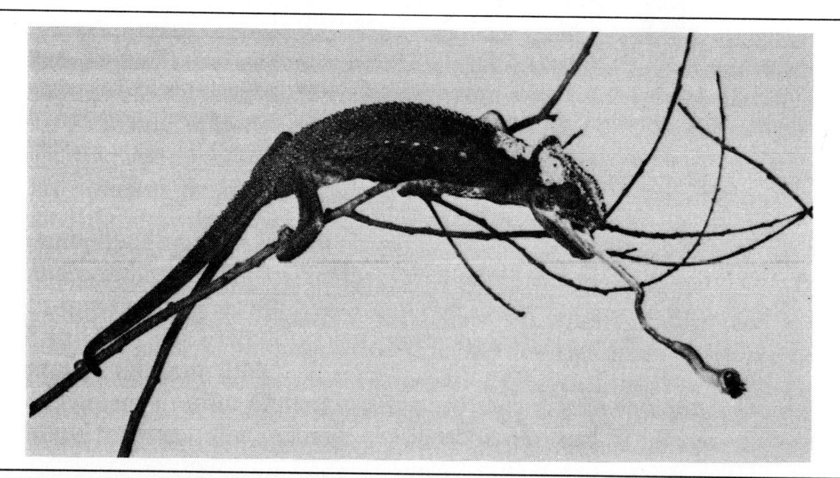

Figure 9–19 A chameleon (*Microsaura pumila*) catching prey. (Photo courtesy of The Zoological Society of San Diego.)

American grasshopper mouse (*Onychomys*), somewhat larger mierkats (three genera of the Old World family Viverridae), raccoons and their relatives, and the monotreme spiny anteaters or echidnas. Bigger insectivorous mammals include the large anteaters and aardvark. Many others eat arthropods as part of a broad diet—some cats (Felidae) and foxes (Canidae), many rodents, viverrids, armadillos (Dasypodidae), the marsupial bandicoots (Peramelidae), possums (Phalangeridae), muskrat-kangaroo (*Hypsiprymnodon*), and many primates, for example. The striped possums (*Dactylopsila* and *Dactylonax*) of New Guinea and northern Australia and the Madagascar primate called the aye-aye (*Daubentonia madagascariensis*) share the habit of feeding on wood-boring insects by tearing open the wood with their incisors and hooking out the insect with a particularly elongated, slender finger.

ADAPTATIONS TO CARNIVORY The patterns of hunting behavior associated with carnivory are varied. Some foragers spend relatively little time searching for the prey; most of the hunting activity involves attack and handling time. Anglerfishes are sedentary ambushers and many of the insectivorous iguanid lizards can be called sit-and-wait foragers. Among the birds, the flycatchers are good examples of sit-and-wait foragers. Sit-and-wait hunting is used more by neotropical than by temperate birds, probably because the large insects on which they prey are more common in tropical regions. Although lions are hardly sedentary, their style of hunting typically involves little searching—they know where to find their quarry. Most of a lion's hunting effort seems to be devoted to a stalk, followed by a short, rapid pursuit and subjugation.

At the other end of the spectrum are animals that spend a great deal of time searching and relatively little time or effort pursuing and subduing their food. Active searchers include the teiid lizards, such as *Cnemidophorus*, and most of the leaf-gleaning birds. Some predators can switch tactics, depending on the prey involved. Wolves harrying a moose would join lions in the category of pursuers, whereas wolves capturing mice are probably closer to the searcher category.

A number of birds and some fishes characteristically forage on insects and other animals flushed by the passage of another animal. Many neotropical birds follow swarms of army ants, snatching up the creatures that scurry to escape the ants. Cattle egrets (*Bubulcus ibis*) follow large herbivorous mammals, or ride on their backs, and capture more insects when associated with these "beaters" than when foraging without them. Other birds ride on bustards or storks or follow monkeys.

Many birds make a living by pirating prey from others (Brockmann and Bernard, 1979). The pirate, or "kleptoparasite," attacks the successful hunter until it relinquishes its prey, which is then collected by the pirate. The success rate of pirates varies with the size and tenacity of the bird that is attacked (Furness, 1978). The effect of kleptoparasites on the fitness of their victims undoubtedly varies from species to species.

The injection of venom is a rather unusual method of subduing prey, but it is found in several vertebrate groups. Vertebrate venoms usually contain agents that attack both the circulatory system and the nervous system. Several kinds of snakes have elaborated this technique (Fig. 9–20). Snakes usually swallow their prey whole, and venom is probably one means of rendering the prey sufficiently inactive that it can be readily swallowed. One group of venomous snakes has grooved rear teeth that permit entry of the venom while the prey is chewed. Many colubrid snakes, including *Hypsiglena* (night snakes) and *Tantilla* (the flat-headed snakes) in North America, belong to this rear-fang group. A second group of highly venomous snakes are the Elapidae—coral snakes, cobras, and mambas. The sea snakes are sometimes classified as elapids and sometimes are placed in their own family, the Hydrophiidae. These reptiles

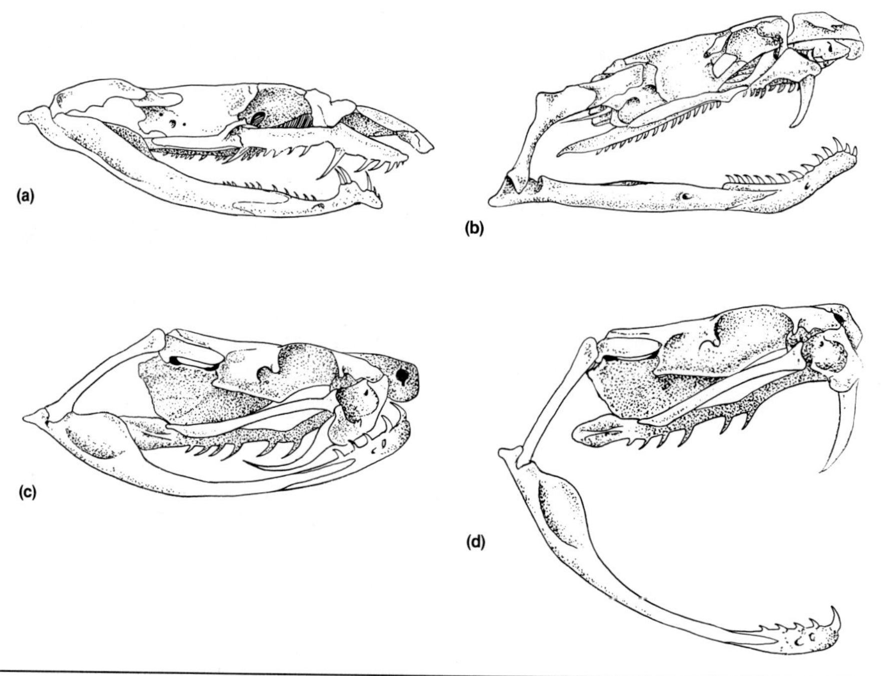

Figure 9–20 Fangs of venomous snakes. (*a*) A rear-fang Asian colubrid, *Dryophis prasinus.* (*b*) A fixed-fang elapid, the African spitting cobra, *Naja nigricollis.* (From *The Life of Reptiles* by Angus Bellairs. Published by Universe Books, New York, 1970.) (*c*) An erectile-fang viperid from Latin America, *Crotalus durissus,* with fangs folded and (*d*) with fangs erect. (*a, c,* and *d* modified from J. Anthony, 1955. Essai sur l'evolution anatomique de l'appareil venimeux des ophidiens. *Ann. Sci. Nat. Zool., 11:*7–53.)

inject venom through fangs fixed at the front of the upper jaw. The mambas (*Dendroaspis*) of Africa and the Australian taipan (*Oxyruanus*) can be highly aggressive and fierce, whereas some coral snakes and some sea snakes tend to be docile. Another group is the Viperidae, which includes the Old World true vipers and the pit vipers of the New World and Asia. Rattlesnakes (*Crotalus* and *Sistrurus*), fer-de-lance and their relatives (*Bothrops*), and bushmasters (*Lachesis*) are some of the American pit vipers; Gaboon vipers (*Bitis gabonica*) and European adders (*Vipera berus*) are examples of true vipers. The vipers have erectile, hollow fangs that lie flat against the roof of the mouth when not in use but are erected to stabbing position during an attack on prey.

Among lizards, only the American *Heloderma* species (gila monster, Mexican beaded lizard) are known to be venomous, although their diet of young vertebrates, eggs, and such hardly seems to require it. The short-tailed shrew has a toxin used to subdue its mammalian and perhaps also its insect prey.

Many vertebrate carnivores have somewhat flexible jaws and skulls; their capacity to change shape and orientation are associated with carnivorous food habits. Numerous teleost fishes have protrusible jaws: The upper and lower jaws are linked in such a way that the upper one moves forward as the lower one moves forward and down (Fig. 9–21). This feature creates a rounded mouth opening with no corners to cause turbulence that might deflect the prey as it is sucked in. Moreover, it allows the jaws to close while the mouth cavity containing the prey remains expanded, thus preventing prey escape. This arrangement in active predators also means that the angle of capture can be the same as the angle of pursuit; that is, the mouth opening is lined up with the trajectory of the attacking fish. The mouth of bottom-feeders can angle downward while the fish's body remains horizontal. Sharks frequently have protrusible upper jaws, which may facilitate grasping and cutting up of prey (Moss, 1972).

A fascinating flexibility of the skull, called cranial kinesis, is well developed in many lizards, birds, and perhaps especially snakes. These groups are predominantly carnivorous. The exact arrangement of bones and the extent and variety of flexion varies greatly, but all forms of kinesis involve at least one hinge on the roof of the skull, allowing the anterior skull elements to rotate on the braincase.

The functional advantages of kinesis have been hotly debated, and the issues have not yet been entirely resolved. It is likely that there are several adaptive functions. In the first place, kinesis may allow both jaws to move toward each other while the animal is grasping its prey, thereby minimizing the risk of deflecting the prey away from the closing jaws. Jaw closure also may be more rapid and the gape may be reoriented so that pursuit and capture trajectories are similar. Lizards have a complex form of cranial kinesis (Fig. 9–22), that allows the jaws to close approximately parallel to each other, rather than scissors-style. This design may provide an advan-

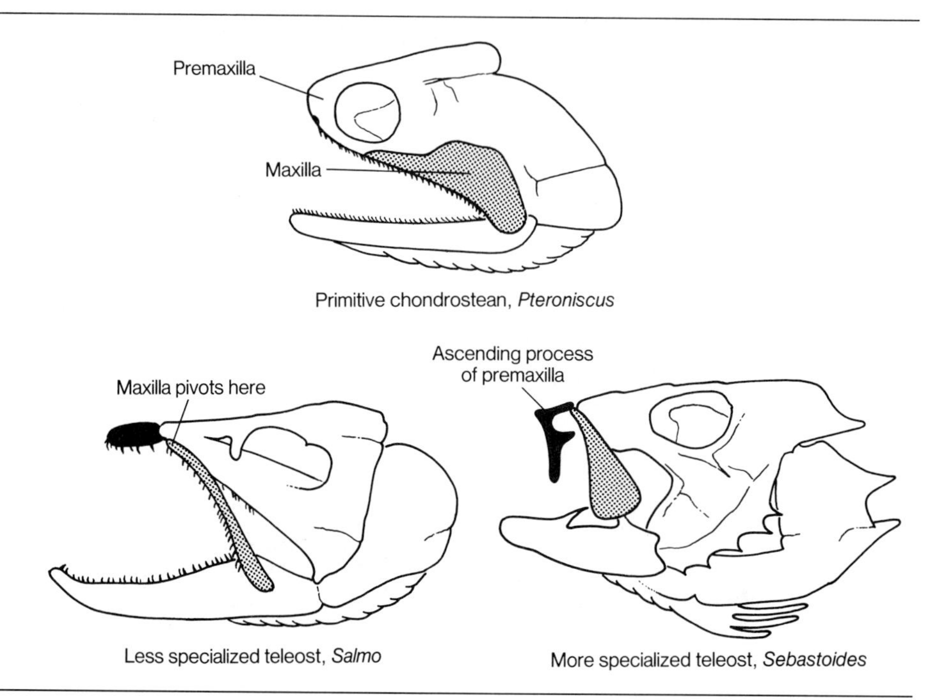

Figure 9–21 The protrusible jaw of some teleosts. Note the major skeletal elements in protrusible and nonprotrusible jaws. (Modified from M. Hildebrand, 1974. *The Analysis of Vertebrate Structure.* John Wiley & Sons, New York.)

tage in grasping prey. Snakes are more frequently spearers than graspers of their prey, at least when feeding on large items. Kinesis here allows the jaws to open so wide that the teeth can be used to transfix the prey and then grasp it (Fig. 9–23). Skulls of many snakes are very flexible, and right and left sides can be moved independently. Each side of the jaw then takes turns in reaching forward to sink into the prey and pull it toward the throat. A fine example of such flexibility is found in rattlesnakes in which each fang can be erected separately. Flexibility of the head and jaws also permits the swallowing of prey much larger than the snake's diameter.

Kinesis in birds is more tightly controlled than in snakes, and the flexibility of the skull is more restricted (Fig. 9–24). Birds, which have no teeth to hold their prey, must depend on precision in aim. Cranial kinesis in birds may have the added advantage of permitting the line of sight to coincide with the line of attack.

Neither mammals nor turtles nor crocodilians have kinetic skulls, although many of them are carnivorous. Other considerations have taken precedence in all these groups. The aquatic reptiles have a secondary palate across the roof of the mouth, separating the nasal passages from the

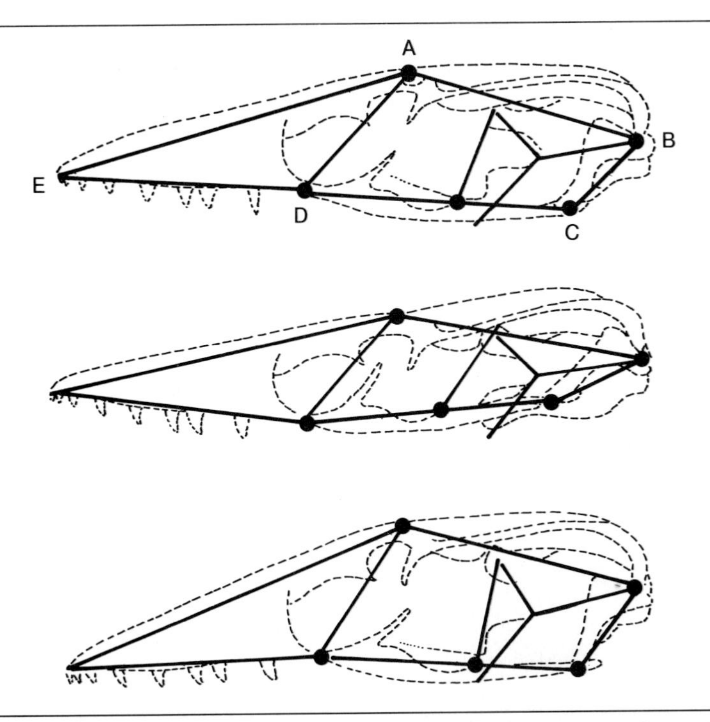

Figure 9–22 Cranial kinesis in a lizard, *Varanus salvator*. The circles indicate the hinges where rotation occurs. The quadrangle labeled *ABCD* changes position as the jaw opens, causing the snout complex (*ADE*) to lift. The remaining elements indicated by black bars constrain movement and link the larger element to the inner braincase. (Modified from T.H. Frazzetta, 1962. A functional consideration of cranial kinesis in lizards. *J. Morphol.*, *111*:287–320.)

feeding apparatus and making kinesis impossible. Mammals have both a secondary palate, which keeps food particles out of the breathing channels, and an enlarged braincase that precludes skull bending (Fig. 9–25).

Many carnivores have teeth that aid in the capture or sectioning of food items. The teeth of invertebrate-eaters, both reptilian and mammalian, are commonly rather simple, conical pegs. Fish-eaters and bird-eaters frequently have long, slender teeth for piercing the scale or feather covering of their prey. The gharials, for instance, have already been mentioned, but tropical bird-eating snakes (*Chondropython* in the Old World, *Corallus* in the New World) have the same feature (Fig. 9–26).

Hinged teeth, which fold back against the jaw or the roof of the mouth, occur in some fishes, amphibians, and reptiles. These teeth are erected during prey capture but can be folded when the prey is swallowed. The folding probably facilitates swallowing and reduces the risk of tooth breakage. This arrangement may be particularly useful for predators feeding on

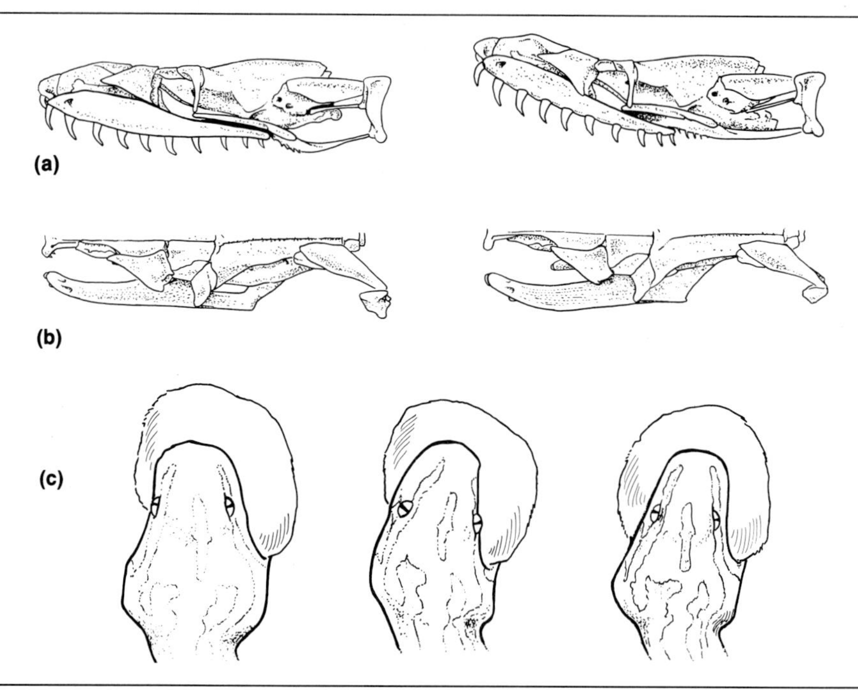

Figure 9–23 Cranial kinesis in a snake, *Python sebae*. (*a*) The skull of *P. sebae* at rest and protracted. (*b*) In addition to protraction and retraction, the tooth-bearing jawbone can rotate to point the teeth outward. (*c*) Moreover, the right and left jaws can move independently to draw the prey into the mouth. (Modified from T.H. Frazzetta, 1966. Studies of the morphology and function of the skull in the Boidae (Serpentes). Part 2. *J. Morphol.*, *118*:217–296.)

hard-bodied prey, such as scaly fishes and well-armored lizards. Lizards shake or pull their larger prey into pieces before swallowing, and snakes swallow their prey whole. Mammalian carnivores, however, usually chew their prey to some extent before they swallow it. Most of these mammals have upper and lower cheek teeth with sharp cutting edges that slice the prey into chunks for swallowing. Neither turtles nor modern birds have teeth. Turtles may shake their prey to pieces, tear it with their claws, or use the edge of their beak to cut it into bits. Birds sometimes have serrated bills that aid in holding prey, but they often render prey inactive and reduce it to pieces by pecking.

Thus far I have concentrated on different foraging methods of different kinds of vertebrates. It is essential to recognize, however, that individual difference in hunting style and ability are common, although rather little studied. Oystercatchers (*Haematopus ostralegus*), for example, have two distinct methods of killing mussels: Some individuals stab through an open-

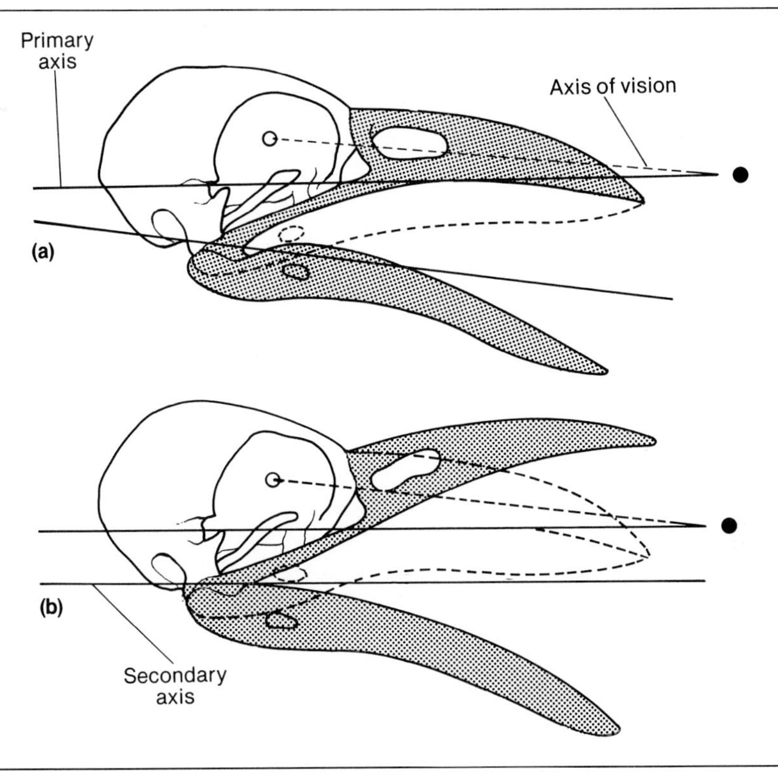

Figure 9–24 Cranial kinesis in a generalized bird. The orientation of the gape in (*a*) a kinetic skull compared to (*b*) a nonkinetic one. (Modified from J. Dorst, 1974. *The Life of Birds,* Vol. 1. Columbia University Press, New York.)

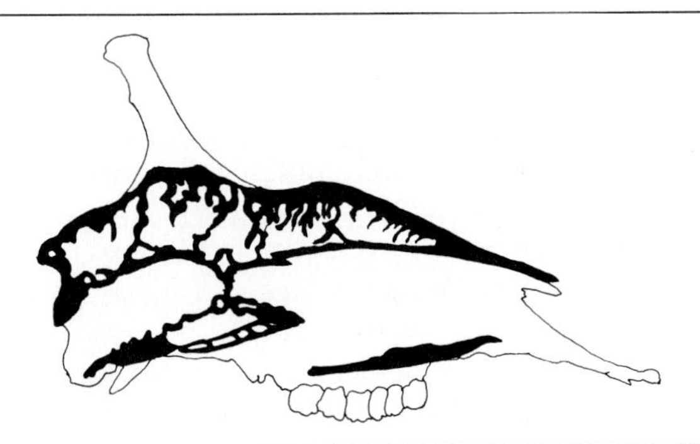

Figure 9–25 Longitudinal section of a giraffe skull, showing the secondary palate and enlarged braincase that make cranial kinesis impracticable in mammals. (From R. Owen, F.R.S., 1886–1888. *On the Anatomy of Vertebrates.* Vol. 2. Birds and Mammals. Longmans, Green, London.)

ing between the shell margins; others hammer their prey to death with a series of blows by the beak. Each young oystercatcher learns its skill from its parents, both of which use the same killing style! Few birds use both methods. Acquisition of these skills takes many months; their perfection

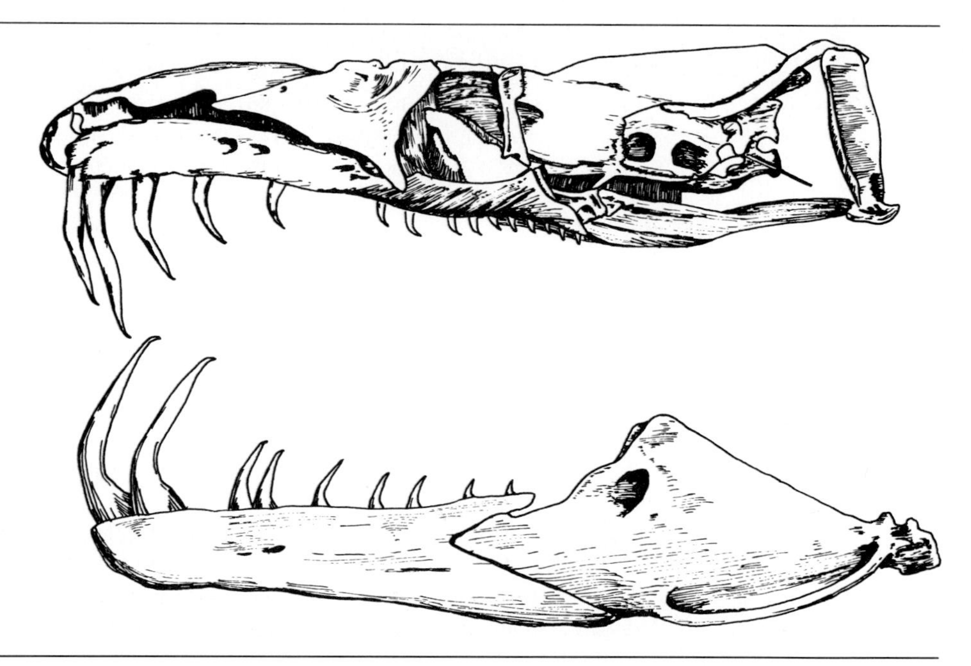

Figure 9–26 The teeth of a bird-eating snake, *Corallus caninus*, from South America. (Redrawn from the original by T.H. Frazzetta.)

may take years. Each technique may be so difficult to do well that individuals need to specialize on one or the other.

Such individual differences must exist in most species, in conjunction with learning experience (as in oystercatchers) or in association with individual differences in morphology and coordination. A sparrow with a large bill can efficiently handle larger seeds than can a conspecific with a small bill. Such individuality in hunting skills is likely to be correlated with individual differences in prey choice, foraging site, habitat selection, and sometimes survival. The relationship of individual foraging skill with prey-capture rates was associated with habitat choice in bluegill sunfish (*Lepomis macrochirus*) (Werner et al., 1981). Furthermore, the ability to handle large, hard seeds enhanced the survival of large-billed individuals of a Galápagos ground finch (*Geospiza fortis*) during a drought, when smaller, softer seeds became rare (Boag and Grant, 1981).

TOOL USE BY CARNIVORES Several vertebrates are known to use tools in obtaining their prey. The chimpanzee inserts a twig or grass stem into termite nests, the termites attach themselves to it; and the predator then retrieves it, covered with prey, to be popped into its mouth. The sea otter floats on its back with a stone on its chest, against which it smashes a mollusk brought up from the bottom. Egyptian vultures (*Neophron percnopterus*) pick up stones and throw them at ostrich eggs in order to get at the contents. Several birds, such as the woodpecker finch (*Camarhynchus pallidus*) of the Galápagos and the Australian orange-winged sitella (*Neositta chrysoptera*), use twigs or cactus spines to probe for insects in crevices. Mollusk-eating birds, such as European thrushes, which feed on land snails, and gulls and crows, which feed on mussels, frequently carry their prey to a rock anvil: The thrushes hammer their prey to bits on their anvils; the gulls and crows drop their shelled food from considerable heights in order to smash it on rocks below. Golden eagles (*Aquila chrysaetos*) reportedly drop desert tortoises (*Gopherus*) to crack open the carapace.

EATING ANIMAL PRODUCTS Animals produce a variety of "products" ranging from eggs and placentas to dung, wax, and honey, and there are vertebrates that eat these products. Placenta-eating is a regular feature at the birth of many eutherian young. The mother, whether carnivore or herbivore, frequently consumes the placenta, probably to reduce the attraction of predators but perhaps also to retrieve some energy and nutrients. Some birds eat feathers, but their function in the digestive process or in providing nutrition does not seem to be established. Egg-eating is an opportunistic indulgence for many vertebrates, but the African egg-eating snakes are highly specialized to this food (Fig. 9–27). The egg is swallowed whole, greatly distending the snake's head and throat. In the snake's throat, the egg is punctured by ventral projections from the vertebrae. Once punctured, the egg collapses from the pressure of the constricting throat muscles and the contents flow into the stomach. The shell is neatly folded,

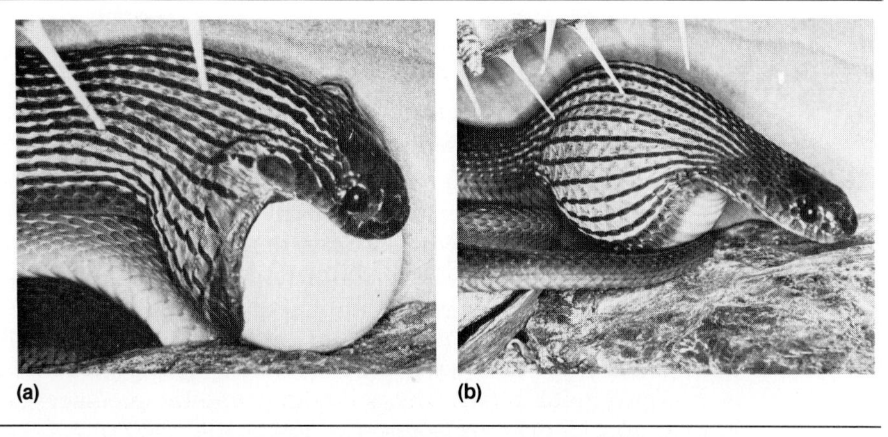

Figure 9–27 The egg-eating snake *Dasypeltis* engulfing an egg. (Photo by C. Gans.)

compressed, and regurgitated. One genus of sea snake, *Emydocephalus*, is said to eat only the eggs of teleost fishes.

The eaters of excrement are numerous. Several fishes, including *Scatophagus* (Scatophagidae) of southeastern Asia, are habitual dung-eaters. Parent birds, especially passerines, sometimes swallow the fecal sacs of their very young offspring. Many female mammals, including American opossums (*Didelphis*) and cats, lick the anal and urinary openings of their young to stimulate excretion and elimination and then consume the products, presumably as a means of nest sanitation. Numerous birds consume the dung of mammals at least upon occasion. Ivory gulls (*Pagophila eburnea*), for instance, eat the dung of polar bears and seals; puffins (Alcidae) and petrels (Procellariidae) eat whale dung. The habit of refection in certain herbivores has already been mentioned. Some bats (such as *Leptonycteris sanborni*) apparently consume their own urine, perhaps as an aid in digesting pollen.

At least one vertebrate consumes saliva. A swiftlet of the genus *Collocalia* in the East Indies constructs its nest with saliva. These nests are made into soup, especially by the Chinese, and are considered a great delicacy.

Honey is considered fine food by many vertebrates, including humans and bears. A specialist on honey (although it eats other things too) is the African ratel or honey-badger (*Mellivora capensis*), a mustelid that digs up bees' nests to obtain the honey (Fig. 9–28*a*). Birds called honeyguides (*Indicator*) eat beeswax and occasionally bee larvae; they are the only animals known to be able to survive on a diet of beeswax for any length of time (Fig. 9–28*b*). Honeyguides are legendary for their interactions with ratels or other large, honey-loving mammals: They make a special call to indicate the location of a bees' nest and reportedly guide the mammals to the nest; the mammal opens the nest, and both creatures share the dividends.

FOOD STORAGE

Food storage is clearly one way to survive a period of food scarcity or unfavorable weather. Large numbers of mammals store food of all sorts. Seeds are stored by kangaroo rats: Unripe seeds in pits are stored outside the burrow, and ripe, dry seeds are stored inside. Pine squirrels (*Tamiasciurus*) in western North America cache heaps of conifer cones with the enclosed seeds in damp places, but in eastern North America, where they coexist with the larger *Sciurus* squirrels in deciduous forest, they do not form caches. The pine squirrels also store mushrooms in dry places. *Sciurus* squirrels and eastern *Tamiasciurus* store acorns and nuts singly in the ground; this behavior is called scatter-hoarding. Neotropical agoutis scatter-hoard fruits in seasons of great fruit abundance. Like their temperate zone counterparts, tropical squirrels scatter-hoard food supplies. Pikas

Figure 9–28 Two specialists on honey: The African honey badger, or ratel, digs up bees' nest; the honeyguide can survive on a diet of beeswax and may lead the ratel to bees' nests. Sometimes these two vertebrates cooperate and share the dividends.

(*Ochotona princeps*) and steppe lemmings or sagebrush voles (*Lagurus*) are among the mammals that store dried herbage or hay. Roots and tubers may be stored by *Spalax* mole rats in the eastern Mediterranean region. European foxes (*Vulpes vulpes*) sometimes store eggs they have raided from gull nests and even cache the carcasses of captured gulls. Leopards or panthers (*Panthera pardus*) sometimes cache the remains of a large prey in a tree for future use.

Food-stashing by birds is not uncommon (Roberts, 1979). Shrikes (*Lanius*) and Australian butcherbirds (*Cracticus*) impale insects and small vertebrates on thorns (Fig. 9–29). These carcasses provide extra food sources during times when food requirements are particularly high or when fresh prey are scarce, and they may also serve other functions, such as marking territory boundaries. Ravens (*Corvus corax*) and pygmy owls (*Glaucidium passerinum*) store prey to feed their young. The acorn woodpecker (*Melanerpes formicivorus*) and, to some extent, the red-headed woodpecker (*M. erythrophthalmus*), both of North America, often store great quantities of acorns in crevices and prepared holes in tree trunks and defend their stores against the intrusions of other animals. Both piñon jays (*Gymnorhinus cyanocephalus*) and nutcrackers (*Nucifraga*) store pine seeds in abundance. These caches are critical to the timing and success of breeding (Ligon, 1978).

Figure 9–29 A loggerhead shrike (*Lanius ludovicianus*) with an impaled insect.

Many organisms store digested and assimilated food in the form of high-energy molecules, especially lipids, at various sites in the body. The amount of stored energy often varies seasonally, or with sex and age. The location of storage also varies greatly, among species, in relation to several kinds of factors, including social and sexual status, thermoregulation, and buoyancy (Pond, 1981). Fat deposits alter the forms of maturing human females as well as the shape of the head and shoulders of males of certain other primates, for instance. The role of localized fat deposits in thermoregulation was mentioned in Chapter 5, and fat contributing to buoyancy in fishes was discussed in Chapter 7.

SUMMARY

Most vertebrates have fairly broad diets compared to many insects, primarily because of the effects of larger body size and the resulting greater food requirements. Nevertheless, a few dietary specialists can be found among the vertebrates. As a rule, herbivores are more specialized in diet than carnivores, whereas carnivores are more likely to specialize on prey size, foraging site, or capture techniques. The difference may be due partly to the many chemical and morphological defenses of plants and their immobility.

The economics of foraging determine both the location and the size of the foraging area. The benefits of a hunting area depend on food density, size of the patch, and the forager's behavior—all of which influence the time and energy expended in foraging. Food choice is also influenced by benefit-cost relationships; time, energy, and nutrients and the demands of other ecological pressures must be weighed in the balance.

A survey of food habits of vertebrates produces examples of species that eat different kinds of prey and indicates the morphological adaptations that accompany various diets. Vertebrates store food of many sorts, usually as a means of providing a supply during times of seasonal shortage and bad weather, and sometimes to facilitate reproduction.

SELECTED REFERENCES

Most of the material on food habits, teeth, and beaks can be augmented by reference to the sources listed at the end of Chapter 2 or in the literature cited in the following studies.

Andersson, M., and J. Krebs, 1978. On the evolution of hoarding behavior. *Anim. Behav., 26*:707–711.

Anthony, J., 1955. Essai sur l'evolution anatomique de l'appareil venimeux des ophidiens. *Ann. Sci. Nat. Zool., 11*:7–53.

Arnold, S.J., 1977. Polymorphism and geographic variation in the feeding behavior of the garter snake *Thamnophis elegans. Science, 197*:676–678.

Baker, H.G., and I. Baker, 1973. Amino-acids in nectar and their evolutionary significance. *Nature, 241*:543–545.

Bearder, S.K., and R.D. Martin, 1980. Acacia gum and its use by bushbabies, *Galago senegalensis* (Primates: Lorisidae). *Int. J. Primatol., 1*:103–128.

Bellairs, A., 1970. *The Life of Reptiles.* Universe Books, New York.

Belovsky, G.E., 1981. Food plant selection by a generalist herbivore: the moose. *Ecol., 62*:1020–1030.

Boag, P.T., and P.R. Grant, 1981. Intense natural selection in a population of Darwin's finches (Geospizinae) in the Galápagos. *Science, 214*:82–85.

Bogert, C.M., 1943. Dentitional phenomena in cobras and other elapids with notes on adaptive modifications of fangs. *Bull. AMNH, 81*:285–360.

Bossema, I., 1979. Jays and oaks: an eco-ethological study of a symbiosis. *Behavior, 70*:1–117.

Brockmann, H.J., and C.J. Barnard, 1979. Kleptoparasitism in birds. *Anim. Behav., 27*:487–514.

Brooks, J.L., 1968. The effects of prey size selection by lake planktivores. *Syst. Zool., 17*:272–291.

Caraco, T., 1981. Energy budget risk and foraging preferences in dark-eyed juncos (*Junco hyemalis*). *Behav. Ecol. Sociobiol., 8*:213–217.

Charles-Dominique, P., 1971. Eco-ethology des prosimiens du Gabon. *Biologica Gabonica, 7*(2):121–228. [With English summary]

Charles-Dominique, P., 1977. *Ecological Behaviour of Nocturnal Primates.* Columbia University Press, New York.

Churchill, D.M., and P. Christensen, 1970. Observations on pollen harvesting by brush-tongued lorikeets. *Aust. J. Zool., 18*:427–437.

Curio, E., 1976. *The Ethology of Predation.* Springer-Verlag, New York.

Dodson, S.I., and D.L. Egger, 1980. Selective feeding of red phalaropes on zooplankton of Arctic ponds. *Ecol., 61*:755–763.

Dorst, J., 1974. *The Life of Birds.* Columbia University Press, New York.

Frazzetta, T.H., 1962. A functional consideration of cranial kinesis in lizards. *J. Morphol., 111*:287–320.

Frazzetta, T.H., 1966. Studies on the morphology and function of the skull in the Boidae (Serpentes). II. *J. Morphol., 118*:217–296.

Furness, R.W., 1978. Kleptoparasitism by great skuas (*Catharacta skua* Brünn) and Arctic skuas (*Stercorarius parasiticus L.*) at a Shetland seabird colony. *Anim. Behav., 26*:1167–1177.

Gans, C., 1961. The feeding mechanism of snakes and its possible evolution. *Am. Zool., 1*:217–227.

Gans, C., 1974. *Biomechanics.* J.B. Lippincott, Philadelphia.

Gill, F.B., and L.L. Wolf, 1977. Nonrandom foraging by sunbirds in a patchy environment.

Ecol., *58*:1284–1296.

Glander, K.E., 1977. Poison in a monkey's Garden of Eden. *Nat. Hist.*, *86*(3):34–41.

Gorlick, D.L., P.D. Atkins, and G.S. Losey, 1978. Cleaning stations as waterholes, garbage dumps, and sites for the evolution of reciprocal altruism? *Am. Nat.*, *112*:341–353.

Grant, P.R., 1981. The feeding of Darwin's finches on *Tribulus cistuides* (L.) seeds. *Anim. Behav.*, *29*:785–792.

Greene, H.W., and G.M. Burghardt, 1978. Behavior and phylogeny: constriction in ancient and modern snakes. *Science*, *200*:74–77.

Harestad, A.S., and F.L. Bunnell, 1979. Home range and body weight—a re-evaluation. *Ecol.*, *60*:389–402.

Hildebrand, M., 1974. *Analysis of Vertebrate Structure.* John Wiley & Sons, New York.

Hofmann, R., 1973. The ruminant stomach. *East. Afr. Monogr. Biol.*, *2*:1–349.

Howell, D.J., 1974. Bats and pollen: physiological aspects of the syndrome of chiropterophily. *Comp. Biochem. Physiol.*, *48A*:263–276.

Jenkins, S.H., 1980. A size-distance relation in food selection by beavers. *Ecol.*, *61*:740–746.

Kardong, K.V., 1979. "Protovipers" and the evolution of snake fangs. *Evol.*, *33*:433–443.

Kenagy, G.J., and D.F. Hoyt, 1980. Reingestion of feces in rodents and its daily rhymicity. *Oecologia*, *44*:403–409.

Krebs, J.R., 1978. Optimal foraging: decision rules for predators. *In* J.R. Krebs and N.B. Davies (Eds.), *Behavioural Ecology: An Evolutionary Approach*, pp. 23–63. Sinauer Associates, Sunderland, MA.

Krebs, J.R., J.T. Erichsen, M.I. Walker, and E.L. Charnov, 1977. Optimal prey selection in the great tit (*Parus major*). *Anim. Behav.*, *25*:30–38.

Krebs, J.R., and N.B. Davies, 1981. *An Introduction to Behavioural Ecology.* Sinauer Associates, Sunderland, MA.

Lagler, K.F., J.E. Bardach, and R.R. Miller, 1962. *Ichthyology.* John Wiley & Sons, New York.

Landry, S.O., 1970. The Rodentia as omnivores. *Q. Rev. Biol.*, *45*:441–454.

Leopold, A.S., 1953. Intestinal morphology of gallinaceous birds in relation to food habits. *J. Wildl. Manag.*, *17*:197–203.

Lesiuk, T.P., and C.C. Linsey, 1978. Morphological peculiarities in the neck-bending Amazonian charocoid fish, *Rhaphiodon vulpinus. Can. J. Zool.*, *56*:991–997.

Ligon, J.D., 1978. Reproductive interdependence of piñon jays and piñon pines. *Ecol. Monogr.*, *48*:111–126.

Lineaweaver, T.H., and R.H. Backus, 1970. *The Natural History of Sharks.* J.B. Lippincott, Philadelphia.

MacArthur, R.H., 1958. Population ecology of some warblers of northeastern coniferous forest. *Ecol.*, *39*:599–619.

MacArthur, R.H., and E.R. Pianka, 1966. On optimal use of a patchy environment. *Am. Nat.*, *100*:603–609.

Maynard Smith, J., and R.J.G. Savage, 1959. The mechanics of mammalian jaws. *School Science Res.*, *40*(141):289–301.

McNab, B.K., 1963. Bioenergetics and the determination of home range size. *Am. Nat.*, *97*:133–140.

Morton, E.S., 1980. Avian arboreal folivores: why not? *In* A. Keast and E.S. Morton (Eds.), *Migrant Birds in the Neotropics*, pp. 123–130. Smithsonian Institution Press, Washington, D.C.

Moss, S.A., 1972. The feeding mechanism of sharks of the family Carcharhinidae. *J. Zool.* (Lond.), *167*:423–436.

Moynihan, M., 1976. Notes on the ecology and behavior of the pygmy marmoset (*Cebuella pygmaea*) in Amazonian Colombia. *In* R.W. Thorington and P.G. Heltne (Eds.), *Neotropical Primates: Field Studies and Conservation*, pp. 79–84. National Academy of Science, Washington, D.C.

Norman, J.R., and P.H. Greenwood, 1963. *A History of Fishes.* Hill & Wang, New York.

Nyberg, D.W., 1971. Prey capture in the largemouth bass. *Am. Midl. Nat.*, *86*:128–144.

Partridge, L., 1976. Individual differences in feeding efficiencies and feeding preferences of captive great tits. *Anim. Behav.*, *24*:230–240.

Pianka, E.R., 1974. *Evolutionary Ecology.* Harper and Row, New York.

Pietsch, T.W., and D.B. Grobecker, 1978. The compleat angler: aggressive mimicry in an antennariid anglerfish. *Science, 201*:368–370.

Pond, C.M., 1981. Storage. *In* C.R. Townsend and P. Calow (Eds.), *Physiological Ecology: An Evolutionary Approach to Resource Use,* pp. 190–219. Blackwell, Oxford.

Pyke, G.H., H.R. Pulliam, and E.L. Charnov, 1977. Optimal foraging: a selective review of theory and tests. *Q. Rev. Biol., 52*:137–154.

Regal, P.J., and C. Gans, 1976. Functional aspects of the evolution of frog tongues. *Evol., 30*:718–734.

Reichman, O.J., 1977. Optimization of diets through food preferences by heteromyid rodents. *Ecol., 58*:454–457.

Rissing, S.W., 1981. Prey preferences in the desert horned lizard: influence of prey foraging method and aggressive behavior. *Ecol., 62*:1031–1040.

Roberts, R.C., 1979. The evolution of avian food-storing behavior. *Am. Nat., 114*:418–438.

Russell, R.E., 1969. Poisons and venoms. *In* W.S. Hoar and D.J. Randall (Eds.), *Fish Physiology.* Vol. 3. pp. 401–449. Academic Press, New York.

Savitsky, A.H., 1981. Hinged teeth in snakes: an adaptation for swallowing hard prey. *Science, 212*:346–349.

Schoener, T.W., 1968. Sizes of feeding territories of birds. *Ecol., 49*:123–141.

Schoener, T.W., 1974. Resource partitioning in ecological communities. *Science, 185*:27–39.

Schoener, T.W., 1979. Generality of the size-distance relation in models of optimal foraging. *Am. Nat., 114*:902–914.

Sibly, R.M., 1981. Strategies of digestion and defecation. *In* C.R. Townsend and P. Calow (Eds.), *Physiological Ecology: An Evolutionary Approach to Resource Use,* pp. 109–139. Blackwell, Oxford.

Smith, A.P., 1982. Diet and feeding strategies of the marsupial sugar glider in temperate Australia. *J. Anim. Ecol., 51*:149–166.

Smith, C.C., and D. Follmer, 1972. Food preferences of squirrels. *Ecol., 53*:82–91.

Smythe, N., 1970. Relationships between fruiting season and seed dispersal methods in a neotropical forest. *Am. Nat., 104*:25–35.

Turner, F.B., R.I. Jennrich, and J.D. Weintraub, 1969. Home ranges and body sizes of lizards. *Ecol., 50*:1076–1081.

Vaughn, T.A., 1978. *Mammalogy.* 2nd ed. Saunders College Publishing, Philadelphia.

Wassersug, R.J., 1975. The adaptive significance of the tadpole stage with comments on the maintenance of complex life cycles in anurans. *Am. Zool., 15*:405–417.

Welty, J.C., 1982. *The Life of Birds.* 3rd ed. Saunders College Publishing, Philadelphia.

Werner, E.E., G.G. Mittelbach, and D.J. Hall, 1981. The role of foraging profitability and experience in habitat use by the bluegill sunfish. *Ecol., 62*:116–125.

Whitford, W.G., and M. Bryant, 1979. Behavior of a predator and its prey: the horned lizard (*Phrynosoma cornutum*) and harvester ants (*Pogonomyrmex* spp.). *Ecol., 60*:686–694.

Wolf, L.L., and F.R. Hainsworth, 1978. Energy: expenditures and intakes. *Chem. Zool., 10*:307–358.

Ziswiler, V., and D.S. Farner, 1975. Digestion and the digestive system. *In* D.S. Farner, and J.R. King (Eds.), *Avian Biology.* Vol. 2, pp. 343–430. Academic Press, New York.

10 *Escape from Predation*

Almost all vertebrates are subject to predation by some other animal and have evolved defenses against attack. Individual defenses have been categorized as primary, those operating even if the prey has detected no predator approaching, and secondary, those brought into play after a prey has detected the approach or attack of a predator (Edmunds, 1974).

Probably no defense is 100 percent successful. Every defense must develop within the limitations resulting from other necessary functions such as eating and reproduction and so might never be perfect; compromises seldom are. Furthermore, at the same time that defenses are evolving, selection favors predators that can beat those defenses. This evolutionary race between eater and eaten is continual and neither may ever win the race permanently. Each new defense selects for new offenses, and new offenses, in turn, select for new defenses.

PRIMARY DEFENSES

A general and obvious primary defense is alertness and the frequent monitoring of the activity of other animals. Virtually all vertebrates survey their surroundings many times during the day, but such behavior is clearly of special importance to those with a high risk of predation. One study suggested that patterns of sleeping in mammals may be correlated with vulnerability to predation (Allison and Ciccheti, 1976). Species considered to be very vulnerable because of large numbers of potential predators or unprotected sleeping sites tend to sleep less deeply and for shorter periods than species that sleep in secure places or are predators themselves. The implication is that vulnerable species have evolved different sleep patterns because they cannot afford to be oblivious of their surroundings for very long.

Retreats

A number of vertebrates have some kind of retreat to which they retire at certain times or in which they live. Burrow- or crevice-dwellers can be found in all vertebrate groups. Protection from predators is often one function of such retreats, although not always the only one. A few vertebrates, such as turtles, armadillos, pangolins, and mailed catfishes (Loricariidae, Callichthyidae), even carry their retreat with them (Fig. 10–1).

Crypsis

Cryptic animals are camouflaged and resemble some part of their environment sufficiently well that predators often fail to distinguish them from their background. Many ground-nesting birds—nighthawks (*Chordeiles minor*), for example, and most female ducks—and copperhead snakes (*Agkistrodon contortrix*) are mottled in shades of brown and often match the dead-leaf litter. Pelagic offspring of fishes are commonly translucent; some arboreal snakes are the color of vines and foliage. A number of green tropical frogs commonly sit on leaves. Several of these frogs reflect near-infrared light (700–900 nm), as do their perches. As a result, they match their background not only in the visible spectrum but in the near-infrared spectrum as well and, in this way, may evade certain predators. Little is known about detection of these wavelengths by vertebrates, however. Local populations of side-blotched lizards and pocket mice (*Perognathus*) have differentiated to match different local backgrounds; individuals on dark soils are dark and those on light soils are pale. Very slow-moving animals such as African chameleons, pottos, and neotropical sloths (*Bradypus, Choloepus*) are likely to be less noticeable to would-be predators because of their extremely deliberate locomotion.

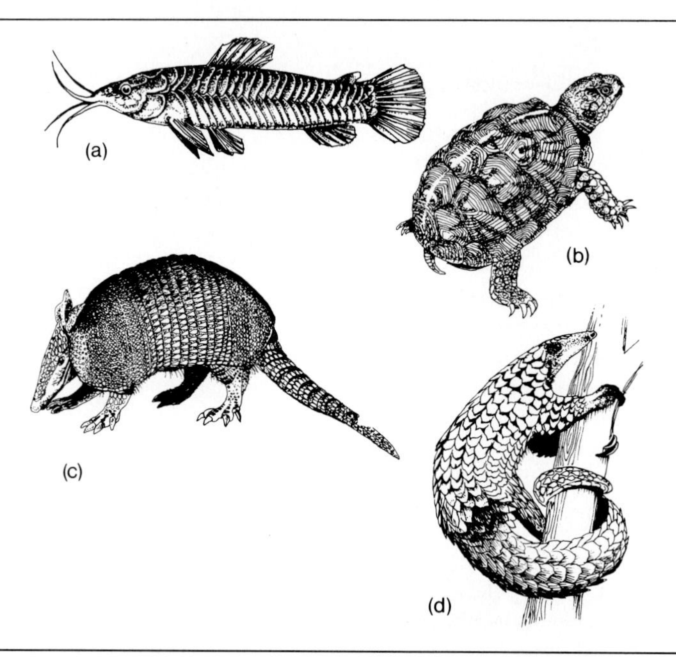

Figure 10–1 The mailed catfish (*Callichthyes*), box turtle (*Terrapene*), armadillo (*Dasypus*), and pangolin (*Manis*). They carry their retreats with them.

Individuals of some species are noted for their ability to change their colors to match a succession of different backgrounds. The change is not generally instantaneous but takes some minutes or even days to be completed. African chameleons are notorious for this ability and several other species, including some of the American *Anolis* lizards and some flounders, also have this ability (Fig. 10–2). Age changes in coloration are notable in deer (such as *Odocoileus*). The fawns are brownish with pale spots that supposedly resemble sunflecks on the ground, thus disguising them when they lie still. Color changes seasonally in some species; for example, certain populations of snowshoe hares (*Lepus canadensis*), ptarmigan (*Lagopus*), and weasels (*Mustela*) are white in winter and brown in summer. The camouflage value of a white covering when snow blankets the ground is evident. Sometimes, however, the seasonal color change may be more closely associated with extremely cold temperatures than with snowfall. Such seemingly unadaptive responses may result from cuing on a proximate factor that no longer relates to the ultimate factor; for example, temperature as a proximate factor may have become inappropriate because of recent changes in snowfall patterns.

Color-matching of the background is commonly accompanied by behavior patterns that enhance the effectiveness of the match. The American

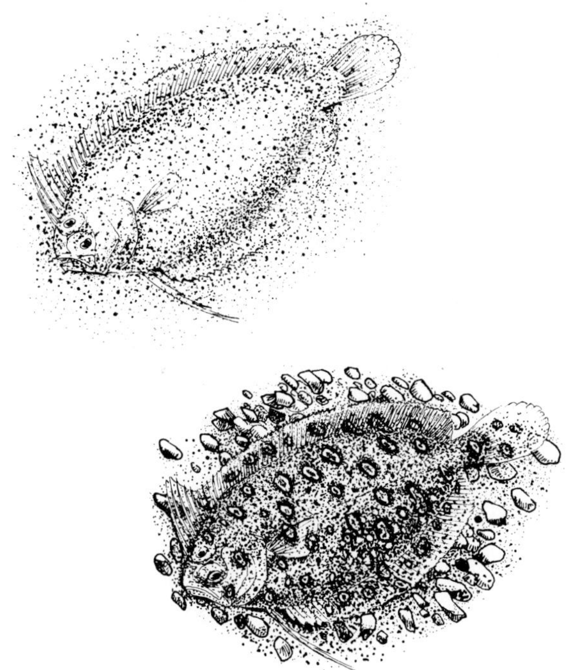

Figure 10–2 Color changes in a flounder on substrates of different graininess and hue.

bittern (*Botaurus lentiginosus*), the cryptic young of many precocial birds, and, in fact, camouflaged species in general usually remain immobile against the background. Some animals select portions of the habitat in which their color-matching is best. In the West Indies, certain species of *Anolis* lizards of different colors live in the same habitat. When startled, each species flees to the portion of the habitat in which it is best concealed—green lizards flee to green vegetation, grey ones to lichen-covered bark, and so on. Australian *Amphibolurus* lizards exhibit a similar response (Gibbons and Lillywhite, 1981). Furthermore, there is evidence for some cryptic *Sceloporus* lizards in Florida and for the green *Iguana iguana* in the neotropics that changes in foraging perches are often timed to coincide with gusts of wind that cause the branches to move. The implication is that movement is less conspicuous when the background is moving too.

The selective advantage of cryptic coloration has been demonstrated frequently in both laboratory and field, mostly with invertebrates. Early experiments with mosquitofishes (*Gambusia patruelis*) showed that a variety of predators (a penguin, a night heron, and a sunfish) preferentially attacked the fishes that were most conspicuous against their background—pale ones in a black tank and dark ones in a white tank (Sumner, 1934) (see Fig. 10–3). Similar experiments have been conducted with mice as prey

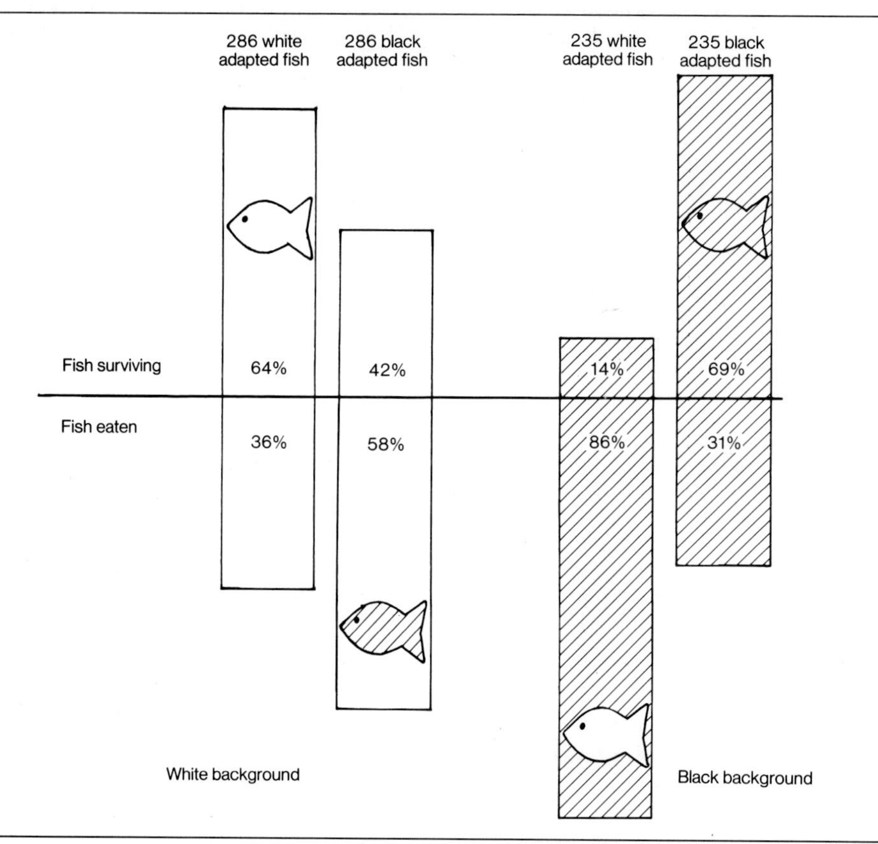

Figure 10–3 Differential predation by penguins on black and white mosquitofishes, *Gambusia patruelis,* seen against dark and light backgrounds. (Modified from M. Edmunds, 1974. *Defence in Animals.* Longman Group Limited, London. Based on data in Sumner, 1934.)

and owls as predators. Kaufman's (1974) work with the oldfield mouse, *Peromyscus polionotus,* which is preyed upon by barn owls (*Tyto alba*) and screech owls (*Otus asio*), will serve as an example. Some populations of the oldfield mouse are composed primarily of pale individuals whereas others have mostly dark individuals; pale individuals are more common in areas of pale soil, and dark individuals are more common on dark soil. Light and dark soil formed the ground covering in two sets of pens containing captive owls. Two mice, similar in age and sex but different in color, were released simultaneously in each of the pens, and the observer recorded which prey the owls captured first. Conspicuous mice were generally caught first, usually within 2 min, although the effects varied with the amount of light available (Table 10–1). More recent work on selective predation by jays on chorus frogs (*Pseudacris triseriata*) and by fishes on gup-

Table 10–1 Numbers of light brown (L) and dark brown (D) oldfield mice captured by owls in experimental enclosures.

*Light class**	*Light soil*	*Dark soil*
1	18L : 19D	12D : 28L†
2	28L : 37D	10D : 32L†
3	11L : 38D†	20D : 26L

* Light class 1 represents nights with half to full moon; class 2 represents nights with less than a half moon; and class 3 represents nights with no moon or heavy cloud cover.
† Statistically significant differences.
(From Kaufman, 1974.)

pies (*Poecilia reticulata*) shows that crypticity is effective in other vertebrates as well, although the interaction may be quite complex (Tordoff, 1980; Endler, 1980).

Color and pattern are also used to disguise the shape of an animal. Many vertebrates are darker dorsally than ventrally. This countershading supposedly reduces the contrast between the shaded undersurface of the animal and the upper surface, which is usually brightly illuminated by light from the sky (Fig. 10–4). An illusory effect of flatness may be created by the countershading, tending to obscure the animal's natural shape. Reverse countershading, with a darkened belly, is found in the Nile catfish (*Synodontis batensoda*), which swims belly-up, but two other fishes that have the habit of swimming upside down have the usual pattern.

The silvery, reflective sides and belly of many fishes combine countershading with camouflage. Crystalline platelets in the scales and skin reflect the light entering the water so effectively that, viewed from below, the fish is virtually invisible against the bright sky above. The platelets are mirrors that reflect light similar in intensity to the light that the observer would receive if the fish were not present. The greatest reflectivity in most silvery fishes is found on the belly, which is viewed against the sky, and gradually diminishes up the sides. The back is usually darker and less reflective because it is normally viewed from above against the background of dark water.

Some fishes of the deep sea produce light, and in many cases, this light is oriented in a direction that contributes to a camouflaging effect. In the hatchetfish (*Argyopelecus aculeatus*), for example, the light-producing organs are aligned along the belly and aimed downward. An elaborate set of reflective surfaces maximizes the downward intensity and diminishes it laterally, so that the fish matches the natural distribution of light in deep water (Fig. 10–5).

Even an apparently uniformly colored animal, however well matched to a background color, may still stand out as a uniform object against a non-

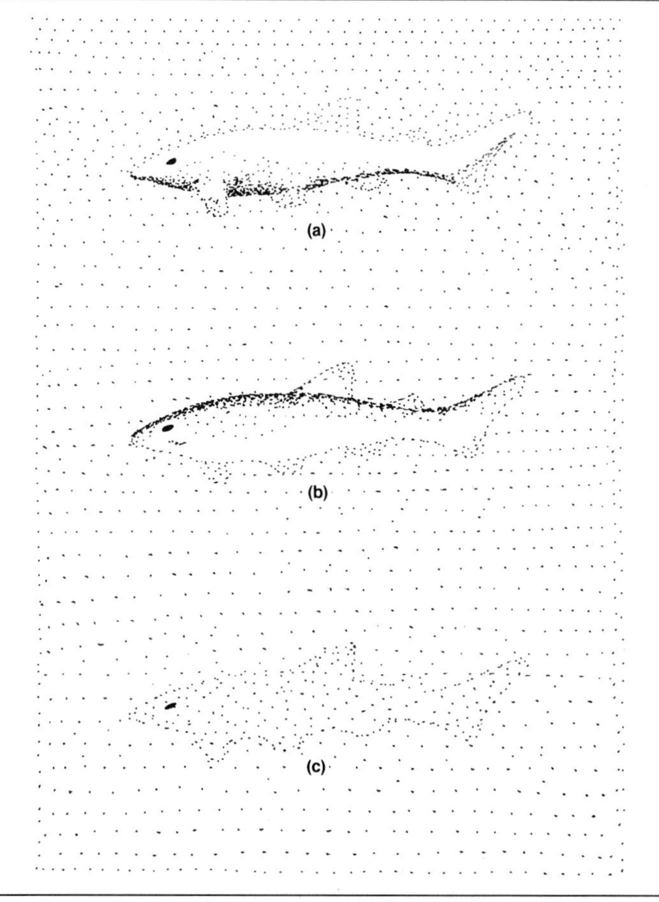

Figure 10–4 The appearance of countershaded and uniformly shaded fishes in different light regimes. (*a*) A uniformly shaded fish with light coming from above. The dorsal surface reflects light and is bright, whereas the venter is shaded. (*b*) A countershaded fish in diffuse light, showing the darker dorsal side. (*c*) A countershaded fish with light from above. The ventral shadow is counterbalanced by the fish's color, rendering the fish inconspicuous against this uniform background. (Modified from M. Edmunds, 1974. *Defence in Animals.* Longman Group Limited, London.)

uniform background. Another modification of color pattern breaks up the outline of an animal; this is called disruptive coloration. The effect of disruptive coloration is greatest when some colors match the background and others contrast sharply, when the different shades do not grade into each other, and when patterns are irregular. Differential predation on young water snakes (*Nerodia sipedon*) in Kansas favored individuals with numerous dark bands and blotches on a paler background color over individuals with fewer marks (Beatson, 1976). In some cases, disruptive color contrasts may give the animal the appearance of some other shape altogether. The body may seem dissected by bright lines on a cryptic surface, or portions of the body may be visually tied together to conceal the true anatomy (Fig. 10–6).

Special color adaptations conceal the eyes in some species. Eyes are essential but rather vulnerable sense organs that may often warrant special protection. Furthermore, because of their distinct shape and shiny surface, they are often quite conspicuous. The eyes may be camouflaged by the presence of other spots around them or by having the body's color patterns repeated on them. Many vertebrates possess black eye masks or stripes; one suggested function for these markings is concealment of the dark pupil.

Warning Coloration

Prey animals that are dangerous or unpalatable to predators are sometimes marked with colors and patterns that advertise the presence and identity of the prey. Predators may avoid such signals automatically or can often learn to avoid such warning colors after a nasty experience. Warning coloration is termed aposematic and usually appears in striking patterns of black, red, and yellow (or white), or black and one of the other colors.

Aposematic coloration is best known in invertebrates, and its adaptive value has been experimentally demonstrated with some insects. In vertebrates, its occurrence is less well documented, although American coral snakes (such as *Micrurus*) and their mimics are classic examples. Skunks carry their black and white colors exposed at all times. Other vertebrates have concealed areas of color that may function as a secondary defense when suddenly exposed. Examples might be the colorful bellies of the North American hognose snakes and mud snakes (*Farancia*).

Mimicry

If two unrelated organisms resemble each other sufficiently, the situation is often one of mimicry. Classically, two forms of mimicry exist. Batesian mimicry is the resemblance of a palatable, innocuous organism (the mimic) to an unpalatable or harmful one (the model). The model is often aposematically colored and, thus, provides warning signals to predators. The mimic is similar in color, but its signal is counterfeit because the mimic is actually edible. Mullerian mimicry involves mutual resemblance between two distasteful or harmful prey; each species in a Mullerian mimicry system mimics the other as a model. In reality, all degrees of intergrade are possible between the two classic extremes. Although the emphasis is on visual resemblance (probably because humans and many predators are visual animals), there is no reason to suppose that mimicry systems using

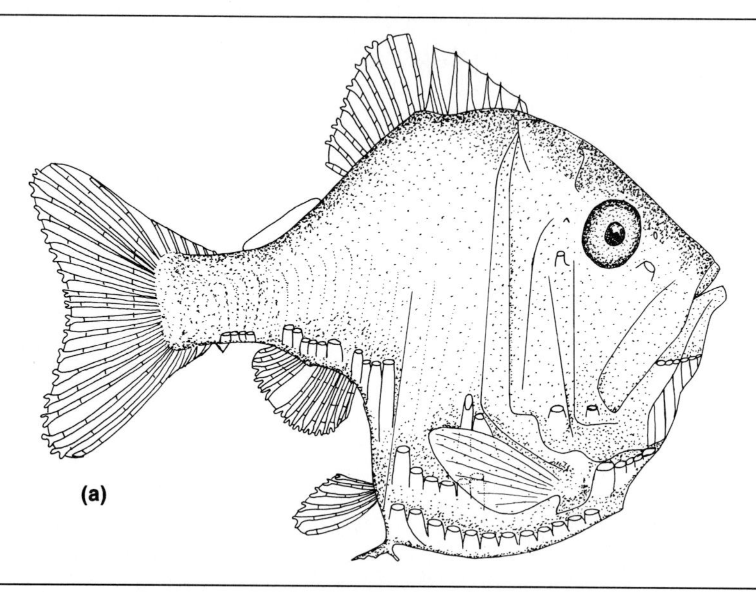

Figure 10–5 The light received by an observer looking at a hatchetfish from different angles. The light reflected by the fish is similar in intensity to that passing by the fish. (Modified from M. Edmunds, 1974. *Defence in Animals.* Longman Group Limited, London.)

other sensory modalities, such as olfaction or audition, do not occur as well. We deal here only with mimicry that functions in escape from predation.

The effectiveness of Batesian mimicry in reducing the risk of predation depends on (1) the ability of the predator to avoid, either innately or through learning, unpleasant prey; (2) the degree of resemblance between model and mimic; (3) the relative unpalatability of the model; and (4), to some extent, the relative abundance of model, mimic, and alternative prey for the predator.

The importance of avoidance by the predator is obvious and has been shown experimentally many times. Most of the experiments have used insect prey; two of these experiments, both from the Browers' work using vertebrate predators, will be mentioned here. The monarch butterfly (*Danaus plexippus*) is mimicked by the viceroy butterfly (*Limenitis archippus*). Adult Florida scrub jays (*Aphelocoma coerulescens*) familiar with the monarch also avoided the viceroy; jays unfamiliar with the monarch readily ate the viceroy. Similarly, toads (*Bufo terrestris*) learned to avoid bumblebees (*Bombus americanorum*) and then also rejected the edible mimic fly, *Mallophora bomboides*.

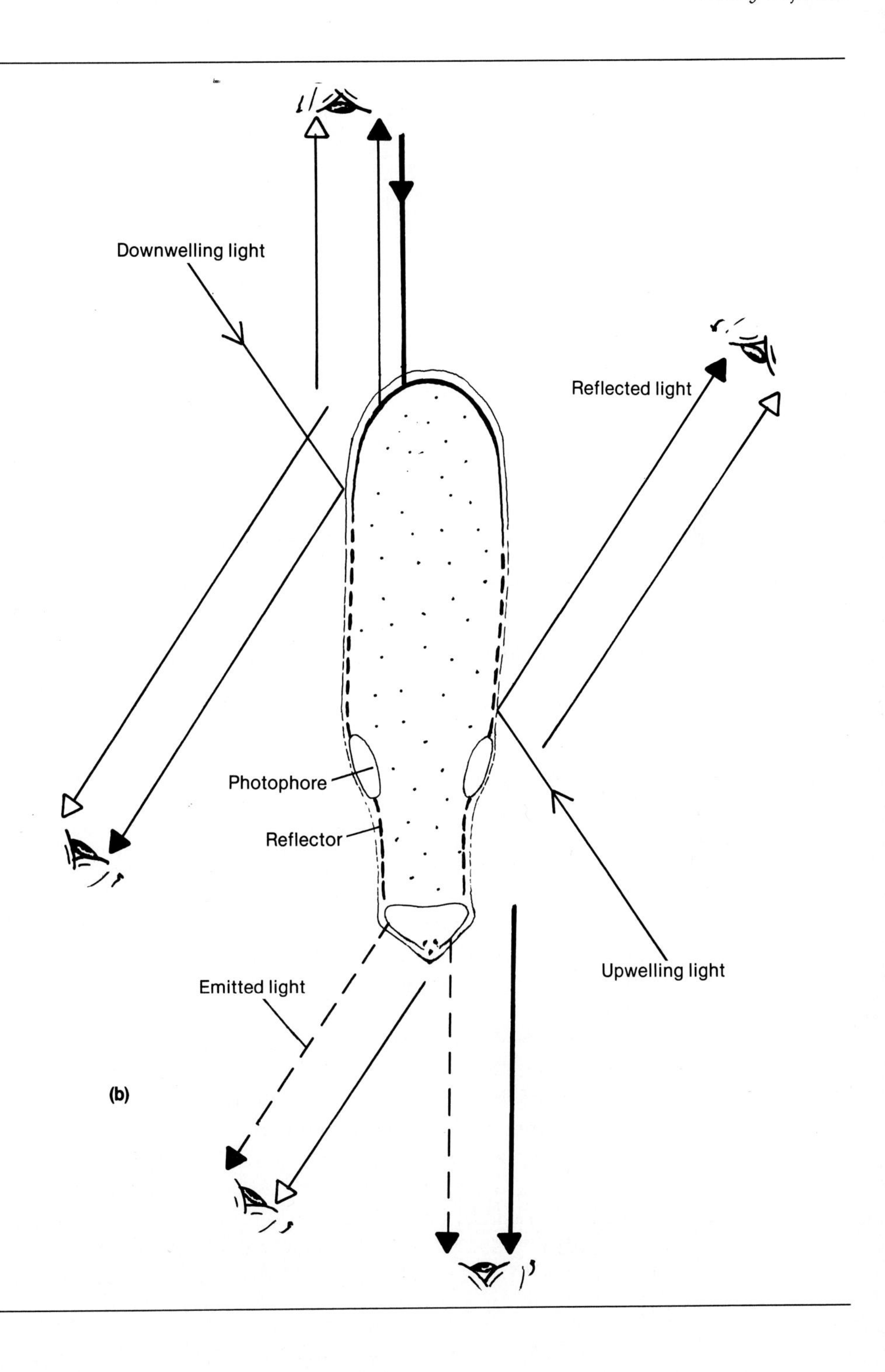
Downwelling light
Reflected light
Photophore
Reflector
Upwelling light
Emitted light
(b)

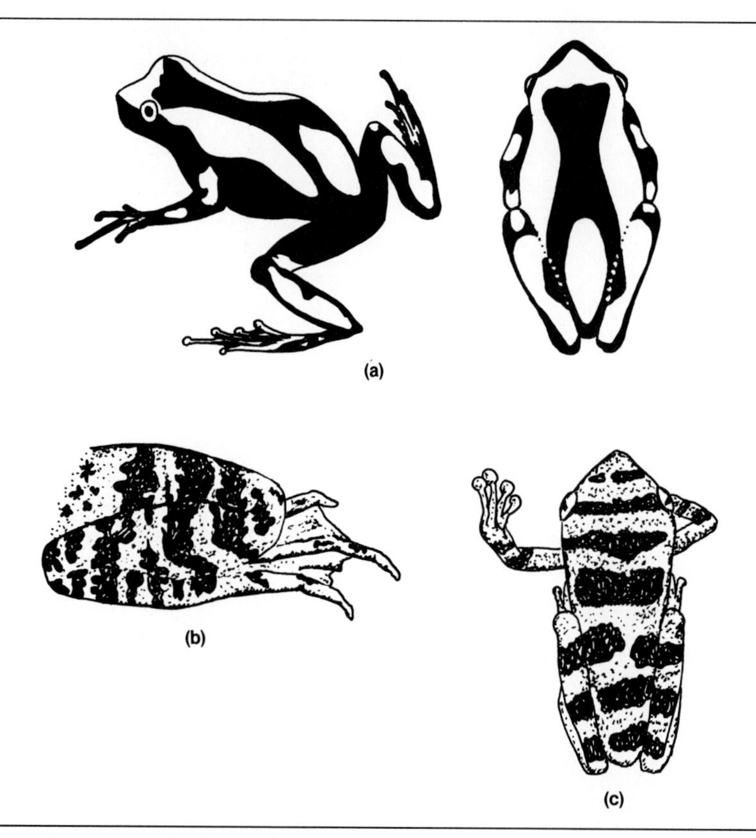

Figure 10–6 Some special effects of disruptive coloration in frogs. Note how body markings can visually link different body parts so that body outlines are obscured. (*a*) *Hyla leucophylla* from South America; (*b*) *Rana temporaria*, from Europe, hind leg; (*c*) *Rhacophorus fasciatus* from the Old World tropics. (Modified from M. Edmunds, 1974. *Defence in Animals*. Longman Group Limited, London.)

Degree of resemblance between model and mimic is also relevant to the effectiveness of mimicry. A number of experiments have shown that any degree of resemblance can render some protection to the mimic, but the amount of protection usually increases with the degree of resemblance. Experiments were done with *Anolis carolinensis* lizards preying on *Photinus* beetles (models), *Tenebrio* larvae (edible controls), and artificial mimics created by gluing *Photinus* wing covers and prothorax to *Tenebrio* (Sexton, 1960). Insects were presented to the lizards in pairs of either mimic plus model or mimic plus control (Fig. 10–7). In all cases, the more complete mimics, with both prothorax and wing covers, were better protected than poorer mimics, with either wing covers or prothorax. In addition, mimics presented at the same time as the real, aposematic model received less protection than mimics presented in the absence of the model. Presumably, *Anolis* could discriminate better when both prey types were seen simultaneously.

Relative unpalatability of the model often influences the degree of protection received by a mimic. The cost to the predator of sampling a totally unpleasant prey is greater than the cost of sampling a somewhat unpleasant one, so predators are less likely to completely avoid prey that are only slightly nasty. The Browers' toads could learn to avoid both bumblebees and honeybees (*Apis*). (Both are models for flies that mimic them.) Bumblebees have more bristles and a more potent sting, however, and may be a more unpleasant prey. Toads that had learned to avoid both models also avoided their mimics, but although bumblebee mimics escaped predation 93 percent of the time, only 46 percent of the *Apis* mimics got away. If we assume that the toads could discriminate equally well between model and mimic in both cases, these data suggest that the nastier effects of attacking bumblebees conferred added protection upon their mimics.

If a model is extremely unpleasant, it remains an effective model even when it is rare relative to the mimic; the predator is so impressed by its noxiousness that the effect needs little reinforcement. If a model is only slightly unpleasant, however, a predator needs numerous encounters to get the message, and mimics then derive little protection unless the model is fairly abundant relative to the mimic. If mimics become common, predators often encounter and sample them and may learn to discriminate between mimic and model. In this way, the predator's behavior creates selection favoring better and better mimicry by differentially eliminating the poorer mimics.

Experiments with mimicry normally present a predator with choices between a limited array of prey types. Wild predators in natural circumstances, however, usually encounter a wide range of possible prey types and, thus, do not necessarily concentrate their feeding efforts on model-mimic systems. As a result, they may encounter both model and mimic less often than in experimental situations. In a natural environment, learning may be slower, more "mistakes" may be made, and other choices are available. Therefore, the degree of protection derived by the mimic may be reduced.

Batesian mimicry systems can be disadvantageous for the model. When a predator has learned to distinguish the edible mimic, it may then begin to attack the model too. Selection then would favor increased repellency of the model and decreased resemblance to the mimic. In short, the model tends to evolve away from the mimic, making good mimicry more difficult to achieve.

Some Batesian mimics are known among vertebrates. The poison-fang blenny (*Meiacanthus nigrolineatus*) is avoided as food by several kinds of predaceous fishes. A different blenny (*Ecsenius gravieri*) in the Red Sea is

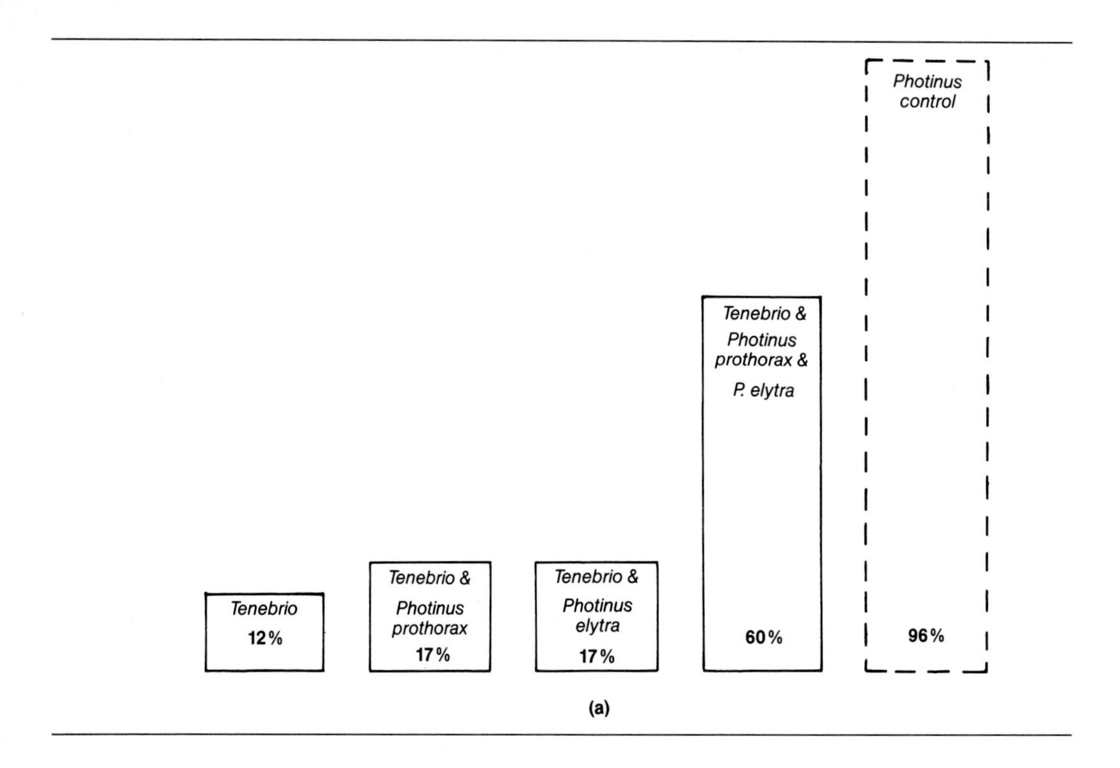

acceptable food and mimics the poison-fang blenny in coloration. Predatory fishes experienced with the poison-fang blenny also reject *E. gravieri.* Other members of these genera in the Pacific and Indian oceans have similar relationships. The weeverfishes (*Trachinus*) have poisonous spines on their fins and gill covers and lie buried in the sand. When a weever is disturbed, the black dorsal fin is erected as a warning to potential predators. The harmless common sole (*Solea*) has a black pectoral fin; it too lies in the sand and erects the black pectoral when disturbed and may be a Batesian mimic of *Trachinus.*

The salamander *Plethodon jordani* produces unpleasant skin secretions that usually deter attacks of experienced avian predators. This salamander, for some reason, has red check patches. Another salamander, *Desmognathus ochrophaeus,* lives in the same area, and individuals of these populations also have red check patches. The mimicry by *D. ochrophaeus* has been shown experimentally to be effective in deterring attacks by birds if those predators have had prior experience with *P. jordani* (Brodie and Howard, 1973).

Red salamanders (*Pseudotriton*) are also thought to mimic certain age classes of the newt *Notophthalmus viridescens* in eastern North America. The life history of the aquatic newt in many locales includes a terrestrial juvenile phase of varying length. The juveniles (called efts) have a noxious skin

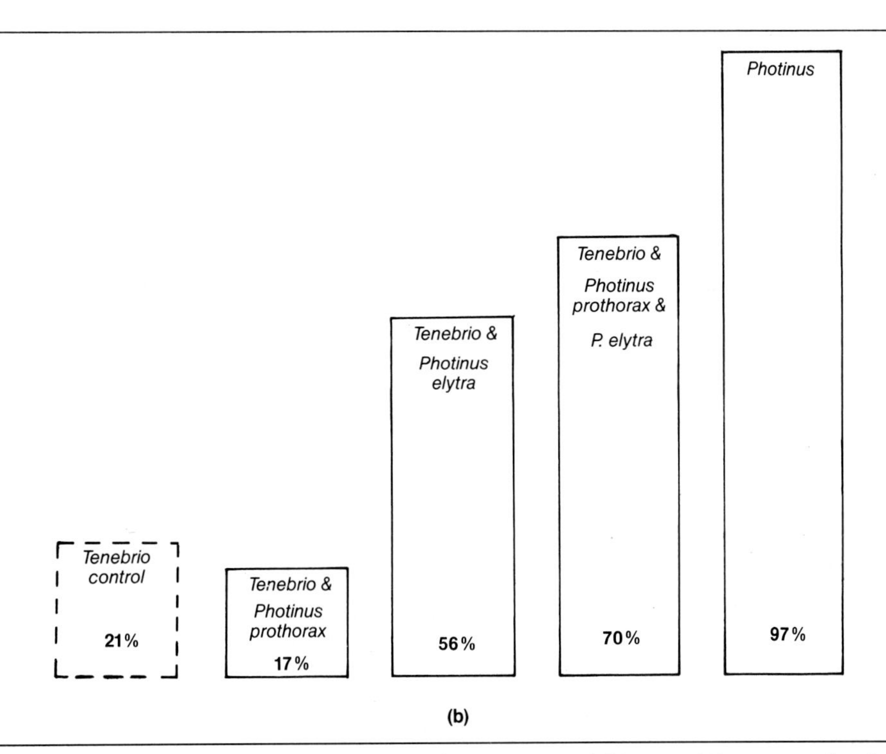

Figure 10–7 Effects of closeness of resemblance of mimic to model in providing protection from predation by *Anolis* lizards. (*a*) Survival when presented with *Photinus* beetles (the model). (*b*) Survival when presented with *Tenebrio* larvae (edible controls). (Modified from M. Edmunds, 1974. *Defence in Animals.* Longman Group Limited, London.)

secretion and are typically red in color. Cool, damp forests provide an ideal habitat for efts, which are often common and active above the ground litter. Red salamanders appear to mimic efts in areas of sympatry, where efts are common and conspicuous. In allopatry or where efts are rarer and spend a smaller proportion of their life in the eft stage, however, red salamanders tend to be cryptically colored (Huheey and Brandon, 1974). More recent work, however, indicates that red salamanders are not fully palatable and, therefore, may be less Batesian than Mullerian mimics. The red morph of *Plethodon cinereus* also seems to mimic efts and is taken less frequently by bird predators than nonmimetic types (Brodie and Brodie, 1980).

Young *Eremias lugubris* lizards (family Lacertidae) in the Kalahari Desert seem to mimic—in color, walking posture, and size—a species of sympatric, noxious beetle (Fig. 10–8). Juvenile *E. lugubris* are black with white stripes and move about with a stiff, arched posture very unlike the lateral

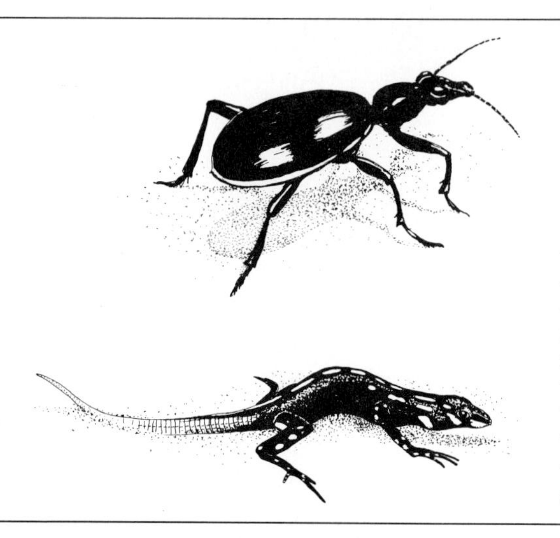

Figure 10–8 *Eremias lugubris* lizard and model beetle. The juvenile lizard may mimic the noxious beetle's size, color, and even walking posture. The lizard's tail, which has no counterpart in the body of the beetle, matches the background color of the sand.

undulation that is typical of quadrupedal lizards. Human observers occasionally mistake these young lizards for beetles; presumably, other visual predators could do so too. Atlhough juvenile *E. lugubris* forage actively in the open, they show fewer visible signs of predator attacks (such as broken tails) than do other young *Eremias* lizards in the region (Huey and Pianka, 1977).

Mullerian mimics are all unpleasant. By resembling each other they reduce the probability of attack by their common predators because the predator only has to learn to avoid one aposomatic color pattern. This form of mimicry is especially well known among insects and its occurrence among vertebrates is uncertain. Probably the best-known mimicry system in vertebrates involves the coral snakes (Elapidae). Just how the system works has been debated, but one suggestion includes Mullerian mimicry.

The coral snakes and coral snake mimics (Colubridae) of the Americas are commonly banded in black, red, and white or yellow (Fig. 10–9); bicolored (black and one of the other colors) forms are also found. True coral snakes such as *Micrurus* are extremely poisonous. They are small and incapable of biting large objects, but they hold on tenaciously in a successful strike. Many of the colubrids are harmless, but some, such as *Erythrolamprus aesculapii,* are mildly toxic.

Although some researchers have supposed that the coral snakes are the models and all the others are mimics, one suggestion is that the mildly venomous species are the models and both the harmless and the very toxic species are mimics. It is argued that the very poisonous coral snakes would probably kill a predator, whose learning process and future avoidance of coral colors would thus be impaired, and the snakes would derive no protection. If a predator attacked the aggressive but less venomous colubrids, however, it might live to exercise its knowledge; therefore, both harmless colubrids and venomous elapids may be protected. The various colubrid models are thus viewed as Mullerian mimics among themselves. Batesian models for the harmless colubrid forms, and models for the dangerous elapid mimics.

The classic view, in which the true coral snakes are viewed as the models and all the others as mimics, has received some support in recent years, however (Greene and McDiarmid, 1981). Smith's (1975, 1977) experiments with great kiskadees (*Pitangus sulfuratus*) and turquoise-browed motmots (*Eumomota superciliosa*) demonstrated that hand-reared young birds that had never been exposed to coral snakes had an apparently inborn aversion to the coral-snake pattern. Adults of both species are known eaters of reptiles. Such unlearned avoidance could result from selection to avoid risky attacks on coral snakes, in which case the colubrids, both harmless and mildly toxic types, would still be Batesian mimics of the elapids.

It is also possible that no aposematic mimicry is involved at all. If avian predators innately avoid bright, ringed patterns simply because the pattern is unusual, convergence on this kind of pattern could occur without mimicry. Furthermore, although it is customary to think of aposematic colors as conspicuous, they may not always be. In the sun-flecked litter on the forest floor, the contrasting pattern may even be concealing (Greene and Pyburn, 1973). A North American colubrid (the milk snake, *Lampropeltis triangulum*)

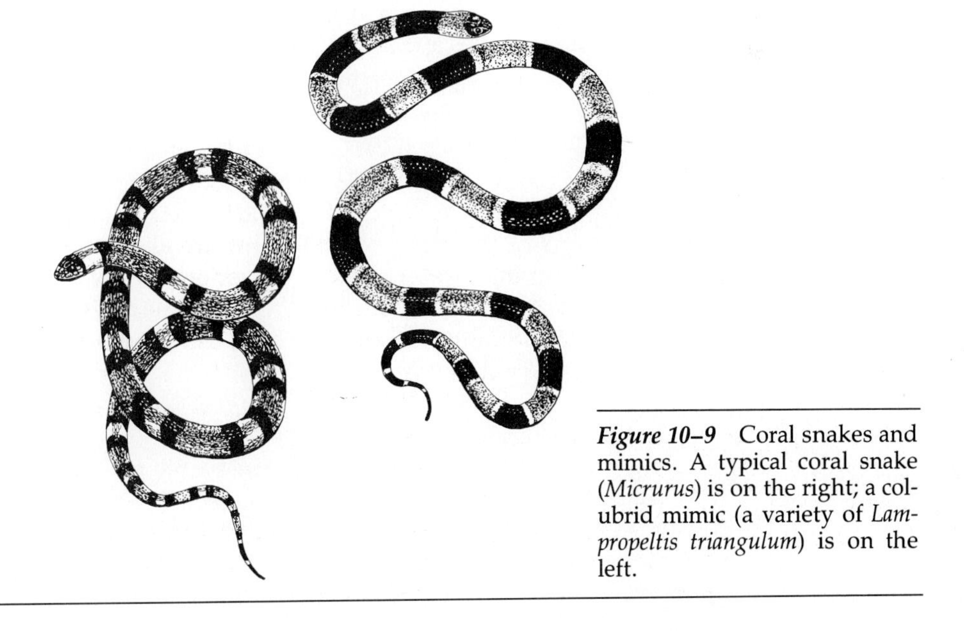

Figure 10–9 Coral snakes and mimics. A typical coral snake (*Micrurus*) is on the right; a colubrid mimic (a variety of *Lampropeltis triangulum*) is on the left.

that has warning patterns but is not sympatric (in the northern part of its range) with coral snakes might be interesting: Its predators would not have been subject to selection for avoidance of coral snakes as dangerous prey. Experiments with these animals might allow us to discriminate between the hypotheses of selection for avoidance of danger and avoidance of novel prey.

No good examples of mimicry in aposomatic colors are documented for birds and mammals, although further study may unearth some. The unpalatable tree-shrews (Tupaiidae) of southeast Asia may be mimicked by squirrels. Edmunds (1974) suggested that African ant-thrushes (*Neocossyphys*) have flesh that smells unpleasant and may be rejected by some predators. They may serve as models for the very similar looking but highly edible rufous flycatchers (*Stizorhina*).

Defensive Associations

Interspecific associations are known to function in the defense of vertebrates. Fishes have perhaps the greatest variety of defensive associations so far reported; a few examples are given here. Several species live in association with sea urchins, down among the spines. *Diademichthyes deversor*, a clingfish, and *Aeoliscus strigatus*, a shrimp fish, inhabit the spiny thicket of *Diadema* urchins and have color patterns and postures that make them relatively inconspicuous there. *Nomeus gronovii* lives among the tentacles of the Portuguese man-of-war (*Physalia*) and is colored with blue and gray stripes that match the blue tentacles of the coelenterate. The fish is partially immune to the stinging cells of the man-of-war, but it seldom bumps against the tentacles and, thus, seldom discharges the stings. Many pomacentrid fishes live in association with sea anemones. Clownfishes (*Amphiprion*) become acclimated to their particular kind of anemone by making tentative approaches and by gradually acquiring a covering of mucus from the anemone; anemone mucus prevents the discharge of stinging cells by the anemone and presumably evolved to prevent the anemone from stinging itself. A Mediterranean goby, *Gobius bucchichii*, is also an anemone fish. Relationships between the fishes and their associates vary greatly, although the fishes probably always derive protection. Some fishes feed their anemones, others nibble their host's tentacles, and still others behave neutrally toward their associates (Mariscal, 1972). (Fig. 10–10).

Defensive associations are found among birds. The amazingly complex case of the giant cowbird and its host caciques and oropendulas is presented in Chapter 15. The sandwich tern (*Sterna sandvichensis*) usually nests in colonies of black-headed gulls (*Larus ridibundus*) or arctic terns (*S. paradisea*) and relies in part on the social defenses of its hosts for protection from certain predators. Some neotropical wrens (Troglodytidae)

Figure 10–10 Some fishes and their invertebrate associates. (*a*) *Nomeus* with a Portuguese man-of-war and (*b*) *Amphiprion* with an anemone.

nest frequently in *Acacia* trees inhabited by protective ants and sometimes wasps. An African weaver (*Ploceus heuglini*) is reported commonly to nest in trees where red weaver ants (*Oecophylla longinoda*) also live: The ants attack nestlings that fall to the ground but not those that remain in the tree. Yellow warblers (*Dendroica petechia*) suffer less nest predation when they nest near a grey-catbird (*Dumetella carolinensis*) activity zone. Proximity to red-winged blackbirds reduces brood parasitism by brown-headed cowbirds (*Molothrus ater*) (Clark and Robertson, 1979). Further study will undoubtedly reveal many more close associations of vertebrates with protective insects or other animals, but few cases are well documented at present.

Escape in Time

Heavy predation at some times of day or year could result in shifts of activity to another time or place. One possible selection factor—but not the only one—that contributes to the evolution of nocturnality in so many vertebrates might be the increased cover provided by the darkness of night. Of course, predators have also become adapted to nocturnal activity. Many nocturnal predators are reported to see better in daylight than at night, however, leading to the inference that they hunt at night because that is when their prey is active. Their hunting would probably be more effective if their prey were diurnal. Evolutionary shifts in seasonal cycles are an obvious possibility too, but I know of no vertebrate data that fit this pattern.

Predation pressures can select for adaptive shifts in the scheduling of life-cycle events such as reproduction. Prey can sometimes avoid heavy predation by being present at only irregular, unpredictable times. Irregularity is more likely to be a successful tactic when the predators are quite specialized to a particular prey species, because specialists can switch to alternate prey less readily than dietary generalists and, therefore, have less chance of surviving a period of prey scarcity. Thus, selection sometimes may favor the evolution of an irregular reproductive schedule as one means for parents to reduce the risk of predation on their young. In conjunction with an unpredictable schedule, it may also be advantageous to produce huge quantities of offspring when reproduction does occur and then sacrifice some young to predators so that they become satiated and allow the remaining young to escape. Production of enormous batches of young usually means that those young cannot be very large, so this escape mechanism brings with it attendant changes in many aspects of reproduction. The hypothesis of predator satiation has received some support from studies of both plants and insects (Janzen, 1971). Some vertebrates may yet be shown to fit the pattern. Synchronized breeding within any population in an area could reduce the risk of predation to the young of any parent, as long as the predators are incapable of raising the predation level proportionately with the increase in prey. Highly unpredictable breeding schedules in vertebrates often seem to be constrained by climatic factors more than predation, but the possibility of predator satiation by vertebrates has been little studied.

Aspect Diversity

Many vertebrate predators are thought to increase their hunting efficiency by means of a search image. A search image implies that the predator has learned how to find and recognize a particular prey type quickly in much the same way as a human in a grocery store can visually sort out a familiar brand of coffee from a shelf containing many brands: One looks for, say, a yellow can with brown lettering; the rest are scarcely seen at all. Predators could form search images by encountering a prey type several times in succession. Becoming more and more familiar with its appearance and where to find it, the predator may gradually come to select that prey out of proportion to its abundance and availability. Clearly, a search image could only be selectively advantageous if foraging this way were profitable.

Croze's (1970) experiments with carrion crows (*Corvus corone*) showed very nicely that search images do function in efficient foraging. The crows foraged over an experimental beach area on which Croze had placed food underneath mussel shells painted in various colors to resemble stones on the beach. Once a crow had found food under a shell of one color, it usually turned over a series of shells of that color, ignoring the others. Once the search image was formed, it could be switched to another color or shell type by eliminating the food reward from the first shells; it was easily restored to the first color by providing occasional rewards. Thus, the search image was persistent as long as some reward was obtained, but new images formed when searching under new shell types was also rewarded.

The use of search images by predators opens a different avenue for escape by the prey. Any prey that looks different from the rest may have an increased chance of escape, as long as the predator is using a search image in foraging. Aspect diversity refers to the different appearances of the prey. Note the contrast between the convergence of morphology and color in mimicry systems, which normally involve warning coloration, and the divergence of the same characteristics in achieving aspect diversity, which usually involves cryptic prey. We might find that cryptic prey species that occur in similar places have been selected to look different from each other. We might even discover that predation pressures have favored the evolution of polymorphic variability within a species.

Such intraspecific variability might span a whole range of sizes, shapes, and colors, but then a predator could more easily generalize from one prey individual to another because it would continually encounter intergrades between variants. We might expect, therefore that intraspecific variability (in the present context of predation) would be disjunct, or discontinuous; that is, several distinct types or morphs would exist within a species with little or no intergradation between them. Distinct morphs would make it more difficult for a predator to form effective search images. For instance, carrion crows hunting for eggs often bypassed experimental eggs that differed in color from the rest. Under these circumstances, any parents capable of producing distinctively different offspring would have an advantage over those that could not. Polymorphism within one locale is not conspicuously common among prey species of vertebrates, but it is known to occur in some fishes such as the three-spined stickleback (Chapter 1). It is difficult to say much more about aspect diversity and polymorphism in prey vertebrates except to note the possibility.

A case of interspecific aspect diversity may be provided by the sandwich tern that nests in colonies of black-headed gulls and may derive some protection of its eggs from aspect diversity. The tern colonies are contained within the gull colonies, and the two species have very different sanitation habits. The gulls usually defecate outside the colony and carry their eggshells away. The terns do not: Their colony is spattered with white. Tern eggs are disruptively marked and, thus, camouflaged against a matching background; gull eggs, however, are cryptic on an unmodified backdrop. Carrion crows are avid predators of gull eggs and appear to form search images not only for the appearance of the eggs but also for the appearance of the place in which they can be found. The eggs of the terns, therefore, are protected by their location in a place that looks different, being decorated with white splotches (Croze, 1970).

Polymorphism is marked in a number of raptorial birds, including the screech owl (*Otus asio*) and many hawks. These polymorphisms usually involve a dark and a pale morph, which may be of either sex. One function of the color difference may be related to predation. Paulson (1973) surveyed many hawk species and found that the polymorphic colors usually occurred on the ventral side, the side exposed to the prey below. Moreover, polymorphism was more common in hawks that preyed on birds and mammals than in those that preyed on fishes, amphibians, or reptiles. Paulson speculated that birds and mammals may be more capable of forming search images than are other vertebrates. Thus, hawks preying on birds and mammals may elude detection by their prey to some degree by their aspect diversity, a nice reversal of the prey avoiding the predator. Arnason (1978) showed that the rarer light-colored morph of the parasitic jaeger (*Stercorarius parasiticus*) in Iceland was more successful than the commoner dark morph in pirating captured fish from puffins. Puffins may lose as much as half their weekly catch to these pirates, primarily to the common dark jaegers, and may recognize the light morph less readily.

SECONDARY DEFENSES

Hiding

Many vertebrates withdraw to their retreats when under attack by a predator. The pearlfishes that live in sea cucumbers dash tail first up the cucumber's anus when frightened by a predator. Turtles pull their heads and appendages into the shell, composed of a dorsal carapace and a ventral plastron. North American box turtles (*Terrapene*), the African side-necked terrapin (*Pelusios*), and tortoises of the African genus *Kinixys* have hinged sections of either the carapace or the plastron that can be closed protectively.

Armored creatures may change their posture to create a retreat of sorts while under attack by a predator. Spiny mammals such as the echidnas and hedgehog (*Erinaceus*) roll up into a ball, spines outward, to protect the vulnerable head and belly. American porcupines (*Erethizon*) turn their backs on the approaching predators and hunch up to expose the spines; they lash their spiny tail in the face of any predators that come too close. Armadillos and pangolins also roll up, presenting an approaching enemy with a wall of armor plates (Fig. 10–11). The fairy armadillo or pichichiego

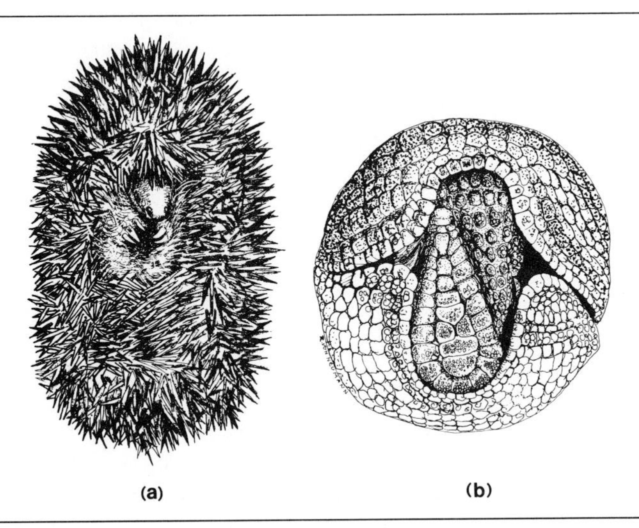

Figure 10–11 Rolled-up defense posture of (*a*) hedgehog and (*b*) armadillo.

(*Zaedyus pichiy*) of Argentina even has bony plates on its rear end; presumably they guard the rear while the armadillo digs.

Of course, these measures are not always effective. Some predators have learned to flip porcupines over on their backs in order to get at the unprotected belly. Others can follow their prey into retreats. Some weasels can follow mice or voles underground, and badgers (*Taxidea taxus* in North America) make their living unearthing ground squirrels and other small, burrowing mammals.

Flight

Straight-line running is a common escape technique, especially for cursorial vertebrates. Moose (*Alces americana*) and gazelles (*Gazella*), for example, frequently try to outrun their predators. Zigzag locomotion, with rapid and erratic changes of directions, is used by many species, especially when pursued by very fast predators. Cottontail rabbits (*Sylvilagus*) are great dodgers, as are other rabbits and hares. Mullet (*Mugil cephalus*) often cruise at speeds greater than those of maximum efficiency, perhaps as a means of reducing predation, and often leap out of the water. Lophotid fishes conceal their getaway in a cloud of inky material.

Both the American kangaroo rats (*Dipodomys*) and the Afro-Asian gerbils (*Meriones*) are saltatory rodents of desert habitats, preyed upon heavily by owls and snakes. These predators are characterized by their ability to make an almost silent approach. Both gerbils and kangaroo rats have greatly enlarged middle ear cavities and extremely sensitive hearing, especially in the low-frequency range, which enables them to detect their stealthy predators (Chapter 4). As the predator nears, the rodents rapidly leap away. Deer mice (*Peromyscus*), which have smaller ear cavities, young gerbils whose ear cavities have not yet reached adult size, and kangaroo rats with experimentally reduced cavity volume are all less successful in evading capture.

Some vertebrates that flee from approaching predators exhibit flash behavior—the exposure of bright colors that are hidden when the animal comes to rest. Predators can become confused by the suddenly alternating appearance and disappearance of color; they may find the color easy to follow, but the prey seems to disappear when it quickly adopts a posture that hides the flash colors and exposes only cryptic colors. The white tail of many rabbits may function in just this way. Gazelles, pronghorn, and some deer display white tails or rump patches while fleeing. Gazelles and a South American rodent, *Dolichotis patagonicum,* indulge in a peculiar, jerky locomotor pattern called stotting. Stotting and the white posterior signal may function as pursuit invitations by coaxing a nearby predator to give chase under conditions favorable to the prey's escape, or as tests of the seriousness of a predator's approach.

Seghers (1974) reported that guppies (*Poecilia reticulata*) in Trinidad occur in populations well isolated from each other and have different responses to predators in different areas. Guppies living where many predatory fishes but few avian predators are found respond to the presence of an aquatic predator by jumping; they respond to an avian predator by rapid scattering and resurfacing. In other areas with fewer predatory fishes but many predatory birds, guppies respond to an object passing overhead by gently sinking and remaining motionless for several seconds. Since piscine predators are few, the risk is apparently small.

Threat

Threat postures may be valid warnings to a predator that a prey is a nasty customer. Such may be the case for the American spotted skunk, *Spilogale putorius,* which stands up on its front legs, exposing its full aposomatic regalia, ready to spray its powerful scent. Other threats are bluffs. The bristling of fur by mammals and feather-fluffing by birds increase the apparent size of the animal without changing their actual defensive ability. Fox snakes (*Elaphe vulpina*) and many other snakes rattle their tails in the ground litter, often producing a rattling sound similar to that of a rattlesnake. Growling, hissing, and snarling are distinctly threatening vocalizations. The egg-eating snake, *Dasypeltis scabra,* mimics three different kinds of deadly vipers: *Causus* spp., the night adders; *Bitis caudalis,* the horned viper; and *Echis carinata,* the carpet viper. These vipers are sometimes

Mullerian mimics of each other, not only in color pattern but also in a rustling, swirling display used as a threat to an intruder.

Animal eyes are known to be threatening organs capable of eliciting avoidance responses by many predators. This phenomenon has been put to good use by numerous insects, which present eyespots to approaching predators with a high frequency of success in deterrence. Threatening eyespots are less well known in vertebrates. However, Edmunds (1974) presents one example of an anuran (*Physalaemus nattereri*) in Brazil that has enormous eyespots on its abdomen that are visible from the rear (Fig. 10–12). At least in some populations, the usual response of this toad to a predator is the presentation of its rump with the eyespots exposed.

Becoming Hard to Handle

Porcupine fishes (Diodontidae), some lizards (*Chamaeleo*), and many anurans quickly puff themselves up to very large sizes when under attack. This not only makes them look bigger but also makes them hard to swallow, although schoolmasters (*Lutianus*) have been found to contain inflated specimens of porcupine fish. The hognose snakes are toad-eaters and puncture inflated prey with large rear teeth mounted on a movable upper jaw. The chuckwalla scuttles to a rocky crevice where it inflates enor-

Figure 10–12 Eyespots on the Brazilian toad *Physalaemus nattereri.* In some populations, the toads readily display these eyespots on the rump when they are molested.

mously, wedging itself in so tightly as to defy extraction. Human predators learned to counter this tactic with a jab of a sharply pointed stick to let the air out.

Another tactic used to become hard to handle is that reported for *Chordylus* lizards in the Old World. These lizards apparently do what the legendary hoop snakes did—namely, seize their tails in their mouths. Unlike the hoop snake, *Chordylus* does not roll away but simply presents no convenient end for a predator to engulf.

Deflection and Diversion

Some fishes, such as *Chaetodon capistratus*, a West Indian butterflyfish, have eyespots on a fin or near the tail, which may serve to deflect the attack of a predator to the "wrong" place (from the predator's point of view). Similarly, certain snakes, such as *Eryx johni*, the Indian sand boa, may misdirect the attack of predators by raising and waving the tail.

Another form of distraction is the autotomy (self-breaking) of tails practiced by many lizards, some salamanders, and some mice. A predator pounces, seizes the tail, and is left with just the tail in its mouth as the now-tailless prey flees for cover. In the case of lizards, the tail often twitches and jerks with an apparent life of its own and probably distracts the predator long enough to give the prey extra escape time. In *Coleonyx variegatus*, a species of gecko from the southwestern United States, experiments have shown that the tail autotomy does indeed reduce the risk of predation significantly. In fact, geckos were observed to flaunt their tails conspicuously in the presence of the predators (night snakes, *Hypsiglena*), and the snakes often oriented their strikes to the waving tail. Gecko tails are enlarged and full of stored fat, increasing their conspicuousness as well as serving as an important energy reservoir when intact (Dial and Fitzpatrick, 1981). Lizards can regenerate their broken tails quite rapidly, often in a matter of weeks. Regeneration is more rapid and more costly of energy in species of lizards that use the tail especially for defense and in species that have high reproductive rates (Vitt et al., 1977). It would be intriguing to discover whether or not predators might exploit lizard tails as a renewable resource and whether the nutrients available in tails result in a significantly decreased number of lizards attacked (Congdon et al., 1974).

Distraction displays of many avian parents (such as the killdeer, *Charadrius vociferus*) are used when a predator approaches the nest or the young. The adult flops about as if it had broken wings or legs, looking like an easy mark but moving away from the nest and always just ahead of the predator. The predator is thus drawn away from the defenseless young. The bowfin (*Amia calva*) reportedly thrashes about as if injured when predators near its young. The intensity of the display should increase in proportion to the cost of replacing the brood.

Feigning Death

American opossums (*Didelphis virginiana*) and the African ground squirrel (*Xerus erythropus*) are examples of mammals that "play dead" when harassed by a predator. Hognose snakes also may lie belly up as if dead. Predators may then relax their attention, allowing the prey to escape, or the killing response may not be elicited by the unmoving prey. A patient predator, however, will often find that the "dead" animal peeks out from under its eyelids to see if the predator has gone or persists in rolling over on its back again when it is righted.

Fighting and Chemical Defense

Claws and hooves, horns and antlers, teeth and beaks, wings and tails are multifarious organs frequently used in defensive fighting. The spines of fishes such as sticklebacks (*Gasterosteus* and *Pygosteus*) are effective feeding deterrents to predators such as pike (*Esox*). *Esox lucius* preferred nonspiny *Phoxinus* minnows over the somewhat spiny *Pygosteus,* which in turn was eaten in preference to the very spiny *Gasterosteus* in an experimental situation (Hoogland et al., 1957).

Many snakes (among them garter snakes, *Thamnophis*) defecate odoriferously when handled, a tactic that tends to discourage predators. American howler monkeys (*Alouatta*) are well known for the hail of branches, fruits, and feces with which they sometimes assail a terrestrial intruder. Black terns (*Chlidonias niger*) nesting in freshwater marshes "dive-bomb" intruders that enter the colony by swooping low and defecating.

Chemical defenses come in many forms. Numerous fishes have venomous spines that deter predators; rabbitfishes (Siganidae), for example, have 13 dorsal, 4 pelvic, and 7 anal spines that deliver poison. Stargazers (Uranoscopidae), stonefishes and scorpion fishes (Scorpaenidae), and toadfishes (Batrachoidididae) are among the venomous teleosts. Chimaeras, piked dogfish (*Squalus acanthias*), stingrays (Dasyatidae), eaglerays (Myliobatidae), and cow-nosed rays (*Rhinoptera*) have venomous spines as well. Many anurans (such as *Dendrobates* and *Rana palustris*) have poisonous or distasteful skin secretions that kill or severely irritate would-be predators. The skin and entrails of pufferfishes (Tetraodontidae) are toxic at least to some predators. Several species, such as the African antthrushes, may have distasteful flesh. Chemical defenses of skunks involve the spraying of a pungent, irritating cloud of fluid in the direction of a predator. The spitting cobras (the ringhals, *Hemachatus haemachatus;* the black-necked cobra, *Naja nigricollis;* and some Indian cobras, *N. naja*) can forcibly eject from the tips of their fangs a substance causing temporary or even permanent blindness if it reaches the eyes of an approaching predator; the drops are said to travel over 2 m.

Social Defenses

Small birds, such as blackbirds (*Agelaius*) and kingbirds (*Tyrannus*), often mob raptors. Mobbing is generally a contagious, group activity involving numbers of attackers. An owl, sitting peacefully in a tree during the day, may be surrounded by large numbers of jays and other birds, all screaming and hopping about, sometimes making feinting attacks. A hawk flying overhead may be pursued by several blackbirds that whiz closely past the hawk, sometimes even striking it. The adaptive nature of this behavior is not always clear (Curio et al., 1978). Often the predaceous creatures under attack are not hunting at the time and are being attacked by smaller animals not known to be prey for that species. Group attacks are known to be effective against predators in some cases, however. Black-headed gulls nest colonially and fly up in groups to assail intruding carrion crows, which prey on gull eggs. The more gulls in the attacking group, the less the hunting success of the crows. Group defense is especially effective against crows but less effective against herring gulls (*Larus argentatus*) and mammalian predators.

Musk oxen (*Ovibos moschatus*) traditionally form a ring, heads outward with the young in the center, when attacked by their natural predators, wolves (*Canis lupus*). This tactic, undoubtedly effective against wolves, yields disaster when the predator is a human with a high-powered rifle.

Schooling of fishes, herding of mammals, flocking of birds—all may reduce predation by localizing the population and reducing the rate of encounter of the predators. Furthermore, the combined motion of all the prey animals may distract the predator. Herds, flocks, and schools may also be a result of each prey animal's putting another individual between itself and the potential predator. Clearly the effectiveness of this strategy varies with both predator and prey. In many cases, it is not clear whether predator attack provided the selective pressure leading to the evolution of this group behavior or whether it is a secondary advantage that resulted once the group had formed for other reasons. The adaptive significance of flocking is still a subject of current research and debate and is discussed at some length in Chapter 11.

Social animals sometimes have specific warning calls that signal the approach of a predator. Marmots (*Marmota*) and prairie dogs (*Cynomys*), for instance, typically live in conspecific colonies and may have particular cries or whistles that specify danger near or far, aerial or terrestrial. Vervet monkeys (*Cercopithecus aethiops*) have specific, learned alarm calls for leopards, eagles, and snakes (Seyfarth et al., 1980). Many birds have warning calls also. The calls of several small birds that have been studied had one characteristic in common: Their structure made them difficult to localize. This feature reduces the risk to the individual giving the call. An obvious question is why the caller should take any risk at all. Several possible answers have been suggested: (1) The caller is warning its offspring or other relatives with shared genes in the vicinity. Female ground squirrels (*Spermophilus beldingi*) live close to their female relatives but males tend to wander. Warning calls are far more frequent in females than in males; furthermore, females without known living relatives nearby call less than females with nearby kin. (2) The caller is being reciprocally altruistic, taking a small risk to warn another animal that may some day return the favor. This is quite a gamble, for there is no guarantee that the receiver of the call will not cheat and fail to reciprocate. Such reciprocity might be most likely to occur between mates, if mate replacement is difficult and calling really deters predation. (3) Calling facilitates flock formation, which reduces the risk of predation. (4) The caller, who can see the predator and perhaps does not need to move from its present location, causes others nearby to dive for cover when the call is heard. Their sudden movement might increase their risk of attack and decrease that of the motionless caller. (5) Warning calls may be signals directed to the predator rather than to conspecifics. Such signals could tell a predator that it has been detected by the potential prey and, therefore, is likely to be less successful in hunting in that spot. It would be advantageous to the caller itself, as well as its neighbors, if the predator is encouraged to go elsewhere.

SUMMARY

Primary defenses generally prevent predator attack and function even if the prey has not detected a predator. These defenses include retreats, crypsis in color and form (camouflage, countershading, disruptive coloration), warning coloration and mimicry, defensive associations, escape in time, and predator satiation. The two basic kinds of mimicry are Batesian, in which a palatable, harmless species mimics an unpalatable or dangerous model, and Mullerian, in which two or more unpleasant prey species mimic each other. Mimicry systems involve convergence in form of conspicuous prey species, but another escape strategy, called aspect diversity, involves divergence of form in cryptic species. Aspect diversity (among species or within a species) may enhance escape from predators, particularly those that form search images.

Secondary defenses include hiding, flight (including possible pursuit invitations), threat, becoming hard to handle, deflection and diversion, feigning death, fighting and chemical defense, and social defenses.

SELECTED REFERENCES

Allison, T., and D.V. Cicchetti, 1976. Sleep in mammals: Ecological and constitutional correlates. *Science, 194*:732–734.

Arnason, E., 1978. Apostatic selection and kleptoparasitism in the parasitic jaeger. *Auk, 95*:377–381.

Beatson, R.R., 1976. Environmental and genetical correlates of disruptive coloration in the water snake, *Natrix s. sipedon*. *Evol.*, *30*:241–252.

Bell, M.A., and T.R. Haglund, 1978. Selective predation of three spine sticklebacks (*Gasterosteus acculeatus*). *Evol.*, *32*:309–317.

Brandon, R.A., G. M. Labanick, and J.E. Huheey, 1979. Relative palatability, defensive behavior, and mimetic relationships of red salamanders (*Pseudotriton ruber*), mud salamanders (*Pseudotriton montanus*), and red efts (*Notophthalmus viridescens*). *Herpetologica*, *35*:289–303.

Brodie, E.D., and R.R. Howard, 1973. Experimental study of Batesian mimicry in the salamanders *Plethodon jordani* and *Desmognathus ochrophaeus*. *Am. Midl. Nat.*, *90*:38–46.

Brodie, E.D., and E.D. Brodie, 1980. Differential avoidance of mimetic salamanders by free-ranging birds. *Science*, *208*:181–182.

Brower, L.P., and J.V.Z. Brower, 1962. Experimental studies of mimicry. VI. The reaction of toads (*Bufo terrestris*) to honeybees (*Apis mellifera*) and their dronefly mimics (*Eristalis vinetorum*). *Am. Nat.*, *96*:297–307.

Charnov, E.L., and J.R. Krebs, 1975. The evolution of alarm calls: Altruism or manipulation? *Am. Nat.*, *109*:107–112.

Clark, K.L., and R.J. Robertson, 1979. Spatial and temporal multi-species nesting aggregations in birds and anti-parasite and anti-predator defenses. *Behav. Ecol. Sociobiol.*, *5*:359–371.

Congdon, J.D., L.J. Vitt, and W.W. King, 1974. Geckos: Adaptive significance and energetics of tail autotomy. *Science*, *184*:1379–1380.

Cott, H.B., 1957. *Adaptive Coloration in Animals.* Methuen, London.

Croze, H., 1970. Searching image in carrion crows. *Z. Tierpsychol.* [Suppl.], *5*:1–85.

Curio, E., U. Ernst, and W. Vieth, 1978. Cultural transmission of enemy recognition: one function of mobbing. *Science*, *202*:899–901.

Denton, E.J., 1970. On the organization of reflecting surfaces in some marine animals. *Philos. Trans. R. Soc. Lond.* (*Biol.*), *258*:285–313.

Denton, E.J., 1971. Reflectors in fishes. *Sci. Am.*, *224*(1):64–72.

Dial, B.E., and L.C. Fitzpatrick, 1981. The energetic costs of tail autotomy to reproduction in the lizard *Coleonyx brevis* (Sauria: Gekkonidae). *Oecologia*, *51*:310–317.

Dice, L.R., 1947. Effectiveness of selection by owls of deer mice (*Peromyscus maniculatus*) which contrast in color with their background. *Contr. Lab. Vert. Biol. Mich.* *34*:1–20.

Edmunds, M., 1974. *Defence in Animals.* Longman, Harlow, England.

Endler, J.A., 1978. A predator's view of animal color patterns. *Evol. Biol.*, *11*:319–364.

Endler, J.A., 1980. Natural selection on color patterns in *Poecilia* reticulata. *Evol.*, *34*:76–91.

Gavish, L., and B. Gavish, 1981. Patterns that conceal a bird's eye. *Z. Tierpsychol.*, *56*:193–204.

Gibbons, J.R.H., and H.B. Lillywhite, 1981. Ecological segregation, color matching, and speciation in lizards of the *Amphibolurus decresii* complex (Lacertilia: Agamidae). *Ecol.*, *62*:1573–1584.

Greene, H.W., 1973. Defensive tail display by snakes and amphisbaenians. *J. Herpetol.*, *7*:143–161.

Greene, H.W., G.M. Burghardt, B.A. Dugan, and A.S. Rand, 1978. Predation and the defensive behavior of green iguanas (Reptilia, Lacertilia, Iguanidae). *J. Herpetol.*, *12*:169–176.

Greene, H.W., and W.F. Pyburn, 1973. Comments on aposematism and mimicry among coral snakes. *Biologist*, *55*:144–148.

Greene, H.W., and R.W. McDiarmid, 1981. Coral snake mimicry: Does it occur? *Science*, *213*:1207–1212.

Harvey, P.H., and P.J. Greenwood, 1978. Antipredator defense strategies: some evolutionary problems. *In* J.R. Krebs and N.B. Davies (Eds.), *Behavioural Ecology: An Evolutionary Approach*, pp. 129–151. Sinauer Associates,

Sunderland, MA.

Harvey, P.H., J.J. Bull, M. Pemberton, and R.J. Paxton, 1982. The evolution of aposematic coloration in distasteful prey: a family model. *Am. Nat.*, *119*:710–719.

Hensel, J.L., and E.D. Brodie, 1976. An experimental study of aposematic coloration in the salamander *Plethodon jordani*. *Copeia*, *1976*:59–65.

Hoogland, R., D. Morris, and N. Tinbergen, 1957. The spines of sticklebacks (*Gasterosteus* and *Pygosteus*) as a means of defence against predators (*Perca* and *Esox*). *Behav.*, *10*:205–236.

Huey, R.B., and E.R. Pianka, 1977. Natural selection for juvenile lizards mimicking noxious beetles. *Science*, *195*:201–203.

Huheey, J.H., and R.A. Brandon, 1974. Studies in warning coloration and mimicry. VI. Comments on the warning coloration of red efts and their presumed mimicry by red salamanders. *Herpetologica*, *30*:149–155.

Janzen, D.H., 1971. Seed predation by animals. *Annu. Rev. Ecol. Syst.*, *2*:465–492.

Kaufman, D.W., 1974. Adaptive coloration in *Peromyscus polionotus:* Experimental selection by owls. *J. Mammal.*, *55*:271–283.

Lay, D.M., 1974. Differential predation on gerbils (*Meriones*) by the little owl, *Athene brahma*. *J. Mammal.*, *55*:608–614.

Mariscal, R.N., 1972. Behavior of symbiotic fishes and sea anemones. *In* H.E. Winn and B.L. Olla (Eds.), *Behavior of Marine Animals.* Vol. 2. *Vertebrates*, pp. 327–360. Plenum Publishing, New York.

McCoskor, J.E., 1977. Fright posture of the plesiopid fish *Calloplesiops altivelis:* an example of Batesian mimicry. *Science*, *197*:400–401.

Neuchterlein, G.L., 1981. "Information parasitism" in mixed colonies of western grebes and Forster's terns. *Anim. Behav.*, *29*:985–989.

Owens, N.W., and J.D. Goss-Custard, 1976. The adaptive significance of alarm calls given by shore birds on their winter feeding grounds. *Evol.*, *30*:397–398.

Paulson, D.R., 1973. Predator polymorphism and apostatic selection. *Evol.*, *27*:269–277.

Peterson, C.H., 1976. Cruising speed during migration of the striped mullet (*Mugil cephalus* L.): an evolutionary response to predation? *Evol.*, *30*:393–396.

Portmann, A., 1959. *Animal Camouflage.* University of Michigan Press, Ann Arbor.

Salmonsen, F., 1972. Zoogeographical and ecological problems in Arctic birds. *Proc. Int. Ornith. Cong.* (IV):25–77.

Schwalm, P.A., P.H. Starrett, and R.W. McDiarmid, 1977. Infrared reflectance in leaf-sitting neotropical frogs. *Science*, *196*:1225–1227.

Seghers, B.H., 1974. Geographic variation in the responses of guppies (*Poecilia reticulata*) to aerial predators. *Oecologia*, *14*:93–98.

Sexton, O.J., 1960. Experimental studies of artificial Batesian mimics. *Behav.*, *15*:244–252.

Seyfarth, R.M., D.L. Cheney, and P. Marler, 1980. Monkey responses to three different alarm calls: evidence of predator classification and semantic communication. *Science*, *210*:801–803.

Sherman, P.W., 1981. Kinship, demography and Belding's ground squirrel nepotism. *Behav. Ecol. Sociobiol.*, *8*:251–159.

Smith, S.M., 1975. Innate recognition of coral snake pattern by a possible avian predator. *Science*, *187*:759–760.

Springer, V.G., and W.F. Smith-Vaniz, 1972. Mimetic relationship involving fishes of the family Blenniidae. *Smithson. Contrib. Zool.* *112*:1–36.

Sumner, F.B., 1934. Does "protective coloration" protect? Results of some experiments with fishes and birds. *PNAS*, *20*:559–564.

Tinbergen, N., 1958. *Curious Naturalists.* Anchor Books, Garden City, NY.

Tordoff, W., 1980. Selective predation of gray jays, *Perisoreus canadensis*, upon boreal chorus frogs, *Pseudacris triseriata*. *Evol.*, *34*:1004–1008.

Trivers, R.L., 1971. The evolution of reciprocal altruism. *Q. Rev. Biol.*, *46*:35–57.

Vitt, L.J., J.D. Congdon, and N.A. Dickson, 1977. Adaptive strategies and energetics of tail autotomy in lizards. *Ecol., 58*:328–337.

Wickler, W., 1968. *Mimicry*. McGraw-Hill Book Co., NY.

Zaret, T.M., 1977. Inhibition of cannibalism in *Cichla ocellaris* and hypothesis of predator mimicry among South American fishes. *Evol., 31*:421–437.

PART 4

SOCIAL AND REPRODUCTIVE PATTERNS

11 Home Range, Territoriality, and Social Groups

Many individual vertebrates disperse from the area where they were born or lived for a while to another area. Large numbers of dispersing animals probably die, but a few establish themselves in new areas and reproduce, often more successfully than if they had stayed "home." Dispersal is particularly common among juvenile vertebrates and occurs most frequently when the original home area is densely occupied relative to the available resources. When resources (space, food, shelter) are adequate or if suitable habitat is predictably unavailable elsewhere, individuals are more likely to remain. Clusters of relatives are found, and when neighbors are related, the operation of kin selection and the importance of inclusive fitness are enhanced. Some of the possible social consequences of kin selection will be discussed in these chapters.

Few vertebrates wander randomly through their environments for their entire lives. At some time or another, usually during their adult lives, each individual of most species settles down in one particular area. Sometimes occupation of the area is temporary (for only a few days or a season) and sometimes it lasts the animal's lifetime. Even nomadic vertebrates settle briefly to breed.

Individuals of some populations live quite solitary existences; others may live in pairs isolated from other such pairs or in "packs" of various sizes and cohesiveness. Spatial relations among conspecific individuals are commonly regulated by social behavior and are fundamentally based on the animal's ecology. The evolution of some forms of social structure can be, in turn, a means of achieving novel ways of exploiting the environment successfully.

Several different patterns of the use of space and their possible adaptiveness are described in this chapter. Emphasis is placed on the adaptive nature of different social organizations rather than on details of behavioral mechanisms. Birds and mammals predominate in the examples because they are the best-studied groups.

HOME RANGE

The area in which an individual or a social group normally lives from day to day and week to week is termed a home range. The home range of one individual or group often overlaps with those of others (Fig. 11–1). Despite the overlap of these activity spaces, the animals using them may mutually avoid each other so that the frequency of encounters is minimal. Home ranges are commonly flexible in size and shape, varying with seasonal changes in resources and activity.

Perhaps the primary adaptive value of remaining in a well-used home range is familiarity with the landscape. The distribution of food resources can be monitored and the location of hiding places can be learned. Foraging may be more effective on familiar ground. Red-backed salamanders (*Plethodon cinereus*), for instance, captured more rewarding food items when hunting in their home area (Jaeger et al., 1981). The importance of the second factor has been demonstrated experimentally with screech owls (*Otus asio*) preying on white-footed mice (*Peromyscus leucopus*) (Metzger, 1967). One set of mice was released into a pen provided with food, water, and cover; they were left there to establish residency. Later, another group of mice, unfamiliar with the pen, also was released, and a screech owl was allowed to hunt the penned prey. Significantly more transients than residents were captured.

Home ranges also permit their owners to learn to recognize their neighbors. Relationships established between known individuals reduce the time and energy spent in investigation of strangers and aggression. Individual recognition and familiarity with neighboring individuals have been demonstrated for several wild animals, including raccoons (*Procyon lotor*), red foxes (*Vulpes vulpes*), and North American ovenbirds (*Seiurus aurocapil-*

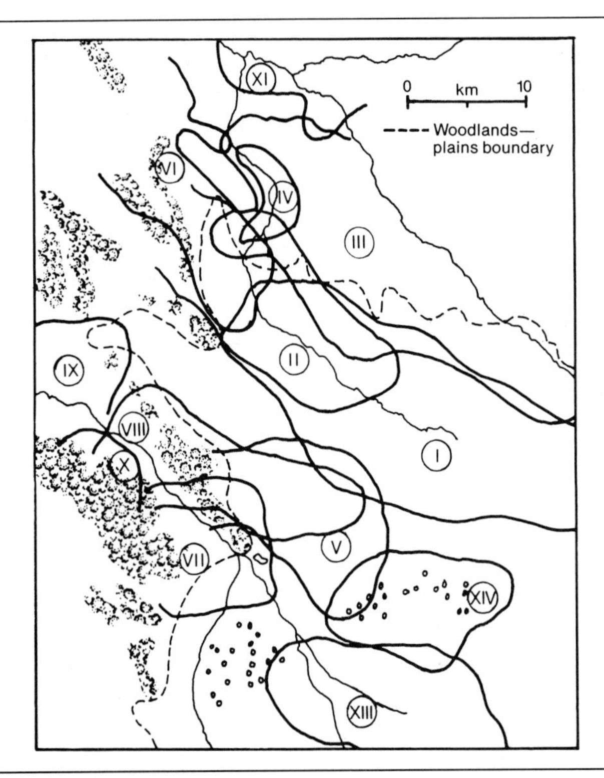

Figure 11–1 Home ranges of lions in the Serengeti Plains of East Africa. Note the extensive overlap of home ranges of different groups, especially groups I and II and groups III and IV. (Reprinted from *The Serengeti Lion* by G. Schaller by permission of The University of Chicago Press. Copyright © 1972 by The University of Chicago Press.)

lus). Male ovenbirds clearly distinguished between the familiar songs of their established neighbors and the unfamiliar songs of strangers and became more agitated and aggressive when they heard a stranger's song (Weeden and Falls, 1959).

The size of an individual's home range varies with many ecological factors, including resource availability and distribution. Home-range size of the desert iguana, *Dipsosaurus dorsalis,* increased markedly after a sandstorm destroyed a large portion of the vegetation that provided foraging sites (Krekorian, 1976). The distribution of nest sites, a critical resource for female yellow-headed blackbirds (*Xanthocephalus xanthocephalus*), may be a major influence on the size of the space claimed by the male blackbirds. Body size and diet, both related to food resources, are often closely related to home-range size for many vertebrates, as we have seen (Chapter 9). The direct relevance of body size to availability of food resources was indicated by Lockie's (1966) study of weasels (*Mustela*) in Great Britain. Two species—*M. nivalis* (called the weasel in Europe and the least weasel in North America) and *M. erminea* (the stoat or ermine)—were found in the same habitat. Both species fed primarily on a vole, *Microtus agrestis.* The body size and average food requirements of the ermine are about twice those of the least weasel, but the ermine's home range is about eight times the size of the least weasel's. This discrepancy can be accounted for by the difference in hunting techniques. The small-bodied least weasel can follow its prey into burrows and therefore has access to larger numbers of voles in an area; the larger ermine hunts primarily above ground and must rely on the voles available there, although it also seems to have a somewhat more varied diet than its smaller congener (King and Moors, 1979).

TERRITORIALITY

Many species of vertebrates defend the home range (or part of it) against conspecific intruders and sometimes against ecologically similar species as well. An area that is defended in such a way that it tends to be the exclusive property of the resident is called a territory (Fig. 11–2). Territory owners often advertise their ownership by various conspicuous displays. Territoriality has been reported from most major groups of vertebrates: fishes (especially osteichthyan fishes), anuran amphibians, lizards (among reptiles), and many birds and mammals. In most cases, defense by a single individual or sometimes by a pair is involved, but sometimes groups defend a shared territory. This is true in lions (*Panthera leo*), several neotropical emballonurid bats (Bradbury and Vehrenkamp, 1976), and many species of communally breeding birds (Gaston, 1978), for example. In addition, sometimes the main territory owner "shares" a territory with "satellite" males that may inherit the territory when the owner dies (see Chapter 13).

The distinction between home range and territory is not a sharp one. Many mammals, including a number of primates and carnivores, mark the borders of their territory with glandular secretions, urine, feces, scratch marks, or vocalizations and are aggressive if a conspecific individual or group approaches that border. Trespassing is common in these same territories, however, when the owners are far away and out of sight, although the behavior of the intruders is frequently somewhat surreptitious. Dominance status in juvenile side-blotched lizards (*Uta stansburiana*) is related to the quality of the home range occupied and probably to their eventual ability to survive (Fox et al., 1981). The importance of dominance implies that some individuals can force others away from their home area and thus implies a degree of exclusivity, although the animals are not called territorial.

Males of other species, including the red-winged blackbird (*Agelaius phoeniceus*), defend strict boundaries against other males in the breeding habitat; but on neutral ground some distance away, they often feed together placidly. Defense of an area and the relative dominance of the residents may vary with the behavior of the intruders, as found in Steller's jay (*Cyanocitta stelleri*). Breeding Steller's jays vigorously defend a certain

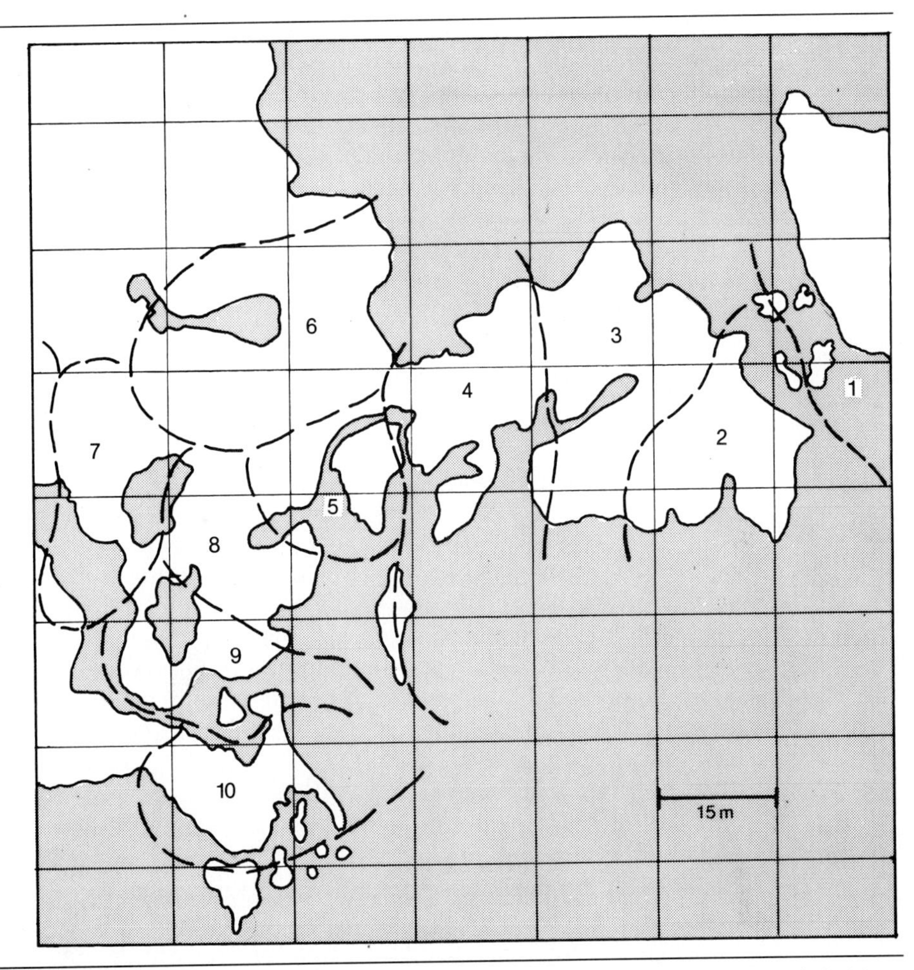

Figure 11–2 Territories of 10 male yellow-headed blackbirds (*Xanthocephalus xanthocephalus*) in a marsh in eastern Washington. (From M.F. Willson, field notes.)

area against other breeders but allow a certain amount of trespassing if the intruders are foraging and behaving in a subordinate fashion. Seasonal variation in the location of territory boundaries is also common.

Defense of a territory necessitates some expenditure of time and energy and therefore must entail some compensating benefits. Brown (1964) emphasized that territoriality will be advantageous if some defendable resource is limited in supply. "Defendable" implies not only that physically territory defense is possible but also that the cost in time and energy is less than the benefits obtained. If the limited resource is food, for instance, the food supply must not be so erratic in its presence or so mobile that its defense diverts too much time or energy from other necessary activities. Many cliff-nesting seabirds feed on schooling fishes in the open ocean; defense of such a food supply presumably is uneconomical because fish schools tend to be unpredictable in time and space and far from the nest. Sunbirds (Nectariniidae) in Africa often defend territories that contain sufficient flowers to supply the nectar needs of an individual. If the flowers are too dispersed, however, they are too far apart to be defendable and no territories are established. Furthermore, if nectar is extremely abundant, the number of intruders is so large that defense is again uneconomical, and the birds do not defend feeding territories (Gill and Wolf, 1975). Similar results were observed in a Hawaiian honeycreeper, *Vestiaria coccinea* (Carpenter and MacMillen, 1976b). In these cases, the costs exceed the benefits at both low and high densities and territoriality is profitable in the intermediate range.

Territories are used for many different purposes. Some species, including most seabirds, defend only the nest site, which is defendable by a bird sitting on the nest. The situation is similar for a number of fishes, including the pupfish (*Cyprinodon*) (Kodric-Brown, 1977) and the bullfrog (*Rana catesbeiana*) (Howard, 1978), in which certain males defend oviposition sites. The area defended by males of some seal and sea lion species is just large enough to contain a male and his harem. On the other hand, at certain seasons some vertebrates defend only a feeding area. Hummingbirds and sunbirds frequently defend a nectar source at any season. Some other species, such as wintering shrikes (*Lanius*) in Africa and northern water thrushes (*Seiurus noveboracensis*) on fall migration through Texas, may hold territories that are apparently used only for feeding. Male vicuñas (*Lama vicugna*) in Peru defend feeding territories in lowland or hillside areas and separate sleeping territories on ridgetops. Vicuña territories are connected by neutral (undefended) corridors by which the male and his family of females and young make daily trips from one area to the other. The adaptive value of such single-purpose territories is generally accepted, but it is not always clear why a particular resource is more limiting for one population than for another.

Multipurpose Territories

Territories of many species are multiple-use property; courting and pair formation, nesting, feeding, and resting occur there. Quite a few controversies have arisen in trying to determine the evolutionary basis for territoriality in these species. There is no reason to assume that all species are alike in the functions and benefits of territorial behavior. Furthermore, it is most unlikely that all uses of a territory have contributed equally to the selective advantage of territory-holding. Often it is difficult to ascertain which uses are primary (related to resources that are limited in supply) and which are secondary (occurring on the territory because the territory is already owned for other, primary, reasons). The population consequences of this behavior must be distinguished from both primary and secondary functions, which have to do with advantages to individuals.

One primary function of territoriality in some populations may be the facilitation of pair formation and pair maintenance. The limited resources in these cases must be (1) places suitable for courtship or copulation or (2) space for uninterrupted interaction between male and female. Male Grevy's zebras (*Equus grevyi*) in Kenya defend territories during the rainy (breeding) season. Territorial borders are marked and maintained at other times, but trespassers are actively excluded only when a female ready to breed is nearby. Would-be trespassers generally yield to the territory owner. This behavior pattern results in the opportunity for the successful territorial stallion to mate without interference, at least temporarily.

Furthermore, breeding territories may reduce the risk to the territory owner of cuckoldry (having a mate that copulates with other individuals outside the established pair bond). The probability of trespassing males is reduced and the female is more easily defended. It is clearly disadvantageous for a male to invest time or energy in offspring fathered by another male.

Another primary function of territoriality can be the defense of food resources. Territorial behavior need not ensure a complete food supply; if territoriality provides some critical food, or perhaps some minimum amount of food, or even readily accessible food at certain times, a primary function of this behavior could be based on food. Predatory arctic birds such as jaegers (*Stercorarius*) and snowy owls (*Nyctea scandiaca*) vary their territorial behavior with the food supply. Jaegers defend feeding territories when their prey is lemmings but have only small nesting territories when they depend primarily on prey pirated from seabirds. Snowy owls have enormous territories when their lemming prey is scarce but small ones, vigorously defended, when lemming populations are high. These correlative studies must be interpreted with care, however, because synchronous establishment can lead to smaller, more densely packed territories than when establishment is staggered through time. If high food levels result in coincident settling of males, the temporal pattern of settling could regulate territory size. Territory size of the ovenbird in eastern North America varies inversely with the density of their invertebrate prey in the ground litter, implying that territory size is probably adjusted to food density (Stenger, 1958). A similar correlation was found for sanderlings (*Calidris alba*) wintering on California beaches; areas of high food density attracted more birds and thus made territorial defense more costly. In this case at least, the inverse correlation of prey density and territory size results from increased costs of defense rather than directly from foraging efficiency (Myers et al., 1979). Red-grouse territory size is inversely related to the nitrogen content of heather shoot-tips, which are the grouses' main food and critical to breeding success (Lance, 1978). Pied wagtails (*Motacilla alba*) in Britain shared their winter feeding territories with other individuals when food resources were at a high level but evicted the interlopers when food abundance was low (Davies and Houston, 1981). Similarly, territory size of the insectivorous iguanid lizard *Sceloporous jarrovi* diminished when extra food was supplied and expanded with the removal of food (Simon, 1975).

Smith's (1968) study of pine squirrels (the red squirrel, *Tamiasciurus hudsonicus,* and the Douglas squirrel, *T. douglasii*) in British Columbia related the probability of winter survival to the size of the crop of conifer cones available on the territory of each individual. Survival was lower on territories with small cone crops than on those with large cone crops, thus clearly indicating that defense of food resources was a critical function of territorial behavior. Foraging was also estimated to be more economical for a single squirrel on an individual territory than it would be for a pair of squirrels on a shared territory.

Nest sites are another critical resource for some species, including many that nest in cavities or on cliff ledges. Males of several bird species mate with several females concurrently and often defend territories containing limited numbers of suitable nesting places. The number of female mates obtained by each male is related to the number of potential nest sites in bobolinks (*Dolichonyx oryzivorus*) and dickcissels (*Spiza americana*), which nest in grassland forbs, some marsh-nesting blackbirds (*Xanthocephalus, Agelaius*) in North America, and savanna-nesting weaverbirds (Ploceidae) in Africa and Asia. Female hillstar hummingbirds (*Oreotrochilus estella*) in the Peruvian Andes defend nesting (and also feeding) territories in a habitat where suitable nesting sites (which are often in rock outcrops) are not common (Carpenter, 1976). Apparently, hollow trees are required for the successful raising of the young of one tropical frugivorous bat (*Artibeus jamaicensis*) in Panama; the hollow trees may be a limited resource that is defended by males and attractive to females. Sites for egg deposition seem to be critical resources for green frogs (*Rana clamitans*) and some other anurans (Wells, 1977), as well as certain fishes.

Greater distance between nests sometimes reduces predation, as shown for great tits (*Parus major*), preyed upon by weasels, and for eider ducks (*Somateria*), preyed upon by arctic foxes (*Alopex lagopus*). Territorial behavior may be one way to achieve larger inter-nest distances, especially when population densities are high and simple mutual avoidance might be inadequate. By increasing the distance between itself and its neighbor, a territorial animal might reduce its risk of predation. Here the resource in question is just the space that provides some protection from predation by reducing the probability that the predator will encounter the nest.

Thus far, we have suggested four factors—mating space, food, nest sites, and space to reduce predation—that may serve as the primary functions of multipurpose territoriality in some populations, although they may be secondary or even irrelevant in others. Only detailed studies will ascertain which functions have been important to the evolution of territorial behavior in particular populations.

Population Consequences

Aspects of territoriality that have given it adaptive value are not to be confused with the population consequences of territorial behavior. As pointed out in Chapter 1, the population consequences are merely by-products of the evolution of territorial systems and are not directly subject to natural selection. Two such consequences are the promotion of dispersal and the provision of a breeding reserve.

Territorial behavior can encourage dispersal and the movement of genes from one population to another. Tompa (1962) studied song sparrows (*Melospiza melodia*) on an island off the coast of British Columbia and found that male song sparrows occupied all the available habitat. Adult males had a fairly low mortality rate and frequently occupied their territories for more than a year. Occasional vacancies were filled by some male offspring of these males, but other sons emigrated to different islands and settled there. Full habitat occupation by territorial males, therefore, clearly promoted dispersal and gene flow. But in no way can the evolution of the territorial habit be explained by the fact that some dispersing individuals intermingled and presumably interbred with individuals in other populations. The evolution of territoriality must be understood as a result of natural selection on individuals. It is advantageous for older males to keep their familiar territories and—because of the lack of suitable space—for certain young males to seek territories elsewhere. The fact that some males have been successful and others have not is a reflection of their individual fitness in terms of acquiring real estate and mates. Fitness and evolutionary success are measured by the individual's genetic contribution to future generations and not by imaginary "benefits" or consequences to the population. If there had been enough room, it is very likely that all of the young male song sparrows would have stayed.

A breeding reserve implies that the loss of individuals from a breeding population can be "made up" numerically by other individuals that replace them. The existence of a breeding reserve in a number of species has been demonstrated experimentally by removing territorial males and observing that other males frequently replace them rather quickly. The replacing males may have been nonbreeders previously or may have tried to mate in other nearby areas. Replacement of territorial individuals means that large-scale male mortality need not result in population decimation. Despite some claims to the contrary, however, this population consequence can not account for the evolution of territorial behavior; breeding reserves are a result, not a cause, of territoriality. Any individual that can successfully breed in an area presumably would do so if it were possible, but some are apparently prevented from breeding there by the territorial behavior of the residents.

Interspecific Territoriality

Territories are usually defended against members of the same species. Occasionally, however, territories of more than one species are mutually exclusive and territoriality is interspecific. Interspecific defense of territories is known in most major vertebrate taxa. The case of the yellow-headed blackbird and the red-winged blackbird was presented in Chapter 3. Another well-documented case involves two congeneric species of squirrels in western North America: the red squirrel and the Douglas squirrel (Smith, 1968). Not only do individuals of both species exclude other conspecifics, but they also actively chase out members of the other species. The result is that space is divided among all individuals of both species (Fig. 11–3).

Ecological pressures favoring interspecific territoriality may shift in time and space. Chaffinches (*Fringilla coelebs*) and great tits (*Parus major*) occupy overlapping territories on the Scottish mainland but are interspecifically territorial on off-shore islands (Reed, 1982). Their nest sites are quite different but their foraging habits are similar and may provide the basis for dividing up the available space. Habitats are probably less diversified on the islands, and interspecific territory defense may replace or reinforce differences in habitat use.

Because defense of territory involves time and energy, it is unlikely that most cases of regular interspecific territoriality are simple misidentifications of one animal by another. Instead, the two species are probably competing for some limited resource that makes such defense worthwhile. Perhaps predation on nests may also select for interspecific spacing of

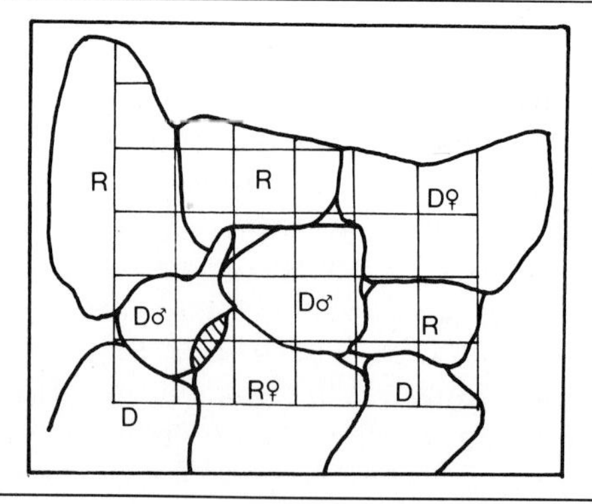

Figure 11–3 Interspecific territoriality in adult red squirrels and Douglas squirrels in a hemlock and mixed conifer forest in British Columbia. The grid is about 33 m on a side. Sex (when known) and species (R or D) of the territory owners are indicated. (Modified from C.C. Smith, 1968. The adaptive nature of social organization in the genus of tree squirrels (*Tamiasciurus*. *Ecol. Monogr.* *38*:31–63. Copyright © 1968, the Ecological Society of America.)

nests (Davies, 1978). Yellowheads and redwings can use similar nest sites and eat similar insect prey; either (or both) resource may provide a basis for spatial segregation. Both species of squirrel harvest and cache conifer cones and are capable of using cones from many of the same tree species. Food resources are probably the ecological basis for exclusive harvesting areas in the squirrels.

A West-Indian coral-reef fish called the Beau Gregory (*Eupomacentrus leucostictus*) defends a small territory against about 40 other species of fishes. Nine of these species are predators on eggs. Of those remaining species, the species most vigorously attacked were those with similar diets and similar energy needs. Such results indicate that defense of food resources and defense of eggs against predation probably are the main factors making interspecific territoriality adaptive for the Beau Gregory (Ebersole, 1977).

Some cases of interspecific territoriality, however, seem to involve unilateral aggression and may not be adaptive to all participants (Murray, 1981). Subordinate individuals may be at a significant disadvantage, if interspecific territoriality excludes them from otherwise usable areas. Sometimes the aggression of the dominant species is only temporary and, eventually, the territories of dominant and subordinate come to overlap.

SOCIAL GROUPS

Many vertebrates regularly occur in groups. A group that forms because of some local attraction such as a salt lick or water hole is called an aggregation. These groups are usually temporary, their members seldom act in concert, and they are not usually included in the definition of true social groups. Social groups are quite cohesive and the animals often act in concert. Individuals may stay together only in certain seasons or they may be essentially permanently grouped. Herds of mammals and schools of fishes may be either breeding or nonbreeding social groups. Groups of birds breeding gregariously are usually called colonies (as is the breeding site), but groups of birds away from the breeding site and nonbreeding social groups of birds customarily are called flocks.

Hypotheses attempting to explain the evolution of social groups tend to fall into three categories: (1) protection against predators; (2) increase of foraging efficiency; and (3) efficiency of locomotion. We should not forget that there are also costs that may be associated with living in groups, including the costs of increased behavioral or competitive interactions among individuals and, in some cases, greater risk of disease or parasitism (Hoogland, 1979). Advantages of group-living are not always easy to discern; for instance, the striped-backed wren (*Campylorhynchus nuchalis*) in Venezuela exhibited no better survival in groups than alone and actually foraged less effectively in groups (Rabenold and Christensen, 1979). Both the costs and the benefits of group-living may differ among individuals in a group, and future research is likely to emphasize such individual variation (Krebs and Barnard, 1980; Baker et al., 1981). Nevertheless, there exists considerable evidence for general benefits in several species, although such benefits are obtained in several ways.

Protection Advantages

Firm evidence to support the antipredation hypothesis is difficult to obtain. Predation on vertebrate prey is rarely observed and, in these days of predator control and habitat destruction, such observations are not likely to become more common. Furthermore, the supposed antipredator advantages to flocking, such as distraction of the predator or increased watchfulness by many eyes, may sometimes be simply secondary effects of living in a group—by-products of the fact that many individuals are found in proximity for other reasons. Indeed, living in a flock might result in selection for increased watchfulness or distraction behavior to counteract the greater conspicuousness of a group.

Some evidence indicates that group-living actually confers some protection from predation. For instance, shorebirds wintering on the California coast (especially dunlins, *Calidris alpina,* and least sandpipers, *C. minutilla*) are preyed upon by merlins (*Falco columbarius*) and other raptors. Single shorebirds were estimated to have a probability of capture over three times greater than that of flocking shorebirds. Clearly, flocking provided some kind of protection (Page and Whitacre, 1975).

A small, tropical freshwater fish, the guppy (*Poecilia reticulata*), is prey to a number of predatory fishes in Trinidad. Seghers (1974) noted that the guppies' tendency to form schools was directly correlated with predation pressure in natural habitats. He also showed in laboratory experiments that the proclivity toward schooling seemed to be heritable and, therefore, subject to natural selection; that is, schooling is not just learned or induced. Furthermore, when schooling and nonschooling individuals were exposed to the same predator (a large cichlid, *Crenicichla alta*) in the laboratory, the predator selectively captured the nonschoolers. Therefore, the adaptive value of schooling behavior in guppies seems to be clearly indicated, although it is difficult to quantify because guppy populations exhibit other differences that may also influence the risk of predation.

Predator protection in groups may work in several ways. One common proposal is that numerous individuals can maintain a better watch for predators than a single animal. Prairie dogs (*Cynomys*) are colonial rodents of the North American plains. The watchfulness of individuals was inversely correlated with the number of other individuals in the vicinity, and potential predators were detected more quickly by greater numbers of watchers, suggesting that collective vigilance may be one benefit of coloniality for some species (Hoogland, 1979, 1981). In contrast, cliff swallows (*Petrochelidon pyrrhonota*) in large colonies did not detect aerial predators more quickly than those in small colonies (Wilkinson and English-Loeb, 1982).

Sometimes a need for increased foraging time may make it advantageous for individuals to group together and share the time spent looking out for predators. Pulliam and colleagues (1974) have shown that aggression among juncos (*Junco*) in Arizona decreased as environmental temperatures decreased and that flocking was then enhanced. They suggested that at low temperatures, especially those below the junco's thermoneutral zone, increased energy is needed to maintain a constant body temperature. To acquire that additional energy, more time is needed for foraging and less time becomes available for aggression. If flocking also increases the time spent in foraging by reducing the time spent by each individual looking out for predators, flocking would become not only possible under such circumstances but also advantageous. Powell (1974) showed for starlings (*Sturnus vulgaris*) that individuals in flocks spent less time on the alert and, thus, could spend more time foraging than when they were alone. The greater number of eyes and ears in the flock apparently reduced the need for constant vigilance by each bird. Consequently, a bird with a limited time budget and the need both to forage and escape predators could balance its budget by sharing the watch for predators and using the time saved to hunt for food.

Hamilton (1971) analyzed the evolution of groups theoretically in terms of "geometry for the selfish herd." Many predators tend to prey especially on peripheral individuals in a group, which provides selective advantage to being in the middle. Each individual has a better chance of not being nearest to an approaching predator if it is closely surrounded by other individuals (Fig. 11–4); thus, each individual tends to seek cover behind the others.

Adélie penguins (*Pygoscelis adeliae*) are very social in most activities. Adélies breed in dense colonies, travel between feeding and breeding areas in flocks, and bathe and feed in flocks. Density of birds in a flock varies predictably. The average distance between adult individuals is two to four body widths in normal circumstances, but when a predator such as a human or a leopard seal (*Hydrurga leptonyx*) is near, the flock bunches together tightly, with only about one and a half body widths between individuals. If the predator approaches, the penguins move even closer to each other until they are touching; when the predator departs, they again spread out (Ainley, 1972). Exactly how this behavior may be successful in reducing predation is not known—it may be a "selfish herd." The observed correlation of behavior changes in the presence of a predator and the observation that avian predation is greater on the periphery of the colony than in the center suggest that predation has been important in the evolution of flocking and colonial breeding in Adélies.

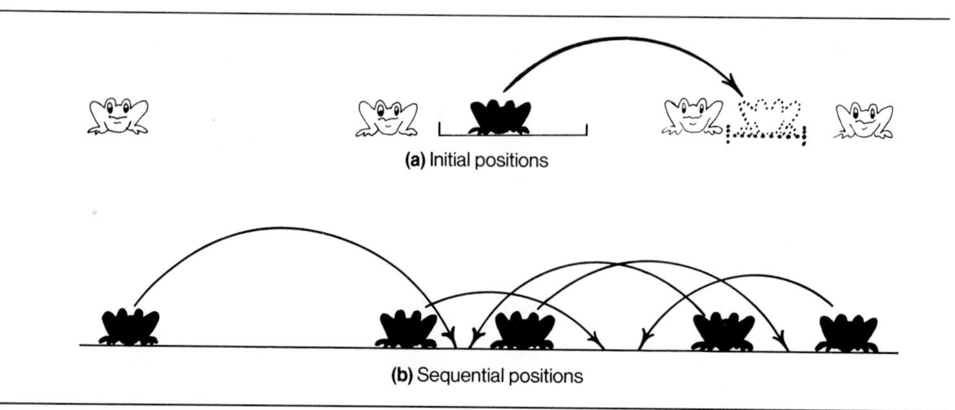

Figure 11–4 Hamilton's (1971) model for a selfish herd represented here by frogs on the perimeter of a pond. Above is shown the position of the frogs as they initially emerge from the water onto the shore. Any frog in a wide open space will have a good chance of being nearest to an approaching predator, such as a snake, and can better its position by hopping to the indicated position. The first frog's neighbors will also be moving to narrower and narrower spaces between frogs as shown below. As a result, the distance between individuals is ever smaller. (From W.D. Hamilton, 1971. Geometry for the selfish herd. *J. Theor. Biol.*, *31*:295–311.)

A similar line of reasoning may apply to some gregariously nesting birds. Many birds nest colonially, each nest located on a small territory sometimes no bigger than the nest itself. Many swallows (Hirundinidae), herons (Ardeidae), seabirds (Laridae, Alcidae), and savannah-nesting weaverbirds nest in colonies and forage over wide areas. Nest sites are patchily distributed and defendable; food resources are widely scattered and not readily defendable. The distribution of food resources in these species may preclude defense of a feeding territory, but that is not sufficient explanation for the evolution of colonial nesting. Possibly colonial nesting confers some protection from predation by means of communal defense and selfish-herd effects, as suggested for the bank swallow (Hoogland and Sherman, 1976).

Communal defense against predators is often successful in such diverse species as musk oxen (*Ovibos moschatus*) and black-headed gulls (*Larus ridibundus*), as discussed in Chapter 10. Fieldfares (*Turdus pilaris*) achieved similar numbers and growth rates of young in both isolated and colonial nests, indicating that foraging success of the parents is similar. A major difference was found, however, in susceptibility of nests to predators; communal defense reduced the probability of nest predation in colonies (Wiklund and Andersson, 1980). Shared defense also appeared to be important in reducing predation on the colonial nests of bluegills (*Lepomis macrochirus*), but selfish-herd effects also were apparent (Gross and Mac-

Millan, 1981). Central nests suffered less egg predation by substrate-level predators such as snails and bullheads because peripheral nests were more likely to be encountered as these predators moved along the bottom. In addition, other *Lepomis* in the water above the nesting colony were deterred by the overlapping defenses of the guardian male bluegills.

Similar protection may be provided by colonial defenses against brood parasites that lay their eggs in the nests of other animals. The yellow-hooded blackbird (*Agelaius icterocephalus*) in Venezuela has a very short nesting season and is parasitized heavily by the shining cowbird (*Molothrus bonariensis*). Small colonies of blackbirds were so severely parasitized that nesting failed altogether, but large colonies were more successful, perhaps in part because the blackbirds built more nests than the local population of cowbirds could exploit but also because of joint defense by numerous male blackbirds (Wiley and Wiley, 1980).

Group-living might also serve to confuse a would-be predator by distraction, or making it difficult to focus on a single prey. Single predators were less successful in capturing schooling prey of two different fish species than in capturing solitary prey in an experimental situation; also, the larger the school, the smaller the percent of individuals captured (Major, 1978). Both guppies and small cyprinid fishes escaped predation by perch (*Perca fluviatilis*) and pike (*Esox lucius*) more often when they were in schools than when they were solitary. Schools disrupted the normal attack behavior of the predators and distracted the predators, which continually switched targets during their pursuits (Neill and Cullen, 1974).

Observations of the wood pigeon, *Columba palumbus*, in England provide some evidence that flocking in this species may have evolved primarily as an antipredation tactic (Murton, 1967; Murton et al., 1971). This is a particularly complex example, which is presented to illustrate the difficulties of disentangling various factors. Wood pigeons live in flocks that have a fairly strong hierarchy of social dominance (Fig. 11–5). Subordinate birds are constantly on the lookout to avoid social conflict and therefore spend less time feeding than do dominant birds. The rate of flock movement is set by the best foragers, and the subordinates sometimes sacrifice feeding to keeping up with the flock. Feeding rates of the subordinates fall below a critical level when food is scarce, and these birds leave the flock and join other flocks. The same process occurs again. As birds move from flock to flock, they lose weight and may actually starve to death, even if adequate food can be found a short distance away.

Why do these birds persist in joining a flock, only to be harassed by their neighbors and subjected to a lowered feeding rate? Field observations indicate that solitary wood pigeons are even more nervous than fugitive subordinates and gather even less food than the lowest-ranking member of a flock. Perhaps selection has favored behavior that reduces predation, despite the consequences to some individuals of flocking at low food densi-

Figure 11–5 A wood pigeon (*Columba palumbus*). These birds may sacrifice food in order to keep up with the flock. Some actually starve to death even though food is available. (Photo by H. Reinhard, courtesy of Bruce Coleman, Inc.)

ties. Large flocks of wood pigeons are, in fact, more successful at detecting and escaping an attacking goshawk (*Accipiter gentilis*).

An alternative explanation might be that the subordinate birds are learning how to choose appropriate foods by observing the behavior of the experienced birds in the flock (Murton, 1971). Thus, solitary birds that spend a lot of time looking around may be watching not only for predators but also for conspecific flocks to join. However, the efforts made to keep up with the flock despite a decrease in feeding suggest that predation initially gave selective advantage to flocking. The strong dependence on observational learning may have evolved subsequently.

Foraging Advantages

Both Davies (1976) and Zahavi (1971) showed that social organization of the European white wagtail could shift dramatically from territorial to flocking in response to changes in dispersion of food during the winter. Davies found that flock birds obtained more insect food per unit time than did territorial birds when the flock found a good foraging place. Success in finding a good place was variable, however, and sometimes food was scarce. Territorial birds, on the other hand, seldom achieved as high a feeding rate, but they had a more dependable food supply. Birds in both social situations could switch to the other. When food was unusually short on the territories, territorial birds would join the flocks (but would return to their territories occasionally to maintain territorial defense). When foraging was unusually good on the territories, flocks moved in and the territories became indefensible.

Schooling in the striped parrotfish (*Scarus croicensis*) in the Caribbean Sea near Panama increases the feeding rates of individuals that do not own territories (Robertson et al., 1976). Striped parrotfishes are herbivorous inhabitants of coral reefs. Some individuals are territorial but are subordinate to a damselfish, *Eupomacentrus planifrons,* which inhibits the feeding of parrotfishes even on their own territories. Aggression by the damselfish apparently prevents some parrotfishes from holding territories. Nonterritorial parrotfishes in schools are attacked by territory owners less often than when they are not in groups, suggesting that schooling helps circumvent the territoriality of competing conspecifics and thus increases the feeding rates of the schoolers.

Flock formation may result in more effective searching for food. Great tits (*Parus major*) found food more quickly when foraging in flocks as a result of the larger number of searching eyes. Furthermore, once one bird was successful, the other flock members promptly began searching in the same or similar areas (Krebs et al., 1972; Baker, 1978). If the "copy-cats" reduce the foraging efficiency of a successful forager, however, there should be selection to disguise foraging success and a tendency for foraging flocks to break up. Cody (1971) suggested that another advantage of flocking might be the thorough exploitation of small patches of food resources at a time, thus maximizing the disparity between gleaned and ungleaned patches and permitting rapid assessment of the food value of the patches. Solitary birds, on the other hand, would move about independently of each other, reduce the overall food abundance gradually and more uniformly, so that favorable feeding sites would become more and more difficult to find. Over a short time period, individual birds sometimes seem to avoid recently exploited areas (Greenwood and Harvey, 1978), but whether flocks exhibit such behavior is apparently unknown.

Great blue herons (*Ardea herodias*) often nest colonially and frequently feed in flocks. Herons often left the nesting colony in groups and exploited different feeding areas from day to day, suggesting that the food resources were ephemeral and patchy and that individuals followed each other. Furthermore, a feeding flock attracted other herons more than a single feeding individual did. Rates of food intake of flocking herons were higher than for solitary birds, and the risk of total failure to capture prey was smaller. Therefore, foraging in flocks seemed advantageous and was facilitated by colonial nesting in these herons (Krebs, 1974). Bank swallows (*Riparia riparia*) also nest colonially, and nesting is often highly synchronized among members of the colony. The adaptive value of synchronization may be that young birds emerging from the nest at or before the peak time of emergence will find many adult birds flying to and from local and ephemeral concentrations of food (Emlen and Demong, 1975). Communal roosts or colonial nesting in a number of other species may have evolved as a means of learning the locations of good foraging places from successful foragers (Ward and Zahavi, 1973). The evolution of such a system of mutual "parasitism" depends on local abundances of food so that food-sharing is not disadvantageous to any individual. The evidence for information-transfer among members of a group is variable: Red-billed weaverbirds (*Quelea quelea*) seem to be able to obtain foraging information from each other (de Groot, 1980), but black-headed gulls apparently do not (Andersson et al., 1981). The extent to which exchange of foraging information has contributed directly to the evolution of group-living is still debated; at least in some cases it seems that it is possible, but in others it is likely to be secondary or irrelevant.

Horn's (1968) study of Brewer's blackbird (*Euphagus cyanocephalus*), conducted partly in the field and partly by means of a model, suggested that colonial nesting in this species may be related to the economical exploitation of unpredictable and patchy food resources (Fig. 11–6). If food for nestlings were evenly distributed in space and time, parent birds would fly the shortest average distance on feeding trips when nests were also evenly dispersed among the food sources. Relatively even dispersion could be accomplished by the establishment of large feeding territories by each pair. If patches of abundant food varied in time and space, however, the parent birds would forage more economically when nesting in a cluster in the center of the foraging area. The average distance flown on a feeding trip would then be less than if the nests were evenly dispersed. In addition, if one bird found a good food patch, neighboring nesters could exploit it by observing the return with a load of food for the nestlings and the flight direction of successful hunting trips. When Brewer's blackbirds feed their young primarily on prey such as aquatic immature dragonflies, they are exploiting a shifting, patchy, continually renewed food resource and frequently nest colonially. When they feed on certain terrestrial insects, which are a less ephemeral food source, however, the blackbirds are more likely to nest solitarily.

Horn's model presumes that a colony can be centrally placed with respect to the food patches, which may not be possible for all colonial species, and that the available foraging patches for the colony all lie in a cluster, isolated from other such clusters (Wittenberger, 1981). This may be true for marsh-nesting Brewer's blackbirds because a single small marsh on a stand of cattails could contain all the foraging patches for a colony.

Some predatory mammals and a few birds live and hunt in conspecific groups. Dwarf mongooses (*Helogale undulatus*) and ground hornbills (*Bucorvus*) in Africa kill large and dangerous snakes by communal attack. Coyotes (*Canis latrans*) sometimes hunt cooperatively, sometimes in pairs, but perhaps also in packs (Bekoff and Wells, 1981).

Kruuk (1972, 1976) showed that African spotted hyenas (*Crocuta crocuta*) often hunt in groups and that their pack size frequently depends on the prey. The small Thompson's gazelle (*Gazella thompsonii*) are usually hunted by solitary hyenas; wildebeest (*Connochaetes taurinus*) adults may be hunted by small packs or solitary individuals. When hunting wildebeest calves, however, packs of two or more hyenas are much more successful

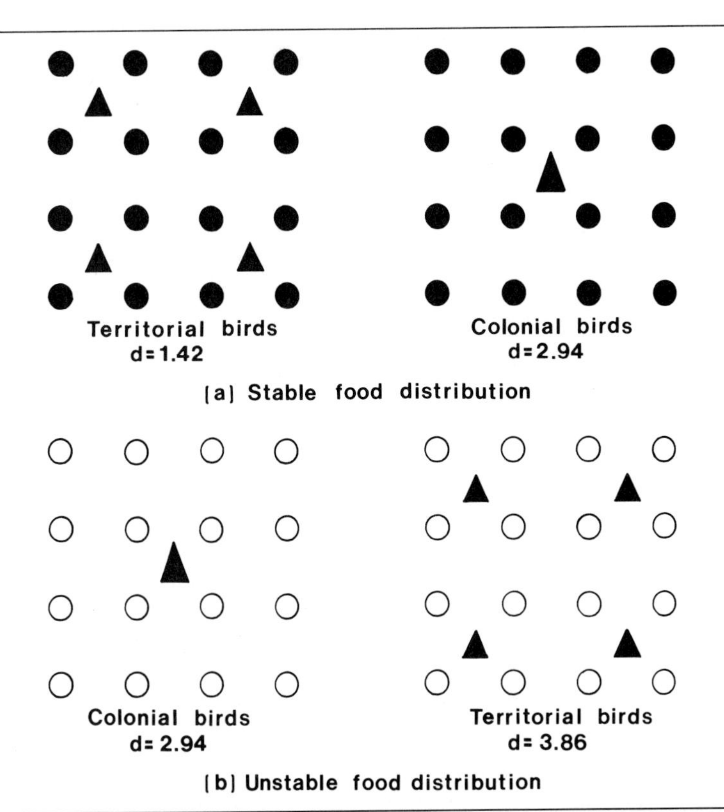

Figure 11–6 Horn's (1968) model for distances flown by adult birds feeding young in the nest (triangles). The dots represent points in the environment at which food may be found; *d* is the round-trip distance. (*a*) In these two situations, food is always found at each of the 16 points. (*b*) In this case, the total quantity of food is the same as before but found unpredictably at only one point at a time. (Modified from H.S. Horn, 1968. The adaptive significance of colonial nesting in the Brewer's Blackbird [*Euphagus cyancephalus*]. *Ecol.*, *49*:682–694. Copyright © 1968, the Ecological Society of America.)

(about 73 percent of the attempts) than single hyenas (15 percent). This difference is due primarily to the effectiveness of several hyenas in overcoming the defenses of the female wildebeest. If the calf simply runs away and does not depend upon its mother, groups of hyenas are not much more effective than single individuals. Hyenas hunt zebra (*Equus burchellii*) in large packs of about 11, but sometimes as many as 25, individuals. Kruuk's hyenas averaged about the same rate of capture (35 to 45 percent of all attempts) in all their adult prey, but pack size grew as the size and defensive ability of the prey increased.

The only social felid, the lion, commonly lives in groups called prides. The typical pride is composed of one or a team of males, several females (usually related to each other) that do most of the hunting for the pride, and their offspring. Caraco and Wolf (1975), using data in the literature, argued that the size of foraging groups of lions may be related to foraging efficiency (Fig. 11–7). When hunting Thompson's gazelle, two or more lions have a higher probability of successful capture than does a single lion (31 percent versus 15 percent). The possible yield of meat per lion (counting only those in the hunting group itself), however, decreases sharply as group size increases because the prey must be divided more ways. The formation of foraging groups of about two lions was estimated to be the best strategy to maximize the food intake per lion feeding on gazelles. Indeed this was observed to be the commonest group size in the Tanzanian study area. Group sizes for lions hunting wildebeest and zebra were larger, averaging four to seven lions, although the estimated size of group for peak efficiency of food intake was again calculated to be only two lions.

Other factors must be involved here, however. For one thing, wildebeest and zebra carcasses, unlike gazelle carcasses, are large and impossible for two lions to consume very rapidly. Hyenas or another group of lions then have ample opportunity to raid the carcass, for two lions cannot continually fend off a pack of scavenging hyenas or a larger group of lions. A larger number of lions is more successful at monopolizing a carcass and may prevent loss of food to the scavengers. Furthermore, in some areas and at some seasons, efficiency and frequency of capture are probably higher than the figure used in these estimates, thus permitting the economical formation of somewhat larger groups. In addition, it can be argued that it is advantageous for a lion to join a group as long as its food intake would be greater with the group than it would be if the lion hunted alone. That is, *any* gain in food intake could favor the joining of a group, even if that gain is not a maximum one. Thus, we should expect to find variation in the size of lion groups; not all groups may be of a size yielding maximum food intake to each lion.

Figure 11–7 (*a*) Capture efficiency and food availability per lion versus group size of foraging lions. Thompson's gazelle is the prey. If the lions chase and capture three gazelles per day, the triangles indicate the average weight of meat per lion in groups of different sizes. The dots indicate the percentage of successful captures by groups of different sizes. Single lions need not share their kill but have a low capture success. Two lions share the kill but have twice the chance of capturing the prey. Three or more lions have a capture success similar to that of two lions but must divide the food more ways. As a result, the food intake per lion is potentially greatest when lions hunt in pairs. (*b*) Average weight of prey per lion per day. Six kg is an estimate of the daily physiological minimum required by an adult lion. Most groups, on average, do not obtain sufficient food to sustain their members. Only groups of two exceed the estimated minimum. (Reprinted from *American Naturalist* by T. Caraco and L.L. Wolf by permission of the University of Chicago Press. Copyright © 1975 by The University of Chicago Press.)

Caraco and Wolf assumed that lion groups of different sizes made an equal number of attempts at prey capture per unit time. More recent evidence suggests that this is not a valid assumption (C. Packer, personal communication); using the observed variation in capture attempts, the peak of foraging efficiency for groups of two disappears, indicating that foraging efficiency does not account very well for group size of hunting

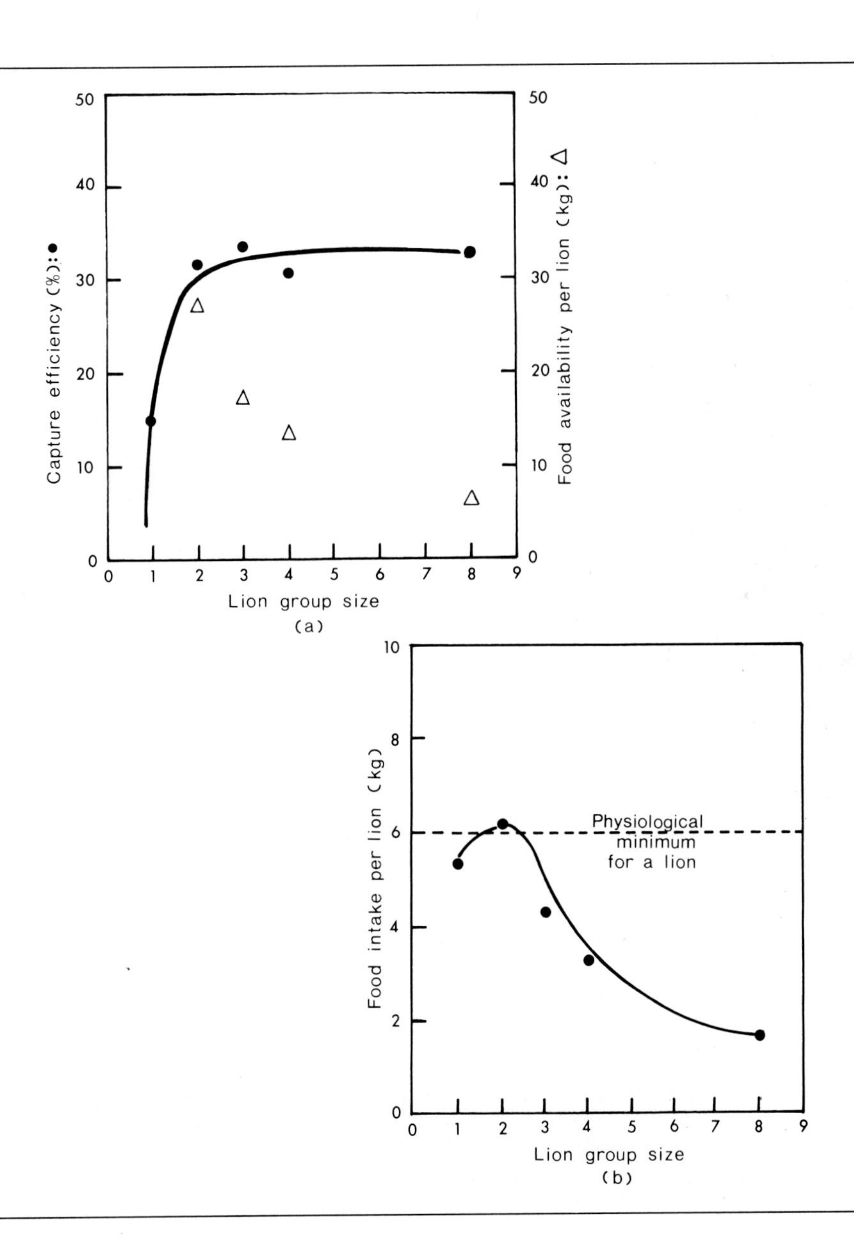

lions. Packer observes that most social carnivores live in open habitats and prey on herding animals and that clumping of the prey may predispose their predators to clumping also. True sociality may then be favored not so much by the putative advantages of group foraging but perhaps through better defense of the carcass, especially against other lions, or better defense of the young against male lions that take over the pride (see Chapter 14). Note that Packer's reassessment shifts the primary basis for lion sociality from foraging to protection, either of food resources or of young.

The estimated optimal size of a hunting group for wolves (*Canis lupus*) was about three individuals when hunting deer but rose to about five when hunting larger ungulates (Nudds, 1978). As in lions, however, the observed sizes of hunting groups was sometimes larger than the estimated optima. Because the members of both lion prides and wolf packs are commonly related to each other, it is possible that kin selection may favor food sharing among a greater than "optimal" number (Rodman, 1981), although this idea is subject to debate.

The cape hunting dog (*Lycaon pictus*) of Africa always hunts socially, usually on smaller prey than that of hyenas or lions. It relies primarily on high-speed chases to tire the quarry and eventually subdue it. Sometimes the chases are run by several dogs; if the prey zigzags, some dogs will cut the corners and, thus, get closer to the fleeing prey. Cape hunting dogs are more specialized group hunters than the others and are more frequently successful in their hunts.

Communal defense of food resources from competitors may be an additional function of living in groups for some species, including lions, as discussed previously. Taiga voles (*Microtus xanthognathus*) in Alaska live in groups of 5 to 10 individuals for much of the year (Wolff and Lidicker, 1981). Each group shares a common food cache and may jointly defend the cache against thieves. The individuals in a group are not especially closely related, so kin selection seems to be of negligible importance. These voles may also conserve body heat by living together. Many primates live in groups in which female relatives remain together for most of their lives. Although the traditional explanation for primate groups is defense against predators, an alternative possibility is that the females cooperate in defense of feeding areas, especially where food is concentrated in patches (Wrangham, 1980).

Locomotor Advantages

The suggestion that flying birds in a flock or swimming fishes in a school may somehow profit by saving locomotor energy is an old one. Until recently, however, little documentation has been available. Moving in a group may facilitate locomotion in several ways. "Drafting" is a common sight on highways: One motor vehicle follows very closely behind another, using it to break the wind and, thus, reducing the energy needed for forward motion. This process may be one reason for the habit of queuing in lobsters; presumably, it could also function in the locomotion of vertebrates. Additionally, the turbulences and vortices created by the fins and undulations of swimming fishes not only create lift that can reduce the locomotor energy of fishes that are heavier than water but may also substantially reduce the effort needed for forward movement. Weihs (1975) showed, in the idealized situation of a mathematical model, that the energetic costs of swimming might be reduced to as little as one fifth that of a solitary fish. Similarly, birds such as geese flying in formation can potentially exploit each other's wing-tip vortices to reduce locomotor costs (Lissaman and Shollenberger, 1970). Such economics may not be achieved by birds in loosely organized, three-dimensional flocks, however (Higdon and Corrsin, 1978).

ADAPTIVE DIFFERENCES IN SOCIAL STRUCTURE

Social structure often changes in response to environmental conditions. The individuals of a population may be sometimes gregarious, sometimes territorial, or sometimes neither, depending on the circumstances. Temporal and spatial changes in social organization are usually related to the availability and defensibility of resources.

Seasonal changes in social organization are very common. Fishes such as basses (*Micropterus*) and sunfishes (*Lepomis*) seasonally defend nesting territories but occupy only individual home ranges the rest of the year. Many north-temperate birds are territorial during the breeding season but gregarious during migration. Some birds, including the Kentucky warbler (*Oporornis formosus*) and the golden-winged warbler (*Dendroica chrysoptera*), are territorial both in their breeding areas and on their tropical wintering grounds. Females of many herding ungulates become temporarily solitary when giving birth but rejoin the herd after a few days.

Geographical or habitat differences in social structure within a species also occur. Song sparrows have been observed to shift from multipurpose territories on islands with dispersed food sources to pairing and nesting territories (but shared feeding zones) on an island with food concentrations. Wildebeest often live in huge, migratory herds in the Serengeti Plains of Tanzania (although a few males are reported to be solitary). In Ngorongoro Crater, however, where permanent water and good grazing are available, many males establish permanent breeding territories, whereas nonbreeding males and females live in separate, small groups. *Cercopithecus aethiops* monkeys (the gray guenons) have a wide distribution in Africa and occupy many kinds of habitat, from rain forest to savanna and scrub forests to grasslands. A population on Lolui Island in Lake Victoria occupied prime habitat of mixed grasslands and thickets composed largely of plant species whose fruit provided food for the monkeys. Groups of monkeys occupied small territories that included both vegetation types. Another population of these monkeys inhabitated disturbed savanna on the mainland. Here few fruits were available and much of the diet was composed of leaves and grasses lower in nutritive value than fruits. Groups of monkeys in this area occupied large home ranges with some overlap between groups.

The marine iguanas (*Amblyrhynchus cristatus*) on Fernandina Island in the Galápagos Islands inhabit a very limited amount of shoreline. Males hold small territories on the shore; females and immatures are crowded inland. The aquatic food supply is seemingly abundant, but basking spots on the shore appear limited. On part of Santa Cruz Island, however, shoreline is extensive and the iguanas reportedly live in isolated family groups, each on its own territory. The side-blotched lizard in Colorado is territorial in open desert but shifts to a hierarchical social structure in adjacent streambeds, where large boulders provide the only suitable habitat. The same species in southern California may defend large multipurpose territories or only small areas around the burrow, depending on the habitat.

The social system of pupfishes in North American desert springs and streams is highly variable (Kodric-Brown, 1981). Breeding males are commonly territorial; this arrangement prevails where the habitat is productive of food but low in predators, so that population densities are high. If the pool is too small to accommodate more than one territorial male, however, a dominance hierarchy will be found among the coexisting males. Where food resources are sparse and the population density is low relative to the available breeding habitat, these fishes consort in pairs, widely spaced in the habitat, without territorial behavior. A fourth social arrangement is found where the habitat forces the costs of territoriality to be excessive: Spawning in groups occurs in dense populations where competition for breeding sites is extremely intense and in streams where strong currents make defense very costly. Individual pupfishes can shift from one behavior pattern to another, depending on the prevailing circumstances.

Changes in social structure have also been demonstrated in captive collections of several kinds of vertebrates. For example, freshwater medaka (*Oryzias latipes*) in a laboratory aquarium shared a localized food source when it was very abundant (Magnuson, 1962). When food was limited (and still localized), however, a dominant individual could set up a territory in the food area and restrict the feeding of subordinate individuals. When food was evenly dispersed throughout the tank, aggressive behavior among the medaka resulted in an even dispersion of the fishes.

Comparative Studies of Related Taxa

Variability of social organizations within a taxonomic group is related to the ecology of the species or populations in that group, just as is true for within-species comparisons. Cross-species comparisons expand the range of contrasts that can be made. Among closely related taxa, such comparisons are useful because many differences can be seen in relation to ecology, and fewer differences are likely to be simply historical or taxonomic than is often the case for distantly related taxa.

Jarman (1974) and Estes (1974) have analyzed the relationship between social organization and ecology for African bovids (antelope and their relatives). Antelope can be grouped into categories of seasonal social structure related to body size and spatial dispersion of males. The social structures are associated with differences in the dispersion of food resources in space and time: Large groups are feasible when food is readily available and evenly spaced; but when food is patchy, antelope tend to be solitary. Furthermore, antelope that are selective feeders on unevenly distributed foods tend to be small, whereas less selective feeders, whose food is common and evenly distributed, tend to be large. Body size is also correlated with tactics for defense against predators. Small antelope are usually cryptic or nocturnal and generally depend on concealment for avoiding predators. Large species, which have fewer potential predators, apparently can afford to be conspicuous and may even utilize communal defense against their predators (Fig. 11–8).

We find that the smallest antelope, such as klipspringers (*Oreotragus oreotragus*) and duikers (*Cephalophus*), feed selectively on a high-quality diet. They eat many fruits, buds, and fresh greens; when they feed, they remove entire food items from the environment and, thus, can affect the food distribution for other individuals. Generally, they are active at night and depend on hiding for avoiding predators. The small species are usually solitary or sometimes found in pairs; they are often territorial, and the

Figure 11–8 Body size among the African bovids is correlated with social organization. The small klipspringer is solitary; the enormous Cape buffalo is gregarious.

spatial dispersion of males and females is similar.

Slightly larger species (reedbuck, *Redunca;* gerenuk, *Litocranius walleri*) are similar in many respects but are somewhat less selective feeders on both grass and herbs. Most adult males are territorial, but subadult males and females occur in small groups of usually three to six animals. The basis for the sexual differences is not explained by the broad correlation among species.

Antelope of the middle size range (Uganda kob, *Kobus kob;* impala, *Aepycores melampus*) are not very selective feeders on browse and grass. Only some adult males succeed in holding territories; nonterritorial males live in bachelor herds, and females also occur in groups. Group size is variable and may include as many as 150 animals.

Large antelope such as the blue wildebeest or gnu and topi (*Damaliscus korrigum*) are quite unselective grazers on low-nutrient foods that are abundant and evenly distributed. Some populations are sedentary and have a social structure similar to the middle-sized species, but other populations are migratory and move about in vast herds. Both this group and the next have few predators; against lions, they employ communal defense tactics.

The largest antelope (African buffalo, *Syncercus caffer;* eland, *Taurotragus*) live in large herds with many males associating with a group of females. A well-developed dominance hierarchy determines the mating privileges of the males. The diet is composed of low-quality foods, and the animals are unselective feeders. They tend to eat only part of a plant, leaving the rest to the individuals that follow, so their foraging has little effect on food dispersion; thus, group-living is facilitated. For these huge antelope, lions are the only serious predators.

The relative costs of dominance encounters is also likely to differ among antelope species; any conditions that sufficiently raise the frequency or danger of dominance battles could favor a shift toward a territorial system (Owen-Smith, 1977). A territorial male antelope seldom wins as many females in one season as a dominant individual in a hierarchy, however, so that the maximum reproductive benefits of a territorial system may be smaller. Owen-Smith calls this a low cost-low benefit system, in which male survivorship is enhanced at the potential sacrifice of a degree of immediate reproductive success.

Studies within a single genus emphasize the importance of the comparative ecology of social organization. Congeners share a phylogenetic history far more extensively than species (such as the antelope) that simply belong to the same family or order. Because historical differences are fewer, differences among congeners indicate the importance of present-day and recent ecology.

Research with the rodent genus *Marmota* provides an example. Barash (1974a) has compared the socioecology of three North American species of marmots (Fig. 11–9). All three species are herbivorous and hibernate when their food is not available. The woodchuck (*M. monax*) of lowland eastern

Figure 11–9 The woodchuck (*Marmota monax*) of eastern North America. (Photo courtesy of the Illinois State Natural History Survey.)

North America lives in a moderate climate with a long growing season (over 150 days a year) for plants. Reproduction is annual and the young mature in one year, presumably thanks to the long period of time for feeding and growth. Adults force the maturing young to disperse well before the next breeding season. Woodchucks are very aggressive among themselves and may be called territorial.

The yellow-bellied marmot (*M. flaviventris*) lives at intermediate elevations in the western part of the continent, where the growing season lasts 70 to 100 days. As a rule, females reproduce annually but may skip a year if conditions are unfavorable. Maturation usually takes 2 years. This species is colonial, but individuals may have separate home ranges. At the highest elevations in its range, the yellow-bellied marmot resembles the following species in many social and ecological respects.

In alpine meadows on the Olympic Peninsula lives the olympic marmot (*M. olympus*). Here the growing season is very short (40–70 days), and the young take 3 years for maturation. Females reproduce every 2 years, and dispersal of the young is delayed. The short season for feeding results in slow growth, and early dispersal would therefore be disadvantageous. As a result, selection has favored decreased aggressiveness among individuals, decreased rates of dispersal of the young, and increased sociality. Olympic marmots are highly colonial and even feed together; individuals do not have separate home ranges. Most colonies consist of an adult male, two adult females (often sisters), a litter of yearlings belonging to one female, and a litter of infants belonging to the other. Interestingly, the alpine species of the Rocky Mountains (*M. caligata*) seems to be similar to the olympic marmot in its social structure.

Studies of the olympic marmot have shown that colony size is greatest in meadow swales in which snow melts away gradually throughout the summer, exposing a series of plant communities on which the marmots feed. Colonies are smaller in marginal habitat on nearby rises, where snow melts rapidly in June, supporting only one or two different plant communities. The sequential release of vegetation from snow in the swales provides marmots with a continual supply of new growth high in calories, protein, and digestibility. Comparisons of social structure among colonies in good and marginal habitats probably would show adaptive differences in the social use of space within this species.

Before we blithely presume that we can explain marmot coloniality, we must attend to the cautionary note sounded by Webb (1981). For yellow-bellied marmots, Webb found that social behavior and levels of antagonism were not correlated with length of the growing season or elevation. Other social and ecological factors that may interact with the degree of coloniality in marmots are discussed in Chapter 13.

SUMMARY

Many vertebrates have home ranges, the size of which is related to resource distribution relative to the animals' requirements. The advantages of having a home range include familiarity with the landscape, which facilitates escape from predators, and familiarity with neighbors, which may reduce overt aggression. Territoriality is found in species whose resources are economically defensible; that is, resources that are limited in supply and therefore advantageous to defend and that are not too patchy or mobile and therefore possible to defend. The selective pressures favoring territoriality differ among species. The most common resources related to territorial behavior are food, nest sites, and perhaps space for uninterrupted pair maintenance or avoidance of predation. Population consequences of territoriality (dispersal, creation of a breeding reserve, population regulation) are not to be confused with the territorial functions that give selective advantage to a territory holder. Interspecific territoriality also typically has an ecological basis in some limited resource.

The evolution of social groups may be related to reduction of predation (by means of increased collective vigilance, decreased individual watchfulness, selfish herds, enemy confusion, communal defense); facilitation of foraging (by means of enhanced food-finding, efficiency of exploitation, cooperative hunting, communal defense of food resources); or locomotor efficiency.

SELECTED REFERENCES

Ainley, D.G., 1972. Flocking in Adélie penguins. *Ibis, 114*:388–390.

Alexander, R.D., 1974. The evolution of social behavior. *Annu. Rev. Ecol. Syst., 5*:325–383.

Andersson, M., F. Götmark, and C.G. Wiklund, 1981. Food information in the black-headed gull, *Larus ridibundus*. *Behav. Ecol. Sociobiol.*, *9*:199–202.

Armitage, K.B., 1974. Male behaviour and territoriality in the yellow-bellied marmot. *J. Zool.* (Lond.), *172*:233–265.

Baker, M.C., 1978. Flocking and feeding in the great tit *Parus major*—an important consideration. *Am. Nat.*, *112*:779–781.

Baker, M.C., C.S. Belcher, L.C. Deutsch, G.L. Sherman, and D.B. Thompson, 1981. Foraging success in junco flocks and the effects of social hierarchy. *Anim. Behav.*, *29*:137–142.

Banks, E.M., and M.F. Willson (Eds.), 1974. Ecology and evolution of social organization. [AAAS symposium.] *Am. Zool.*, *14*:1–264.

Barash, D.P., 1974a. The evolution of marmot society: a general theory. *Science*, *185*:415–420.

Barash, D.P., 1974b. Neighbor recognition in two "solitary" carnivores: The raccoon (*Procyon lotor*) and the red fox (*Vulpes fulva*). *Science*, *185*:794–796.

Barash, D.P., 1977. *Sociobiology and Behavior*. Elsevier Scientific, Amsterdam.

Bekoff, M., and M.C. Wells, 1981. Behavioural budgeting by wild coyotes: The influence of food resources and social organization. *Anim. Behav.*, *29*:794–801.

Berry, K.H., 1974. The ecology and social behavior of the chuckwalla, *Sauromalus obesus*. *U. Calif. Publ. Zool.*, *101*:1–60.

Bertram, B.C.R., 1978. Living in groups: predator and prey. *In* J.R. Krebs and N.B. Davies (Eds.), *Behavioural Ecology: An Evolutionary Approach*, pp. 64–96. Sinauer Associates, Sunderland, MA.

Bradbury, J.W., 1977. Social organization and communication. *In* W.A. Wimsatt (Ed.), *Biology of Bats*. Vol. 3, pp. 1–72. Academic Press, New York.

Bradbury, J.W., and S.L. Vehrenkamp, 1976. Social organization and foraging in emballonurid bats. I. Field studies. *Behav. Ecol. Sociobiol.*, *1*:337–381.

Brown, J.L., 1963. Aggressiveness, dominance and social organization in the Stellar jay. *Condor*, *65*:460–484.

Brown, J.L., 1964. The evolution of diversity in avian territorial systems. *Wils. Bull.*, *76*:160–169.

Brown, J.L., and G.H. Orians, 1970. Spacing patterns in mobile animals. *Annu. Rev. Ecol. Syst.*, *1*:239–262.

Burghardt, G.M., H.W. Greene, and A.S. Rand, 1977. Social behavior in hatchling green iguanas: Life at a reptile rookery. *Science*, *195*:689–691.

Buskirk, W.H., 1976. Social systems in a tropical forest avifauna. *Am. Nat.*, *110*:293–310.

Caraco, T., and L.L. Wolf, 1975. Ecological determinants of group sizes of foraging lions. *Am. Nat.*, *109*:343–352.

Carpenter, F.L., 1976. Ecology and evolution of an Andean hummingbird (*Oreotrochilus estella*). *U. Calif. Publ. Zool.*, *106*:1–74.

Carpenter, F.L., and R.E. MacMillen, 1976a. Energetic cost of feeding territories in a Hawaiian honeycreeper. *Oecologia*, *26*:213–223.

Carpenter, F.L., and R.E. MacMillen, 1976b. Threshold model of feeding territoriality and test with a Hawaiian honeycreeper. *Science*, *194*:639–642.

Clutton-Brock, T.H., and P.H. Harvey, 1977. Primate ecology and social organization. *J. Zool.* (Lond.), *183*:1–39.

Cody, M.L., 1971. Finch flocks in the Mohave Desert. *Theor. Popul. Biol.*, *2*:142–158.

Collias, N.E., and E.C. Collias, 1977. Weaverbird nest aggregation and evolution of the compound nest. *Auk*, *94*:50–64.

Crook, J.H. (Ed.), 1970. *Social Behavior in Birds and Mammals*. Academic Press, New York.

Crook, J.H., J.E. Ellis, and J.D. Goss-Custard, 1976. Mammalian social systems: Structure and function. *Anim. Behav.*, *24*:261–274.

Curio, E., 1976. The ethology of predation. *Zoophysiol. Ecol.*, *7*:1–250.

Davies, N.B., 1976. Food, flocking and territorial behaviour of the pied wagtail (*Motacilla alba yarrelli* Gould) in winter. *J. Anim. Ecol.*, *45*:235–253.

Davies, N.B., and A.I. Houston, 1981. Owners and satellites: The economics of territory defenses in the pied wagtail, *Motacilla alba*. *J. Anim. Ecol.*, *50*:157–180.

de Groot, P., 1980. Information transfer in a socially roosting weaverbird (*Quelea quelea*; Ploceinae): an experimental study: *Anim. Behav.*, *28*:1249–1254.

Ebersole, J.P., 1980. Food density and territory size: an alternative model and a test on the reef fish *Eupomacentrus leucostictus*. *Am. Nat.*, *115*:492–509.

Eisenberg, J.F., 1966. The social organization of mammals. *Handbuch der Zool.*, *8*(39):1–92.

Emlen J.M., 1978. Territoriality: a fitness set-adaptive function approach. *Am. Nat.*, *112*:234–241.

Emlen, S.T., and N.J. Demong, 1975. Adaptive significance of synchronized breeding in a colonial bird: a new hypothesis. *Science*, *188*:1029–1031.

Estes, R.D., 1974. Social organization of the African Bovidae. *In* V. Geist and F. Walther (Eds.), *The Behaviour of Ungulates and Its Relation to Management*, pp. 166–205. IUCN, Morges, Switzerland.

Fisler, G.F., 1969. Mammalian organizational systems. *Los Angeles County Museum Contrib. Sci.*, *167*:1–32.

Fox, S.F., E. Rose, and R. Myers, 1981. Dominance and the acquisition of superior home ranges in the lizard *Uta stansburiana*. *Ecol.*, *62*:888–893.

Franklin, W.J., 1974. The social behavior of the vicuña. *In* V. Geist and F. Walther (Eds.), *The Behaviour of Ungulates and Its Relation to Management*, pp. 477–487. IUCN, Morges, Switzerland.

Gartlan, J.S., and C.K. Brain, 1968. Ecology and social variability in *Cercopithecus aethiops* and *C. mitis*. *In* P.C. Jay (Ed.), *Primates: Studies in Adaptation and Variability*, pp. 253 292. Holt, Rinehart and Winston, New York.

Gaston, A.J., 1978. The evolution of group territorial behavior and cooperative breeding. *Am. Nat.*, *112*:1091–1100.

Geist, V., and F. Walther (Eds.), 1974. *The Behaviour of Ungulates and Its Relation to Management*. IUCN, Morges, Switzerland, Publ. 24, Vols. 1, 2.

Gill, F.B., and L.L. Wolf, 1975. Economics of feeding territoriality in the golden-winged sunbird. *Ecol.*, *56*:333–345.

Goss-Custard, J.D., 1970. Feeding dispersion in some over-wintering wading birds. *In* J.H. Crook (Ed.), *Social Behaviour in Birds and Mammals*, pp. 3–35. Academic Press, New York.

Goss-Custard, J.D., et al., 1972. Survival, mating and rearing strategies in the evolution of primate social structure. *Folia Primatol.*, *17*:1–19.

Greenwood, P.J., and P.H. Harvey, 1976. The adaptive significance of variation in breeding area fidelity of the blackbird (*Turdus merula* L.). *J. Anim. Ecol.*, *45*:887–898.

Greenwood, P.J., and P.H. Harvey, 1978. Foraging and territory utilization of blackbirds (*Turdus merula*) and song thrushes (*Turdus philomelos*). *Anim. Behav.*, *26*:1222–1236.

Gross, M.R., and A.M. MacMillan, 1981. Predation and the evolution of colonial nesting in bluegill sunfish (*Lepomis macrochirus*). *Behav. Ecol. Sociobiol.*, *8*:163–174.

Hamilton, W.D., 1971. Geometry for the selfish herd. *J. Theor. Biol.*, *31*:295–311.

Hamilton, W.J., III, R.E. Buskirk, and W.H. Buskirk, 1976. Defense of space and resources by chacma (*Papio ursinus*) baboon troops in an African desert and swamp. *Ecol.*, *57*:1264–1272.

Higdon, J.J.L., and S. Corrsin, 1978. Induced drag of a bird flock. *Am. Nat.*, *112*:727–744.

Hoogland, J.L., 1979a. Aggression, ectoparasitism, and other possible costs of prairie dog (Sciuridae, *Cynomys* spp.) coloniality. *Behaviour*, *69*:1–35.

Hoogland, J.L., 1979b. The effect of colony size on individual alertness of prairie dogs (Sciuridae: *Cynomys* spp.). *Anim. Behav.*, *27*:394–407.

Hoogland, J.L., 1981. The evolution of coloniality in white-tailed and black-tailed prairie dogs (Sciuridae: *Cynomys leucurus* and *C. ludovicianus*). *Ecol.*, *62*:252–272.

Hoogland, J.L., and P.W. Sherman, 1976. Advantages and disadvatanges of bank swallow (*Riparia riparia*) coloniality. *Ecol. Monogr. 46*:33–58.

Horn, H.S., 1968. The adaptive significance of colonial nesting in the Brewer's blackbird (*Euphagus cyanocephalus*). *Ecol., 49*:682–694.

Howard, R.D., 1978. The influence of male-defended oviposition sites on early embryo mortality in bull frogs. *Ecol., 59*:789–798.

Hrdy, S.B., and D.B. Hrdy, 1976. Hierarchical relations among female hanuman langurs (Primates: Colobinae, *Presbytis entellus*). *Science, 193*:913–915.

Hunsaker, D., and B.R. Burrage, 1969. The significance of interspecific social dominance in iguanid lizards. *Am. Midl. Nat., 81*:500–511.

Jaeger, R.G., R.G. Joseph, and D.E. Barnard, 1981. Foraging tactics of a terrestrial salamander: sustained yield in territories. *Anim. Behav., 27*:1100–1105.

Jarman, P.J., 1974. The social organization of antelope in relation to their ecology. *Behaviour, 48*:215–267.

Jewell, P.A., and C. Loizos, 1966. Play, exploration and territory in mammals. *Symp. Zool. Soc. Lond., 18*:1–280.

Kenward, R.E., 1978. Hawks and doves: attack success and selection in goshawk flights at wood-pigeons. *J. Anim. Ecol., 47*:449–460.

King, C.M., 1975. The home range of the weasel (*Mustela nivalis*) in an English woodland. *J. Anim. Ecol., 44*:639–665.

King, C.M., and P.J. Moors, 1979. On co-existence, foraging strategy and the biogeography of weasels and stoats (*Mustela nivalis* and *M. erminea*) in Britain. *Oecologia, 39*:129–150.

Kleiman, D.G., and J.F. Eisenberg, 1973. Comparisons of canid and felid social systems from an evolutionary perspective. *Anim. Behav., 21*:637–659.

Klingel, H., 1974. A comparison of the social behaviour of the Equidae. *In* V. Geist and F. Walther (Eds.), *The Behaviour of Ungulates and Its Relation to Management*, pp. 124–131. IUCN, Morges, Switzerland.

Klomp, H., 1972. Regulation of the size of bird populations by means of territorial behaviour. *Netherlands J. Zool., 22*:456–488.

Kodric-Brown, A., 1977. Reproductive success and the evolution of breeding territories in pupfish (*Cyprinodon*). *Evol., 31*:750–766.

Krebs, J.R., 1971. Territory and breeding density in the great tit *Parus major* L. *Ecol., 52*:2–22.

Krebs, J.R., 1974. Colonial nesting and social feeding as strategies for exploiting food resources in the great blue heron (*Ardea herodias*). *Behav., 51*:99–131.

Krebs, J.R., and C.J. Barnard, 1980. Comments on the function of flocking in birds. *Proc. Int. Ornith. Cong., 17*:795–799.

Krebs, J.R., and N.B. Davies, 1981. *An Introduction to Behavioural Ecology*. Sinauer Associates, Sunderland, MA.

Krebs, J.R., M.H. MacRoberts, and J.M. Cullen, 1972. Flocking and feeding in the great tit *Parus major*—an experimental study. *Ibis, 114*:507–530.

Krekorian, C.O., 1976. Home range size and overlap and their relationship to food abundance in the desert iguana, *Dipsosaurus dorsalis*. *Herpetologica, 32*:405–412.

Kruuk, H., 1964. Predators and antipredator behaviour of the black-headed gull (*Larus ridibundus* L.). *Behav.* [*Suppl.*], *11*:1–129.

Kruuk, H., 1972. *The Spotted Hyena*. University of Chicago Press, Chicago.

Kruuk, H., 1976. Functional aspects of social hunting by carnivores. *In* G. Baerends, C. Beer, and A. Manning (Eds.), *Function and Evolution in Behaviour*, pp. 119–141. Oxford University Press, Oxford.

Lance, A.N., 1978. Territories and the food plant of individual red grouse. II. Territory size compared with an index of nutrient in heather. *J. Anim. Ecol., 47*:307–313.

Lazarus, J., 1972. Natural selection and the functions of flocking in birds: a reply to Murton. *Ibis, 114*:556–558.

Lissaman, P.B.S., and C.A. Shollenberger, 1970. Formation flight of birds. *Science, 168*:103–105.

Lockie, J.D., 1966. Territory in small carnivores. *Symp. Zool. Soc. Lond., 18*:143–165.

Lyon, D.L., J. Crandall, and M. McKone, 1977. A test of the adaptiveness of interspecific territoriality in the blue-throated hummingbird. *Auk, 94*:448–454.

Magnuson, J.J., 1962. An analysis of aggressive behavior, growth and competition for food and space in medaka (*Oryzias latipes* [Pisces, Cyprinodontidae]). *Can. J. Zool., 40*:313–363.

Major, P.F., 1978. Predator-prey interactions in two schooling fishes, *Caranx ignobilis* and *Stolephorus purpureus. Anim. Behav., 26*:760–777.

Metzger, L.H., 1967. An experimental comparison of screech owl predation on resident and transient white-footed mice (*Peromyscus leucopus*). *J. Mammal., 48*:387–391.

Morse, D.H., 1970. Ecological aspects of some mixed-species foraging flocks of birds. *Ecol. Monogr., 40*:119–168.

Murray, B.G., 1981. The origins of adaptive interspecific territorialism. *Biol. Rev., 56*:1–22.

Murton, R.K., 1967. The significance of endocrine stress in population control. *Ibis, 109*:622–623.

Murton, R.K., 1971. Why do some birds feed in flocks? *Ibis, 113*:534–536.

Murton, R.K., A.J. Isaacson, and N.J. Westwood, 1971. The significance of gregarious feeding behaviour and adrenal stress in a population of wood-pigeons, *Columba palumbus. J. Zool.* (Lond.), *165*:53–84.

Myers, J.P., P.G. Connors, and F.A. Pitelka, 1979. Territory size in wintering sanderlings: The effect of prey abundance and intruder density. *Auk, 96*:551–561.

Neill, S.R., and J.M. Cullen, 1974. Experiments on whether schooling by their prey affects the hunting behaviour of cephalopods and fish predators. *J. Zool.* (Lond.), *172*:549–569.

Nudds, T.D., 1978. Convergence of group size strategies by mammalian social carnivores. *Am. Nat., 112*:957–960.

Orians, G.H., and M.F. Willson, 1964. Interspecific territories of birds. *Ecol., 45*:736–745.

Owen-Smith, N., 1977. On territoriality in ungulates and an evolutionary model. *Q. Rev. Biol., 52*:1–38.

Page, G., and D.F. Whitacre, 1975. Raptor predation on wintering shorebirds. *Condor, 77*:73–83.

Parker, G.A., 1974. Assessment strategy and the evolution of fighting behaviour. *J. Theor. Biol., 47*:223–243.

Powell, G.V.N., 1974. Experimental analysis of the social value of flocking by starlings (*Sturnus vulgaris*) in relation to predation and foraging. *Anim. Behav., 22*:501–505.

Pulliam, H.R., K.A. Anderson, A. Misztal, and N. Moore, 1974. Temperature-dependent social behaviour in juncos. *Ibis, 116*:360–364.

Rabenold, K.N., and C.R. Christensen, 1979. Effects of aggregation on feeding and survival in a communal wren. *Behav. Ecol. Sociobiol., 6*:39–44.

Radadov, D.V., 1973. *Schooling in the Ecology of Fish.* [Translation from Russian.] John Wiley & Sons, New York.

Rand, A.S., 1967. The adaptive significance of territoriality in iguanid lizards. *In* W.W. Milstead (Ed.), *Lizard Ecology: A Symposium,* pp. 109–115. University of Missouri Press, Columbia.

Rappole, J.H., and D.W. Warner, 1976. Relationships between behavior, physiology and weather in avian transients at a migration stopover site. *Oecologia, 26*:193–212.

Reed, T.M., 1982. Interspecific territoriality in the chaffinch and great tit on islands and the mainland of Scotland: playback and removal experiments. *Anim. Behav., 30*:171–181.

Robertson, D.R., H.P.A. Sweatman, E.A. Fletcher, and M.G. Cleland, 1976. Schooling as a mechanism for circumventing the territoriality of competitors. *Ecol., 57*:1208–1220.

Rodman, P.S., 1981. Inclusive fitness and group size with a reconsideration of group size in lions and wolves. *Am. Nat., 118*:275–283.

Rothstein, S.I., 1979. Gene frequencies and selection for inhibitory traits, with special emphasis on the adaptiveness of territoriality.

Am. Nat., 113:317–331.

Schaller, G.B., 1972. *The Serengeti Lion.* University of Chicago Press, Chicago.

Seghers, B.H., 1974. Schooling behavior in the guppy (*Poecilia reticulata*): an evolutionary response to predation. *Evol., 28*:486–489.

Simon, C.A., 1975. The influence of food abundance on territory size in the iguanid lizard, *Sceloporus jarrovi. Ecol., 56*:993–998.

Smith, C.C., 1968. The adaptive nature of social organization in the genus of tree squirrels *Tamiasciurus. Ecol. Monogr., 38*:31–63.

Snapp, B.D., 1976. Colonial breeding in the barn swallow (*Hirundo rustica*) and its adaptive significance. *Condor, 78*:471–480.

Southwick, C.H., and M.F. Siddiqi, 1974. Contrasts in primate social behavior. *Bioscience, 24*:398–406.

Spinage, C.A., 1969. Territoriality and social organization of the Uganda defassa waterbuck, *Kobus defassa ugandae. J. Zool.* (Lond.), *159*:329–361.

Stamps, J.A., 1977. Social behavior and spacing patterns of lizards. *In* C. Gans and D.W. Tinkle (Eds.), *Biology of the Reptilia.* Vol. 7, pp. 265–334, Academic Press, New York.

Stamps, J.A., 1977. The relationship between resource competition, risk, and aggression in a tropical territorial lizard. *Ecol., 58*:349–358.

Stenger, J., 1958. Food habits and available food of ovenbirds in relation to territory size. *Auk, 75*:335–346.

Stokes, A.W. (Ed.), 1974. *Territory.* Dowden, Hutchinson & Ross, Stroudsburg, PA.

Tenaza, R., 1971. Behavior and nesting success relative to nest placement in adélie penguins (*Pygoscelis adeliae*). *Condor, 73*:81–92.

Thompson, W.A., I. Vertinsky, and J.R. Krebs, 1974. The survival value of flocking in birds: a simulation model. *J. Anim. Ecol., 43*:785–820.

Tompa, F.S., 1962. Territorial behavior: The main controlling factor of a local song sparrow population. *Auk, 79*:687–697.

Tompa, F.S., 1963. Behavioral response of song sparrows to differential environmental conditions. *Proc. Int. Ornith. Congr* (XIII):729–739.

Tullock, G., 1979. On the adaptive significance of territoriality: comment. *Am. Nat., 113*:772–775.

Walta, E.C., 1982. Resource characteristics and the evolution of information centers. *Am. Nat., 119*:73–90.

Ward, P., and A. Zahavi, 1973. The importance of certain assemblages of birds as "information-centres" for food-finding. *Ibis, 115*:517–534.

Waser, P.M., 1981. Sociality or territorial defense? The influence of resource renewal. *Behav. Ecol. Sociobiol., 8*:231–237.

Waser, P.M., and R.W. Wiley, 1979. Mechanisms and evolution of spacing in animals. *In* P. Marker and J.G. Vandenbergh (Eds.), *Handbook of Behavioral Neurobiology,* pp. 159–223. Plenum Publishing, New York.

Webb, D.R., 1981. Macro-habitat patch structure, environmental harshness, and *Marmota flaviventris. Behav. Ecol. Sociobiol., 8*:175–182.

Weeden, J.S., and J.B. Falls, 1959. Differential response of male ovenbirds to recorded songs of neighboring and more distant individuals. *Auk, 76*:343–351.

Weihs, D., 1975. Some hydrodynamical aspects of fish schooling. *In* T.Y.T. Wu, C.J. Brokaw, and C. Brennen (Eds.), *Swimming and Flying in Nature.* Vol. 2, pp. 703–718. Plenum Publishing, New York.

Wells, K.D., 1977. The social behavior of anuran amphibians. *Anim. Behav., 25*:666–693.

Wells, K.D., 1980. Behavioral ecology and social organization of a dendrobatid frog (*Colostethus inguinalis*). *Behav. Ecol. Sociobiol., 6*:199–209.

Wiklund, C.G., and M. Andersson, 1980. Nest predation selects for colonial breeding among fieldfares (*Turdus pilaris*). *Ibis, 122*:363–366.

Wiley, R.H., and M.S. Wiley, 1980. Spacing and timing in the nesting ecology of a tropical blackbird: comparison of populations in different environments. *Ecol. Monogr.,*

50:153–178.
Wilkinson, G.S. and G.M. English-Loeb, 1982. Predation and coloniality in cliff swallows (*Petrochelidon pyrrhonota*). *Auk, 99*:459–467.
Wilson, E.O., 1975. *Sociobiology.* Belknap Press, Cambridge, MA.
Wittenberger, J.F., 1981. *Animal Social Behavior.* Duxbury, Boston.
Wolff, J.O., and W.Z. Lidicker, 1981. Communal winter nesting and food sharing in taiga voles. *Behav. Ecol. Sociobiol., 9*:237–240.
Wrangham, R.W., 1980. An ecological model of female-bonded primate groups. *Behav., 75*:263–300.
Zahavi, A., 1971. The social behaviour of the white wagtail (*Motocilla alba alba*), wintering in Israel. *Ibis, 113*:203–211.

12 *Sex*

SEXUAL SYSTEMS

Vertebrates typically reproduce sexually: a male and a female each contribute a haploid gamete to form a diploid zygote. Just a few populations reproduce asexually: females produce diploid female offspring, genetically identical to themselves and to each other, with no collaboration by a male. Why is sexual reproduction so common in vertebrates and in many other groups of organisms? The evolution of sex has intrigued biologists for years, and controversy is still raging in the literature on the subject. Although the final word is not yet in, a few important aspects of this subject can be presented here.

Asexual reproduction has some potential advantages. First, there is no need to produce structures or behaviors designed to attract or find a mate. In addition, no time is spent in searching for mates, nor is there any risk of not finding one. Therefore, there is no "cost of mating." Second, there is no "cost of making males." The offspring of a sexual female usually are only partly females (the rest are males), but all the offspring of an asexual female are female. If both kinds of females produce equal total numbers of offspring, the proportion of sexually reproducing females in a population should get smaller with each generation, and sexual reproduction might eventually disappear. Third, asexual reproduction usually maintains the parental genotype, so that successful genotypes are not disassembled in each generation. Meiosis and recombination in sexual reproduction can break up successful gene combinations. Fourth, asexual parents pass on to future generations all of their genes, not just (approximately) half of them, which is the case for sexually reproducing individuals. This is called the "cost of meiosis." The importance of this cost to the evolution of sex has been debated; some researchers believe that this cost is not paid by individuals but by genes and that it is a less important obstacle to the evolution of sex than some of the other costs.

On the other hand, we can also readily imagine some advantage to sexual reproduction. Despite the potentially high costs of mating and the other possible costs, sex offers the benefit of producing offspring that differ among themselves (as well as from the parents). When environments vary in space and time, parents increase the likelihood of having some successful offspring if they produce a varied array. If all the young were alike, then clearly they could succeed in only one kind of spatio-temporal "patch," which might not be available at that time or place. Not only does recombination generate a diversified progeny, but it also increases the probability that some novel genetic combination will have unusually high fitness even in a constant environment. Although the same process also leads to novel combinations that are unusually poor, the value of the highly fit may outweigh the price of producing some unfit offspring. In fact, the production of such low-quality young might help eliminate deleterious genes from the lineage.

We know that many environments are highly variable. Some winters are colder than others, some summers are drier, and blizzards and hailstorms occur irregularly; all of these short-term variations are superimposed on long-term climatic and geologic cycles. Spatial variation is instantly apparent to even a casual observer of fields and forests, mountains and deserts, or, on a smaller scale, sandbars and mud flats, or upper and lower sides of a fallen log. Another kind of variation is provided by other organisms: the biological rather than the physical environment. Competitors, predators, and prey or host organisms are all capable of changes that can affect a given species. In fact, coevolutionary interactions among species may provide a continually changing environment favoring reproductive systems that foster variability of the progeny. The importance of the biological environment to the evolution of sexual systems has been emphasized both for animals and for plants. Glesener and Tilman (1978) present evidence that sexual reproduction in animals is most prevalent in areas inhabited by large numbers of potentially interacting species and that asexual reproduction occurs most commonly in less biologically diverse areas.

Most animals are capable of phenotypic flexibility in responding to environmental changes: behavior may change, physiological state may alter, and so on. Because this is one means of coping with a variable environment, we could ask why sex might be adaptive to variable environments if phenotypic plasticity is present. If an animal has responded phenotypically to one kind of stress, its ability to react phenotypically to a second stress is often impaired. A physiological state that enhances tolerance of moderate heat loads might decrease resistance to severe heat stress, for example, or an animal subjected to excessive physical stress may be more subject to disease. Bateson (1963) argued that it is therefore advantageous for certain characteristics to be controlled genetically—especially those characteristics relevant to regular and predictable environmental stresses—whereas phenotypic plasticity should be a more common way of coping with unpredictable stresses. Sexual reproduction may offer the best of both worlds: genetic control of many characteristics in each individual with new genetic possibilities in subsequent generations, plus a degree of phenotypic flexibility.

However plausible may be the association of variable progeny with variable environments, there are some problems to be solved before accepting it as an explanation for the evolution of sex. First, there is little or no empirical evidence on the fitness of vertebrate offspring with different genotypes in different environments. Such information is extremely difficult to obtain, but it is nevertheless necessary as direct support for the variability hypothesis. Second, there are large numbers of asexual species that seem to be successful; many of these coexist, in the same environments, with related sexual species. They may occupy relatively invariant environments or they may have very generalized genotypes capable of success in a variety of environments. They also may have previously underrated means of producing variant offspring. Third, recombination may have other possible functions, such as the repair of DNA, that could contribute to the evolution of sexual processes.

Sexual reproduction may have several consequences at the population level as well as at the individual level. Often it maintains genetic variation, which may permit a rapid phylogenetic change or prevent a population from going extinct if the environment changes drastically. If the environment changes from one generation to the next, there is a greater probability that some genotype in the population will be suited to the changed conditions than if all offspring were identical to their parents and thus suited to the previous conditions. The population would then be less likely to become extinct and might be able to evolve more rapidly. The extent to which such population consequences have contributed to the prevalence of sexual reproduction is still debated.

Separate Sexes or Hermaphroditism?

Most kinds of vertebrates have separate males and females, with each individual exercising a single sexual role. A few vertebrates are hermaphroditic, however: The occurrence of functional gonads in both sexes in a single individual is known as androgyny, or, more commonly, hermaphroditism. Hermaphroditus was the mythological son of Hermes and Aphrodite. His body was joined to that of a water nymph when she pleaded with the Greek gods that he never be separated from her!

The typical form of hermaphroditism is simultaneous—functionally male and female at the same time. This sexual arrangement is known to occur in a number of families of fishes; it is particularly common in certain families such as the Serranidae and Labridae (Smith, 1975). Although it occurs especially in coral-reef fishes, it also occurs in deep-water types, such as the aleposauroids. Simultaneous hermaphrodites in vertebrates usually cross-fertilize each other, one individual donating sperm to and receiving sperm from another. All other mature individuals are potential mates, not just half of them, as is the case when sexes are separate, so mate-finding is facilitated. Whether mate-finding is, indeed, a problem for these fishes can be argued, however (Fischer, 1981).

Hermaphroditism may also allow an increase of reproductive success at a relatively low cost when the gain in fitness from a single sexual role is very uncertain or when a small expenditure on sperm production increases the number of offspring without incurring the considerable costs of egg production. If animals live in groups of unpredictable sexual composition, a paucity of males could make sperm production (by an hermaphrodite) very profitable, because the hermaphrodite could then father many offspring. Because sperm are thought to be much less expensive to produce than eggs (which contain nutrition-rich yolk), the expenditure of resources on sperm production could be small, detracting little from the production of eggs, while greatly increasing the number of offspring from the single parent. Likewise, rarity of females would make egg production advantageous to an hermaphrodite because its success as a male would probably be low in the face of competition from males; egg production, however, would have a good chance of being successful. The ability to shift the balance of sexual function might make hermaphroditism advantageous (see also Borgia and Blick, 1980); however, we do not know if existing simultaneous hermaphrodites actually encounter the conditions favoring such flexibility.

Fischer (1981) and others have suggested that a small expenditure on sperm production, in addition to a large expenditure on egg production, could greatly enhance the total reproductive output of each individual at small cost. We are then left with the question of why simultaneous hermaphroditism is not more common in vertebrates. Constraints on its evolution may occur if the costs of successful male function are high in some species. Success as a male may require elaborate behaviors or morphological devices whose development would detract markedly from reproductive success of the female role, so that the overall reproductive success of the hermaphrodite would be diminished. The underlying conditions making possible a significant gain in reproductive success at low cost are poorly known. Various aspects of the social system may facilitate simultaneous hermaphroditism in some cases (Fischer 1980, 1981, for the black hamlet, *Hypoplectrus nigricans*). Ecological constraints may limit the advantages of specializing on a male role: The harlequin bass (*Serranus tigrinus*), for example, spawns only near sunset, perhaps as a means of reducing predation on eggs; the short time available for spawning may limit the number of females a male can court (Pressley, 1981).

Self-fertilizing (simultaneous) hermaphrodites are rare. The Caribbean cyprinodontid *Rivulus marmoratus* inhabits coastal waters, where it is subject to high mortality from violent storms. Not all populations of this species are entirely hermaphroditic; males are known from some populations. The ability of hermaphrodites to self-fertilize permits reproduction without a mate and may thus enhance the ability to recolonize a repeatedly disturbed habitat. Self-fertilization also means that the progeny will not be as varied as is possible with cross-fertilization, so the putative advantages of sexual reproduction have been largely relinquished.

Sequential hermaphroditism refers to functioning first as one sex and then as the other during an individual's lifetime. When the initial sexual state is male, the condition is known as protandry (first-male); when the initial state is female, it is called protogyny (first-female). This system can reduce inbreeding by preventing sibling matings, since siblings will normally be the same age and, hence, of the same sex at that time. In addition, sequential hermaphroditism may be selected when one sex gets a greater age- or size-dependent advantage than the other. For instance, if large females can lay more eggs than small ones when small males are at no disadvantage to larger males, protandry may be favored. When large males have some mating advantage over small ones—such as dominance in competition for mates or territories—but size makes less difference to females, protogyny may be favored. Whichever sex acquires less increase of fitness from having a large size would occur first in the sequence. Populations of *Pseudolabrus celidotus*, a protogynous New Zealand wrasse, differ in the timing of sex change (Jones, 1980). Individuals in certain populations grow fast and change sex at an earlier age than those in other populations, indicating that size, more than age, is important to success in a male role.

A fascinating case of sequential hermaphroditism is reported for a coral-reef "cleaner" fish, *Labroides dimidiatus*, in the wrasse family (Robertson, 1972). This species is protogynous and the change from female to male is controlled by social interaction. These fishes live in groups composed of a territorial male with his harem of three to six mature females and several immature individuals. A strong, often linear, dominance hierarchy exists: Larger, older individuals are dominant over those that are smaller. The largest and oldest individual is a male that actively dominates all others in

Figure 12–1 *Labroides dimidiatus.* An intriguing example of sequential hermaphroditism. (Photo by J. Burton, courtesy of Bruce Coleman, Inc.)

the group. If the male dies, the next largest individual becomes functionally male, unless neighboring males interfere and occupy the dead male's territory. The change in sex begins about two hours after the dominant male's death and is complete two to four days later.

The advantage of this system remains hypothetical at present. Some limitation of suitable habitat and feeding sites may make it disadvantageous for the subordinate individuals to move away, or large size may be advantageous in aggressive encounters. If aggression is sufficiently severe, selection might favor large size and greater aggressiveness in the sex whose gametes are cheapest to produce, namely the male, whereas smaller females can shunt energy not spent in aggression toward the production of their relatively expensive eggs. Neither the specific benefits nor the possible costs of sex-changing are well understood at this time.

Sequential hermaphroditism is also known or thought to be socially controlled in another reef fish, *Anthias squamipinnis,* in some scarids, and in clownfishes (*Amphiprion*). *Anthias* males are greatly outnumbered by the smaller females (Fishelson, 1975), which may indicate that hiding places are fewer and predation is more severe on larger individuals; thus, there may be a cost that accompanies the attainment of large size and reproductive success as a male. The clownfishes, or anemone fishes, typically live in pairs, but in this case, the system is protandrous. Females are larger than males and defend the home anemone against intruders. One species, *A. akallapisos,* lives in colonial anemones, and several small males may live with a large female, although only one apparently breeds (Fricke, 1979). We do not know precisely why the clownfishes are protandrous but the other species discussed are protogynous.

Why is hermaphroditism so uncommon in vertebrates, although it occurs regularly in certain invertebrates and in most plants? If separation of sexes is indeed adaptive, it must be true that sexual specialization—maleness or femaleness—results in greater reproductive success than hermaphroditism. Several factors may be involved. First, animal reproductive organs seldom serve both male and female functions, but in plants, floral parts such as petals may do double-duty. This may mean that the cost of maintaining two separate sexual systems, with associated behaviors, is greater for most animals than it is for a plant. Second, adult vertebrates are typically mobile, so that mate-finding is less of a problem than it is for many sessile invertebrates or plants. Third, mobility and the potential for complex behavior patterns makes possible elaborate interactions among individuals. Males may fight with each other for access to females and they may coax reluctant females by fancy courtship procedures. Once these interactions are possible, the advantages of sexual specialization are potentially greater. A doubling of the energy devoted to sperm production and male behavior, for instance, may more than double reproductive success, even though less energy can then be devoted to femaleness. When the payoff is sufficiently great, allocation of resources to expression of a single sex becomes advantageous.

Warner (1978, 1980) suggested that terrestrial living may make the evolution of hermaphroditism unlikely. Life on land apparently has selected for heightened investment by females in their young. Terrestrial vertebrates tend to produce larger young than aquatic forms, in part because very tiny food items on which a young one could feed are less available on land. Furthermore, the risks of desiccation are greater and additional embryonic membranes provide some protection against drying. Terrestrial young are often borne within the mother's body for some time and, in general, the female role seems to have become more specialized. The greater the investment and role specialization, the less likely is a switch of sexual roles.

Parthenogenesis

Asexual reproduction occurs regularly in a few kinds of vertebrates. Parthenogenesis ("virgin birth") refers to the production of offspring by females without participation by male gametes. Regular parthenogenesis is known among fishes, amphibians, and reptiles. No males are even known for at least certain populations of a variety of lizards, *Typhlops* snakes, *Ambystoma* salamanders, and several species of fishes. Among homeotherms, regular, spontaneous parthenogenesis is reported only for turkeys; parthenogenetically produced turkeys are always male.

Ambystoma tremblayi is closely related to *A. laterale,* and they coexist in southeastern Michigan. *Tremblayi* is composed entirely of females that use spermatophores (packages of sperm) from male *laterale* to stimulate development of their eggs. The sperm make no genetic contribution at all, so, in a sense, the *tremblayi* females are "parasitizing" the male *laterale* "hosts." The use of nonconspecific sperm to stimulate egg development is called gynogenesis. Females of *A. tremblayi* are sometimes courted by *laterale* males and so obtain spermatophores that way, but they apparently also pick up spermatophores left on the bottom of the pond in other courting bouts. Therefore, even if *laterale* males somehow discriminated against *tremblayi* females, breeding is not prevented.

Gynogenetic populations of *Ambystoma tremblayi* are triploid (there are three chromosome sets instead of the normal two) and probably originated from hybridization of *laterale* with *A. jeffersonianum*. Fecundity of the gynogenetic females is lower than for normal females, larval development is slower, and larval survivorship is less. Furthermore, diploid larvae have more severe competitive effects on the body size and survivorship of triploid larvae than vice versa. Present coexistence of *A. tremblayi* and *A. laterale* is probably transitory; it is headed toward extinction of the triploid form or toward the reproductive independence of the triploids by the evolution of total independence from males or of a tetraploid form (for example, a $3n$ egg with a $1n$ sperm from a diploid male) capable of bisexual reproduction (Wilbur, 1971).

Unisexuality also occurs in the atherinid fish, *Menidia,* in Texas, in the cyprinid goldfish, *Carassius auratus,* in Russia, and in several populations of live-bearing poeciliid fishes (including at least five species of *Poeciliopsis* in Mexico and one of *Poecilia* in Texas and Mexico). All the unisexual populations of *Poeciliopsis* appear to have arisen by hybridization between normal species and seem to be concentrated in transitional habitats. Two of the known unisexual populations are gynogenetic and triploid. Three others are diploid and have a very unusual form of meiosis. Sperm of the "host" males not only stimulate development of the eggs but penetrate the egg, and the paternal chromosomes contribute to the phenotype of the offspring. When the offspring mature and form gametes, however, only the maternal chromosomes are contained in the eggs; all the paternal ones are discarded during meiosis. Thus, even though the offspring exhibit both paternal and maternal characteristics, only the maternal ones are passed on. Unisexual reproduction is maintained.

The poeciliids are commonly found in isolated pools in dry streambeds during the dry season, but isolates may become confluent during the rainy season when the stream is full. The coexistence of normal and all-female populations of *Poeciliopsis* appears to be maintained as a result of certain behavioral characteristics of males of the "host" populations. When density of the males is high, males interact with each other and form dominance hierarchies. Dominant males control access to conspecific females; subordinate males, although they prefer conspecific females if given a choice, then court and mate with the gynogenetic females. A feedback mechanism is postulated (Fig. 12–2) in which the success of gynogenetic females at high male densities results in a relative increase of gynogenetic females and a concomitant decrease of normals, which leads to a decrease in the number of subordinate males and a subsequent lowered success of gynogenetic females. The relative frequency of diploids increases as a

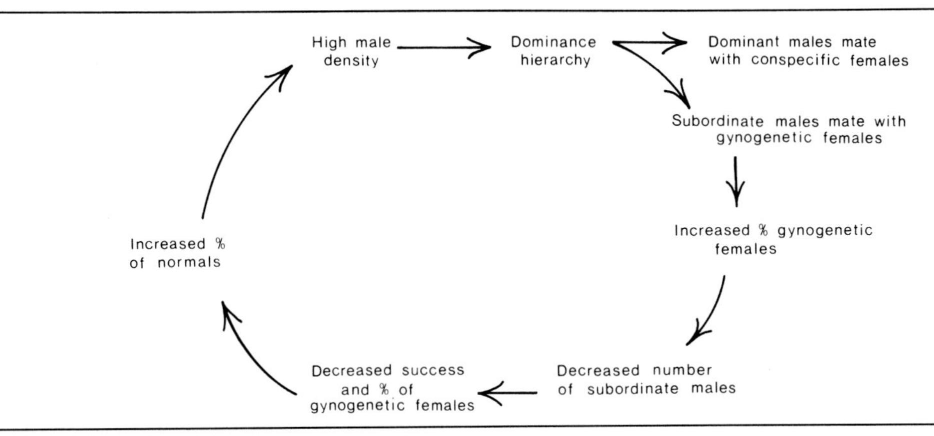

Figure 12–2 In this proposed feedback mechanism, the success of gynogenetic *Poeciliopsis* changes with the density of normal males of the "host" population. (From McKay, F.E., 1971. Behavioral aspects of population dynamics in unisexual-bisexual *Poeciliopsis* (Pisces: Poeciliidae). *Ecol.*, 52:778–790.)

result, and the cycle begins anew. This system reportedly can be maintained as long as random extinction of one of the forms does not occur and as long as gynogenetic females average at least half the fitness of normal females (McKay, 1971; Moore and McKay, 1971). Furthermore, the fitness of the unisexual females appears to be particularly high in habitats intermediate between the habitats of the normal species that formed the hybrid unisexual populations; the fitness of the normal species is probably low in these intermediate habitats (Moore, 1976). Seasonal changes in fitness of diploid and triploid forms may also contribute to maintenance of the polymorphism (Thibault, 1978).

The situation in parthenogenetic species of *Cnemidophorus* lizards is rather similar to that of parthenogenetic fishes and salamanders. The nine known all-female populations in the arid southwest of North America all seem to have a hybrid origin from normal species. Some populations are diploid, some are triploid, and some are mixed. They all occupy woodland or grassland habitats that are in some way marginal for this genus and for which normal *Cnemidophorus* species are not well adapted. The unisexual populations are characteristic of habitats at the periphery of the habitat spectrum (rocky areas, higher altitudes) of normal species. Perpetuation of the parthenogenetic species may depend on the availability of unoccupied, marginal habitats, although parthenogenetic and sexual populations apparently intermingle in some cases (Cuellar, 1979). Unisexual *Cnemidophorus* in Brazil are found in disturbed habitat near human settlement.

Asexual reproductive patterns may be adaptive to individuals of species exploiting ephemeral habitats that are continually destroyed and recreated.

Individuals that can reproduce alone could have an initial advantage over those that must find mates. Problems of mate-finding may be particularly severe at extremely low population densities, as might characterize populations invading a new habitat. Or perhaps asexual species are really "fugitive species" found only in transitory habitats that they occupy rapidly before more competitive species can establish themselves. Theoretically, exceptionally stable environments might also select for parthenogenesis by providing no advantage for a varied progeny and by releasing time or energy needed for mate-finding to be used directly in production of young. Evidence from vertebrates does not provide much support for this alternative hypothesis, although examples may be found in invertebrates and plants.

It is also possible that parthenogens possess broadly adapted genotypes that enable them to be successful in a variety of ecological situations (M. Lynch, ms.). Whether they arise by hybridization or by fortuitous genetic changes, parthenogens may persist if they are ecologically somewhat different from the parental types and if they are cytogenetically prevented from successfully backcrossing with the parents. The ecological and evolutionary basis for the persistence of vertebrate parthenogens in a family tree of sexually reproducing forms is still controversial.

SEX DETERMINATION

Most vertebrates seem to have genetic sex determination. The particular sex chromosomes received by each zygote determine its sex. Sex chromosomes often are morphologically distinguishable, but there is no obvious reason why sex chromosomes must look different; genetic sex determination may occur without differentiated chromosomes. Mammalian females are the homogametic sex, producing gametes that contain the same sex chromosome (called the X chromosome); thus, females have two X chromosomes. Male mammals are XY and heterogametic, producing some gametes with an X chromosome and some with a Y. Sex of offspring is determined by the sex chromosomes in the gametes that fuse to form the zygote: Two Xs (or the absence of a Y) produce a female, and an X and a Y produce a male. This situation need not mean, however, that males determine the sex of their offspring, for females may be able to influence the competitive ability and acceptance of X- and Y-bearing sperm.

In contrast, birds have homogametic males and heterogametic females. The situation in the other classes is more complicated, however. Lizards tend to have male heterogamety, but snakes generally have female heterogamety. Fishes and amphibia have both patterns; in fact, both male and female heterogamety have been reported within a single species of fish (*Xiphophorus maculatus*) and in one mammal (the wood lemming, *Myopus schisticolor*). It seems likely that sex chromosomes have originated repeatedly on the course of vertebrate evolution; it is not clear that it makes any adaptive differences which sex is heterogametic.

Environmental sex determination is known for many vertebrates; in some cases, it can override genetic sex determination (Bull et al., 1982). Environmental determination of sex occurs in alligators (*Alligator mississippiensis*); male offspring predominate in wild populations, but females are more common in "farm" populations, in which food resources are unusually abundant (Nichols and Chabreck, 1980). The microclimate of the nest is important in determining sex of the developing young. Environmental sex determination is quite common (but not universal) in turtles in which nest temperature is known to have significant effects (Bull and Vogt, 1979; Morreale et al., 1982). Low nest temperatures often tend to favor males, whereas high temperatures may favor females. Thus, female turtles may be able to control the sex of their offspring by choosing particular sites for egg deposition. Temperature-dependent sex determination is also known in some lizards (Bull, 1980).

The Atlantic silverside (*Menidia menidia*) has both genetic and environmental sex determination; sex is determined during a specific interval in larval development (Conover and Kynard, 1981). As also found for reptiles, temperature is an important environmental factor in this system: Warm temperatures favor females and cold ones favor males. The progeny of different mothers differed greatly in their sensitivity to temperature. Females predominate early in the season in natural populations when the population is low because of mortality over the winter. These females, then, are older and larger than most of the males in the annual breeding population. Perhaps the temperature-dependent sexual lability is related to the advantages of large size for females (as suggested by the investigators) or to the great reproductive success to be achieved by mated females early in the season (males might not achieve much early success if males outnumbered females early on).

Environmental sex determination is also known in other fishes. Some instances of socially controlled sex changes have already been discussed. Clones of the hermaphroditic fish *Rivulus marmoratus* exhibited temperature effects on sex expression in the laboratory, although such sex shifts are not known in the wild (Harrington, 1975).

Sex determination in mammals and birds is normally genetically determined. Pathological shifts can be induced, however, by the injection of DDT into eggs; this treatment feminized male California gulls (*Larus californicus*), which were then unable to breed at all as adults (Fry and Toone, 1981). Other pathologies include "free-martins" in cattle, in which a female calf that is a twin to a male calf is masculinized by hormone transfer from her twin brother. Such females are also infertile because their gonads have active male complements of sex chromosomes (Ohno et al., 1976).

Possible adaptive advantages of genetic or environmental sex determination are not well known. It is reasonable to conjecture that environmental sex determination can be adaptive when the sex-dependent fitness of an offspring is not predictable in advance and that genetic sex determination may be advantageous when sexual roles are so specialized that their development must begin as early as possible. But how unpredictable must the success be and what conditions make it so? And can sexual role specialization really be so constraining that sex shifts early in development (which is the case in the fishes and reptiles discussed) would be detrimental to proper execution of the roles? A degree of parental control over sex expression of offspring is sometimes possible in both kinds of sex-determining systems, so this alone is unlikely to affect the evolutionary dichotomy between the two. Phylogenetic constraints may limit the possibilities for certain taxa; either sex-determining system may be difficult to attain in some taxa simply because the right preconditions for its evolution are lacking. The possible costs attendant upon each method of sex determination are also unknown.

SEX RATIOS

A sex ratio is defined here as the ratio of the number of males to the number of females. In this section, we are concerned particularly with the sex ratio of the offspring produced by a set of parents. Offspring sex ratios may have strong effects on the fitness of their parents. The evolution of sex ratios has been subjected to a variety of theoretical treatments, sometimes with divergent conclusions. A summary of the most basic theoretical considerations is presented here. The evolution of sex ratios is a problem rife with confusion and debate; it is the subject of much contemporary research; a recent review is provided by Charnov (1982).

Sex ratios have been labeled at three times in a life history. The primary sex ratio is that at conception; the secondary sex ratio is that at birth or hatching; and the tertiary sex ratio is that at the age of reproduction. Obviously, these ratios can be different if there is differential mortality of males and females during development. For instance, we know that young human males have a higher probability of death than females and that the primary sex ratio favors males; the tertiary ratio in early adulthood is about 1 : 1, however.

The classical model for explaining sex ratios is that of Fisher (1958). The model assumes random mating in the population, genetic determination of sex by autosomal genes (not on sex chromosomes), sex-ratio genes affecting only sex ratio and no other traits (if they did, advantages accruing through the other traits could compensate for any disadvantages relating to sex ratio), constant average fitness (if fitness varies, selection could favor shifting sex ratios), and equilibrium conditions in general. Fisher's model originally applied to sex ratios at the level of an entire population, but modifications of the original model suggested that it could be applied to individuals as well (MacArthur, 1965; Verner, 1965). The basic argument is framed, not in terms of sex ratio directly, but in terms of the total parental expenditure on offspring of each sex. In general, the collective costs of producing male offspring will tend to equal the collective costs of producing female offspring. For example, whenever the costs of making sons are less than the costs of making daughters, parents with an inherent tendency to make sons will have a temporary advantage; thus raising the total son-making costs until they again equal those for daughters. Fisher argued that this must be true because all sexually reproducing animals have a parent of each sex; also, of necessity, the genetic contributions of all males must equal that of all females. Because all sons are worth as much to parental fitness as all daughters, selection should favor equal parental expenditures on both sexes. The primary sex ratio is then determined by the per-capita cost of individuals of each sex. If young males require more food than young females because they are larger or have higher metabolic rates, for example, the cost per male may be higher than the cost per female; consequently, fewer males can be reared. This appears to be true for grackles (*Quiscalus quiscula*) and red-winged blackbirds (*Agelaius phoeniceus*) in which males are heavier than females—even as nestlings—and the nestling sex ratio favors females (Howe, 1976; J.H. Myers, 1978; Fiala, 1981). In European sparrowhawks (*Accipiter nisus*), however, males are larger than females, but the apparent costs of raising males are no higher; therefore, size is not necessarily a good indicator of cost (Newton and Marquiss, 1979). The cost of raising males seems to be higher in red deer (*Cervus elaphus*) and a number of other polygynous mammals, although the sex ratio at birth favors males. The explanation for this deviation from expectations is still uncertain (Clutton-Brock et al., 1981). Most snakes have a 1 : 1 sex ratio at birth, little parental care, and no known differences in the cost of raising males and females, but a few species seem to have skewed sex ratios, for reasons still unknown (Shine and Bull, 1977). If juveniles of one sex—males, for example—have a higher probability of death than the other sex and the parents will replace a dead male with another male, the primary sex ratio favors males in order to maintain equal collective costs for each sex. This seems to be the case for the human sex ratio, which was mentioned previously. If per-capita costs are equal, the primary sex ratio should be 1 : 1.

Within the Fisherian model as applied to individuals, we would expect that parents might shift the sex ratios of their progenies in favor of whichever sex is rarer or whichever one is receiving the lesser collective expenditure. There is some evidence that certain vertebrates may do this to some degree and in some circumstances: guppies (*Poecilia reticulata*), zebra finches (*Taeniopygia guttata*), laboratory strains of the house mouse (*Mus musculus*), and woodchucks (*Marmota monax*).

It is becoming clear that the assumptions of the Fisherian model are not always valid and that progeny sex ratios may be adjusted irrespective of the balance of expenditure. One of the more important exceptions to the classical model occurs when average fitnesses are not equal (Bull, 1981; Charnov et al., 1981). Trivers and Willard (1973) noted that the prospects of success for male and female offspring might differ greatly depending on the physical condition of the mother. Mothers in good condition should produce mainly whichever sex can profit most from a good start in life, whereas mothers in poor condition should produce the other sex. This argument is based on three notions that will be discussed in the following paragraphs.

First, the health of the mother is correlated with the condition of the young. This is known to be true for several species of mammals, including ungulates, rodents, and primates, especially those mammals with small litter sizes. The birth weight of mule deer (*Odocoileus hemionus*) fawns in Utah was 18.3 kg when they were born in the spring following a mild winter; however, it was only 16.3 kg when they were born after a severe winter when food for the mothers was scarce (Robinette et al., 1957).

Second, differences in the condition of the young when they achieve independence of their parent tend to persist into adulthood. Although some compensatory growth may be possible, this assumption of enduring differences is supported by data from domesticated animals, humans, deer, and salmon. Body size of young Atlantic salmon (*Salmo salar*) was weakly correlated with their sizes two years earlier (Ryman, 1972). Growth during the first year of life in male red deer is related to mature body size and combat prowess (Clutton-Brock et al., 1981).

Third, differences in maternal condition affect future male reproductive success more than they affect female reproductive success. In species with little male parental care, male reproductive success usually varies greatly because some males breed frequently and others not at all when they lose in the competition for mates. Female reproductive success, in contrast, varies less in such species. Slight differences in physical condition, then, can have greater effects on male reproductive success than on female success. Great variation in male reproductive success, as compared to the reproductive success of females, is known in ungulates, carnivores, pinnipeds, humans, and birds.

Under these circumstances, then, it could be to the advantage of a female to produce mainly male offspring when she is in good condition because these males will then go on to be successful in competition for females. On the other hand, the production of female offspring when the female parent is in poor condition could be advantageous, because even puny females can have some reproductive success. Data from numerous species demonstrate that adverse environmental conditions, stress, and perhaps age are sometimes associated with deviations from expected sex ratios; these deviations are not always in favor of females (Trivers and Willard, 1973; Clutton-Brock et al., 1981; J.H. Myers, 1978), however, perhaps because other factors are also involved. Mature female white-tailed deer (*Odocoileus virginianus*) usually produce female-biassed sex ratios (an average of 43 percent males, at birth) when they are poorly fed (Verme, 1969). However, sex ratios also vary with litter size: Litters of one were 67 percent male, litters of two or three only 42 to 45 percent males. The tendency to produce multiple young is, in deer, usually associated with good nutrition. Thus, nutrition levels influence sex ratio and litter size in directions that produce conflicting effects on sex ratio. Wolves (*Canis lupus*) produced 66 percent male pups in parts of Minnesota with high wolf densities but only 38 to 50 percent males in areas of lower density (Mech, 1975). The sex ratio of moose (*Alces alces*) calves in Sweden varied both temporally and spatially (Reuterwall, 1981). On the other hand, female wood rats (*Neotoma floridana*) typically invest equally in sons and daughters but, on a restricted diet, differentially care for female young during lactation (McClure, 1981). Body size in this species varies with nutrition during growth and influences the reproductive success of males more than females. Thus, wood rats seem to conform to the expectations of the Trivers-Willard model.

Local competition among relatives may also influence sex ratios. If populations are localized and males compete for the chance to mate, selection might favor female-biassed sex ratios, with just enough males to accomplish fertilization (Hamilton, 1967; Alexander and Sherman, 1977). If females compete for resources, sex ratios might favor males (Clark, 1978; see also Hoogland, 1981). In general, sex ratios might favor whichever sex disperses farther from the natal area, because individuals of that sex are less affected by local competition and can contribute more to parental fitness (Taylor and Bulmer, 1980; Bulmer and Taylor, 1980). It is important to remember, however, that such a mechanism could only operate if competitive effects are sufficiently deleterious and if the costs of dispersal are not excessive. That conditions may not always produce local competitive effects on sex ratios is indicated by observing that dispersal is accomplished chiefly by males in mammals but by females in birds (Greenwood, 1980); yet, we have (so far) noted no general tendency for male-biassed sex ratios in mammals and female-biassed ones in birds. Interestingly, the sex ratio of pups in the Cape hunting dog is male-biased, but it is females that usually disperse (Frame et al., 1979); this example provides a pattern opposite to that predicted by the local competition hypothesis.

Male Richardson's ground squirrels (*Spermophilus richardsonii*) in Alberta disperse from the natal area and are subject to high mortality; females stay on or near their mother's home range (Michener, 1980). Two of ten females for whom detailed histories were available were the progenitors of most of the local population in the next two years. These two females were also the only ones to produce female-biassed litters. Unfortunately, we do not know if their success was due to their sex-biassed litters or to some other trait.

Nevertheless, the basic importance of the relative fitness gain to be achieved through offspring of a particular sex seems to be accepted widely. All-female litters in certain small mammals may be related to demographic conditions, and the costs and benefits of producing a given sex may vary with population density (Oksanen, 1981). The secondary sex ratio of yellow baboons (*Papio cynocephalus*) is correlated with the social rank of the mother: High-ranking females produce mainly daughters, which tend to inherit the mother's rank, whereas low-ranking females produce more sons (Altmann, 1980). Offspring sex ratios in captive bonnet macaques (*Macaca radiata*) favored males, but high-ranking mothers produced a higher proportion of females than did low-ranking mothers (Silk et al., 1981). In addition, the macaques produced more male offspring the year following a season of good survival of male young and more females following a season of good survival of female young. Thus, the secondary sex ratio tracked survival patterns rather than compensating for high mortality. Zebra finches seem to be able to shift the sex ratio of their young to favor the sex represented by whichever parent is most attractive to members of the other sex (Burley, 1981). Nestling sex ratios in broods of primary and secondary females of polygynous yellow-headed blackbirds favor males in primary broods (Patterson et al., 1980). Many other factors could influence the costs and benefits of skewed sex ratios; we have only begun to explore this evolutionary problem.

Skewed sex ratios may result from nonequilibrium conditions, in which selection has not yet eliminated a bias. Sometimes sex ratios depend on the particular parents involved: Certain males of a fish, *Platypoecilus maculatus*, produced more sons than daughters when mated with a given female; likewise, some females produced fewer sons with some males than with others (Kosswig, 1964). Whether this is merely the result of polygenic sex determination or is, in some way, related to the ability of certain parents to produce *better* sons (or daughters) is not entirely clear.

Sex ratio in clutches of the snow goose (*Chen caerulescens*) differs among individuals, and the sex of each egg depends greatly on the order of egg-laying by the female: The first eggs laid are mostly male, whereas later ones are mostly females (Ankney, 1982). In this case, sex determination seems not to be entirely random, but the mechanism of sex determination, the possible adaptive value of a sequential shift in sex ratio, and the consequences of individual differences in sex ratios are still unknown.

There is considerable evidence that vertebrate progeny sex ratios are sometimes biassed, although the direction of the bias often cannot be explained by any existing adaptive rationale. Williams (1979) found little evidence that vertebrate sex ratios were adaptive at all and argued instead that genetically determined sex ratios are determined chiefly by the random assortment of chromosomes during meiosis and the random union of gametes at fertilization; an example is provided by the eastern bluebird

(*Sialia sialis*) (Lombardo, 1982). Williams may be right, but it is too soon to give up the search for adaptive patterns when so little is known about such a complex matter.

There seems to be rather little evidence for genetic variance in primary sex ratios of vertebrates, at least in the much-studied populations of humans and their domesticated animals, although a few cases are known (Charnov,1982; Maynard Smith, 1980). If, in fact, the primary sex ratio is commonly fixed at 1:1 (and thus selection cannot alter the primary sex ratio itself), parents may modify the sex ratio of their progeny during the period of parental care by investing differentially in sons and daughters (as wood rats are known to do, for example). Parents must be able to recognize the sex of their offspring, which may be possible, after birth, at least for some birds and mammals, and, for live-bearing vertebrates of any class, a sort of physiological sex-recognition might be possible even before birth of the young. Differential investment can be advantageous if the fitnesses of sons and daughters vary differently with the level of investment received (Maynard Smith, 1980). As we have seen, both differential investment and differential fitness-gains are possible, but whether they come together in the way proposed by Maynard Smith is still unknown.

To summarize briefly: If sex and sex ratio are genetically determined, shifting selection pressures may alter primary sex ratios, as in the basic model of Fisher. There may also be selection for the ability to alter secondary sex ratios. In contrast, if primary sex ratios have become fixed, all changes in sex ratios result, not from shifts in the frequencies of sex-ratio genes in the population, but from phenotyic adjustments, as in Maynard Smith's model; in this case it is only the ability to modify secondary and subsequent sex ratios that is subject to selection. A number of factors determine the relative advantages (and costs) of particular sex ratios and investment patterns, including mating patterns in the population, the amount of fitness-gain per unit of parental investment, the capacities of the individual parents, the total costs of sons vs. daughters, and so on, and sometimes these factors may produce conflicting selection pressures. Furthermore, it remains possible that sex ratios are random and nonadaptive. A further complication derives from species with environmental sex determination, discussed earlier in this chapter. The progeny sex ratio in these species is determined by the outcome of the sex determination process. Parents may be able to control the environment of the developing young, at least in some cases, and therefore they may be able to determine the sex ratio of the progeny (for example, if female turtles select nest sites according to the temperature regime of the site). The costs and consequences of sex ratio variation when sex is not determined genetically are still unresolved. The evolution of sex determination mechanisms and sex ratios may be closely intertwined in complex and fascinating ways, but this interaction has only begun to be explored.

SEXUAL SELECTION

Sexual selection refers particularly to the evolution of reproductive traits, especially those involving the process of mate attraction and assessment. Because such traits have little to do, directly, with survival, some biologists have considered sexual selection to be distinct from natural selection. However, because success in mating is intimately involved with an individual's relative contribution to future generations (its fitness), obviously sexual selection is best seen as one facet of natural selection.

There are two essential ingredients of sexual selection, as originally set forth by Darwin and later developed by Bateman (1948), Trivers (1972), and others. Members of one sex vie with each other for the chance to mate with the other sex, and members of the other sex choose among them. Thus, one aspect of the process involves intrasexual competition, whereas the other consists of intersexual preferences. Typically, it is the males that compete among themselves and the females that choose among the males. The reproductive success of females is commonly limited by food resources; female gametes are more costly than male gametes because of the nutritional endowment of the egg, which is sometimes very great, and females usually spend more time and energy in parental care. Female reproductive success is seldom limited by the number of males with which she mates. Males, on the other hand, generally have little to lose and much to gain by mating with a number of females, because they usually invest relatively little in gametes and in parental care. Females, then, are a limiting resource for males, and males usually compete for females. Occasionally, these patterns of parental investment and intrasexual competition are reversed (Chapter 15) or the cost of gamete production by males may be quite high (Nakatsuru and Kramer, 1982), but the principles remain the same.

The variation in reproductive success of members of the competing sex (usually males) generally should be greater than the variation in reproductive success of the other sex (usually females) (Bateman, 1948; Wade and Arnold, 1980). Some males will court and service many females, whereas others win none; but most females will mate and produce some offspring. High variability of male reproductive success is known for *Drosophila melanogaster* in the laboratory (Bateman, 1948), *Sceloporus jarrovi* (Ruby, 1981), red-winged blackbirds (Payne, 1979), and red deer, for example. It is also evident in the sample of breeding yellow-headed blackbirds presented later. A survey of male mating success in various species of birds showed that the variation of male success was, as expected, greater in species in which male competition was intense and in which males did not provide parental care than in species in which there were lesser levels of male-male competition (Payne and Payne, 1977).

Male-male competition may entail actual combat or aggressive displays. Whatever attributes stand witness to a male's prowess should be favored by females. The reproductive success of "good" males then exceeds that of poorer males, and genotypes producing "good" features are perpetuated differentially. In some cases, females seem to judge males by the same characteristics that help determine male status among males, but in other cases, female choice may be based on traits that are not clearly related to male competitive ability.

The large horns of bighorn rams (*Ovis canadensis*) are used primarily for intermale aggression in butting contests of strength and, subsequently, for display. Rams with larger, heavier horns commonly dominate less well endowed (often younger) males, and these marks of dominance are recognized by both males and females. Any male genetically constituted to produce small horns at maturity would be at a clear disadvantage. In some African antelope, too, horns are used as weapons against other males as much as for display.

Females seem to exercise an active preference for large males in several species, such as the bicolor damselfish (*Eupomacentrus partitus*) (Schmale, 1981), the mottled sculpin (*Cottus bairdi*) (Brown, 1981), and some anurans. The ornamental tail feathers of the long-tailed widowbird (*Euplectes progne*) of eastern Africa are important elements in female choice (Andersson, 1982). Females prefer males with long tails, and long-tailed males mate with more females than short-tailed males. Tail length did not affect the ability of males to defend their territories, and therefore these observations suggest that the long tail length of males is maintained by female choice rather than intermale competition.

Displays of male red-winged blackbirds (*Agelaius phoeniceus*) are important in male-male interactions. Some of the major displays emphasize the red epaulets. Field experiments showed that most male redwings with experimentally blackened or whitened epaulets lost their territories to normal males, but few of the normal males did (Peek, 1972; Smith, 1969, 1972). The small proportion of black males that kept their territories occupied poorer sections of marsh where few males attempted to establish themselves. The ability of a male blackbird to claim and hold a good territory is central to his eventual success in winning females. The possible importance of red epaulets to female choice was not clearly established in these experiments, which only demonstrate the importance of the character in male–male competition.

Female preference may be based on an active choice of a mate, in which case females usually inspect several males before choosing one. In some cases, however, female choice is rather passive: Any male that has bested his competition is accepted. Male elephant seals contend among themselves for the possession of a stretch of beach; females usually mate with the beach owner. Here the males have sorted themselves and it is apparently advantageous for females to take the winner. Females have not relinquished all choice in the matter, however; they protest more when accosted by a subordinate male, which often elicits interference by the dominant. In general, both actively and passively choosing females presumably improve their own fitness by making the choices they do.

The property owned by the male also may influence in his mating success. A preliminary study of yellow-headed blackbirds illustrates the importance of the male's real estate (Willson, 1966). Yellowheads are marsh-nesters of western North America, and males usually mate with several females. A good territory for a male yellowhead contains a number of suitable nest sites for females to use; in fact, a male's success in acquiring females can be correlated with amount of habitat the male possesses that is suitable for nesting. Females almost always place their nests at the edge of a clump of bulrush or cattail, so "edge" is a desirable feature of a territory. In marsh areas with broken stands of vegetation, a small territory sufficed to acquire a large amount of edge, but in stands of continuous vegetation, large territories were the only way to acquire edge.

Table 12–1 shows pairing and breeding success of males (and females) in different parts of the habitat. Clearly, males owning much edge acquired more females and, hence, produced more young than males with little edge; nesting success of individual females on these territories did not differ, however. Furthermore, nesting success of females was highest in small territories with broken stands of vegetation, where production of insects was probably greater. Hence, males on these territories had higher production of young than males with large territories in continuous vegetation, even though they acquired similar numbers of females. The very best kind of territory in this marsh, for both males and females, was small and in clumped vegetation with much edge.

Possession of a suitable territory also seems to be closely related to mating success in green frogs (*Rana clamitans*). Males that spend the most time in densely vegetated territories obtain the most mates. Female green frogs use territories for oviposition sites. Large males were more successful than small ones in one year, but the effect of body size was less than that of territory quality. Defense of oviposition sites may be particularly advantageous in frogs that have a long breeding season. Arrival of females is spread out in time and direct conflict among males for possession of females is low. Territory defense then becomes economical (Wells, 1977). Similar patterns are found in the bullfrog.

The property of a male long-billed marsh wren (*Cistothorus palustris*) consists of his territory and clusters of woven, domed nests. The male builds the nests and uses them in advertising for females, which select and put the finishing touches on one of them for use as a brood nest. Bowerbirds (Ptilonorhynchidae) build elaborate courtship "bowers" that are used in courtship (Chapter 13); there is a general tendency for males with the fanciest plumage to build the simplest bowers and for males with the plainest feathering to build the most intricate bowers. This broad correlation across species suggests that the signs of male distinctiveness can be quite separate from morphology.

Table 12–1 Territory features and nesting success of yellow-headed blackbirds.

	Short edge length	*Long edge length*	*Totals and grand means*
Small territories			
Number of territories	4	4	8
Number of nests	10	18	28
Number of females per male	2.25	4.00	3.13
Number of fledged per nest	1.10	0.83	0.93[1]
Number of fledged per territory	2.75	3.75	3.25[2]
Large territories			
Number of territories	2	3	5
Number of nests	4	12	10
Number of females per male	2.00	3.67	3.00
Number of fledged per nest	0.25	0.82	0.67[1]
Number of fledged per territory	0.50	3.00	2.00[2]
Totals and grand mean			
Number of territories	6	7	
Number of nests	14	30	
Number of females per male	2.17[3]	3.86[3]	
Number of fledged per nest	0.86	0.83	
Number of fledged per territory	2.00[4]	3.42[4]	

[1–4] Statistically significant (or almost significant) differences between each pair of indicated values.

Sexual selection (male–male competition, female choice, or both) can produce an evolutionary tendency toward increasing male size or colorfulness. At some point, that tendency will be countered by other factors. Males that are too big or too bright may be subject to greater risks of predation or energetic limitations (Searcy, 1979), so we do not expect the trend fostered by sexual selection to continue indefinitely with respect to any single trait. As it happens, it is difficult to demonstrate that male mortality increases as a result of greater male-female difference, but a cross-species comparison among blackbirds (Icteridae) showed that greater sexual difference was associated with greater risks of male mortality (Searcy and Yasukawa, 1981). The problem is that we have not yet ascertained whether the degree of difference itself is the cause of greater mortality or whether some associated trait (such as the degree of polygamy) is the cause.

SEXUAL DIMORPHISM

Males and females are sometimes very different in size, color, or pattern; this condition is called sexual dimorphism ("two forms"). Very commonly, males are larger or more highly colored than females, but numerous cases exist in which the reverse is true. Many of the same general factors are involved regardless of which sex is more conspicuous, so we can first discuss sexual dimorphism as a general case, pointing out whenever possible the selective factors that may have tipped the balance one way or the other.

Sexual dimorphism has evolved in response to a variety of selective pressures (Clutton-Brock et al., 1977), but because many cases do not fit comfortably into existing explanations, the following discussion is not exhaustive. A survey of sexual dimorphism among vertebrates would not be very instructive in the absence of ecological and behavioral correlations; broad taxonomic correlations are almost nonexistent. Therefore, let us review the available hypotheses regarding the evolution of sexual dimorphism by choosing examples from any convenient taxon. These hypotheses fall into three major (and sometimes overlapping) categories: One deals with mating, another with maternal care, and the other with foraging. It is important to realize that many species are not dimorphic and that countervailing selection pressures may favor monomorphism (Burley, 1981).

Mating

One function of sexual dimorphism may be reproductive isolation of different species. Male dabbling ducks of the genus *Anas* are commonly very different from each other; females of this genus, however, exhibit fewer differences—sometimes to the point of being indistinguishable in the field (Fig. 12–3). Females choose their mates and would be selected against if they chose a male of the wrong species: Hybrid offspring are usually inviable or sterile. Mate choice in most vertebrate species with nonrandom mating is exercised primarily by females; females usually have more energy invested in eggs than males do in sperm, so "mistakes" are more costly to them. Distinctive plumage in male *Anas* makes recognition easier and facilitates mate selection by females. Natural selection has probably favored cryptic coloration in females because they usually nest and incubate eggs on the ground and, hence, are exposed to predation. Distinctive male plumage is particularly important in continental areas where many species are found together. Pair formation in many species of *Anas* begins on the wintering grounds, where six or eight species may occur together. On islands such as Hawaii where there are few or no congeners, males often resemble females and usually are more cryptically dressed. Other factors must be involved too, because some continental congeners also lack dimorphism (e.g., *Anas rubripes*); the fact that these males become cryptic

Figure 12–3 Males and females of two species of *Anas* ducks, the pintail (*A. acuta*) on the left and the American wigeon (*A. americana*) on the right. Note the distinctive plumages of the males and the more similar and cryptic plumage of the females. Females of these two species, although more similar than the males, are still distinguishable by a birdwatcher. Females of the blue-winged and cinnamon teal (*A. discors* and *A. cyanoptera*) appear identical, although the males differ vividly.

instead of females becoming colorful suggests that there are costs associated with the production or wearing of gaudy dress. Reproductive isolation could be a primary function of sexual dimorphism in many species, including darters (*Etheostoma*) in freshwater streams and neotropical *Anolis* lizards.

Some species may use sexual dimorphism to speed pair formation. For instance, pair formation in some migratory species of birds takes place after arrival on the breeding grounds. The short breeding season necessitates rapid initiation of nesting and, hence, speedy pair formation. Therefore, males that are easily recognized may be favored. This explanation has been used to account for the evolution of sexual dimorphism in North American migratory warblers (Parulidae), tanagers (*Piranga*), and orioles (*Icterus*). Many migratory species are monomorphic, however, including the ovenbird (*Seiurus aurocapillus*), most of the thrushes (Turdidae), virtually all the vireos (*Vireo*) in North America, and many European songbirds. Thus, this hypothesis does not seem to be very general.

Sexual dimorphism in size (Fig. 12–4) tends to be especially marked in species that are promiscuous or polygamous and in which competition for mates is great. Greater size is one way (but not the only way) to enhance dominance in encounters with other males. In polygamous and promiscuous species, males compete intensely with each other for control of females. Elephant seals (*Mirounga*), elk (*Cervus*), some antelope, muscovy ducks (*Cairina*), and red-winged and yellow-headed blackbirds are a few examples of species in which competition for mates has probably enhanced sexual size dimorphism.

Figure 12–4 Some examples of sexual size dimorphism. (*a*) Northern elephant seals (*Mirounga angustirostris*). (*b*) *Anolis garmani*. (*c*) Red-winged blackbird (*Agelaius phoeniceus*). (*d*) An anglerfish (*Cryptopsaras couesi*).

In the case of *Anolis garmani*, the frequency of copulation—an index of reproductive success—increased with increasing body size for both males and females, but the increase was more marked in males (Trivers, 1976). Male anoles compete for territories. Large males usually hold large territories, and large territories usually hold more females. Success in intermale competition results in reproductive advantages, and Trivers suggested that the more than twofold size difference between males and females can be ascribed to the increase in fitness obtained through success in intermale competition.

Sexual dimorphism in size is rather well developed in some anurans; females are generally larger than males. Among the North American *Rana* frogs, species in which males are territorial are often less dimorphic than nonterritorial species in which the breeding season is very short (Wells, 1978). Large size is thought to be advantageous in territory defense, but large size is probably also advantageous to nonterritorial males that fight over females directly. If large body size is beneficial to males in both territorial and direct combat systems, what can account for the greater dimorphism in the nonterritorial species? One answer may come from differences in the life histories of males and females. Male wood frogs (*Rana sylvatica*) would be better at displacing others if they attained large size, but high male mortality in this species could make deferred maturity disadvantageous (Howard, 1980). If male mortality were great enough, selection could favor breeding at small sizes despite the loss of combat ability. Even in some populations of the territorial bullfrog (*R. catesbeiana*), males mature earlier and begin to breed at smaller sizes than females (Howard, 1981). Males suffer heavy predation when they are large, so again deferred maturity would be likely to lower lifetime reproductive success. Growth rates of males and females are the same; sex differences in size are attributed to differences in age at breeding. Thus, while male-male contests may contribute to sexual differences in body size in some way, the benefits of breeding early and the costs of waiting can differ between the sexes, thereby producing differences in the average sizes of breeding individuals of each sex. If differential male mortality were greater in combative than in territorial anurans, the difference in degree of sexual dimorphism could be attributed to life-history patterns.

Size differences between the sexes in two species of sandpipers appear to facilitate the formation of new pairs (Jehl, 1970). Females of the stilt sandpiper (*Micropalama himantopus*) and the least sandpiper (*Calidris minutilla*) are somewhat larger than males. Pairs are formed on the breeding ground in the tundra region very soon after the birds arrive. Among pairs that are forming for the first time (the individuals were not mated to each other the previous year), pair formation and initiation of nesting is most rapid in pairs with the greater sexual difference in size. Early nesting may be advantageous because early broods have access to extensive feeding areas while the ground is still wet; as the ground dries up during the summer, rich feeding areas are much less available. Exactly how size differences affect the speed of pair formation is not known. Sex recognition can be accomplished by behavioral differences, so perhaps size differences may be related to the establishment of behavioral dominance of one individual over the other—a condition that is often suggested to enhance stability of the pair bond. Why the female rather than the male is larger is also not clear.

An unusual condition is found in many species of anglerfishes. There are almost 100 species of ceratioid anglers in 11 families; they are deep-sea fishes with enormous size differences between males and females. The males are considered to be dwarfed. Males are always parasitic on females in the families Ceratiidae, Linophrynidae, and in perhaps one or two other families. In several other families, the sexual parasitism is probably facultative, and in still others, the males may attach nonparasitically and temporarily to females while spawning (Pietsch, 1976). Parasitic males have reduced appendages and modified teeth for holding onto the female's body. In some species, they are nourished by the circulatory system of the female. Reduction of nonsexual body parts in the males perhaps permits increased allocation of nutrients to gamete production. If that is true, then small size and simple construction may be viewed as the results of selection for increased reproductive output by small fathers. The reproductive success of parasitized females must also have risen; otherwise, their expenditure of energy in feeding the male and in dragging him around would have no compensation. It is also possible that simultaneous selection for large females increased the disparity in sizes.

Patterns of sexual dimorphism in wrasses of the Caribbean are complex and intriguing. Males of many species are dimorphic in color; some are colored like females, but older, larger individuals sometimes lose female coloration and acquire new hues. Female-colored males tend to spawn in groups, presumably in direct competition with each other, and sometimes they "poach" females from territorial, colorful males. Their testes are larger than those of male-colored individuals, which usually are territorial and spawn singly with a female. Intermale competition apparently has selected for increased sperm production and larger gonads in the group spawners (Warner and Robertson, 1978). Large populations of *Thalassoma bifasciatum*, on large coral reefs, contain high numbers of female-colored males. When many of these males are present, the reproductive success of the territorial males decreases, because aggression and interference decrease the time for mating. Most spawnings are accomplished by female-colored males in these populations. In contrast, males in smaller populations are more successful when they are large and territorial than when they are small and female-colored. In this case, we would expect a higher proportion of the males to be territorial because their success is greater than when they are in large populations (Warner and Hoffman, 1980a,b). This system is particularly complex because these wrasses are also sex-changers; large individuals can shift from female to territorial male, given the proper social conditions.

Maternal Care

Ralls (1976, 1977) suggests that, among mammals (and possibly some fishes, herptiles, and birds), large females may be better mothers. Large female mammals often bear large babies, which may survive better than small ones. Large mothers may also be better able to carry the offspring or to defend them. When litter size is variable, large females are often able to produce large litters. Large females may also be able to produce young more frequently. Thus, big mothers may often be better mothers in a variety of ways. Ralls' "big-mother hypothesis" certainly helps explain cases of sexual size dimorphism in which females are larger than males. What remains to be explained is why large mothers have been selectively favored in some species but not in others.

Bats are the only volant mammals. Female bats carry their young not only during pregnancy but sometimes after birth. Among the vespertilionids, females of species that bear large offspring and/or large litters are large compared to males of the same species. The bigger the load of offspring, the bigger the female relative to the male. Furthermore, even when body size is held constant statistically, wings of female bats are often larger than males' wings. Big mothers may have been favored in bats as a means of reducing the relative load of offspring and of reducing wing-loading (P. Myers, 1978). Both factors are probably important in maintenance of the female's ability to forage while carrying her young. Large size may also allow females to pack more insects in their stomachs; consumption of larger meals may be particularly useful when supporting the high costs of flying with a load and of lactation.

The energetics of maternal care may contribute to size limitations in females of some species. Male anubis baboons (*Papio anubis*) are considerably larger than females, a trait commonly explained in terms of competition for mates. The daily energy expenditure of a female has been estimated at about 70 percent of a male's—except during lactation. Milk production is an expensive activity for mammals; it raises the daily energy requirement of a lactating female baboon to the male's level. If some factor sets an upper limit to energy requirements, the smaller body size of females can be interpreted as a means of allowing the addition of the lactation burden without exceeding that limit (T. Demment, personal communication). The ecology of parental care and sexual dimorphism has only begun to be explored.

Foraging

If food resources are limited, selection may sometimes favor sexual size dimorphism as a means of partitioning food resources between the sexes.

Food can be partitioned in a number of ways, including size of food item and location. Foragers with large bodies or large jaws frequently choose larger food items and may also forage in different places than conspecifics of smaller body size.

An example of sexual differences in foraging is provided by Schoener's (1967) study of *Anolis conspersus* on Grand Cayman Island in the Caribbean. This lizard is the only anole on the island. Adult males are significantly larger than adult females; maximum snout-vent length for males is about 70 mm; for females, it is about 50 mm. Subadult males have the same body sizes as females but tend to have larger heads and longer jaws; head length of adult males is about 1.4 times that of females. These sex and size differences are associated with differences in microhabitat occupation and insect prey captured. Adult males tend to occupy larger and higher arboreal perches than females, and subadult males are intermediate. Adult males captured somewhat larger prey than females; prey of subadult males was more similar in size to that of adult males than that of females. These sexual differences are especially well developed in anoles living on islands that lack congeneric competitors.

Ladder-backed woodpeckers (now part of the genus *Melanerpes*) occur both on the North and Central American mainland and on Caribbean islands (Fig. 12–5). Mainland males in North America usually are larger than their mates. On islands, however, which have fewer species that forage in woodpecker fashion, sexual differences are greater than on the mainland. This difference is interpreted as evidence that intersexual com-

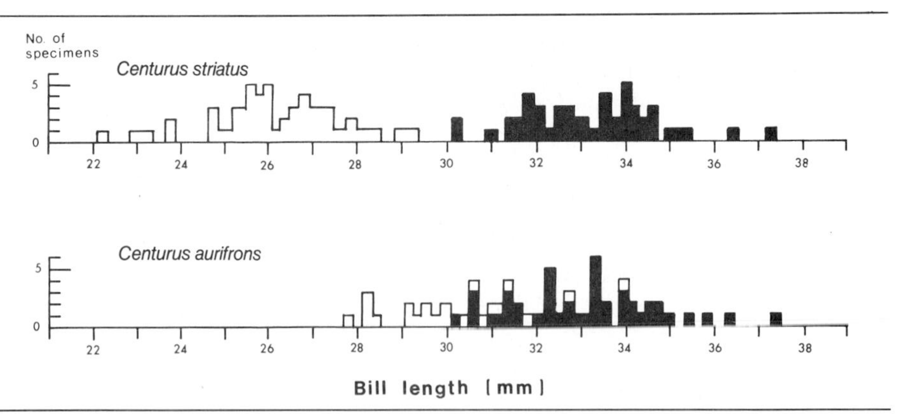

Figure 12–5 Variation in bill lengths of *Centurus* woodpeckers on the mainland (*C. aurifrons*) and on a Caribbean island (*C. striatus*). Sexual differences in bill size are much more pronounced on the island. Solid histograms, males; open histograms, females. (Modified from R.K. Selander, 1966. Sexual dimorphism and differential niche utilization in birds. *Condor, 68*:113–151. Reprinted with permission of author.)

petition favors dimorphism on both island and mainland; interspecific competition may restrict divergence on the mainland (Selander, 1966). Ideally, we also need to know that bill-size differences cannot be accounted for simply by allometric association with body size; differences in body size might evolve in response to other ecological pressures as well.

Size differences between the sexes in raptorial birds have generated considerable discussion. Although the arguments continue, it seems likely that these size differences can be related to different strategies of resource exploitation (Reynolds, 1972). A study of North American hawks and owls (Snyder and Wiley, 1976) relates the degree of size dimorphism to the degree of food shortage as the breeding season progresses. The energy demands of the raptors' offspring rise throughout the nestling period, so that foraging by the adults becomes increasingly strenuous. Sexual size differences are most marked in species that feed on birds, which, in north temperate regions, typically reach peak population sizes in spring and early summer. In contrast, mammal and insect populations usually continue to rise through the summer in the latter part of the raptor breeding cycle, and raptors preying on mammals or insects are less dimorphic than those preying on birds. Furthermore, most nestling mortality in bird predators occurs late in the nesting cycle when feeding the young is thought to be most difficult for the adults. Food shortages for bird-feeding hawks may have selected for a division of the food resources by males and females that share a hunting range.

Snyder and Wiley note that one European bird hawk, Eleanora's falcon (*Falco eleanorae*), is exceptional. This colonial falcon nests on Mediterranean islands and is not strongly dimorphic. It breeds in the autumn and feeds primarily on fall migrant birds. Unlike the breeding birds that are prey to North American hawks during their nesting season, fall migrants are a continually renewed resource that is not depleted by falcon predation. As a result, apparently the food shortages seen in North American hawks are not encountered by the falcon and dimorphism is not marked.

Interspecific comparisons in the genus *Accipiter* are in line with the idea that the intensity of intersexual competition for food may vary with the diet. Three species of *Accipiter* in western North America are *A. striatus* (the sharp-shinned hawk), *A. cooperii* (the Cooper's hawk), and *A. gentilis* (the goshawk). Sexual differences in size are greatest in sharpshins and least in goshawks. All accipiters feed primarily on other birds, but range of prey size and the proportion of mammals in the diet increase greatly with body size of the predator. Expansion of the resource base increases the probability of encountering suitable prey and decreases the importance of sexual size differentials.

Andersson and Norberg (1981) suggest that the use of other birds as prey provides strong selection for agility and selection for small size. Small size is favored particularly for males, because large females may be better able to carry the developing eggs and to lay larger clutches. Because so few species of predators specialize on birds, most of the potential competition for food will be with conspecifics. The possible effects of each sex on the food supply for the mate should favor divergence of body size. Species with a broader resource base share their food supply with numerous other predators, and intraspecific competition has relatively less impact on available resource levels, so then there is less selection favoring size differences.

Body size is closely related to the ability of an animal to store energy; large individuals can store more fat than small ones. Some sexual differences in body size may be related to differences in the ecology and behavior of males and females (Downhower, 1976). The female finback whale (*Balaenoptera physalus*), for instance, is larger than the male and stores fat for the long migration to the breeding ground, where foraging is poor. Male pinnipeds are often larger than females, perhaps in part because they must spend long fasting periods on land fighting for females (Chapter 13). Thus, the entire life history may impinge upon the evolution of body-size differences.

Clearly, sexual dimorphism in size favored by natural selection related to mating and reproduction may subsequently influence sexual differences in foraging. If body sizes or jaw sizes are different for any reason, this variation may affect prey size and foraging habits, even though foraging was not the original evolutionary reason for size differences. Therefore, if dimorphic males and females differ in diet, foraging techniques and sites, or nonbreeding geographic range, we cannot, in the absence of more information, decide which differences came first.

SUMMARY

Sexual reproduction in vertebrates is probably adaptive because a variable progeny has a higher probability of finding suitable living space in environments that vary both temporally and spatially. A few fishes, salamanders, and lizards reproduce parthenogenetically. These appear to be of hybrid origin, tend to occupy marginal or shifting habitats, and are sometimes outcompeted by their sexual relatives. Hermaphroditism among the vertebrates is recorded only from fishes. Simultaneous hermaphroditism may be advantageous in colonizing ephemeral habitats; sequential hermaphroditism is favored when relative fitness in each sex depends on age or size.

Sex ratios are based partly on equal parental expenditure on sons and daughters, each taken collectively. The individual cost of each male or female determines how many of each can be produced with the appropriate 50 percent share of the total parental expenditure on reproduction. Sex ratios may be modified, in a potentially adaptive way, in directions that reduce the total cost of rearing the brood or that capitalize on temporary shifts in the average fitness of males and females.

Sexual selection has two aspects: male-male interactions and male-female interactions. Interactions among males determine their status and acceptability to females, which choose "good" males as mates. Mate choice may be based on morphology, behavior, or property.

Sexual dimorphism can evolve in response to selective pressures favoring reproductive isolation, male-male aggression and female recognition of winning males, rapid pair formation, large mothers or small fathers, and foraging ecology of the sexes.

SELECTED REFERENCES

Alexander, R.D., and P.W. Sherman, 1977. Local mate competition and parental investment in social insects. *Science, 196*:494–500.

Altmann, J., 1980. Baboon Mothers and Infants. Harvard University Press, Cambridge, MA.

Andersson, M., 1982. Female choice selects for extreme tail length in a widowbird. *Nature, 299*:818–820.

Andersson, M., and R.A. Norberg, 1981. Evolution of reversed sexual size dimorphism and role partitioning among predatory birds, with a size scaling of flight performance. *Biol. J. Linn. Soc., 15*:105–130.

Ankney, C.D., 1982. Sex ratio varies with egg sequences in lesser snow geese. *Auk, 99*:662–666.

Arnold, S.J., 1977. The evolution of courtship behavior in New World salamanders with some comments on Old World salamandrids. *In* D.H. Taylor and S.I. Guttman (Eds.), *The Reproductive Biology of Amphibians*, pp. 141–183. Plenum Publishing, New York.

Bateman, A.J., 1948. Intrasexual selection in *Drosophila. Heredity, 2*:349–368.

Bateson, G., 1963. The role of somatic change in evolution. *Evol., 17*:529–539.

Beatty, R.A., 1967. Parthenogenesis in vertebrates. *In* C.B. Metz and A. Monroy (Eds.), *Fertilization*. Vol. 1, pp. 413–440. Academic Press, New York.

Borgia, G., and J. Blick, 1980. Sexual competition and the evolution of hermaphroditism. *J. Theor. Biol., 89*:523–532.

Breder, C.M., and D.E. Rosen, 1966. *Modes of Reproduction in Fishes*. Natural History Press, Garden City, NY

Bremermann, H.J., 1980. Sex and polymorphism as strategies in host-pathogen interactions. *J. Theor. Biol., 87*:671–702.

Brown, L., 1981. Patterns of female choice in mottled sculpins. (Cottidae, Teleostei). *Anim. Behav., 29*:375–382.

Bull, J.J., 1980. Sex determination in reptiles. *Q. Rev. Biol., 55*:3–21.

Bull, J.J., 1981. Sex ratio evolution when fitness varies. *Heredity, 46*:9–26.

Bull, J.J., and R.C. Vogt, 1979. Temperature dependent sex determination in turtles. *Science, 206*:1186–1188.

Bull, J.J., R.C. Vogt, and M.G. Bulmer, 1982. Heritability of sex ratio in turtles with environmental sex determination. *Evol., 36*:333–341.

Bulmer, M.G., and P.D. Taylor, 1980. Dispersal and the sex ratio. *Nature, 280*:448–449.

Burley, N., 1981. The evolution of sexual indistinguishability. *In* R.D. Alexander and D.W. Tinkle (Eds.), *Natural Selection and Social Behavior*, pp. 121–137. Chiron, New York.

Burley, N., 1982. Facultative sex-ratio manipulation. *Am. Nat., 120*:81–107.

Campbell, B. (Ed.), 1972. *Sexual Selection and the Descent of Man*. Aldine, Chicago.

Charnov, E.L., 1982. *The Theory of Sex Allocation*. Princeton University Press, Princeton, NJ.

Charnov, E.L., and J. Bull, 1977. When is sex environmentally determined? *Nature, 266*:828–830.

Charnov, E.L., R.L. Los-den Hartogh, W.T. Jones, and J. van den Assem, 1981. Sex ratio evolution in a variable environment. *Nature, 289*:27–33.

Choat, J.H., and D.R. Robertson, 1975. Protogynous hermaphroditism in fishes of the family Scaridae. *In* R. Reinboth (Ed.), *Intersexuality in the Animal Kingdom*, pp. 263–283. Springer-Verlag, Berlin.

Clark, A.B., 1978. Sex ratio and local resource competition in a prosimian primate. *Science, 201*:163–165.

Clutton-Brock, T.H., P.H. Harvey, and B. Rudder, 1977. Sexual dimorphism, socionomic sex ratio and body weight in primates. *Nature, 269*:797–800.

Clutton-Brock, T.H., S.D. Albon, and F.E. Guinness, 1981. Parental investment in male and female offspring in polygynous mammals. *Nature, 289*:487–489.

Cole, C.J., 1975. Evolution of parthenogenetic species of reptiles. *In* R. Reinboth (Ed.), *Intersexuality in the Animal Kingdom*, pp. 340–385. Springer-Verlag, Berlin.

Conover, D.O., and B.E. Kynard, 1981. Environmental sex determination: interaction of temperature and genotype in a fish. *Science, 213*:577–579.

Cox, C.R., and B.J. Le Boeuf, 1977. Female incitation of male competition: A mechanism in sexual selection. *Am. Nat., 111*:317–335.

Cuellar, O., 1977a. Genetic homogeneity and speciation in the parthenogenetic lizards *Cnemidophorus velox* and *C. neomexicanus:* evidence from intraspecific histocompatibility. *Evol., 31*:24–31.

Cuellar, O., 1977b. Animal parthenogenesis. *Science, 197*:837–843.

Cuellar, O., 1979. On the ecology of coexistence in parthenogenetic and bisexual lizards of the genus *Cnemidophorus. Am. Zool., 19*:773–786.

Darwin, C., 1859. *The Origin of Species.* Murray, London.

Darwin, C., 1871. *The Descent of Man, and Selection in Relation to Sex.* Murray, London.

Downhower, J.F., 1976. Darwin's finches and the evolution of sexual dimorphism in body size. *Nature, 263*:558–563.

Echelle, A.A., and D.T. Mosier, 1981. All-female fish: a cryptic species of *Menidia* (Atherinidae). *Science, 212*:1411–1413.

Emlen, J.M., 1968a. A note on natural selection and the sex ratio. *Am. Nat., 102*:94–95.

Emlen, J.M., 1968b. Selection for the sex ratio. *Am. Nat., 102*:589–591.

Emlen, S.T., and L.W. Oring, 1977. Ecology, sexual selection, and the evolution of mating systems. *Science, 197*:215–223.

Farr, J.A., 1977. Male rarity or novelty, female choice behavior, and sexual selection in the guppy, *Poecilia reticulata* Peters (Pisces: Poeciliidae). *Evol., 31*:162–168.

Ficken, M.S., and R.W. Ficken, 1967. Age-specific differences in the breeding behavior and ecology of the American redstart. *Wils. Bull., 79*:188–199.

Fischer, E.A., 1980. The relationship between mating system and simultaneous hermaphroditism in the coral reef fish, *Hypoplectrus nigricans* (Serranidae). *Anim. Behav., 28*:620–633.

Fischer, E.A., 1981. Sexual allocation in a simultaneously hermaphroditic coral reef fish. *Am. Nat., 117*:69–82.

Fishelson, L., 1975. Ecology and physiology of sex reversal in *Anthias squamipennis* (Peters), (Teleostei: Anthiidae). *In* R. Reinboth (Ed.), *Intersexuality in the Animal Kingdom*, pp. 284–294. Springer-Verlag, Berlin.

Fisher, R.A., 1958. *The Genetical Theory of Natural Selection.* Dover, New York. [Originally published in 1929.]

Frame, L.H., J.R. Malcolm, G.W. Frame, and H. van Lawick, 1979. Social organization of African wild dogs (*Lycaon pictus*) on the Serengeti Plains, Tanzania 1967–1978. *Z. Tierpsychol., 50*:225–249.

Fricke, H.W., 1979. Mating system, resource defense and sex change in the anemonefish *Amphiprion akallopisos. Z. Tierpsychol., 50*:313–326.

Fry, D.M., and C.K. Toone, 1981. DDT-induced feminization of gull embryos. *Science, 213*:922–924.

Geist, V., 1977. A comparison of social adaptations in relation to ecology in gallinaceous bird and ungulate societies. *Annu. Rev. Ecol.*

Syst., *8*:193–207.

Ghiselin, M.T., 1969. The evolution of hermaphroditism among animals. *Q. Rev. Biol.*, *44*:189–208.

Ghiselin, M.T., 1974. *The Economy of Nature and the Evolution of Sex.* University of California Press, Berkeley.

Glesener, R.R., and D. Tilman, 1978. Sexuality and the components of environmental heterogeneity: Clues from geographic parthenogenesis in terrestrial animals. *Am. Nat.*, *112*:659–673.

Greenwood, P.J., 1980. Mating systems, philopatry and dispersal in birds and mammals. *Anim. Behav.*, *28*:1140–1162.

Halliday, T.R., 1978. Sexual selection and mate choice. *In* J.R. Krebs and N.B. Davies (Eds.), *Behavioural Ecology: An Evolutionary Approach*, pp. 180–213. Sinauer Associates, Sunderland, MA.

Hamilton, W.D., 1967. Extraordinary sex ratios. *Science*, *156*:477–488.

Hamilton, W.D., 1980. Sex versus non-sex versus parasite. *Oikos*, *35*:282–290.

Hamilton, W.D., P.A. Henderson, and N.A. Moran, 1981. Fluctuation of environment and coevolved antagonist polymorphism as factors in the maintenance of sex. *In* R.D. Alexander and D.W. Tinkle (Eds.), *Natural Selection and Social Behavior*, pp. 363–381. Chivon, New York.

Harrington, R.W., 1975. Sex determination and differentiation among uniparental homozygotes of the hermaphroditic fish *Rivulus marmoratus* (Cyprinodontidae: Atheriniformes). *In* R. Reinboth (Ed.), *Intersexuality in the Animal Kingdom*, pp. 249–262. Springer-Verlag, Berlin.

Hoogland, J.L., 1981. Sex ratio and local resource competition. *Am. Nat.*, *117*:796–797.

Howard, R.D., 1980. Mating behaviour and mating success in wood frogs, *Rana sylvatica. Anim. Behav.*, *28*:705–716.

Howard, R.D., 1981. Sexual dimorphism in bullfrogs. *Ecol.*, *62*:303–310.

Howe, H.F., 1976. Egg size, hatching asynchrony, sex, and brood reduction in the common grackle. *Ecol.*, *57*:1195–1207.

Howe, H.F., 1977. Sex ratio adjustment in the common grackle. *Science*, *198*:744–746.

Jehl, J.R., 1970. Sexual selection for size differences in two species of sandpipers. *Evol.*, *24*:311–319.

Jones, G.P., 1980. Growth and reproduction in the protogynous hermaphrodite *Pseudolabrus celidotus* (Pisces: Labridae) in New Zealand. *Copeia*, *1980*:660–675.

King, J.R., 1973. Energetics of reproduction in birds. *In* D.S. Farner (Ed.), *Breeding Biology of Birds*, pp. 78–117. National Academy Press, Washington, D.C.

Kosswig, C., 1964. Polygenic sex determination. *Experientia*, *20*:190–199.

Leigh, E.G., 1970. Sex ratio and differential mortality between the sexes. *Am. Nat.*, *104*:205–210.

Lombardo, M.P., 1982. Sex ratios in the eastern bluebird. *Evol.*, *36*:615–617.

Lowther, P.E., 1975. Geographic and ecological variation in the family Icteridae. *Wils. Bull.*, *87*:481–495.

Lynch, M., ms. Genomic incompatibility, general-purpose genotypes, and geographic parthenogenesis.

MacArthur, R.H., 1965. Ecological consequences of natural selection. *In* T.H. Waterman and H.J. Morowitz (Eds.), *Theoretical and Mathematical Biology*, pp. 288–297. Blaisdell, New York.

Maynard Smith, J., 1978. *The Evolution of Sex.* Cambridge University Press, Cambridge, England.

Maynard Smith, J., 1980. A new theory of sexual investment. *Behav. Ecol. Sociobiol.*, *7*:247–251.

McClure, P.A., 1980. Sex-biased litter reduction in food-restricted wood rats (*Neotoma floridana*). *Science*, *211*:1058–1060.

McKay, F.E., 1971. Behavioral aspects of population dynamics in unisexual-bisexual *Poeciliopsis* (Pisces: Poeciliidae). *Ecol.*, *52*:778–790.

Mech, D.L., 1975. Disproportionate sex ratios of wolf pups. *J. Wildl. Mgt.*, *39*:737–740.

Michener, G.R., 1980. Differential reproduction among female Richardson's ground squirrels and its relation to sex ratio. *Behav. Ecol.*

Sociobiol., *7*:173–178.

Mittwoch, U., 1975. Chromosomes and sex differentiation. *In* R. Reinboth (Ed.), *Intersexuality in the Animal Kingdom*, pp. 438–446. Springer-Verlag, Berlin.

Moore, W.S., 1976. Components of fitness in the unisexual fish *Poeciliopsis monacha-occidentalis*. *Evol.*, *30*:564–578.

Moore, W.S., and F.E. McKay, 1971. Coexistence in unisexual-bisexual species complexes of *Poeciliopsis* (Pisces: Poeciliidae). *Ecol.*, *52*:791–799.

Moore, W.S., R.R. Miller, and R.J. Schultz, 1970. Distribution, adaptation and probable origin of an all-female form of *Poeciliopsis* (Pisces: Poeciliidae) in northwestern Mexico. *Evol.*, *24*:789–795.

Morreale, S.J., G.J. Ruiz, J.R. Spotila, and E.A. Standora, 1982. Temperature-dependent sex determination: current practices threaten conservation of sea turtles. *Science*, *216*:1245–1247.

Murton, R.K., and N.J. Westwood, 1977. *Avian Breeding Cycles*. Clarendon, Oxford.

Myers, J.H., 1978. Sex ratio adjustment under food stress: maximization of quality or numbers of offspring? *Am. Nat.*, *112*:381–388.

Myers, P., 1978. Sexual dimorphism in size of vespertilionid bats. *Am. Nat.*, *112*:701–711.

Nakatsuru, K., and D.L. Kramer, 1982. Is sperm cheap? Limited male fertility and female choice in the lemon tetra (Pisces, Characidae). *Science*, *216*:753–755.

Newton, I., and M. Marquiss, 1979. Sex ratio among nestlings of the European sparrowhawk. *Am. Nat.*, *113*:309–315.

Nichols, J.D., and R.H. Chabreck, 1980. On the variability of alligator sex ratios. *Am. Nat.*, *116*:125–137.

Ohno, S., L.C. Christian, S.S. Wachtel, and G.C. Koo, 1976. Hormone-like role of H-Y antigen in bovine freemartin gonad. *Nature*, *261*:597–598.

Oksanen, L., 1981. All-female litters as a reproductive strategy: defense and generalization of the Trivers-Willard hypothesis. *Am. Nat.*, *117*:109–111.

Patterson, C.B., W.J. Erekmann, and G.H. Orians, 1980. An experimental study of parental investment and polygyny in male blackbirds. *Am. Nat.*, *116*:757–769.

Payne, R.B., 1979. Sexual selection and intersexual differences in variance of breeding success. *Am. Nat.*, *114*:447–452.

Payne, R.B., and K. Payne, 1977. Social organization and mating success in local song populations of village indigobirds, *Vidua chalybeata*. *Z. Tierpsychol.*, *45*:113–173.

Peek, F.W., 1972. An experimental study of the territorial function of vocal and visual display in the male red-winged blackbird. *Anim. Behav.*, *20*:112–118.

Pieau, C., 1975. Temperature and sex differentiation in embryos of two chelonians, *Emys orbicularis* L. and *Testudo graeca* L. *In* R. Reinboth (Ed.), *Intersexuality in the Animal Kingdom*, pp. 332–339. Springer-Verlag, Berlin.

Pietsch, T.W., 1976. Dimorphism, parasitism and sex: reproductive strategies among deepsea ceratioid anglerfishes. *Copeia*, *1976*:781–793.

Pressley, P.H., 1981. Pair formation and joint territoriality in a simultaneous hermaphrodite: the coral reef fish *Serranus tigrinus*. *Z. Tierpsychol.*, *56*:33–46.

Ralls, K., 1976. Mammals in which females are larger than males. *Q. Rev. Biol.*, *51*:245–276.

Ralls, K., 1977. Sexual dimorphism in mammals: avian models and unanswered questions. *Am. Nat.*, *111*:917–938.

Reuterwall, C., 1981. Temporal and spatial variability of the calf sex ratio in Scandinavian moose *Alces alces*. *Oikos*, *37*:39–45.

Reynolds, R.T., 1972. Sexual dimorphism in hawks: A new hypothesis. *Condor*, *74*:191–197.

Robertson, D.R., 1972. Social control of sex reversal in coral-reef fish. *Science*, *177*:1007–1009.

Robertson, D.R., and S.G. Hoffman, 1977. The roles of female mate choice and predation in the mating systems of some tropical labroid fishes. *Z. Tierpsychol.*, *45*:298–320.

Robertson, D.R., and R.R. Warner, 1978. Sexual patterns in the labroid fishes of the western Caribbean. II. The parrotfishes (Scaridae). *Smithson. Contrib. Zool., 255*:1–26.

Robinette, W.L., et al., 1957. Differential mortality by sex and age among muledeer. *J. Wildl. Mgt., 21*:1–16.

Ruby, D.E., 1981. Phenotypic correlates of male reproductive success in the lizard, *Sceloporus jarrovi*. *In* R.D. Alexander and D.W. Tinkle (Eds.), *Natural Selection and Social Behavior*, pp. 96–107. Chiron Press, New York.

Ryman, N., 1972. An analysis of growth capability in full sib families of salmon (*Salmo salar* L.). *Hereditas, 70*:119–128.

Schmale, M.C., 1981. Sexual selection and reproductive success in males of the bicolor damselfish, *Eupomacentrus partitus* (Pisces: Pomacentridae). *Anim. Behav., 29*:1172–1184.

Schoener, T.W., 1967. The ecological significance of sexual dimorphism in size in the lizard *Anolis conspersus*. *Science, 155*:474–477.

Searcy, W.A., 1979. Sexual selection and body size in male red-winged blackbirds. *Evol., 32*:649–661.

Searcy, W.A., and K. Yasukawa, 1981. Sexual size dimorphism and survival of male and female blackbirds (Icteridae). *Auk, 98*:457–465.

Selander, R.K., 1966. Sexual dimorphism and differential niche utilization in birds. *Condor, 68*:113–151.

Selander, R.K., 1972. Sexual selection and dimorphism in birds. *In* B. Campbell (Ed.), *Sexual Selection and the Descent of Man*, pp. 180–230. Aldine, Chicago.

Shapiro, D.Y., 1980. Serial female sex changes after simultaneous removal of males from social groups of a coral reef fish. *Science, 209*:1136–1137.

Shapiro, D.Y., and R. Lubbock, 1980. Group sex ratio and sex reversal. *J. Theor. Biol., 83*:411–426.

Shine, R., and J.J. Bull, 1977. Skewed sex ratios in snakes. *Copeia, 1977*:228–234.

Sibley, G.C., 1957. The evolutionary and taxonomic significance of sexual dimorphism and hybridization in birds. *Condor, 59*:166–191.

Silk, J.B., C.B. Clark-Wheatley, P.S. Rodman, and A. Samuels, 1981. Differential reproductive success and facultative adjustment of sex ratios among captive female bonnet macaques (*Macaca radiata*). *Anim. Behav., 29*:1106–1120.

Smith, C.L., 1975. The evolution of hermaphroditism in fishes. *In* R. Reinboth (Ed.), *Intersexuality in the Animal Kingdom*, pp. 295–310. Springer-Verlag, Berlin.

Smith, C.L., and E.H. Atz, 1973. Hermaphroditism in the mesopelagic fishes *Omosudis lowei* and *Alepisaurus ferox*. *Copeia, 1973*:41–44.

Smith, D.G., 1969. Role of the epaulet in red-winged blackbird territorial maintenance. *Am. Zool., 9*:1064.

Smith, D.G., 1972. The red badge of rivalry. *Nat. Hist., 81*(3):44–51.

Snyder, N.F.R., and J.W. Wiley, 1976. Sexual size dimorphism in hawks and owls of North America. *Ornith. Monogr., 20*:1–96.

Taylor, P.D., and M.G. Bulmer, 1980. Local mate competition and the sex ratio. *J. Theor. Biol., 86*:409–419.

Templeton, A.R., 1982. The prophecies of parthenogenesis. *In* H. Dingle and J.P. Hegmann (Eds.), *Evolution and Genetics of Life Histories*, pp. 75–101. Springer-Verlag, New York.

Thibault, R.E., 1978. Ecological and evolutionary relationships among diploid and triploid unisexual fishes associated with the bisexual species, *Poeciliopsis lucida* (Cyprinodontiformes: Poeciliidae). *Evol., 32*:613–623.

Treisman, M., 1976. The evolution of sexual reproduction: a model which assumes individual selection. *J. Theor. Biol., 60*:421–431.

Trivers, R.L., 1972. Parental investment and sexual selection. *In* B. Campbell (Ed.), *Sexual Selection and the Descent of Man*, pp. 136–179. Aldine, Chicago.

Trivers, R.L., 1974. Parent-offspring conflict. *Am. Zool., 14*:249–264.

Trivers, R.L., 1976. Sexual selection and resource-accruing abilities in *Anolis garmani*. *Evol., 30*:253–269.

Trivers, R.L., and D.E. Willard, 1973. Natural selection of parental ability to vary the sex ratio of offspring. *Science, 179*:90–92.

Uzzell, T., and I.S. Darevsky, 1975. Biochemical evidence for the hybrid origin of the parthenogenetic species of *Lacerta saxicola* complex (Sauria: Lacertidae) with a discussion of some ecological and evolutionary implications. *Copeia, 1975*:204–222.

Verme, L., 1969. Reproductive patterns of white-tailed deer related to nutritional plane. *J. Wildl. Mgt., 33*:881–887.

Verner, J., 1965. Selection for sex ratio. *Am. Nat., 99*:419–421.

Vrijenhoek, R.C., 1979. Factors affecting clonal diversity and coexistence. *Am. Zool., 19*:787–797.

Wade, M.J., and S.J. Arnold, 1980. The intensity of sexual selection in relation to male sexual behaviour, female choice, and sperm precedence. *Anim. Behav., 28*:446–461.

Warner, R.R., 1975. The adaptive significance of sequential hermaphroditism in animals. *Am. Nat., 109*:61–82.

Warner, R.R., 1978. The evolution of hermaphroditism and unisexuality in aquatic and terrestrial vertebrates. *In* E.S. Reese and F.J. Lighter (Eds.), *Contrasts in Behavior*, pp. 77–101. John Wiley & Sons, New York.

Warner, R.R., 1980. The coevolution of behavioral and life-history characteristics. *In* G.W. Barlow and J. Silverberg (Eds.), *Sociobiology: Beyond Nature/Nurture?* pp. 151–188. AAAS, Washington, D.C.

Warner, R.R., and D.R. Robertson, 1978. Sexual patterns in the labroid fishes of the western Caribbean. I. The wrasses (Labridae). *Smithson. Contrib. Zool., 254*:1–27.

Warner, R.R., D.R. Robertson, and E.G. Leigh, 1975. Sex change and sexual selection. *Science, 190*:633–638.

Warner, R.R., and S.G. Hoffman, 1980a. Population density and the economics of territorial defenses in a coral reef fish. *Ecol., 61*:772–780.

Warner, R.R., and S.G. Hoffman, 1980b. Local population size as a determinant of mating system of sexual composition in two tropical marine fishes (*Thalassoma* spp.). *Evol., 34*:508–518.

Wells, K.D., 1977. Territoriality and male mating success in the green frog (*Rana clamitans*). *Ecol., 58*:750–762.

Wells, K.D., 1978. Territoriality in the green frog (*Rana clamitans*): vocalizations and agonistic behaviour. *Anim. Behav., 26*:1051–1063.

Werner, Y.L., 1980. Apparent homosexual behaviour in an all-female population of a lizard, *Lepidodactylus lugubris* and its probable interpretation. *Z. Tierpsychol., 54*:144–150.

Wilbur, H.M., 1971. The cological relationship of the salamander *Ambystoma laterale* to its all-female, gynogenetic associate. *Evol., 25*:168–179.

Wilbur, H.M., D.I. Rubenstein, and L. Fairchild, 1978. Sexual selection in toads: the role of female choice and male body size. *Ecol., 32*:264–270.

Williams, G.C., 1975. *Sex and Evolution.* Princeton University Press, Princeton, NJ.

Williams, G.C., 1979. The question of adaptive sex ratio in outcrossed vertebrates. *Proc. R. Soc. Lond.* (Biol.), *205*:567–580.

Williams, G.C., 1980. Kin selection and the paradox of sexuality. *In* G.W. Barlow and J. Silverberg (Eds.), *Sociobiology: Beyond Nature/Nurture?* pp. 371–384. AAAS, Washington, D.C.

Willson, M.F., 1966. Breeding ecology of the yellow-headed blackbird. *Ecol. Monogr., 36*:51–77.

Wright, J.W., and C.H. Lowe, 1968. Weeds, polyploids, parthenogenesis and the geographical and ecological distribution of all-female species of *Cnemidophorus. Copeia, 1968*:128–138.

Yntema, C.L., and N. Mrosousky, 1980. Sexual differentiation in hatchling loggerheads (*Caretta caretta*) incubated at different controlled temperatures. *Herpetologica, 36*:33–36.

13 Courtship and Mating Systems

COURTSHIP

Functions of Courtship

Vertebrate males and females usually court each other as a prelude to mating. Courtship potentially serves several functions: species identification, sex recognition, synchronization of male and female reproductive cycles, and assessment of individual characteristics. The relative importance of these functions may vary in different species. For example, the need for correct species identification is great when a number of similar species are found sympatrically at the time of pair formation. The length of time available for pair formation may place a premium on quick sex recognition and individual assessment in some species.

The assessment of an individual's suitability as a mate is an important element of courtship in many species. The choice of a mate can have important consequences for an individual's reproductive success. The quantity and quality of resources provided by a territory owner can affect the ability to rear offspring. The ability of any individual to claim a nest site successfully and, in some cases, provide parental care may be associated with experience, so reproductive experience may be a relevant criterion (Burley and Moran, 1979; Yasukawa et al., 1980; Yasukawa, 1981). Preferences may be exercised for previous mates, especially if the previous reproductive attempts were successful (Coulson, 1966). Body size is important in many anurans: Large males can best smaller ones and are commonly preferred by females (Berven, 1981; Gatz, 1981; Ryan, 1980; Davies and Halliday, 1979). Females may prefer males that are from a different sibling group and yet not too different in appearance, possibly as a means of avoiding inbreeding and achieving a suitable level of outcrossing (Bateson, 1978). In other words, females may favor matings that will yield offspring with "optimal" gene combinations. Perhaps females can choose males that have, in some sense, superior genotypes, so that their offspring—especially their sons—will have these superior traits. The "sexy son" hypothesis is hotly debated, in terms of both theory and available evidence (Wittenberger, 1981c; Heisler, 1981; Weatherhead and Robertson, 1981; Searcy and Yasukawa, 1981), and it is fair to say that controversy will continue. Many features, morphological or behavioral, could provide signals that are of use in mate choice. Song repertoires are a clue to male quality in red-winged blackbirds (Yasukawa et al., 1980), and song rates are correlated with later male parental care in stonechats (*Saxicola torquata*), although the basis of the association is unclear (Greig-Smith, 1982).

The opportunity to exercise a choice of mates can vary with circumstances. Pupfish females have considerable opportunity for mate choice in habitats where males defend breeding territories but much less opportunity where males are not territorial but form hierarchies, spawning groups, or mate only in pairs (Kodric-Brown, 1981). The sex ratio of available adults may differ in time and space, as it does in the wood frog (Berven, 1981). When males are common, male-male competition is intense and males may differ greatly in mating success, perhaps more as a result of intermale conflict than of direct female choice. Where males are few, however, females have a relatively greater chance of encountering an inferior male and should exercise more active choice. Rather simple preferences by females for large males may be obscured by other effects (Howard, 1981).

Female choice of mates is generally thought to be stronger and more prevalent than male choice (Chapter 12; Burley, 1977), but males also often exercise some choice. Male wood frogs and male mottled sculpins strongly prefer large females (Berven, 1981; Downhower and Brown, 1981), presumably because of their greater fecundity. In addition, sometimes adult females outnumber males in a population. If the number of females accommodated by each male is limited, as it is in the monogamous gulls (*Larus*), females may compete for males (Burger and Gochfeld, 1981). Males too should be expected to prefer some females to others (Dewsbury, 1982;

Nakatsuru and Kramer, 1982). Some birds have mating systems in which male choice (and female competition for mates) is the rule, as we shall see.

One would expect patterns of courtship to reflect the importance of choice to both sexes. The greater the selectivity of one sex, especially of personal traits (rather than property), the more elaborate should be the displays of the other.

Another function of courtship, especially of prolonged courtship, may be insurance for the mated male against cuckoldry (Trivers, 1972). When the participation of the male in parental care is extensive, he stands to lose a great deal by raising another male's offspring instead of his own. Laboratory experiments with ring doves (*Streptopelia risoria*) suggested that males are less likely to court females whose behavior indicates that they have recently associated with another male (Erickson and Zenone, 1976). When males do pair with such females, their pre-mating behavior is aggressive, which may delay the female's reproductive processes long enough (several days) that any sperm stored from previous matings with other males become ineffective (Zenone et al., 1979). Male ring doves spend considerable time and effort in all aspects of parental care; thus, avoidance of cuckoldry is very important.

Male eastern bluebirds (*Sialia sialis*) defended their mates against conspecific male intruders but also became aggressive toward their mates when an intruder was present (Gowaty, 1981). The intensity of this response declined through the season from a peak at the time of egg-laying. The defensive behavior pattern might be a form of nest-site defense, because it was also exhibited against house sparrows, which use similar nest sites. Aggression against the mate was shown only in the presence of a conspecific intruder, which suggests that protection of parentage may be involved. Male bluebirds provide considerable parental care for their young.

Male mountain bluebirds (*S. currucoides*) are also aggressive against intruding conspecific males and escort their mates very closely during the egg-laying period (Power and Doner, 1980). Such behavior reduces the risk of cuckoldry, although it may have other functions as well. Male starlings (*Sturnus vulgaris*) share incubation duties with their mates but do not begin incubating until all the eggs are laid; the female begins earlier (Power et al., 1981). This gives the male times to guard his mate throughout the egg-laying period, a time when he monitors her activities more closely than after the last egg is laid. These observations do not conclusively establish that the behavior evolved to prevent cuckoldry, but they are consistent with that idea. Mate-guarding is recorded for a number of avian species (Birkhead, 1982); it typically occurs principally during the egg-laying period, which is the usual period of copulation and, hence, of susceptibility to cuckoldry.

Males are not always successful in preventing their females from wandering. Territorial male red-winged blackbirds frequently mate with several females. A number of males were surgically sterilized and released on their territories, where they resumed territorial and mating activities. The eggs produced by their females were frequently infertile, as expected, but a surprising number of fertile eggs were laid: As many as 69 percent of the females produced viable eggs. Females whose sterilized males held territories farthest from normal males produced the most infertile eggs. Apparently, the females copulated not only with their own males but with neighboring males (Bray et al., 1975). The cost of cuckoldry to a male redwing is less than to a male ring dove, because male redwings usually spend less time and energy in caring for offspring. Therefore, we would expect redwing males to be more careless about their female's extracurricular activities than would males of species in which male parental investment is high. (At this point, however, the possibility still exists that females sought other males because their own could not sire offspring; thus, it is possible that their wandering behavior was not a normal occurrence.)

Courtship Patterns

Courting animals of any species commonly exhibit behavior that is distinct from the courtship behavior of all other species. Visual behavior patterns in courtship are often rather stereotyped and consistent among individuals of the same species. Separate actions are combined to form a total display. Often the sequence of events is predictable; one such sequence is illustrated in Figure 13–1.

When it is especially important to avoid mistakes in identification, natural selection is likely to favor displays composed of redundant signals. When the major message of a display concerns a variable behavioral state, however, natural selection may favor displays composed of components that can vary independently of each other, thus increasing the amount of information conveyed by a single display. The song spread display of carib grackles (*Quiscalus lugubris*) in Trinidad is composed of several elements, including bill and wing elevation. The degree of elevation in displaying males depends on the presence and sex of the recipient of the display. Bill elevation, a sign of aggressiveness, is more pronounced in the presence of another male than when the displaying male is alone or with a female. Higher wing elevations occur in the presence of a female. The same basic display can also be used to convey a series of related behavioral states by means of independently varying components of the display (Wiley, 1975).

Sexual dimorphism in color, pattern, and ornamentation is common in visually oriented vertebrates, and courtship displays usually emphasize distinctive visual features (Fig. 13–2). However, courting individuals are sometimes very dull in color; the source of visual stimulation is often transferred to other objects. The courting nests of male long-billed marsh wrens are one example (Chapter 12).

Figure 13–1 The courtship pattern of the 10-spined stickleback (*Pygosteus pungitius*). (*a*) Male dances to female; (*b*) female turns to male; (*c*) male dances to nest and female follows; (*d*) male points to nest entrance and female watches; (*e*) female has entered the nest, and male quivers on female's tail; and (*f*) female leaves, having deposited the eggs, and male enters to fertilize the eggs. (Modified from D. Morris, 1958. The reproductive behavior of the ten-spined stickleback (*Pygosteus pungitius.*) *Behav. Suppl.*, *6*:1–154.)

More dramatic are the bowerbirds (Ptilonorhynchidae) of New Guinea and northern Australia. They are closely related to the gaudy birds of paradise but have much less exotic plumage, and males of most of the 19 species court their mates in prepared areas called bowers (Fig. 13–3). Males of some species clear a courtship stage and pave it with fresh leaves or ferns, sometimes adding snail shells, beetle wings, and strips of grass as decoration. Others build conical towers out of sticks, usually about 1½ m high, and decorate the walls, floor, and surrounding area with fresh flowers, berries, and mosses, which are replaced when they fade. The most specialized members of this group are "avenue builders": Males lay down a mat of twigs into which they insert two parallel rows of interwoven sticks to form a parade ground. Males of two avenue-building species paint the walls of the avenue with saliva-mixed pigments, using a wad of bark or leaves.

Levels of overt aggression are also a common means of sex identification. Usually male vertebrates are more active and aggressive then females, and passive "female-like" behavior will shut off male aggression in many species. An anesthetized male banded gecko, *Coleonyx variegatus*, of southwestern North America was treated by other males as a female.

Voice may be a major adjunct of visual displays, many of which are accompanied by characteristic sounds. The importance of auditory stimuli is probably greatest in anurans and birds but occurs in other vertebrates as well. The songs of male birds and anurans may be related to intermale aggression, but, in most cases, they also attract females. Voice in anurans is also a means of close-range sex recognition. Male frogs and toads have been found attempting to mate with other males, stones, sticks, and other objects. If the assaulted object is another male, he emits a special croak of protest that identifies him as a male and discourages the mating attempt.

Figure 13–2 A display of the Asian mandarin duck, *Aix galericulata.* When mated birds meet after a separation, both male and female often drink. Then the male "mock-preens," touching the banner feathers of the wing on the side nearest to the female. (*a*) In the drinking part of the greeting display, the head plumes are raised. (*b*) With head plumes still raised, the male points to the conspicuous banner feathers. (*c*) Head plumes are sleeked after the display. In this series, the male's behavior further emphasizes certain conspicuous plumage feathers, such as the banner feathers. (From M. Bastock, 1967. *Courtship: An Ethological Study.* Chicago, Heinemann Educational Books.

Chemical and tactile cues are often of prime importance in the courtship of mammals, reptiles, amphibians, and some fishes. They often function together in courtship displays. Long-range chemical signals certainly occur in insects, but, in vertebrates, olfactory cues apparently work primarily at close range. Male mammals often sniff at female genitals or urine; in some cases, such as the North American grasshopper mouse (*Onychomys leucogaster*), genital sniffing is reciprocal. Courtship behavior in colubrid snakes such as *Thamnophis* (the garter snakes) and *Storeria* (DeKay's and red-bellied snakes) appears to be quite dependent on the sense of smell. Occlusion of olfactory organs normally results in immediate cessation of courtship procedures. Mutual entwining is also an important element in their courtship. Certain members of the boid snakes have vestigial hind limbs with which the male scratches the female's flanks, presumably in a stimulatory way. Male red-eared slider turtles (*Pseudemys scripta*) swim back-

(a)
(b)
(c)

Figure 13–3 Some display bowers of bowerbirds. (*a*) The stage of a tooth-billed bowerbird, *Scenopoectes dendrirostris*, of northeastern Australia; (*b*) The conical "maypole" of the crestless gardener, *Amblyornis inornatus*, of western New Guinea. (*c*) The avenue of the great gray bowerbird, *Chlamydera nuchalis*, of eastern Australia. (Drawings based on illustrations from *Birds of Paradise & Bower Birds* by E. Thomas Gilliard. Copyright © 1969 by Margaret Gilliard Person. Reproduced by permission of Doubleday & Company, Inc.)

ward in front of the female, tickling the female's face with long foreclaws. In the related *P. floridana*, however, the male is positioned over the female's back and reaches forward to scratch her face from that position.

The phenomenon of "courtship feeding" in birds serves not only as a possible element of the courtship ritual but, more importantly, as a source of nutrition for egg-laying females. Female titmice (*Parus*) are unable to acquire enough food by their own efforts: The male's input is necessary for successful egg production and incubation (Royama, 1966). Females may also use courtship feeding as an indicator of a male's future performance in parental care. The number of fish delivered by a male common tern (*Sterna hirundo*) during courtship was correlated with the number he brought to his chicks and the fledging success of the brood. Females of this species also seem to prefer males that bring many fish, but it is not yet established that their choice is based on fish delivery itself or some other feature of the male.

Certain fishes, such as the three-spined stickleback (*Gasterosteus aculeatus*), consume some of their own eggs and steal, but do not eat, the eggs of others. Males of these fishes guard nests in which several females deposit eggs. Occasional egg cannibalism by the male may provide energy to prolong the time he can defend his nest. Furthermore, males sneak into the nests of other males, sometimes spawning with the resident male's female. Rohwer (1978) speculates, however, that the main function of these trespasses may be egg thievery. The stolen eggs are deposited in the thief's nest, where they may help convince courted females that the male and his nest are suitable, since other females apparently have chosen them. Moreover, the effects of egg-eating by the attendant male and of egg thievery by other males will be diluted by the presence of other eggs. Perhaps egg-stealing is an important part of male courtship in these species. Simultaneously, there should be selection on nest-owners to deter thieves: Concealment of the nest might reduce the success of would-be thieves, and hatching success is greater in concealed nests (Sargent and Gebler, 1980).

Courtship may last for weeks or months, as it often does in *Homo sapiens*. More commonly, a few hours or days are spent in courtship, as is probably the case for many birds. Some species have essentially no courtship processes, however. Male hellbenders (*Cryptobranchus alleganiensis*) simply follow females that are ready to lay their eggs and fertilize the eggs as they are extruded from the female's cloaca. The mating situation among wild muscovy ducks (*Cairina moschata*) living in the forested swamps of Latin America is also fairly simple. Males are about twice as heavy as females, and they fight violently among themselves in the establishment of dominance hierarchies. Mating efforts are described as violent attacks upon the female: A furious chase is followed by the male biting the female's head or back feathers until she is quiescent. The male then leaps vigorously onto her back several times, frequently falling off, and finally copulates. In African elephants (*Loxodonta africana*), a male reportedly approaches a female and lays his trunk on her back. If she is receptive, she permits him to mount; if not, she moves away.

The form of the courtship displays and the sensory modalities (visual, olfactory, and so on) used are influenced by the location of the display, including the habitat and objects in the environment. Ecological limitations, such as energetic cost and risk of predation, also restrict the options for courtship patterns. Therefore, the evolution and adaptive aspects of courtship behavior are best understood in the context of a species' environment and not in isolation.

McKinney (1976) has emphasized that the nature of courtship displays in ducks varies with the social environment of the species. Factors such as selection for prolonged pair-bond maintenance, opportunity for stolen copulation with other females, and the intensity of competition for mates by both males and females have influenced the orientation of a display, the intended receiver of the signal, the distance at which the display is usually given, and the form of the display. Territorial ducks such as the shoveler (*Anas clypeata*) have long-distance threat signals that are absent in nonterritorial species such as the North American green-wing teal (*Anas crecca*). Because ducks often court in groups, both males and females need a display to indicate the individual in which they are interested; a number of courtship displays have a precise orientation of head and body with respect to the courted individual. Certain male displays are given only in the presence of both a female and another male nearby. Such displays probably convey both threat to the other male and pair-bond affirmation to the female.

Habitat may also have a profound influence on visual and vocal displays. Woodland birds commonly have greater vocal repertories than related species of open habitats. Songs form a large portion of the courtship process in such visually obstructed habitats, whereas flight displays are emphasized by birds of open areas. Social systems are also relevant to the form of courtship. Extended flight displays are more common in territorial than in colonial bird species, which tend to display on or near the nest. Territorial fishes often have elaborate courtship patterns and frequently use the territory for nesting; schooling fishes usually have a fairly simple courtship and tend to shed their gametes in no special place.

Risk of predation can be a selective pressure in courtship patterns. We could expect that individuals engaged in active courtship have reduced awareness of events around them and might be unusually susceptible to predation. This possibility has been applied specifically to the evolution of courtship in eider ducks in the North Pacific (McKinney, 1965) (although it undoubtedly is also of more general importance). The small Steller eider (*Polysticta stelleri*) tends to feed in shallow water; the large common eider (*Somateria mollissima*) usually feeds in deeper water. Courtship displays in the large species are significantly longer, more conspicuous, and composed of less rigidly predictable sequences than in the small species. One explanation for this difference is that larger birds in deep water can dive to elude a surprise attack by an aerial predator and, thus, do not need to be constantly alert. The smaller birds have little refuge in shallow water, so they must not be distracted for long by the courtship process. *Polysticta stelleri* is also more gregarious and more readily takes wing when disturbed, which again suggests the importance of predator pressure.

Furthermore, bat predation may have influenced the evolution of mate-attraction calls of some neotropical anurans (Tuttle and Ryan, 1981). Fringe-lipped bats (*Trachops cirrhosus*) in Panama can discriminate among the calls of many species of frogs, choosing small and edible species in preference to large and unpalatable ones. The bats even seem to use the same elements of the calls that female frogs do. Thus, there is a cost to male advertisement, at least in small and edible species. One would then expect that the calling of such frogs would be somehow less conspicuous and locatable than the calling of large and unpalatable frogs, and apparently it is. Because the risk of predation to a male frog in a large chorus is less than it is in a small chorus (and the chance of mating is greater), predation pressure may have been one evolutionary pressure favoring the formation of large choruses (Ryan et al., 1981).

Communal Courtship

Vertebrates of many species engage in group displays at the time of courtship. A number of temperate-zone colubrid and crotalid snakes, which often hibernate in clusters, frequently court soon after emergence, before the individuals disperse. Many males may court each female, but usually only one is accepted at that time. Many dabbling ducks also form courting groups: Several males display to each other and to a female, who chooses one male for a mate. During the spring in woodlands of eastern North America, courting parties of brown-headed cowbirds (*Molothrus ater*)—several males and a female—are a common sight.

LEKS The epitomes of communal courtship are display arenas and leks. The lek is a traditional display ground, used year after year, on which each of a number of males holds a small territory as a court for display and mating. Females come to the lek to mate and then go off to rear the young by themselves. The males go on displaying and courting other females. In some species, the display area of each male is very small (a few square meters or less) and many such areas can be packed close together. But in others, each male's court may be rather large, and they are aggregated in what is sometimes called an "exploded arena." Lek behavior is known for some birds, mammals, anurans, and a few fishes (as well as some insects).

Manakins (Pipridae) are small fruit-eating neotropical passerines that are usually very sexually dimorphic. Females are typically greenish but males of most species have conspicuous plumage patterns. Males of the lek-forming species display actively for much of the year on leks of various sizes; sometimes the leks are used by as many as 50 or 60 males, but sometimes as few as 6. Each male tends to remain associated with the same lek for many years and often keeps the same display court. Females, on the other hand, often visit more than one lek.

Males of the white-bearded manakin (*Manacus manacus*) in Tinidad are rather long-lived for small birds; some marked individuals were known to be alive and sexually active at ages as great as 11 years (Lill, 1974a,b). They occupy their display courts for most of the year, departing only during the molt and if food becomes scarce. Courts in this species are small, round areas of bare ground with small saplings on which the display occurs. A typical display sequence is shown in Figure 13–4.

Female white-bearded manakins visit the lek mainly between March and August and exhibit definite preferences among the lek males. Mating success is decidedly nonrandom among males—in an extreme case, one male on a lek with nine other males performed 73 percent of all observed copulations at that lek, and three of the males accomplished 95 percent of all copulations. Males that are most aggressive against other males early in the season usually are most favored by females later in the season. The cue used by the females might be the court itself—if certain courts are occupied by particular kinds of males—or perhaps some subtle behavioral trait of the male.

Other lek-forming birds include some neotropical hummingbirds (such as the so-called hermits, *Phaethornis*, and related species), many species of grouse, the ruff (*Philomachus pugnax*) and a few other shorebirds, one genus of cotingid (*Rupicola*), two birds of paradise, two bustards (*Otis*), and one ploceid weaver (*Euplectes jacksonii*). A number of studies, both in North America and Europe, have shown that grouse leks resemble in many ways the leks of manakins. Display areas of the males are usually small and tightly clustered. Males engage in vigorous competition for display areas and not all males are successful in claiming one. Furthermore, of the territorial males, some are far more successful than others in mating females. In the sage grouse (*Centrocercus urophasianus*) of the western United States, for example, fewer than 10 percent of the males accomplished more than 75 percent of the copulations. In the greater prairie chicken (*Tympanuchus cupido*) of the central United States, 89 percent of the copulations were by the two most dominant males on a lek.

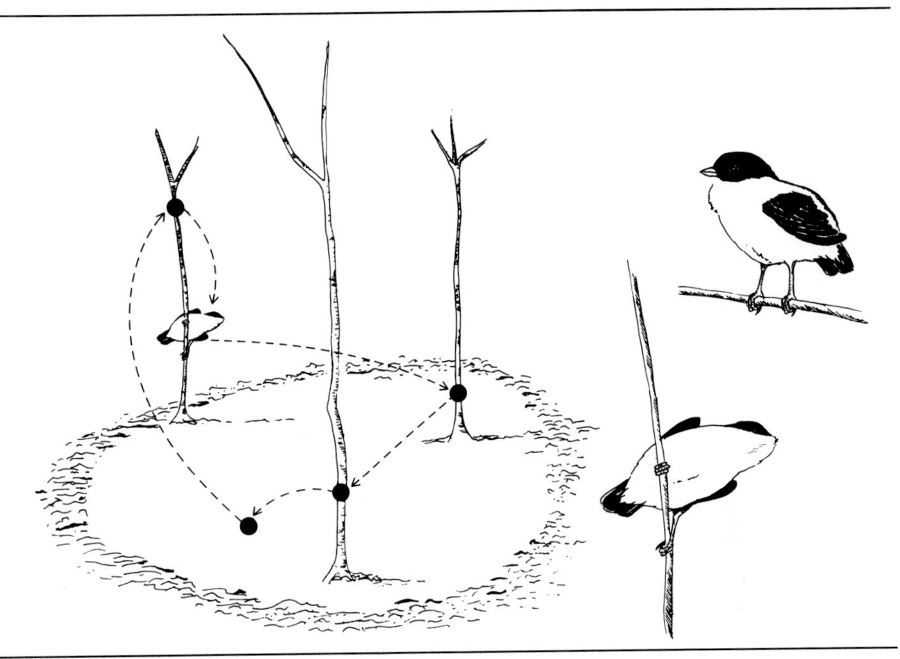

Figure 13–4 Snapping display of a male white-bearded manakin on his display arena. In a horizontal position with his "beard" fluffed up, he jumps from sapling to sapling, sometimes to the ground, and back to his starting position, giving a loud snap with each jump. When a female visits the court, she jumps back and forth with the male, but in the opposite direction, so that they cross in midair. If she is ready to mate, the male jumps rapidly first to the ground and then, with a grunting noise, to a position higher on the sapling than the female. He slips quickly down the stem and copulates. (From H. Sick, 1967. Courtship behavior in the manakins (Pipridae): A review. *Living Bird.* 6:5–22.)

Leks of the ruff deserve more mention, particularly because of the highly developed polymorphism among males. Not only are males more colorful than females, but male color patterns are highly varied. Males tend to fall into two main behavioral categories, each characterized by high frequencies of different color patterns. Some males successfully hold a display court and occupy it day after day, whereas others of the same morphs drift around the margin of the lek and may eventually hold a display court of their own. Morphs of different colors are usually in the "satellite" category. These are males that sometimes share display courts of the territorial males (not always the same ones and not always on the same lek); and they seldom own a territory. The presence of a satellite sometimes increases the number of females that visit a territory. Satellite males are sometimes quite successful in mating females, especially when relationships among neighboring territorial males are unsettled. Satellites copulate most frequently early in the season, on newly established leks, and after turnovers in territory occupancy.

Maintenance of this complex morphological and behavioral polymorphism depends on unknown conditions. Van Rhijn (1973) suggested that, in some years, there is an influx to the population of large numbers of territorial males. These males establish new leks where satellite males can copulate frequently so that the frequency of satellites in the next generation is increased. In other years when the recruitment of territorial males is less, opportunities for successful satellites are also less; consequently, the frequency of satellites in the next generation is lower. Alternating selection pressures may be a plausible explanation, but to date we have no information on what controls the number of territorial or satellite males entering the population.

Leks are also known, more rarely, among mammals. Lekking is well-known in the Uganda kob (*Kobus kob*). Males of certain populations hold small display territories that are clustered on a traditional site (Fig. 13–5). Here they court and mate females, but little feeding is possible on these small trampled spaces. Therefore, after several days of sexual activity, males are easily ousted by challengers and go off to feed, usually with a bachelor herd of nonbreeding males. Eventually, they return to try to reclaim a territory. Lek formation in the kob appears to be best developed in good habitat where populations are dense and male-male competition is high. Males of less dense populations establish large territories on which they feed and court passing females. More study is needed on the distribution and defendability of food resources as well as on the intensity of

Figure 13–5 A male kob on the lek. (Photo by L.L. Rue, courtesy of Bruce Coleman, Inc.)

intermale competition before this system can be well understood. But in the kob, as in avian lek-formers, differential mating success among males is great.

Lek formation also occurs in other large African antelope, including some populations of the topi (*Damaliscus korrigum*) and the lechwe (*Kobus leche*), and among the bats. Some bats defend feeding territories, roosting places, or groups of females. Others, including the West African hammerhead bat (*Hypsignathus monstrosus*), court females on leks (Bradbury, 1977). These bats live in equatorial African forests, feeding on fruit, and are perhaps among the homeliest of bats (Fig. 13–6). Males have enormous larynxes, which occupy more than half the body cavity, and huge muzzles, which are used in producing calls that attract females to the lek.

Some surgeonfishes (Acanthuridae) may also form leks, but detailed information is not available (Barlow, 1974). Another coral-reef fish, the blue-headed wrasse, also courts on leks; males of some populations move as much as half a kilometer to join spawning groups. There they establish temporary territories, and individuals often return repeatedly to the same territory (Robertson and Hoffman, 1977). Females usually choose to mate with the largest males. Some choruses of frogs can be viewed as leks as well.

Figure 13–6 Male hammer-headed bat. Males gather in leks to attract females with vocalizations. (Courtesy of Centre National de la Recherche Scientifique.)

No general explanation for the evolution of leks in all these vertebrates is available. In fact, a universal explanation may be impossible. There is really no reason to suppose that the selective pressures were the same in all cases. In general, emancipation of males from parental care may have been a major predisposing factor. Nest predation in manakins, and perhaps in other species as well, is very high. This danger might favor parental care by a single adult, thus reducing the amount of activity around the nest. Furthermore, at least some manakin species can obtain their adult daily food requirement in about 10 percent of the daylight period, so it is probably easy for a single parent to feed itself and the small broods of offspring adequately. The ability to get enough food in such a short time each day certainly facilitates male emancipation and may pertain to a number of fruit-eating and nectar-feeding species. Grouse are ground-nesters; perhaps nest predation is also high in this species, especially in open habitats. Furthermore, the young are hatched in an advanced condition and are able to leave the nest almost immediately to feed themselves; two attentive parents may not be of any advantage. Emancipation of males is not a sufficient explanation for lek formation, however, since emancipated males of many species (and most mammals) do not form leks.

A second precondition for the evolution of leks is an absence of defendable food or nesting resources. Defense of a feeding/nesting territory is then disadvantageous, and an alternative system is required. Males might display and advertise in isolation, as argus pheasants (*Argusianus argus*) do (Davison, 1981), but there are some potential advantages to group courtship, at least for some species.

What then accounts for the evolution of leks? Several suggestions have been made, but the situation is still only dimly understood. The advantages of lek behavior can be viewed either from a male or a female perspective or from both points of view (Wittenberger, 1981b).

Predation can select for formation of groups (Chapter 11) and may have contributed to the evolution of group courtship. Reduction of the risk of predation could be advantageous both to the competing males and to the choosing females while they are on the lek. There is some evidence that favors this interpretation. Species of grouse that form leks are found in open country, whereas woodland species have scattered display sites (Wittenberger, 1978; Bradbury, 1981). Male blue grouse (*Dendragapus obscurus*) shift from group display in the open to solitary display in dense brush. It is thought that the effect of group vigilance in reducing vulnerability to predator attack is greater in open grassland habitats than in forests, although this needs to be documented. Some lekking species, however, including hermits and manakins, display communally in dense cover; others, including the hammer-headed bat, lek in both open and wooded habitats. Either the behavior of the predators of these species is different than for grouse, or some other factor must be at work.

Males might make themselves more conspicuous to females by displaying together and, indeed, this seems to be the case for the neotropical frog, *Physalaemus pustulosus* (Ryan et al., 1981). Although mating success in greater prairie chickens increased with lek size up to a point, the correlation disappeared when leks were larger than about 15 males; furthermore, such correlations are not found in several other species of lekking birds (Wittenberger, 1981b). Some researchers doubt that grouped males, in general, have a significantly greater attractive effect than solitaries (Bradbury, 1981). Even when males mate more successfully in large groups, it remains to be seen whether the greater success can be attributed to lek size itself or to other factors such as differences in the availability of females (which may be controlled by resources critical to their survival and nesting success, for instance).

Males in a displaying group may have a greater stimulatory effect on females (Snow, 1962, 1976). When sexes meet only for copulation, there is little opportunity for gradual reduction of fearful or hostile responses and no long period of adjustment of the mates to each other. The importance of this in the absence of male parental care is debatable. In any case, neither male conspicuousness nor male stimulatory effects seem to provide a sufficient explanation for the evolution of leks, because neither idea explains why the effect should be found in certain species but not in others faced with seemingly similar situations.

If male attentions are disruptive to female foraging and nesting, females might prefer males that wait to be found; the cost of searching for males might be less than the cost of male harassment, especially when population densities are high (Wrangham, 1980). It is not clear, however, why the waiting males should cluster together, unless an additional pressure favored group formation.

The ecology of females might exert significant pressures on the mating system (Bradbury, 1981). If the important food resources are patchy, widely scattered, or difficult to find, females must move over large home ranges to garner enough food, both for survival and for reproduction. Neither resource-based territories nor females themselves are likely to be economically defended by a male, and males can attract females only by their personal traits and self-advertisement. In this way, a necessary precondition for the evolution of leks is met. Because the average success of males is not necessarily greater in a group, Bradbury suggested that female preferences may make lekking advantageous. Females might prefer clusters of males because this allows them to assess a greater number of potential mates or perhaps because females can then influence the mate choice of other females. The evidence for either of these processes is small. Although male mating success is often highly uneven, as we have seen, the phenotypic traits preferred by females are not at all obvious in most cases (Wittenberger, 1981b).

The great variation in male reproductive success in many lekking species raises a significant question. Some individuals are highly successful because, for whatever reason, they seem to be preferred by many females. For these individuals, lek behavior has no disadvantages sufficient to provide counter-selection pressures. What about the subordinate males that seldom attract females? Why should they stay in areas where they seem to be such failures? There are two kinds of answers. One is that the advantages of group living (such as reduction of predation) are sufficient to keep them there. The other is that, once a lek system has evolved, the subordinate males have little choice. If they went off by themselves to display, their chances of mating would be even less than on the fringes of a lek. By remaining at the lek, they may be able to obtain some females when the favored males are occupied or worn out, and they might even move up (eventually) to a favored position themselves.

Satellite Males

The occurrence of satellite males has already been mentioned with respect to the polymorphic ruff, in which a dominant territory owner may share his display area with a satellite. This curious phenomenon turns out to be moderately common in gregariously breeding species, including several anurans, fishes, lizards, and mammals. Satellites often achieve some success in mating, although seldom as much as dominants. Satellite-male behavior is most likely to be a successful strategem when intermale competition is intense and dominant males are occasionally preoccupied with fighting or courting. The behavior of satellites is described as sneaky and sometimes disruptive to the dominant male's activities. About 16 percent of the calling males of the green tree frog (*Hyla cinerea*) in a pond in Georgia were attended by satellites. These "sexual parasites" were successful in intercepting over 40 percent of the approaching females and, in this species, satellites were about as successful as callers (Perrill et al., 1978, 1982). Individuals shift status in both directions. Occasionally, satellites may assist the dominant individual, either by increasing male attraction, as in the ruff, or by sharing territory defense, as in the waterbuck (*Kobus ellipsiprymnus*) (Wirtz, 1981).

Satellite males in the pupfish are smaller than territorial males. They mate with almost as many, or sometimes even more, females than territory owners, but their reproductive success is lower because they mate with smaller females that lay fewer eggs. Still other sexually mature males in a New Mexico population do not breed at all and so are even less successful than the satellites; in fact, over 90 percent of the males are reproductively inactive in a given season (Kodric-Brown, 1977). Male bluegills (*Lepomis macrochirus*) also pursue alternative strategies (Gross and Charnov, 1980). Some males are parental, defending nesting territories and tending broods.

These individuals seem not to become sexually mature until about seven years of age. Other males, however, mature at age two, are not parental, and are called cuckolders. They sneak in on a spawning pair and release sperm. Although this gambit is not terribly successful, accounting for only an average of 14 percent of the eggs fertilized, most males in these populations adopted this behavior pattern, perhaps because nesting sites were very limited, and their other options were severely restricted.

COOPERATIVE DISPLAYS Manakins of some genera engage in very unusual joint displays in which two or three males "dance" together. Two males of the blue-backed manakin, *Chiroxiphia pareola*, on the Caribbean island of Tobago perform a cartwheel dance in the presence of a female or sometimes a juvenile male in female-like plumage. As one male jumps up and hovers, the second moves up to its place on the perch. The first male then lands, and the second one leaps into the air as the first shuffles up the perch to the takeoff point (Fig. 13–7). According to some observers, one of the males is usually younger than his partner. In contrast to the white-bearded manakins, displaying blue-backed manakins do not appear to be territorial but may use several display courts in turn. Males of the closely related long-tailed manakin (*C. linearis*) in Costa Rica form long-term bonds with other males as regular display partners. One male is dominant over the other and performs most, if not all, of the copulations with females attracted to the display court (Foster, 1977).

Even more complex and exotic is the performance of the blue or swallow-tailed manakin (*C. caudata*). This species lives in the forests of Paraguay and adjacent areas of Argentina and Brazil. The males are powder blue, with black wings, tail, and head, and a brilliant red cap. Three males crouch in a row on a low twig near a female, calling in chorus. The male nearest the female then flies up to hover, display his glowing red cap, and calls in front of her. That male then moves to the end of the row. The other males, crouching and quivering, sidle along the perch toward the female. As the first male lands, the next male repeats his performance, the entire sequence occurring rather rapidly and rigidly (Fig. 13–8). Only the dominant one of the courting males gives the precopulatory display to the female and mates with her after a dancing session (Snow, 1976).

Some aspects of manakin cooperative displays can be considered in the same framework as the evolution of leks (Foster, 1981; Bradbury, 1981); that is, clustering of males and patterns of female choice apply to these manakins as well as to other animals with communal display behavior. Manakins with cooperative displays, however, exhibit the additional feature that several males are needed to perform the courtship display at all. It seems likely that one can imagine a conceptual gradient from the traditional lekking situation, with each male on a separate territory, through systems with satellite males (recall that satellite males seem to aid the

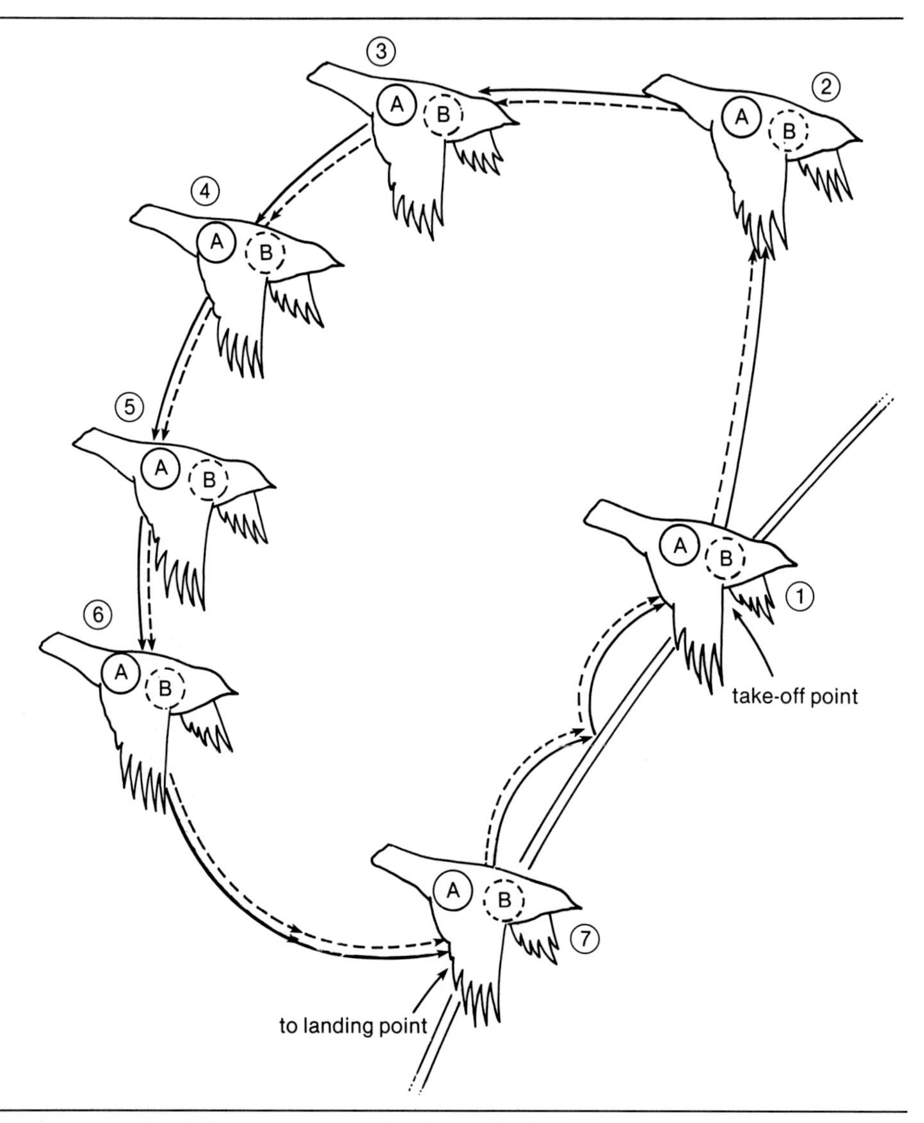

Figure 13–7 The cartwheel dance of two male blue-backed manakins. When bird *A* is at the takeoff point, bird *B* is in position 3. When bird *A* reaches position 3, bird *B* is just landing. When *A* has landed, *B* is in position 2, just after takeoff. (Courtesy of the American Museum of Natural History.)

territory owner in some cases), to cooperatively displaying males. A cooperative display could be advantageous if practice improved the display performance of low-ranking males or if a joint display somehow increased the likelihood that a female returns to that display ground (Foster, 1981).

Figure 13–8 The communal display of the swallow-tailed manakin. Each male is identified by a letter. As the first male leaps up to display before the female, the others sidle along the branch until the next is in position to begin his leap. The first male goes to the end of the line when his display is finished. The jumps get lower and faster as the dance proceeds. (From H. Sick, 1967. Courtship behavior in the manakins (Pipridae): A review. *Living Bird*. *6*:5–22.)

Low-ranking males can move up the hierarchy and eventually breed (Foster, 1981). Relationships among the males seem to be unknown, so far, but Foster (1977) argued that kin selection is unlikely for *C. linearis*, given the promiscuous mating system, small clutch size, and high mortality of young.

The American turkey (*Meleagris gallopavo*) has a complex and variable social organization. Harem formation and association of a male with several females is customary in eastern forest habitats, but in southwestern arid brush, the situation is sometimes rather different. Groups of males vie with each other for dominance status, especially at high densities and in relatively open habitat. A strict dominance relationship also exists within each group. During the mating season, most of the copulations are performed by the dominant individual of the dominant group. Subordinate males have much lower reproductive success, so it is appropriate to ask why they persist in these groups. Although brothers are found in such groups, so are unrelated individuals, so a kin-selection argument is probably not appropriate (Balph et al., 1980). It seems likely that the fitness of subordinates in groups is higher than that of solitaries, for some reason. The modified social structure reported for the Texas population is no longer much in evidence because that population now exists at a much lower density and their reproductive activity is low.

Teams of male lions compete for access to groups of females. These males are sometimes related, coming from the same natal pride, and kin selection might operate in these associations. However, in many cases, the males of a breeding coalition are not related and compete actively among themselves for access to females that are ready to mate. Therefore it seems that kinship is probably not a primary factor regulating the interactions of lion males in breeding coalitions (Packer and Pusey, 1982).

Notice that kin selection may be suggested as an explanation for acceptance of subordinate status or for cooperative behavior in many of the cases discussed here. However, other explanations must be invoked (instead or in addition), and ecological/behavioral constraints and opportunities probably contribute more to explaining cooperative courtship than does kinship. Whatever role kinship may play in the evolution of vertebrate cooperation, it is important to recognize that these diploid animals are always more closely related to their own parents and offspring than to other relatives (unlike the haplo-diploid insects). Therefore, when siblings cooperate, it is likely that ecological conditions prevent them from being successful alone and, given those limitations, kin selection might favor cooperation with relatives (rather than nonrelatives) when possible. Lacking cooperative relatives, an individual might still profit from collaboration with nonrelatives. Thus kin selection in vertebrates operates in a framework of ecological constraints and availability of relatives.

MATING SYSTEMS

The term "mating system" refers to relationships between mated males and females. If a male and a female are associated with each other only for copulation and either may mate with more than one individual, the arrangement is called promiscuous. Monogamy is a pair bond between one male and one female to the exclusion of all others. In some cases, a pair bond may last the lifetimes of the mates; in others, a new pair is formed for each litter or each season. A sequence of monogamous matings by an individual may be called serial monogamy.

Polygamy refers to a bond between a single individual of one sex and a harem composed of two or more of the other sex. The commonest form of polygamy is polygyny, in which one male is mated to several females during the same time span. In some cases, bonds are formed with all the females at about the same time; in others, a distinct lag occurs between initiation of successive pair bondings. The alternative form of polygamy is polyandry, in which one female is mated to several males during the same time period. As in polygyny, bonds may be formed more or less simultaneously or quite sequentially. Serial monogamy sometimes may be called polygamy by some researchers; since many behavioral and ecological aspects of serial monogamy are different from those of polygamy (as it is defined here), however, the use of "polygamy" will be restricted to multiple bonds overlapping in time.

Clearly it is possible for members of one sex to behave as if they were monogamous while members of the other behave promiscuously. Copulations outside the pair-bond are quite common among some mammals, waterfowl, and colonial, monogamously mated birds (Gladstone, 1979; Barash, 1981). Males in these species often subdue and mate with females paired with other males. Typically, the assaulted female protests, often vigorously. Although these incidents may not be disadvantageous to the female if she obtains better offspring, she runs the risk of disrupting her existing pair-bond. If she has difficulties rearing offspring by herself, a broken pair-bond could result in the loss of her investment in the eggs. If this investment is high, the cost of cuckoldry to the female could be high; however, it is also advantageous to the cuckolding male that the female's pair-bond not be broken, because the reproductive success to be gained by "extra-marital" copulations depends on successful rearing of the young. Such behavior, therefore, should be most prevalent in species in which the cost of egg-laying is relatively low and in which paternal care of offspring is poorly developed, or in species in which, for some reason, cuckoldry is difficult to detect.

In general, the evolution of mating systems involves several basic considerations: the relative investment of the sexes in parental care, the degree of competition for mates, the opportunity for multiple mating, and the ecological relationships of the species. A potential conflict of interest exists between males and females. The opportunities for mating more than once in a season often differ between the sexes, and so do the costs and benefits of doing so. If the initial cost of producing young is high, as it often is for females, an individual that deserts the first brood to start another may incur excessive resource demands. If such costs are relatively small, as they often are for males, and other mates are available, then it often should be advantageous for the individual to desert its first mate and find another. If mate desertion means that no young can be reared successfully, however, then it is to the advantage of neither sex to desert the other. The conflict arises when male attention (or more of it) could improve female survival or the size of her brood, but male fitness is enhanced by multiple mating (and less paternal care).

The resolution of the potential conflict will depend on the relative magnitudes of the costs and benefits, which are likely to vary among, and even within, species. Some environments may present limited options either for multiple matings by one sex or for simple pairwise mating by the other. Generally, we would not expect any individual to enter into a mating arrangement that would lower its fitness. Sometimes the choice will not be between good and poor reproductive success but, rather, between poor and none at all, in which case an individual might well settle for the better of the available options. Furthermore, the ability of an individual to make decisions is surely imperfect and mistakes are inevitable. This is true especially when the environment is temporally unpredictable, so that the cues used to make a decision early in the season may not be accurate indicators of things to come (Wittenberg, 1981a; Lenington, 1980).

General Considerations

Most vertebrates are promiscuous. This mating system prevails in all vertebrate classes except birds. Male promiscuity may be favored if males cannot contribute to care of the young. This might be the case in a number of herbivorous mammals, such as the ungulates, for example. Females of these species are equipped to nourish the young, and protection of the young from predators is usually by flight or concealment. If males cannot contribute to care of the young, they are obviously free to pursue other activities, including the search for other mates. If female reproductive success cannot be increased by male help, this implies that food resources, predator protection, or other factors do not limit reproduction in ways that male attendance would alleviate. Of course, this situation raises the question of what factors have favored the evolution of males incapable of helping. For instance, why don't male mammals lactate (Daly, 1979)?

Male promiscuity may also be favored if the benefits of deserting the female outweigh any decrease in her eventual success in rearing the brood (except when success decreases to zero). For this to occur, a male must be able to find enough new mates, which also will be at least partially successful, so that his total number of offspring from all mates will exceed the number attainable with a single mate. Among the ecological factors facilitating this condition are offspring that enter the world in a relatively advanced stage of development (which reduces the cost of promiscuity to the female by enhancing her likelihood of success in rearing them), the nondefendability of resources required for successful production of offspring, and the indefensibility of females themselves (Elliott, 1980). These factors do not make promiscuity advantageous—they merely make it easier; not all species with advanced young and indefensible resources are promiscuous, nor are all species with helpless young nonpromiscuous. Specific ecological conditions that create high probabilities of finding additional mates might include long breeding seasons with frequent opportunities for new matings (Knowlton, 1979) or an adult sex ratio that unpredictably favors females. The factors producing these conditions remain to be explored, however.

Notice that females in this situation do not benefit; the intrasexual conflict seems to have been resolved in favor of males. However, some might argue that these females do not lose, because their sons can do the same thing as their fathers, and enhanced maternal fitness might be seen if the grandchildren were enumerated.

Correspondingly, female promiscuity might evolve if females could achieve greater reproductive success by deserting a male (Wittenberger, 1981b). Regular female promiscuity seems to be less common than male promiscuity. A female might do better alone if the male's activities attract predators, interfere with the female's activities, or reduce critical resources for the female. Then selection could favor mate desertion by females. In addition, once male parental care has evolved, and if resources are sufficiently abundant, the female of a few species may leave the male in charge of the first clutch while she produces another one. The male may also father the second clutch, and the monogamous pair will raise two broods, thus raising their total reproductive output (e.g., *Alectoris rufa*). From this arrangement, a next step might involve a female laying more than two clutches in close succession, but to do so, she needs to find another male to care for each additional brood. If resources allow the formation of multiple clutches, and if parental males can be found, it may be advantageous to the female to go on producing eggs and changing mates. There may be a cost to the males, if they sire only a single brood instead of two, but they may have little choice in the matter. Multiple-clutch mate-switching systems are uncommon, indicating that few species both possess the proper preconditions and encounter suitable ecological conditions for the evolution of this arrangement.

Of course, both males and females may behave promiscuously; this is the common condition in many promiscuous species. One example is Belding's ground squirrel (*Spermophilus beldingi*): Male ground squirrels mate multiply but so do the females; two or three males usually sire the offspring of a single litter (Hanken and Sherman, 1981). Some birds and mammals live in more or less restricted groups composed of several males and several females; all individuals may mate with group members of the other sex. Although such arrangements can be called polyandrous, polygynous, or both, in a restricted sense, they are promiscuous within the group. This situation is common in some primate groups, in lions, and in rheas (*Rhea americana*), which are large cursorial birds of the South American plains. Flocks of female rheas lay a collective clutch of eggs for one male and then for a series of others (see Chapter 15).

Polygyny is the second most common mating system in mammals and in birds; it also occurs in some reptiles and fishes. Emancipation of the male from parental care may facilitate polygyny in some cases, but it is clearly irrelevant where polygynous males provide paternal care. Thus, male emancipation cannot provide a general condition for the evolution of polygyny. There seem to be two fundamental kinds of polygyny. The first type is based on male territorial defense of some limited resource. Females may choose a male in part by using characteristics of the male's property, or males may simply control access to resources limiting for females. This seems to be a common occurrence in both birds and mammals. The second type is common particularly in polygynous mammals. Males often defend the females themselves, rather than food supplies or nest sites. This will be the case when resources are difficult to defend but females are not, because they group together (for whatever reasons, Chapter 11) (Emlen and Oring, 1977).

Resource-defense polygyny sometimes involves only male control of certain essential resources. Resource-defense polygyny probably describes the mating arrangements of the chuckwalla (*Sauromalus obesus*). Dominant males own territories from which they chase intruding subordinates. Several females live within the territory of each dominant male, with whom they mate (Berry, 1974). Harem size in the South American lizard *Liolaemus tenuis* is related to the size of the tree trunk defended by the male; the tree presumably provides some limited resource (Manzur and Fuentes, 1979).

Males of the northern elephant seal (*Mirounga angustirostris*) defend the beaches that are required by the females for giving birth to their young. Although suitable beaches may be a limited resource, males also defend the females themselves from the attentions of subordinate males. Thus, to some degree, these animals may represent both types of polygyny. Male Weddell seals (*Leptonychotes weddelli*) defend breathing holes in the ice; these are used by females to emerge onto the ice to give birth. Only a limited number of females use each hole; unfortunately, it is uncertain that the male defender is indeed the father of the pups.

Many birds exhibit resource-defense polygyny. When males engage in parental care and nesting cycles of polygynously mated females overlap in time, males often do not distribute their paternal attentions equally among broods. If they feed the nestlings chiefly at the nest of their primary female, whose nestlings hatch first, any subsequently mating females must rear their young with less paternal help. The later-nesting females may have a choice between mating with an unmated male, and being a primary female, or mating with an already-mated male, and becoming a secondary or tertiary female. Everything else being equal, we would expect females not to pay this potential cost of polygyny and to prefer the unmated male, if his paternal care would increase the success of her brood or significantly reduce her workload and thus enhance her prospects of survival.

As usual, however, all factors are seldom equal. Male parental contributions do not vary with harem size in all species. A female may be unable to assess accurately the prospects of male assistance. Territories of the males may be so unequal in quality that mating polygynously, even as a secondary female, may be better than mating with an unmated male and having primary status. Sometimes, harem females may even cooperate in such activities as nest defense, providing another advantage to settling in a harem. The territorial male may control several resources, whose importance shifts as the season progresses. Females might then find it difficult to evaluate properly the quality of each territory. Furthermore, not all options may be open for all females. To add to the complexity, members of the same species sometimes do things differently in different locations, and the mating system of the species may operate differently even though polygyny prevails throughout. Advantages to males of this system are generally clear; the situation for females seems to be more variable. We will survey some case studies that exemplify some of these problems.

Female-defense polygyny occurs when females themselves can be defended more economically than resources. Male parental care is generally irrelevant here; males control the harem during the mating season, or part of it, and have little direct involvement with their females and young after copulation and conception. Females may obtain no special benefit from this arrangement, or perhaps they derive some benefit from producing "sexy sons." In general, females form groups for their own advantage, and males can then defend a number of females economically (Wittenberger, 1980a).

It is sometimes argued that a sex ratio skewed in favor of females would favor the evolution of polygyny. However, most evidence supports the hypothesis that the reverse is usually true—polygyny may result in a skewed sex ratio. Adult males in sexually dimorphic species are sometimes fewer than females because their morphology or behavior makes them more susceptible to mortality. Males of the great-tailed grackle (*Quiscalus mexicanus*) seem to have greater difficulty flying and less agility in locomotion than females because of their enormous tails—they may be great display accoutrements but are aerodynamic disasters in storms and strong winds. In addition, males of polygynous species often mature more slowly than females, so that the adult sex ratio may be further skewed. This too, is probably a result (not a cause) of the intense intermale competition characteristic of polygyny, in which first-year males are at a real disadvantage in trying to breed. On the other hand, the monomorphic and polygynous long-billed marsh wren apparently has a 1 : 1 adult sex ratio.

However, it is certainly not impossible for an unbalanced sex ratio to foster polygyny. A case in point may be provided by Tengmalm's owl (Chapter 11): Males perennially defend their nest sites, but females are reported to settle in densities related to annual availability of their rodent prey. As a result, males mate polygynously in good mouse years but not at all in poor ones. Moreover, experimental removal of male song sparrows (*Melospiza melodia*), which are usually monogamous, produced an adult sex ratio favoring females, which then tended to enter polygynous matings (Smith et al., 1982).

Monogamy is the prevailing mating system among birds, but it also occurs in certain mammals and fishes. Two general conditions favoring the evolution of monogamy may include the importance of both parents for feeding or protecting young and the inability of males to defend a territory with enough resources to attract more than one female. Each general condition can be split into subsets of more specific conditions whose relative importance is likely to vary among taxa (Wittenberger and Tilson, 1980). Some of these differences will be discussed in the following paragraphs, as we survey selected studies of mating systems, especially in birds and mammals.

Mammals

Mammals engage in both resource-defense and female-defense polygyny. Defense of females occurs in a number of ungulates, many of which often live in groups. For instance, plains zebra (*Equus quagga*) commonly live in groups of up to 16 individuals. Some of these groups are composed of bachelor stallions; others are composed of a stallion, his harem of several females, and their foals. These small units sometimes aggregate into herds, especially at night, which suggests that predation is a major selective pressure favoring gregariousness. Females would be expected to form some kind of cluster whether or not a male were in attendance. Harem stallions often defend their families against predators such as spotted hyenas (*Crocuta crocuta*). This defense is not very effective, however, and more mares than stallions are killed. The protective value of the male, therefore, may be unlikely to select very strongly for the formation of female groups. It seems more probable that males compete for and attach themselves to existing bunches of females.

Primate females may also gather together and form groups, which may form a basis for female-defense polygyny. Some bats, including *Phyllostomus hastatus* in Trinidad, seem to exhibit preformed female groups defended by a male (McCracken and Bradbury, 1981). Females of this bat species roost in caves. They form clusters of individuals that are stable from year to year, although they are not especially closely related. The basis for the formation of female groups is uncertain but may lie in the benefits of cooperative foraging (as suggested also for many primates). The clusters average about 18 females, and each cluster is tended by a male who fathers most or all of the young.

Northern elephant seals have been studied on island beaches off the coast of California (LeBoeuf, 1974). These animals feed at sea but haul out on land two times a year: once to breed and once to molt. Breeding activity lasts from about December to March. Females give birth when they arrive on the beaches and then are ready to mate. Males arrive first on the breeding beaches and display and fight among themselves, with the result that a strong dominance hierarchy is established. Males of the highest social ranks in the male hierarchy do most of the copulating with females. Adult males are about three times as big as females and mature more slowly. They are seldom able to maintain a dominant rank (if they ever achieve one) for more than three years and often for only a year. Subordinate males usually have very low reproductive success. Their success increases, however, if the number of females on the beach grows so large that they cannot be controlled by the dominant male or if access to the beach is possible from several directions, which would make defense of a harem more difficult by the dominant bull. Competition among males is intense. A male who is dominant very early in the season may be worn out and replaced by the time most of the females are ready to breed. Male-male aggression and copulation attempts with females rather frequently result in the crushing of pups: Males pay no attention to pups or their mothers in the course of status conflicts with males or when mating with other females. The probability of a male killing one of his own offspring sired the previous year is rather small, however, due to the large annual turnover rate of dominant males. Therefore, a modest increase in number of successful copulations is all that is needed to compensate for this accidental destruction.

Bartholomew's (1970) model for the evolution of polygyny in pinnipeds (Fig. 13–9) at large is based on two general mammalian traits (male aggressiveness and large size of aquatic species) and on two characteristics of many pinnipeds (offshore aquatic foraging and terrestrial parturition). These features are seen as preadaptations to pinniped polygyny that make possible the great gregariousness during the breeding season and harem formation by males. It is not altogether clear, however, just why certain pinnipeds mate on land as well as giving birth there—terrestrial mating makes polygyny possible (Wittenberger, 1981b). Polygyny is less well developed among pinnipeds that breed on ice instead of land (Pierotti and Pierotti, 1980). These species breed when air temperatures are very low, which may create thermoregulatory problems that prevent males from

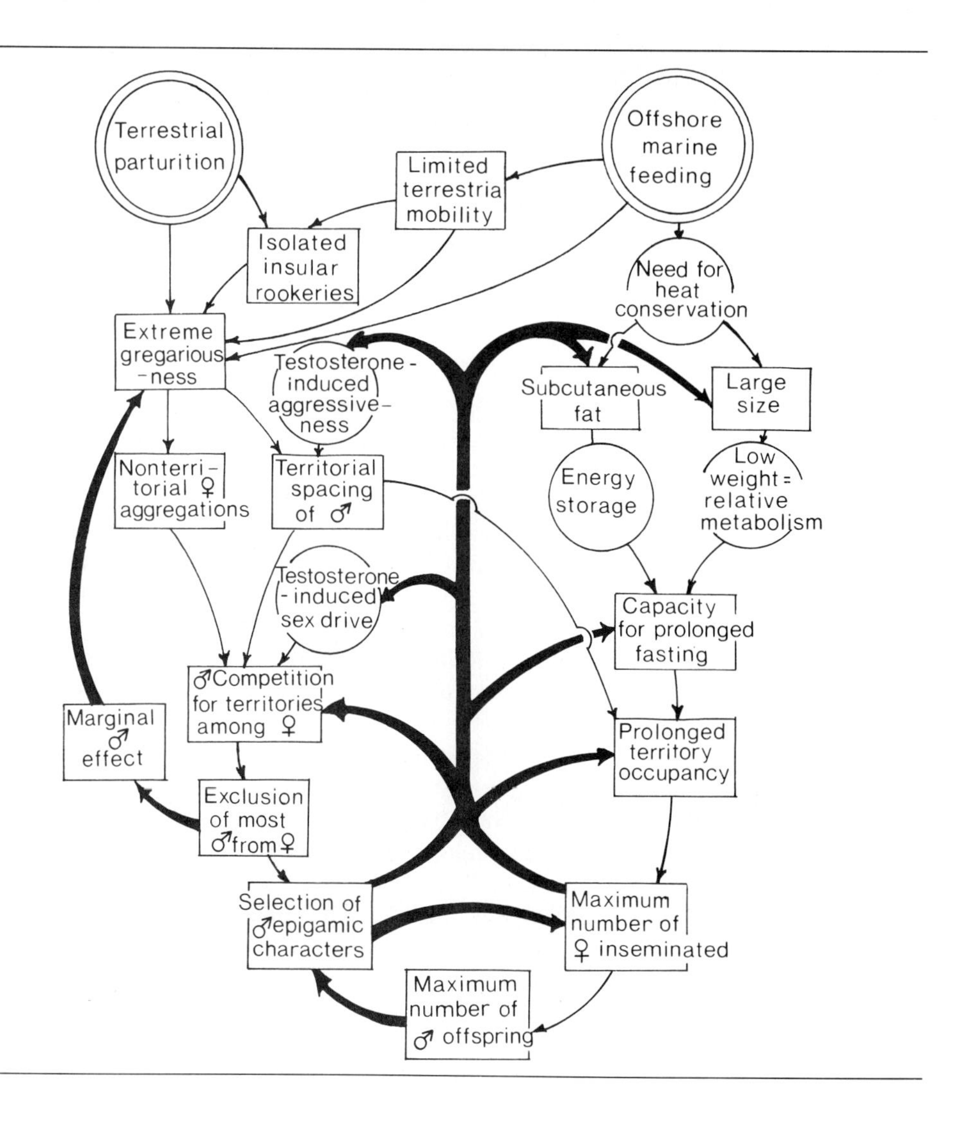

remaining out of the less cold water for long periods. Short emergence times for males could prevent the establishment of elaborate dominance hierarchies and intense intermale competition, thereby, reducing the potential for the development of very large harems.

Males of some species control access to limited resources that are necessary for successful breeding. For example, certain neotropical bat species such as *Artibeus jamaicensis* roost in hollow trees. Males of these species can obtain possession of a harem by "owning" a roost tree. In a sense, this is a parental investment by the male, since the roost may be used as a brood chamber for the young, but the parental investment is an indirect result of competition among males for mates. Resource-defense polygyny is also

Figure 13–9 Bartholomew's (1970) model for pinniped polygyny. The double circles indicate two important features of most pinnipeds. The small, single circles indicate characteristics of most mammals. The rectangles mark factors typical of polygynous pinnipeds. The model is interpreted as follows.

Most pinnipeds feed offshore, which produces the need for good heat conservation and, because of their extreme adaptations for aquatic locomotion, limited mobility on land. The females of some species give birth on land, however, on isolated beaches on islands. Because of their limited mobility and the small number of suitable beaches, the pinnipeds are highly gregarious on these islands. Males in the dense aggregations of animals tend to be aggressive and territorial, although the females are not. Males compete among themselves for territories and females, with the result that some males are excluded and hang around the margins of the gregarious group.

Returning to the right-hand side of the model, the need for heat conservation produced selection for subcutaneous fat deposits and large body size. These two features mean that the animals have a low metabolic rate (per gram) and improved capacity for storage of energy, which in turn permits them to go without food for a long time. As a result, they can stay on land for long periods. Thus, males can occupy a territory for some time, which improves their chances of fertilizing females. Males that are successful in this way have greater fitness: Prolonged territory occupancy, coupled with intermale competition and exclusion of some males from breeding, increases the fitness of some males. Because these males are likely to be larger than others, an enormous feedback loop to earlier elements of the model is created, enhancing each of the indicated features. (Modified from G.A. Bartholomew, 1970. A model for the evolution of pinniped polygyny. *Evol.*, 24:546–559.)

reported for Kenyan populations of the insectivorous bat, *Pipistrellus nanus*, for which roosting sites (in and about human habitations) are a limited resource defended by males (O'Shea, 1980). South African populations apparently have abundant roost sites in rolled-up leaves, no evident resource competition, no male defense, and no harems. Other factors must also be involved in structuring the system, however, because certain other South African bats also roost in rolled-up leaves and exhibit resource-defense polygyny.

A fascinating case of polygyny occurs in yellow-bellied marmots (*Marmota flaviventris*), which live at middle elevations in mountainous regions of western North America. The animals are active outside their burrows from May to September. Some females lived alone, without a regular male, and had a much lower reproductive success than females who were regularly associated with males (Downhower and Armitage, 1971). Solitary females bore litters less frequently than group females and may also have had lower survival of offspring. Group living was clearly advantageous for females, compared to a solitary life. Males associated with one to four females. If the number of surviving young is plotted against harem size, the resulting curve for females decreases steeply, demonstrating that polygynous matings were less advantageous for females than monogamy (Fig. 13–10). For a male, however, a harem size of two or three was best; the sum of his females' reproductive output was then greatest. Polygyny was common in this population, despite its relative disadvantage (compared to monogamy) to females. Downhower and Armitage (1971) argued that selection had maximized the mean reproductive success of male and female and that the observed mating system was, perhaps, an adaptive compromise.

There are at least two possible alternatives to the compromise hypothesis. One is that females survive better when they live in a harem and that their enhanced survival compensates for lowered reproduction (Elliott, 1975; Wittenberger, 1980). Group females may have a lower risk of predation (Elliott, 1975). However valid this suggestion may be, it does not

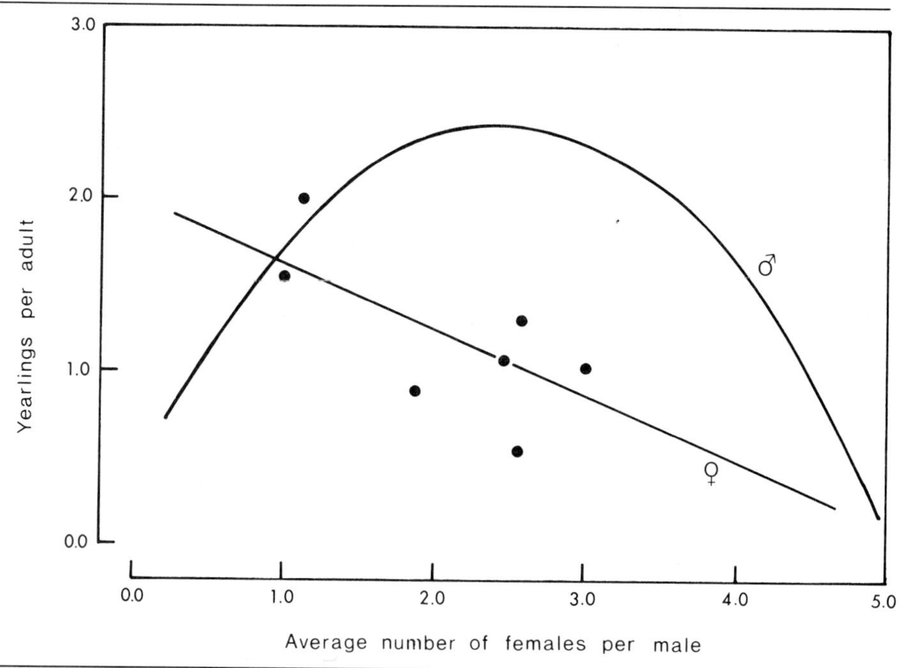

Figure 13–10 Changes in the number of surviving offspring (yearlings) per male and per female, as harem size increases in the yellow-bellied marmot. The data points for females are averages over a 5- to 7-year time span for seven harems. Note that the reproductive success of females tends to decrease with increasing harem size. The curve for male reproductive success is the sum of the reproductive successes of all the females in a harem. This curve peaks at a harem size of two to three females. Thus, females and males cannot simultaneously maximize their reproductive output in these conditions. (Reprinted from *American Naturalist* by J.F. Downhower and K.B. Armitage by permission of The University of Chicago Press. Copyright © 1971 by The University of Chicago Press.)

explain the observed variation in harem size, so additional explanations are necessary.

A second alternative is that females may have restricted options. Environmental limitations may make their optimum choice (monogamy) unavailable, and polygyny may be the only alternative. The time of beginning of the growing season for plants was correlated with the reproductive success of the herbivorous marmots: In years when the snow stayed on the ground for a long time and the growing season began late, harems were larger and average litter sizes were smaller because fewer females had successful pregnancies. When snow covers much of the habitat, some of the females have little choice but to join other females in a polygynous arrangement. Late growing seasons also might lower reproductive success directly, by reducing the supplies of suitable food for gestating females. Females without young are less aggressive than pregnant and lactating females, so in years with late growing seasons, there is less aggression among females and more can fit into a male's territory. Thus, it is possible to suggest that polygyny itself induces lowered reproduction and that females are forced into polygyny by environmental constraints, or that environmental limitations reduce reproduction and facilitate the clustering of females. In either case, female choices are limited in this kind of environment. Variation in harem size could result from variation in suitability of the male's territory for feeding.

Moreover, females often share burrows and conserve thermoregulatory energy, which might enhance survival (Anderson et al., 1976). If hibernation burrows are limited in supply, the advantages of burrow-sharing might help account for the formation of female alliances and for variation in harem size.

Monogamy in mammals is quite widespread, occurring in a number of taxa (Kleiman, 1977; Wittenberg and Tilson, 1980). Among the Carnivora, it is found especially among the canids (jackals, wolves, foxes) but also in some viverrids (mongooses, meerkat) and mustelids (otters, European badger). Some pinnipeds are monogamous, as are some bats, ungulates, insectivores, primates, and rodents.

Monogamous mammals generally produce helpless offspring, in contrast to the more advanced young of most other mammals (Zeveloff and Boyce, 1980). The offspring are usually small, although litter size may be relatively large, and gestation periods tend to be short. These authors suggest that maternal investment in such offspring is low compared to nonmonogamous species. As a result, the opportunities for paternal investment are considerable, and parental males in monogamous pair-bonds should be more common in these cases. In contrast, when maternal investment at birth is high, there is less opportunity for paternal investment and less selection for monogamy. Kleiman (1977), too, notes that male parental care is typical of most monogamous mammals. Such care may take the form of feeding young, defending or carrying them, or perhaps teaching them. Thus, it is tempting to suppose that the importance of paternal care in rearing the young may have been crucial in the evolution of a monogamous mating system. However, other factors could be involved (Kleiman and Malcolm, 1981). Territory quality may not be sufficiently varied to permit polygyny (Wittenberger and Tilson, 1980). Mated females are often aggressive against other females, which would reinforce a monogamous association, but the basis for the aggression is likely to lie in resource defense (if the "extra" females are not part of a larger social group) or in monopolizing the male's paternal care (Kleiman, 1981). Furthermore, the ecological conditions that favored low maternal investment in the first place certainly require elucidation.

Polyandry appears to be recorded only in certain societies of *Homo sapiens* and usually takes the form of wife-sharing by brothers. These societies, mainly in Asia, are commonly characterized by subsistence agriculture on small family holdings, rights of primogeniture (the first son inherits property and the right to marry), female infanticide, and a sex ratio skewed in favor of males (Alexander, 1974). Landholdings are so small that further subdivision is impossible. Numerous male offspring increase the labor force and farm productivity, but they have no place to go when they mature since all available land is owned. Younger brothers may work as hired hands for their older brothers, and older ones may use wife-sharing as an inducement to cooperation. Inclusive fitness, even of younger brothers, may be enhanced by this arrangement, since they are not likely to be successful if they leave home to wander and remain landless. In a number of these societies, the poorest families are polyandrous, middle-class families tend to be monogamous, and rich ones may be polygynous!

Birds

Polygyny is relatively rare in birds, occurring in about 2 percent of the species (Lack, 1968). Almost all cases seem to be resource-defense polygyny. If some limited resource is distributed in a patchy way and a few males can maintain exclusive ownership of a disproportionate share of those resources, females may be obliged to enter polygynous matings to obtain access to the required resources. The resources defended by a territorial male may be food, nest sites, or protective spaces. Spatial variability of the environment is an essential prerequisite to the evolution of the system. If the spatial variability and the population of males are sufficiently large, some males will occupy the good patches while others are forced to claim second-rate patches. Females may then have a choice between good and poor territories. Most males that hold a territory may get at least one mate, but some may hold such poor patches that no females settle there. The number of unmated but territorial males will depend on how well males can judge female preferences and how well they can assess territory quality, which may vary seasonally.

Males of polygynous species usually contribute less parental care than do males of related monogamous species. They spend their time and energy courting new females. In times of food scarcity (when cold, wet weather reduces the availability of insects, for example), the help of a male in obtaining food for the nestlings may be very important. Such appeared to be true for the great reed warbler (*Acrocephalus arundinaceus*), a polygynous marsh-nesting passerine in Europe. Nestlings frequently starved to death in one of the two nests of bigamous males, presumably because the male's attention was focused primarily on the nest of only one of his mates (Dyrcz, 1977). If a female mates with a polygynous male, she runs a high risk of obtaining less help (or maybe none at all) in brood-rearing. Ideally, then, the superiority of the male or his territory must be great enough to compensate for the possible loss of the mate's help.

Several studies of polygyny in territorial birds have demonstrated that territories really do differ, and females can respond to these differences by entering a polygynous mating.

Dickcissels (*Spiza americana*) preferentially established territories in weedy oldfields, and prairies were a second choice in Kansas (Zimmerman, 1982). Females reproduced with equal success in both habitats and seemed to distribute themselves in each habitat according to the availability of suitable nest sites. The patchy habitat of oldfields allowed some males to claim territories with several good nest sites whereas others had few or none at all. Polygynously mated females were as successful as monogamously mated ones, but males in oldfields were more polygynous and had more offspring than those in prairies.

Many studies of polygyny have concerned the blackbirds and their relatives (Icteridae) because they are relatively easy to study and provide good comparative situations. The case of the yellow-headed blackbird was presented in Chapter 12. Another example is another marsh-nester: the red-winged blackbird. Table 13–1 shows the reproductive success, as estimated by survival of young up to the time of departing from the nest, of male and female redwings in eastern Washington in two different years. As expected, male reproductive success increased markedly with increasing harem size in both years. Female reproductive success tended to increase also, but the trend was more pronounced in the second year of the study as a result of differences in the weather. Stormy weather in the second year increased differentials in territory characteristics: The availability of food near the nest and the strength of the supporting vegetation available for nest sites probably were critical in bad weather.

In the mild year, primary females had higher nesting success (here measured not as young raised per female but as percentage of nests that raised at least one young); females of lower status tended to have lower chances of some success (Orians, 1980). In the severe season of the second year, however, primary females had a lower likelihood of success, presumably because of the inclement weather early in the season. Thus, the un-

Table 13–1 Reproductive success of male and female redwings in eastern Washington.

Number of females per male	*Number of males in sample*	*Number of young per male*	*Number of young per female*
1966			
1	8	0.88	0.88
2	14	1.57	0.78
3	11	4.45	1.48
4	7	4.14	1.03
5	9	6.33	1.27
6	3	9.66	1.61
1967			
1	7	0.43	0.43
2	15	1.40	0.70
3	14	2.43	0.81
4	7	3.57	0.89
5	5	6.20	1.24
6	1	11.00	1.83

(From Holm, 1973.)

predictability of weather conditions seems to affect the relative reproductive success of females nesting at different times. Therefore, it must be difficult for females to predict the consequences of choosing to join existing harems of different sizes.

Redwings in northern Illinois, however, apparently encountered a different situation (Lenington, 1980). Females in large harems raised fewer young than those in small harems, indicating a possible real cost to joining a larger harem. The values for estimated reproductive success were based only on nests that produced *some* young, however, and total failures were omitted, so it is difficult to evaluate the actual cost.

The paternal roles of male redwings in eastern and western United States are different: Eastern males provide more care of the young than western males. Furthermore, there are regional differences in the nesting synchrony of females sharing a male's territory. Nesting synchrony among the females on a single territory in Washington was similar to that of females on different territories, but in Indiana, the temporal spacing was somewhat greater within a harem than in different harems (Yasukawa and Searcy, 1981). These results are interpreted to mean that there may be competition among eastern females for male attention, which could be expected if male parental care is important to reproductive success.

A comparison of western redwings with yellowheads is complicated by several factors. Male yellowheads are significantly more active in parental care than are western redwings, but they allocate their attentions unevenly among their broods. Primary females receive significantly more male help in feeding young than do secondary females, and their offspring are often somewhat heavier. Therefore, female yellowheads may incur a greater cost in mating polygynously than do western redwing females because they sacrifice more paternal care. It remains to be seen, however, if yellowhead territories differ enough to compensate for their loss or if females are constrained to accept secondary status in order to mate at all.

A further complexity results from age differences in reproductive success of females of both species: Female at least two years old lay more eggs and raise more and larger nestlings than yearling females (Crawford, 1977). Older females were usually the first nesters on a male's territory; yearlings usually had secondary status. On the few occasions when yearling females were first-nesters on a territory, their nesting success was greater than that of secondary yearlings. In yellowheads, their nesting success was equal to that of older females. These unusually successful young females were usually mated monogamously, however, and in yellowheads, they presumably had the full attention of the male. These differences in reproductive success are summarized in Table 13–2.

Table 13–2 Mean clutch size and fledging success in relation to pairing status in female red-winged and yellowheaded blackbirds.

	First female			*Second female*		
	N	Number of eggs per nest	Number of young fledged per female	*N*	Number of eggs per nest	Number of young fledged per female
Redwing yearling	6	4.0[1]	1.2[3,6]	29	2.9[1]	0.6[4,6]
Redwing older adult	24	4.2	1.8[3,7]	11	4.1	1.3[4,7]
Yellowhead yearling	7	4.0[2]	2.0[8]	17	2.8[2]	0.8[5,8]
Yellowhead older adult	13	4.1	2.0[9]	4	4.0	1.2[5,9]

(From Crawford, 1977.)
[1–9] Statistically significant differences between each pair of indicated values.

Clearly, it is impossible to discern just what is regulating the occurrence and degree of polygyny in these blackbirds without studying each area in great detail. Not only must we know age, status, settling times, and success of each female on a study area, but the territories of the males must be examined carefully throughout the season for ecological features that may determine territory quality and its predictability. Such measures should not only include vegetation cover, nest sites, and food supply but should also include the risk of predation and brood parasitism. Picman (1980) has suggested that redwing females may cluster their nests and cooperate in nest defense against the depredations of the long-billed marsh wren. Furthermore, the possible cost to a secondary female must be measured not only by her reproductive success (perhaps including the sex ratio of her young) for that season but by her likelihood of survival and successful reproduction in subsequent seasons.

Other studies of polygynous passerines reinforce the importance of assessing the entire ecological picture. The bobolink (*Dolichonyx oryzivorus*) is a grassland-nesting icterid. Harem size reflected vegetation structure to some degree (Martin, 1974) but varied closely with the availability of the major prey (caterpillars) for nestlings (Wittenberger, 1980b). The correlation of harem size was especially clear in certain kinds of grasslands. Secondary female bobolinks were usually young and unexperienced or were females whose first nesting attempts elsewhere failed. As in the marsh-nesting blackbirds, age complicates the possible costs of lower status.

Lark buntings (*Calamospiza melanocorys*) are also polygynous; harem size is apparently determined by the number of suitable nest sites present on a male's territory (Plezczynska, 1978). Although secondary females on average produce about as many young as monogamously mated females nesting at the same time, some females mate monogamously even when their reproductive success for the season might be higher if they were secondary females on very good territories. One explanation for this choice might be that male assistance for primary females means that females do not have to work so hard and, consequently, reach the end of the breeding season in good condition, with good prospects of survival (Plezczynska and Hansell, 1980).

The long-billed marsh wren has evolved yet another solution to the problem of polygyny and male parental care (Verner, 1964, 1965). Males in eastern Washington may have two or three females with overlapping nesting cycles and will feed only the last brood. In western Washington, however, nesting cycles of females are usually staggered so that there is little overlap between nestling periods, and males attend to each of their broods in sequence. Eastern Washington has a continental climate with cold winters and hot summers, in contrast to the mild climate of the western part of the state. A strong seasonal alternation often means that production of insects is concentrated in the spring and summer months, probably resulting in a rich burst of insect food for the wrens during the breeding season.

As a result, the female can feed a nest full of young. The male cannot significantly augment the food provided by the female and is thus free to court additional females with no delay for parental duties.

Paternal feeding of nestling Cetti's warbler (*Cettia cetti*) in Britain diminished sharply with increasing harem size (Bibby 1982), but, in this case, the growth and survival of young was equally good in large and small harems. Highly polygynous males were larger than monogamous ones and may have held better territories. Both male and female Cetti's warblers had higher reproductive success in polygyny than in monogamy, but just how the evolutionary ecology of this species might differ from that of most of the other birds discussed here is not known.

Several studies have sought broad ecological correlations with the occurrence of polygyny in birds. Verner and Willson (1966) surveyed North American passerines and found that polygyny (and a few cases of promiscuity) was most prevalent in marshes, grasslands, and savannas. (Since that study, the short-billed marsh wren, *Cistothorus platensis*, dickcissel, indigo bunting, *Passerina cyanea*, and several other species in several families have been shown to be regularly polygynous, so the correlation does not depend on any single taxonomic family. Several of these additional species nest in marsh or grassland, and one or two nest in forest edge.) We suggested that the high productivity of marshes and grasslands, where the sun's energy is utilized within a fairly narrow vertical belt, may make possible the variability in food and nest-site density necessary for polygyny to prevail. Crook (1962, 1963) found a similar correlation in weaverbirds (Ploceidae): Forest species are usually monogamous, whereas savanna and grassland species are often polygynous. These correlations are not perfect, of course, and in both studies, some species inhabiting grassland and marshes are monogamous. Differences in how these species exploit the habitat are clearly important to an understanding of why some are monogamous and others are polygynous, but these differences have yet to be analyzed.

In contrast to these studies, von Haartman's (1969) found no correlation of habitat and mating system for European passerines. Perhaps the productive marsh and grassland habitats found extensively in North America and Africa have been less widespread in Europe; consequently, there would have been less adaptive radiation in birds specializing in their exploitation. Furthermore, extensive habitat modification by humans is of very long standing in Europe and may have disrupted relationships that were originally present. von Haartman found that most polygynous European passerines are hole- or dome-nesters. Covered nests may permit more flexibility in the female's time budget. Protection from the elements could mean that a female needs to spend less time in the nest warming the young, allowing her to spend more time foraging. Also, improved insulation in a covered nest might reduce heat loss and food requirements of the young. Protective from predators may reduce selection against short nestling periods, permitting slower nestling growth and lower rates of food delivery. This argument assumes that food-delivery rate is not critical in regulating brood size and that an increase in feeding rate due to male participation would not increase the female's reproductive success. Otherwise, male help might permit an even larger clutch size. Therefore, we infer that brood size is limited by other factors or that good nest sites are so patchy that one male can defend more than one and so limited that females must claim them despite the cost of polygyny.

Polyandry is very rare in birds, occurring regularly in less than 1 percent of the species. Several known cases fall into the category of sequential polyandry: A female lays clutches of eggs for two or more males, and the males take over parental duties. In some species, the female incubates the last clutch herself. This is the case for one species of button quail (Turnicidae) and several shorebirds (Jenni, 1974; Howe, 1975; Emlen and Oring, 1977; Schamel and Tracy, 1977; Maxson and Oring, 1980). Clearly these cases can also be called serial monogamy or even promiscuity. Their most interesting feature is the unusual role of the male in parental care.

Simultaneous polyandry is documented for at least three or four bird species and may occur in others. The Tasmanian native hen (*Tribonyx mortierii*) lives in small groups that are usually composed of siblings; apparently, brothers indulge in wife-sharing. We know virtually none of the ecological pressures on this species.

Jenni and Collier (1972) have described an intriguing arrangement in the American jacana (*Jacana spinosa*). Male jacanas in a marsh in Costa Rica held small territories, and females defended larger areas that often included several male territories. Some females were monogamous, some had four males, and the average was about two males per female. Females defended their territories against other females and their males' holdings against other males. Females copulated frequently with their mates, often with all their mates on the same day, and mate bonds with all males were maintained simultaneously. Only males incubate the eggs and care for the young. Female jacanas are about 75 percent heavier than males and are dominant over males in all situations. A male reportedly accepts any female whose territory includes his. Because nesting success is low and the birds spend a lot of time threatening and attacking predators (other birds and reptiles), Jenni (1974) suggests that predation on jacana eggs and chicks may be very high. Perhaps large size in female jacanas evolved as a means of producing many eggs and thereby replacing lost clutches quickly. Eggs are smaller relative to female body size than eggs of other shorebirds, facilitating egg replacement. Males would then increase their fitness by mating with large females that could both produce more eggs and successfully defend a territory. Jacana ecology and behavior are indeed fascinating subjects for future research.

Galápagos hawks (*Buteo galapagoensis*) appear to be simultaneously polyandrous (Faaborg et al., 1980). Up to four males attend a single female and her nest. They raise more young than when only one male is present, although there are fewer young per male. Unfortunately, paternity of the brood is unknown, so we cannot be sure that more than one male in fact fathered the young. Survival of males in a mating group in a territory is greater than when they are nonterritorial, which might compensate for the lowered individual reproductive success. Harris' hawk (*Parabuteo unicinctus*) may also be polyandrous (Mader, 1979), with two or more males helping rear the young with a single female.

Avian monogamy is thought to have evolved when male help is essential to reproductive success (Wittenberger and Tilson, 1980). Males might be particularly important when competition for resources is intense and two birds can defend those resources better than one, when two parents are required to feed even a single chick, or when it is necessary to protect the brood from predators or climatic extremes. The importance of male parental care can be assessed by removing males after the eggs are laid and monitoring the success of the young and the female's workload. Ideally, such experiments should be done for a variety of species in different ecological circumstances. At least for the cardinal (*Cardinalis cardinalis*), temporary removal of males had little effect (Richmond, 1978). It is also possible that the environment is sufficiently uniform that a few males cannot monopolize a limited resource; that is, monogamy evolves by default when polygyny is impossible. This possibility is difficult to assess, because environmental patchiness must be viewed through the ecology of each population of birds. For example, redwings and field sparrows (*Spizella pusilla*) often nest in the same field in Illinois. The redwing is often polygynous but field sparrows are typically monogamous. What are the differences in ecology that might foster the differences in mating system? At this point, we have no idea.

Another important consideration for the evolution of monogamy might be the relative costs of other aspects of the life history. Sandpipers exhibit a range of mating systems, including monogamy. Nonmonogamous species typically have parental care by only one parent, while monogamous species have biparental care. Monogamous species do not migrate as far as nonmonogamous ones. Perhaps an early departure (and desertion of the mate and brood) reduces the risks of long migration (Myers, 1981). This suggestion is important because it recognizes that the entire life history must be taken into account in weighing costs and benefits, but it does not permit us to predict which sex will actually desert.

Other Vertebrates

Monogamy is the typical arrangement in Nile crocodiles (*Crocodylus niloticus*). Both sexes guard the young for several weeks (Pooley and Gans, 1976). The evolutionary basis for monogamy in this species is unknown.

Good examples of pair-bonding and monogamy in fishes are found in the jewelfish (*Hemichromis bimaculata*) and in other tropical freshwater cichlids. This family is almost unique in that male and female often have an extended courtship period and guard their developing young together. The guarding period suggests that predation pressure may be high for many cichlids (Wittenberger and Tilson, 1980). Guarding of the young by both parents (and hence temporary monogamy) is also reported for an African characin (*Hepsetus odoe*), which builds a froth nest in shoreline vegetation, and in the pomacentrid genus *Amphiprion*, which place their young in the tentacles of sea anemones. Some surgeonfishes (Acanthuridae) of coral reefs and a few other species also practice biparental care.

Polyandry seems to occur in some populations of the clownfish *Amphiprion clarkii* (Keenleyside, 1979). Where large colonies of anemones are available, several males may hold subterritories within the female's territory and all may breed.

SUMMARY

The functions of courtship may include species identification, sex recognition, synchronization of male and female cycles, assessment of individual characteristics important in mate choice, and protection against cuckoldry. Courtship patterns are influenced by the functions of courtship, ecological limitations involving costs and risks, sexual dimorphism and sexual differences in behavior, usable modes of sensory perception, and the nature of the message conveyed. Certain vertebrates court in groups; some of these species display on traditional leks and a few others engage in joint display. In both these cases, marked differentials in fitness occur among males.

Three major mating systems are known among sexually reproducing vertebrates: promiscuity, monogamy, and polygamy (most commonly polygyny, but sometimes polyandry). Monogamy may have evolved in response to selection for biparental care of offspring and lack of opportunity for multiple mating. Polygamous systems may involve competition among members of one sex (usually male) for control of female groups, which is common among mammals. They might also entail competition among members of one sex for resources in patchy environments that attract members of the other sex, which is more common among birds. Female choices among males and territories in polygamous avian systems are subject to many pressures and constraints, which vary temporally, spatially, and taxonomically.

SELECTED REFERENCES

Alexander, R.D., 1974. The evolution of social behavior. *Annu. Rev. Ecol. Syst.*, *5*:325–383.

Altmann, S.A., S.S. Wagner, and S. Lenington, 1977. Two models for the evolution of polygyny. *Behav. Ecol. Sociobiol.*, *2*:397–410.

Andersen, D.C., K.B. Armitage, and R.S. Hoffmann, 1976. Socioecology of marmots: female reproductive strategies. *Ecol.*, *57*:552–560.

Balph, D.F., G.S. Innis, and M.H. Balph, 1980. Kin selection in Rio Grande turkeys: a critical assessment. *Auk*, *97*:854–860.

Barash, D.P., 1977. *Sociobiology and Behavior.* Elsevier, Amsterdam.

Barash, D.P., 1981. Mate guarding and gallivanting by male hoary marmots (*Marmota caligata*). *Behav. Ecol. Sociobiol., 9*:187–193.

Barlow, G.W., 1974. Contrasts in social behavior between Central American cichlid fishes and coral-reef surgeon fishes. *Am. Zool., 14*:9–34.

Bartholomew, G.A., 1970. A model for the evolution of pinniped polygyny. *Evol., 24*:546–559.

Bastock, M., 1967. *Courtship: An Ethological Study.* Aldine, Chicago.

Bateson, P., 1978. Sexual imprinting and optimal outcrossing. *Nature, 273*:659–660.

Beecher, M.D., and I.M. Beecher, 1979. Sociobiology of bank swallows: reproductive strategy of the male. *Science, 205*:1282–1285.

Berry, K.H., 1974. The ecology and social behavior of the chuckwalla, *Sauromalus obesus. U. Calif. Publ. Zool., 101*:1–60.

Berven, K.A., 1981. Mate choice in the wood frog, *Rana sylvatica. Evol., 35*:707–722.

Bibby, C.J., 1982. Polygyny and breeding ecology of the Cetti's warbler *Cettia cetti. Ibis, 124*:288–301.

Birkhead, T.R., 1982. Timing and duration of mate guarding in magpies, *Pica pica. Anim. Behav., 30*:277–283.

Bradbury, J.W., 1977. Lek mating behavior of hammer-headed bats. *Z. Tierpsych., 45*: 225–255.

Bradbury, J.W., 1980. Foraging, social dispersion, and mating systems, pp. 189–207. *In* G.W. Barlow and J. Silverberg (Eds.). *Sociobiology: Beyond Nature/Nurture?* AAAS, Washington, D.C.

Bradbury, J.W., 1981. The evolution of leks. *In* R.D. Alexander and D.W. Tinkle (Eds.), *Natural Selection and Social Behavior*, pp. 138–169. Chiron, New York.

Bradbury, J.W., and S.L. Vehrenkamp, 1977. Social organization and foraging in emballonurid bats. III. Mating systems. *Behav. Ecol. Sociobiol., 2*:1–17.

Bray, O.E., J.J. Kennelly, and J.L. Guarino, 1975. Fertility of eggs produced on territories of vasectomized red-winged blackbirds. *Wils. Bull., 87*:187–195.

Breder, C.M., and D.E. Rosen, 1966. *Modes of Reproduction in Fishes.* Natural History Press, Garden City, NY

Brunning, D.F., 1974. Social structure to reproductive behavior in the greater rhea. *Living Bird, 13*:251–294.

Buechner, H.K., and H.D. Roth, 1974. The lek system in Uganda kob antelope. *Am. Zool., 14*:145–162.

Burley, N., 1977. Parental investment, mate choice, and mate quality. *Proc. Nat. Acad. Sci.* (USA), *74*:3476–3479.

Burley, N., and N. Moran, 1979. The significance of age and reproductive experience in the mate preferences of feral pigeons, *Columba livia. Anim. Behav., 27*:686–698.

Burger, J., and M. Gochfeld, 1981. Unequal sex ratios and their consequences in herring gulls (*Larus argenteus*). *Behav. Ecol. Sociobiol., 8*:125–128.

Carey, M., and V. Nolan, 1975. Polygyny in indigo buntings: a hypothesis tested. *Science, 190*:1296–1297.

Coulson, J.C., 1966. The influence of the pair-bond and age on the breeding biology of the kittiwake gull *Rissa tridactyla. J. Anim. Ecol., 35*:269–279.

Cox, C.R., and B.J. LeBoeuf, 1977. Female incitation of male competition: a mechanism in sexual selection. *Am. Nat., 111*:317–335.

Crawford, R.D., 1977. Breeding biology of year-old and older female red-winged and yellow-headed blackbirds. *Wils. Bull., 89*:73–80.

Cronin, E.W., and P.W. Sherman, 1976. A resource-based mating system: the orange-rumped honeyguide. *Living Bird, 15*:5–32.

Crook, J.H., 1962. The adaptive significance of pair formation types in weaver birds. *Symp. Zool. Soc. Lond., 8*:57–70.

Crook, J.H., 1963. Monogamy, polygamy, and food supply. *Discovery, 24*:35–41.

Daly, M., 1979. Why don't male mammals lactate? *J. Theor. Biol., 78*:325–346.

Davies, N.B., and T.R. Halliday, 1979. Competitive mate searching in male common toads, *Bufo bufo. Anim. Behav., 27*:1253–1267.

Davison, G.W.H., 1981. Sexual selection and

the mating system of *Argusianus argus* (Aves: Phasianidae). *Biol. J. Linn. Soc., 15*:91–104.

Dewsbury, D.A., 1982. Ejaculate cost and mate choice. *Am. Nat., 119*:601–610.

Downhower, J.F., and K.B. Armitage, 1971. The yellow-bellied marmot and the evolution of polygamy. *Am. Nat., 105*:355–370.

Downhower, J.F., and L. Brown, 1981. The timing of reproduction and its behavioral consequences for mottled sculpins, *Cottus bairdi*. *In* R.D. Alexander and D.W. Tinkle (Eds.), *Natural Selection and Social Behavior,* pp. 78–95. Chiron, New York.

Dudley, D., 1974. Paternal behavior in the California mouse, *Peromyscus californicus*. *Behav. Biol., 11*:247–252.

Dyrcz, A., 1977. Polygamy and breeding success among great reed warblers *Acrocephalus arundinaceus* at Milicz, Poland. *Ibis, 119*:73–77.

Eisenberg, J.F., 1966. The social organization of mammals. *Handbuch der Zoologie,* Band 8, Lief., *39*:1–92.

Elliott, P.F., 1975. Longevity and the evolution of polygamy. *Am. Nat., 109*:281–287.

Elliott, P.F., 1980. Evolution of promiscuity in the brown-headed cowbird. *Condor, 82*:138–141.

Emlen, S.T., and L.W. Oring, 1977. Ecology, sexual selection, and the evolution of mating systems. *Science, 197*:215–223.

Erickson, C.J., and P.G. Zenone, 1976. Courtship differences in male ring doves: avoidance of cuckoldry? *Science, 192*:1353–1354.

Ewer, R.F., 1968. *Ethology of Mammals.* Plenum Publishing, New York.

Faaborg, J., T. deVries, C.B. Patterson, and C.R. Griffin, 1980. Preliminary observations on the occurrence and evolution of polyandry in the Galápagos hawk (*Buteo galapagoensis*). *Auk, 97*:581–590.

Foster, M.S., 1977. Odd couples in manakins. A study of social organization and cooperative breeding in *Chrioxiphia linearis*. *Am. Nat., 111*:845–853.

Foster, M.S., 1981. Cooperative behavior and social organization of the swallow-tailed manakin (*Chiroxiphia caudata*). *Behav. Ecol. Sociobiol., 9*:167–177.

Gatz, A.J., 1981. Size selective mating in *Hyla versicolor* and *Hyla crucifer*. *J. Herpetol., 15*:114–116.

Gilliard, E.T., 1959. Notes on the courtship behavior of the blue-backed manakin. *Am. Mus. Novit., 1942*:1–20.

Gilliard, E.T., 1969. *Birds of Paradise and Bower Birds.* Natural History Press, Garden City, NY.

Gladstone, D.E., 1979. Promiscuity in monogamous colonial birds. *Am. Nat., 114*:545–557.

Gowaty, P.A., 1981. Aggression of breeding eastern bluebirds (*Sialia sialis*) toward their mates and models of intra- and interspecific intruders. *Anim. Behav., 29*:1013–1027.

Graul, W.D., 1973. Adaptive aspects of the mountain plover social system. *Living Bird, 12*:69–94.

Greig-Smith, P.W., 1982. Song-rates and parental care by individual male stonechats (*Saxicola torquata*). *Anim. Behav., 30*:245–252.

Gross, M.R., and E.L. Charnov, 1980. Alternative male life histories in bluegill sunfish. *Proc. Nat. Acad. Sci.* (USA), *77*:6937–6940.

Halliday, T.R., 1978. Sexual selection and mate choice. *In* J.R. Krebs and N.B. Davies (Eds.), *Behavioural Ecology: An Evolutionary Approach,* pp. 180–213. Sinauer Associates, Sunderland, MA.

Hanken, J., and P.W. Sherman, 1981. Multiple paternity in Belding's ground squirrel litters. *Science, 212*:351–353.

Hartshorne, C., 1973. *Born to Sing.* Indiana University Press, Bloomington.

Heisler, I.L., 1981. Offspring quality and the polygyny threshold: a new model for the "sexy son" hypothesis. *Am. Nat., 117*:316–328.

Holm, C.H., 1973. Breeding sex ratios, territoriality, and reproductive success in the red-winged blackbird (*Agelaius phoeniceus*). *Ecol., 54*:356–365.

Howard, R.D., 1981. Male age-size distribution and male mating success in bull frogs. *In* R.D. Alexander and D.W. Tinkle (Eds.), *Natural Selection and Social Behavior,* pp. 61–77. Chiron, New York.

Howe, M.A., 1975. Behavioral aspects of the pair bond in Wilson's phalarope. *Wils. Bull.,*

87:248–270.

Jenni, D.A., 1974. Evolution of polyandry in birds. *Am. Zool., 14*:129–244.

Jenni, D.A., and G. Collier, 1972. Polyandry in the American jacana (*Jacana spinosa*). *Auk, 89*:743–765.

Keenleyside, M.H.A., 1979. *Diversity and Adaptation in Fish Behaviour.* Springer-Verlag, Berlin.

Kleiman, D.G., and J.R. Malcolm, 1981. The evolution of male parental investment in mammals. *In* D.J. Gubernick and P.H. Klopfer (Eds.), *Parental Care in Mammals,* pp. 347–387. Plenum Publishing, New York.

Klingel, H., 1974. A comparison of the social behaviour of the Equidae. *In* V. Geist and F. Walther (Eds.), *The Behaviour of Ungulates and Its Relation to Management.* Vols. 1, 2, pp. 124–132. IUCN (24), Morges, Switzerland.

Knowlton, N., 1979. Reproductive synchrony, parental investment, and the evolutionary dynamics of sexual selection. *Anim. Behav., 27*:1022–1033.

Kodric-Brown, A., 1977. Reproductive success and the evolution of breeding territories in the pupfish (*Cyprinodon*). *Evol., 31*:750–766.

Kodric-Brown, A., 1981. Variable breeding systems in pupfishes (Genus *Cyprinodon*): adaptations to changing environments. *In* R.J. Naiman and D.L. Soltz (Eds.), *Fishes in North American Deserts,* pp. 205–235. Wiley-Interscience, New York.

Krebs, J.R., and N.B. Davies, 1981. *An Introduction to Behavioural Ecology.* Sinauer Associates, Sutherland, MA.

Lack, D., 1968. *Ecological Adaptations for Breeding in Birds.* Methuen, London.

LeBoeuf, B.J., 1974. Male-male competition and reproductive success in elephant seals. *Am. Zool., 14*:163–176.

Lenington, S., 1980. Female choice and polygyny in red-winged blackbirds. *Anim. Behav., 28*:347–361.

Lill, A., 1974a. Sexual behavior of the lek-forming white-bearded manakin (*Manacus manacus trinitatus* Hartert). *Z. Tierpsychol., 36*:1–36.

Lill, A., 1974b. Social organization and space utilization in the lek-forming white-bearded manakin, *M. manacus trinitatus* Hartert. *Z. Tierpsychol., 36*:513–530.

Loiselle, P.V., and G.W. Barlow, 1978. Do fishes lek like birds? *In* E.S. Reese and F.J. Lighter (Eds.), *Contrasts in Behavior,* pp. 31–75. John Wiley & Sons, New York.

Mader, W.J., 1979. Breeding behavior of a polyandrous trio of Harris' hawk in southern Arizona. *Auk, 96*:776–788.

Manzur, M.I., and E.R. Fuentes, 1979. Polygyny and agonistic behavior in the tree-dwelling lizard *Liolaemus tenuis* (Iguanidae). *Behav. Ecol. Sociobiol., 6*:22–28.

Martin, S.G., 1974. Adaptations for polygynous breeding in the bobolink, *Dolichonyx oryzivorus. Am. Zool., 14*:109–119.

Maxon, S.J., and L.W. Oring, 1980. Breeding season time and energy budgets of the polyandrous spotted sandpiper. *Behav., 74*:200–263.

McCracken, G.F., and J.W. Bradbury, 1981. Social organization and kinship in the polygynous bat *Phyllostomus hastatus. Behav. Ecol. Sociobiol., 8*:11–34.

McKinney, F., 1965. The spring behavior of wild Steller eiders. *Condor, 67*:273–290.

McKinney, F., 1976. The evolution of duck displays. *In* G. Baerends, C. Beer, and A. Manning (Eds.), *Function and Evolution in Behaviour,* pp. 331–357. Oxford University Press, New York.

Morris, d., 1958. The reproductive behavior of the ten-spined stickleback (*Pygosteus pungitius* L.). *Behav. Suppl., 6*:1–154.

Morton, E.S., 1975. Ecological sources of selection on avian sounds. *Am. Nat., 109*:17–34.

Myers, J.P., 1981. Cross-seasonal interactions in the evolution of sandpiper social systems. *Behav. Ecol. Sociobiol., 8*:195–202.

Nakatsuru, K., and D.L. Kramer, 1982. Is sperm cheap? Limited male fertility and female choice in the lemon tetra (Pisces: Characidae). *Science, 216*:753–755.

Orians, G.H., 1969. On the evolution of mating systems in birds and mammals. *Am. Nat., 103*:589–603.

Orians, G.H., 1972. The adaptive significance of mating systems in the Icteridae. *Proc. Int. Ornith. Cong., 15*:389–398.

Orians, G.H., 1980. Some adaptations of marsh-nesting blackbirds. Princeton Monographs in Population Biology, *14*:1–295.

O'Shea, T.J., 1980. Roosting, social organization and the annual cycle in a Kenya population of the bat *Pipistrellus nanus*. *Z. Tierpsychol.*, *53*:171–195.

Payne, R.B., and K. Payne, 1977. Social organization and mating success in local song populations of village indigo birds, *Vidua chalybeata*. *Z. Tierpsych.*, *45*:113–173.

Perrill, S.A., H.C. Gerhardt, and R. Daniel, 1978. Sexual parasitism in the green tree frog (*Hyla cinerea*). *Science*, *200*:1179–1180.

Perrill, S.A., H.C. Gerhardt, and R.E. Daniel, 1982. Mating strategy in male green tree frogs (*Hyla cinerea*): an experimental study. *Anim. Behav.*, *30*:43–48.

Picman, J., 1980. Behavioral interactions between red-winged blackbirds and long-billed marsh wrens and their role in the evolution of the redwing polygynous mating systems. Ph.D. thesis, University of British Columbia.

Pierotti, R., and D. Pierotti, 1980. Effects of cold climate on the evolution of pinniped breeding systems. *Evol.*, *34*:494–507.

Pleszczynska, W.K., 1978. Microgeographic prediction of polygyny in the lark bunting. *Science*, *201*:935–937.

Pleszczynska, W.K., and R.I.C. Hansell, 1980. Polygyny and decision theory: testing of a model in lark buntings (*Calamospiza melanocorys*). *Am. Nat.*, *116*:821–830.

Pooley, A.C., and C. Gans, 1976. The Nile crocodile. *Sci. Am.*, *234*:114–124.

Power, H.W., and C.G.P. Doner, 1980. Experiments on cuckoldry in the mountain bluebird. *Am. Nat.*, *116*:689–704.

Power, H.W., E. Litovich, and M.P. Lombardo, 1981. Male starlings delay incubation to avoid being cockolded. *Auk*, *98*:386–389.

Richmond, A.R., 1978. An experimental study of advantages of monogamy in the cardinal. Ph.D. thesis, Indiana University.

Robertson, D.R., and S.G. Hoffman, 1977. The rules of female mate choice and predation in the mating systems of some tropical labroid fishes. *Z. Tierpsychol.*, *45*:298–320.

Rohwer, S., 1978. Parent cannibalism of offspring and egg-raiding as a courtship strategy. *Am. Nat.*, *112*:429–440.

Royama, T., 1966. A re-interpretation of courtship feeding. *Bird Study*, *13*:116–129.

Ryan, M.J., 1980. Female mate choice in a neotropical frog. *Science*, *209*:523–525.

Ryan, M.J., M.D. Tuttle, and L.K. Taft, 1981. The costs and benefits of frog chorusing behavior. *Behav. Ecol. Sociobiol.*, *8*:273–278.

Ryan, M.J., M.D. Tuttle, and A.S. Rand, 1982. Bat predation and sexual advertisement in a neotropical anuran. *Am. Nat.*, *119*:136–139.

Sargent, R.C., and J.B. Gebler, 1980. Effects of nest site concealment on hatching success, reproductive success, and paternal behavior of the three-spine stickleback, *Gasterosteus aculeatus*. *Behav. Ecol. Sociobiol.*, *7*:137–142.

Schuster, R.H., 1976. Lekking behavior in Kafue lechwe. *Science*, *192*:1240–1241.

Searcy, W.A., 1979a. Female choice of mates: a general model for birds and its application to red-winged blackbirds (*Agelaius phoeniceus*). *Am. Nat.*, *114*:77–100.

Searcy, W.A., 1979b. Male characteristics and pairing success in red-winged blackbirds. *Auk*, *96*:353–363.

Searcy, W.A., and K. Yasukawa, 1981. Does the "sexy son" hypothesis apply to mate choice in red-winged blackbirds? *Am. Nat.*, *117*:343–348.

Selander, R.K., 1972. Sexual selection and sexual dimorphism in birds. *In* B. Campbell (Ed.), *Sexual Selection and the Descent of Man*, pp. 180–230. Aldine, Chicago.

Shepard, J.M., 1975. Factors influencing female choice in the lek mating system of the ruff. *Living Bird*, *14*:87–111.

Sick, H., 1967. Courtship behavior in the manakins (Pipridae): a review. *Living Bird*, *6*:5–22.

Smith, J.N.M., Y. Yom-Tov, and R. Moses, 1982. Polygyny, male parental care, and sex ratio in song sparrows: an experimental analysis. *Auk*, *99*:555–564.

Snow, D.W., 1962. A field study of the black and white manakin, *Manacus manacus*, in Trinidad. *Zoologica*, *47*:65–104.

Snow, D.W., 1976. *The Web of Adaptation*. Collins, London.

Thomas, J.A., and E.C. Birney, 1979. Parental care and mating system in the prairie vole, *Microtus ochrogaster*. *Behav. Ecol. Sociobiol.*, *5*:171–186.

Trivers, R.L., 1972. Parental investment and sexual selection. *In* B. Campbell (Ed.), *Sexual Selection and the Descent of Man*, pp. 136–179. Aldine, Chicago.

Tuttle, M.D., and M.J. Ryan, 1981. Bat predation and the evolution of frog vocalizations in the neotropics. *Science*, *214*:677–678.

van Rhijn, J.G., 1973. Behavioral dimorphism in male ruffs, *Philomachus pugnax* (L.). *Behav.*, *47*:153–229.

Verner, J., 1964. Evolution of polygyny in the long-billed marsh wren. *Evol.*, *18*:252–261.

Verner, J., 1965. Breeding biology of the long-billed marsh wren. *Condor*, *67*:6–30.

Verner, J., 1976. Complex song repertoire of male long-billed marsh wrens in eastern Washington. *Living Bird*, *14*:263–300.

Verner, J., and M.F. Willson, 1966. The influence of habitats on mating systems of North American passerine birds. *Ecol.*, *47*:143–147.

von Haartman, L., 1969. Nest-site and the evolution of polygamy in European passerine birds. *Ornis Fenn.*, *46*:1–2.

Watts, C.R., and A.W. Stokes, 1971. The social order of turkeys. *Sci. Am.*, *224*:112–118.

Weatherhead, P.J., 1979. Ecological correlates of monogamy in tundra-breeding savannah sparrows. *Auk*, *96*:391–401.

Weatherhead, P.J., and R.J. Robertson, 1981. In defense of the "sexy son" hypothesis. *Am. Nat.*, *117*:349–356.

Wiley, R.H., 1973. Territoriality and non-random mating in sage grouse, *Centrocercus urophasianus*. *Anim. Behav. Monogr.*, *6*:87–169.

Wiley, R.H., 1975. Multidimensional variation in an avian display: implications for social communication. *Science*, *190*:482–483.

Wilson, E.O., 1975. *Sociobiology*. Belknap Press, Cambridge, MA.

Wirtz, P., 1981. Territorial defense and territory take-over by satellite males in the waterbuck *Kobus ellipsiprymnus* (Bovidae). *Behav. Ecol. Sociobiol.*, *8*:161–162.

Wittenberger, J.F., 1976. The ecological factors selecting for polygyny in altricial birds. *Am. Nat.*, *110*:779–799.

Wittenberger, J.F., 1978. The evolution of mating systems in grouse. *Condor*, *80*:126–137.

Wittenberger, J.F., 1980a. Group size and polygamy in social mammals. *Am. Nat.*, *115*:197–222.

Wittenberger, J.F., 1980b. Vegetation structure, food supply, and polygyny in bobolinks (*Dolichonyx oryzivorus*). *Ecol.*, *61*:140–150.

Wittenberger, J.F., 1981a. Time: a hidden dimension in the polygyny threshold model. *Am. Nat.*, *118*:803–822.

Wittenberger, J.F., 1981b. *Animal Social Behavior*. Duxbury, Boston.

Wittenberger, J.F., 1981c. Male quality and polygyny: the "sexy son" hypothesis revisited. *Am. Nat.*, *117*:329–342.

Wittenberger, J.F., and R.L. Tilson, 1980. The evolution of monogamy. *Annu. Rev. Ecol. Syst.*, *11*:197–232.

Wolf, L.L., and J.S. Wolf, 1976. Mating system and reproductive biology of malachite sunbirds. *Condor*, *78*:27–39.

Wrangham, R.W., 1980. Female choice of least costly males: a possible factor in the evolution of leks. *Z. Tierpsychol.*, *54*:357–367.

Yasukawa, K., 1981. Male quality and female choice of mate in the red-winged blackbird (*Agelaius phoeniceus*). *Ecol.*, *62*:922–929.

Yasukawa, K., and W.A. Searcy, 1981. Nesting sychrony and dispersion in red-winged blackbirds: Is the harem competitive or cooperative? *Auk*, *98*:659–668.

Yasukawa, K., J.L. Blank, and C.B. Patterson, 1980. Song repertoires and sexual selection in the red-winged blackbird. *Behav. Ecol. Sociobiol.*, *7*:233–238.

Zenone, P.G., M.E. Sims, and C.J. Erickson, 1979. Male ring dove behavior and the defense of genetic paternity. *Am. Nat.*, *114*:615–626.

Zeveloff, S.I., and M.S. Boyce, 1980. Parental investment and mating systems in mammals. *Evol.*, *34*:973–982.

Zimmerman, J.L., 1982. Nesting success of dickcissels (*Spiza americana*) in preferred and less preferred habitats. *Auk*, *99*:292–298.

14 Life History Patterns and Reproduction

LIFE CYCLES

Most vertebrates have a simple life cycle. A newborn individual develops into an adult directly, with relatively few morphological or ecological changes. But developing individuals of some species undergo such dramatic changes in morphology and ecology that their life cycles are called complex (Fig. 14–1). The best examples of complex life cycles among the vertebrates are provided by amphibians whose larval or tadpole stage is radically unlike the adult in numerous ways. Anuran tadpoles, for instance, typically change their appendages, locomotor style, jaw and gut structure, foraging and dietary habits, and often feeding habitat as well, when they metamorphose into adults. Complex life cycles also occur in some fishes (Chapter 2) and in lampreys, in which the detritus-feeding ammocoete larva is quite different from the adult.

Two (or more) distinct stages of a life cycle are partially independent of each other because each stage is characterized by different behavioral, morphological, and physiological adaptations to different environments. Yet, at the same time, they are intimately linked through development and

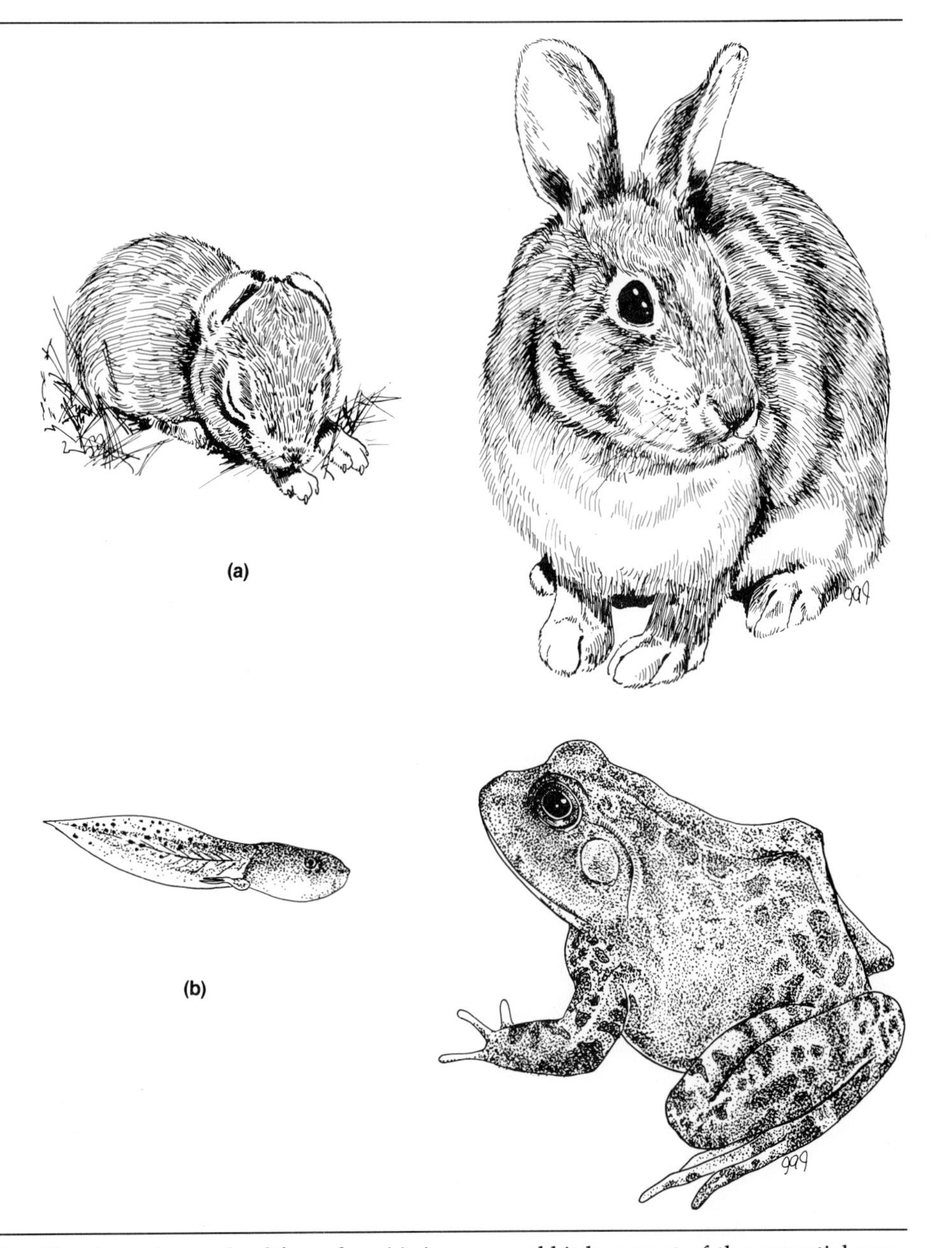

Figure 14–1 Simple and complex life cycles. (*a*) A young rabbit has most of the essential morphological features of an adult. In this simple life cycle, development is fairly direct. (*b*) A young frog (tadpole) is very different from the adult, however. The complex life cycle entails a dramatic metamorphosis.

reproduction. Evolutionary changes in one stage may proceed more slowly than those of the other, and that stage may thus be less well adapted to its environment than the other one. The adult stage may reproduce poorly and supply the larval stage with very few individuals, or the larvae may metamorphose too slowly to maintain the adult population and a spiral of diminishing numbers may begin. Istock (1967) suggested that complex life cycles therefore may be inherently unstable in the long run and that selection has often favored the reduction of one stage or the other.

Nevertheless, the existence of well-developed complex life cycles in amphibians (not to mention myriads of invertebrates and plants) argues strongly for the existence of selective pressures favoring the maintenance of complicated life histories. Some of the assumptions of Istock's model may be unrealistic. In fact, anuran tadpoles probably represent an evolutionary specialization that permits young anurans to exploit rich, but temporary, bodies of water (Wassersug, 1975). These bodies of water may offer a food supply of organic particles sufficiently abundant to allow more rapid growth than would be possible for small froglets feeding on aerial and terrestrial insects.

Larval development for some anurans is indeed rapid. The spadefoot toads (*Scaphiopus*) of the North American deserts lay their eggs in highly ephemeral rain pools; larval life may be as short as 1½ to 2 weeks and is commonly less than 4 weeks. Even the longer-lived tadpoles of the ranid frogs such as the bullfrog (*Rana catesbeiana*), which may spend a year or more as larvae, may be successful because they are able to exploit seasonal bursts of food resources.

Some anurans and salamanders have lost the larval stage. Wassersug reported that these species typically occur in relatively constant, nonseasonal environments that are either uniformly productive of food or consistently lacking surface water. Omission of a larval phase may also result from selection to escape the numerous environmental uncertainties typical of some aquatic habitats (Wilbur and Collins, 1973). One frog species, *Breviceps parvus* of southern Africa, reproduces independently of free water (the adults supposedly cannot even swim): The entire larval development occurs within the egg capsule in a mucus-lined burrow or under a stone. Salamanders in the genera *Plethodon* and *Aneides* and *Eleutherodactylus* frogs also have larvae that complete their transformation to adults within the egg. One species of salamander (*Salamandra atra*) and one genus of anuran (*Nectophrynoides*) bear live young that emerge with adult morphology. In general, amphibians without larval stages are thought to occupy environments that lack the ephemerally rich patches exploited by tadpoles.

Adult amphibians often are more terrestrial than the larvae and may be considered as the dispersal phase of the life cycle. An adult phase is adaptive where larval habitats become unsuitable; dispersal then becomes essential. Salamanders are particularly interesting because a typical adult stage is sometimes absent and reproduction may be accomplished by sexually mature individuals with the external morphology of larvae. Reproduction by sexually mature "larvae" is not associated with a short generation time and early maturation. Instead, there is some evidence that larval stages are prolonged where the aquatic environment is favorable and the terrestrial environment is unfavorable (Wilbur and Collins, 1973). The incredibly varied life histories found within a single salamander species, the tiger salamander (*Ambystoma tigrinum*), may exemplify these ideas. Individuals of some populations in Texas inhabit permanent waters and breed in early spring. They are sexually mature while retaining external larval morphology and achieve a large body size. Individuals of other populations metamorphize rapidly into small adults, inhabit ephemeral pools, and breed whenever suitable conditions prevail (Rose and Armentrout, 1976). Curiously, the adult phase is not known to be lost from anuran life cycles.

The newt, *Notophthalmus viridescens*, is a salamander in which both larvae and adults are aquatic, but many populations have an intervening terrestrial "eft" stage, thus creating a three-stage life cycle (Fig. 14–2). Some coastal populations in Massachussetts lack the terrestrial phase. The aquatic larvae in these populations feed nearly year-round, growing rapidly and maturing at a younger age than do individuals in other populations possessing the eft stage. Efts can feed and grow mainly during the summer months, primarily when rain increases the activity of their invertebrate food and creates suitable (moist) foraging conditions. Consequently, their growth is slower and more erratic than that for fully aquatic individuals (Healy, 1973) and maturation is delayed. This unique stage of the life history may be an adaptation for dispersal, a means of colonizing new ponds (Gill, 1978). Presumably the advantages of dispersal outweigh the disadvantages of slower growth in these populations. Why adult newts are so aquatic and a novel life-history phase was necessary to accomplish dispersal is unexplained.

Some amphibians, including the tiger salamander, have still another variation in the life cycle. In addition to the metamorphosing and nonmetamorphosing individuals, some individuals may be cannibalistic. These individuals are typically large-bodied and may have different body proportions as well. Cannibalistic and scavenging morphs occur in spadefoot toads (*Scaphiopus bombifrons*), whose tadpoles are typically herbivorous. These dietary changes may permit more rapid growth and earlier transformation. The occurrence of carnivorous morphs may be particularly advantageous when breeding pools dry up with unusual rapidity (Curio, 1976). Note that the cannibalistic toad larvae eventually metamorphose and that the carnivorous diet is seen as a means of exploiting a temporary resource.

In contrast to many of the amphibians, most fishes and all reptiles, birds, and mammals possess simple life cycles with direct development. Although body proportions, diet, physiology, and behavior all may change

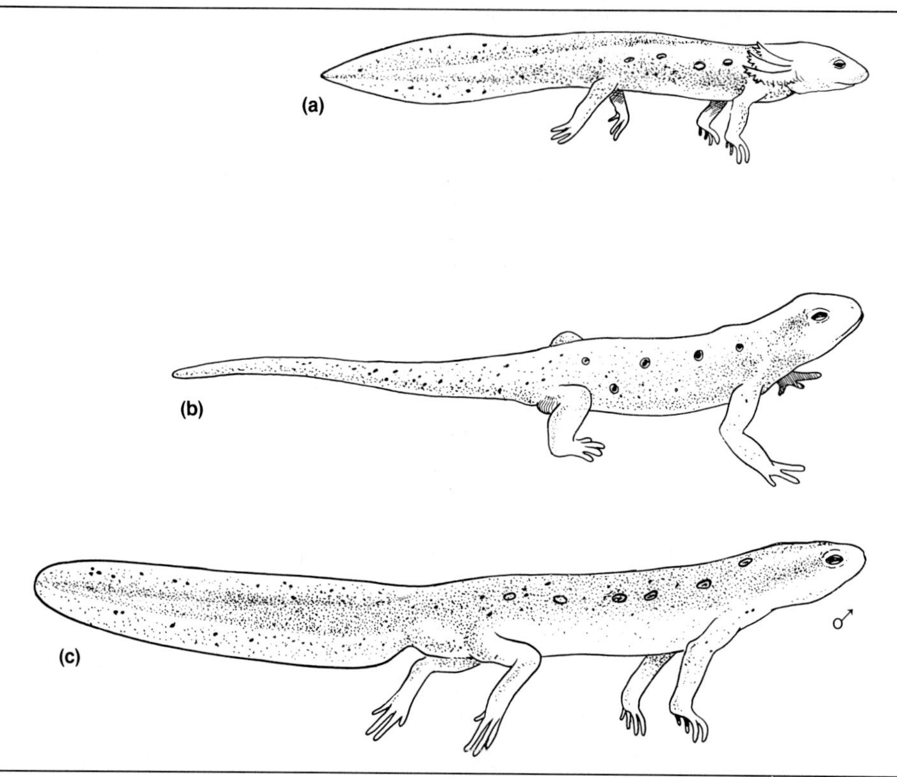

Figure 14–2 A three-stage life cycle represented by many individuals of the newt, *Notophthalmus viridescens.* The aquatic juvenile phase (*a*) precedes the terrestrial eft (*b*), which in turn precedes the aquatic adult (*c*).

during development, changes tend to be gradual and quite continuous over a period of time rather than occurring during a single, disjunct episode of metamorphic reorganization. The ecology of a young mammal or bird, especially during parental care, is obviously different from that of the adults—the habitat may be a nest or den, the diet milk or selected high-protein items, and so on. Nevertheless, no one seriously calls these young "larvae." These differences are smaller and the development changes are less radical than those exhibited by organisms with complex life cycles.

LIFE HISTORY PATTERNS

The three major life-history activities of all organisms are growth, maintenance of the individual (avoiding predation, finding resources, and so on), and reproduction. The first two activities are important evolutionarily because of their relationship to successful reproduction, which provides our measure of fitness. If the time and resources available to an organism are limited, an increase in one of these three activities is likely to occur at the expense of the others. Thus, major periods of growth may be achieved during nonreproductive periods, and maintenance activities may suffer during peaks of reproductive activity. For instance, male *Ranidella* frogs in Australia invest heavily in reproductive activities, depleting their energy reserves, but allocate little to growth (MacNally, 1981). If minimum maintenance requirements are high, then the proportion of the time and resource budget to be allocated to reproduction may be low. The concept of "trade-offs" among various portions of the time and resource budget is sometimes called the principle of allocation (Cody, 1966). The resources in question may be energy or minerals (calcium, phosphorus; and so on), or even vitamins or water.

It is by no means true, however, that resource budgets of organisms are always constant and limited. Expensive activities often take place during periods of food-resource abundance so that some or all of the added costs of reproduction may be paid, not by reallocation within the previously existing budget, but by addition of new resources (Hirschfield, 1980). When this is the case, trade-offs between portions of the budget need not be expected. Similar considerations apply to comparisons of different species (Dunn, 1980). If individuals of one population produce more or larger offspring than those of another, two interpretations are possible: Perhaps selection has favored a different pattern of budgetary allocation or perhaps resources are more available and higher costs of reproduction can be met without alteration of the basic budget. In addition, if reproductive allocation necessitates a compensatory shift in allocation to maintenance, we would expect that the magnitude of reproductive investment would be inversely correlated with subsequent survival, but this is observed in only some cases (Smith, 1981).

If augmented resources are spent entirely on reproduction and, as a result, none of the increase is expended for growth or maintenance, then it still can be argued that reproduction occurs at some cost to the other two activities (even though they do not decrease), because they are not increased. The allocational principle is still relevant, although its interpretation and evidence of it are slightly different. The question to be answered then becomes "Why is all the additional resource allocated unequally among the various activities?" instead of "Why does an increase of one activity necessitate a decrease of the others?"

The distinction between the constraints imposed by trade-offs in the basic allocation pattern and those of resource availability can be important in interpreting variations in life histories. Within the limits set by minimum requirements for growth, maintenance, and reproduction, what might be evolutionarily advantageous is constrained by what is ecologically possible (as well as by phylogenetic limitations, as discussed in Chapter 1). It is important to know whether observed variations of life histories are the result of selection for different balancing of the allocation pattern or of differences in resource availability and other ecological conditions. For example, if development of young is faster in one place than in another, it could be because of differences in selection pressures or because of differences in ecological constraints. For instance, temperature has a direct effect on developmental rate and larval size in the green frog (*Rana clamitans*); thus, site differences in larval developmental times may be related to prevailing temperatures as much as, or more than, to selection for life history and allocational differences (Berven et al., 1979).

Two additional examples may clarify this point. Several species of birds breed both on the North American mainland and on the Atlantic island of Bermuda. The insular populations of some of these occur at markedly higher densities than mainland populations, and the number of eggs laid in each nest (clutch size) is significantly lower (Crowell and Rothstein, 1981). It is possible that higher densities result in greater allocation of time and energy to intraspecific competition and that selection has favored a shift in the allocation pattern toward greater maintenance and lower reproductive costs. It is also possible that higher densities mean that fewer food resources are available to each set of parents and, consequently, that fewer young can be produced. Concrete evidence to distinguish these alternatives is still lacking.

In a similar vein, the number of eggs laid by each set of parents is known to vary among mainland populations of many species (Ricklefs, 1980). One general trend is for larger clutches to occur at higher latitudes. Clutch size is reported to be related directly to a ratio between an estimate of summer food availability and the density of breeding adults; adult density (the denominator in the ratio) has the greater effect. Although there may be some selection for greater allocation to reproduction at high latitudes, Ricklefs argues that the direct effect of density is much greater. In short, the ecological possibilities are sufficiently limited at lower latitudes that, even though selection might favor increased reproductive allocation there, it cannot be achieved, simply because resources are in limited supply. Ricklefs goes on to suggest that mortality of adults over the winter has a strong effect on the density of breeding adults; thus, he interprets geographic variation in clutch size as a function of winter mortality.

Ecological conditions may control "reproductive effort" directly in many cases, without a concomitant decrease in growth and maintenance. Only careful study of each case will tell us if basic allocation patterns underlie differences in life histories. We will return to this point later in this chapter.

Many life-history features are closely linked to body size. As a general rule, at least for mammals, as the average body size of a species increases, the life span, the age of sexual maturity, and the total weight of a litter increase; however, the average number of young produced in a lifetime declines (Western, 1979). The weight of the litter relative to body size decreases in association with the lower relative metabolic efficiency of larger mammals (Tuomi, 1980; Robbins and Robbins, 1979; Bekoff et al., 1981). All these features are therefore related to each other and selection on one feature may necessitate concomitant changes in the others. The relationships are allometric, so that the life history traits change less rapidly than does body size on a logarithmic plot. (The exponent is less than 1; see Chapter 1.) Although the absolute values often differ (as indicated by the coefficient of W in the equation and the intercept of the line in Figure 14–3), the relative relationship (indicated by the exponent and the slope of the line) is quite similar in three ecologically and phylogenetically very different orders of mammals. Litter size is related to body size in ways that vary with taxon and with body size itself, so the constraints of body size on litter size seem to be complex (Tuomi, 1980). It seems likely that comparable relationships can be described for other vertebrate taxa as well (Western and Ssemakula, 1982).

Body size (and metabolic rate), therefore, may constrain the evolution of life histories within certain bounds because several important traits are linked together. Such restrictions must be borne in mind when considering possible allocational trade-offs among life-history features. It is possible that many life-history differences can be explained in terms of selection on body size rather than selection on allocation patterns. The effect of body size discussed here pertains specifically to comparisons among species. Within-species comparisons (individual variation) may not be affected in the same way.

Reproductive Rates

Because reproduction is the life-history feature most directly related to fitness, the evolution and ecology of reproductive rates are central to an understanding of life histories. Three elements constitute the major features of reproductive rates: (1) the number of young produced during each reproductive episode, which is called litter size (in mammals and other live-bearers) or clutch size (in birds and other egg-layers); (2) the frequency of reproduction, both within each season and throughout a lifetime; and (3) the age of first reproduction.

Litter sizes and clutch sizes range from one (as in many primates and ungulates) to the millions (as in some oceanic fishes). A general inverse correlation exists between litter size and the probability of offspring survival: Animals with large litters commonly have low juvenile survival rates, and those with small litters enjoy higher offspring survival. One cannot argue legitimately, however, that small litter sizes have evolved because of greater survival or that large ones have evolved to counter high mortality. Clearly, any individual whose young survived well but who

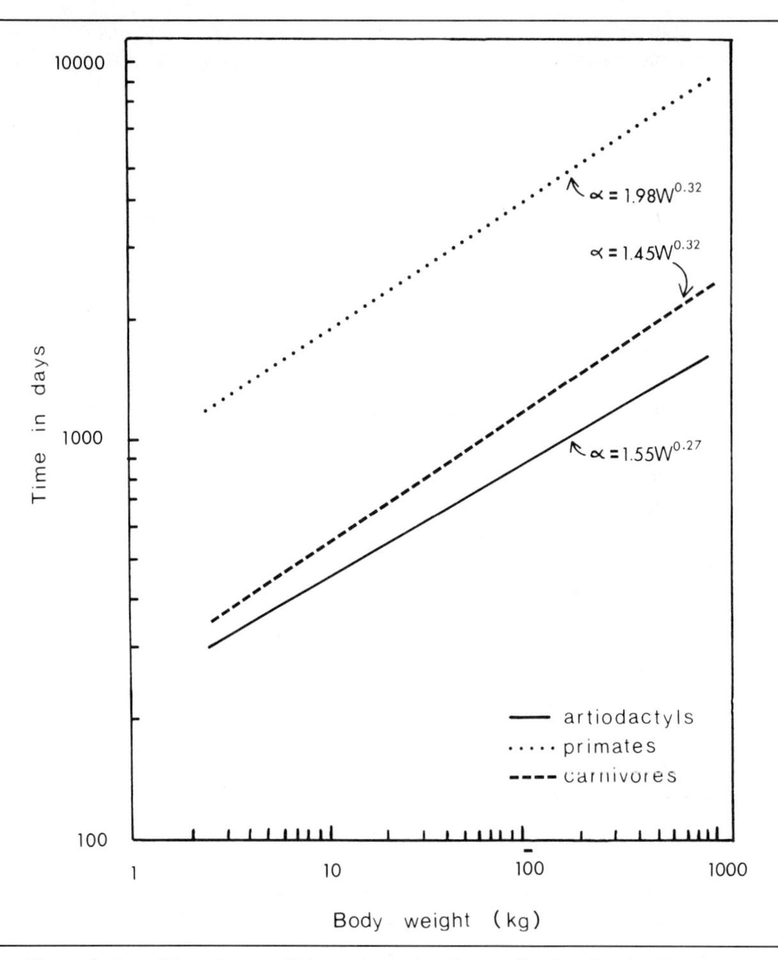

Figure 14–3 The relationship of age of first reproduction to body size in three groups of mammals. Note the similarity of the slopes (exponents in the equation) and the differences in the intercepts of the lines. (From D. Western, 1979. Size, life history and ecology in mammals. *Afr. J. Ecol., 17*:185–204.)

nevertheless managed to produce a large litter would have a greater fitness than an individual that, in the same circumstances, produced a small litter. Instead, evolution of small litters may come about through the costs of reducing juvenile mortality. Greater parental investment in each offspring is one means of reducing mortality risks, but more protection or nutrition for each young may necessitate the production of fewer young and, thus, a reduction in litter size. In fact, a trade-off between offspring quality and offspring quantity is commonly suggested (Smith and Fretwell, 1974; Salthe and Mecham, 1974; Pianka, 1976). Among the fishes, for instance, those whose eggs contain large quantities of nutritive yolk often have smaller clutches than those with small eggs. Such trade-offs are sometimes apparent even with a single species such as herring (*Clupea harengus*). In addition, baby herrings hatched from large eggs are bigger and survive food shortages better than those from small eggs (Mann and Mills, 1979). In some populations of the lizard *Gerrhonotus coeruleus,* litter size and size of the newborn offspring are inversely correlated (Stewart, 1979). Such trade-offs are not universal, however. There is some minimum egg or offspring size below which survival would be impossible, and this sets bounds on the flexibility of the trade-off. Some species may produce eggs or offspring near the minimum size and adjust only their numbers. Changes in resource levels may permit both size and number to increase simultaneously.

Frequency of reproduction interacts with litter size to determine the potential total number of young per unit of time. The number of litters produced within a season is related to duration and cost of producing and rearing the first litter and the length of the breeding season. Perhaps the ultimate means of increasing litter frequency, while spreading out the costs of maturing the young, is overlapping litters. Female poeciliid fish simultaneously carry as many as five batches of young, each at a different developmental stage, and the interval between litters is very short. Kangaroos may have one offspring free-ranging but still nursing, one in the pouch, and one embryo. Overlapping litters may be an important means of increasing offspring numbers for some species (Burley, 1980; Miller, 1979). The number of litters produced within a lifetime varies from one to many. A few vertebrates characteristically reproduce only once; this pattern is typical of Pacific salmon (*Oncorhynchus*), eels (*Anguilla*), and a few other fishes. It is also reported for male (but not female) marsupial mice of the genus *Antechinus* (Braithwaite and Lee, 1979). Once-in-a-lifetime reproduction is called semelparity (the term is derived from the Greek myth of Semele, who went up in flames at the sight of her lover Zeus in his full celestial glory; Zeus rescued and raised the single offspring of their union).

Most vertebrates are potentially iteroparous (repeated births) to some degree; long-lived species may reproduce for many years. Iteroparity obviously depends on the ability to survive from one breeding season to the next and, thus, on the risks of mortality from environmental factors. An increase in the risk of mortality at a certain age tends to select for an increased reproductive allocation at earlier ages and a diminished allocation after that age (Schaffer, 1979; Michod, 1979). In addition, the costs intrinsic to the reproductive process itself may influence survival (Goodman, 1974; Wooller and Coulson, 1977). If the extrinsic risks of mortality are sufficiently high, selection may even favor an increase of reproductive effort (a very large litter, for example) to the point that intrinsic costs raise the probability of death to certainty, which seems to happen in Pacific salmon. We also might expect an increase in reproductive effort with increasing age in iteroparous species, in part because the expectation of future reproduction declines (Williams, 1966); thus, the cost of present reproduction diminishes. This seems to be true for the California gull (*Larus californicus*) in Wyoming: Older parents fed their young more often than younger (but not inexperienced) birds, and their reproductive success was correspondingly greater (Pugesek, 1981). If the extrinsic risks are low, however, selection might sometimes (but not necessarily) favor lower intrinsic costs, so that the probability of survival remains good. Iteroparity thus depends not only on extrinsic sources of mortality but also on relatively low intrinsic costs of reproduction. Semelparity may evolve when extrinsic risks are unavoidably high or when there is some advantage to risking everything on a single reproductive episode. These ideas are developed further in the next section.

Intrinsic costs of reproduction can be considered as two types. One is dependent on the size of the brood—the more young, the higher the cost; the other is independent of fecundity—a fixed amount of effort is expended per reproductive attempt. Fecundity-independent costs might include such things as the expenditure of energy in setting up and maintaining a territory, an arduous prebreeding migration, and intensive forms of parental care, including incubation and brooding or live-bearing. When such costs are high compared to those that vary with brood size, selection may often favor the evolution of large broods and, concomitantly, lower frequencies of reproduction. Such a life-history pattern means that the fixed costs are encountered less often and that the fixed cost per offspring is less than it would be for smaller, more frequent, broods. In the absence of other limitations on brood size, we would then expect to observe low frequencies of reproduction and large broods in species with high fixed reproductive costs; evidence for a number of poikilothermic species supports the hypothesis (Bull and Shine, 1979).

The energetic costs of feeding young can be three to four times that of self-maintenance for some birds (Hails and Bryant, 1979). Adults may lose weight, especially when food is scarce and resource requirements of the young are at a peak. Although this was true for the house martin (*Delichon urbica*) (Bryant, 1979), it may not pertain to all parental birds. Female house wrens (*Troglodytes aedon*) lose weight, but mostly during incubation before the time of peak nestling demand, and the loss is independent of brood size. In such cases, the weight loss may be adaptive in that it reduces significantly the energy needed for flying and thus may improve the female's own chances of survival or even allow her to rear more young (Freed, 1981; Norberg, 1981). Furthermore, although male mortality seemed to be related to brood size and parental effort in the pied flycatcher (*Ficedula hypoleuea*) (Askenmo, 1979), mortality was not related to effort for parent song sparrows (*Melospiza melodia*) or female tree swallows (*Iridoprocne bicolor*) (DeSteven, 1980). So it is difficult to generalize concerning the costs of reproduction relative to brood size.

The energetic costs of reproduction for mammals can also be high, although the major costs to a female occur, not during gestation, but during lactation. The amount of food ingested by female cotton rats (*Sigmodon hispidus*) increased 19 to 22 percent during pregnancy (compared to the ingestion rate when nonreproductive), but 70 percent (for small Texas females) to 111 percent (for large Kansas females) while the young were nursing (Mattingly and McClure, 1982). Kansas females even withdrew additional resources from their bodies to aid in feeding their young. Lactational costs were related directly to litter size.

Fishes (and some birds) commonly deplete lipid reserves in the body when eggs are produced and, in some cases, muscle protein is depleted as well (Wootton, 1979). Growth is often inversely correlated with levels of egg production. Because body size and clutch size in fishes are often directly related, reduction of growth means that present fecundity tends to reduce future fecundity. Still another cost is a greater risk of dying; at least for some fishes, reproduction is associated with a lower probability of survival (Mann and Mills, 1979).

Because males and females often have different patterns of reproductive investment, the intrinsic costs may also differ. Bertram (1975) noted, for instance, that the effective reproductive life of a male lion (*Panthera leo*) averaged 2 or 3 years (a team of males usually can maintain ownership of a pride longer than a single male), whereas the reproductive life of a female lion is about 12 years. House martins increase their chance of raising two broods a season by starting early, but the cost of doing so is greater for females than for males (Bryant, 1979). Furthermore, extrinsic mortality risks may sometimes also differ between the sexes because of different nonreproductive behavior patterns. As a result, life history schedules of males and females may differ and one sex may mature earlier, reproduce more often, or live longer than the other.

The age of first reproduction is a very important feature of reproductive life histories. Fundamentally, an individual that successfully reproduces at an earlier age than others of its age class contributes to the population's gene pool faster than one that delays reproduction (Cole, 1954; Lewontin, 1965). This effect is especially marked in growing populations. Suppose that a cohort of females of the same age will eventually produce the same total number of young in their lifetimes. The ones that start turning out successful young the soonest will contribute genes to future generations at the highest rate, assuming that their own offspring will survive and do likewise (Fig. 14–4). Even if the females that delayed reproduction produce larger litters when they finally do reproduce, the advantage of successful, early reproduction may not be offset unless those delayed litters are very large and the survival of delayed breeders is improved. Even iteroparity may not overcome the advantage of early reproduction, all else being equal.

Other things are not always equal, however, and there are cases of

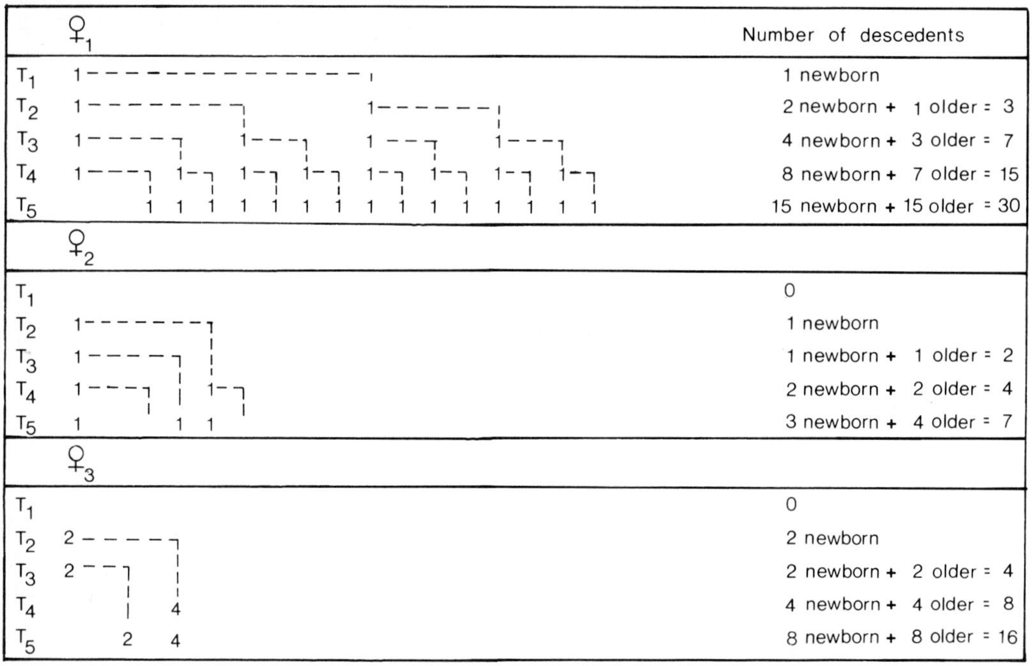

♀1		Number of descedents
T_1	1	1 newborn
T_2	1 1	2 newborn + 1 older = 3
T_3	1 1 1 1	4 newborn + 3 older = 7
T_4	1 1 1 1 1 1 1 1	8 newborn + 7 older = 15
T_5	1 1 1 1 1 1 1 1 1 1 1 1 1 1 1	15 newborn + 15 older = 30
♀2		
T_1		0
T_2	1	1 newborn
T_3	1	1 newborn + 1 older = 2
T_4	1 1	2 newborn + 2 older = 4
T_5	1 1 1	3 newborn + 4 older = 7
♀3		
T_1		0
T_2	2	2 newborn
T_3	2	2 newborn + 2 older = 4
T_4	4	4 newborn + 4 older = 8
T_5	2 4	8 newborn + 8 older = 16

Figure 14–4 The importance of breeding early. Imagine three females. One (♀$_1$) produces one (female) offspring per year for 4 years, and her daughters do likewise. If each reproductive time is indicated by T, a summary of the reproductive history of ♀$_1$ could be given as $1T_1 + 1T_2 + 1T_3 + 1T_4$. A second female (♀$_2$) delays reproduction until her second adult year but then performs as the first female; her reproductive history is given as $1T_2 + 1T_3 + 1T_4 + 1T_5$. A third female (♀$_3$) delays reproduction in the manner of ♀$_2$ but produces a litter of two females in each of 2 years and so do her daughters; her reproductive history is $2T_2 + 2T_3$. The consequences of these three life-history patterns are easily indicated in a numerical example. In each case, we enumerate the number of descendants of the original female present in each time period. Clearly, at any given time, ♀$_1$ has more descendants than either of the others (including ♀$_3$ with a larger litter) and ♀$_2$ has the lowest rate of contribution of genes to future generations.

deferred reproduction. Many wild sheep and goats live gregariously in societies organized in dominance hierarchies. Competition for access to females is keen, and only high-ranking, older males are very successful. Young males are neither fiercely competitive with older males nor reproductively effective. Males of such species commonly achieve full adult stature and horn size more gradually, over a longer period of time, than those of territorial species. The low probability of successful competition and mating by young males, with a good chance of surviving for several years, may have selected for deferred breeding and protracted growth. In this way, males may avoid the stress of severe intermale conflict and futile reproductive attempts. The accoutrements of adulthood are only assumed when age confers greater likelihood of success (Geist, 1966; Schaller, 1977). Deferred reproduction is also characteristic of males, but not females, in red-winged and yellow-headed blackbirds.

Young adults often have small litters (Coulson and Horobin, 1976; Reiter et al., 1981; DeSteven, 1978), which may be associated in part with their greater inexperience. Young individuals, even if they are of adult size, often have less success in hunting for themselves than do full adults; this has been observed for brown pelicans (*Pelecanus occidentalis*) and little blue herons (*Florida caerulea*), for instance. Hence small litters of first-time reproducers are probably a result, at least partly, of the inferior hunting ability and the lower probability of success in rearing a large litter. A lower survival of offspring of young female great tits in England may also be related to their inexperience at parental tasks, although breeding experience as yearlings did not seem to enhance the success of subsequent breeding attempts in the European sparrowhawk (*Accipiter nisus*) (Newton et al., 1981).

Reproductive success of young adults of some species would be so low (we infer) that it is disadvantageous for them to try to breed at all at an early age. Breeding in many species is deferred well beyond the age at which adult size is reached. In fact, breeding may be deferred even when the young adults are physiologically competent to breed. Both males and females of some species (many seabirds, for instance) have prolonged prebreeding stages (about five years in Manx shearwaters, *Puffinus puffinus*). For these species, we can guess that the acquisition of well-developed foraging skills may be particularly important and difficult. In the case of shearwaters and many other oceanic birds, the prey are difficult to locate, require special hunting techniques, and are often far from land. In view of the potential impact of early reproduction on fitness, the factors selecting for deferred breeding must be quite commanding.

Many ectothermic vertebrates have indeterminate growth; they increase in size throughout most of their lives. Age is correlated with body size, and body size is correlated with litter or clutch size in many fishes and herps (Mann and Mills, 1979; Wootton, 1979; Salthe and Mecham, 1974). An expected increase in body size and litter size sometimes may permit a delay in achieving sexual maturity, provided that the likelihood of surviving the delay and the expected increase in fecundity are great enough. In effect, the increase in future fitness must be great enough to compensate for the delay.

Despite a great deal of thought and effort, the evolution of life-history patterns is only beginning to be understood. Some basic concepts are emerging, but perhaps more impressive are the many different kinds of life histories that seem to work and the differing importance of various factors in separate populations.

Allocation Adjustments

Many people have assumed that reproductive rates have evolved to balance mortality rates; that is, if a population has a given death rate, the birth rate is supposed to compensate so that the population size remains the same. Most vertebrates that are long-lived (having low adult mortality rates) produce small broods. It cannot be true, however, that small broods have evolved in compensation for low adult death rate and, thus, to avoid overpopulation. Any individual whose genotype allowed it to rear larger numbers of successful offspring would obviously have an advantage over other individuals, and its genotype would then become more frequent in the population. Nevertheless, birth rates and death rates can influence each other. Over short time periods we may observe that high birth rates may produce large populations and high densities, with resulting high mortality. Conversely, high mortality may reduce population size so much that more resources are left to the survivors, whose reproductive output may then increase. These responses are generally phenotypic responses and need not involve evolutionary change.

The evolutionary effects of mortality on allocation patterns depend on the age-specific probability of death, as mentioned in the previous section. The time schedule of the risk of mortality during a typical lifetime determines the adaptive value of different means of maximizing an individual's reproductive output. For simplicity, imagine a population with only two age classes: juveniles and adults. A high or unpredictably variable risk of death to juveniles lends selective value to adult longevity and iteroparity; adults that live long and reproduce repeatedly are clearly more likely to be represented in the next generation than are adults that gamble on the success of a single brood. Thus, a sculpin (*Cottus gobio*) produces many clutches a season in the warm streams in southern Britain but only lives two or three years; in the north, however, the breeding season is so short and the risk of juvenile death so high that selection seems to have favored a greater longevity of 8 to 10 years (Mann and Mills, 1979). High juvenile mortality also will select for any means of improving juvenile survival. On the other hand, a high mortality risk for adults tends to favor the evolution of large litters, several litters per season, reproduction early in life, or combinations of these. The probability of reproducing more than once is small, in this case, and the genotypes producing big litters, several litters, or reproducing early in life will leave more offspring than those that do not. Semelparous reproductive schedules represent extreme cases of this pattern. Ideally, age classes would be more finely divided, but the idea is adequately illustrated by this general example.

Correlations of mortality schedules and reproductive patterns in natural populations can be found. A complex example is provided by a comparison of certain salamander life histories. Salamanders of the family Plethodontidae are terrestrial and lungless. Most species live in moist habitats. The males breed annually, the females biennially. Several plethodontids occur in southern and central California, where the climate is less predictable, especially with respect to rainfall, than it is farther north. The dry season is longer, and the onset of the rainy season and the amount of rain during the wet season more variable. This environment creates potentially severe problems for terrestrial amphibians. Two species of salamanders (*Aneides lugubris* and *Ensatina eschscholtzii*) appear to have typical plethodontid life histories, and both are relatively large. Another species, *Batrachoseps attenuatus,* is considerably smaller—about one fifth to one tenth the size of the other two—and its life history is quite different from the typical plethodontid pattern. This small salamander usually breeds annually, but it refrains from breeding altogether if the individuals are unusually thin. Sexual maturity is also delayed—from an average of 2.5 to 3.5 years in females—if conditions are poor. The number of eggs laid is highly variable, and females having lower energy reserves (less fat stored in the tail) often resorb at least some of their eggs. In contrast to many other plethodontids, *B. attenuatus* females do not tend the eggs and, therefore, do not enter a two- to three-month period of semifasting that is required of attentive females. *Batrachoseps attenuatus* deposits its eggs at the beginning of the rainy season (in the autumn), although other plethodontids typically breed in the spring. Because the duration and wetness of the rainy season vary from year to year, however, juvenile mortality tends to be high and variable. Maiorana (1976) suggested that these characteristics of juvenile mortality have produced selection for long-lived adults that reproduce frequently. Climatic unpredictability may have prevented the evolution of consistently large sets of eggs or early maturity. The differences in life history observed between *B. attenuatus* and plethodontids of other genera may be related to body size: Larger salamanders can store more fat and consume it more slowly than can small ones. Thus, seasonal unpredictabilities probably have less serious consequences for these species, and life-history modifications are not favored (Maiorana, 1976).

Furthermore, geographic variation in life-history patterns occur among populations of some North American lizard species (*Cnemidophorus tigris, Uta stansburiana*). Some populations are composed of individuals with a short life span (high probability of adult mortality), early maturation, and multiple broods. Individuals of other populations are characterized by long lives (low risk of death to adults), late maturation, and single broods each season (Tinkle, 1969; Derickson, 1976). Yet, in many instances, such correlations are not clear-cut and may appear only in restricted conditions. The prairie swift (*Sceloporus undulatus*) ranges widely in North America, occurring in three principal habitat types: eastern woodlands, central and southwestern grasslands and deserts, and western rocky plateaus and canyonlands (Ferguson et al., 1980). There was rather little consistency among life-history traits for animals in the same habitat class, however. Three grassland-desert populations were all characterized by an early age of maturity and relatively low adult survival; two of these populations had a relatively high allocation to reproduction (the mass of the clutch was large, relative to female mass), whereas one had a relatively low reproductive allocation. Thus, it seems that life-history traits do not necessarily vary in unison but rather can vary independently. The lowest annual adult survival for the prairie swift was recorded for the grassland populations; these swift also matured the earliest, but early maturity also occurred in a woodland population with good adult survivorship. The expected associations do not always hold, therefore, indicating that other factors must be involved. It seems likely that direct ecological constraints differ among the populations; for instance, resource levels are likely to differ, but these have not been assessed.

Ecological Constraints

Whatever the life-history pattern favored by selection operating through the extrinsic risks and intrinsic costs, it depends on ecological conditions for its achievement. A variety of ecological factors may impose direct constraints on reproductive rates.

A study of Atlantic salmon has demonstrated a positive correlation of the average age of first spawning with the estimated difficulty and energetic cost of upstream migration (Schaffer and Elson, 1975). Note that, unlike Pacific salmon of the genus *Oncorhynchus,* Atlantic salmon do not typically die after their first spawning. The average age of sexual maturation on short, easy rivers in Quebec was 1.6 years, but on a longer, more difficult river, it was 2.6 years (counting only the years spent at sea and not those as juveniles in fresh water) (Fig. 14–5). In addition, the greater the age of first spawning, the more likely the salmon were to skip at least one year before breeding again, apparently devoting this time to substantial additional growth. Furthermore, Atlantic salmon that spawned in a short river after two years in the ocean lost 44 percent of their body weight, compared to individuals of equivalent age that did not breed (Schaffer and Elson, 1975). On a much longer river, very few individuals spawned after only two years at sea, but those few showed an average loss of 64 percent of their body weight. Loss of body weight could easily affect postbreeding survival of parents. Large (and older) individuals have two advantages over small ones: Their litter sizes are usually larger and they are stronger swimmers (and, hence, better able to negotiate difficult streams). These correlations indicate a difference in resource allocation patterns of fish that use different rivers. Schaffer and Elson suggested that selection has probably favored delayed maturation and delayed repeat-breeding for individuals that spawn in difficult rivers because of their greater probability of success attendant upon the achievement of larger size, but it remains to be

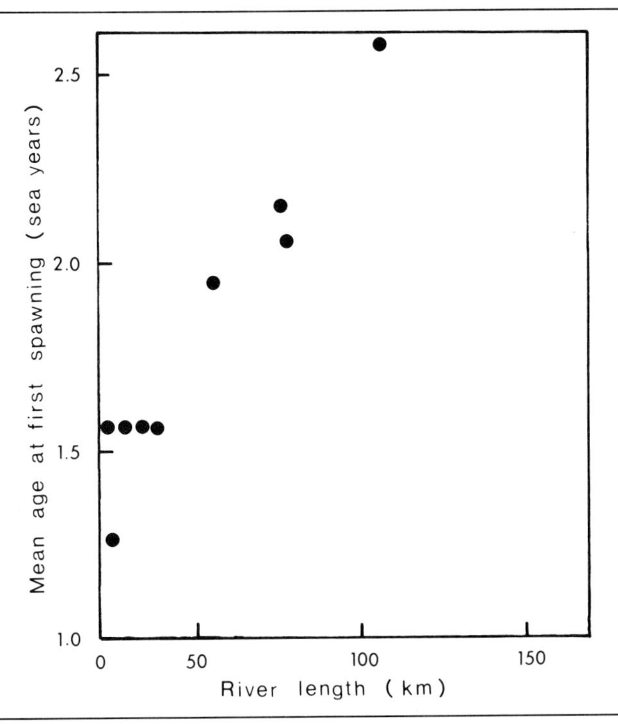

Figure 14–5 Mean age of first spawning in *Salmo salar* versus length of river in which the salmon breed. Data for Quebec show a strong positive correlation of age and river length. (Modified from W.M. Schaffer and P.F. Elson, 1975. The adaptive significance of variations of life history among local populations of Atlantic salmon in North America. *Ecol., 56*:577–590. Copyright © 1975, the Ecological Society of America. Also from W.M. Schaffer, 1982: Personal communication.)

seen if the life-history differences are genetically controlled or imposed by ecological restrictions directly.

Intense harvesting of Atlantic salmon (and other salmon species as well) removes large adults from the population, with the interesting consequence that more and more male salmon mature early and return to their breeding rivers when young and small. Furthermore, apparently removal of large adults also allows some young males to breed precociously, without ever going out to sea. The effect of river length on age of breeding is, not surprisingly, greater where fishing pressure is least. Clearly, behavioral interactions among these fish contribute additional variability to the life-history. Similarly, seal and whale populations subject to heavy hunting pressure are known to have a lower age of first reproduction for females, because of reduced interaction among individuals (Reiter et al., 1981). It will be interesting to learn to what extent the correlations discussed by

Schaffer and Elson might be explained by factors other than river length itself.

Some vertebrates seem to reproduce as much as is permitted by their food supply, whereas others do so sometimes, and still others perhaps never. In the first case, litter size or frequency usually mirrors changes in food availability; this evidence is reviewed in following paragraphs. In the second case, other constraints may prevent a great rise in litter size or frequency even when food levels are high, but low food availability results in lowered reproduction. In the third case, other constraints are paramount and reproductive effort is held below the ability of the animals to garner food. If the brood size is experimentally increased, the parents then are able to rear additional young successfully at no significant cost to their own survival.

Geographic variation in litter size is common among vertebrates; a common trend is an increase of litter size with increasing altitude and latitude. Clutch size of passerine birds at the latitude of Central Europe averages about twice as large as in tropical Africa. Even in a single species, such as the European robin (*Erithacus rubecula*), a marked increase in clutch size with increasing latitude is observed (Fig. 14–6). Many North American mammals have a corresponding trend, which is most marked in nonpredaceous genera such as *Microtus* (voles), *Peromyscus* (deer mice), *Sylvilagus* (cottontail rabbits), and *Eutamias* (chipmunks). Furthermore, among closely related species living in forest and savanna, the savanna species often have larger litters than the forest species. The usual clutch size for African forest ploceids (weaver finches) is two eggs, for example, whereas that of savanna and grassland species is sometimes two but more often three. These trends probably are related to the energetics of parental care. The seasonal fluctuations in climate are more marked at high altitudes and latitudes and in savannas, so that the seasonal burst of food productivity is also marked. As a result, more resources per unit of time are available for individuals living in these areas and more young can be successfully reared. Also, as Ricklefs suggested, perhaps seasonal climates induce greater adult mortality, so that resource availability per individual is increased.

Marked annual or seasonal differences in clutch or litter size are well known for some vertebrates. Arctic foxes (*Alopex lagopus*) of interior Greenland feed mainly on lemmings; they produce large litters in years of high lemming populations and small ones in years of low lemming populations. In contrast, coast-dwelling foxes, which feed on shore animals and carrion that vary little in annual abundance, produce smaller and less variable litter sizes. Similarly, many raptorial birds produce large broods in years of high prey populations and small ones in bad prey years. Unusually bad times of food availability may be marked by failure to breed at all, and sometimes the female resorbs eggs or embryos already present. This response also conserves parental energy.

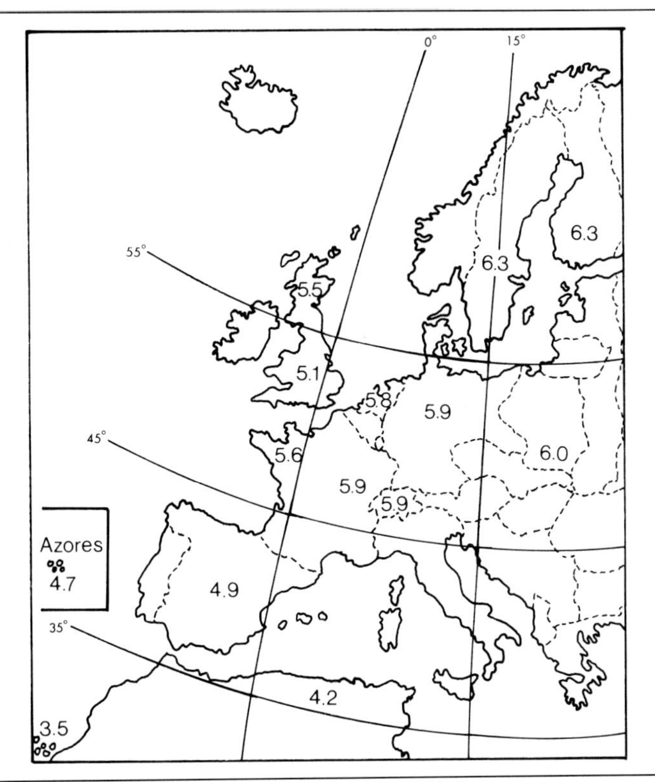

Figure 14–6 Average clutch size of the European robin (*Erithacus rubecula*) in different parts of Europe and North Africa. Note the tendency for larger clutches to be produced at higher latitudes. Although clutch sizes on Atlantic islands are smaller, on average, than on the mainland at the same latitude, a similar trend—increasing clutch size with increasing latitude—prevails on islands. (From D. Lack, 1954. *The Natural Regulation of Animal Numbers.* Clarendon, Oxford University Press.)

Swifts (*Apus apus*) in England may lay clutches of two eggs or three (rarely one). These fast-flying birds forage on the wing for flying insects. In years of good weather, when aerial insects are abundant, the clutches of three produce more young; but in years of rainy weather, when such insects are few, clutches of two leave more offspring (Table 14–1). Clutches of three were more common in the population as a whole and were clearly advantageous much of the time. The ability to produce smaller clutches was maintained in the population because in years of inclement weather the smaller broods had the advantage. However, we do not know if all the variation in clutch size was simply a phenotypic response by individuals or if, in addition, some individuals were genetically constituted to produce only broods of two or three and selection alternately favored different

Table 14–1 Nestling survival and brood size of the swift (*Apus apus*) in England.

A. In rainy weather

Brood size	N	Number fledged	Percentage fledged	Young fledged per brood
1	7	6	86	0.9
2	48	24	50	1.0
3	36	11	31	0.9

B. In fair or average weather

Brood size	N	Number fledged	Percentage fledged	Young fledged per brood
1	24	20	83	0.8
2	118	112	95	1.9
3	15	12	80	2.4

(From Lack and Lack, 1951.)

genotypes. Studies of another species, the European great tit (*Parus major*), indicated that the genetic contribution to variability in clutch size was considerable, but the directions of selection on clutch size seemed to fluctuate (von Noordwijk et al., 1981).

Protein reserves in the muscles of individual female red-billed queleas (*Quelea quelea*), an African weaver finch, are highest at the onset of breeding and are correlated with the number of eggs laid (Jones and Ward, 1976). Feeding conditions for adult females contribute to the protein reserves and thus indirectly affect clutch size. The clutch size of many tundra-nesting geese is similarly limited, although female geese acquire most of their food before migrating northward (Newton, 1977). Clutch size of mallard ducks (*Anas platyrhynchos*) was correlated with the female's reserves of lipids, at least for the first clutch of the season (Krapu, 1981). Production of the second clutch, however, relied entirely on availability of food at the time.

The tree lizard, *Urosaurus ornatus,* inhabits the deserts and mountains of southwestern North America. It is insectivorous, and both clutch size and number of clutches are lower in dry years, when insects are scarce. Females stored less fat before reproducing when food availability was low, and clutch size declined from an average of 11 eggs to 7 eggs. Most females produce two large clutches in wet years, but in dry years most females produce only one small clutch (Ballinger, 1977).

The ability of parents to feed their brood can limit brood size. As brood size of the great tit increases, parents increase their total feeding effort, but the amount of food per nestling decreases despite this behavioral adjustment (Fig. 14–7). Although heat conservation in large broods slightly reduces individual food requirements, offspring from large broods often leave the nest at a smaller size than those from small broods. Small fledglings apparently have a decreased chance of surviving after leaving the nest (Perrins, 1980). Furthermore, field experiments with the European starling (*Sturnus vulgaris*) in New Jersey showed that artificial increases in food supply permitted parent starlings to rear more nestlings of normal weight than could parents relying on a natural food supply (Crossner, 1977). Both sets of observations show that food resources can place a limit on the success of raising large broods. To produce fewer young than are possible obviously results in a loss of fitness. To produce extra-large broods is also ineffective in increasing fitness because parental energy is wasted on young that cannot be fed adequately.

Habitat differences in brood size can also be correlated with differences in food supply. Great tits nesting in deciduous (mainly oak) woods in the Netherlands have an average clutch size of 10 or 11 eggs, but those that nest in nearby pine woods lay an average of only 8 or 9 eggs each (Van Balen, 1973). In addition, survival of nestlings was roughly 75 percent in pine woods but ranged from 80 percent to 98 percent in oak woods. The abundance of caterpillars in oak woods is generally at least 20 to 100 times as great as it is in pine woods. One might wonder why great tits nest in

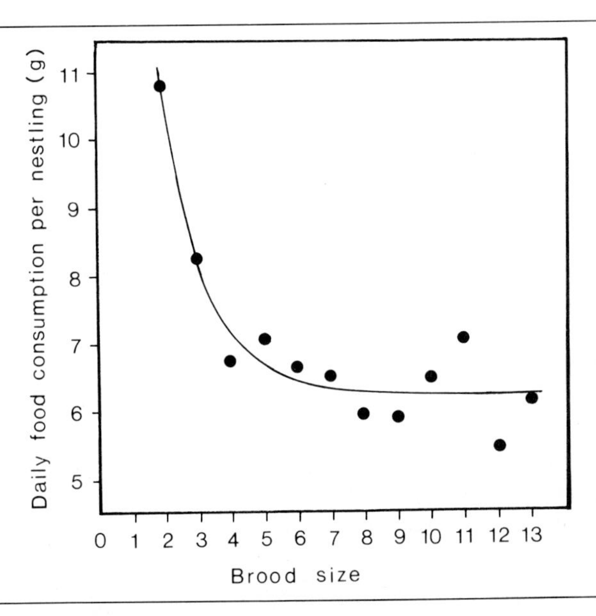

Figure 14–7 Food consumption (dry weight of food per day) per *Parus major* nestling versus brood size. (Modified from J.H. Van Balen, 1973. A comparative study of the breeding ecology of the Great Tit [*Parus major*] in different habitats. *Ardea, 61*:1–93.

pine woods at all, given their relatively low reproductive success in this habitat, or why at least the clutch size has not been reduced to a more efficient size by individuals that nest in pine woods. The use of pine woods by great tits may be a recent phenomenon, a result of the planting of extensive pine plantations and the introduction of nest boxes in these marginal habitats. Pine woods seem to be occupied by the overflow from the much more desirable oak woods, and these individuals maintain their oakwood adaptations. In all likelihood, they would move back to oak woods if space somehow became available.

Individual parents differ in their ability to raise large broods. This may be related to age and experience (see earlier discussion), to quality of the home range or territory (Högstedt, 1980), or to inherent abilities to gather food. Individual variation in parental ability is known for hooded crows (*Corvus corone*), great tits (*Parus major*), and several other avian species (Perrins, 1980; Loman, 1980). The effects of competition at high densities vary among individuals—some are much better competitors than others. Successful competitors of the Everglades pygmy sunfish (*Elassoma evergladei*) had larger ovaries and more eggs. (Similarly, male pygmy sunfish successful in competition for food often exhibited a higher reproductive effort than did comparable males at lower densities.)

Many researchers have suggested that predation on eggs and young may limit clutch size, in various ways discussed in the next paragraphs Whether or not there are shifts in allocation patterns as a result is not known. The clutch size of long-tailed skua (*Stercorarius longicaudus*) seems to be limited by the number an adult can incubate. This predatory bird makes no nest and incubates only two eggs—one on each foot (Andersson, 1981). If the skua made a nest to permit a larger clutch, the rate of egg predation reportedly would increase. Here it might be argued that predation risks determine the nesting habits, which in turn affect the clutch size.

Similarly, one hypothesis for the evolution of small clutches (one or two eggs, usually) in bellbirds, manakins, and some other tropical birds considers space in the nest as a limited resource (Snow, 1970; Lill, 1975). Predation rates on eggs and nestlings are often quite high (as much as 80 or 90 percent of the nests) in tropical forests, and many of the predators may search visually for nests. Such predation may select for small inconspicuous nests that cannot contain many eggs. Fruit-eating by some of these species facilitated brood care by a single parent (the female); hence, brood size is probably not limited by food resources.

For vertebrates whose young do not remain in a nest or den after birth or hatching, a major task of parents is warning and guarding the young. The more young in the brood, the higher the probability of their spreading out over a large area while foraging and the more difficult the task of the attending parent. Furthermore, once a predator has found a brood, it may often return to that area to hunt again, and much or all of the brood may be lost. Experiments with broods of semipalmated sandpipers (*Calidris pusilla*) showed that, although growth rates of the young in normal and experimentally enlarged broods were similar, predation was higher on the larger broods (Safriel, 1975).

The risk of predation may be higher for gravid females weighted down by a large clutch or litter. The running speeds of female skinks are reduced 20 to 30 percent, and individuals with large clutches tend to be slower than those with small clutches (Shine, 1980). Furthermore, Vitt and Congdon (1978) showed that lizards (such as *Cnemidophorus*) that forage actively and flee from predators often carry a lower weight of eggs, relative to body size, than do more sedentary, cryptic lizards (such as *Phrynosoma*). However, other lizards may merely shift their escape tactics when carrying eggs or young: *Lacerta vivipara* females flee less readily and rely more on crypsis when they are gravid, but they seem not to be more subject to predation as a result (Bauwens and Thoen, 1981). This lizard is an active forager but carries a relatively large mass of eggs, so it does not fit the Vitt-Congdon model.

Perrins (1977) suggested that predation may sometimes limit clutch sizes below the level set by food resources. Large numbers of nestlings not only require more feeding trips by the parents but also may clamor more loudly when hungry. Perhaps even more important is the amount of time the eggs and young spend in the nest. Suppose a bird lays an egg a day and the nest has a constant daily predation risk during the egg phase. A bird that lays 5 eggs will, on average, suffer less from predation than one that lays 10 eggs. If the incubation period is 15 days, the smaller clutch is at risk for 20 days but the larger one for 25 days. The larger clutch thus has a predation risk 25 percent greater than the smaller one, due simply to the duration of exposure to predation. Too large a clutch may decrease the probability of survival and, thus, make large clutches disadvantageous. The higher the daily risk of predation, the lower the most adaptive clutch size. Such effects should be negligible, however, when the egg-laying period is short relative to the total length of time eggs are in the nest—that is, when the incubation period is very long compared to the laying period. Perrins emphasizes that data to substantiate this theoretically possible effect of predation are virtually nonexistent.

Many birds begin incubating before the last egg of a clutch is laid. As a result, the eggs commonly hatch asynchronously and siblings may differ considerably in age and size. Especially if food is scarce, younger, smaller nestlings may starve and perhaps be eaten by their older siblings. Brood reduction may adjust brood size to the available food supply, at least in some cases, and might be viewed as further evidence of the importance of food supply in regulating brood size. However, one cannot presume that early incubation and hatching asynchrony are adaptations specifically to allow subsequent brood reduction. In fact, there is reason to believe that asynchronous hatching is often adaptive in hastening the time of hatching of some of the clutch to take advantage of peaks in food abundance (possibly at the expense of later-hatching offspring). Furthermore, if nestling growth rates can vary such that early periods of slow growth can be compensated by later periods of rapid growth (as may be true of some hole-nesting species with long nestling periods), asynchronous hatching may allow parents to raise more young than would otherwise be possible. Some could be stinted temporarily while others were fed; thus, the total feeding effort of the parents for a large brood would be less than normally expected. Synchronous hatching is apparently less common than asynchronous hatching. Synchrony is favored when there is a high risk of nestling predation compared to egg predation, because synchrony reduces the total length of time that nestlings are present in the nest (Clark and Wilson, 1981). Conversely, asynchronous hatching is associated with relatively low rates of nestling mortality.

Senescence

The age distribution of mortality not only influences age-specific allocations to reproduction but also is responsible for senescence (the process of aging) (Medawar, 1957; Williams, 1957; Hamilton, 1966). Even if we assume that the probability of dying from some random accident is constant for each time interval, the total probability of death is greater over many such intervals than over a few intervals. If the risk of death during one year is 10 percent, the risk of death over two years is 10 percent plus 10 percent of the remainder for a total of 19 percent. For three years, the risk is 19 percent plus 8 percent (which is 10 percent of the survival after two years), or 27 percent. The increasing cumulative probability of death means that, even without senescence, an individual is less likely to be alive to reproduce at the greater ages. The expectation of future reproduction is usually highest sometime near the age of sexual maturation and declines thereafter. Furthermore, litters produced later in life often have smaller effects on the rate of genetic contribution to future generations than do earlier litters. Therefore, any deleterious expression of a gene during the period of high reproductive expectation will have greater effects on fitness than a similar expression later in life. As a result, selection favors postponement of the expression of detrimental genetic traits. The consequence of this postponement is the accumulation of deleterious characteristics later in life, which produces an increasing condition of decrepitude, which we call senescence. Total elimination of the negative effects is virtually impossible if the same genes also produce beneficial effects at other times, especially early in the reproductive life of the individual. Furthermore, selection may be less effective in cohorts of greater age because mortality (from any cause) has already reduced the amount of genetic variation on which selection can act.

Senescence, therefore, is characteristic of all adults. It proceeds at different rates at different ages, however, depending on the expectation of reproduction. It is often thought that older animals cease reproducing because they are senescent; instead, the reverse is true: They senesce because their expectation of reproduction drops. In fact, postreproductive vertebrates generally do not occur, except in modern humans and their domesticated animals. Williams (1957) calls this an "artifact of civilization"—a result of improved medical practices. Even in humans it must be remembered that an individual is not really postreproductive until its offspring are self-sufficient. Even then, the role of grandparent may contribute to fitness.

BREEDING SEASONS

Well-marked annual breeding seasons are a familiar phenomenon among vertebrates. Most species court, mate, and rear their young only during certain portions of the year; the breeding season of many species is a predictable annual event. Temperate-zone and polar animals generally breed in spring and summer (the large penguins are notable exceptions). Clear-cut breeding periodicity is common among tropical vertebrates too, especially where the wet-dry cycle is most pronounced, but even in tropical rain forests, most species have a breeding season.

Some species have prolonged periods of parental care: The young stay with the parent(s) for over a year. Vertebrates whose breeding cycles are longer than a year also commonly begin each new cycle at a particular time of year. However, parents with immature young will not generally start a new brood until the first one has approached or reached the age of independence from intensive parental care.

Well-marked seasonality of breeding is notably absent in some circumstances. A few species may start a breeding cycle at any time of year and seemingly have no regular nonbreeding season; the sooty tern (*Sterna fuscata*) on Christmas Island in the Pacific and tropicbirds (*Phaethon*) and the black noddy (*Anous minutus*) on Ascension Island in the Atlantic are examples. Yet another schedule is followed by a number of irregular, opportunistically breeding species, including several bird species of interior Australia that live in unpredictable habitats. These species breed whenever conditions are right, no matter what month it is. To understand the adaptive nature of these differences, we must look at the ecological conditions surrounding each species.

Ecological Conditions

The evolutionary determinants of breeding times are ecological conditions. Individuals that try to breed when conditions are not suitable leave fewer descendants than others and thus are gradually eliminated from the population by natural selection. We have called these conditions ultimate factors.

Each species may have different sets of ecological factors that are of prime importance in determining breeding time. For some, the presence of a food supply for the dependent young or for newly independent offspring may be critical; for others, it may be the availability of sufficient food for egg construction or gestation. Suitable habitat, or cover, or nesting sites and material may be central in some cases. Sometimes several such factors together may influence breeding time for a species. Thus, the breeding season may be a compromise among selection pressures from several factors.

A good example is provided by the well-studied *Parus major.* Breeding primarily in deciduous woods, but sometimes also in coniferous woods and in hedgerows, the great tit feeds largely on lepidopteran caterpillars during the breeding season. Ecologists have long supposed that the breeding season was timed to correspond with the availability of caterpillars. At first researchers emphasized the food supply for nestlings, but peak nestling periods did not coincide especially well with peak caterpillar abundances. More recent work, however, has shown that the food supply for advanced nestlings and fledglings, which have recently left the nest, is of great importance because food requirements at this time are very high. Other critical considerations are the energy supply for the female when she is producing eggs and incubating. Egg production can be a costly process for female birds (from 10 to 70 percent of daily energy requirements, depending on the species). Great tit males alleviate the problem somewhat by feeding the female during egg-laying and incubation, but they do so primarily away from the nest. As incubation proceeds, the female leaves the nest less and less often and loses a considerable amount of weight, especially at the end of incubation and during the hatching period.

Therefore, the breeding of *Parus major* appears to face several critical times in terms of the available food resources. The peak of caterpillar abundance lasts two to three weeks, but the possible critical periods of a breeding cycle span at least five weeks. It is obviously impossible, therefore, for breeding to be timed in such a way that each critical period matches the peak of food abundance. Observed breeding times may be a compromise among these selective factors, or future research may show that certain of the so-called critical periods are actually more crucial than others. In any event, there is a broad correlation of the timing of the nesting cycle with caterpillar abundance (Fig. 14–8). We can safely assert that great tits' breeding is, in a general way, timed to occur when their major food resource is abundant, even though the details of the association are not clear.

Tundra-nesting geese depend on food reserves that are stored up primarily before their long northward migration. These stores are used both

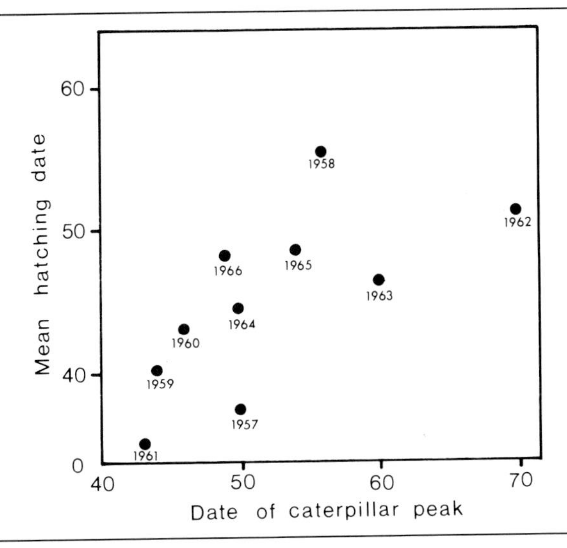

Figure 14–8 Relation between the mean hatching date of the first *Parus major* brood and the date of the peak of caterpillar abundance in an oak woods in the Netherlands, 1957 to 1966. Dates are counted from 1 April = 1. (Modified from J.H. Van Balen, 1973. A comparative study of the breeding ecology of the Great Tit [*Parus major*] in different habitats. *Ardea, 61*:1–93.)

to provision the eggs and to sustain the birds during incubation. They arrive on their breeding grounds at about the same time every year, although the date of snow-melt varies by as much as three weeks. Pairs have already formed and eggs have been fertilized before arrival. If the birds arrive before the ground is bare, they wait. The longer they wait before laying eggs, the more their reserves are depleted. If breeding is delayed for more than about two weeks, no breeding is attempted that year, because food reserves are inadequate by then. Late breeding is selected against because it has a low probability of success. Not only are late clutches small, but the young may not have enough time to acquire their flight feathers before winter closes in. Late-hatching brent geese (*Branta bernicla*) have been found frozen in the ice early the following spring, perfectly preserved, and with no defects except that their feather development was four or five days short of making them airborne. Under these circumstances, adult geese that are greatly delayed in breeding would do better not to try and to save their energies for another year (Newton, 1977). In the case of these geese, initiation of breeding depends on the level of adult food reserves and the availability of snowless nest sites.

Still other factors may well provide selection for particular breeding schedules in highly social vertebrates such as gulls (*Larus*). Many gulls, including the herring gull (*L. argentatus*), are cannibalistic, eating eggs and chicks of their own species (often their neighbors in the colony). Other species, such as the great black-backed gull (*L. marinus*), also prey on herring gull chicks. As long as the number of eggs and chicks killed remains similar throughout the breeding season, any eggs or young produced during the peak of breeding will have a lower risk of predation than those occurring in the company of fewer others. This produces selection for synchronized breeding among members of a colony, as shown for the sooty tern (*Sterna fuscata*) in the Seychelles Islands (Feare, 1976), and may modify the time schedule induced by food availability or other factors.

On the other hand, a study of the glaucous-winged gull (*L. glaucescens*) in British Columbia suggested that offspring mortality is not always constant through the season (Fig. 14–9), (Hunt and Hunt, 1976). As a result, timing patterns in this species are likely to be different from those of herring gulls. Glaucous-winged gulls frequently killed conspecific chicks that wandered into their territories from neighboring territories in the colony. Gull chicks wandered about more extensively when the food supply was lower; hence, they trespassed more frequently and were killed more often. Gulls of this species are particularly aggressive against trespassing chicks when they have chicks of their own; therefore, murder by the neighbors tended

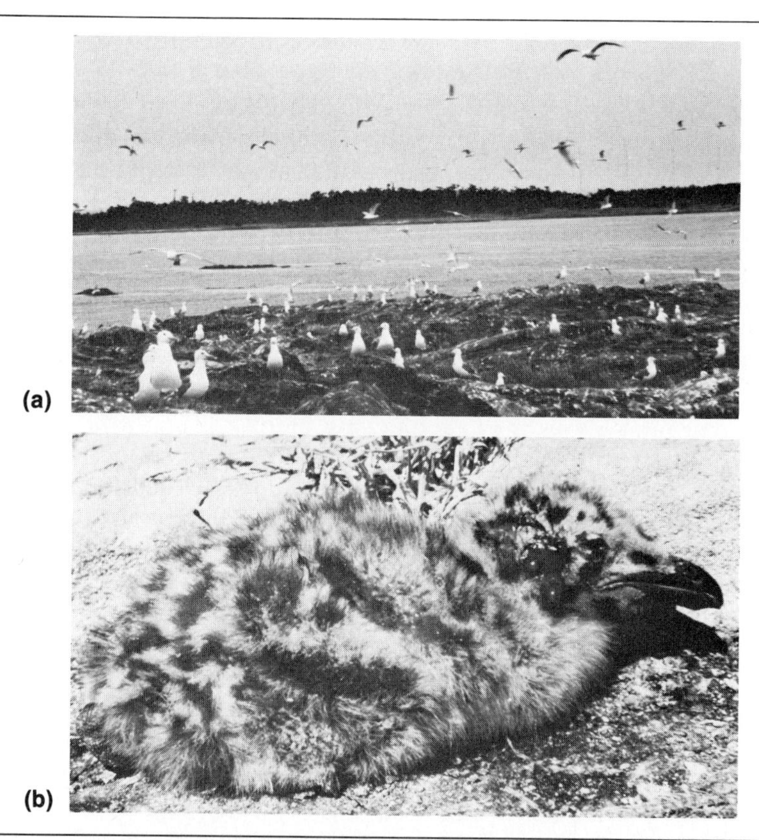

Figure 14–9 (*a*) Glaucous-winged gull colony. In this species, asynchronous breeding helps to reduce infanticide. (*b*) A gull chick that has been attacked by an adult. (*a* and *b* courtesy of Wayne Campbell.)

to be least when fewer of the neighbor's eggs had hatched. Thus, asynchronous breeding in glaucous-winged gulls seemed to have some advantages—at least in some years. Because gull species differ considerably in habitat, predatory or aggressive inclinations, and colony organization, these studies open up numerous avenues of research on ecological-behavioral interactions that may influence the timing of breeding.

Social factors may be major regulators of reproduction in lions. Females in a lion pride may produce young at any time of year, but all females of a group tend to give birth at approximately the same time. As a result, cubs of the pride are generally about the same age; in fact, cub survival is better then than if cubs 7 to 12 months older are present. Adult females share the feeding of the cubs and may be more attentive when more cubs require attention. Furthermore, the absence of older cubs reduces competition for food among the young lions, which is especially important in seasons when food is scarce. Behavioral interactions among the females may contribute to reproductive synchrony.

Another significant factor sometimes operates when a new male lion or team of males takes over the pride. Cub mortality is unusually high at these times, perhaps because of social stress induced by the presence of new males and cub-killing by the new arrivals. Infanticide by new males is presumably adaptive to them in that it removes offspring sired by the previous males and causes the females to be ready to mate both sooner and at about the same time as each other. The new males may gain by the early fertilization of the females and also by the enhanced synchrony of litter production. This process may not be as disadvantageous to females as it might seem. The gestation period of lions is short, cubs are small, and cub mortality is high even in the absence of infanticide. Therefore, replacement of lost cubs is not very costly and may have happened even without cub-killing. Furthermore, because the improved breeding synchrony for the litters fathered by the new males may result in better cub survival, the disadvantages to females are probably small (Bertram, 1975). Nevertheless, we should expect to find behavioral traits in females that further reduce their losses (Packer and Pusey, 1983). Infanticide by new males entering a social group is also known among primates, rodents, and some other species. Whether or not infanticide by incoming males is really an adaptation for improving male fitness (and just how it might be adaptive) or whether it is merely a result of stress that may accidentally influence fitness is a matter of some debate (Hrdy, 1979; Boggess, 1979; Packer and Pusey, in press; Sherman, 1981). Here, the social environment may provide both proximate cues and suitable conditions for successful reproduction.

Vertebrates mate at a time that enables them to produce young at an advantageous season, and development of the young usually begins immediately. Development is delayed or slowed in a number of species, however: a number of bats, bears, mustelids, rodents, insectivores, armadillos, seals, and a deer, among mammals (Renfree and Calaby, 1981), and several reptiles and salamanders. Females may copulate and store the sperm, producing delayed fertilization, or implantation of the embryo in the uterus may be delayed. These mechanisms all have a common result: the extended temporal separation of copulation and birth.

The selective advantages of these tactics are not well understood. The difficulty that males and females have in finding each other may provide the selective pressure in some cases. If females can reliably be found at certain places when they lay eggs or give birth and are scattered far and wide at other times, one solution to the problem may be for mating also to occur at those places. If the gestation period were short, it would be advantageous to delay the development of offspring so that the young could be produced at the appropriate season. Delayed development could also be adaptive if suitable conditions for growth and maturation of the young occur unpredictably in time or space; then females carrying sperm or delayed zygotes might be able to produce young rapidly when suitable conditions are encountered. However, these suggestions may not cover all the cases of delayed development, and more study is needed.

Stimuli for Breeding

To accomplish the appropriate timing, various cues or proximate factors may be used. Changing day length (photoperiod) is a major trigger of breeding readiness for many species. The physiological details of this reaction are quite well known: Visual perception of changing photoperiod is transmitted to the pituitary gland, which then secretes hormones that stimulate development of the gonads. Photoperiod is a cue that varies in a highly predictable way during each year and is thus a reliable signal of the appropriate season. As we all know, however, spring weather is quite variable, and a given date in one year may be much colder or warmer than the same date in another year. Consequently, it is not surprising to learn that temperature frequently exerts a powerful modifying effect on the general response to photoperiod. In the great tit, for example, spring temperature may even be the most important proximate factor determining the onset of breeding. Animals would be predicted to use whatever environmental cues prove to be reliable in stimulating breeding at an ecologically appropriate time. Therefore, we find that different cues are used in different circumstances.

An annual rain/dry cycle may be used when those climatic events override the importance of changes in day length. Most desert anurans can respond quickly to rain, and the filling of temporary streams and pools, by emerging from their burrows and mating promptly. The breeding of desert birds typically is closely linked to rainfall patterns. Particularly in the Old World, where deserts are of great antiquity, avian species are very flexible in their breeding. Some species such as the crimson chat (*Epthianura tri-*

color) and certain wood swallow (*Artamus*) of Australia are nomadic and move into areas of recent rain. The pink-eared duck (*Malacorhynchus membranaceus*) in the semiarid parts of interior Australia feeds primarily on plankton and is highly dependent on food-rich floodwaters for breeding. Both this species and the less specialized gray teal (*Anas gibberifrons*) are extremely nomadic. Some birds, such as the Australian apostle bird (*Struthidea cinerea*) and desert chat (*Ashbyia lovensis*), are stimulated to breed by the mere sight or sound of rain or by seeing the green vegetation that sprouts after a rain. Some other birds, such as the crimson chat and the magpie of southwestern Australia, *Gymnorhina dorsalis*, rear more young if rains are unusually heavy, thereby capitalizing on occasional bonus conditions.

Behavioral interactions among individuals, especially between males and females, often provide additional regulation of breeding time. One function of courtship is to synchronize male and female readiness. A series of experiments with the iguanid lizard *Anolis carolinensis* (Fig. 14–10) has documented the importance of such interactions (Crews, 1975). These lizards live in the southeastern United States and spend the fall and winter in dormancy, emerging in late January (males) or late February (females). Dormant females were brought into the laboratory in fall and winter and exposed to controlled environmental conditions of photoperiod and temperature normally associated with increasing sexual activity in spring. Most females subjected to this treatment gradually entered breeding readiness and produced mature, yolky eggs (Fig. 14–11). Successful egg production was clearly influenced also by interactions with male anoles. If the

Figure 14–10 *Anolis carolinensis.* This iguanid lizard furnishes a classic example of how courtship can synchronize male and female sexual readiness. (Photo by A.C. Echternacht.)

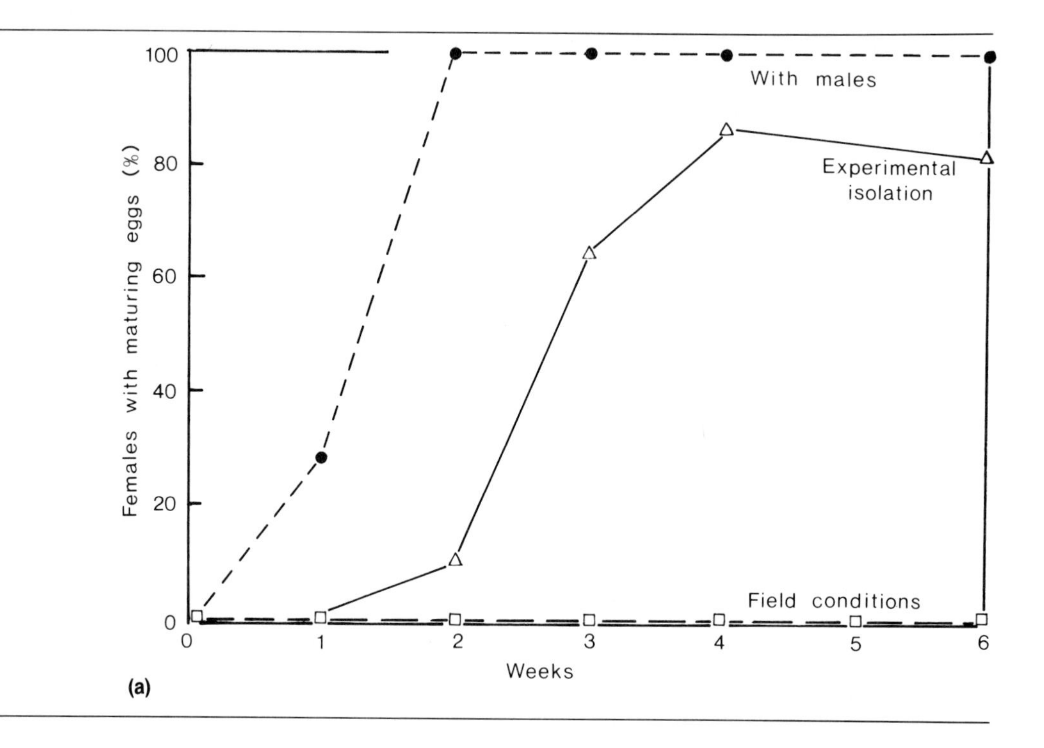

males were already sexually active when placed with the dormant females, intermale aggression reduced the frequency of successful courtship to zero and many fewer females produced mature eggs. If both males and females began in the dormant condition and were exposed to the same schedule of environmental events at the same time, intermale aggression had ended and a dominant male emerged from the male group by the time the females were ready. The dominant male then began courting females as they became sexually ready, and successful mating ensued. Further experiments clearly demonstrated that the amount of male courtship to which females were exposed profoundly influenced the females' gonadal development. Moreover, they showed that intermale aggression inhibited such development.

Once the breeding season has begun, the length of time until the next cycle varies greatly among species. Many small birds and rodents may rear several broods or litters each year. If an early brood is lost, replacement broods are often attempted. Other species, including most larger mammals and birds, rear only one batch of offspring per season. Still others, such as elephants, many seabirds, and whales, do not breed so often as once a year. Female African elephants, at least in some populations, normally produce an offspring only at four- to nine-year intervals. The number of breeding cycles within each breeding season ultimately depends on the

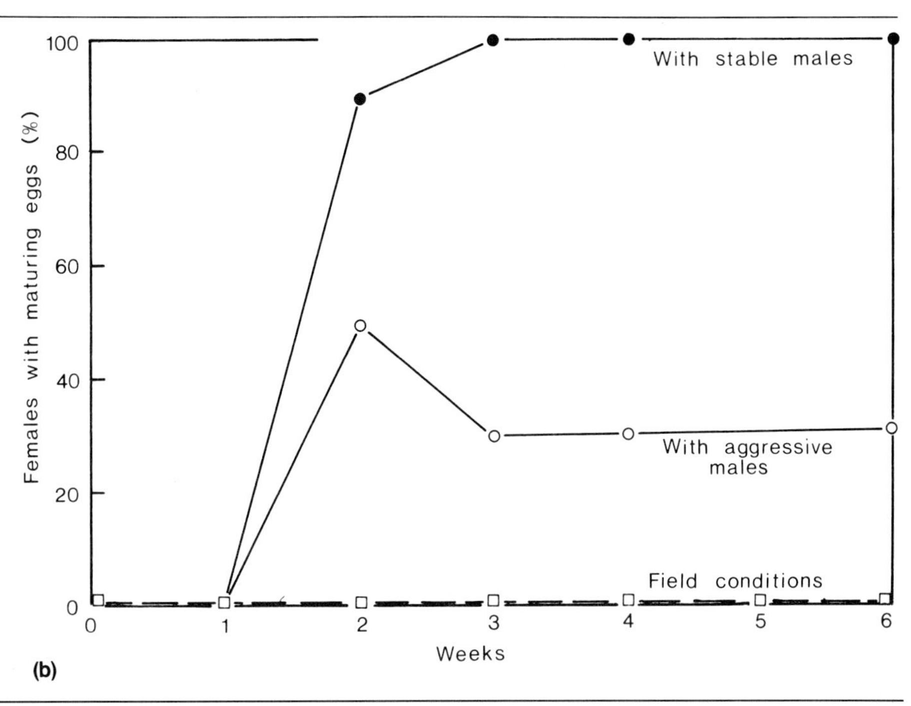

Figure 14–11 (*a*) Patterns of ovarian growth in female *Anolis carolinensis* in the field in winter, in isolation in the experimental situation, and with males in the experimental conditions. The presence of males results in increased egg production. (*b*) Ovarian growth in females in the field that were exposed to active, aggressive males and exposed to males whose dominance relationships were already established by the time females were ready to breed. (From D. Crews et al., 1974. Effects of unseasonal environmental regime, group composition and males' physiological state on ovarian recrudescence in the lizard, *Anolis carolinensis. Endocrinol.*, *94*:541–547.)

length of the suitable season, the length of the period of parental care, and the necessity of performing other activities (such as molting or preparing for migration) that might utilize resources otherwise available for breeding. The frequency of breeding is regulated physiologically by a gonadal refractory period (the gonads do not respond to the usual stimuli). Some minimum refractory period may be necessary to avoid depletion of energy reserves, but, in many cases, natural selection seems to have increased the length of the refractory period beyond this minimum, thereby ensuring that breeding does not recommence until the proper season and that accidents of weather or behavior do not trigger nonadaptive attempts to breed.

Male sexual cycles in most vertebrates tend not to be as sharply defined as those of females (as in *Anolis carolinensis*, Fig. 14–12). The energetic cost of producing sperm is generally far less than that of producing eggs (less

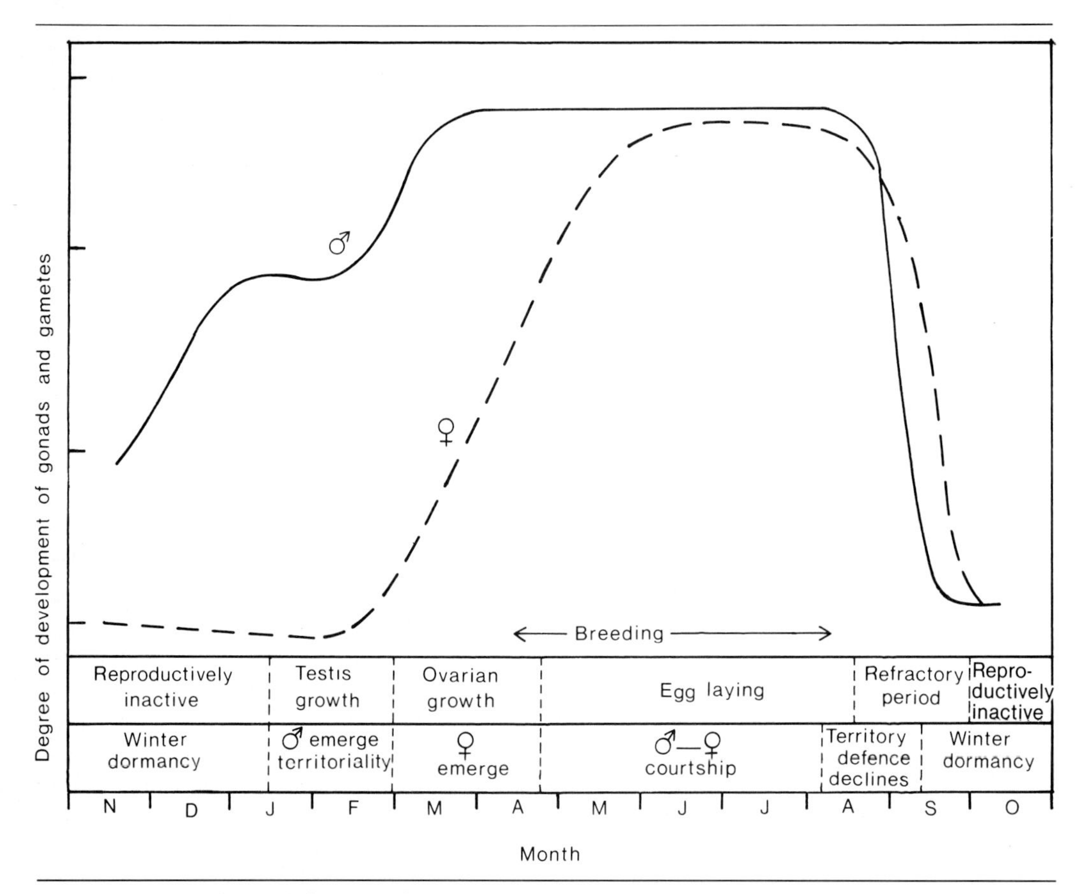

Figure 14–12 Relative definition of the reproductive cycles in male and female *Anolis carolinensis.* Males have functional gonads markedly earlier than females, and they are ready to breed for a longer period. (Modified from D. Crews, 1975. Psychobiology of reptilian reproduction. *Science, 189:*1059–1065. Copyright © 1975 by the American Association for the Advancement of Science.)

than 1 percent of the daily energy budget as opposed to more than 10 percent for females in most of the species studied so far). Hence, the male's physiological cost of being ready to breed is less than that of a female, and a male can afford to stay in reproductive condition longer. As a result, he is often able to copulate with other females that are somewhat out of synchrony with each other. Furthermore, male aggression is often closely tied to success in breeding and is physiologically linked with the sexual cycle. Therefore, when male aggression precedes breeding, it is not surprising that the male cycle begins before the female cycle.

SUMMARY

Although complex life cycles may appear to be inherently unstable in evolutionary time, they are probably adaptive specializations to the exploitation of ephemeral habitats.

The schedule of life-history events is broadly defined by the evolution of patterns of resource allocation to various activities. When animals have time and energy budgets that are limited, the allocation of these commodities to various activities can have an important influence in molding life histories. Trade-offs may occur between present and future reproduction and between quantity and quality of offspring. Litter size, frequency of reproduction, and age of first reproduction are all central to reproductive rates of individuals and are influenced by allocational patterns. A major selective factor determining allocational patterns is the age-specific risk of death, which broadly sets the adaptive values of various reproductive schedules. Within the framework of a life-history pattern, reproductive rates are often constrained by limited and variable resources. Considerable evidence shows that variability of food may often influence litter size, success of those litters, frequency of reproduction, and age of first breeding. The risk of predation may also limit litter size.

Breeding seasons are generally timed to occur when ecological conditions are suitable. Optimum conditions for each phase of the breeding cycle may differ and may often be unattainable for all phases of the cycle. Therefore, the precise time of breeding probably is often a compromise between conflicting requirements of different phases, although the importance (to the time of breeding) of some parts of the cycle is undoubtedly greater than that of others. Food resources for developing eggs and young, conditions for newly independent offspring, availability of suitable nest materials and sites, interaction with conspecifics (especially in gregariously breeding species), and other demands on parental time and energy—all contribute to defining the most suitable breeding season. Proximate factors that stimulate the onset of breeding are those that are successful predictors of suitable ecological conditions. Environmental cues such as photoperiod, rainfall, and temperature often provide major stimuli, and the "fine tuning" of breeding time is often produced by behavioral interactions among individuals.

SELECTED REFERENCES

Andersson, M., 1981. Reproductive tactics of the long-tailed skua *Stercorarius longicaudus*. *Oikos*, *37*:287–294.

Askenmo, C., 1979. Reproductive effort and return rate of male pied flycatchers. *Am. Nat.*, *114*:748–753.

Ballinger, R.E., 1977. Reproductive strategies: food availability as a source of proximal variation in a lizard. *Ecol.*, *58*:628–635.

Bauwens, D., and C. Thoen, 1981. Escape tactics and vulnerability to predation associated with reproduction in the lizard *Lacerta vivipara. J. Anim. Ecol., 50*:733–743.

Beckoff, M., J. Diamond, and J.B. Mitton, 1981. Life-history patterns and sociality in canids: body size, reproduction, and behavior. *Oecologia, 50*:386–390.

Bertram, B.C.R., 1975. Social factors influencing reproduction in wild lions. *J. Zool.* (Lond.), *177*:463–482.

Berven, K.A., D.E. Gill, and S.J. Smith-Gill, 1979. Counter gradient selection in the green frog, *Rana clamitans. Evol., 33*:609–623.

Boggess, J., 1979. Troop male membership changes and infant killing in langurs (*Presbytis entellus*). *Folia Primatol., 32*:65–107.

Braithwaite, R.W., and A.K. Lee, 1979. A mammalian example of semelparity. *Am. Nat., 113*:151–155.

Bryant, D.M., 1979. Reproductive costs in the house martin (*Delichon urbica*). *J. Anim. Ecol., 48*:655–675.

Bryant, E.H., 1971. Life history consequences of natural selection: Cole's result. *Am. Nat., 105*:75–76.

Bull, J.J., and R. Shine, 1979. Interoparous animals that skip opportunities for reproduction. *Am. Nat., 114*:296–303.

Burley, N., 1980. Clutch overlap and clutch size: alternative and complementary reproductive tactics. *Am. Nat., 115*:223–246.

Charnov, E.L., and J.R. Krebs, 1974. On clutch-size and fitness. *Am. Nat. 116*:217–219.

Charnov, E.L., and W.M. Schaffer, 1973. Life-history consequences of natural selection: Cole's result revisited. *Am. Nat., 107*:791–793.

Clark, A.B., and D.S. Wilson, 1981. Avian breeding adaptations: hatching asynchrony, brood reduction, and nest failure. *Q. Rev. Biol., 56*:253–277.

Cody, M.L., 1966. A general theory for the evolution of clutch size. *Evol., 20*:174–184.

Cody, M.L., 1971. Ecological aspects of reproduction. *In* D.S. Farner and J.R. King (Eds.), *Avian Biology.* Vol. 1, pp. 461–512. Academic Press, New York.

Cohen, D., 1976. The optimal timing of reproduction. *Am. Nat., 110*:801–807.

Cole, L.C., 1954. The population consequences of life-history phenomena. *Q. Rev. Biol., 29*:103–137.

Coulson, J.C., and J. Horobin. 1976. The influence of age on the breeding biology and survival of the Arctic tern *Sterna paradisea. J. Zool.* (Lond.), *178*:247–260.

Crews, D., 1974. Effects of group stability, male-male aggression, and male courtship behaviour on environmentally-induced ovarian recrudescence in the lizard *Anolis carolinensis. J. Zool.* (Lond.), *172*:419–441.

Crews, D., 1975. Psychobiology of reptilian reproduction. *Science, 189*:1059–1065.

Crews, D., J.S. Rosenblatt, and D.S. Lehrman, 1974. Effects of unseasonal environmental regime, group composition and males' physiological state on ovarian recrudescence in the lizard, *Anolis carolinensis. Endocrinology, 94*:541–547.

Crossner, K.A., 1977. Natural selection and clutch size in the European starling. *Ecol., 58*:885–892.

Crowell, K.L., and S.I. Rothstein, 1981. Clutch sizes and breeding strategies among Bermuda and North American passerines. *Ibis, 123*:42–50.

Curio, E., 1976. *The Ethology of Predation. Zoophysiol. Ecol., 7*:1–250.

Derickson, W.K., 1976. Ecological and physiological aspects of reproductive strategies in two lizards. *Ecol., 57*:445–458.

DeSteven, D., 1978. The influence of age on the breeding biology of the tree swallow *Iridoprocne bicolor. Ibis., 120*:516–523.

DeSteven, D., 1980. Clutch size, breeding success, and parental survival in the tree swallow (*Iridoprocne bicolor*). *Evol., 34*:278–291.

Dunn, E.H., 1980. On the variability in energy allocations of nestling birds. *Auk, 97*:19–27.

Feare, C.J., 1976. The breeding of the sooty tern *Sterna fuscata* in the Seychelles and the effects of experimental removal of its eggs. *J. Zool.* (Lond.), *179*:317–360.

Ferguson, G.W., C.H. Bohlen, and H.P. Woolley, 1980. *Sceloporus undulatus:* comparative life history and regulation of a Kansas population. *Ecol., 61*:313–322.

Foster, M.S., 1975. Temporal patterns of resource allocation and life history phenomena. *Florida Scientist, 38*:129–139.

Freed, L.A., 1981. Loss of mass in breeding wrens: stress or adaptation? *Ecol., 62*:1179–1186.

Geist, V., 1966. The evolution of horn-like organs. *Behav., 27*:177–214.

Giesel, J.T., 1976. Reproductive strategies as adaptation to life in temporally heterogeneous environments. *Annu. Rev. Ecol. Syst., 7*:57–79.

Gill, D.E., 1978. The metapopulation ecology of the red-spotted newt, *Notophthalmus viridescens* (Rafinesque). *Ecol. Monogr., 48*:145–166.

Goodman, D., 1974. Natural selection and a cost ceiling on reproductive effort. *Am. Nat., 108*:247–268.

Hails, C.J., and D.M. Bryant, 1979. Reproductive energetics of a free-living bird. *J. Anim. Ecol., 48*:471–482.

Hamilton, W.J., 1966. The moulding of senescence by natural selection. *J. Theor. Biol., 12*:12–45.

Healy, W.R., 1973. Life history variation and the growth of juvenile *Notophthalmus viridescens* from Massachusetts. *Copeia, 1973*:641–647.

Higuchi, H., 1976. Comparative study on the breeding of mainland and island subspecies of the varied tit, *Parus varius. Tori, 25*:11–20.

Hirshfield, M.F., 1980. An experimental analysis of reproductive effort and cost in the Japanese medaka, *Oryzias latipes. Ecol., 61*:282–292.

Hirshfield, M.F., and D.W. Tinkle, 1975. Natural selection and the evolution of reproductive effort. *PNAS, 72*:2227–2231.

Högstedt, G., 1980. Evolution of clutch size in birds: adaptive variation in relation to territory quality. *Science, 210*:1148–1150.

Hrdy, S.B., 1977. Infanticide as a primate reproductive strategy. *Am. Sci., 65*:40–49.

Hrdy, S.B., 1979. Infanticide among animals: a review, classification, and examination of the implications for the reproductive strategies of females. *Ethol. Sociobiol., 1*:13–40.

Hunt, G.L., and M.W., Hunt. 1976. Gull chick survival: The significance of growth rates, timing of breeding and territory size. *Ecol., 57*:62–75.

Hussell, D.J.T., 1972. Factors affecting clutch size in arctic passerines. *Ecol. Monogr., 42*:317–364.

Immelmann, K., 1971. Ecological aspects of periodic reproduction in birds. *In* D.S. Farner (Ed.), *Avian Biology*. Vol. 1, pp. 342–389. Academic Press, New York.

Istock, C.A., 1967. The evolution of complex life cycle phenomena: an ecological perspective. *Evol., 21*:592–605.

Jones, P.J., and P. Ward, 1976. The level of reserve protein as the proximate factor controlling the timing of breeding and clutch-size in red-billed quelea *Quelea quelea. Ibis, 118*:547–574.

King, J.R., 1973. Energetics in reproduction of birds. *In* D.S. Farner (Ed.), *Breeding Biology of Birds*, pp. 78–107. National Academy Press, Washington, D.C.

Klomp, H., 1970. The determination of clutch-size in birds. *Ardea, 58*:1–124.

Krapu, G.L., 1981. The role of nutrient reserves in mallard reproduction. *Auk, 98*:29–38.

Lack, D., 1946. Clutch size and brood size in the robin. *Brit. Birds, 39*:98–109, 130–135.

Lack, D., 1948. Further notes on clutch size and brood size in the robin. *Brit. Birds, 41*:98–104, 130–137.

Lack, D., 1954. *The Natural Regulation of Animal Numbers*. Clarendon, Oxford.

Lack, D., 1968. *Ecological Adaptations for Breeding in Birds*. Methuen, London.

Lack, D., and E. Lack, 1951. The breeding biology of the swift *Apus apus. Ibis, 93*:501–546.

Larson, S., 1960. On the influence of the arctic fox *Alopex lagopus* on the distribution of Arctic birds. *Oikos, 11*:276–305.

Lewontin, R.C., 1965. Selection for colonizing ability. *In* H.G. Baker and G.L. Stebbins (Eds.), *The Genetics of Colonizing Species*, pp. 77–94. Academic Press, New York.

Lill, A., 1975. The evolution of clutch size and male "chauvinism" in the white-bearded manakin. *Living Bird, 13*:211–231.

Loman, J., 1980. Brood size optimization and adaptation among hooded crows *Corvus corone. Ibis, 122*:494–500.

Low, B.S., 1976. The evolution of amphibian life histories in the desert. *In* D.W. Goodall (Ed.), *Evolution of Desert Biota,* pp. 149–195. University of Texas Press, Austin.

MacNally, R.C., 1981. On the reproductive energetics of chorusing males: energy depletion profiles, restoration and growth in two sympatric species of *Ranidella* (Anura). *Oecologia, 51*:181–188.

Maiorana, V.C., 1976. Size and environmental predictability for salamanders. *Evol., 30*:599–613.

Mann, R.H.K., and C.A. Mills, 1979. Demographic aspects of fish fecundity. *Symp. Zool. Soc. Lond., 44*:161–177.

Mattingly, D.K., and P.A. McClure, 1982. Energetics of reproduction in large-littered cotton rats (*Sigmodon hispidus*). *Ecol., 63*:183–195.

Medawar, P.B., 1957. *The Uniqueness of the Individual.* Methuen, London.

Michod, R.E., 1979. Evolution of life histories in response to age-specific mortality factors. *Am. Nat., 113*:531–550.

Millar, J.S., 1977. Adaptive features of mammalian reproduction. *Evol., 31*:370–386.

Miller, P.J., 1979. Adaptiveness and implications of small size in teleosts. *Symp. Zool. Soc. Lond., 44*:263–306.

Morton, E.S., 1971. Nest predation affecting the breeding season of the clay-colored robin. *Science, 171*:920–921.

Newton, I., 1977. Timing and success of breeding in tundra-nesting geese. *In* B. Stonehouse and C. Perrins (Eds.). *Evolutionary Ecology,* pp. 113–126. University Park Press, Baltimore.

Newton, I., M. Marquiss, and D. Moss, 1981. Age and breeding in sparrowhawks. *J. Anim. Ecol., 50*:839–853.

Norberg, R.A., 1981. Temporary weight decrease in breeding birds may result in more fledge young. *Am. Nat., 118*:838–850.

O'Connor, R.J., 1978. Brood reduction in birds: selection for fratricide, infanticide and suicide? *Anim. Behav., 26*:79–96.

O'Connor, R.J., 1979. Egg weights and brood reduction in the European swift (*Apus apus*). *Condor, 81*:133–145.

Packer, C., and A.E. Pusey, 1983. Adaptations of female lions to infanticide by incoming males. *Am. Nat., 121*:91–113.

Packer, C., and A.E. Pusey, in press. Infanticide in carnivores. *In* G. Hausfater and S. Blaffer Hrdy (Eds.), *Infanticide in Animals and Man.* Aldine, Hawthorne, New York, NY.

Parsons, J., 1975. Seasonal variation in the breeding success of the herring gull: an experimental approach to pre-fledgling success. *J. Anim. Ecol., 44*:553–573.

Perrins, C.M., 1970. The timing of birds' breeding seasons. *Ibis, 112*:242–255.

Perrins, C.M., 1977. The role of predation in the evolution of clutch size. *In* B. Stonehouse and C. Perrins (Eds.), *Evolutionary Ecology,* pp. 181–191. University Park Press, Baltimore.

Perrins, C.M., 1980. Survival of young great tits, *Parus major. Proc. Intern. Ornith. Cong., 17*:159–174.

Perrins, C.M., and D. Moss, 1974. Survival of young great tits in relation to age of female parent. *Ibis, 116*:220–224.

Perrins, C.M., and D. Moss, 1975. Reproductive rates in the great tit. *J. Anim. Ecol., 44*:695–706.

Pianka, E.R., 1976. Natural selection of optimal reproductive tactics. *Am. Zool., 16*:775–784.

Pianka, E.R., and W.S. Parker, 1975. Age-specific reproductive tactics. *Am. Nat., 109*:453–464.

Pugesek, B.H., 1980. Increased reproductive effort with age in the California gull (*Larus californicus*). *Science, 212*:822–823.

Reiter, J., K.J. Panken, and B.J. LeBoeuf, 1981. Female competition and reproductive success in northern elephant seals. *Anim. Behav., 29*:670–687.

Renfree, M.B., and J.H. Calaby, 1981. Background to delayed implantation and embryonic diapause. *J. Reprod. Fertil., 29*:1–9.

Ricklefs, R.E., 1973. Fecundity, mortality, and avian demography. *In* D.S. Farner (Ed.),

Breeding Biology of Birds, pp. 366–435. National Academy of Sciences, Washington, D.C.

Ricklefs, R.E., 1977. On the evolution of reproductive strategies in birds: reproductive effort. *Am. Nat.*, *111*:453–478.

Ricklefs, R.E., 1980. Geographic variation in clutch size among passerine birds: Ashmole's hypothesis. *Auk*, *97*:38–49.

Robbins, C.T., and B.L. Robbins. Fetal and neonatal growth patterns and maternal reproductive effort in ungulates and subungulates. *Am. Nat.*, *114*:101–116.

Rose, F.L., and D. Armentrout, 1976. Adaptive strategies of *Ambystoma tigrinum* Green inhabiting the Llano Estacado of West Texas. *J. Anim. Ecol.*, *45*:713–739.

Royama, T., 1969. A model for the global variation in clutch size. *Oikos*, *20*:562–567.

Rubenstein, D.I., 1981. Individual variation and competition in the Everglades pygmy sunfish. *J. Anim. Ecol.*, *50*:337–350.

Sadleir, R.M.F.S., 1969. *The Ecology of Reproduction in Wild and Domestic Mammals.* Methuen, London.

Safriel, U.N., 1975. On the significance of clutch size in nidifugous birds. *Ecol.*, *56*:703–708.

Salthe, S.N., and J.S. Mecham, 1974. Reproductive and courtship patterns. *In* B. Lofts (Ed.), *Physiology of the Amphibia*, pp. 309–521. Academic Press, New York.

Schaffer, W.M., 1974. Selection for optimal life histories: the effects of age structure. *Ecol.*, *55*:291–303.

Schaffer, W.M., 1979. The theory of life-history evolution and its application to Atlantic salmon. *Symp. Zool. Soc. Lond.*, *44*:307–326.

Schaffer, W.M., and P.F. Elson, 1975. The adaptive significance of variations of life history among local populations of Atlantic salmon in North America. *Ecol.*, *56*:577–590.

Schaffer, W.M., and M.L. Rosenzweig, 1977. Selection for optimal life histories. II. Multiple equilibria and the evolution of alternate reproductive strategies. *Ecol.*, *58*:60–72.

Schaller, G.B., 1977. *Mountain Monarchs.* University of Chicago Press, Chicago.

Sherman, P.W., 1981. Reproductive competition and infanticide in Belding's ground squirrel and other animals. *In* R.D. Alexander and D.W. Tinkle (Eds.). *Natural Selection and Social Behavior*, pp. 311–331. Chiron Press, New York.

Shine, R., 1980. "Costs" of reproduction in reptiles. *Oecologia*, *46*:92–100.

Skutch, A.F., 1976. *Parent Birds and Their Young.* University of Texas Press, Austin.

Slade, N.A., and R.J. Wassersug, 1975. On the evolution of complex life cycles. *Evol.*, *29*:568–571.

Smith, C.C., and S.D. Fretwell, 1974. The optimal balance between size and number of offspring. *Am. Nat.*, *108*:499–506.

Smith, J.N.M., 1981. Does high fecundity reduce survival in song sparrows? *Evol.*, *35*:1142–1148.

Snow, B.K., 1970. A field study of the bearded bellbird in Trinidad. *Ibis*, *112*:299–329.

Sohn, J.J., 1977. Socially induced inhibition of genetically determined maturation in the platyfish, *Xiphophorus maculatus. Science*, *195*:199–201.

Stearns, S.C., 1976. Life-history tactics: a review of the ideas. *Q. Rev. Biol.*, *51*:3–47.

Stearns, S.C., 1977. The evolution of life history traits: a critique of the theory and a review of the data. *Annu. Rev. Ecol. Syst.*, *8*:145–171.

Stewart, J.R., 1979. The balance between number and size of young in the live bearing lizard *Gerrhonotus coeruleus. Herpetol.*, *35*:342–350.

Tinkle, D.W., 1969. The concept of reproductive effort and its relation to the evolution of life histories of lizards. *Am. Nat.*, *103*:501–516.

Tuomi, J., 1980. Mammalian reproductive strategies: a generalized relation of litter size to body size. *Oecologia*, *45*:39–44.

Van Balen, J.H., 1973. A comparative study of the breeding ecology of the great tit *Parus major* in different habitats. *Ardea*, *61*:1–93.

von Noordwijk, A.J., J.A. von Balen, and W. Scharloo, 1981. Genetic and environmental variation in clutch size of the great tit (*Parus major*). *Neth. J. Zool.*, *31*:342–372.

Wassersug, R.J., 1975. The adaptive significance of the tadpole stage with comments on the maintenance of complex life cycles in anurans. *Am. Zool., 15*:405–417.

Western, D., 1979. Size, life history and ecology in mammals. *Afr. J. Ecol., 17*:185–204.

Western, D., and J. Ssemakalu, 1982. Life history patterns in birds and mammals and their evolutionary interpretation. *Oecologia, 54*:281–290.

Wilbur, H.M., and J.P. Collins, 1973. Ecological aspects of amphibian metamorphosis. *Science, 182*:1305–1314.

Wilbur, H.M., D.W. Tinkle, and J.P. Collins, 1974. Environmental certainty, trophic level and resource availability in life history evolution. *Am. Nat., 108*:805–817.

Wiley, R.H., 1974. Effects of delayed reproduction on survival, fecundity, and the rate of population increase. *Am. Nat., 108*:705–709.

Williams, G.C., 1957. Pleiotropy, natural selection, and the evolution of senescence. *Evol., 11*:398–411.

Williams, G.C., 1966. Natural selection, the cost of reproduction, and a refinement of Lack's principle. *Am. Nat., 100*:687–692.

Wooller, R.D., and J.C. Coulson, 1977. Factors affecting the age of first breeding of the kittiwake *Rissa tridactyla*. *Ibis, 119*:339–349.

Wootton, R.J., 1979. Energy costs of egg production and environmental determinants of fecundity in teleost fishes. *Symp. Zool. Soc. Lond., 44*:133–159.

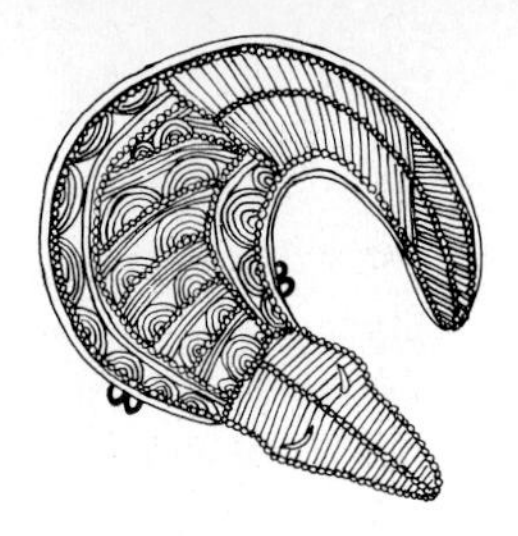

15 Parental Care

INTERNAL PARENTAL CARE

Females of all vertebrates invest each offspring with energy and nutrients before that offspring leaves the mother's body. Different species expend different amounts of nutritive material in producing young, and condition of the young at birth or hatching varies widely. Each offspring enters the world either as an egg or as a more or less free-living creature unencumbered by a shell. In general, offspring invested with much energy are produced at a more advanced stage than those that are not so endowed. Rather little is understood about the adaptive basis of different kinds of maternal care while the young are still internal, but some generalizations are emerging.

Forms of internal parental care range from ovipary (egg-laying) through various grades of ovoviviparity (the egg hatches inside the mother) to true viviparity (live-bearing). Two important changes occur along this gradient. One is that the offspring generally is retained within the mother's body for longer periods. In addition, the primary source of embryo nutrition shifts from the egg yolk to direct maternal supply.

The laying of eggs means that the offspring will receive no further nutrition until it hatches, so the developing embryo must rely entirely on its private store of nourishment until hatching. All the resources in the yolk obviously must be provided before the egg is laid. To do this, a female must either store up resources for egg production or exploit temporary abundances of resources while the eggs are developing. When eggs hatch internally, and in true live-bearers, embryos may then be nourished directly by the mother to some degree. If embryo development is continuous, a steady influx of nutrients is required; if the abundance of resources varies, maternal metabolism may compensate for temporary shortages. At least in mammals, the cost of a developing embryo is small compared to the cost of lactation after birth, however, so the cost of continuous development may not be great.

Because egg-layers usually retain their young in the body for shorter periods than do live-bearers of either type, oviparous females commonly are less encumbered by the weight and bulk of the offspring. Pregnant mammals and snakes are often rather ungainly and slow, indicating that the carrying of young may impose locomotor difficulties and perhaps increase the risk of predation. However, this is certainly not true for some viviparous species (such as marsupials and bears) in which the tiny offspring are born in a very undeveloped state. Some oviparous fishes and amphibians carry their eggs around and guard them for a period of time; presumably the costs of encumbrance are less than those of predation.

At least among snakes (but less so among lizards), the breeding season is short in many areas outside the tropics, and many species of nontropical snakes rear only one brood a year. Single-broodedness might make the extra brood protection provided by viviparity more advantageous, because the option of a second brood is not available.

Viviparous ectotherms can easily carry their unborn young from place to place, perhaps seeking sites with temperatures favorable to development (Tinkle and Gibbons, 1977; Shine and Bull, 1979). Viviparity (and ovoviviparity) occurs in a high proportion of reptiles at high latitudes and high elevations, suggesting that viviparity might be an adaptation to cold climates, in which thermally favorable spots might hasten embryo development. The largest numbers of viviparous (in the broad sense) reptiles occurs at middle latitudes, however, where variable and unpredictable climates (rather than simply cold ones) might favor viviparity by favoring egg-retention until conditions are most propitious for birth and survival of young.

Viviparous young of ectotherms are usually born in a relatively advanced stage of development compared with the young of related species that hatch from eggs. It could be argued, therefore, that viviparity is favored whenever a suitable environment for small and undeveloped young is not available. Climatic factors need not be the only ones involved. Internal care of the young also might be advantageous in lowering certain kinds of predation or when appropriate small food items for very small young are uncommon. However, it is important to remember that existing developmental patterns may make difficult a shift from oviparity to viviparity. For instance, calcium in the eggshell seems to be important to the development of young turtles and crocodilians (Packard et al., 1977). Thus, the necessity for calcium and the mechanism of its retrieval might preclude the achievement of live-bearing in these kinds of reptiles.

Among mammals, the generalization that live-bearers produce better developed young than egg-layers has some exceptions. A number of mammals have very short gestation periods and bear very small young that are in a relatively early stage of development. Although marsupial and placental mammals are all viviparous, the length of the gestation period varies greatly. Young marsupials remain inside the mother for 13 to 35 days, depending on the species. The tiny young (less than a gram in total weight) are born in an undeveloped state: The hind limbs are no more than limb buds. When the young are born, they crawl with their front limbs to the mammary glands and cling there during the next stage of development. The gestation period in placental mammals ranges from about three weeks in shrews and mice to almost two years in elephants. The young are larger, relative to the size of the mother, and more fully developed than marsupial young.

Low (1978) and others have suggested that such developmental differences between marsupials and placentals, far from being historical accidents or phylogenetic limitations, are likely to be adaptive. Early emergence of offspring in marsupials may be adaptive in unpredictable environments in which conditions suitable for completion of parental care may be ephemeral. If maturation of offspring is therefore uncertain and cannot be significantly improved by greater parental investment, it may be advantageous for the parent to unencumber itself by getting rid of the young, thereby increasing its own chances of survival and its chance to reproduce again. Short gestation periods with tiny offspring (relative to the size of the mother) might reduce the dangers of abortion, which increase with increasing offspring size. Many small marsupials are so short-lived, however, that the expectations of future reproduction are minute (Russell, 1982). Also, the ecological conditions that might have favored the production of very diminutive offspring by some placental mammals, such as bears, do not seem to have received much speculation, not to mention close examination. Moreover, eutherians commonly abort fetuses, especially in times of environmental stress, and, thus, may be able to adjust their litter sizes in much the same way as marsupials do (Morton et al., 1982). Gestation periods and offspring growth rates in many eutherians are correlated with basal metabolic rate, and the same may be true for marsupials. Marsupial metabolic rates are generally lower than those of eutherians, and it is possible that the developmental differences between the two groups may be linked to their respective metabolic rates (Morton et al., 1982). Metabolic rates, then, may constrain the evolutionary possibilities available for reproductive patterns, and those rates themselves may be determined both by phylogenetic limitations and selection pressures that are important outside the breeding season as well as within it.

Although rationales for prolonged egg-retention have been suggested, it is more difficult to account for the variation in the source of embryo nutrition. Ovoviviparity sometimes is considered as a phylogenetically limited attempt at true viviparity, a suggestion that implies a selective advantage for viviparity that cannot be realized because suitable genetic variation is lacking. However, although this may be true, the ecological conditions that might favor egg-provisioning versus direct maternal supply to the embryo have not been investigated sufficiently.

Most vertebrates are oviparous: a few sharks and all skates, most bony fishes, most amphibians and reptiles, all birds, and monotreme mammals. Ovoviviparity is characteristic of most sharks and all rays, a number of teleosts (such as some poeciliids, brotulids, and hemirhamphids), and the coelacanth (*Latimeria*) (Wourms, 1981). Some amphibians are ovoviviparous, including many caecilians, a few anurans (*Nectophrynoides*, a species of *Eleutherodactylus*), and a salamander (*Salamandra atra*). A number of reptiles also reproduce by means of internally hatched eggs.

The hatched embryos of ovoviviparous species are nourished by the mother in a variety of ingenious ways (Wourms, 1981). A "uterine milk" is secreted by villi on the oviduct wall of some sharks and absorbed by the yolk sac or the embryo's gill filaments. Butterfly-ray (*Gymnura*) embryos are nourished by extra-long villi that pass through the spiracles and reach the foregut. Growing embryos of sand sharks (*Odontaspis*) feed on unfertilized eggs and small embryos in the oviduct. Similar variety can be found among the amphibians.

True viviparity is found, not only in most mammals, but also in a few sharks (such as the blue shark, *Prionace glauca*) and a number of teleosts (including some poeciliids, anablepids, and embiotocids, for example). The classic case of advanced development at birth is the surf perch (*Cymatogaster aggregatus*), an embiotocid, in which males are sexually mature at birth. (However, this achievement pales in comparison with certain invertebrates, whose young become sexual mature and mate with their siblings before birth!) A placenta connecting mother and offspring is developed to varying degrees in viviparous reptiles; a well-developed placental link occurs in some skinks (*Chalcides, Lygosoma*).

As a broad generalization, viviparity (including ovoviviparity) is associated with internal fertilization, in which sperm from the male is placed in the female's reproductive tract. Oviparity in fishes and amphibians is commonly associated with external fertilization, in which males and females shed their gametes and the zygote is formed outside the parent's body. A number of oviparous species, however, are known to have internal fertilization, including *Ascaphus truei*, in which the male's tail serves as an intromittent organ, and *Eleutherodactylus coqui* (Townsend et al., 1981). Furthermore, a different sort of internal fertilization occurs among salamanders, in which the male deposits a spermatophore to be picked up by a female from the bottom of a pond, and in some seahorses, in which the female deposits her eggs in the male's brood pouch, where he fertilizes them. Certain cichlid fishes in which the mother broods the young in her mouth exhibit unusual devices that may encourage a facsimile of internal fertilization. The males of these species develop special morphological devices, often brightly colored, that are exhibited near the genital opening during courtship. The female is attracted to this device and tries to pick it up, but gets her own eggs and the male's sperm in the process. Although these various appendages and tassels have been interpreted as devices to induce the female to retrieve her eggs (Keenleyside, 1979), it seems likely that they also foster success of the displaying male's sperm. Finally, all birds are oviparous with internal fertilization. All in all, viviparity presumably requires internal fertilization, but oviparous animals often fertilize their eggs internally also. Thus, more factors than just the condition of the young must determine the type of fertilization mechanism used. One obvious trend is that external fertilization is far more likely to be successful in animals that mate in water, where the gametes have no risk of desiccation and sperm can swim through the medium to reach the eggs.

Conditions of young mammals at birth or of birds at hatching can be classified further into two broad categories. Altricial young are very undeveloped: They are small, usually naked, and neither eyes nor limbs function very well. At the other end of the spectrum, precocial young are born active, open-eyed, and furred or feathered. Case (1978a) pointed out that condition of the young at birth is not associated with body size in mammals; rather, there seems to be a good correlation with foraging style and risks of predation. Altricial young are particularly advantageous in active foragers, which spend large amounts of time in the search for food or which must vigorously pursue and subdue their prey. The young of such species would probably require a long juvenile period anyway, just to learn how to hunt successfully. Mammals that consume food that is readily found and easily handled can carry the additional weight with less handicap. Furthermore, their young can quickly acquire adult food habits and do not require a long dependent period for learning techniques of the hunt. Altricial young, therefore, may be expected more commonly among carnivores than among grazers and browsers.

Species whose young are well protected from predation by a safe den are likely to be more altricial than those whose young have no safe retreat. Thus, beavers (*Castor canadensis*) produce highly altricial young, which are protected in the lodge or burrow, whereas the porcupine (*Erethizon dorsatum*) bears more advanced—and spiny—youngsters that can protect themselves to some degree. Similarly, spiny insectivores (hedgehogs, tenrecs) are more precocial than those without spines. Ungulates build no nests and their young are precocial. In addition, if the risk of predation on adults is high and extended pregnancy reduces the chance of escape, selection may favor shorter gestation and highly altricial young. Similar considerations can be applied to precocial and altricial birds. However, none of these factors seem to explain very well why the young of hares (*Lepus*) are precocial but those of cottontail rabbits (*Sylvitagus*) are altricial.

The condition of young that hatch from eggs is related to the size of the egg, particularly the yolk (Kaplan, 1980; O'Connor, 1980; and many others); for birds, it is egg size relative to female body size that seems to be critical (Lack, 1968). Relatively large eggs have longer developmental times and the young are quite advanced at hatching. Large and advanced young often survive better than small ones. Furthermore, large eggs and advanced young at hatching would be adaptive if food for very small offspring is hard to obtain, if predation on small offspring is unusually severe, and if carrying a load of large eggs is not too difficult for the female. (Recall that some of these same factors were mentioned with respect to viviparity in other classes of vertebrates, but, for whatever reason, birds have not evolved live-bearing habits.) Large eggs might also be adaptive if the egg itself is subject to metabolic stresses that can be surmounted by the use of energy reserves. Thus, egg size and endowment of the embryo are probably directly related to offspring fitness in many cases. Nevertheless, there are obviously limits to egg sizes. To some degree, there is a trade-off between egg size and egg number, and parental fitness is often enhanced not by making one enormous egg but by making several eggs of lesser size (Chapter 14). The nature of the nest site and body size (and size of the pelvic opening through which the egg passes) of the female provide additional constraints; the selection pressures on these features, at some point, may interact with those on egg size. When the time for provisioning an egg is short, it may be advantageous to lay smaller eggs, despite the smaller hatchling, rather than waiting to accumulate enough resources for a larger egg (Birkhead and Nettleship, 1982). Variation in egg size often occurs within and between clutches of the same female as well as among females, and the patterns of this variation do not yet have a general explanation.

EXTERNAL PARENTAL CARE

After the eggs or young have left the female's body, the amount of external parental care provided varies greatly among the vertebrates. Because the range of variation is so different in different classes, a taxonomic survey of this variability is one reasonable way to organize the welter of detail. No attempt at a catalog of all variants is either possible or desirable here; the aim is to describe in brief the kinds of parental behavior exhibited and to outline its variety.

Parental behavior can be effective in reducing mortality from certain factors and is an evolutionary means of increasing offspring survival. As a general rule, increased offspring care necessitates a reduction in offspring number, so we often find a broad inverse correlation of litter size and parental investment per young.

Fishes

Most marine and some freshwater species of fishes simply abandon their offspring to the elements at spawning; more than half of all fish families have no species that exhibit any parental care. Some strew their eggs into open water and others place them in various hiding places, but no further attention is paid by the parents to the young. In general, these species breed in well-oxygenated water; alternatively, the young have very well-developed respiratory organs. They often produce huge batches of eggs (*Mola mola*, the ocean sunfish, is reported to produce an estimated 3 million eggs). In contrast, some species deposit their eggs in the gill chambers of mussels and crabs. In this group, egg number is relatively low; examples include the cyprinids *Rhodeus* and *Acheilognathus* and the liparid *Careproc-teus*.

The eggs or young of many other species are guarded by one or both parents. More than 80 families of bony fishes contain species with parental care; in most of these, care is provided only by males. Adults of many of these species not only guard the offspring but commonly also fan the eggs. Fanning increases oxygen supply, decreases the settling of silt, and may help prevent fungal infection. Male and female of the neotropical characin, *Copeina arnoldi*, leap in unison from the water and cling upside down on an overhanging leaf or stone. After egg deposition, the male splashes water on the eggs several times an hour for three days until the larvae emerge and drop into the water.

Construction of a nest represents another form of parental investment, and fishes make many kinds of nests (Fig. 15–1). Sunfishes and many cichlids scoop out nests in the bottom substrate. Other cichlids, such as the angelfishes (*Pterophyllum*) in Amazonia, clean off "nursery areas" on leaf surfaces; they attach their eggs to one and transfer the newly hatched larvae to another. If a young falls off, the attending parent picks it up by mouth and returns it to the leaf nursery. *Nannacara taenia*, another cichlid, performs similarly and, in addition, has a danger display that sends the young into hiding. *Aequidens coeruleo-punctatus*, a Panamanian stream fish, lays its eggs on sodden leaves; the female reportedly pulls the leaf under the bank or into shallow water when danger threatens. Nests of plant material are constructed by the mormyrid *Gymnarchus niloticus* and by sticklebacks (*Gasterosteus*). Bubble or froth nests are built by certain inhabit-ants of still and stagnant tropical waters such as the Siamese fighting fish (*Betta splendens*), some gouramis (*Colisia, Trichopodus*), and a catfish (*Callich-*

thyes callichthyes). Members of several genera (*Ancistrus, Symbranchus, Fluta,* and *Protopterus,* for example) build deep burrows in which the eggs are guarded.

A complex form of parental behavior is the brooding of eggs or young in contact with a parent's body (Fig. 15–1). *Kurtus* males carry bunches of eggs on special hooks on their foreheads. Other species incubate the eggs in special areas of the skin (females of certain South American catfishes,

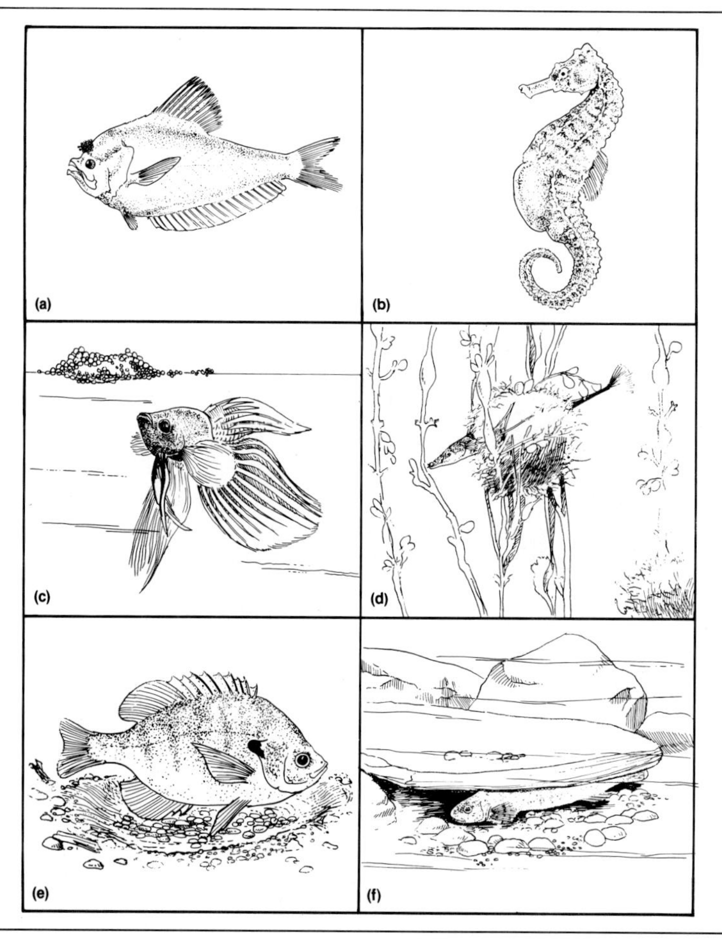

Figure 15–1 Some modes of egg-tending in fishes. (*a*) The egg mass on a forehead hook of male *Kurtus*. (*b*) Brood pouch of a male sea horse. (*c*) Bubble nest of the Siamese fighting fish, *Betta splendens*. (*d*) Grass nest of a 15-spine stickleback (*Spinachia spinachia*). (*e*) Nest scoop of a sunfish on the bottom of a pond. (*f*) Eggs placed on the undersides of rocks by the johnny darter, *Etheostoma nigrum*.

including *Bunocephalus* and *Aspredo*, and males of certain pipefishes including *Phyllopteryx* and *Nerophis*). Special brood pouches form on the lower lip of some loricariid catfishes, on the ventral fins of female *Solenostomus*, and on the venter of male sea horses and pipefishes (*Syngnathus, Siphistoma*). Still others brood the eggs or young in the mouth (marine cardinalfishes *Apogon* and a number of cichlids) or in the gill chambers (North American cavefishes of the genera *Amblyopsis* and *Chologaster*). Mouth-brooding adults are usually able to distinguish between their sometimes errant young and prey.

No fishes are known to bring food to their young, but adults of several Central American species of *Cichlasoma* (Cichlidae) reportedly turn over leaves for their foraging offspring. The young of some cichlids also graze on mucus on their parent's skin.

The high frequency of paternal care in fishes, as compared with other kinds of vertebrates, has generated much discussion and the end of the story has not yet been told. At this point, we can suggest that several factors may have contributed to the evolution of parental care by males alone (Gross and Shine, 1981; Kennleyside, 1979, 1981; Ridley, 1978; Perrone and Zaret, 1979; Maynard Smith, 1977; Blumer, 1979; Werren et al., 1980). Several separate, but related questions can be asked. First, why is there any parental care at all? What factors make it both possible and advantageous? Because fishes so seldom feed their young, the most likely factors favoring care may be the risk of predation and habitat conditions that limit aeration or foster pathogens of the eggs. Use of a nest for protection clearly depends on a safe site for the nest itself, which is no doubt related to the absence of nests in pelagic oceanic fishes. Although aeration and guarding are the commonest functions of piscine parental care, there have been no studies showing that predation of eggs and young is a particularly great risk for the species that guard, compared to those that don't.

Next, one might ask what determines whether one or both parents participate in parental care. There must be some conditions under which offspring success is enhanced by the attentions of both parents. Keenleyside (1979) suggested that perhaps the uniparental care that predominates among temperate-zone fishes that exhibit parental care may be sufficient to guard against the relatively few kinds of predators in that area. In contrast, biparental care is known particularly in tropical species (the cichlids), many of which inhabit waters where there are many kinds of potential predators. The actual levels of predation remain to be documented, however. One possible difficulty with this idea is that a number of other tropical species have uniparental care. Among cichlids, these are usually species in which the female broods the offspring in her mouth rather than species that use some sort of nest site for at least part of the period of parental care. Thus, perhaps one could argue that strict mouth-brooders are somehow exempt from the impact of all those potential predators. Brooding females may, for instance, move far from the spawning site to special nursery areas where perhaps the risk of predation is reduced. However, it is interesting that the mouth-brooders have specialized predators all their own. For example, certain species of *Haplochromis* have means of forcing congeneric mouth-brooders to disgorge their offspring in order to be consumed by the attacker (Keenleyside, 1979).

A factor facilitating biparental care is the lack of opportunity for finding another mate. If desertion of one mate is accompanied by a good chance of breeding soon with another individual, and if the deserted mate can be successful in rearing the brood alone, biparental care may be less advantageous than multiple matings and uniparental care. However, a lack of further mating opportunities is not sufficient to select for biparental care; it merely makes it easier.

Finally, what determines which sex is the parental one in uniparental species? If a male can be certain that the young are his, this could predispose him to care for them. However, patterns of paternal care do not parallel estimates of the certainty of paternity, because paternal care is recorded for species in which satellite or cuckolding males are common as well as for those in which a male has high confidence in his paternity. Sometimes males tend young that are not their own, although such behavior may be incidental to other events (Constantz, 1979). Furthermore, if the certainty of paternity does not improve predictably when a male deserts one female for another, he would do as well to stay with the first female. Whether he stays or deserts will then be determined largely by the opportunities for other matings; whether he helps or not will be determined chiefly by the utility of doing so. In either case, the certainty of paternity itself does not explain the behavior. Paternal confidence can be no more than a predisposing factor facilitating paternal care.

The ability of many kinds of male fishes to defend a nest site without sacrificing additional matings may contribute to the evolution of paternal care. Because care commonly centers on protection from predators and not on feeding the young, a male may be able to guard the eggs and young of several females. If sites for depositing eggs are limited, males might defend them territorially. On the other hand, if selection favors parental care of some sort and site-attachment is a suitable way to provide care, then the male may become territorial (reversing the order of causation just described). Males, rather than females, are more commonly territory-defenders, probably because their other reproductive costs are often lower (Chapter 11). The male's site-attachment enables him to care for a brood at the same time that he advertises for more females. Again, the defense of oviposition sites does not necessitate male parental care; it merely facilitates it.

Solitary maternal care of external young is no easier to explain. Prolonged maternal care is recorded for the Antarctic plunder fish (*Harpagifer bispinis*), although males will substitute for an absent female (Daniels, 1979), and attentive mothers are also known in other species. Maternal care in cichlids is found principally in the mouth-brooders (*Sarotherodon*), perhaps those species in which the males display on leks and in which the nursery areas are at some distance from the spawning site. Thus, presumably the advantages of multiple matings by the male and of paternal care cannot be simultaneously realized. (One might inquire why the males don't display in the nurseries.) Furthermore, such arguments can only suggest why males of these species *don't* care for the young, not why females *do*. If females are larger than males, they might be better equipped than males for offspring defense, but such size differentials certainly do not characterize all guardian mothers. Are some females merely stuck with the job because some parental care is essential and males obtain greater reproductive success in other ways?

Sometimes even two parents seem to be insufficient for adequate protection of a brood. Communal care of a group of offspring by several sets of parents occurs in several cichlids and is probably adaptive in improving protection against predation. Furthermore, parents adopt strays and orphans and even kidnap young from other broods. This unusual behavior may reduce the risk of predation to the parents' own young by increasing the size of the school (McKaye and McKaye, 1977).

Amphibians

Parental care in salamanders is little developed and typically involves no more than guarding a batch of eggs. Female salamanders of most plethodontids, of *Ambystoma opacum*, and of the aquatic *Necturus* and *Amphiuma* are usually engaged in guarding the eggs. The highly aquatic *Cryptobranchus* and a related species, *Hynobius nebulosus*, have external fertilization and paternal care of the eggs (Salthe and Mecham, 1974). Adults of *Hemidactylium* and *Aneides* (both North American plethodontid genera) may briefly tend the young after hatching; their presence may help control water balance and fungal infections of the offspring.

Parental care of some form is widespread among anurans, occurring in about 14 families (but in perhaps only 10 percent of all the species) (Wells, 1981). Egg attendance is the most common form of care, but tadpole care occurs in several families, especially dendrobatids.

Anurans demonstrate rather complicated parental behavior, particularly associated with transporting young (Fig. 15–2). Eggs of the European midwife toad (*Alytes obstetricans*) are carried wrapped about the rear end of the male. Adults of several mainly tropical genera (*Pipa, Dendrobates, Gastrotheca*) tote the eggs and young about in special pockets on their backs. An Australian leptodactylid frog, *Rheobatrachus silus*, has a unique form of parental care—females apparently brood the young in their stomachs and

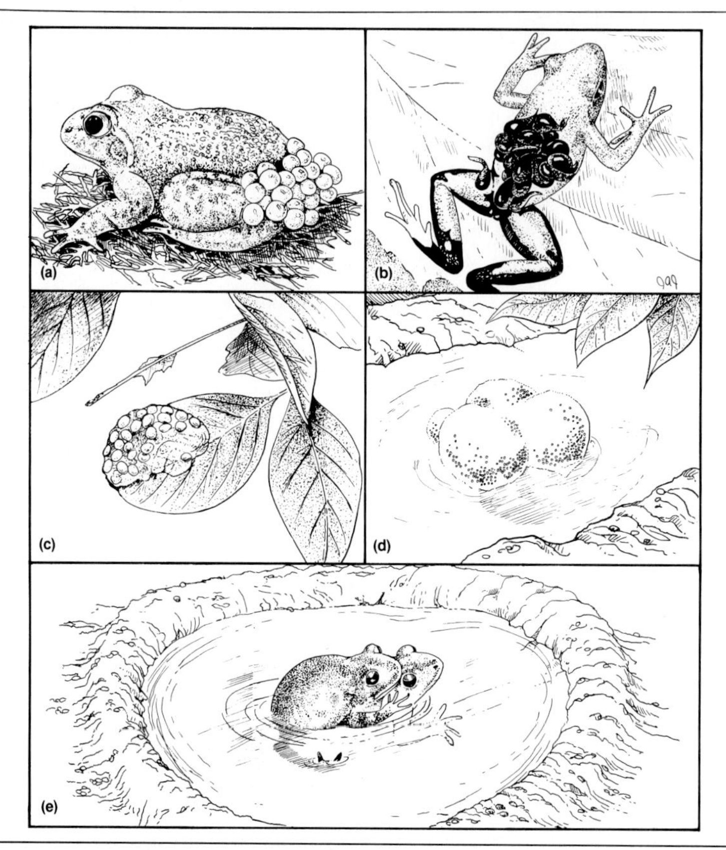

Figure 15–2 Some modes of egg-tending in anuran amphibians. (*a*) The European midwife toad (*Alytes obstetricans*). (*b*) The piggyback young of *Phyllobates bicolor.* (*c*) Egg mass of *Agalychnis callidryas* on a leaf. (*d*) A froth nest of *Engystomops pustulosus.* (*e*) The breeding pool of *Hyla faber.*

somehow avoid digesting them. The so-called mouth-breeding frog, *Rhinoderma darwinii,* lives in southern South America. Several males of this species guard a clutch of eggs from a single female. When the eggs are about to hatch, each male picks up 5 to 15 of them with its tongue and inserts them into a huge vocal pouch that extends all the way down the back and up the belly. The young hatch and go through metamorphosis there.

Other anurans build nests to house the eggs. Froth or bubble nests on the surface of water, on the ground, or on vegetation are characteristic of a number of species, including some species of *Leptodactylus* and *Rhacophorus.* A burrow provides a nest of sorts for the eggs of numerous genera. Male *Hyla faber* build small pools with clay walls in which the eggs are placed.

Paternal and maternal care both occur in amphibians with about equal frequency; biparental care is relatively rare (Gross and Shine, 1981). Little seems to be understood about the evolution of parental care in amphibians, and few patterns are even evident. Territorial behavior is known for only some of the species with paternal care (Gross and Shine, 1981). Male *Hynobius nebulosus* salamanders guard the eggs of one female at a time (Ridley, 1978) and, in so doing, relinquish the opportunity to mate concurrently with several females. *Hyla rosenbergi* males also cannot mate multiply, because a second female in the mating pool disturbs the eggs, which sink and die (Wells, 1981). Thus, many of the factors suggested to be important for parental-care patterns among fishes seem not to pertain to amphibians, and much more work is needed first to ascertain the distribution of various forms of parental care and then to discern what processes were likely to be important in producing the observed pattern.

Reptiles

Nests of some sort are built by turtles, crocodilians, many lizards, and, reportedly, the king cobra (*Ophiohagus hannah*). Among the crocodilians, gharials (*Gavialis*) and some crocodiles make hole nests in riverbanks or on lake or ocean shores. Other crocodiles, alligators, and caimans make mound nests by piling up debris and placing the eggs in a depression in the mound. Mound-nesting may be an adaptation to cooler, shady habitats; the heat produced by decaying vegetation in the mound could be advantageous in speeding development. Mounds in marshes may provide dry nest sites above the waterline. Mound-nest builders usually guard the nest and may open the mound when the young hatch. Choice of nest site may also affect water balance and embryo growth, at least in some turtles (Packard et al., 1981).

Parental care is uncommon in lizards; females of a few genera, including *Eumeces* skinks, *Ophisaurus*, and *Gerrhonotus* in North America, guard the batch of eggs. Females of these species were thought to warm their bodies periodically in sunny spots and return to the nest where this heat is transmitted to eggs, but the efficacy of this behavior is not well documented. Brooding behavior in snakes does not appear to be common, but females of *Python molurus* incubate their eggs by producing heat endogenously as they coil around them. Parental care after hatching or birth seems to be almost unknown in most reptiles; in *Eumeces obsoletus*, the great plains skink, however, the female tends and grooms the young after hatching. The female Nile crocodile both guards the nest and opens it when the young emit their hatching calls. She often carries the brood from the nest in the riverbank to the river by picking them up gently in her formidable jaws and tucking them into a gular pouch (Pooley, 1976). Similar behavior is known in at least three other crocodilians.

Birds

Most birds build a nest of some sort. Exceptions include the brood parasites, the emperor and king penguins (*Aptenodytes forsteri* and *A. patagonicus*), which incubate their single egg on top of their webbed feet, and the fairy tern (*Gygis alba*), which usually places its egg on a bare tree branch. Many seabirds, shorebirds, vultures, and goatsuckers simply place their eggs on the ground with little or no preparation of the site. Constructed nests vary from flimsy platforms of a few sticks to elaborate hanging baskets and from simple domed structures to enormous multifamily apartment houses (Fig. 15–3).

Once the eggs are laid, all birds, except brood parasites and the megapodes (Megapodiidae), incubate them with body heat. Megapodes are large ground birds of the Australian region that bury their eggs in mounds of decaying vegetation and warm earth. Heat from this nest material incubates the eggs. Parents of some megapodes leave their clutch entirely; those of other species, such as the mallee fowl (*Leipoa ocellata*), attend the nest and regulate its interior temperature by adding or removing nest material (Fig. 15–4).

Incubation periods vary greatly in length, but the factors that select for different incubation times are complex and not well understood. There is evidence that incubation periods are shorter in birds whose nests are open or accessible to predators, suggesting that predation is one factor selecting for shorter incubation times.

When avian eggs have hatched, nestlings may leave the nest almost immediately or remain for months, depending on the species. Precocial young hatch with downy covering and well-developed legs and eyes and are able to move about freely soon after hatching (Fig. 15–5). Some leave the nest within a few hours. Young may be completely independent of their parents, as in megapodes, or dependent on parents for food, as in grebes (Podicipedidae) and rails (Rallidae); or they may obtain their own food while accompanied by a parent, as in most ducks (Anatidae) and grouse (Tetraonidae). Other precocial young stay in the nest for several days at least and are fed there by the parents; gulls and terns (Laridae) are good examples.

In contrast, the young of many species are altricial at hatching. Development of the eyes and feathers at hatching varies among species, but, in all cases, altricial nestlings are fed and sheltered from severe weather by at least one parent. Nestling periods of nest-dwelling young birds vary from just over a week, as in the field sparrow, to over five months in some albatrosses (Diomedeidae). Parental care continues during the period after the young have left the nest, sometimes only for a few weeks (many passerines). Sometimes, however, the parents are in attendance for several months (royal tern, *Thalasseus maximus*, and some other seabirds), while the young are acquiring sophisticated hunting techniques (Davies, 1976).

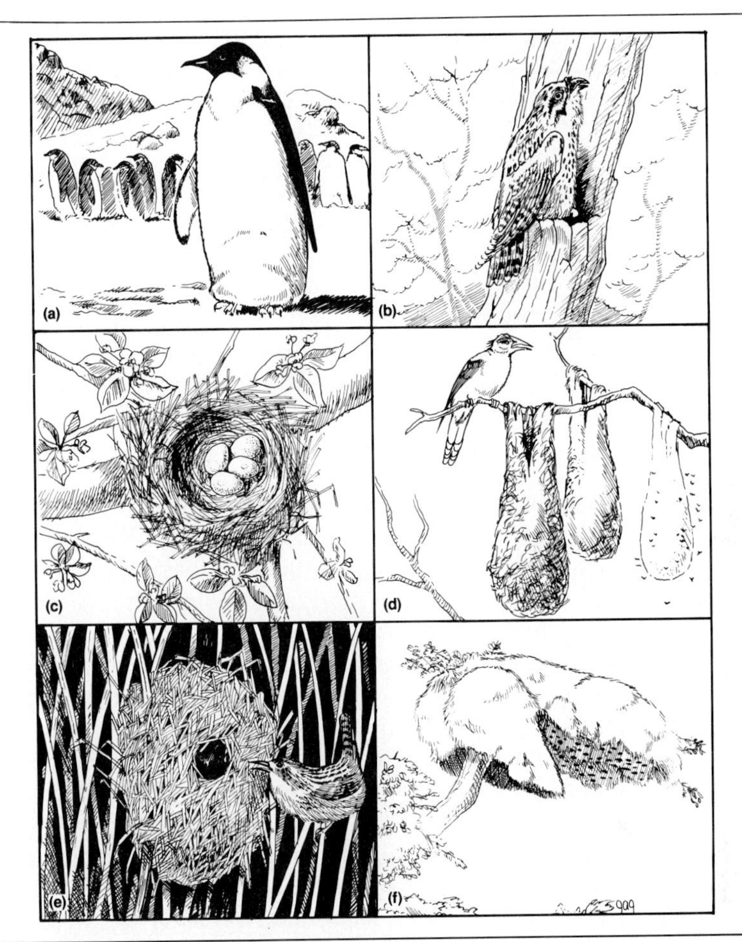

Figure 15–3 Some modes of egg-tending in birds. (*a*) A male emperor penguin with its egg between feet and abdomen—the father can walk without dropping the egg. (*b*) The placement of the egg of the common potoo (*Nyctibius griseus*) of the neotropics and the position of the incubating bird. This unusual pose is actually cryptic—the bird resembles a broken tree branch. (*c*) A simple open nest of an American robin, *Turdus migratorius*. (*d*) The deep pendant nest of the chestnut-headed oropendula (*Zarhynchus wagleri*). (*e*) A covered nest of a long-billed marsh wren (*Cistothorus palustris*). (*f*) The complex multiple dwelling of the sociable weaver (*Philetairus socius*).

Most birds feed their young on food with a high protein content (often other vertebrates, insects, or sometimes seeds). Birds that eat large quantities of seeds during at least part of the year usually feed their young a high proportion of insects. The cardueline finches (American goldfinches, European goldfinches and greenfinches, *Carduelis*) are unusual in that their

Figure 15–4 Mallee fowl and nest. The parents regulate nest temperature by adding or removing material. (Photo by J. Brownlie, courtesy of Bruce Coleman, Inc.)

young receive considerable quantities of regurgitated seeds as well. Fruit-eating birds such as the Central American quetzal (*Pharomachrus mocinno*) reportedly feed their very young nestlings largely on insects, but other frugivores, including manakins (Pipridae), commonly deliver many fruits to their offspring in the nest. Fruits usually contain less protein than meat or seeds. (Many seeds contain as much as 35 percent protein, for instance, but 14 percent is a high protein value for a fruit.) Low protein concentrations in fruits may contribute to slow growth in species such as the South American bearded bellbird (*Procnias averano*) and the oilbird (*Steatornis caripensis*), whose nestlings are apparently dependent totally on fruit, although fruit selected by these species is often somewhat higher in protein and oils than other fruit. In partial frugivores, it is possible that slow growth rates evolved as a means of reducing parental activity near the nest, thus reducing the risk of a predator noticing the nest. These slow

Figure 15–5 An altricial hatchling of the bohemian waxwing (*Bombycilla garrulus*) and a precocial hatchling of Gambel's quail (*Lophortyx pictus*).

rates then perhaps permitted use of the low-protein diet, which provides enough protein to account for observed growth rates.

Adults of several species feed their nestlings at least partially on material produced on their own digestive tract. Pigeons and doves (Columbidae) regurgitate "pigeon's milk", which is actually quite similar to rabbit's milk. Flamingos (Phoenicopteridae), emperor penguins, and petrels (in the Procellariiformes) all regurgitate oily substances to their young. Material fed to young emperor penguins is also unusually high in protein: The dry weight composition includes 59 percent protein and 29 percent fats. We might expect that these dietary adaptations would be most likely in species that would otherwise have difficulty obtaining suitable nestling food. In this connection, we note that petrels are small seabirds that forage widely over the ocean and may have difficulty carrying bulky meals to their land-bound offspring. Pigeons and doves commonly feed on hard seeds and fruits that their very young nestlings cannot handle; older nestlings are fed on a mixture of "milk" and regurgitated, softened seeds. Similar reasoning may apply to other birds that supplement the nestling diet, but the suggestion remains hypothetical. Moreover, not all widely foraging seabirds supplement the nestling diet as do petrels, so other, unknown factors are no doubt involved.

Some birds are able to carry their young from place to place. Grebes often transport their young on the parents' backs. The woodcock (*Philohela minor*) can fly with its young carried between its feet and body. Males of the neotropical finfoot (*Heliornis fulica*) carry the altricial young in special pockets under the wing; the young can travel in this way while the father flies or swims. A number of other species sometimes carry their offspring in their beaks or feet.

Both parents of most bird species have some role in parental care: Some investigators have thought that, at least in some cases, biparental care may be essential to reproductive success (Chapter 13). Sometimes both sexes perform the same functions; sometimes one sex broods while the other feeds the young. The emperor penguin male is the sole incubator of the female's single egg. He holds the egg between feet and abdomen for two entire winter months without a break until the female returns from the sea to care for the chick. Incubation in most of the Old World hornbills (Bucerotidae) is performed by the female, who has sealed herself into her nesting cavity by plastering the entrance with mud. The male feeds her through a small slit in the plaster. This division of labor persists into the nestling period. However, males do no parental care at all in promiscuous species such as most hummingbirds and grouse.

Perhaps the most unusual situation is that in which the male has become the parental sex and the female's parental role has been reduced to egg-laying. It is not unusual for the female of some species, including the house wren (*Troglodytes aedon*), to leave care of the first brood of fledglings to the male while she begins a second clutch. In members of 13 families, however, solitary male care is recorded (Ridley, 1978). Shorebirds (Charadriiformes) provide most of the examples, but tinamous, button-quail, kiwis, cassowaries, emus, and several other birds exhibit paternal care. In some cases, a male cares for the eggs of several females (e.g., some tinamous, emus); in others, he tends only the young of a single female. The females, on the other hand, may produce eggs for only one male (monogamy), for several males in succession (serial monogamy or even female promiscuity), or for several males simultaneously (polyandry). Birds that are serially monogamous (or "sequentially polyandrous") with well-developed paternal care sometimes employ biparental care for the last clutch of the season or in especially inclement weather (as in the spotted sandpiper, *Actitis macularia*). Other shorebirds produce two clutches, one tended by the male, the other by the female; sometimes there are monogamous pairs, but sometimes the female changes mates for the subsequent clutch (examples include the mountain plover, *Charadrius montanus,* Temminck's stint, *Calidris temminckii,* and probably some other *Calidris* species). It is not always easy to tell if a mate-changing female has, in fact, deserted her first male, because such females sometimes reassociate with their previous mates and may replace lost clutches (Maxson and Oring, 1980). Thus, some of the cases of apparent serial monogamy may really be nonobvious or partial forms of simultaneous polyandry; the spotted sandpiper may be an example.

Paternal care in birds is generally considered to have evolved from the very prevalent condition of biparental care, but the factors favoring emancipation of females are anything but clear. Production of two (or more) clutches in rapid succession may be favored if clutch size is somehow limited but females are able to respond quickly to high food abundance by producing more clutches (Parmalee and Payne, 1973; Graul, 1976; Maxson and Oring, 1980). This is clearly advantageous to females if males are capable of rearing broods alone; it is equally advantageous to the male if he is the father of the subsequent broods. The advantage to the male if he is not the father of subsequent broods is less clear. However, when clutches are frequently destroyed by predation or bad weather, selection might favor females that replace the lost clutches. This ability could be enhanced by the evolution of large female body size and/or of clutch weights that are small in relation to female size (Ross, 1979, and others). Still, obtaining replacement clutches would be advantageous to the male chiefly if his own opportunities of getting another female were small, which could be true if breeding of females were very synchronous.

It seems likely that mate-changing by females originated after paternal care in these birds. Once paternal care occurs in a pair that produces two clutches (each with uniparental care), female territoriality becomes more possible. Once that occurs, behavioral dominance of females might prevent their mates from excluding additional males, whereupon the first male's best option is to return to incubate the eggs (Maxson and Oring, 1980). Mate-changing by the female may occur if she is, in fact, attempting to mate with several males in sequence and, thus, to produce several clutches (Pienkowski and Greenwood, 1979). Such an attempt may often fail if the season is very short or if males are few and far between; therefore, the would-be series of males is stopped at two and a third clutch can only be cared for by the female herself.

Male parental care in species that are single-brooded and monogamous, or in which males tend the eggs of several females is even more mysterious. If the cost to the female of producing a clutch is very high (if egg size is necessarily large, for instance, but clutch size cannot be altered), the depleted female might tend the eggs poorly, making a parental role advantageous to the male. Here one supposes that females are essentially unable to incubate or perhaps even to care for the young later on. This scheme might be applied to kiwis, which have small clutches of very large eggs (relative to female size) and seem to be monogamous (Ridley, 1978). Whether such a speculation could be applied to rheas or to some other birds seems dubious, for female rheas that mate polygynously with one male (who tends their eggs) go on to mate with others, suggesting that the cost of egg production in rheas may not be particularly great. At least there is no lack of opportunity for productive research on these questions!

As in fishes, parents of some birds merge their broods after hatching and the combined brood is tended jointly by the parents. In the common eider (*Somateria mollissima*), brood mixing reportedly improves survival of females and young. However, in the shelduck (*Tadorna tadorna*), it seems to lower the survival of the young and has been interpreted as a maladaptive response to uncommonly high nesting densities (Patterson et al., 1982). Some penguins, pinyon jays (*Gymnorhinus cyanocephalus*), and some other birds pool their broods into conspecific creches of dozens to hundreds of young birds, possibly as a means of reducing predation. Parents commonly feed their own young in these enormous groups, but at least in the pinyon jay, adults sometimes feed young that are not their own (Balda and Balda, 1978).

Discussion of the complex situations of cooperative or communal breeding is deferred to a later section of this chapter.

Mammals

All young mammals are fed by their mothers on milk, and parental care is often simply maternal care. As we have seen (Chapter 14), the cost of lactation can be very high.

The monotremes lay eggs, which are incubated in an abdominal pouch by the spiny echidnas (*Tachyglossus aculeatus* in Australia and *Zaglossus bruijni* in New Guinea) and in a nesting burrow by the platypus (*Or-*

nithorhynchus anatinus). Although a female monotreme does not have nipples, the milk is sucked by the young from her fur in the mammary area.

Most marsupials brood their highly altricial young in a pouch or in ventral folds of skin that contain the mammary glands. The offspring of the large red kangaroo (*Megaleia rufa*) remains continually in the pouch for about six months; then it begins to make outside expeditions. The joey leaves the pouch permanently after about six more weeks, although it continues to nurse for several more months, even when the mother is harboring a new baby. Young koala bears (*Phascolarctos cinereus*), after leaving the mother's pouch, feed on half-digested *Eucalyptus* leaves from her anus. In marsupial species that have litter sizes greater than one, the young emerge from the pouch at an earlier stage and may be carried by the female or sometimes left in a nest.

Placentals exhibit a spectrum of offspring dependency from fully altricial to highly precocial. Young of most rodents and carnivores as well as rabbits (such as *Sylvilagus*) are born helpless and with closed eyes. Those of most artiodactyls, hares (*Lepus*), and domesticated guinea pigs (*Cavia porcellus*) are born open-eyed, active, and precocial.

Lactation and nursing may continue for a long time after the young have begun to eat solid food, as is the case in many precocial ungulates. The length of time over which the young nurse depends on more than the degree of development at birth; in terrestrial mammals it tends to increase with increasing body size. Lactation periods are short in most pinnipeds, however; during this time, the young gain weight and the females lose it rapidly. After lactation, the young leave the rookery with their mothers and go to sea. Whales also have short lactation times for their body size. Rapid growth is permitted by the production of milk that is unusually high in fat and protein compared to that of most mammals, although reindeer and some rabbits also produce a similarly nutritious milk. The relatively short lactation periods of these aquatic mammals seem to be unexplained.

The end of lactation signals the beginning of independence for the young of some species, including many rodents. In many species, however, the offspring remain with their mother well beyond this time. This is true of many primates, carnivores, and ungulates. The extended period of juvenile dependence provides protection from predators and sometimes also from starvation while the young learn to hunt for their own food. The juvenile may also acquire familiarity with the species' social behavior. Young male gibbons (*Hylobates klossi*) in Indonesia often receive parental assistance in establishing a territory near the home territory (Tilson, 1981).

Although maternal care prevails, paternal care does occur. The largest number of genera with some paternal care fall into three orders: Primates, Carnivora, and Rodentia (Kleiman and Malcolm, 1981). Paternal care in mammals may involve provisioning the young, especially in carnivores, in which males often regurgitate meat for the litter. Males may brood the young, as in the California deer mouse (*Peromyscus californicus*); the presence of the father apparently contributes to thermoregulation and, by so doing, improves the growth rate of the young (Dudley, 1974a, b). Males may guard their offspring against predators of various sorts, including conspecific males, or teach them various skills. Several factors may predispose the males of certain species to provide care of the young, including an inability to find additional females or the high costs of doing so, a social system with close interaction of individuals and strong social ties, and, of course, the ability of the male to contribute anything at all (Kleiman and Malcolm, 1981).

Male ability to provide care must depend in part on the nature of the risks to which the young are exposed and on whether the male is competent to counter those particular risks. The warm body of any male mammal can provide heat to reduce the risk or cost of chilling, but the ability to protect against predators depends on the size and ferocity of the male relative to the predators. A male's capacity to provision the young depends partly on diet; food must be portable in sufficient quantities. (Further, carnivorous mammals may be more commonly territorial than herbivorous ones, so that carnivores are more likely to remain associated with the female and her litter [Gross and Shine, 1981].) The fact that male mammals do not lactate may have at least two interacting explanations (Daly, 1978). Biparental care in mammals is clearly derived phylogenetically from a condition of maternal care. The fundamental sexual dimorphism of mammals (females equipped to lactate but males not) may be so well-developed, and the physiological mechanisms involved in female lactation so complex, that the requisite evolutionary shift in male physiology is very unlikely to be achieved. Ecological constraints may be added to the phylogenetic ones. The high costs of lactation raises the probability that the milk supply limits litter size, so male lactation might augment litter size. This might be true especially in species that produce many large litters, but males of these species also may have good opportunities for multiple mating, which may preclude intensive paternal investment.

Communal care of young is found in certain carnivorous mammals, including the cape hunting dog and lion. Lionesses often nurse the cubs of other females in the pride, and adult wolves and hunting dogs bring meat to young of the pack. These social carnivores commonly are characterized by a high degree of relatedness among group members (especially the females in lions) and kin selection has probably contributed to the evolution of this behavior. Other factors must be at work too, for if everything else were equal, it is evolutionarily better to care for your own young than for those of even a near relative. Communal living by carnivores, in itself, is an insufficient explanation, since spotted hyenas (*Crocuta crocuta*) are highly gregarious but do not share parental chores. One might argue that if the major sources of mortality for cubs are predation or disease, provisioning the young is seldom limiting to reproductive success. Then adults could have the ability to feed both their own cubs and those of their relatives. Perhaps, in addition, reciprocal arrangements may be made to the benefit of both adults: one watches/feeds the young while the other is off hunting, turn and turn about. Reciprocity could be at work even when provisioning is difficult, although the door may then be open to cheating.

Furthermore, in lions at least, males of similar age are likely to remain together as a corporate unit most of their lives, so a female may benefit her own young by helping to provide him with a companion. Female lions, too, commonly remain together in the pride and cooperate both in hunting and rearing of offspring. So a lioness helping rear her nieces, half-sisters, cousins may assist her own daughters by providing future cooperators.

The brown hyena (*Hyaena brunnea*) in the Kalahari is also a communal breeder, females helping feed each other's offspring. Although kin relationships among clan members are apparently less close than in lions, reciprocal benefits may be obtained. This hyena is primarily a scavenger, spending long periods away from the den in search of food. Perhaps when food is found, there is often enough for more than one litter of cubs (or the female's milk supply is more than adequate for her own litter), and reciprocal sharing enhances survival of all the cubs. Since adults are so often away from the den, joint protection of the communal den may not be of major importance (Owens and Owens, 1979).

COOPERATIVE BREEDING

Cooperative or communal breeding comes in a variety of forms. Perhaps the commonest is the assistance of a breeding pair by one or more relatives. Sometimes unrelated individuals serve as helpers, too. Several breeders sometimes share a nest and its brood; there may be more than one female, more than one male, or both, and the mating system may be polygynous, polyandrous, monogamous, or promiscuous. Cooperative breeding in one form or another is known to occur in at least 10 families and more than 150 species of birds (Brown, 1978; Emlen, 1982a), in more than 25 species of mammals (Emlen, 1982a), and in a few fishes (Taborsky and Limberger, 1981). As study of this phenomenon increases, the numbers of species known to exhibit some form of this behavior is growing rapidly.

A small sample of the many known instances will illustrate some of the variety. Helpers occur fairly frequently among carnivorous mammals, in which biparental care is not uncommon. Packs of wolves normally have a single breeding pair; the other adults are commonly siblings of the breeding pair or offspring of a previous season. These nonbreeding adults generally help care for the pups born to the mated pair. Older pups often stay with their parents and help rear the next litters in jackals. Similarly, among birds, helpers are often the young of previous matings, and, in some instances, they stay with their parents for several years. Unrelated, nonbreeding helpers occur frequently in the African pied kingfisher (*Ceryle rudis*) and the white-fronted bee-eater (*Merops bullockoides*) in eastern Africa (Reyer, 1980; Emlen, 1981) and are known to occur at least occasionally in several other species.

Several males may breed with a single female; Harris' hawk and the Galápagos hawk and the Tasmanian native hen are examples (Chapter 13). The naked mole-rat (*Heterocephalus glaber*) is a colonial, fossorial rodent from eastern Africa; only one female in a colony breeds, perhaps with several males. The other members of the colony are often related and perform all the work of burrowing, collecting food, nest-building, and tending (but not feeding) the young (Jarvis, 1981). Two or more females may share a nest, mating promiscuously with several males; this occurs in the pukeko (*Porphyrio porphyrio*) in New Zealand and in some populations of the acorn woodpecker (*Melanerpes formicivorus*), for instance. Lions have a comparable arrangement. The groove-billed ani (*Crotaphaga sulcirostris*) in Central America mates monogamously but nests communally, with several pairs to a nest (Vehrenkamp, 1977). Several female ostriches (*Struthio camelus*) in East Africa lay eggs in the nest of each male, but only one female engages in parental care, and females seem to mate with several males (Bertram, 1980) at least during the course of the season if not for each clutch.

How can this remarkable variety be accounted for? When helpers are relatives, they may be increasing their inclusive fitness by increasing the reproductive success of their kin. Kin selection can only facilitate the evolution of helping, however. It is not a sufficient explanation; nor is it even a necessary element of the explanation in all cases, because nonrelatives are known to help in some species.

Ecological pressures and constraints must mold these systems, but in ways that are only beginning to be understood. Emlen (1982a, b) has provided one promising scheme, based on assessment of the costs and benefits to all the participants. There are two features of cooperative breeding that need explanation: the failure of certain sexually mature adults to breed and their participation in brood-rearing.

Failure to breed presumably occurs when the costs of breeding are prohibitively great. If suitable habitat is limited, and the adult population long-lived, sites for the establishment of territories by young individuals are rare. The chance of successful dispersal from the natal area is low in such stable environments (Brown, 1978). Under these circumstances, selection could favor stay-at-homes that continue to reside with their parents. This limitation is thought to apply to a number of species, such as the Mexican jay (*Aphelocoma ultramarina*) (Brown, 1978). The acorn woodpecker in California seems to conform to this explanation, although populations in New Mexico and Arizona seem less crowded. Emlen, pulling together the work of several others, pointed out that California acorn woodpeckers live at high density on territories that seldom become vacant; the average number of adults per group was about five, almost half the yearlings stayed on their home territory, and 70 percent of all groups had helpers. At certain locations in New Mexico and Arizona, density was lower and territories fell vacant occasionally. The average group size was about three adults, perhaps a third of the juveniles stayed home, and less than 60 percent of the groups had helpers. The woodpeckers in another Arizona locale usually occupied their breeding range only seasonally, and over 90 percent of the territories were vacated. Virtually all birds dispersed, the average group size was just over two, and only 16 percent of the groups had helpers. A positive correlation of density with low territory turnover, large group size, low juvenile dispersal, and a high frequency of helpers is apparent.

Breeding failure can be frequent in fluctuating, unpredictable environments, such as the arid regions of Africa and Australia. Australian desert birds are often nomadic and a number of cooperative species on both continents are colonial. There is no evidence that the suitable habitat is fully occupied. Instead, it seems that the probability of successful reproduction varies enormously from year to year, and perhaps from site to site. The seasonal rains in East Africa (and the irregular rains in arid Australia) vary greatly in amount and the timing. The frequency of helpers at the nest in the white-fronted bee-eater (Fig. 15–6) in Kenya increased in years when the rainfall was low in the month before breeding began and the insect populations were correspondingly reduced (Emlen, 1981, 1982a). This inverse correlation suggests that the birds assess the prospective difficulty of breeding and, when faced with improbable success, opt for helping instead of breeding themselves. Obviously, not all adults respond to seasonal uncertainties in the same way, since some usually breed (with the aid of the helpers). Some helpers come from nesting attempts that failed, but others no doubt elected not to try to breed at all. It would be interesting to learn if certain categories of individuals characteristically try to breed in rainy seasons of any sort whereas others are characteristically inclined to give up

Figure 15–6 A white-fronted bee-eater, which nests colonially and cooperatively. Helpers at the nest are often close relatives of the breeders, but other individuals sometimes help, too.

easily. Thus, cooperative breeders in unpredictable environments face the same problem of limited breeding opportunities that those of stable environments encounter, but the ecological factors restricting the chance of breeding are different.

Because our knowledge of which species have helpers is still expanding rapidly, it is premature to attempt a contrast of species, with and without helpers, that live in the same environment. It seems very likely that not all birds of stable and saturated habitat and not all birds of uncertain environments have helpers at the nest. Eventually, it will be enlightening to examine the ecological differences that might result in different responses to the same environment.

There are obvious costs to breeding failure. If a breeding attempt fails, the cost of mating, egg-laying, and so on, is wasted. A reproductive season may be skipped entirely, lowering the number of broods produced in a lifetime. Helpers that remain on their natal territory and postpone their first breeding effort have a sometimes greatly delayed age of first reproduction. The costs of attempting to breed are presumably greater than the costs of not breeding, which accounts for the failure of young adults to disperse and establish their own nest. This constraint does not explain why the nonbreeders participate in helping the breeding effort of others, however. Here the interpretation becomes complicated.

First, it is useful to consider the benefits of helping; there are at least two types. The breeding pair may benefit by obtaining a greater success in fledging young or a greater frequency of brood production. White-fronted bee-eaters raised larger broods as the size of the cooperative group increased (at least up to four adults) and the probability of successful fledging also was greater (Emlen, 1981). In environments where hunting was poor, helpers reduced the rate of starvation of young nestling pied kingfishers, and unrelated helpers were accepted only when parents could not rear the young alone (Reyer, 1980). Helpers increased the reproductive success of inexperienced females in the splendid wren (*Malurus splendens*) (Rowley, 1981). Gray-crowned babblers (*Pomatostomus temporalis*) in Australia are territorial; there is one breeding pair per group and the helpers are usually young of previous broods. Reproductive success increased with group size, and experimental removal of helpers decreased reproductive success (Brown and Brown, 1981; Brown et al., 1982). Thus, the greater success of larger groups could be attributed directly to the effect of helpers and not to the quality of the territory. (This factor of territory quality is not relevant for the bee-eaters, which are colonial.) Helpers in the babblers do not, however, increase the feeding rate of the young. Instead, the foraging rates of all adults diminishes, presumably lowering the breeding costs and allowing more frequent breeding attempts (Brown and Brown, 1981). Similar benefits may accrue in Florida scrub jays (*Aphelocoma coerulescens;* Stallcup and Woolfenden, 1978) and in experienced females of the splendid wren (Rowley, 1981).

The helpers themselves may benefit, reducing the cost of not breeding. If the reproductive success of the breeders is enhanced and the helpers are closely related to the breeders, the inclusive fitness of the helpers will be increased. Although the helpers are producing no direct descendants, their contribution to relatives augments their fitness. Fitness gain through relatives may seldom compensate completely for a lack of personal fitness, but given that helpers have opted not to breed, some fitness gain is better than none. Clearly any costs involved in helping must be lower than the possible benefits.

Helpers, at least in some species, have a good chance of inheriting the territory of the breeding pair. Both male and female helpers in the green woodhoopoe (*Phoeniculus purpureus*) in Kenya have high probabilities of moving up to breeding status because of rather high adult mortality (Ligon, 1981). Florida scrub jays can do the same, or at least use the home territory as a base to establish a territory next door (Woolfenden and Fitzpatrick, 1978). Helpers gain experience in all stages of the nesting cycle, which probably enhances their own probability of success once they become established independently. Helpers in the neotropical brown jay (*Psilorhinus morio*) improved their helping ability, even within a single season, and fledging success of the flock was best correlated with the number of experienced helpers (Lawton and Guidon, 1981).

If helpers are not related to the breeders, kin selection is clearly irrelevant. Nonkin may function as helpers if the association allows them to move up to breeding status, as seems to occur in the green woodhoopoe (Ligon, 1981) and may occur in pied kingfishers (Reyer, 1980). Association with a group of offspring also may improve the chances of obtaining helpers for future breeding efforts. In the white-fronted bee-eater, social ties among apparently unrelated individuals may foster reciprocal helping; "friends" are more likely to give and to receive help than strangers (Emlen, 1981).

It is possible that conflict may arise between breeder and helper, particularly if it is advantageous for the helper to leave but beneficial to the breeder that the helper stays (Emlen 1982b). When this happens, breeders might share their mates with helpers, thus potentially increasing the direct personal fitness of the helper at the cost of their own. Copulation by helper males is recorded for a number of birds and mammals. Note here that it is the breeding male that endures the fitness loss, not the female, and that further conflicts may then occur between male and female concerning the role of the auxiliary male. Regular mate-sharing by males is known in the polyandrous hawks and the Tasmanian native hen discussed in Chapter 13, and in the Australian noisy miner (*Manorina meloncephala*), in which several seemingly unrelated males mate with a female and tend the young (Dow, 1979).

Another compromise between breeder and helper could take the form of nest-sharing by several pairs, and sharing of parental care by several adults. Communal rearing of young in lions and brown hyenas has already been described, and it is also reported for some mongooses and a few others.

Among birds, the pukeko in New Zealand has a variable social order: some pairs nest alone, some pairs have nonbreeding helpers, and some mate promiscuously and rear young in a group (Craig, 1980). Paired females lay more eggs and raise more young than group-breeders, but group formation may be necessitated, despite the reproductive loss, when many birds crowd into prime habitat and more than two birds are needed to defend a territory (Craig, 1979). Just what determines why some group females have helpers but do not share the nest is not clear.

The monogamous groove-billed ani nests preferentially in marshes in Costa Rica but also nests in pastures. Marsh nests were shared, on average, by more pairs than pasture nests, which thus had fewer adults in attendance. The average number of juveniles per pair per year was similar for all group sizes in each habitat (Vehrencamp, 1978). Breeding success is not distributed equally among all the pairs sharing a nest, however, (Vehrencamp, 1977). Females roll each other's eggs out of the nest (perhaps suggesting conflict among them). Older females begin laying later than others in the breeding group and toss out more of their eggs, so that most of the eggs that are incubated and hatched are those of the older females. Most of the incubation and parental care is performed by the dominant male, who is mated to the older, dominant female; this female does less work than her mate and less than other females, with possible consequences for future survival. The number of incubated eggs of dominant females increased with group size (from 3.9 in simple pairs to 5.8 in groups of three pairs), whereas the size of incubated clutches of subordinate females was similar to or less than those of females nesting in simple pairs (and more of the eggs they laid were tossed) (Vehrencamp, 1977). Nestling success probably was independent of parentage, so that dominant females had both the reproductive advantage with the present clutch and the potential survival advantage. Subordinate male survival increased with group size, although the dominant male in all groups was subject to high predation while incubating. Thus, advantages accruing to the dominant female from nest-sharing are clearly indicated. Less clear is any reason for the subordinate females to engage in this behavior, although their mates did enjoy enhanced survival. Since gains can be shown for both dominants and subordinates, perhaps the situation does not really entail a resolution of conflict, although the egg-tossing behavior suggests that conflict exists. There is plenty of room for further study.

BROOD PARASITISM

Brood parasites have abdicated all parental responsibilities; they lay their eggs in the nests of other animals and parental care is provided by the host parents. Some brood parasites are facultative parasites, utilizing foster parents only occasionally, but many appear to be obligatory parasites. Some parasitize their own species, and others use hosts of different species.

Intraspecific brood parasitism is reported for 55 species of birds, mostly ducks (Yom-Tov, 1980); it is likely that detailed study will discover the habit in a number of other species. The redhead duck (*Aythya americana*) is one of the species in which intraspecific parasitism seem to be common (Weller, 1959). Unfortunately, little seems to be known about the success of the parasitic eggs. The extra eggs usually reduce the nesting success of the host (Yom-Tov, 1980); female goldeneyes (*Bucephala clangula*) with experimentally parasitized broods responded by reducing their own clutch size (Andersson and Eriksson, 1982). The habit of depositing some eggs in the nests of conspecifics seems most likely to be successful when brood size is not limited by parental care, because otherwise, the extra eggs and young could mean reduced feeding rates for all the young, if they are developing simultaneously. However, if the parasitic eggs hatched ahead of the host (and, thus, if the parasitic female deposited the egg early, in the nest of a host that starts incubating well before her clutch is complete), the success of the young parasite might be achieved, but perhaps at a cost to the host's brood.

Interspecific brood care occurs among fishes, but its ecology and behavior have been little studied. At least four North American cyprinid fishes—*Notropis cornutus, N. chrysocephalus, N. rubellus* (shiners), and *Phoxinus erythrogaster* (red-bellied dace)—reportedly often spawn in the nests of more than seven other species, including *Nocomis biguttatus, Micropterus dolomieui, Campostoma anomalus, Semotilus atromaculatus,* and various *Lepomis*. The host species usually provides parental care. It is not clear, however, to what extent the male parent may also contribute to nest-guarding. Another North American stream fish, the orange-throated darter (*Etheostoma spectabile*), reportedly has a strong but facultative affinity for the nests of *M. dolomieui*, the smallmouth bass. Young darters appear to seek out nests of the bass when the male is guarding and, thus, obtain some protection from predation. South African freshwater fishes of three species (*Haplochromis polystigma, H. macrostoma,* and *Serranochromis robustus*), in which the females are mouth-brooders, are hosts to the young of *H. chrysonotus.*

McKaye (1981) favors the idea that interspecific brood care in fishes is, in fact, adaptive to the host parents. One advantage might be that the dangers of predation are spread out over a larger brood, and the risk to any one offspring are reduced. Some fishes even kidnap the young of others and herd them together with their own brood. It remains to be seen, however, whether the possible advantage to the host is relevant to all cases of interspecific brood care in fishes. Some cases of interspecific brood care may turn out to be parasitic.

Interspecific brood parasitism occurs in five avian families (Payne, 1977); the Anatidae (ducks), Cuculidae (cuckoos), Icteridae (cowbirds), Indicatoridae (honeyguides), and Ploceidae (whydahs and the cuckoo-weaver). Brood parasites commonly depress the nesting success of their hosts. For example, the pied crested cuckoo (*Clamator jacobinus*) is parasitic on several other species in Africa and Asia. Laughing thrushes (*Garrulax*) were the chief hosts in the hills of India, whereas babblers (*Turdoides*) were the primary hosts in the lowlands. Presence of a young cuckoo in a babbler nest reduced the breeding success from an average of about 2.5 host young per nest in unparasitized nests to only 1.1 young (Gaston, 1976). Because host individuals have lower reproductive success than individuals that have not been parasitized, selection normally should favor the avoidance of parasitism. A common means of avoidance is rejection of the parasite egg: The host may throw out the strange egg or perhaps build a new nest. Many brood parasites have evolved means of diminishing rejection, however, as we will see in the following paragraphs.

Some ducks are facultative brood parasites. These include the redhead duck and the ruddy duck (*Oxyura jamaicensis*). Female redheads vary in their parasitic tendencies; some seem to be chiefly parasitic, but others both lay in other birds' nests and build one of their own (Weller, 1959). The reasons for this variation are not known. The success rate of nestlings in alien broods seems to be rather low, but it is presumably high enough that the parasitic females usually gain some reproductive success. How much success and under what circumstances is still to be discovered.

The black-headed duck (*Heteronetta atricapilla*) of southern South America is the only known obligate brood parasite in this family. A marsh-dweller, it parasitizes the nests of many other bird species in its habitat, including two species of coots (*Fulica*), the white-faced ibis (*Plegadis falcinellus*), and the rosybill duck (*Netta peposaca*). Host eggs and nests are varied in color and form. The red-fronted coot (*F. rufifrons*) was the most common host species in Weller's (1968) study, and black-headed duck eggs hatched most successfully in nests of this species. Ducklings of this brood parasite remain in the host nest only a day or two, receiving no food from the host female, and can feed themselves on aquatic crustaceans. Unlike the occasionally parasitic redheads and ruddy ducks, young black-headed ducks do not remain with the brood of the host and become independent when they leave the nest. Perhaps because of the extraordinary precocity and independence of the duckling, brood parasitism by black-headed ducks seems to have little effect on the nesting success of the host (Weller, 1968).

The parasitic honeyguides are more dully colored than their nesting relatives. Duller coloration may make them less conspicuous and less likely to incur aggression by their hosts. Honeyguide nestlings have a hook at the end of the bill, which is used to kill foster siblings.

Many cuckoos of the Old World have a well-developed system of egg mimicry, host specificity, and habitat segregation (where habitats are relatively undisturbed). Even within a single species in the genus *Cuculus*, evolution sometimes has resulted in local races whose eggs commonly match, in color and size, those of the local host (Fig. 15–7). Each race has a characteristic set of closely related host species, and each is found in the habitat occupied by that set of hosts. Egg mimicry is an excellent way to reduce rejection behavior by the host. Where habitats have been much disturbed by human activities, habitat differentiation is obscured; several parasitic forms may now coexist and the egg mimicry system tends to be less finely adjusted (Southern, 1954). This appears to be the case in western Europe, where destruction of natural habitats is widespread and European cuckoo (*C. canorus*) eggs of many colors may be found in the nests of a single host species. The best remaining case of good mimicry in Europe occurs in the great marshes of Hungary, which are relatively undisturbed and where cuckoo eggs match those of the great reed warbler quite well. Similar effects of habitat and host specificity are known from India (Southern, 1954).

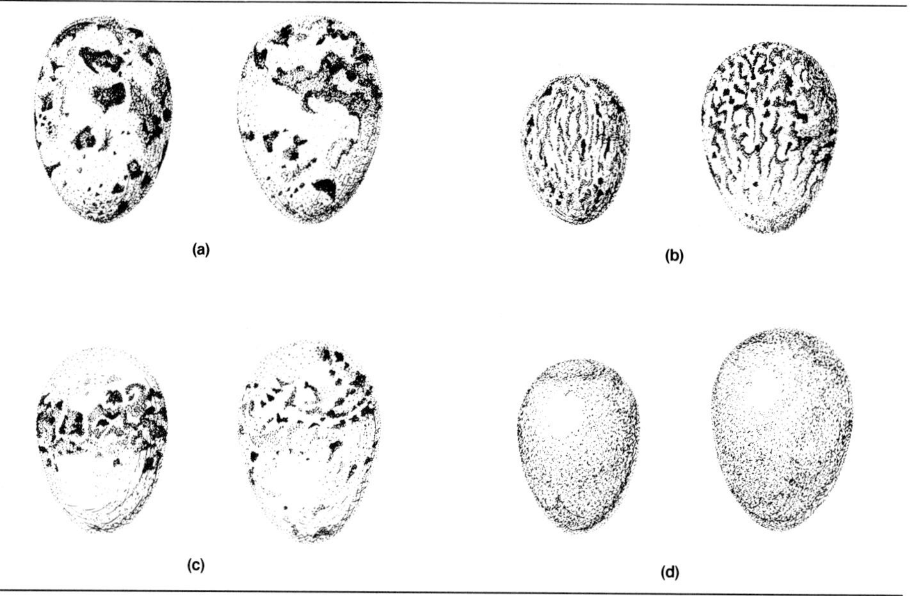

Figure 15–7 Eggs of the brood-parasitic European cuckoo (*Cuculus canorus*) and some of its hosts. In each pair, the host egg is on the left and the cuckoo egg on the right. (*a*) Great reed warbler, *Acrocephalus arundinaceus.* The ground color is greenish blue with brown splotches. (*b*) Yellow wagtail, *Motacilla flava.* A tan background is freckled with yellow and brown. (*c*) Red-backed shrike, *Lanius collurio.* A grayish background has a light mottling of brown and tan. (*d*) European redstart, *Phoenicurus phoenicurus.* The eggs are blue. Note the difference in egg size as well as color: Although the cuckoo eggs are typically larger than the host eggs, the size of the cuckoo egg also varies in parallel to that of the host. (Modified from W. Wickler, 1968. *Mimicry in Plants and Animals.* Weidenfeld & Nicolson Ltd., London.)

Cuckoo species that parasitize hosts no larger than themselves usually remove at least one host egg or nestling from the host nest. Sometimes the adult cuckoo does this. In other species, the young cuckoo, having a hollow place in its back, simply backs up to its foster siblings and heaves some or all of them over the side. This behavior obviously results in more food for the remaining young parasite, which grows very quickly. Furthermore, the young cuckoo is often larger than its nest mates at hatching and, thus, has a head start in growth. As a result, host nestlings often starve while their parents stuff the gaping maw of the young parasite, for the biggest and most demanding offspring is usually fed first. Species parasitic on hosts larger than themselves (such as crows) do not remove host offspring. The nestlings of these cuckoo species tend to mimic nestlings of their host. This mimicry usually takes the form of a similarity in color and pattern of the mouth lining, which is exposed when the young gape for food, but it is not as well developed as egg mimicry of other species. The neotropical striped cuckoo (*Tapera naevia*) is sometimes parasitic on the smaller rufous-and-white wren (*Thryothorus rufalbus*). The cuckoo nestling can kill its nest mates by biting them with its hooked bill. Furthermore, the mouth lining of the nestling parasite matches that of the host (Morton and Farabaugh, 1979). Thus, the striped cuckoo exhibits both nestling removal and nestling mimicry.

Only two genera of New World cuckoos, both in South America, are regularly parasitic. In addition, occasional facultative parasitism is well documented, as in the yellow-billed and black-billed cuckoos, *Cuculus americanus* and *C. erythrophthalmus* of North America. An important food of these cuckoos is cicadas, large insects that emerge in great bursts of abundance, some species annually, but others periodically, at intervals of up to 17 years. A long-term survey of cuckoo reproduction in Indiana provided evidence that times of high cicada abundance were marked by increased reproductive effort by yellow-billed cuckoos; more eggs were laid and some of these were laid in black-billed cuckoo nests (Nolan and Thompson, 1975). The North American cuckoos seem to be able to respond rapidly to high levels of food abundance; both clutch size and breeding times shifted when cicadas were unusually abundant. These interesting observations suggest that the mechanisms permitting quick exploitation of flushes of prey may be so sensitive that some females lose the normal timing sequence of nest-building and egg-laying and so deposit their eggs in other nests. Any female whose foundlings were raised by other birds and were successful would have, at least potentially, a higher reproductive output than females that only raised their own brood. This may have been the route by which obligate brood parasitism evolved.

The whydahs or indigobirds (*Vidua*) of Africa parasitize estrildid fire finches (*Lagonosticta*). All these species lay white eggs, but each kind of indigobird is a parasite of a specific form of fire finch. Many vocalizations of male indigobirds mimic those of male fire finches of the particular host species. Because indigobirds probably learn the song of their foster father while they are still in the nest, use of the host male song for mate attraction results in proper mate and host selection (Payne, 1973). Furthermore, the nestling brood parasites mimic host nestlings in the markings inside the mouth, the markings on the edge of the bill, and the pattern of food begging.

The North American brown-headed cowbird (*Molothrus ater*) is an obligate brood parasite. Some of its hosts have evolved behavior that rejects the cowbird egg, but others have not, perhaps because cowbirds have begun parasitizing these hosts only relatively recently. Furthermore, discrimination that takes the form of deserting the parasitized nest will only be favored if renesting by the host is likely to be successful (Rothstein, 1975a, b). Other American cowbirds range from occasional through facultative to obligate parasites. Perhaps the most complex and intriguing case of avian brood parasitism is Smith's (1968) report of the neotropical giant cowbird, *Scaphidura oryzivora*, an icterid whose hosts are the colonial-nesting oropendulas and caciques (other icterids of at least four genera). The full story is very complex, but an outline is presented here.

Smith studied a series of oropendula and cacique colonies in Panama. Some of these colonies were lightly parasitized by giant cowbirds; egg mimicry by the cowbirds was well developed, and adult oropendulas and caciques chased intruding *Scaphidura*. Other colonies were heavily parasitized; egg mimicry was not so well developed, and aggression against cowbird intruders was much less. Ordinarily, we would predict that individuals in heavily parasitized colonies would be at a great disadvantage and have greatly lowered fitness. This was not the case, however. In these colonies, the nesting success of the hosts was greater with nestling cowbirds than without! For instance, a nest with two host nestlings and one cowbird nestling had an average fledging success of 0.53 host young per nest, but those with two or three hosts and no cowbirds averaged 0.19 host young per nest. On the other hand, in the lightly parasitized colonies with discriminating hosts, fledging success tended to be higher in unparasitized nests, as expected: Nests with two or three host chicks and no cowbirds produced an average of 0.54 young, but those with two hosts and one cowbird chick produced an average of only 0.20 host young. Fledging success of nests with cowbirds in heavily parasitized colonies was similar to that of nests without cowbirds in lightly parasitized colonies. What accounts for this difference among colonies? How can cowbird parasitism be advantageous to hosts in any colonies?

Smith observed that the lightly parasitized colonies were always located in a tree that was also inhabited by certain wasps and bees. The heavily parasitized colonies did not have this association. One of the major enemies of oropendula and cacique nestlings is a botfly whose eggs or larvae are placed on the young birds and whose larvae burrow into the chicks' bodies. Chicks with several botfly larvae often die. By some means, the wasps and bees reduce the level of botfly infestation of host nestlings. Colonies unprotected by wasps or bees are subject to severe attack by botflies in nests lacking cowbird young. Nests containing cowbird offspring had much lower botfly infestations and better survival of host chicks because the active cowbird young preened their nest mates and removed the botflies!

The evolution of brood parasitism is still something of a mystery in many respects. The habit may have originated by an accidental lack of synchrony between the appearance of the egg and the building of the nest, as sometimes happens in nonparasitic species. The ecological conditions that gave selective advantage to this habit and allowed its continuation are not certain, however. Hamilton and Orians (1965) suggested that parasitic species use foods that are very erratic in spatial or temporal distribution or unusual in some way; thus, being tied to one place for nesting would be difficult. Honeyguides, for instance, feed on the larvae and wax in bees' nests; cuckoos ingest quantities of hairy caterpillars (unusual for birds) and exploit cicada emergences; and cowbirds perhaps originally evolved to forage by following ungulate herds. Yet other species of birds do at least some of these things and are not parasitic. Moreover, the dependence of cowbirds on herds is certainly not obligatory at present.

Payne (1977) showed that brood parasitism may improve the chances that at least some offspring will escape predation. By scattering its eggs among the nests of several host individuals, a brood parasite increases the probability that at least one young will survive to fledging. The advantages of not "putting all the eggs in one basket" are greater at higher levels of nest predation. Still, only certain birds have adopted this tactic of reducing predation. Abdication of parental responsibilities (and consumption of sporadically rich sources of prey) might also allow female brood parasites to lay more eggs per season than if they invested parental care in nestlings. The success of those eggs obviously depends on the exploitation of hosts whose diet is suitable for young brood parasites (which may not be difficult, since so many bird species are insectivorous) and of hosts with low rejection rates.

SUMMARY

Oviparity is more common than viviparity among vertebrates, but viviparity has evolved several times, in different classes. The advantages and disadvantages of each mode of reproduction are not altogether clear. Viviparous ectotherms in regions with cold or variable weather can move their broods to favorable sites. Viviparity sometimes means that the young are carried by the mother for longer times than eggs would be. Live-bearers only sometimes produce young that are developmentally more advanced than those of egg-layers and only sometimes are more encumbered by the weight of their burden.

Parental care devoted to young after hatching or birth is highly variable in kind and amount. One or both parents may be involved. There is considerable speculation concerning the evolution of uniparental (paternal or maternal) and biparental care; central factors include the ability of the parent(s) to give care and the opportunity to mate with additional individuals.

Cooperative breeding is rather widespread. One factor important in its evolution is probably the low probability of successful breeding by the helpers, which could explain why mature juveniles remain with their parents. Improved future reproductive success (achieved in various ways) of the helpers and improved inclusive fitness of the helpers may explain why they provide help.

Brood parasitism may occur in fishes and is well known in birds. Commonly disadvantageous to the host, this habit offers advantages to the parasite, which may include the ability to lay more eggs (since it is not constrained to care for the young) and perhaps a dilution of the effects of egg predation. It is not clear why only some birds have evolved this habit, however.

SELECTED REFERENCES

Alexander, R.D., 1974. The evolution of social behavior. *Ann. Rev. Ecol. Syst., 5*:325–383.

Alvarez, H., 1975. The social system of the green jay in Colombia. *Living Bird, 14*:5–44.

Balda, R.P., and J.H. Balda, 1978. The care of young piñon jays (*Gymnorhinus cyanocephalus*) and their integration into the flock. *J. F. Ornith., 119*:146–171.

Balon, E.K., 1975. Reproductive guilds of fishes: a proposal and definition. *J. Fisheries Research Board of Canada, 32*:821–864.

Barash, D.P., 1977. *Sociobiology and Behavior.* Elsevier Scientific, Amsterdam.

Bertram, B.C.R., 1978. Living in groups: predators and prey. *In* J.R. Krebs and N.B. Davies (Eds.), *Behavioral Ecology: An Evolutionary Approach,* pp. 64–96. Sinauer Associates, Sunderland, MA.

Bertram, B.C.R., 1980. Breeding system and strategies of ostriches. *Proc. Intern. Ornith. Cong., 17*:890–894.

Birkhead, T.R., and D.N. Nettleship, 1982. The adaptive significance of egg size and laying date in thick-billed murres *Uria lomvia. Ecol., 63*:300–306.

Blumer, L.S., 1979. Male parental care in the bony fishes. *Q. Rev. Biol., 54*:149–161.

Boucher, D.H., 1977. On wasting parental investment. *Am. Nat., 111*:786–788.

Bradbury, J.W., and S.L. Vehrenkamp, 1977. Social organization and foraging in emballonurid bats. IV. Parental investment patterns. *Behav. Ecol. Sociobiol., 2*:19–29.

Breder, C.M., and D.E. Rosen, 1966. *Modes of Reproduction in Fishes.* Natural History Press, Garden City, NY.

Brown, J.L., 1978. Avian communal breeding systems. *Annu. Rev. Ecol. Syst., 9*:123–155.

Brown, J.L., and R.P. Balda, 1977. The relationship of habitat quality to group size in Hall's babbler (*Pomatostomus halli*). *Condor, 79*:312–320.

Brown, J.L., and E.R. Brown, 1981. Kin selection and individual selection in babblers. *In*

R.D. Alexander and D.W. Tinkle (Eds.). *Natural Selection and Social Behavior*, pp. 244–256. Chiron Press, New York.

Brown, J.L., E.R. Brown, S.D. Brown, and D.D. Dow, 1982. Helpers: effects of experimental removal on reproductive success. *Science, 215*:421–422.

Case, T.J., 1978a. Endothermy and parental care in the terrestrial vertebrates. *Am. Nat., 112*:861–874.

Case, T.J., 1978b. On the evolutionary and adaptive significance of postnatal growth rates in the terrestrial vertebrates. *Q. Rev. Biol., 53*:243–282.

Constantz, G.D., 1979. Social dynamics and parental care in the tessellated darter (Pisces: Percidae). *Proc. Acad. Nat. Sci. Phila., 131*:131–138.

Craig, J.L., 1979. Habitat variation in the social organization of a communal gallinule, the pukeko, *Porphyrio porphyrio melanotus. Behav. Ecol. Sociobiol., 5*:331–358.

Craig, J.L., 1980. Breeding success of a communal gallinule. *Behav. Ecol. Sociobiol., 6*:289–295.

Daly, M., 1979. Why don't male mammals lactate? *J. Theor. Biol., 78*:325–345.

Daniels, R.A., 1979. Nest guard replacement in the Antarctic plunder fish *Harpagifer bispinis:* possible altruistic behavior. *Science, 205*:831–833.

Davies, N.B., 1976. Parental care and the transition to independent feeding in the young spotted flycatcher (*Muscicapa striata*). *Behaviour, 59*:280–295.

Dow, D.D., 1979. Agonistic and spacing behaviour of the noisy miner *Manorina melanocephala*, a communally breeding honeyeater. *Ibis, 121*:423–436.

Dudley, D., 1974a. Contributions of paternal care to the growth development of the young in *Peromyscus californicus. Behav. Biol., 11*:155–166.

Dudley, D., 1974b. Paternal behavior of the California mouse, *Peromyscus californicus. Behav. Biol., 11*:247–252.

Emlen, S.T., 1978. The evolution of cooperative breeding in birds. *In* J.R. Krebs and N.B. Davies (Eds.), *Behavioural Ecology: An Evolutionary Approach*, pp. 245–281. Sinauer Associates, Sunderland, MA.

Emlen, S.T., 1981. Altruism, kinship, and reciprocity in the white-fronted bee-eater. *In* R.D. Alexander and D.W. Tinkle (Eds.), *Natural Selection and Social Behavior*, pp. 217–230. Chiron Press, New York.

Emlen, S.T., 1982a. The evolution of helping. I. An ecological constraints model. *Am. Nat., 119*:29–39.

Emlen, S.T., 1982b. The evolution of helping. II. The role of behavioral conflict. *Am. Nat., 119*:40–53.

Ewer, R.F., 1968. *Ethology of Mammals.* Plenum Publishing, New York.

Gaston, A.J., 1976. Brood parasitism by the pied crested cuckoo *Clamator jacobinus. J. Anim. Ecol., 45*:331–348.

Gaston, A.J., 1978. The evolution of group territorial behavior and cooperative breeding. *Am. Nat., 112*:1091–1100.

Graul, W.D., 1976. Food fluctuations and multiple clutches in the mountain plover. *Auk, 93*:166–167.

Graul, W.D., S.R. Derrickson, and D.W. Mock, 1977. The evolution of avian polyandry. *Am. Nat., 111*:812–816.

Gross, M.R., and R. Shine, 1981. Parental care and mode of fertilization in ectothermic vertebrates. *Evol., 35*:775–793.

Hamilton, W.J., and G.H. Orians, 1965. Evolution of brood parasitism in altricial birds. *Condor, 67*:361–382.

Hogarth, P.J., 1976. *Viviparity.* Arnold, London.

Hopson, J.A., 1973. Endothermy, small size, and the origin of mammalian reproduction. *Am. Nat., 107*:446–452.

Jarvis, J.U.M., 1981. Eusociality in a mammal: cooperative breeding in naked mole-rate colonies. *Science, 212*:571–573.

Jensen, R.A.C., and C.F. Clinning, 1974. Breeding biology of two cuckoos and their hosts in South West Africa. *Living Bird, 13*:5–50.

Kaplan, R.H., 1980. The implications of ovum size variability for offspring fitness and

clutch size within several populations of salamanders (*Ambystoma*). *Evol.*, *34*:51–64.
Keenleyside, M.H.A., 1979. *Diversity and Adaptation in Fish Behaviour.* Springer-Verlag, Berlin.
Keenleyside, M.H.A., 1981. Parental care patterns in fishes. *Am. Nat.*, *117*:1019–1022.
Kleiman, D.G., and J.R. Malcolm, 1981. The evolution of male parental investment in mammals. *In* D.J. Gubernick and P.H. Klopfer (Eds.). *Parental Care in Mammals*, pp. 347–387. Plenum Publishing, New York.
Lack, D., 1968. *Ecological Adaptations for Breeding in Birds.* Methuen, London.
Lawton, M.F., and C.F. Guidon, 1981. Flock composition, breeding success, and learning in the brown jay. *Condor*, *83*:27–33.
Ligon, J.D., 1981. Demographic patterns and communal breeding in the green woodhoopoe, *Phoeniculus purpureus*. *In* R.D. Alexander and D.W. Tinkle (Eds.), *Natural Selection and Social Behavior*, pp. 231–243. Chiron Press, New York.
Low, B.S., 1978. Environmental uncertainty and the parental strategies of marsupials and placentals. *Am. Nat.*, *112*:197–213.
Maxson, S.J., and L.W. Oring, 1980. Breeding season time and energy budgets of the polyandrous spotted sandpiper. *Behav.*, *74*:200–263.
Maynard Smith, J., 1977. Parental investment: a prospective analysis. *Anim. Behav.*, *25*:1–9.
McKaye, K.R., 1981. Natural selection and the evolution of brood care in fishes. *In* R.D. Alexander and D.W. Tinkle (Eds.), *Natural Selection and Social Behavior*, pp. 173–183. Chiron Press, New York.
McKaye, K.R., and N.M. McKaye, 1977. Communal care and kidnapping of young by parental cichlids. *Evol.*, *31*:674–681.
Miller, R.J., 1964. Behavior and ecology of some North American cyprinid fishes. *Am. Midl. Nat.*, *72*:313–357.
Morton, E.S., and S.M. Farabaugh, 1979. Infanticide and other adaptations of the nestling striped cuckoo *Tapera naevia*. *Ibis*, *12*:212–213.
Morton, S.R., H.F. Recher, S.D. Thompson, and R.W. Braithwaite, 1982. Comments on the relative advantages of marsupial and eutherian reproduction. *Am. Nat.*, *120*:128–134.
Nolan, V., and C.F. Thompson, 1975. The occurrence and significance of anomalous reproductive activities in two North American non-parasitic cuckoos *Coccyzus* spp. *Ibis*, *117*:496–503.
O'Connor, R.J., 1975. Growth, metabolism in nestling passerines. *Symp. Zool. Soc. Lond.*, *35*:277–306.
O'Connor, R.J., 1980. Energetics of reproduction in birds. *Proc. Intern. Ornith. Cong.*, *17*:306–311.
Owens, D.D., and M.J. Owens, 1979. Communal denning and clan associations in brown hyenas (*Hyaena brunnea*, Thunberg) of the central Kalahari Desert. *Afr. J. Ecol.*, *17*:35–44.
Packard, G.C., C.R. Tracy, and J.J. Roth, 1977. The physiological ecology of reptilian eggs and embryos, and the evolution of viviparity within the class Reptilia. *Biol. Rev.*, *52*:71–105.
Packard, G.C., M.J. Packard, T.J. Boardman, and M.D. Ashen, 1981. Possible adaptive value of water exchanges in flexible-shelled eggs of turtles. *Science*, *213*:471–473.
Parmalee, D.F., and R.B. Payne, 1973. On multiple broods and the breeding strategy of Arctic sanderlings. *Ibis*, *115*:218–226.
Patterson, I.J., A. Gilboa, and D.J. Tozer, 1981. Rearing other people's young: brood-mixing in the shelduck *Tadorna tadorna*. *Anim. Behav.*, *30*:199–202.
Payne, R.B., 1973. Behavior, mimetic songs and song dialects, and relationships of the parasitic indigobirds (*Vidua*) of Africa. *Ornith. Monogr.*, *11*:1–333.
Payne, R.B., 1977. The ecology of brood parasitism in birds. *Annu. Rev. Ecol. Syst.*, *8*:1–28.
Pienkowski, M.W., and J.J.D. Greenwood, 1979. Why change mates? *Biol. J. Linn. Soc.*, *12*:85–94.
Pooley, A.C., 1976. Mother's day in the crocodile pool. *Animal Kingdom*, *79*(1):7–13.

Reyer, H-U., 1980. Flexible helper structure as an ecological adaptation in the pied kingfisher (*Ceryle rudis rudis* L.). *Behav. Ecol. Sociobiol., 6*:219–227.

Ridley, M., 1978. Paternal care. *Anim. Behav., 26*:904–932.

Ross, H.A., 1979. Multiple clutches and shorebird egg and body weight. *Am. Nat., 113*:618–622.

Rothstein, S.I., 1975a. Evolutionary rates and host defenses against avian brood parasitism. *Am. Nat., 109*:161–176.

Rothstein, S.I., 1975b. An experimental and teleonomic investigation of avian brood parasitism. *Condor, 77*:250–271.

Rowley, I., 1981. The communal way of life in the splendid wren, *Malurus splendens. Z. Tierpsychol., 55*:228–267.

Russell, E.M., 1982. Parental investment and desertion of young in marsupials. *Am. Nat., 119*:744–748.

Sadleir, R.M.F.S., 1973. *The Reproduction of Vertebrates.* Academic Press, New York.

Salthe, S.N., and J.S. Mecham, 1974. Reproductive and courtship patterns, pp. 309–521. *In* B. Lofts (Ed.), *Physiology of the Amphibia.* Vol. 2. Academic Press, New York.

Sherman, P.W., 1977. Nepotism and the evolution of alarm calls. *Science, 197*:1246–1253.

Shine, R., and J.J. Bull, 1979. The evolution of live-bearing in lizards and snakes. *Am. Nat., 113*:905–923.

Skutch, A.F., 1976. *Parent Birds and Their Young.* University of Texas Press, Austin.

Smith, C.L., C.S. Rand, B. Schaeffer, and J.W. Atz, 1975. *Latimeria,* the living coelacanth, is ovoviviparous. *Science, 190*:1105–1106.

Smith, N.G., 1968. The advantages of being parasitized. *Nature, 219*:690–694.

Southern, H.N., 1954. Mimicry in cuckoo's eggs. *In* J. Huxley, A.C. Hardy, and E.B. Ford (Eds.), *Evolution as a Process,* pp. 219–232. Allen & Unwin, London.

Stacey, P.B., 1982. Female promiscuity and male reproductive success in social birds and mammals. *Am. Nat., 120*:51–64.

Stallcup, J.A., and G.E. Woolfenden, 1978. Family status and contributions to breeding by Florida scrub jays. *Anim. Behav., 26*:1144–1156.

Taborsky, M., and D. Limberger, 1981. Helpers in fish. *Behav. Ecol. Sociobiol., 8*:143–145.

Thibault, R.E., and R.J. Schultz, 1978. Reproductive adaptations among viviparous fishes (Cyprinodontiformes, Poeciliidae). *Evol., 32*:320–333.

Tilson, R.L., 1981. Family formation strategies of Kloss's gibbon. *Folia Primatol., 35*:259–287.

Tinkle, D.W., and J.W. Gibbons, 1977. The distribution and evolution of viviparity in reptiles. *University of Michigan Museum of Zoology (Misc. Publ.), 154*:1–55.

Townsend, D.S., M.M. Stewart, F.H. Pough, and P.F. Brussard, 1981. Internal fertilization in an oviparous frog. *Science, 212*:469–471.

Trivers, R.L., 1972. Parental investment and sexual selection. *In* B. Campbell (Ed.), *Sexual Selection and the Descent of Man,* pp. 136–179. Aldine, Chicago.

Tyler, M.J., and D.B. Carter, 1981. Oral birth of the young of the gastric brooding frog *Rheobatrachus silus. Anim. Behav., 29*:280–282.

Vehrencamp, S.L., 1977. Relative fecundity and parental efforts in communally nesting anis, *Crotophaga sulcirostris. Science, 197*:403–405.

Vehrencamp, S.L., 1978. The adaptive significance of communal nesting in groove-billed anis (*Crotophaga sulcirostris*). *Behav. Ecol. Sociobiol., 4*:1–33.

Wake, M.H., 1977a. Fetal maintenance and its evolutionary significance in the Amphibia: Gymnophiona. *J. Herpetol., 11*:379–386.

Wake, M.H., 1977b. The reproductive biology of caecilians: an evolutionary perspective. *In* D.H. Taylor and S.I. Guttman (Eds.), *The Reproductive Biology of Amphibians,* pp. 73–101. Plenum Publishing, New York.

Wake, M.H., 1978. The reproductive biology of *Eleutherodactylus jasperi* (Amphibia, Anura, Leptodactylidae), with comments on the evolution of live-bearing systems. *J. Herpetol., 12*:121–133.

Weir, B.J., and I.W. Rowlands, 1973. Reproductive strategies of mammals. *Annu. Rev. Ecol. Syst.*, *4*:139–163.

Weller, M.W., 1959. Parasitic egg laying in the redhead (*Aythya americana*) and other North American Anatidae. *Ecol. Monogr.* *29*:333–365.

Weller, M.W., 1968. The breeding biology of the parasitic black-headed duck. *Living Bird*, *7*:169–207.

Wells, K.D., 1981. Parental behavior of male and female frogs. *In* R.D. Alexander and D.W. Tinkle (Eds.), *Natural Selection and Social Behavior*, pp. 184–197. Chiron Press, New York.

Werren, J.H., M.R. Gross, and R. Shine, 1980. Paternity and the evolution of male parental care. *J. Theor. Biol.*, *82*:619–631.

West Eberhard, M.J., 1975. The evolution of social behavior by kin selection. *Q. Rev. Biol.*, *50*:1–33.

White, F.N., and J.C. Kinney, 1974. Avian incubation. *Science*, *186*:107–115.

Wickler, W., 1968. *Mimicry in Plants and Animals.* McGraw-Hill Book Co., New York.

Woodward, B.D., 1982. Local intraspecific variation in clutch parameters in the spotted salamander (*Ambystoma maculatum*). *Copeia*, *1982*:157–160.

Woolfenden, G.E., and J.W. Fitzpatrick, 1978. The inheritance of territory in group-breeding birds. *Bioscience*, *28*:104–108.

Wourms, J.P., 1981. Viviparity: the maternal-fetal relationship in fishes. *Am. Zool.*, *21*:473–515.

Wourms, J.P., and O. Bayne, 1973. Development of the viviparous brotulid fish, *Dinematichthyes ilucoeteoides*. *Copeia*, *1973*:32–40.

Yom-Tov, Y., 1980. Intraspecific nest parasitism in birds. *Biol. Rev.*, *55*:93–108.

Index